PHONONS 89

PHONONS 89

Heidelberg, 21–25 August 1989
Federal Republic of Germany

Proceedings of the
Third International Conference on
Phonon Physics
and the
Sixth International Conference on
Phonon Scattering
in Condensed Matter

Volume 2

Editors

S Hunklinger   W Ludwig   G Weiss

World Scientific
Singapore • New Jersey • London • Hong Kong

0383 1632

PHYSICS

Published by

World Scientific Publishing Co. Pte. Ltd.,
P O Box 128, Farrer Road, Singapore 9128
USA office: 687 Hartwell Street, Teaneck, NJ 07666
UK office: 73 Lynton Mead, Totteridge, London N20 8DH

Library of Congress Cataloging-in-Publication data is available.

PHONONS '89

ISBN 981-02-0034-X

Printed in Singapore by JBW Printers & Binders Pte. Ltd.

v

CONTENT

VOLUME 1

Chapter 3: Phonons and Low-Energy Excitations in Amorphous Solids

VOLUME 2

Chapter 5: Phonons in Superlattices and Quantum Wells   701

Chapter 9: Excitations in Quantum Matter   1017

Chapter 13: Nonequilibrium Phonons   1223

Chapter 14: Transport Phenomena   1303

5

Phonons in Superlattices and Quantum Wells

PHONON TRANSMISSION THROUGH PERIODIC, QUASIPERIODIC AND RANDOM SUPERLATTICES

S. Tamura

Department of Engineering Science, Hokkaido University, Sapporo 060, Japan

I. INTRODUCTION

The recent development of techniques of growing thin films enables the realization of high-quality crystalline and amorphous multilayer structures, i.e., superlattices (SL's).[1] In addition to the important effects on the electronic states and transport properties,[2] the periodic alternation of layers with different acoustic properties yields interesting physical effects on the vibrational properties.[3] The most fundamental effect on the propagation of acoustic phonons arises from the folding of the phonon dispersion relation along the direction perpendicular to the layers and the resulting opening of frequency gaps (stop bands).[4] This can be viewed as Bragg reflection of long wavelength phonons due to an artificial periodicity much longer than the atomic lattice spacing. Experimentally these zone-folding effects and associated phonon stop bands are observed in the Raman spectrum [5,6] and phonon spectroscopy of both crystalline[7] and amorphous[8] SL's. More recently, phonon imaging has made it possible to observe the anisotropic spatial distribution of phonon stop bands in periodic SL's.[9]

Evidently periodicity is not the only possible layering scheme for SL's. In their pioneering work, Merlin et al.[10] fabricated a semiconducting layered structure based on the Fibonacci sequence, called a Fibonacci SL. The study of phonon propagation in this quasi-periodic system has also been developed both theoretically[11] and experimentally[12]. This is motivated by the special characteristics which are expected for wave propagation in the systems where the Bloch theorem is not applied. We shall also discuss briefly the numerical results for the phonon transmission through random SL's .

II. PERIODIC SL's

Because growth conditions require only a small acoustic mismatch between constituent layers of SL's, it might be naively expected that no significant effect arises when phonons propagate through such layered structures. Actually, no unusual features are observed in the phonon transmission through single interface of, say, GaAs/AlAs heterostructures. However, a dramatic effect appears when the number of layers is increased. A typical result is shown in Fig.1. Very significant dips in transmission occur locally and periodically in frequency.[9] These sharp features are caused by Bragg reflection of phonons in this artificially modulated periodic system.

Analogous to the case for photons, we can write the Bragg condition for phonons in a SL with periodicity D as

$$2D\cos\theta = m\lambda \ , \ (m=1,2,\cdots\cdots), \tag{1}$$

704

where λ is the phonon wavelength, θ is the angle the phonon wave vector makes with the normal of interfaces, and m stands for the order of the reflection. Since the basic reciprocal SL vector $\vec{G}_0$ has a magnitude $|\vec{G}_0| = G_0 = 2\pi/D$, an equivalent form of the Bragg condition is

$$2q - mG_0, \qquad (2)$$

where $q = 2\pi\cos\theta/\lambda$ is the component of the wave vector normal to the SL interfaces. Hence, phonons with $q = q_m = m\pi/D = mG_0/2$ would be reflected strongly in SL's. These wave numbers q_m correspond to the center and edge of the mini (folded) Brillouin zone with width G_0. The transmission dips in Fig.1 should correspond to the frequencies satisfying this equation at which the phonon dispersion curves have forbidden gaps (stop bands).

The first experimental observation of phonon Bragg reflection and associated zone-folding effect in SL's was made by Narayanamurti et al.[7] in a phonon spectroscopy experiment with GaAs/AlGaAs multilayers. They generated phonons by a Sn tunneling junction ($2\Delta_{Sn} = 1.2$meV = 290GHz) and detected by an Al tunnel junction ($2\Delta_{Al} = 0.54$meV = 130GHz). The voltage-tunable phonon spectroscopy makes it possible to scan the energy region $2\Delta_{Al}$ to $2\Delta_{Sn}$. Figure 2 exhibits the transmission rate of longitudinal (L) phonons propagating normal to the interfaces of a (111)-GaAs/Al$_{0.5}$Ga$_{0.5}$As SL.[7] The phonon energy at which the transmission dip occurs coincides well with 0.93meV (225GHz) corresponding to the first-order (m=1) phonon-Bragg reflection predicted in this SL structure. The profile of the transmission dip excellently agrees with the calculation. More recently, the same kind of phonon spectroscopy method was applied to amorphous SL's.[8] Similar transmission dips attributed to the stop bands at the center and boundary of the folded zone were observed for transverse (T) phonons also propagating normal to the layer interfaces.

Now we shall consider more general case where phonons propagate oblique to the interfaces. In this case phonon-polarization vectors are not parallel or perpendicular to the layer interfaces and hence mode conversion occurs by the transmission and reflection at interfaces. This gives an interesting possibility of intermode-phonon Bragg reflection

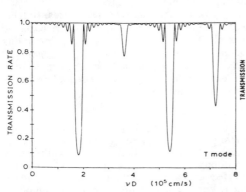

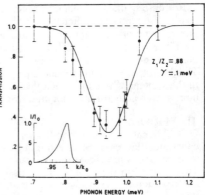

Fig.1 Transmission rate vs frequency for T phonons at normal incidence in a (001) GaAs/AlAs SL with 15 periods.

Fig.2 Normalized L-phonon transmission for the (111) GaAs/Al$_{0.5}$Ga$_{0.5}$As SL observed by Narayanamurti et al. [7]

in addition to ordinary intramode Bragg reflection.[9] For the intermode Bragg reflection between L and T phonons, the condition (2) should be generalized as

$$q_L + q_T = mG_0, \tag{3}$$

where q_L and q_T are the normal components of the wave vectors of the L- and T-phonons. In the frequency (ν) versus q plane, Eq.(3) is satisfied at points where the folded dispersion curves intersect within the minizone, whereas Eq.(2) for the intramode Bragg reflection is satisfied at the center and edges of the minizone. Accordingly, the "anticrossing", or "intrazone stop band" of the dispersion relations evidences the occurrence of the intermode-phonon Bragg reflection in SL's.

The direct check of this happening is to calculate the dispersion relations.[13] This is done by introducing the transfer matrix T which relates the acoustic fields across the unit period of a SL, or

$$U_{n+1} = TU_n, \tag{4}$$

where U_n (n is an index specifying the location of the unit cell) consists of six components describing the amplitudes of transmitted and reflected waves for three phonon modes. Now the Floquet theorem implies $U_{n+1} = e^{iqD}U_n$. Hence, the eigenvalues of the transfer matrix give $e^{\pm iqD}$ for three different modes. Because the transfer matrix depends on phonon frequency, these equations establish the dispersion relations of phonons for an arbitrary propagation direction. (For normal propagation, they lead to the dispersion relation derived by Rytov many years ago.)[14]

An example of the phonon dispersion relations for oblique propagation is shown in Fig.3(a).[9] As expected, intrazone gaps can be seen in addition to the gaps at the zone center and boundary. The former frequency gaps (coupled-mode stop bands) occur exactly in the region of ν-q plane where the L and slow transverse (ST) dispersion curves would otherwise intersect. The existence of these stop bands in the phonon dispersion

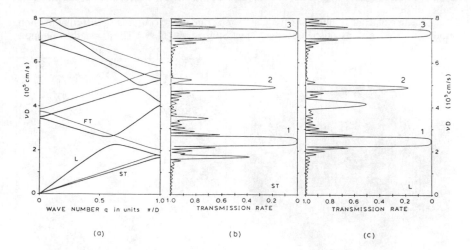

Fig.3 (a) Dispersion relation for phonons propagating in the (110) plane of a (001) GaAs/AlAs SL. Incident angle is 40° for L phonons. (b) Transmission rate of ST phonons in the same SL. Incident angle is 21.20°. (c) Transmission rate of L phonons.[9]

relation is undoubtedly the origin of the prominent dips in transmission.

Figures 3(b) and 3(c) show the phonon transmission rate of ST and L phonons, where the SL parameters and phonon-propagation directions are the same as those in Fig.3(a).[9] All of the calculated dips in transmission have one-to-one correspondences with the frequency gaps in the dispersion curves. In particular, the dips labeled 1-3 are common to L and ST modes, originating from intermode Bragg reflections between these modes.

The phonon dispersion relation in SL's depends sensitively on the propagation direction. As the angle of incidence increases, the dispersion curves become steeper. This means that the frequency at which stop bands appear becomes larger for a large angle of incidence. Hence, for a selected phonon frequency, stop bands due to different Bragg processes will occur at different angles of incidence.

The first experimental verification of intermode-phonon-Bragg reflection in SL's was made by using the phonon-imaging technique.[15] Figure 4(a) is a constant-velocity image for the L phonons transmitted through an $In_xGa_{1-x}As/AlAs$ SL (x=0.15) with $d_A=d_B=20Å$ ($D=d_A+d_B$) and 40 periods.[9] Phonons are detected by a PbBi tunnel junction with onset frequency v_c=850 GHz. We find very striking features. One is the narrow nearly circular "dip" in transmission and the other is a diamond-shaped structure. The calculated 2D-stop-band distribution in this SL is shown in Fig.4(b), where a narrow frequency range, 845 to 855 GHz, is assumed. Three different stop-band structures are apparent and they are quite similar to the experimental observation. The identification of the origins of these structures was made by calculating the dispersion relations for L phonons propagating in the three different directions A, B and C indicated in Fig.4(b). Seeking the frequency gaps in which the selected frequencies are involved, we find that the "ring" corresponds to the lowest zone-boundary stop band associated with the first-order Bragg reflection, while the "diamond-shaped structure" corresponds to the intrazone stop band due to intermode Bragg reflection of L phonons into FT phonons. The stop band at the direction C is due to the coupling of L phonons with ST phonons. However, this appears at large angle of incidence, making the observation of the structure rather difficult.

The existence of intrazone frequency gaps due to the coupling of different phonon modes in SL's was also observed by Santos et al.[16] They made phonon

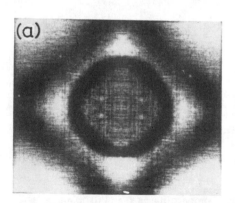
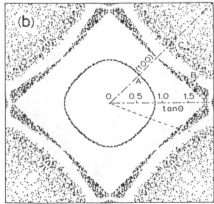

Fig.4 Phonon images for a (001) $In_{0.15}Ga_{0.85}As/AlAs$ SL. (a) Experimental image of L phonons. (b) Calculated stop-band distribution.[9]

transmission-spectroscopy measurements in a-Si:H/a-SiN$_x$:H amorphous SL's with Sn tunneling junctions as the phonon detectors. Transmission dips which cannot be attributed to the zone-center and zone-boundary gaps of T and L phonons were observed at frequencies for which the predicted dispersion curves of L and T phonons cross inside the minizone.

III. FIBONACCI SL's

The discovery of quasicrystals[17] has stimulated interest in the physical properties of systems with quasiperiodic order. Specifically, Merlin et al.[10] realized a quasiperiodic Fibonacci sequence of semiconducting layers. The SL consists of double-layer building blocks A[(17-Å AlAs)/(42-Å GaAs)] and B[(17-Å AlAs)/(20-Å GaAs)]. The sequence of A and B blocks in the SL is determined by the Fibonacci generation number N. A SL of the Nth generation G_N is created by concatenating the two prior generations: $G_N = G_{N-1}G_{N-2}$, where $G_1 = \{A\}$ and $G_2 = \{AB\}$. Thus the higher generation starts with the sequence $\{ABAABABA \cdots\}$. This structure can be regarded as the realization of a 1D model showing quasiperiodic order, or quasicrystals.[18]

The interference effects of phonons reflected at the interfaces of constituent layers are described by the structure factor

$$S(Q) = \sum_j r_j \exp(iQz_j), \qquad (5)$$

where Q is the sum of the normal components of the wave vectors of the incident and reflected phonons (for instance, $Q=2q$ for a reflection in which the mode of phonon does not change) and r_j is the amplitude-reflection coefficient at the jth interface located at $z=z_j$. The calculation of Eq.(5) for a periodic SL consisting of alternating A and B layers with thicknesses d_A and d_B yields[19]

$$S(Q) = S_P(Q) = \frac{2\pi}{D} \sum_m f^P(Q) \, \delta(Q-Q_m), \qquad (6)$$

where $Q_m = 2\pi m/D$. The δ-function peaks correspond to the Bragg reflection of phonons and the modulation factor f^P determines the relative strength of the Bragg reflection, and hence the width of frequency gaps in the dispersion relation. The structure factor for Fibonacci SL's are calculated by using the projection method,[20-22] yielding[23]

$$S(Q) = S_F(Q) = \frac{2\pi}{d} \sum_{m,n} f^F_{m,n}(Q) \, \delta(Q-Q_{m,n}), \qquad (7)$$

where $d = \tau d_A + d_B$, $Q_{m,n} = 2\pi(m+n\tau)/d$, $\tau = (\sqrt{5}+1)/2$ is the golden mean, $f^F_{m,n}$ is proportional to $\sin \phi_{m,n}/\phi_{m,n}$ with $\phi_{m,n} = \pi\tau^2(md_A - nd_B)/d$. Thus the structure factor of Fibonacci SL's consists also of δ peaks, indicating the presence of Bragg-like reflection of phonons with wave number $Q=Q_{m,n}$. The form of the modulation factor $f^F_{m,n}$ implies that the largest reflection takes place for Q satisfying $\phi_{m,n} = 0$. For $d_A/d_B = \tau$ (which is approximately satisfied by the Fibonacci SL made by Merlin et al.[10]), we recognize that the major Bragg-like reflection occurs for those $Q_{m,n}$ with n/m close to τ, i.e., m and n are successive Fibonacci numbers, $(m,n) = (F_{p-1}, F_p)$, where $F_{p+1} = F_p + F_{p-1}$ and $(F_0, F_1) = (1,1)$. For these values of (m,n) $Q_{m,n}$ becomes $2\pi\tau^{p+1}/d$. The "order" of the major reflection, therefore, is classified by a single integer p and these reflections are commonly labeled τ^p.

These basic characteristics of phonon reflection in Fibonacci SL's can be confirmed by calculating numerically the transmission rate of phonons. This calculation is also made by using the transfer-matrix method. Let T_A and T_B be the transfer matrices associated with A and B blocks. Then, after the transmission of phonons through the Nth generation of the blocks, the acoustic field U_D to be detected becomes $U_D = T_N U_S$, where $T_N = \cdots T_B T_A T_A T_B T_A$ is the product of F_N matrices T_A and T_B ordered in a Fibonacci manner and U_S is the acoustic field in the substrate of the SL. (For periodic SL's the sequence of T_A and T_B in T_N is simply alternating.) Now U_D consists of only the transmitted wave, while U_s consists of both incident and reflected waves. Hence, this equation is used to calculate the transmission rate of phonons in Fibonacci SL's.

Figure 5(a) exhibits the calculated frequency dependence of T-phonon transmission for normal propagation.[11] The SL system assumed is the same as the Fibonacci SL originally fabricated by Merlin et al.[10], but a smaller generation of the Fibonacci sequence with thirteen A and eight B blocks is assumed. The similar transmission rate was also calculated by McDonald and Aers[24] for GaAs/AlAs SL in which both A(GaAs) and B(AlAs) are composed of single layers. The distribution of transmission dips is obviously much more complicated than that in the periodic SL. However, significant dips are easily identified according to the discussion given above and several frequencies $v = v_{m,n} \equiv (m,n)$ predicted are indicated by arrows in this figure. The main dips are observed at $v_{m,n}$ for which m and n are neighbouring Fibonacci numbers, as indicated by τ^p.

For comparison, we have also shown in Fig.5(b) the transmission rate in the

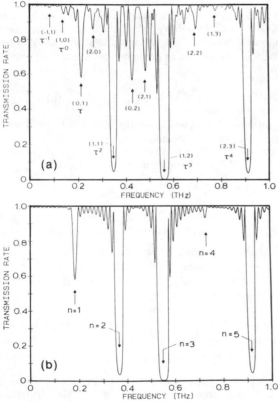

Fig.5 Frequency dependence of T phonon transmission. (a) Fibonacci SL. (b) Corresponding periodic SL.

periodic SL consisting of the sequence ABABAB···. (Transmission dips are labeled by the order of reflection.) Surprisingly, all of the main dips in transmission for the Fibonacci SL are very close in width, magnitude, and frequency to the periodic Bragg dips. The reason for these correlations, which also exist in the angular dependenc, has recently been established theoretically by analyzing the structure factors of both periodic and quasiperiodic SL's.[23]

Another interesting feature observed in Figs.5(a) and 5(b) is the fact that the transmission dips are small in the frequency range 0.7 to 0.8THz (and also at 0THz). The relative magnitudes of dips are basically related to the interference effects of reflected phonons and described by the structure factors given above. For the Fibonacci SL, it can also be explained in terms of the scaling properties of the quasiperiodic system.[25] The phonon spectrum in Fibonacci SL's is obtained by imposing the periodic boundary condition after the Nth generation (in the limit of $N \to \infty$). This leads to the dispersion relation for normal propagation

$$x_N \equiv \mathrm{tr}\,(T_N)\,/\,2 = \cos(q d_N), \qquad (8)$$

where $d_N = F_{N-1} d_A + F_{N-2} d_B$ is the thickness of the Fibonacci SL of the Nth generation. Thus the phonon frequency is in the spectrum if $|x_N| \le 1$. A recursion relation for the x_N's can be obtained by noting that $T_{N+1} = T_{N-1} T_N$. The result is

$$x_{N+1} = 2 x_N x_{N-1} - 2 x_{N-2}, \qquad (N \ge 1). \qquad (9)$$

With the use of this relation it is shown that

$$I = x_{N+1}{}^2 + x_N{}^2 + x_{N-1}{}^2 - 2 x_{N+1} x_N x_{N-1} - 1 \qquad (10)$$

is an invariant independent of N.[25] This invariant is related to the scaling index which classifies the spectrum of the system. It has been shown that the larger the invariant I, the more strongly the quasiperiodicity will affect the properties of SL's, i.e., diminishing of band widths etc. In a periodic system I is equal to zero and the spectrum is absolutely continuous. For any finite I, the scaling index is smaller than unity and the spectrum becomes singular continuous.(The corresponding eigenstates are critical.)

The invariant quantity I for SL's was first calculated by McDonald and Aers for single-layer systems.[24] In the Fibonacci SL we are considering this invariant becomes

$$I = \frac{1}{4}\left(\frac{Z_2 - Z_1}{Z_2 + Z_1}\right)^2 \sin^2\alpha \left(\sin^2\beta + \sin^2\delta - 2\sin\beta \sin\delta \cos(\beta-\delta) \right), \qquad (11)$$

where Z is the acoustic impedance (1 and 2 denote AlAs and GaAs, respectively), $\alpha = k_1 d_1$, $\beta = k_2 d_{2A}$ and $\delta = k_2 d_{2B}$. In Fig.6 I versus frequency for T phonons is plotted. Note that I vanishes like ν^4 as the frequency goes to zero. In addition I vanishes at finite frequencies for which α or $\beta - \delta$ are multiples of π (resonance condition). The scaling index approaches unity for frequencies near any zero of I, and the spectrum behaves like that in periodic systems. Thus, the phonon transmission becomes almost perfect at frequencies exactly satisfying I=0. The smallness of I (<0.03) in the entire frequency region suggests that the Fibonacci SL is rather close to the periodic system, which is reflected in the fact that the gap regions are small compared to the allowed bands.

The experimental study of the phonon transmission in Fibonacci SL's was made by phonon imaging to measure transmission versus propagation angle rather than the frequency dependence.[12] The sample identical to the one specified by Merlin et al.[10], i.e., the 13th-generation SL with 377 separate A and B blocks,was used. The experiments were performed in the same manner to those described for periodic SL's, using a tunneling junction detector with phonon cut-on frequency ν_c=850GHz. The experimental result for L phonons is shown in Fig.7(a). This sharp image indicates that ballistic reflected in the fact that the gap regions are small compared to the allowed bands.

710

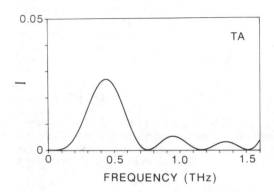

Fig.6 Invariant quantity I versus frequency for T phonons.

The experimental study of the phonon transmission in Fibonacci SL's was made by phonon imaging to measure transmission versus propagation angle rather than the frequency dependence.[12] The sample identical to the one specified by Merlin et al.[10], i.e., the 13th-generation SL with 377 separate A and B blocks,was used. The experiments were performed in the same manner to those described for periodic SL's, using a tunneling junction detector with phonon cut-on frequency $v_c=850GHz$. The experimental result for L phonons is shown in Fig.7(a). This sharp image indicates that ballistic transmission of high-frequency ($v>850GHz$) phonons through a large number (>750) of interfaces is detected, attesting to the quality of this layered structure. In the image, dark areas again indicate attenuation by the SL and their spatial distribution agrees well with Fig.7(b) which plots the directions along which L phonons, $v=850\text{-}900GHz$, are predicted to have a transmission rate smaller than 0.5. However, the observed phonon attenuations have been restricted to those due to large transmission dips which significantly correlate with Bragg dips in the corresponding periodic system. The frequency resolution of current phonon detectors is not enough to resolve the fine dips characteristic of the quasiperiodicity of the underlying sequence.

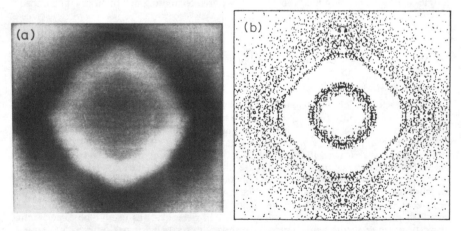

Fig.7 (a) Experimental phonon image of L phonons in a Fibonacci SL. (b) Calculated directions of L phonons in which the transmission rate is smaller than 0.5.[12]

IV. RANDOM SL's

The close correlation of the main transmission dips of phonons in double-layer periodic and quasiperiodic SL's inspires us to study further the phonon transmission in random SL's consisting of the same A and B blocks assumed in Sec.III. The random sequences of fifty A and fifty B blocks are generated by flipping coins. Figure 8 exhibits the frequency dependence of transmission averaged over the results for ten such different random sequences. Remarkably, the distinct structures are still clearly observed against our naive expectation for random systems. Especially, the transmission dips are sharply defined at frequencies around 0.92THz and 1.47THz. The existence of the latter dip in the random SL is related to the Saxon-Hutner theorem,[26] i.e., any frequency region that is a spectral gap for both pure A-type and B-type SL's is also a gap for any mixed SL consisting of A and B cells. Actually, the pure A- and B-type SL's have common frequency region around 1.47THz in which the frequency gaps overlap each other. It is also important to notice that there exist frequency ranges in which the transmission rate is close to unity. These frequencies correspond to ones at which the resonance condition is satisfied for constituent blocks of SL's. All these features are important for the experimental verification of this theory.

The above results suggest that we might also find sharp features in the angular dependence of the phonon transmission in random SL's. In Fig.9 we have plotted the angular dependence for the L mode in the (110) plane of the same random SL systems. This figure again exhibits the averaged results for five different random arrays of the blocks. Remarkably, we still find the local angular region in which the transmission rate is enhanced up to 0.8 to 0.9. Although the random SL samples have already been grown by the molecular-beam-epitaxy method,[27] no experiment is reported on the phonon-transmission spectrum. The results of Figs.8 and 9 suggest that both phonon spectroscopy and imaging experiments are promising methods to verify the spectral structures predicted for the random SL's.

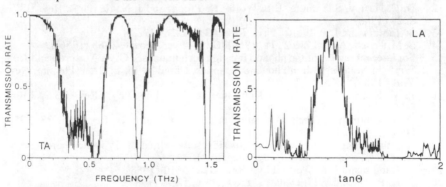

Fig.8 Frequency dependence of T phonon transmission in a random SL.

Fig.9 Angular dependence of L phonon transmission in a random SL.

V. CONCLUSIONS

Theory and experiment are in good accord on phonon transmission in periodic and quasiperiodic SL's as far as the major features are concerned. The future improvement of frequency and angular resolution of phonon detectors will make it possible to observe the predicted fine structures of the spectra characteristic of quasiperiodic and many nonperiodic (but deterministic) SL's.[28] The theory can further be extended to the problem of transmission and reflection of ballistic phonons in more general SL's consisting of multicomponents in a unit cell and in disordered SL's with a few defect cells or thickness fluctuations - a subject of continuing interest.

712

ACKNOWLEDGMENTS
This paper is based on work in close collaboration with J.P.Wolfe, D.C. Hurley, F.Nori, H.Morkoc, J.Nagle, K.Ploog and T.Watanabe. The author is indebted to M.Ramsbey for a critical reading of the manuscript. Financial support by the Foundation For C & C Promotion is gratefully acknowledged.

1. L. Esaki and R. Tsu, IBM J. Res. & Dev. 14, 61 (1970).
2. See, for example, Heterojunctions and Semiconductor Superlattices, edited by G. Allan et al. (Springer Verlag, Berlin, 1986).
3. For a recent review, see J. Sapriel and B. Djafari Rouhani, "Vibrations in Superlattices" to appear in Surface Science Reports.
4. A. S. Barker, Jr., J. L. Merz, and A. C. Gossard, Phys. Rev. B17, 3181 (1978).
5. J. Sapriel, J. C. Toledano, R. Vacher, J. Kervarec, and A. Regreny, Phys. Rev. B28, 2007 (1983).
6. C. Colvard, R. Merlin, M. V. Klein, and A. C. Gossard, Phys. Rev. Lett. 45, 298 (1980); C. Colvard, T. A. Gant, M. V. Klein, R. Merlin, R. Fischer, H. Morkoc, and A. C. Gossard, Phys. Rev. B31, 2080 (1985).
7. V. Narayanamurti, H. L. Stormer, M. A. Chin, A. C. Gossard, and W. Wiegmann, Phys. Rev. Lett. 43, 2012 (1979).
8. O. Koblinger, J. Mebert, E. Dittrich, S. Dottinger, W. Eisenmenger, P. V. Santos, and L. Ley, Phys. Rev. B35, 9375 (1987).
9. D. C. Hurley, S. Tamura, J. P. Wolfe, and H. Morkoc, Phys. Rev. Lett. 58, 2446 (1987); S. Tamura, D. C. Hurley, and J. P. Wolfe, Phys. Rev. B38, 1427 (1988).
10. R. Merlin, K. Bajema, R. Clarke, F. -Y. Juang, and P. K. Bhattacharya, Phys. Rev. Lett. 55, 1768 (1985).
11. S. Tamura and J. P. Wolfe, Phys. Rev. B36, 3461 (1987).
12. D. C. Hurley, S. Tamura, J. P. Wolfe, K. Ploog, and J. Nagle, Phys. Rev. B37, 8829 (1988).
13. S. Tamura and J. P. Wolfe, Phys. Rev. B35, 2528 (1987).
14. S. M. Rytov, Akust. Zh. 2, 71 (1956) [Sov. Phys. Acoust. 2, 68 (1956)].
15. For a recent review of the phonon imaging technique, see G. A. Northrop and J. P. Wolfe in Nonequilibrium Phonon Dynamics, edited by W. E. Bron (Plenum, New York, 1985), p.65.
16. P. V. Santos, J. Mebert, O. Koblinger, and L. Ley, Phys. Rev. B36, 1306 (1987).
17. D. Schechtman, I. Blech, D. Gratias, and J. W. Chan, Phys. Rev. Lett. 53, 1951 (1984).
18. See, for example, The Physics of Quasicrystals, edited by P. J. Steinhardt and S. Ostlund (World Scientific, Singapore, 1987).
19. S. Tamura and J. P. Wolfe, Phys. Rev. B38, 5610 (1988).
20. R. K. P. Zia and W. J. Dallas, J. Phys. A18, L341 (1985).
21. V. Elser, Phys. Rev. B32, 4892 (1985).
22. M. W. C. Dharma-wardana, A. H. MacDonald, D. J. Lockwood, J. -M. Barbibeau, and D. C. Houghton, Phys. Rev. Lett. 58, 1761 (1987).
23. S. Tamura and T. Watanabe, Phys. Rev. B39, 5349 (1989).
24. A. H. MacDonald and G. C. Aers, Phys. Rev. B36, 9142 (1987).
25. M. Kohmoto, L. P. Kadanoff and C. Tang, Phys. Rev. Lett. 50, 1870 (1983); M. Kohmoto and Y. Oono, Phys. Rev. Lett. 102A, 145 (1984); M. Kohmoto, B. Sutherland and C. Tang, Phys. Rev. B35, 1020 (1987).
26. See, for example, J. M. Ziman, Models of Disorder (Cambridge Univ. Press, London, 1979).
27. R. Merlin, K. Bajema, J. Nagle and K. Ploog, J. Phys (Paris) C5, 503 (1987).
28. S. Tamura and F. Nori, Phys. Rev. B (to be published).

CONFINEMENT EFFECTS ON FREE CARRIER SCATTERING FROM OPTICAL PHONONS IN SEMICONDUCTOR QUANTUM WELLS

R. Haupt, L. Wendler, M. Fiedler and F. Bechstedt

Sektion Physik der Friedrich-Schiller-Universität Jena, Max-Wien-Pl. 1, Jena, GDR - 6900

Above 40K the scattering of quasi-free electrons by optical phonons is the most important process in III-V semiconductors and their micro-structures, such as quantum wells (QW). Therefore, the mobility of car-riers in high-speed devices is mainly limited by this scattering pro-cess. Several authors have been treated the problem of free-carrier scattering in microstructures.

Most of the work was done in considering free-carrier scattering des-cribing the optical phonons and their interaction as in 3D bulk materials. Experiments indicate that this simplification is not suffi-cient. Therefore a careful theoretical treatment of the influence of the actual geometric configuration on both electrons and lattice vibra-tions is necessary.

This concept is realized in the present paper for a symmetrical QW and polar Fröhlich interaction.

There are two basic differences between a layered structure and bulk material concerning carrier scattering by optical phonons:

(i) Conduction and valence band discontinuities at the heterointer-faces lead to a confinement of the carrier motion perpendicular to the interfaces. In the QW considered here subbands occur.

(ii) The existence of the interfaces is accompanied by an altering of the spectrum of the optical phonons. Instead of the ordinary 3D-dispersion-free LO phonons confined modes with vanishing electro-static potential at the heterointerfaces and interface phonons, accompanying fields of which are localized at the interfaces, appear.

Electrons and phonons allow different approximations.
The electron energies $E_K(\vec{k}_{\shortparallel})$ and the corresponding wave functions are calculated with respect to the finite discontinuity of the conduction band and of the band masses at the interfaces in parabolic approximation. We model continuum states by a synthetic infinite barrier QW with a thickness much broader than a real QW.
Calculations concerning the lattice vibrational problem are based on the continuum approximation. Recent studies comparing microscopic and macroscopic models show, that the electrodynamic continuum model yield sufficient results also for very low well thicknesses greater than a = 2nm. Three types of long-wavelength optical phonons are included in our calculations:

(i)   the dispersive interface phonons with field maxima at the inter-
       faces. There exist two symmetric and two antisymmetric modes;
(ii)   the well-material-like LO_1-phonons with a electrostatic potential,
       which is localized in the QW and vanishes at the interfaces. The
       z-component of the wavevector is discrete;
(iii) and the LO_2-phonons of the QW barriers, which contribute to the
       scattering process due to the nonvanishing electron probability
       outside of the QW.

The scattering rate of quasi-free electrons in the K-th subband by long-wavelength optical phonons in a QW can be described by

$$W_K(\vec{k}_{\shortparallel}) = \frac{2\pi}{\hbar} \sum_j \sum_{K'} \sum_{\eta=\pm 1} \sum_{\vec{q}_{\shortparallel}} (n_B(\omega_j(\vec{q}_{\shortparallel})) + \frac{1}{2} + \eta\frac{1}{2} - \eta n_F(E_{K'}(\vec{k}_{\shortparallel}-\eta\vec{q}_{\shortparallel})-\mu)) \times \tag{1}$$

$$\times |M_{K'K}(\vec{q}_{\shortparallel})|^2 \, \delta(E_K(\vec{k}_{\shortparallel}) - E_{K'}(\vec{k}_{\shortparallel}-\eta\vec{q}_{\shortparallel}) - \eta\hbar\,\omega_j(\vec{q}_{\shortparallel})).$$

Herein $E_K(\vec{k}_{\shortparallel})$ is the parabolic energy of an electron in the K-th sub-band with the kinetic energy $\hbar^2 k_{\shortparallel}^2/2m_1$. j labels the different types of long-wavelength optical phonons with the dispersion $\omega_j(\vec{q}_{\shortparallel})$. $\eta = \pm 1$ corresponds to the emission/absorption of an optical phonon and $\vec{k}_{\shortparallel}$ and $\vec{q}_{\shortparallel}$ are the 2D wave vectors of the electrons and phonons in the plane of the interfaces. $n_F(x)$ and $n_B(x)$ are the ordinary Fermi-Dirac and Bose distribution functions. $M_{K'K}(\vec{q}_{\shortparallel})$ is the matrix element of the

Fröhlich-type electron-phonon interaction.

We discuss the results for the scattering rate for an electron in the lowest subband of a QW. This QW consists of a smaller gap semiconductor (GaAs) embedded symmetrically between a wider gap semiconductor ($Ga_{0.75}Al_{0.25}As$).

At first we model the potential of the electron by an infinite barrier QW. The scattering into all possible subbands (intra- and intersubband scattering) and the emission or absorption of all possible long-wavelength optical phonons (according to the correct selection rules) are considered. We get the following results:

(i)   In comparison with the scattering rate applying the incorrect ordinary 3D bulk LO phonons a reduction of the scattering rate is observed. This deviation increases with decreasing layer thickness.

(ii)   In contrast to the case of large thicknesses for small layer thicknesses (a = 2.5nm) the contribution of the interface phonons to the scattering rate is the dominant one.

(iii) The total scattering rate (sum over all contributions) is only weakly dependent on the well width.

If we include the correct height of the potential barriers (206 meV) and the discontinuity of the effective mass of the electrons at the heterointerfaces the scattering rates are somewhat changed:

(i)   The contribution of the confined LO_1 phonons is lowered. A weak coupling to the LO_2 phonons appears. These effects are especially pronounced for very small layer thicknesses (a = 2.5nm).

(ii)   The scattering rate concerning the interface phonons is only weakly dependent on the barrier height.

The consideration of higher electron densities leads to a drastical reduction of the scattering rates for lower kinetic energies of the electrons. The sharp onset of the scattering rate for the emission is changed to a more or less smooth increase of the scattering (in dependence on the temperature) up to a maximum. Higher electron densities reduce the scattering rate.

LATTICE DYNAMICS OF GaSb/InAs SUPERLATTICES
IN A BOND-CHARGE-MODEL APPROACH

L. COLOMBO

Institut Romand de Recherche Numérique en Physique des Matériaux
(IRRMA) - PH-Ecublens - CH 1015 LAUSANNE (Switzerland)

L. MIGLIO

Dipartimento di Fisica dell' Universitá di Milano
via Celoria 16 - I-20133 MILANO (Italy)

ABSTRACT. By using a planar configuration for the evaluation of long-range Coulomb interactions, within a bond-charge-model approach we made a three-dimensional calculation for the phonon dispersion relations in a (001) $(GaSb)_3(InAs)_3$ superlattice. Here we present a preliminar study of the effects of interface modelling on the lattice dynamics of such a superlattice.

1. INTRODUCTION

Lattice dynamics of semiconductor superlattices has been investigated by several Authors both from the theoretical and experimental point of view[1]. Most of the work has been devoted to the $GaAs/AlAs$ system, grown along the (001) direction. The typical features of the superlattice phonon dispersion relations along the growth direction (namely, the folding of acoustic branches, the confinement of the optical ones and the possible appearance of interface-localized vibrations) have been described using simple 1-dimensional model, as well as more realistic three dimensional ones (rigid-ion, shell or bond charge models). On the other hand, the phonon dynamics in the planes parallel to the interfaces has been considered only very recently[2,3] and no calculations are present for superlattices other than $GaAs/AlAs$.

In the present work we compute the phonon dispersion relations for a (001) $(GaSb)_3(InAs)_3$ superlattice along the growth axis as well as along directions normal to it. The short range repulsive force constants have been described within the bond charge model (BCM)[4] which has proved to be one of the best dynamical models for lattice dynamics of semiconductors (either for bulk, for surfaces and for superlattices). The long range Coulomb interactions are taken into account very carefully in order to correctly describe the macroscopic electric field effects on optical phonons. Here we make use of an interplanar approach to the calculation of such interactions presented elsewhere[3] and already applied to $GaAs/AlAs$ and $GaAs/Ge$ systems.

2. RESULTS AND DISCUSSION

The $GaSb/InAs$ superlattice is not a common-anion system and consequently has two different interfaces: an heavy In-Sb and a light Ga-As one. The BCM parameters can be evaluated for the bulk crystals either by a fitting procedure or by an interpolation between cations and ions in different crystals (for details see our forthcoming paper of reference 5), while the interface force constants (IFC) must be guessed in some way. Here we adopted two different interface modellings: *(i)* the IFCs are taken as the average between the two correspondig bulk values ; *(ii)* the IFCs are taken as *InSb*-like and *GaAs*-like for heavy and light interface, respectively. In case *(i)* the two interfaces display equivalent bondings, the only difference being the reduced mass of the interface unit cell. On the contrary, in case *(ii)* we can describe the local interface bonding as indipendent on the microscopic environment by considering the *In-Sb (Ga-As)* bonds as bulk-like.

As mentioned before, the long range Coulomb interactions have been introduced in the dynamical matrix with no truncation along the growth direction of the superlattice. We used a planar approch that essentially consists in evaluating as a first step the in-plane and inter-plane Coulomb force constants and afterwards to sum them along the superperiodicity direction. This allows us to describe in a proper way the effects of the macroscopic electric field associated to optical phonons. Because of their intrinsic uniaxial symmetry, the superlattices are anisotropic systems in which the macroscopic electric field affects in different ways the optical phonons propagating along the growth direction or normally to it, causing the discontinuity of some optical branches at zone center. This feature is known as the *anysotropy of optical phonons* and has been both predicted theoretically and observed experimentally[1].

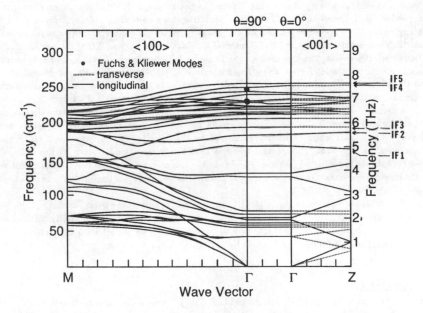

Fig.1 - *Phonon dispersion relations for a (001) (GaSb)₃(InAs)₃ superlattice along the growth direction (right panel) and along the < 100 > direction (left panel), parallel to the interfaces. Along the high-symmetry growth direction transverse (longitudinal) modes are plotted as dotted (full line) curves. Central panel shows the angular dispersion and the frequency of Fuchs and Kliewer modes (see text).*

In Fig.1 we present the resulting dispersion relations for interface modelling type *(i)* for both growth < 001 > (right panel) and and normal-to-it < 100 > (left panel) directions. Along < 001 > direction we plotted transverse branches as dotted lines and marked the interface phonons as IF_n,

$n = 1, .., 5$. An analysis of the displacement patterns lets us to attribute $IF_{1,2,3}$ to the heavy $In\text{-}Sb$ interface, while $IF_{4,5}$ are localized at the $Ga\text{-}As$ light one. This result agrees with the difference in reduced mass of the interface unit cell. Moreover the frequency of IF modes compares quite well to that one obtained by Fasolino et al.[6] within a ab-initio linear-chain model for a $GaSb/InAs$ superlattice, the only difference being the bigger $IF_1 - IF_2$ splitting in the present calculation (as a matter of fact, these two branches result almost degenerate at zone center and splitted by an amount of some cm^{-1} at zone boundary in the calculation of Ref.6). Anyway, a model calculation by Valence Overlap Shell Model[7], presented at this Conference, displays a quantitative agreement to our results for the $IF_1 - IF_2$ gap.

Turning now to the dispersions along the $< 100 >$ direction, we observe the usual distribution of phonon branches showing crossings and anti-crossings of normally polarized and same polarized curves, respectively. Along this direction the IF_n modes show the same interface character as for $< 001 >$ direction, even if less sharp (mainly the IF_2 branch). On the contrary, we do not observe any acustic interface phonon (i.e. Stoneley wave).

Finally, in the central panel we report the angular dispersion obtained rotating the wavevector (fixed in modulus and very close to the Γ point) in the xz plane. Most of the phonon branches are dispersionless, i.e. continuos in Γ, while two high-frequency pairs show a θ-dependence (here θ is the angle that the wavevector forms with the growth z axis). These are the anisotropic optical modes: their frequencies fall in the LO-TO interval for $\theta = 0°$ (the LO and TO splittings for $GaSb$ and $InAs$ perfectly overlap) and merge into the Fuchs and Kliewer ones as obtained in a dielectric continuun model calculation for a $GaSb - InAs$ interface. This behaviour can be understood comparing the multi-slab (superlattice) and single-slab dynamics in the limit of vanishing wavevector: this comparison is quite a complex argument and here we refer to the original paper[3].

A useful comparison on the role played by the interface modelling on the phonon dispersion relations along the growth direction can be made looking at case (i) vs. case (ii) (here not displayed). The modes which result more affected by the interface modellig turn out to be the interface ones, as aspected. The general trend is that the frequencies localized at the $In - Sb$ ($Ga - As$) interface are softened (stiffened) and that the transverse vibrations result more sensitive than the longidutinal ones. On the contrary, the acoustic folded and optic confined branches change only very little.

In conclusion, once more the lattice dynamics turns out to be the most sensitive probe for local modifications of the atomic environment and, in this particular case, the tunability of interface modes with the interface modelling would be a very efficient tool for superlattice characterization, as soon as suitable experimental probes will be available.

3. REFERENCES

1) For a very recent review see M.G.Cottam and D.R.Tilley, "Introduction to surface and superlattice excitations", Cambridge University Press (Cambridge, 1988)

2) E.Richter and D.Strauch, Solid State Comm. **64**, 867 (1987); S.Ren, H.Chu and Y.Chang, Phys. Rev. **37**, 8899 (1988); S.Ren, H.Chu and Y.Chang, Phys. Rev. Lett. **59**, 1841 (1988)

3) L.Miglio and L.Colombo, Surface Science (1989), in press

4) W.Weber, Phys. Rev. **B15**, 4789 (1977)

5) L.Colombo and L.Miglio, to be published

6) A.Fasolino, E.Molinari and J.C.Maan, Phys. Rev. **B39**, 3923 (1989); A.Fasolino, E.Molinari and J.C.Maan, Phys. Rev. **B33**, 8889 (1986)

7) D.Berdekas and G.Kanellis, present volume.

FREQUENCY GAPS IN ACOUSTIC PHONON TRANSMISSION IN PERIODIC, QUASIPERIODIC AND REAL SUPERLATTICES FOR NORMAL INCIDENCE

R. HOTZ, Theoretische Physik, Universität des Saarlandes

D-6600 Saarbrücken, Federal Republic of Germany

We compare acoustic phonon transmission through ideal periodic, quasiperiodic and realistic 2-component SLs. The layer thicknesses of the realistic SLs are distributed statistically about a periodic sequence. The Fibonacci SL is built up from 2 layer initial building blocks $S_1 \equiv A$ and $S_0 \equiv B$ by the well known inflation rule $S_r = S_{r-1} S_{r-2}$ [1].

For normal incidence the transfer matrix disintegrates into 2x2 matrices for the decoupled modes.

TRACE OF TRANSFER MATRIX

In an **infinite periodic** SL the wave number k of a propagating decoupled mode is given by the trace of the transfer matrix $2x(\Omega) = 2\cos(kd)$. Thus frequency gaps are defined by $|x(\Omega)| > 1$.

To investigate the Fibonacci SL, we treat the sequence of layer systems (SL_r), $r = 0, 1, 2, \ldots$ consisting of a periodic repetition of the $2F_r$ layers S_r. From the recursion relation for the traces of transfer matrices, $x_{r+1}(\Omega) = 2x_r(\Omega)x_{r-1}(\Omega) - x_{r-2}(\Omega)$ [2], we obtain convergence theorems for the frequency gaps of the Fibonacci SL:

1. The common gap frequencies for 2 successive superlattices (SL_r) and (SL_{r+1}) are gap frequencies for all $(SL_{r'})$ with $r' > r$, if $|x_{r-1}| \le |x_r x_{r+1}|$.

2. A remaining gap can only grow with r', but it is limited by the zeros of x_{r+1} nearest to the gap of (SL_{r+1}).

3. The widths of new gaps arising at (SL_r) are limited by $\pi/(F_{r-1}t_A + F_{r-2}t_B) \to 0$, for $r \to \infty$, $t_A = t_1 + t_2$, $t_B = t_3 + t_4$ with the ratios $t_n = d_n/c_n$ of layer thicknesses d_n and sound velocities c_n.

The infinite Fibonacci SL exhibits a hierarchy of **continuous** frequency gaps. Their location is determined only by the sequence t_n, moreover their widths depend on a common factor $E=(\rho_1 c_1/\rho_2 c_2 + \rho_2 c_2/\rho_1 c_1 - 2)/2$.

As an **example** we investigate in this paper the incommensurate structure, defined in (Fig. 1,a,b) with t_n forming a Fibonacci sequence. Theorem 1 holds then without the last condition. The trace $2x_7(\Omega)$ shows all the Fibonacci frequency gaps which we can distinguish by eye.

NUMBER OF STATES

Counting the number of states of a finite layer stack below any frequency Ω and dividing it by the number of layers in the stack we get a function $Z(\Omega)$ which shows the frequency gaps as plateaux. [3,4]

In the **ideal case** (Fig.1,a,b) $Z(\Omega)$ is proportional to the wave number

$$Z(\Omega) = \frac{k}{\pi} \cdot \begin{cases} (d_A + d_B)/2 & \text{for the periodic 4-layer system } (SL_2) \\ (d_A\tau + d_B)/(\tau+1) & \text{for the Fibonacci SL, } \tau=(\sqrt{5}+1)/2. \end{cases}$$

Gaps are located at $Z=s/2$ in the periodic case and at $Z=(p+q\tau)/(\tau+1)$ in the Fibonacci case, where s,p, and q are integers. The most pronounced Fibonacci gaps start as $Z=n\tau^m$ [1], but that is not the rule. In our example with $t_A/t_B=\tau$ we found a main gap at $Z=3\tau+1$ (Fig. 1,b). If we require $d_A/d_B=\tau$ instead, we can find a main gap at $Z=2+\tau$ (for certain sound velocities).

At low frequencies $Z(\Omega)$ shows almost the same gap pattern for **realistic layer systems** (Fig. 1,c,d) as for the periodic system. Buth with increasing frequency the plateaux gradually disappear. Above a certain frequency, depending on the variance, the number of states grows within the ideal gaps as fast as within the ideal bands.

TRANSMISSION RATE

To decide whether transmission experiments could show frequency gaps, we calculated the transmission rate (TR) of stacks with different numbers of layers (NL):

With increasing NL the TR of **periodic SLs** developes all gaps successively, first the bigger ones (e.g. for NL$\cong$40 at $\nu\cong$840GHz) and finally the smallest (for NL$\cong$700); a further increase of NL does not change the TR.

The TR of **Fibonacci SLs** developes infinitely new gaps with increasing NL: the bands become more and more fragmented while the already developed gaps do not change visibly.

The TR of **realistic SLs**, imperfect (SL_2), differs more and more from that of the ideal SL, the higher the frequency. With increasing NL the TR vanishes gradually at the (former) gaps, at the band edges and within some bands. E.g. in the imperfect SL (Fig. 1,c) the $\nu \cong 840GHz$ gap arises and the following band disappears for $NL \cong 100$. In this case transmission experiments cannot show the gap structure above 800GHz.

REFERENCES

1. Tamura, S., Wolfe, J.P., Phys. Rev. B 36, 3491 (1987)

2. Kohmoto, M., Kadanoff, L.P., Tang, C., Phys.Rev.Lett., 50, 1870 (1983)

3. Würtz, D., Schneider, T., Soerensen, M.P., Physica 148A, 343 (1988)

4. Hotz, R., Thesis, Saarbrücken (1988)

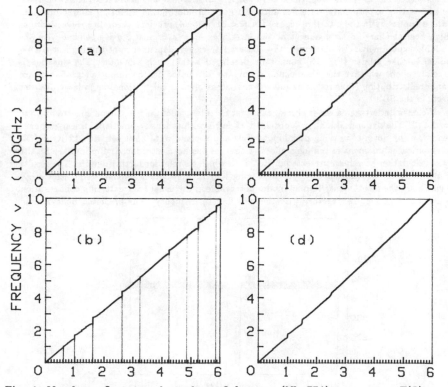

Fig. 1: **Number of states / number of layers**, (NL=754) $Z(\Omega) \longrightarrow$

(a): Per. 4-layer system (SL_2) with $d_1,..,d_4$=12, 5, 6, 5 [nm], ρ_2/ρ_1=1.16 c_1,c_2=4.3, 5.8 [km/s]. (b): Fibonacci superlattice: $\lim(SL_r)$ for $r \to \infty$.

(c), (d): Realistic layer systems with thicknesses distributed about that of (SL_2) with a variance of 4% and 7% respectively.

722

DISORDER EFFECTS ON THE PHONON SPECTRA OF ULTRATHIN
SEMICONDUCTOR SUPERLATTICES: FIRST PRINCIPLES RESULTS

S. Baroni, SISSA, Strada Costiera 11, I-34014 Trieste, Italy;

P. Giannozzi, IRRMA, PHB-Ecublens, CH-1015 Lausanne, Switzerland;

E. Molinari, CNR – Istituto "Corbino", Via Cassia 1216, I-00189 Roma, Italy.

The phonon spectra of ultrathin $(GaAs)_n(AlAs)_n$ (001) superlattices are calculated by means of an *ab initio* linear-response density-functional approach. For $n = 1, 2, 3$, we find that results for perfectly ordered superlattices are not consistent with the available Raman experimental results. By studying prototype supercells aimed at simulating completely or partially disordered superlattices, we show that the experimental data can be explained if cationic mixing affecting also the inner planes is taken into account.

Phonon spectra of $(GaAs)_m(AlAs)_n$ (001) superlattices (SL's) have been studied rather extensively in the last years. For "thick" SL's $(m, n > 3)$, the comparison of theoretical results and light scattering experimental data has allowed to obtain information useful for characterization of the samples[1,2]. In the ultrathin regime $(m, n < 3)$, instead, previous theoretical investigations which neglect charge rearrangement at the interfaces show a dramatic disagreement with the available experiments. In particular, the observed thickness dependence of the highest $GaAs$- and $AlAs$- like modes (Fig. 1b) cannot be described within such schemes. The question is therefore open, whether the discrepancy should be attributed to the importance of interface charge redistribution (thus requiring full self-consistent phonon calculations) or to large disorder effects in the SL's.

We have undertaken a series of self-consistent calculations of the SL phonon spectra for $m = n = 1, 2, 3$. The dynamical matrix is obtained *ab initio* within density-functional linear-response theory[3], using norm-conserving pseudopotentials and a plane wave basis set with 12 Ry cutoff. No empirical adjustment of the ionic effective charges has been introduced. Preliminary results of the calculation have been presented in Ref. 4, to which we refer for further details. The results for the thickness dependence of the highest-lying longitudinal optical phonon frequencies ω_{LO1} — which are the most intense Raman modes observable in the usual backscattering configuration — are shown in Fig. 1a, together with the bulk LO(Γ) frequency. The predicted behaviour in

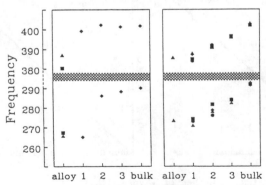

Fig. 1. Frequencies of the $GaAs$-like and $AlAs$-like LO_1 modes of $(GaAs)_n(AlAs)_n$ SL's, as a function of n. Note the break in the energy scale: the dashed band separates the $GaAs$ and $AlAs$ energy ranges. *Left*: theory, present work; diamonds: perfectly ordered superlattices; triangles: alloy from a 16-atom supercell calculation; squares: chalcopyrite structure. *Right*: experiments; diamonds: from Ref. 6a; squares: from Ref. 6b; triangles: from Ref. 6c; circles: from Ref. 6d.

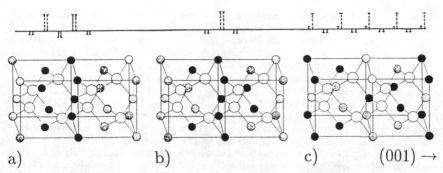

Fig. 2. *Bottom:* Prototype structures used to study the effect of increasing cationic mixing in the $(GaAs)_2(AlAs)_2$ superlattice (the correspondig optical phonon spectra are shown in Fig. 3). Black, dashed and empty circles represent Al, Ga, and As atoms respectively. *a)* perfect $n = 2$ superlattice; *b)* $n = 2$ superlattice with "interface cationic intermixing": the unit cell contains one pure-Al plane, one pure-Ga plane, and two mixed $Ga - Al$ planes; *c)* chalcopyrite structure: all cationic planes are mixed $Ga - Al$ planes. *Top:* Amplitude of longitudinal ionic displacements for the topmost LO_1 $AlAs$-like mode. A bar is shown for the displacement of each of the two atoms of the unit cell in the (001) plane; in order to distinguish the displacements of the different cations lying in the mixed planes, the displacements of Al are shown by broken bars. (See also Ref. 7).

the $AlAs$- and $GaAs$- like ranges is very different: in the first case ω_{LO1} is very slightly sensitive to variations in the layer thickness n, while in the second a jump occurs between the $n = 1$ and the $n = 2$ case. This result can be understood by inspection of the bulk dispersion along the (001) SL growth direction, as reported in Refs. 4 and 5: While the width of our calculated bulk GaAs LO branch is rather large ($50 cm^{-1}$), in good agreement with inelastic neutron scattering experiments, for AlAs we find that the LO branch is almost dispersionless ($8 cm^{-1}$); this is also in agreement with the available experimental data, which are however very scarce for this material. For perfectly ordered superlattices, it is indeed expected[1,2] that the ω_{LO1} modes should fall within the bulk bandwidth.

The comparison between our theoretical results (Fig. 1a) and experimental data by several groups[6] (Fig. 1b) shows that fully accounting for interface charge rearrangements — as it is done in our calculations — is not enough to reconcile theory and experiments. In particular the existence of $AlAs$-like experimental points falling out of the bulk $AlAs$ (001) bandwidth can be by no means reproduced by calculations for perfect superlattices.

In order to understand whether such a discrepancy is due to cationic intermixing, we have performed lattice dynamical calculations for selected supercell systems aimed at simulating different situations where interfaces are not abrupt. Let us consider the three geometries shown in the lower part of Fig. 2. The first case (Fig. 2a) is the ideal $n = 2$ SL; the second and the third cases, having a doubled periodicity parallel to the interface, contain mixed $Ga - Al$ planes. The structure of Fig. 2b simulates a situation where intermixing takes place only at the interfaces of the $n = 2$ SL, leaving one pure Ga and one pure Al cationic plane per cell; the chalcopyrite structure of Fig. 2c is taken as the prototype of a situation where cationic intermixing affects all the planes. The Γ-point spectrum for these geometries is shown in Fig. 3 for phonons propagating along the growth axis and polarized alike. Disorder shifts ω_{LO1} downwards. When mixed cationic planes are present only at the interface (Fig. 2b), the shift turns out to be appreciable in the $GaAs$-like range, but negligible in the $AlAs$-like range. This is due to the fact that the highest frequencies are still related to vibrations of the pure-Ga or pure-Al cationic planes, and therefore their shift is only due to further confinement (to one pure-Ga or pure-Al monolayer instead of two). The shift is therefore related to the (001) LO bulk bandwidth, much bigger in $GaAs$ than in $AlAs$. A relevant shift of both ω_{LO1} modes can be obtained only if no pure-cation plane is left in the structure. This is confirmed by the calculation for the structure of Fig. 2c; here all the modes are related to vibrations of the mixed cationic planes (Al atoms of the mixed planes in the $AlAs$-like range, Ga atoms of the mixed planes in the $GaAs$-like range), and they no longer fall within the (001) LO bulk bandwidth. To illustrate the nature of such modes, in the topmost part of Fig. 2 we display the amplitude of longitudinal displacements associated to the $AlAs$-like ω_{LO1} frequencies in the three cases[7]. A more disordered situation, with intermixing in all the cationic planes, has been simulated by a 16-atom FCC supercell where 4 Al and 4 Ga atoms occupy the the cationic sites at random. The result for the ω_{LO1} modes are shown in the first column of Fig. 1a. Also in this case, it is clear that full cationic intermixing has a very strong effect on the highest $AlAs$-like mode, which is pulled out of the bulk (001) dispersion. These results strongly support the idea that structural intermixing — rather than significant charge rearrangements — are responsible of the observed spectra of ultrathin superlattices.

The sensitivity of $AlAs$-like modes to structural disorder makes them particularly suitable for characterization purposes. In fact, confinement-induced and disorder-induced shifts can be easily distinguished in this case: the flatness of the bulk $AlAs$ dispersion along (001) makes confinement-induced shifts very small; disorder-induced shifts, instead, turn out to be very large. This last fact can be attributed to two effects: On one hand, disorder can produce mixing with bulk states with wavevector q not lying along (001). Such states are more dispersed[8] than those along (001): for example $\omega_{LO}(\Gamma) - \omega_{LO}(L) = 27 cm^{-1}$, to be compared with $\omega_{LO}(\Gamma) - \omega_{LO}(X) = 8 cm^{-1}$. On the other hand, the effective polarization responsible for the LO-TO splitting is reduced by alloying, which therefore brings LO modes closer to the TO branch; the importance of such contribution for $AlAs$ is indicated by its large LO-TO splitting: $\omega_{LO}(\Gamma) - \omega_{TO}(\Gamma) = 40 cm^{-1}$. A similar identification of disorder-induced from confinement-induced shifts is much harder for $GaAs$-like modes, due the absence of a strong anisotropy in the bulk $GaAs$ dispersion and to the smaller LO-TO splitting.

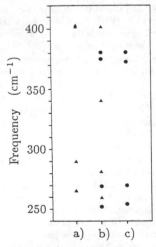

a) b) c)

Fig. 3. Frequencies of the $GaAs$-like and $AlAs$-like LO modes for the three structures of Fig. 2. Triangles indicate modes of pure-Ga or pure-Al (001) atomic planes; circles indicate modes either of Ga or of Al ions of the mixed (001) planes. The displacement patterns of the topmost $AlAs$-like modes are shown in the upper part of Fig. 2.

References

1. For a recent review, see e.g.: B. Jusserand and M. Cardona, in *Light Scattering in Solids V*, edited by M. Cardona and G. Güntherodt (Springer, Berlin, 1989), p. 49.
2. See: A. Fasolino and E. Molinari, *Proceedings of the IV Internat. Conf. on Modulated Semiconductor Structures*, Ann Arbor 1989 (Surf. Sci., in press), and references therein.
3. S. Baroni, P. Giannozzi and A. Testa, Phys. Rev. Lett. 58, 1861 (1987).
4. S. Baroni, P. Giannozzi and E. Molinari, in *Proceedings of the XIX Internat. Conf. on the Physics of Semiconductors*, edited by W. Zawadzki (Institute of Physics, Polish Academy of Sciences, Wroclaw, 1988), p. 795.
5. S. Baroni, P. Giannozzi and E. Molinari, to be published.
6. a) M. Cardona, T. Suemoto, N.E. Christensen, T. Isu and K. Ploog, Phys. Rev. B36, 5906 (1987); b) A. Ishibashi, M. Itabashi, Y. Mori, K. Kawado and N. Watanabe, Phys. Rev. B33, 2887 (1986); c) M. Nakayama, K. Kubota and N. Sano, Solid State Commun. 53, 493 (1985); d) T. Toriyama, N. Kobayashi and Y. Horikoshi, Jpn. Journ. Appl. Phys. 25, 1895 (1986).
7. Notice that while the displacements of the cationic sublattice are purely longitudinal, in general some degree of transverse vibrational amplitude is present for the As sublattice in the case of the geometries of Figs. 2b and 2c. [The longitudinal character of the modes plotted in Fig. 3 was indeed defined in terms of the cationic displacements only.] However for the highest LO_1 modes, which are Γ-derived, such transverse amplitude is always very small and is not shown in the figure.
8. P. Pavone, S. De Gironcoli, P. Giannozzi and S. Baroni, to be published.

RELATIONSHIP OF MICROSCOPIC AND MACROSCOPIC THEORIES FOR OPTICAL PHONONS IN GaAs-AlAs SUPERLATTICES

F. Bechstedt, H. Gerecke and L. Wendler

Friedrich-Schiller-Universität, Max-Wien-Platz 1, 6900 Jena, GDR

The interpretation of recent Raman scattering studies[1] of optical phonons in $(GaAs)_{N_1}(AlAs)_{N_2}$ (001) superlattices indicates discrepancies to the treatment of Fröhlich as well as deformation-potential electron-phonon interactions in the framework of the conventional dielectric continuum theory[2,3]. For systems involving thin layers this theory seems to be beyond its legitimate limit. The atomic displacements do not fulfil the continuity condition at the interfaces.

For that reason, we critically examine the continuum theory for long-wavelength optical phonons by comparison with a parallel micros-copic model. Starting point is a rigid-ion model[4] in which the elastic interaction of the atoms is described by one radial force constant. The microscopic electric field produced by the vibrating ion charges is spatially averaged with respect to the Wigner-Seitz cell of the under-lying fcc structure to describe the same field as in the macroscopic theory. Despite of the simplicity the resulting equations of motion must be solved numerically.

The formal transition of the microscopic theory to the continuum approach can be done taking the limit of vanishing atomic distances un-der conservation of the thicknesses $d_b = N_b a/2$ (lattice constant $a \to 0$) of GaAs (b=1) and AlAs (b=2) in the $(GaAs)_{N_1}(AlAs)_{N_2}$ (001) superlattice. Defining the continuous distance z $(0 \leqslant z \leqslant d_1 + d_2 = d)$ of the molecules with reduced masses μ_b and atomic charges e_b to the interfaces and neg-lecting the bulk phonon dispersion one gets for the polarization vectors $\bar{e}_{bj}(z,\theta)$ of a long-wavelength optical phonon mode j with the frequency $w_j(\theta)$ and the propagation direction $\bar{e}_Q = \sin\theta \bar{e}_y + \cos\theta \bar{e}_z$ the equations

$$\left[\omega_j^2(\theta) - \omega_{TOb}^2 (\delta_{\alpha x} + \delta_{\alpha y}) - \omega_{LOb}^2 \delta_{\alpha z} \right] e_{bj\alpha}(z,\theta) = -\frac{e_b}{\sqrt{\mu_b}} \left[E_{j\gamma}(\theta)\delta_{\alpha y} + D_{jz}(\theta)\delta_{\alpha z} \right] \tag{1}$$

where the macroscopic electric field components

$$\begin{pmatrix} E_{j\gamma}(\Theta) \\ D_{jz}(\Theta) \end{pmatrix} = -\sin\Theta \frac{16\pi}{a^3} \sum_{b=1}^{2} \frac{e_b}{\sqrt{\mu_b}} \frac{1}{d} \int_{d_1\delta_{b2}}^{d_1\delta_{b2}+d_b} dz \begin{pmatrix} \sin\Theta & \cos\Theta \\ \cos\Theta & -\sin\Theta \end{pmatrix} \begin{pmatrix} e_{bj\gamma}(z,\Theta) \\ e_{bjz}(z,\Theta) \end{pmatrix} \tag{2}$$

and the bulk-zone-center optical phonons w_{TOb} and w_{LOb} are introduced.

The first type of solutions, characterized by $E_{jy}(\Theta)=D_{jz}(\Theta)=0$, yields the confined optical phonons with frequencies $w_j(\Theta)=w_{TOb}$ or w_{LOb} for GaAs-like (b=1) and AlAs-like (b=2) modes and polarization directions parallel to the Cartesian direction $\alpha =x,y,z$. Because of the high degeneracy of each vibronic level and the automatically fulfilled boundary conditions of the macroscopic electrodynamics there is an arbitrariness in the z-dependence of the solutions. Starting from the idea of standing waves and the continuity of the displacements at the interfaces we write in agreement with the results of the microscopic theory[4] $(0 \le z-d_1\delta_{b2} \le d_b)$

$$e_{bj\alpha}(z,\Theta) = \sqrt{\frac{2}{d_b}} \begin{cases} \dfrac{\cos(q_n^b(\Theta)(z-d_1\delta_{b2}-d_b/2))-\cos(q_n^b(\Theta)d_b/2)}{\sin(q_n^b(\Theta)d_b/2)} & n=1,3,5... \\ \sin(q_n^b(\Theta)(z-d_1\delta_{b2}-d_b/2)) & n=2,4,6,..., \end{cases} \tag{3}$$

where the confinement wavevector $q_n^b(\Theta)$ satisfies the equation

$$q_n^b(\Theta)d_b/2 = \tan(q_n^b(\Theta)d_b/2) \tag{4}$$

for symmetric p-polarized TOn and LOn phonons and propagation directions $\Theta > 0$. For all other branches – s-polarized TOn (arbitrary Θ and n), antisymmetric p-polarized TOn/LOn (arbitrary Θ, even n) and p-polarized TOn/LOn ($\Theta=0$, arbitrary n) – it holds $q_n^b(\Theta)=\pi n/d_b$ in (3).

The second type of solutions of the integral equations (1) for $\Theta > 0$ is characterized by $E_{jy}(\Theta)$, $D_{jz}(\Theta) \neq 0$. These four p-polarized solutions are the well-known macroscopic interface phonons of the Fuchs-Kliewer type[2,3]. The polarization vectors are constant in each material. Their directions vary between y and z in dependence on Θ and j. One interface phonon has to be arranged in the four p-polarized GaAs-like and AlAs-like LOn and TOn series. Thereby, for $\Theta > 0$ TO1 and the last appearing symmetric LON_b (or $LO(N_b-1)$) have to be replaced by a corresponding interface phonon if the vanishing bulk phonon dispersion and the real limitation $n \le N_b$ is taken into account.

A strict comparison of the results from the macroscopic continuum theory (3) and atomic displacement patterns in the framework of the microscopic rigid-ion model is possible for cations and anions. Such a comparison is only problematic for propagation directions $\Theta > 0$ since in this case all p-polarized symmetric LOn and TOn modes of the microscopic theory are accompanied by long-range electric fields of the type (2) whereas in the dispersionless macroscopic treatment only the interface phonons produce such fields. More in detail we state the following similarities and discrepancies between the two theories:

(i)   The spectrum resulting within a dispersionless continuum theory is rather poor. Apart from the four dispersive interface phonon branches the only and therefore highly degenerated vibronic levels are given by w_{LOb} and w_{TOb}.

(ii) For all modes which are also within the microscopic treatment not accompanied by long-range electric fields $E_{jy}(\Theta)$ and $D_{jz}(\Theta)$ the line shapes of the displacement patterns resulting within the continuum theory and the parallel microscopic model agree widely if the appropriate solutions (3) are taken into account.

(iii) Small discrepancies between the two theories mainly concern the boundary regions. Firstly, the envelope functions vanish at the interfaces whereas the confinement in the microscopic theory is not complete due to the finite bulk phonon dispersion. Secondly, the effective material layer thicknesses are increased in the microscopic theory according to the chemical equivalence of the two interfacial As layers.

(iv) The comparison in the case of the p-polarized symmetric LOn and TOn modes ($\Theta > 0$) exhibits some problems due to the influence of the long-range electric fields. A strict identification of macroscopic and microscopic modes is only possible if the spatial dispersion is also taken in the continuum approach.

1) Sood, A.K., Menendez, J., Cardona, M., and Ploog, K., Phys. Rev. Lett. _54_, 2111 + 2115 (1985).

2) Wendler, L. and Haupt, R., phys. stat. sol. (b) _143_, 487 (1987).

3) Enderlein, R., Bechstedt, F., and Gerecke, H., phys. stat. sol. (b) _148_, 173 (1988).

4) Bechstedt, F. and Gerecke, H., phys. stat. sol. (b) _154_,Nr.2 (1989).

THEORY OF PHONON-POLARITON MODES IN LOW-DIMENSIONAL STRUCTURES

N. Raj,[1] D.R. Tilley,[1] T. Dumelow[2] and T.J. Parker[2]

[1]   Department of Physics, University of Essex, Colchester CO4 3SQ, UK

[2]   Department of Physics, Royal Holloway and Bedford New College, Egham, Surrey TW20 0EX, UK

The availability of high-quality layered specimens has renewed interest in electromagnetic waves, or polaritons, in such structures. For a film on a substrate the theoretical expressions are well known, but need careful application for a case like CdTe on GaAs where the reststrahlen bands occupy different frequency intervals. For a periodic structure (superlattice) a complete theoretical description can be developed by a transfer-matrix formalism. When the polariton wavelength is much greater than the superlattice period the full expressions simplify into an effective-medium description.[1-3] This characterises the superlattice as a uniaxial medium and in reststrahlen bands the principal dielectric-tensor components, which are parallel and perpendicular to the layers, vary with frequency in quite different ways. The dispersion curves of both bulk and surface phonon-polaritons then contain a number of branches.

For an isotropic film of thickness L bounded by isotropic semi-infinite media, Maxwell's equations together with boundary conditions in s-polarisation yield[4]

$$(q_{1z}+q_{2z})(q_{2z}-q_{1z})^{-1} = (q_{3z}+q_{2z})(q_{2z}-q_{3z})^{-1}\exp(2iq_{2z}L) \tag{1}$$

where the media are 1 (semi-∞), 2 (film) and 3 (semi-∞), $q_{iz} = (\varepsilon_i\omega^2/c^2-q_x^2)^{1/2}$, $i = 1,2,3$, ε_i being the dielectric functions, and all field components contain the factor $\exp(iq_xx-i\omega t)$, so that x is the direction of propagation. The dispersion equation for p-polarisation is obtained from (1) by the replacement $q_{i+1z}/q_{iz} \rightarrow \varepsilon_{i+1}q_{iz}/\varepsilon_iq_{i+1z}$ for $i = 1,2$ except in the exponential.

For a mode confined within the film, q_{1z} and q_{3z} are pure imaginary provided there is no energy transfer into the substrate (we ignore damping). Eqn. (1) and the p-analogue then describe guided modes if q_{2z} is real; this requires $\varepsilon_2 > \varepsilon_1$ and $\varepsilon_2 > \varepsilon_3$. The p equation can also describe modes of surface type if q_{2z} is imaginary; such modes do not occur in s-polarisation. For a CdTe film on a GaAs substrate the main spectral features, as observed in attenuated total reflection (ATR) occur in the reststrahlen bands, which lie between 141 cm^{-1} and 169 cm^{-1} for CdTe and between 270 cm^{-1} and 294 cm^{-1} for GaAs. In the former region, a surface mode is seen in p-polarisation;[5] it is essentially the surface polariton on a CdTe/vacuum interface, perturbed and made leaky by the proximity of the GaAs substrate. This corresponds to a real value of q_{3z}. In the latter region, guided modes occur in both p and s polarisation. Typical dispersion curves for a number of CdTe thicknesses are shown in Fig. 1. Also shown is a scan line for ATR with 20° angle of incidence in a Si prism. The frequencies at which this scan line intersects the dispersion curves correspond to the dips in the ATR spectra.[5]

The simplest description of a superlattice, i.e. a periodic medium in which thicknesses d_1 and d_2 of materials 1 and 2 alternate, is to ascribe to each layer the dielectric function ε_1 or ε_2 of the bulk medium. Within this bulk-slab model, optical propagation is described by a 2 x 2 transfer matrix T.[4,6,7] In the far infrared the inequality $\lambda \gg (d_1 + d_2)$ applies, and the dispersion equation simplifies[2] to that of a uniaxial medium with $\varepsilon_{xx} = (\varepsilon_1 d_1 + \varepsilon_2 d_2)/(d_1+d_2)$ and $\varepsilon_{zz}^{-1} = (\varepsilon_1^{-1} d_1 + \varepsilon_2^{-1} d_2)/(d_1+d_2)$, the z direction being normal to the layers. This result can also be obtained by electromagnetic field continuity arguments.[1] Although these equations are very simple, they can lead to striking anisotropy in the frequency dependences of ε_{xx} and ε_{zz}. As an example, we cite the surface polariton on a semi-infinite superlattice, which has the dispersion equation[4,7]

$$q_x^2 = (\omega^2/c^2)\varepsilon_m \, \varepsilon_{zz}(\varepsilon_{xx}-\varepsilon_m)/(\varepsilon_{xx}\varepsilon_{zz}-\varepsilon_m^2) \qquad (2)$$

where ε_m is the dielectric constant of the bounding medium. For the vacuum interface of a GaAs/Al$_x$Ga$_{1-x}$As superlattice, (2) predicts 4 branches, as shown in Fig. 2. A necessary condition for the existence of the surface mode is $\varepsilon_{xx} < 0$. We distinguish in Fig. 2 between real surface modes, with $\varepsilon_{zz} < 0$ and virtual surface modes, with $\varepsilon_{zz} > 0$. The former have $q_x \to \infty$ for some frequency, and therefore have an electrostatic counterpart, whereas dispersion curves for the latter terminate at a finite point of the (q_x,ω) plane. All 4 modes shown in Fig. 4 have been observed by ATR.[8]

The bulk-slab model is inadequate for short-period superlattices, in which d_1 and d_2 each correspond to only a few monolayers, since we must then take account of the confinement of optic phonons in calculating the dielectric function. To first order a correction to the bulk phonon frequencies will suffice, but a complete description requries the inclusion of higher-order (folded) confined modes.[9]

730

REFERENCES

1.	Agranovich, V.M. and Kravstov, V.E., Sol. State Comms 55, 85 (1985).

2.	Raj, N. and Tilley, D.R., Sol. State Comms 55, 373 (1985).

3.	Liu, W.M., Eliasson, G. and Quinn, J.J., Sol. State Comms 55, 533 (1985)

4.	Cottam, M.G. and Tilley, D.R., "Introduction to Surface and Superlattice Excitations", Cambridge University Press (1989)

5.	Dumelow, T., El-Gohary, A.R., Maslin, K.A., Parker, T.J., Tilley, D.R. and Ershov, S.N., submitted to Semicond. Sci. Technol.

6.	Yeh, P., Yariv, A. and Hong, C-S, J. Opt. Soc. Am. 67, 423 (1977).

7.	Raj, N. and Tilley, D.R., "The Electrodynamics of Superlattices", Chapter 7 of "The Dielectric Function", ed. D.A. Khirznitz, L.V. Keldysh and A.A. Maradudin, Elsevier (1989).

8.	El-Gohary, A.R., Parker, T.J., Raj, N., Tilley, D.R., Dobson, P.J., Hilton, D., and Foxon, C.T.B., Semicond. Sci. Technol. 4, 388 (1989)

9.	Dumelow, T., Hamilton, A., Parker, T.J., Samson, B., Smith, S.R.P., Tilley, D.R., Hilton, D., Moore, K. and Foxon, C.T.B. These Proceedings.

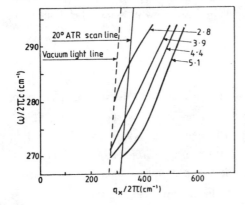

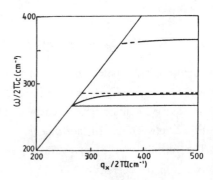

Fig. 1 Dispersion curves ω versus q_x for s-polarised guided modes of a CdTe film on a GaAs substrate in the GaAs reststrahlen band. CdTe thicknesses (in μm) are marked. Also shown are the vacuum light line $\omega = cq_x$ and the ATR scan line for $20°$ incidence in Si (refractive index 3.413)

Fig. 2 Surface phonon-polariton disperison curve for a semi-infinite GaAs/Al$_{0.14}$Ga$_{0.86}$As superlattice with $d_1 = (^1/_2)d_z$. Continuous lines are real modes, and broken lines, virtual modes.

ANGULAR DEPENDENCE OF ACOUSTIC PHONON STOP BANDS IN PERIODIC AND QUASIPERIODIC a-Si:H/a-SiN$_x$:H SUPERLATTICES

R. HOTZ, Theoretische Physik, Universität des Saarlandes,

D-6600 Saarbrücken, Federal Republic of Germany

To investigate a Fibonacci superlattice (SL) [1] we deal with a sequence of layer systems (SL$_r$), r=0, 1, 2, ... consisting of a periodic repetition of blocks S$_r$ with 2F$_r$ layers. The Fibonacci building blocks are defined recursively: S$_r$=S$_{r-1}$S$_{r-2}$. S$_1$ and S$_0$ consist in each case of one a-Si:H layer and one a-SiN$_x$:H layer with thicknesses d$_1$, d$_2$ and d$_3$, d$_4$ respectively.

PROPAGATING MODES IN PERIODIC LAYER SYSTEMS

The elastic vibrations of systems of isotropic layers are characterized by the plane of propagation, the frequency Ω and the component K of the wave vector parallel to the layers. In transmission experiments phonons pass through a substrate and arrive at the SL with an angle of incidence Θ. Which can be calculated from the experimental arrangement and the elastic properties of the substrate. The ratio $K/\Omega=\sin\Theta/c_S$ remains constant within the SL (c_S is the phase velocity of the substrate). There are 6 possible plane waves in each layer with K and Ω given. We denote the coefficients of their stimulation in the n-th layer with the vector $\underline{Y}_n(K,\Omega)$. The coherence conditions at the interfaces yield matrices connecting $\underline{Y}_n$ and $\underline{Y}_{n+1}$.

For a periodic repetition of F layers the transfer matrix $M(K,\Omega)$ connects the coefficient vectors for the first layers of adjacent periodicity blocks

$$\underline{Y}_{F+1}(K,\Omega) = M(K,\Omega)\,\underline{Y}_1(K,\Omega). \tag{1}$$

The 6 eigenvalues of $M(K,\Omega)$ are $\lambda_1,..,\lambda_3$, $1/\lambda_1,...,1/\lambda_3$. Eigenvalues of modulus 1 correspond to propagating waves (SL-modes) with wave number k

$$\lambda(k,\Omega) = \exp(ikd), \tag{2}$$

where d is the thickness of the periodicity block. There are 2 independent polarizations: **a-modes** (pure transverse, vibrating normal to the plane of propagation) and **p-modes** (coupled transverse and longitudinal, vibrating parallel to it).[2]

ANGULAR DEPENDENCE OF STOP BANDS

The dependence of the frequencies of propagating modes on the angle of incidence Θ, i.e. Ω/K is shown in Fig. 1. The regions with propagating **a-modes** are the single shaded areas (left); the empty areas in between are their stop bands. The regions with one propagating **p-mode** are the single shaded areas (right) and the dark areas are the regions with two propagating p-modes. The complementary regions indicate the stop bands: empty regions are total stop bands for both p-modes and single shaded areas are stop bands only for one of the two p-modes.

The relation (2) between K, Ω and k yield the dispersion surfaces $\Omega(k)$ (we use the abbreviation $\underline{k} \equiv (k, K)$). For a periodic 2-layer system[3] we derived from these the group velocity $\underline{v} = \text{grad } \Omega(\underline{k})$. By means of the magnitude of group velocity we could almost completely classify the p-modes as quasi-transverse (slow) and quasi-longitudinal (fast) ones. The angular dependence of their stop bands is shown in Fig. 5 of (Ref. 4).

CONCLUSIONS

We compared the angular dependent stop bands of (SL_r) for r=2,3,6,7. All (SL_r) stop bands appear slightly shifted and deformed at (SL_{r+1}). The new (SL_{r+1}) stop bands are generally smaller than those of (SL_r). With increasing r the widths of new stop bands and the differences to the old ones decrease rapidly. In our example the differences are scarcely visible if r is 6 or 7. For $\Theta=0$ we proved this convergence theorem analytically.

REFERENCES

1) Merlin, R., Bajema, K., Clarke, R., Juang, F.Y., Bhattacharya, R.K., Phys. Rev. Lett. 55, 1768 (1985)

2) Hotz, R., Siems, R., Solid State Comm., 51, 793 (1984)

3) Santos, P.V., Ley, L., Mebert, J., Koblinger, O., Phys. Rev. B, 36, 4858 (1987)

4) Hotz, R., Siems, R., Superlattices and Microstructures, 6, 139 (1989)

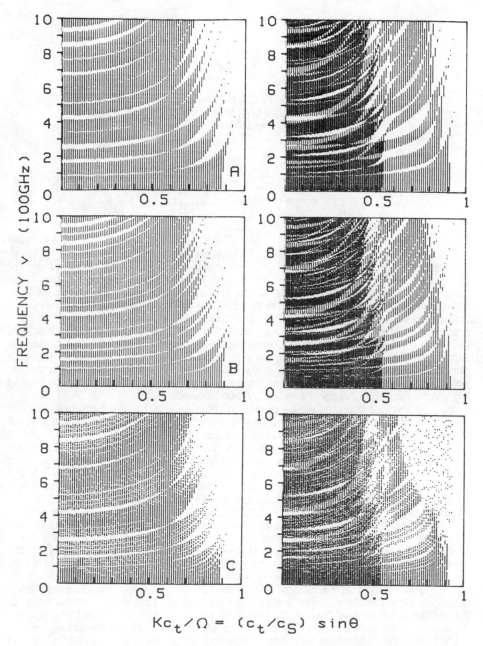

Fig. 1: **Angular dependence of stop bands** (left: a–modes, right: p–modes)
in a periodic sequence of $2F_r$-layer blocks (SL_r): r=2, 3, 6 (A,B,C)
The thicknesses are $d_1,...,d_4$ = 12, 5, 6, 5 [nm]

PHONON CALCULATIONS IN InAs/GaSb HETEROSTRUCTURES

D. BERDEKAS and G. KANELLIS
Physics Department 313-1, University of Thessaloniki
540 06 Thessaloniki, Greece.

InAs/GaSb superlattices (SL's) of high crystalline quality have been grown[1], and their electronic properties have been extensively studied in view of interesting applications. Their dynamical properties are also of interest although only a Linear-chain model[2] has been applied to a 15x14 SL of this kind, tracing roughly the basic features of the normal modes of vibration in this system. The unit cell masses of InAs and GaSb are almost equal, resulting in overlapping ranges of phonon frequencies in these compounds. Therefore extented modes may appear in the SL, in contrast to the confined modes in the GaAs/AlAs systems[3], while the interface modes, may belong to the well separated optical phonon frequency ranges of InSb and GaAs. Moreover, the bulk phonon dispersion is experimentally known for all four above mentioned, binary compounds allowing for a comparison of the interatomic interactions between the same ions, in the SL and in the coresponding bulk compounds.

In this communication, we present a more realistic three-dimensional calculation of the lattice dynamics of these SL's using a ten-parameter Valence Overlap Shell Model[4] (VOSM) with modified short-range interactions[5].

The above VOSM parameters have been fitted to the phonon dispersion curves of InAs[6] and GaSb[7] and they are found to be for InAs (GaSb) as follows: ionic and shell charges, $Z=2.00$ (kept constant for both), $Y_1 = 4.930$ (4.277) and $Y_2 = -2.513$ (-3.502) in proton charges, core-sheli coupling constants $k_1 = 13.790$ (15.111) and $k_2 = 4.829$ (4.448), short range interaction parameters $\lambda = 1.718$ (1.717), $kr_1\theta = 0.116$ (0.171) $k'r_1\theta = -0.179$ (-0.195), $kr_2\theta = -0.078$ (-0.052) and $k'r_2\theta = -0.008$ (-0.003) in 10^5dyn/cm.

The short range interactions between the pairs of ions In-Ga and As-Sb are approximeted by the average value of the coresponding In-In and Ga-Ga, and As-As and Sb-Sb interactions respectively, and those of the pairs In-Sb and Ga-As by the same average of the In-As and Ga-Sb interactions. The remaining interactions in the SL are taken to be the same as in the two bulk binary compounds.

The method of calculation of the phonon dispersion in the SL's has been described elswhere[8]. We should only notice here that, in order to account correctly for the self-terms of the final Dymamical matrix for the SL, the method has to be applied separetely to the matrices describing short range and long range interactions respectively before the degrees of freedom concerning the shells are eliminated. The so resulting four matrices are then combined by the elimination of the "electronic" degrees of freedom to give the final Dymamical matrix for the SL. The matrix $\underline{N}$ which transforms the primitive translation vectors of the zinc blende structure to the ones of the SL, the SL primitive cell (the supercell) and the coresponding first Brillouin zone (BZ) are the same as for Ge/Si SL's[9] with simple (P-type) crystallographic unit cell. The Ox' and Oy' axes, referenced bellow, are in the $x_0 + y_0$ and $y_0 - x_0$ directions of the underlying zinc blende structure respectively, which is assumed to have a lattice constant of 6.07A.

Both, 1x1 and 3x3 SL's show the C^1_{2v} (Pmm2) space group symmetry and the SL primitive cells contain four and twelve atoms respectively. They occupy the following four Wyckoff sites all having the mm symmetry: In and Ga atoms at a and d sites, As and Sb at b and c. Each one of the above sites contributes one of A_1, B_1 and B_2 irreducible representations. Since the point group of the wavevector k, for k along the SL axis, is the same as for the center of the zone, the entire phonon branches for $k = (0, 0, k_z)$ have the same symmetry properties as their end points at the zone center. The point group of the wavevector along the Ox'-axis is C_s. The same point group holds also valid for k in the zOx' plane. The symmetry of the A_1 (longitudinal) and B_1 (transverse) modes becomes Δ_1 and the B_2 (transverse) modes turn into Δ_2 symmetry as k developes a nonzero k_x component.

In Fig.1 the phonon dispersion is shown for an 1x1 SL, for k parallel to the SL axis (right panel) and perpendicular to it (left panel). In the central part of the figure we give the frequency of each mode as the direction of the k changes from the one to other direction. The highest frequency A_1 mode lies above the LO branche of InAs and falls into the LO band of GaAs, the mode involving mainly vibrations of the Ga and As atoms. As k changes direction from parallel to perpendicular to the SL axis (remaining very small), the mode remains longitudinal with decreasing frequency and when k becomes parallel to the Ox'-axis the displacement pattern becomes (almost) identical to that of the higher frequency B_1 mode for k parallel to the SL axis. Hence, the anisotropy of the optical phonons[10] is followed by an anticrossing of the modes of the same (Δ_1) symmetry as k changes direction. The remaining Δ_1 (from A_1 and B_1) modes show little dispersion while Δ_2 (from B_2) modes show no dispersion for the same change of k.

In Fig.2 the phonon dispersion is shown for a 3x3 SL for the same directions of the wavevector as in Fig.1. Optic modes of all symmetries for k parallel to the SL axis resulting from bulk optic modes, show almost no dispersion while for k along Ox' the dispersion is larger. Several optic modes resulting from bulk acoustic branches show large dispersion mainly for k along Ox' where many anticrossings occur between modes of Δ_1 symmetry.

The above calculated frequencies are in good agreement with extrapolated into the miscibility gap, far-infrared experimental data for InAs/GaSb quaternary alloys[11], for frequencies above 180cm^{-1}, while for the lower frequency range the available data are not suitable for comparison.

We thank Dr. S. Ves for many heipfull disscusions. This work has been financially supported in part, by the Greek Ministry of Research and Technology.

REFERENCES

1. L.L.Chang and L.Esaki, Surf.Sci. 98,70 (1980)
2. A.Fasolino,E.Molinari and J.C.Maan, Phys.Rev. B33,8889 (1986)
3. E.Richter and D.Strauch, Sol.St.Commun. 64,867 (1987)
4. K.Kunc and O.Nielsen, Comp.Phys.Commun. 17,413 (1979)
5. G.Kanellis,W.Kress and H.Bilz, Phys.Rev. B33,8724 (1986)
6. N.S.Orlova, Phys Stat.Sol. b119,541 (1983)
7. M.K.Farr,J.G.Traylor and S.K.Sinha, Phys.Rev. B11,1587 (1975)
8. G.Kanellis, Phys.Rev. B35,746 (1987)
9. G.Kanellis,to be published in the Proceedings of the NATO-ASI on "Spectroscopy of Semiconductor Microstrucres", 9-13 May 1989, Venezia (Italy), eds. G.Fasol, A. Fasolino and P. Lugli, Plenum Press, London (1989).
10. S.F.Ren,H.Chu and Y.C.Chang, Phys.Rev.Lett., 59,1841 (1987)
11. G.Pickering, J.Electron.Materials 15,51 (1986)

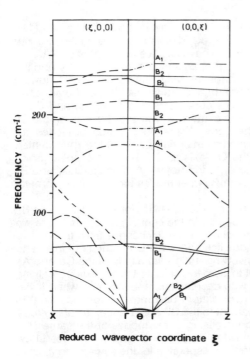

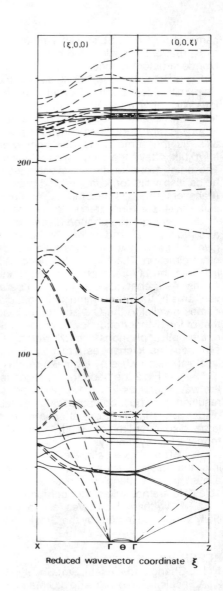

FIG. 1: Phonon dispersion for an 1x1 InAs/GaSb SL for wavevectors papallel (right panel), perpendicular (left panel) to the SL axis and for small wavevectors of intermediate directions (central part). In the right panel dashed lines represent longitudinal modes of A_1 symmetry, while in the left panel they represent modes of mixed polarization and Δ_1 symmetry (vibrations along Ox' and Oz). Solid lines represent transverse modes of B_1 and B_2 symmetry in the right panel, while in the left panel they represent modes of Δ_2 symmetry. Modes of B_2 and Δ_2 symmetry are polarized along Oy'.

FIG. 2: Phonon dispersion for a 3x3 InAs/GaSb SL for wavevectors parallel and perpendicular to the SL axis (right and left panel respectively) and for small wavevectors of intermediate directions (central part). Conventions and symmetries as in Fig.1.

GENERALIZED CONTINUUM APPROACH TO PHONON
MODES IN HETEROJUNCTIONS

A. Greiner, G. Mahler
Universität Stuttgart
Pfaffenwaldring 57
7000 Stuttgart 80

A. Kriman, D. K. Ferry
CSSER, Arizona State University
Tempe, Arizona

1 Introduction

Continuum theory is a well known approximation scheme for the discrete atomic displacement fields of condensed matter. In its original form it is restricted to acoustic fields with constant velocity of sound. Generalizations include "multi-phase" displacement fields which can be transformed to optical and acoustic fields, their respective dispersive corrections, anharmonicity and couplings to other fields.

In this contribution we use such a generalized scheme for an investigation of the reflection coefficient R of a heterojunction as it applies to acoustic and optical fields including dispersive and near-field corrections around the interface.

2 Local Field Theory

It is assumed that there exists a Lagrangian-density $\mathcal{L}$ in terms of the vector field Ψ_i and its first-order derivative with respect to time, $\dot{\Psi}_i$, and finite-order derivatives with respect to the spatial coordinates r_j, $\Psi_{i,j}$ etc.
Then the equations of motion read

$$\frac{d}{dt}\frac{\partial \mathcal{L}}{\partial \dot{\Psi}_\nu} + \frac{d}{dr_i}\frac{\partial \mathcal{L}}{\partial \Psi_{\nu,i}} - \frac{d}{dr_i}\frac{d}{dr_j}\frac{\partial \mathcal{L}}{\partial \Psi_{\nu,ij}} = \frac{\partial \mathcal{L}}{\partial \Psi_\nu} \qquad (1)$$

For a linear theory the Lagrangian is a bilinear form containig a set of parameters constrained by the respective point symmetry. Parameter field patterns appear by allowing for spatial dependencies: They constitute a phenomenological structure model.

As an important special case we consider a heterojunction, defined by parameter fields which are constant in each half-space (1), (2), separated by the interface $r_3 = 0$. The solution of the (linear) equation of motion requires appropriate boundary conditions, which have been derived.

3 Multiphase Displacement Fields

Starting from a biphase displacement-field $\mathbf{u}^{\nu}$ ($\nu=1,2$) and imposing material invariance properties with respect to infinitesimal translations and rotations, we may transform the corresponding Lagrangian with respect to the center-of-mass displacement field, $\mathbf{s}$ and relative displacement-field, $\mathbf{w}$. In the long-wavelength limit the $\mathbf{s}$ and $\mathbf{w}$-fields decouple and approach the acoustic and optical mode, respectively.

For an isotropic medium the equations of motion according to eq. (1) decouple into longitudinal (L) and transversal (T) parts:

$$\rho_o \ddot{\mathbf{s}}_L = (\lambda + 2\mu)\triangle \mathbf{s}_L - (F_A + F_B)\triangle^2 \mathbf{s}_L \tag{2a}$$

$$\rho_o \ddot{\mathbf{s}}_T = \mu\triangle \mathbf{s}_T - F_A\triangle^2 \mathbf{s}_T \tag{2b}$$

$$\rho_{eff} \ddot{\mathbf{w}}_L = -B_L^{ww} \mathbf{w}_L + (\tilde{\lambda} + 2\tilde{\mu})\triangle \mathbf{w}_L \tag{2c}$$

$$\rho_{eff} \ddot{\mathbf{w}}_T = -B^{ww} \mathbf{w}_T + \tilde{\mu}\triangle \mathbf{w}_T \tag{2d}$$

$\lambda^{s(w)}$, $\mu^{s(w)}$ are the Lamé-coefficients, $F_{A,B}$ dispersive corrections, ω_{LO} the optical long-wavelength frequencies, ρ_{eff} = reduced mass density. For a non-polar medium $B_L^{ww} = B^{ww}$. The resulting bulk dispersion-relations $\omega^2(q)$ are obvious expansions in terms of q^2.

4 Acoustic Modes in Heterojunctions

We consider a longitudinal mode, $s_3(r_3)$ impinging perpendicularly on the interface from medium (1). We first assume $(F_A + F_B) \geq 0$, so that $\omega^2(q) = const$ has only two real solutions, but also two imaginary ones. While in the harmonic theory the reflection coefficient of such an interface is constant, we get corrections due to the modified dispersion relation depending on even powers in q. The imaginary solutions are the reasons for near field corrections. Neglecting the imaginary solutions compared to small real solutions q the reflection coeffient may increase or decrease (see Fig.1).

5 Optical Modes in Heterojunctions

The bulk dispersion-relation $\omega^2(q)$ = const. has two solutions in this case. If the optical phonon frequencies ω_{LO} in either bulk materilas equal one another, the reflection coefficient is constant. In the case where the ω_{LO} differ the reflection coefficient equals unity for q=0 and the expansion of the square of it contains all powers in q starting with linear decreasing (see Fig.2).

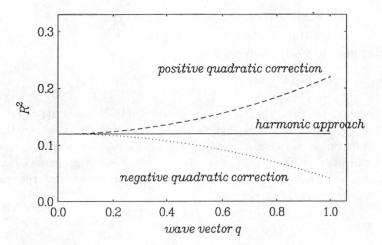

Fig. 1: *LA reflection coefficient*

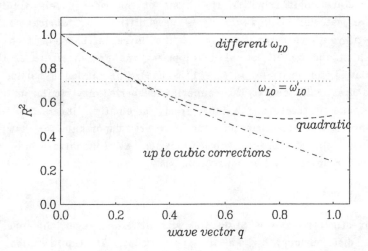

Fig. 2: *LO reflection coefficient*

References

Mahler, G., Kriman, A. and Ferry, D. K., Proc. Picosecond Electronics and Optoelectronics, Salt Lake City 1989

CONFINED OPTICAL MODES IN SI/GE STRAINED LAYER SUPERLATTICES

E. FRIESS, K. EBERL, U. MENCZIGAR AND G. ABSTREITER
Walter Schottky Institut, Technische Universität München
Am Coulombwall, D-8046 Garching, FRG

ABSTRACT

Optical phonons in short period Si/Ge strained layer superlattices (sls) are studied by Raman spectroscopy. Samples grown with lattice constants in the whole range between the Si and Ge bulk values and individual layer thicknesses between 2 and 18 monolayers are investigated. Longitudinal and transverse confined modes are observed in (100) and (110) orientation respectively.

INTRODUCTION

Si/Ge short period sls are of considerable interest due to the possibility of band structure engineering on the basis of the most widely applied semiconductor material Si. However, it is very difficult to fabricate high quality multilayer structures because of the large lattice mismatch of 4 % between Si and Ge. Historical development and present state of the art are reviewed, for example, in Ref. 1) and 2). The lattice dynamics is also altered drastically due to the superlattice periodicity similar to the electronic band structure. Aside from scientific interests the understanding of superlattice phonons is very important for sample characterization. In the present communication we concentrate on the optical phonon properties in various Si/Ge sls.

EXPERIMENTAL RESULTS

We have investigated several Si/Ge sls with different layer thicknesses and strain distributions by Raman spectroscopy at liquid nitrogen temperature. Fig. 1 shows a typical spectrum of a Si_4Ge_{12} sls grown on (100) Ge substrate. Three main features are evident. At about 480 cm^{-1} the Si optical phonon mode appears, which originates from Si–Si bond vibrations within the Si slabs. In this sample the full lattice mismatch of 4 % is accomodated by biaxial strain within the Si layers giving rise to the major contribution of the downward shift of about 45 cm^{-1} from the Si bulk value[3]. On the other hand, the Ge layers are unstrained, therefore the series of Ge modes appears close to the bulk value of about 300 cm^{-1}. At around 400 cm^{-1} a small signal is observed which is caused by Si-Ge vibrations near the interfaces[3]. As already mentioned, optical modes are expected to be confined within the corresponding layers and evanescent in the neighbouring layers. This is due to the fact that Si and Ge LO bulk phonon dispersions do not overlap in energy[4]. Therefore a series of modes similar to standing waves in each slab builds up. These vibrations are influenced only by properties of their host layers and nearly independent from the thickness and properties of the adjacent layers. This is demonstrated for example in Fig. 2 where Si mode energy is plotted versus Ge layer thickness for

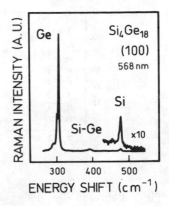

Fig.1 Raman spectrum of optical sls phonons in a Si_4Ge_{18} structure.

constant Si layer thickness of 4 monolayers. We have obtained similar results for the inverse case although Ge modes are only quasi-confined[5] because of the overlap of the Ge LO phonon modes with the Si LA bulk branches. This is expected to lead to small dispersion of the Ge modes.

The dispersive nature, however, is not observed in the experiments. The Ge modes behave as if they are well confined to the Ge layers. With this simple picture of standing waves an effective wave vector in the bulk Brillouin zone $k_{eff} = m \cdot \pi / d_{eff}$ can be assigned to each confined mode where m is the number of half waves which have to

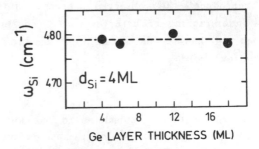

Fig.2 Si mode energy versus Ge layer thickness. The dashed line simply labels constant energy.

fit into the slab and $d_{eff} = d_0 \cdot (n + \delta)$ is the confinement length where d_0 is the thickness of one monolayer, n is the number of monolayers of a slab and δ accounts for penetration into adjacent layers. The mode energy is then expected to be $\omega = \Omega(k_{eff})$ with $\Omega(k)$ being the bulk phonon dispersion. Plotting measured energies versus k_{eff} allows comparison with bulk dispersion relations. Results for Ge optical phonons

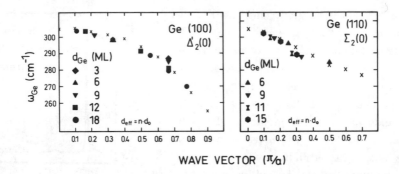

WAVE VECTOR (π/a)

Fig.3 Confined Ge mode energies versus effective wave vector (δ = 0) compared with bulk neutron scattering (crosses)[6] for (100) and (110) sls

Fig. 3 for are presented in sls grown on (110) and (100) Ge subtrates. No strain shifts have to be taken into account because all Ge layers are unstrained in these samples. The Raman data are in excellent agreement with neutron scattering results both for LO (Ge(100)) and TO (Ge(110)) modes. The application of this simple model is, however, limited especially when π/k comes close to d_o. Full three-dimensional calculations[7],[8] are able to account for the details and thus can reproduce the observed spectra quite well. For a quantitative understanding of both Ge and Si confined modes the effect of built-in strain has to be taken into account additionally. This is described elsewhere[3].

REFERENCES

1)   Pearsall, T.P., submitted to Crit. Rev. of Solid State and Material Sci.
2)   Abstreiter, G., III. Int. Symposium on Si MBE, Strasbourg, May 1989, to be published in Thin Solid Films.
3)   Friess, E., Menczigar, U., Eberl, K., and Abstreiter, G., to be publ.
4)   Fasolino, A., Molinari, E., Proc. 3rd Int. Conf. on Modulated Semiconductor Structures, J.Physique Coll. 48 C5 Suppl.11 569 (1988)
5)   Fasolino, A., Molinari, E., Maan, J.C., Phys. Rev. B39, 3923 (1989).
6)   Nielsson G., and Nelin, G., Phys. Rev. B3, 364 (1971).
7)   Alonso, M.I., Cerdeira, F., Niles, D., Cardona, M., Kasper, E., and Kibbel, H., preprint.
8)   Ghanbari, R.A., Fasol, G., Solid State Communication 70, 1025, (1989).

RAMAN SCATTERING FROM ACOUSTIC, OPTICAL AND INTERFACE VIBRATIONAL MODES IN STRAINED SHORT–PERIOD (001) Si/Ge SUPERLATTICES

W. Bacsa, M. Ospelt, H. von Känel and P. Wachter

Laboratorium für Festkörperphysik, ETH Hönggerberg,

CH–8093 Zürich, Switzerland

Abstract: We have investigated strained short–period (8–10 ML) Si/Ge superlattices (SL) grown on a laterally graded (001) Si_xGe_{1-x} alloy buffer layer. The effect of the interfacial structure on the vibrational modes has been studied on SL's with Ge layers ranging in thickness from 4 down to 1 monolayer (ML). Various excitations between 410–490 cm^{-1} appear to be associated with the interfacial structure. In addition to folded LA and optical modes, we have found evidence for a confined Si–like longitudinal acoustic mode.

In the last two years, it has become possible to grow strained Si/Ge superlattices (SL's)[1] by molecular beam epitaxy (MBE). SL's with periods ranging from 50 down to a few monolayers (ML) have been grown on different substrates (e.g. Si(001), Ge(001)/Si, Si_xGe_{1-x}(001)/Si) confining the strain either to the Ge or Si layers, or distributing it symmetrically to both [2]. Lattice vibrational properties have been investigated by Raman spectroscopy showing folded LA phonons, the effect of strain on confined and quasi–confined LO phonons and a number of interface excitations [3].

In the following we shall discuss first order Raman spectra of short period Si/Ge SL's with different layer thicknesses grown by MBE using a non–rotated 3 inch Si(001) substrate. After a Si_xGe_{1-x} buffer layer of approximately 1000 Å thickness, a SL with 490 periods was grown. The average composition of the SL was the same as the one of the buffer. The SL's obtained in this way exhibit a continuously varying Ge layer thickness (1–4 ML) and period (8–10 ML). Furthermore on the Ge deficient side, the total thickness of SL and buffer was just below its critical value [4]. Rutherford backscattering, He ion channeling, transmission electron microscopy (TEM) and Raman spectroscopy (below) show a gradual transition to a fully relaxed state towards the Ge–rich side. The Raman spectra have been measured in backscattering geometry $(z(x+y,all)\bar{z})$ at 295 K, 10 K and in a helium atmosphere using an argon laser with 60–100 mW power at 514 nm. The resolution of the spectrometer was set to 4–8 cm^{-1}.

Fig. 1 shows, in the optical phonon region (250–520 cm^{-1}) a quasi–confined Ge LO mode (291–302 cm^{-1}), a interface excitation (410 cm^{-1}) and a confined Si LO mode (503–518 cm^{-1}). Strain and confinement affect these modes by shifting them from their bulk energies [2]. In the acoustic phonon region (0–250 cm^{-1}) one observes a folded LA phonon (FLA) (155–230 cm^{-1}) which increases strongly in energy as the thickness of Ge layers decreases from 4 to 1 ML. It thus follows more and more the folded LA Si dispersion with dropping Ge content and merges eventually into an additional broader structure appearing at 250 cm^{-1} (Fig.1, top) due to disorder induced LA and second–order scattering [7]. Moreover as the Ge layer thickness decreases below 3 ML the FLA intensity drops sizably, indicating that the phonons experience less and less the SL periodicity. Some of the investigated SL's in this period range show a weak excitation at 350 cm^{-1} (Fig. 2). This excitation can be attributed to a confined LA Si–like mode derived from the part of the LA Si phonon branch extending above the phonon spectrum of Ge. This is consistent with lattice dynamical calculations of SL with similar structures[5].

Short period Si/Ge SL's show further excitations between 410 and 520 cm^{-1}. Some of them have been shown to be confined modes of higher order [3]. Similar excitations in SL's with 1–4 ML thick Ge layers appear to be of different origin (fig. 2.) [3]. Broad structures in the same energy range have been found in alloy crystals [7]. But the SL excitations in Fig. 2 are distinct and show no comparable broadening. A recent TEM investigation [6] has revealed an ordering of interfacial Si and Ge atoms in double layers along (111) directions instead of a statistical distribution. This results in a doubling of the unit cell in the <111> directions. In the Raman spectra of fig. 3 one observes three additional excitations (436, 460, 491 cm^{-1}). It is interesting to note that taking into account the doubling of the unit cell by folding the L points of the Si optical phonon branches into Γ leads to new modes at 428 cm^{-1} (TO) and 497 cm^{-1} (LO). In this regard we can associate the excitations at 436 cm^{-1} and 491 cm^{-1} with the interface ordering and the weak excitation at 460 cm^{-1} with random disorder. This seems to agree with recent lattice dynamical calculations of Alonso et al. [7]. A detailed annealing study of these excitations in Si/Ge SL's with different strain distributions will be given elsewhere.

References:
1. Applied Sciences, Series E: Vol. 160, ed. Y.I. Nissim et al. (1989).
2. M. Ospelt, W. Bacsa, J. Henz, K. A. Mäder, H.von Känel, Superlattices and Microstructures, 4, 717 (1988), 5, 71 (1989); E. Kasper, H. Kibbel, H. Jorke, H. Brugger, E. Friess, G. Abstreiter, Phys. Rev. B, 38, 3599 (1988).

3. E. Friess, H. Brugger, K. Eberl, G. Kötz and G. Abstreiter, Solid State Comm., 69, 899 (1989); W. Bacsa, M. Ospelt, J. Henz, H. von Känel, P. Wachter, E–MRS proceedings (Strasbourg, 1989), Thin Solid Films in press.

4. R. People, J.C. Bean, Appl. Phys. Lett. 49, 229 (1986).

5. M.I. Alonso, M. Cardona, G. Kanellis, Solid State Comm., 69, 479 (1989); J. White, G. Fasol, R. Ghanbari, E–MRS proceedings (Strasbourg, 1989), Thin Solid Films in press; E. Molinari et al. Appl. Phys. Lett., 54, 1220 (1989).

6. E. Müller, H.U. Nissen, M. Ospelt, H. von Känel E–MRS proceedings (Strasbourg, 1989), Thin Solid Films in press.

7. M. I. Alonso and K. Winer, Phys. Rev. B, 39, 10056 (1989).

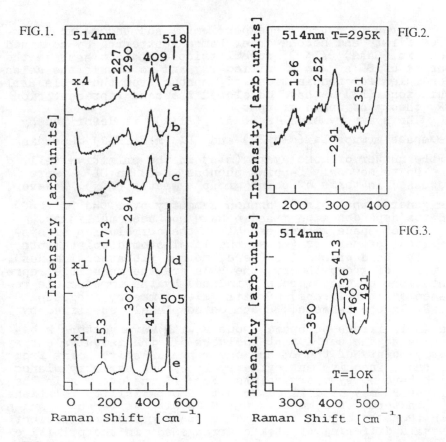

FIG.1: Unpolarised Raman spectra of SL 507/x at 295 K: a) $Si_7Ge_{1.2}$ (/1) b) $Si_{6.8}Ge_2$ (/4) c) $Si_{6.4}Ge_{2.5}$ (/2) d) $Si_{6.4}Ge_3$ (/5) e) $Si_{6.4}Ge_4$ (/3) (Si_nGe_m: n,m [ML]).

FIG.2&3: Raman spectra of SL 507/2 ($Si_{6.4}Ge_{2.5}$, fig. 1.c) between 150–500 cm^{-1}.

PHONONS IN SINGLE CRYSTAL SUPERLATTICE FAMILY $(GaAs)_m(AlAs)_n$

B.H.BAIRAMOV, M.DELANEY[*], R.A.EVARESTOV, T.A.GANT[*],
YU.E.KITAEV, M.V.KLEIN[*], D.LEVI[*], H.MORKOÇ[*]

A.F.Ioffe Physical-Technical Institute, Leningrad,USSR
[*]University of Illinois at Urbana-Champaign,
IL 61801, USA

We report extensive experimental and theoretical study of first- and second-order Raman scattering by confined modes in $(GaAs)_m(AlAs)_n$ superlattices (SL). To analyze the spectra of SL's in unified way and to obtain the selection rules for combinations of phonons from the whole Brillouin zone (BZ) we have developed the band representation (BR) theory of space groups.

The symmetry of SL's $(GaAs)_m(AlAs)_n$ is described by the space groups $D_{2d}^5 (m+n=2k)$ and $D_{2d}^9 (m+n=2k+1)$ depending on the number of monolayers $(m+n)$ in the primitive cell.

Up to now only Γ-point phonons $(k=0)$ in SL's were discussed in terms of point group symmetry (D_{2d}). However, the general analysis of phonon symmetry over the BZ $(k\neq 0)$ and its dependence on m and n have not been published.

BR of space groups establish the correlation of system local properties (here, the local atom displacements) with its band properties (here, normal vibrational modes). In terms of group theory, the BR's are the reducible representations of space group F induced by the irreducible representations (irreps) of its site symmetry subgroups $M_q \subset F$. In the k-basis, BR can be completely specified by the small irreps of space group F with wave vectors k belonging to the set K that includes all the inequivalent symmetry points of the BZ and one representative point from all those inequivalent symmetry lines and symmetry planes which have no symmetry points. For all the other k in the BZ, the BR may be derived from the compatibility relations.

In terms of symmetry, the SL's with different m and n even those belonging to the same space group are distinct crystals differing by atomic arrangement in the primitive cell.

The phonon symmetry analysis with well-known method of constructing the vibrational representation of the cry-

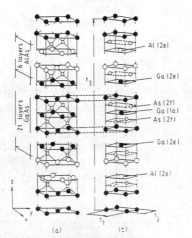

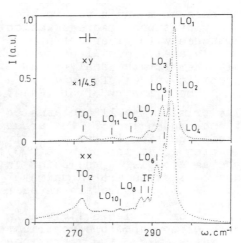

Fig.1 The structure of $(GaAs)_{21}(AlAs)_6$ SL and the positions of the atoms in Wyckoff notation showing cubic (a) and centred tetragonal (b) unit cells.

Fig.2 First-order Raman scattering spectra of confined GaAs LO_1 phonons of $(GaAs)_{21}(AlAs)_6$ SL for depolarized $z(xy)\bar{z}$ (Γ_2) and polarized $z(xx)\bar{z}$ (Γ_1) scattering configurations. T=10K, λ_i=5145Å.

stal and its subsequent decomposition into irreps of crystal factor-group as well as the analysis with so called "site symmetry" method do not allow to solve this problem in general way for arbitrary m and n but require to perform the calculations for each given SL all over again. Moreover, the above methods are very cumbersome especially for the crystals with the large number of atoms in the primitive cell to which SL's also belong.

In the approach used in the present work one needs to construct the BR's only for two space groups D_{2d}^5 and D_{2d}^9 and then to derive the formula for arrangements of atoms in the primitive cell over the Wyckoff positions for arbitrary m and n. Combining them one can easily perform the phonon symmetry analysis of any given SL's.

We have established that there are eight non-equivalent cases of atomic arrangement corresponding to different sets of m and n and constructed the BR's of space groups D_{2d}^5 and D_{2d}^9.

Then we performed the phonon symmetry analysis and established that the types of phonon symmetry for all the SL's belonging to the same space group do not depend on the specific values of m and n. However, the contribution of specific atoms to the phonons with definite symmetry depends on m and n due to the change of atomic arrangement. The vibrational representation, i.e. the number of phonon branches with given symmetry changes as well. The results of the analysis will be published elsewhere. Below, we apply them for the analysis of experiments on the SL with a specific values of m and n.

The symmetry of SL $(GaAs)_{21}(AlAs)_6$ (m+n=27) is D_{2d}^9. The primitive cell contains 54 atoms. The number of vibrational modes is 162. The arrangement of atoms over the Wyckoff position in the primitive cell is the following: the Ga atom is at the point a(000), the site symmetry is $\bar{4}2m$; the As atom is at the point 1c(01/2 1/4)-$\bar{4}2m$, 20 Ga atoms and 6 Al atoms are in pairs at the position 2e(00z)(00$\bar{z}$)- mm and 26 As atoms are in pairs at the position 2f(0 1/2z) (1/2 0$\bar{z}$)-mm (Fig.1).

From the Table 1 given in [1] it is seen, that definite groups of atoms contribute to the phonons of given symmetry. E.g., the vibrational representation at Γ-point is

$26\Gamma_1 (Ga_e^z ; Al_e^z ; As_f^z) + 28\Gamma_2 (Ga_{a,e}^z ; Al_e^z ; As_{c,f}^z) + 54\Gamma_5 (Ga_{a,e}^{xy} ; Al_e^{xy} ; As_{c,f}^{xy})$, where the

superscripts are the components of atomic displacements, and subscripts - are the Wyckoff positions occupied by the atoms contributing to the mode of a given symmetry. We see, for example, that only z-components of Ga and Al atoms at e positions as well as of As at f-positions contribute to Γ_1-mode. Fig.2 shows the first-order Raman spectra of $(GaAs)_{21}$-$(AlAs)_6$ SL in the region of confined LO_1 modes.

The analysis of phonons from the other points of BZ whose combinations could be seen in second-order Raman [2,3] scattering considerably increase the number of independent atomic groups.

In conclusion, it is shown that the experimental study of first- and second-order Raman scattering in SL's and their analysis with BR's theory of space groups potentially increases the analitical applications of the Raman scattering and opens up the new possibilities to obtain broad information on atomic structure of SL's. Morever, the monocrystal nearly lattice matched SL's-being the series of distinct man -made crystals-are the good model system to enlarge the field of applications of BR's theory itself.

REFERENCES

1 Bairamov, B.H., et al. Superlattices and Microstructures, 5, (1989) (in press).
2 Sood, A.K., Menendez, J., Cardona, M., Ploog, K., Phys.Rev. B, 32, 2, 1412-1415 (1985).
3 Bairamov, B.H., et al. Zh.Eksp.Teor.Fiz. 95, 2200-2209 (1989).

STUDY OF DIELECTRIC RESPONSE FUNCTIONS OF SUPERLATTICES USING RAMAN AND FAR INFRARED SPECTROSCOPY

T Dumelow[1,2], A Hamilton[1], T J Parker[1], B Samson[2], S R P Smith[2], D R Tilley[2], D Hilton[3], K Moore[3], and C T B Foxon[3].

1)   Department of Physics, Royal Holloway and Bedford New College, Egham, Surrey TW20 0EX, UK.
2)   Department of Physics, University of Essex, Wivenhoe Park, Colchester, Essex CO4 3SQ, UK.
3)   Philips Research Laboratories, Cross Oak Lane, Redhill, Surrey RH1 5HA, UK.

INTRODUCTION

In the optic phonon region, the dielectric properties of a superlattice follow the bulk properties of its constituent layers provided the superlattice period is sufficiently large. It then behaves as a single uniaxial medium[1]. In short period superlattices, phonon confinement shifts the LO and TO frequencies from their bulk values, gives rise to additional modes, and generally complicates the optical properties. Chu and Chang[2] describe a model for the resulting dielectric function which relates it simply to the confined phonon frequencies. Thus the in-plane component of the dielectric tensor has poles at the confined TO frequencies and the out-of-plane component has zeroes at the confined LO frequencies.

In this paper we use confined optic phonon frequencies, measured by Raman spectroscopy, in the above model to calculate theoretical far infrared oblique incidence spectra. These are compared with experimental data.

EXPERIMENTAL MEASUREMENTS

Each sample had 1 μm of superlattice of the type $(GaAs)_n(AlAs)_n$ (n is the number of monolayers per superlattice period) bounded by 0.1 μm GaAs cladding layers and grown by molecular beam epitaxy on a GaAs substrate. Using 514.5 nm exciting radiation, Raman spectra were recorded, with the top cladding layer removed, in near back-scattering geometry at ~325 K as previously reported[3]. Far infrared power reflectivity measurements were made at 77 K and a 45° angle of incidence using a Michelson interferometer.

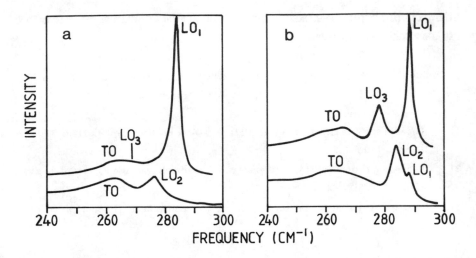

Fig. 1: Room temperature Raman spectra for (a) n=4 and (b) n=6 samples. The polarisations are $z(x,y)\bar{z}$ (top curves) and $z(x,x)\bar{z}$ (bottom curves).

Fig. 1 shows Raman spectra, in the GaAs optic phonon region, for the n=4 and n=6 samples in $z(x,x)\bar{z}$ and $z(x,y)\bar{z}$ polarisations (where x=(100), y=(010), and z=(001)), showing B_2 (LO_1 and LO_3 modes) and A_1 (LO_2 modes) symmetries respectively. Some TO features are also seen, although they are forbidden in strict back scattering geometry. The observed frequencies[3] are in agreement with a linear chain model calculation.

Fig. 2 shows s- and p-polarised far infrared reflection spectra in the same region. We also show calculated spectra, using first order oscillator strengths derived from the LO_1-TO_1 splittings and temperature corrected Raman frequencies wherever available[3]. We have observed a series of confined mode features whose frequencies are in agreement with those observed using Raman spectroscopy, despite some discrepancy in the n=2 LO_1 value. Good overall agreement between experimental and theoretical curves is obtained, showing that the dielectric function must take the form described by Chu and Chang[2].

ACKNOWLEDGEMENTS

The authors wish to thank the UK Science and Engineering Research Council for supporting this work and research studentships for AH and BS.

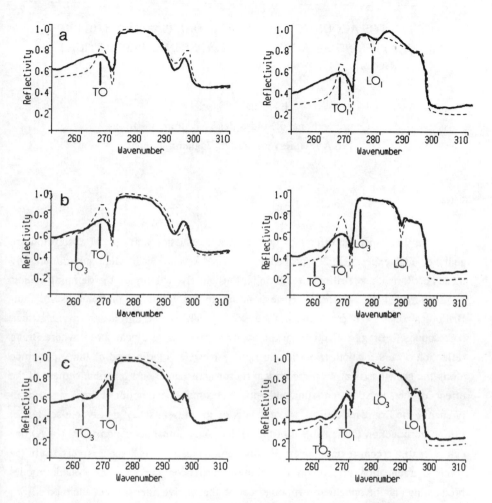

Fig. 2: *Oblique incidence experimental 77 K far infrared reflectivity spectra for samples with (a) n=2 (b) n=4 and (c) n=6. Experimental data are shown by solid lines, and calculated spectra by dashed lines. Left hand curves show s-polarised spectra and right hand curves show p-polarised spectra. Marked frequencies are the confined phonon values used in the calculation.*

REFERENCES

1)   Raj N. and Tilley D.R., Solid State Communications <u>55</u>, 373 (1985).

2)   Chu H. and Chang Y-C, Phys. Rev. <u>B38</u>, 12369 (1988).

3)   Dumelow T., El Gohary A.R., Hamilton A., Maslin K.A.,, Parker T.J., Raj N.,
      Samson B., Smith S.R.P., Tilley D.R.T., Dobson P.J, Foxon C.T.B., Hilton D.,,
      and Moore K., Materials Science and Engineering <u>B</u>, in press.

EFFECTS OF ACOUSTIC AND OPTICAL DAMPINGS ON THE LIGHT SCATTERING BY ACOUSTIC PHONONS IN SUPERLATTICES

J. He and J. Sapriel

Centre National d'Etudes des Télécommunications

196 Av. H. Ravera, 92220 Bagneux, France.

In this paper the effects of the acoustic attenuation and optical absorption are analyzed and experimentally probed in GaAl/AlAs and $Si/Si_{0.5}Ge_{0.5}$ superlattices by means of Raman and Brillouin scatterings. Particularly striking are the damping effects for scattering wave vectors q corresponding to the limits of the successive mini-Brillouin zones. We particularly focus on q values ranging around π/D, which corresponds to Bragg reflection for the acoustic phonons, and near $2\pi/D$, where Bragg reflection occurs for both acoustic and light waves (D is the period of the superlattice along the growth axis z). In these critical regions one can expect an enhancement of the effects of the acoustic attenuation and optical absorption since both of them significantly change the Bragg reflections. More details on the present study are given elsewhere [1]. For an introduction to the subject of the folded longitudinal acoustic modes (FLA) one can refer to a recent review paper[2]. In this communication we'll successively study the acoustic attenuation effects on the phonon dispersion curves, the acoustic modes broadening and the optical absorption effect on the relative intensities of the modes.

The energies of the acoustic modes are measured by light scattering at different laser wavelengths in a $Si-Si_{0.5}Ge_{0.5}$ superlattice ($d_{(Si)} = d_{(SiGe)} = 140$ Å) and are refered by crosses on Fig1. The solid lines correspond to the values calculated in the case without acoustic attenuation. At low frequencies ($\omega < 6cm^{-1}$) the crosses are in good agreement with the solid line and a gap appears[3] (~1 cm^{-1}). For $\omega > 12cm^{-1}$ we found experimental points within "the gap". Actually at these frequencies the acoustic attenuation can no longer be neglected . The acoustic damping makes the modes at the Brillouin-zone edge shift into the "forbidden gaps". In Fig 2 are represented the calculated dispersion curve (Fig 2-a) and attenuation-frequency relation of the acoustic phonons in the considered $Si-Si_{0.5}Ge_{0.5}$ superlattice. The increase of the attenuation at light frequency changes

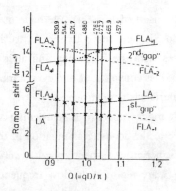

Fig 1

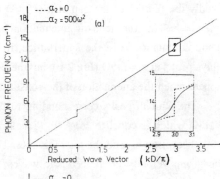

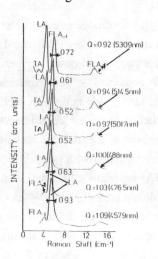

Fig3

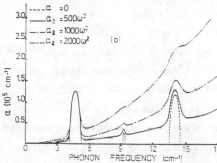

Fig 2

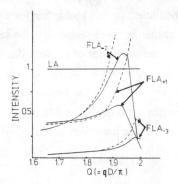

Fig 4

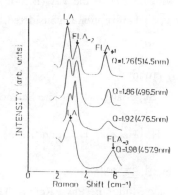

Fig 5

markedly the feature of the gap at k = 3π/D (see the insert of Fig 2-a). We have used $\alpha_1 = 0$ and $\alpha_2 = A\omega^2$ for the acoustic attenuation in Si and $Si_{0.5}Ge_{0.5}$ respectively. As to the acoustic attenuation α in the superlattice, it increases with the frequency and always displays peaks at k = nπ/D (Fig 2-b), but the peaks are less conspicuous and even almost disappear when the attenuation of the constitutive materials is very strong.

The Raman peaks corresponding to the FLA modes undergo a broadening Δω (FWHM) which is equal to $2\alpha V_g$, $4\alpha_{op}V_g$ and $2\pi V_g/l$, due to the acoustic attenuation α, the optical absorption α_{op} and the total thickness l of the superlattice respectively; $V_g = d\omega/dk$ is the group velocity of the acoustic waves. In Fig 3 are reported the low frequency spectra of the superlattice with different excitation laser wavelengths. For low-frequency modes the linewidth displays a minimum at Q = 1 due to the flatness of the acoustic dispersion curves. This is particularly clear on the upper mode of the doublet.

In Ref 4 the relative intensities of the different modes have been calculated by taking into account the modulation of the photoelastic, acoustic, and optical properties in the superlattice, but the optical absorption has been neglected. A good agreement between the theory and the experiments has been obtained for a wide range of scattering wave vectors, not too close to the Brillouin-zone boundaries[5]. However one can expect a significant effect of the optical absorption on the relative intensities for scattering wave vectors near Q = 2, 4, 6. Fig 4 shows the intensity calculated with (solid lines) or without (dashed lines) optical absorption for a GaAs/AlAs superlattice ($d_1 = 167$ Å, $d_2 = 458$ Å) for a scattering wave vector near the second-Brillouin edge (Q ~ 2). When Q is very close to 2, the intensities of the folded modes with respect to the Brillouin line are much reduced by the optical absorption. The Raman spectra corresponding to this Q range are shown in Fig 5. One can notice on Fig 5 that when Q tends to 2, FLA_{-2} increases, reaches a maximum and then decreases with respect to the LA (Brillouin line). This is indeed the behaviour predicted when the optical absorption is considered.

References

[1] He,J., Sapriel,J., and Azoulay, R., Phys. Rev. B, 40, (1989).

[2] Sapriel, J. and Djafari-Rouhani, B., "Vibrations in Superlattices" Surface Science Reports 10, 189 (1989).

[3] Brugger, H., Reiner, H., Abstreiter, G., Jorke, H., Herzog, H.J. and Kasper, Superlattices and Microstructures, 2, 451 (1986).

[4] He, J., Djafari-Rouhani, B. and Sapriel, J., Phys. Rev. B, 37, 4086 (1988).

[5] Sapriel, J., He, J., Djafari-Rouhani, B., Azoulay, R. and Mollot, F., Phys. Rev. B, 37, 4099 (1988).

BANDGAPS IN THE PHONON EMISSION SPECTRA OF METALLIC SUPERLATTICES

H. Waelzel, P.Berberich, J. M. Triscone[1] and M. G. Karkut[1]
Physik–Department E10 TU München, D–8046 Garching, FRG
[1]Section de Physique, Université de Genève 4, CH–1211 Genève 4, Switzerland.

Phonons in superlattices (SL) have interesting anomalies of their dispersion curves. Bragg–scattering at the superstructure leads to mini–zones within the Brillouin–zone according to the Bragg condition $q = n\pi/\Lambda$ (n = 1,2,...), where Λ is the modulation wavelength of the SL. The phonon branches split at the zone boundaries and band gaps open up. These bandgaps have been studied in recent years by phonon transmission and by Raman scattering spectroscopy.[1] Here we present measurements of the phonon emission spectra of Joule heated metallic superlattices, which also show up these characteristic bandgaps.

The Pt/Pd– and V/Mo–superlattices were made by magnetron sputtering in an UHV chamber. The native oxide layer of the (111) Si substrate has not been removed before evaporation. The substrate was held at about 700 °C being the optimum temperature for coherent growth of a V/Mo–SL on (100) MgO.[2] X–ray diffractograms revealed that the V/Mo–SL was primarily (100) textured with $\Lambda = 75$ Å, while the Pt/Pd–SL turned out to be polycrystalline with $\Lambda = 85$ Å. The experimental setup for measuring the phonon emission spectra at 1 K is shown in the inset of Fig. 1. The metallic SL (area 1x1 mm^2) is Joule heated by voltage pulses of 100 ns duration. Longitudinal (L) as well as transverse (T) phonons are detected by a superconducting Al–I–Al tunnel junction on the opposite site of the crystal. The Si:B crystal acts as a phonon filter which is tuned by applying uniaxial stress in $(1\bar{1}0)$ direction.[3]

The solid lines in Fig. 1a and b show the signal decrease ΔS of the longitudinal phonons for two different input powers, which gives a qualitative picture of the emitted spectrum of the Pt/Pd–(6x42.5 Å/42.5 Å)–SL. Above the $2\Delta_D$–threshold at 110 GHz there are three acoustic thickness resonances of the

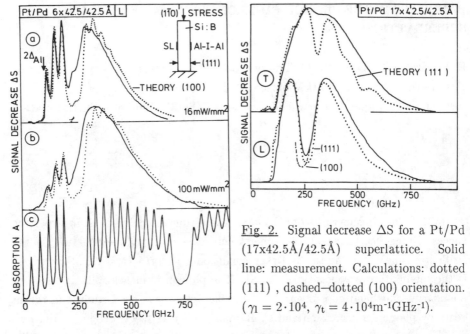

Fig. 2. Signal decrease ΔS for a Pt/Pd (17x42.5Å/42.5Å) superlattice. Solid line: measurement. Calculations: dotted (111), dashed–dotted (100) orientation. ($\gamma_l = 2 \cdot 10^4$, $\gamma_t = 4 \cdot 10^4 \text{m}^{-1}\text{GHz}^{-1}$).

Fig. 1. (a) and (b): Signal decrease ΔS for a Pt/Pd (6x42.5Å/42.5Å) superlattice for 2 different input powers. Solid line: measurement. Dotted line: calculation for (100) orientation of SL. (c) Calculated absorption A. Inset: Experimental setup.

total multilayer. The first bandgap is clearly resolved at 250 GHz. Note that there is a small resonance within the bandgap which represents a surface mode. We calculated the spectra by using a transfer matrix technique analogous to optical multilayer filters[4]. Assuming a lossy medium due electron–phonon interaction we calculated the transmission T, the reflexion R, and from this the absorption A = 1–R–T of the phonon intensity as a function of frequency f. For the amplitude attenuation coefficient of the phonons by electrons we used $\alpha = \gamma \cdot f$, γ being an adjustable parameter. A(f) is shown in Fig. 1c for $\gamma = 2 \cdot 10^4$ m^{-1}GHz^{-1} with sound velocities for the (100) direction of the constituent metals. According to detailed balance the absorptivity and the emissivity of the film should be the same. Then the emission spectrum of the SL at a temperature T_1 into the substrate at T_0 should be proportional to $A(f) \cdot f^3 \cdot [n(f,T_1) - n(f,T_0)]$, n(f,T) being the Bose–Einstein distribution function. This spectrum was folded with the spectral response of the spectrometer[3] to obtain $\Delta S(f)$. The resulting curves are shown in Fig. 1 as

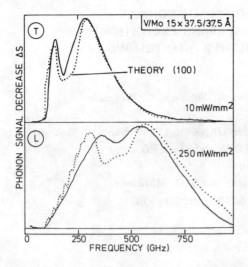

Fig. 3: Signal decrease ΔS for a (100)–V/Mo–(15x37.5Å/37.5Å)–SL. Solid line: measurement. Dotted line: calculation.

($\eta = 2 \cdot 10^4$, $\gamma_t = 4 \cdot 10^4 \text{m}^{-1}\text{GHz}^{-1}$).

dotted lines. T_1 was treated as an adjustable parameter. The position of the band-gap is smaller by 4% than expected. For (111) orientation the difference is 12%.

Fig. 2 shows corresponding measurements (solid lines) for a Pt/Pd–SL with the same modulation wavelength as in Fig.1 but with more layers. Here, the calculated positions of the bandgaps agree better with the measurements if (111) orientation (dotted line) of the superlattice is assumed than for (100) orientation (dashed dotted line). We could resolve two bandgaps in the spectrum of the transverse phonons at 100 GHz and 320 GHz, respectively, the positions being in fair agreement with calculated ones. In comparison with the calculation the bandgaps are getting smaller and smaller with increasing frequency indicating that the $\alpha(f)$ has a stronger frequency dependence than assumed. Indeed, a f^3–dependence is more appropriate.

Fig.3 shows the emission spectra of a (100) V/Mo (15x37.5 Å/37.5 Å) SL. The first bandgap was resolved for both phonon modes. Calculated spectra are shown as dotted curves. While the positions of the bandgaps agree within 10%, their shape is inconsistent with the calculations indicating the presence of structural defects, e.g. dislocations and interface roughness by alloying of the metals.

References
1. For a recent review, see Tamura S., Phonon transmission through periodic, quasiperiodic and random superlattices, these conference proceedings.
2. Karkut M.G., Ariosa D., Triscone J.M., Fischer O., Phys. Rev. B32, 4800 (19
3. Berberich P. and Schwarte M., Z. Phys. B64, 1 (1986).
4. J.M. Eastman: In Physics of Thin Films, Vol 10, ed. by G. Haas and M.H. Francombe (Academic Press, New York, 1978), p. 167.

RAMAN SCATTERING FROM A COUPLED
LO-PHONON-HOLE-INTERSUBBAND EXCITATION IN
GaAs-Ga$_{0.57}$Al$_{0.43}$As MULTIPLE QUANTUM WELLS

M. Dahl, J. Kraus, B. Müller, G. Schaack, G. Weimann[*]

Physikalisches Institut der Universität Würzburg,
Am Hubland, 8700 Würzburg, FRG
and
[*]Walter-Schottky-Institut der T.U. München,
Am Coulombwall, 8046 Garching, FRG

INTRODUCTION

From the very first investigations of semiconductors by inelastic light scattering the interaction of phonons and electronic transitions was one of the prominent topics of research[1]. In polar semiconductors as GaAs the macroscopic electric field of longitudinal optical (LO) phonons normally results in a strong coupling to electronic excitations of collective character. Another, but much weaker interaction mechanism is of deformation potential (polaron) type and active in both polar and non-polar materials. From Raman scattering experiments in non-polar semiconductors[2] and also from the investigation of the line shape of the TO-phonon in GaAs[3] (both heavily doped) it is well-known, that such a weak coupling of a discrete phonon line to a quasicontinuum of electronic excitations results in Fano-interference phenomena. The interaction manifests itself as a small renormalisation of frequency (usually a softening) and lifetime (broadening) of the phonon.

EXPERIMENT

Our sample was a p-type modulation-doped GaAs-Ga$_{0.57}$Al$_{0.43}$As multiple quantum well heterostructure of well-width L_z = 10 nm (10 periods). Be-doped layers of 3.5 nm in the center of the barriers are separated from the GaAs-wells by spacers of 24.5 nm. The high-mobility two-dimensional hole gas has a density of $2.4 \cdot 10^{11}$ cm^{-2} (T = 77 K).

The resonant Raman scattering experiments were performed in backscattering geometry at 2 K and with magnetic fields up to B = 7.5 T oriented normal to the heterostructure layers.

RESULTS

With the laser tuned in resonance with the first excited conduction subband state we could reproduce well the results found by Heiman et al.[4] for the $h_0 \to h_1$-heavy hole intersubband transition of an equivalent sample. Under resonance conditions for the second excited conduction subband we observed another electronic excitation, which we identify as $h_0 \to h_2$.

In the case of hole intersubband transitions a frequency difference (depolarization shift) between excitations of collective character ("polarized spectra") and single particle transitions ("depolarized spectra") is almost not existent or - depending on the carrier density - small[5]. This agrees well with our observations, which indicate, that for the coupling between the LO-phonon and transitions in the valence subband system depolarization field effects are of low importance. Therefore the interaction is of nearly pure polaron type and we were able to observe coupling phenomena just as a consequence of an accidental coincidence of the sharp phonon line and the broad $h_0 \to h_2$- transition (line width of about 20 cm^{-1}). This is in conspicuous contrast to results found for conduction subband excitations in n-doped heterostructures[6], where depolarization field effects are dominating.

The observed $h_0 \to h_2$-transition energy is tunable by an external magnetic field and decreases with growing B. In our sample the maximum of the electronic excitation agrees with the LO-phonon frequency at $B \approx 7$ T.

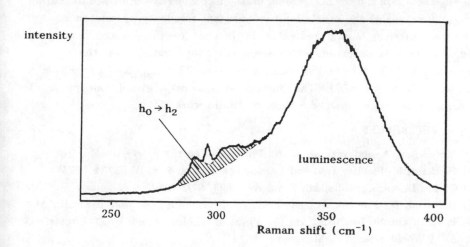

Fig. 1: Raman spectrum obtained with an excitation frequency of 14158 cm^{-1} (1.7553 eV) at B = 6.5 T; scattering geometry $z(xx)\bar{z}$ ("polarized spectrum")

Fig.1 shows a typical spectrum as obtained for B = 6.5 T using a laser frequency of about 60 cm^{-1} above resonance condition. The strong line in the right part of Fig.1 is due to luminescence from the second excited conduction subband to the h_0-valence subband. On the left flank of this line scattering intensity from the $h_0 \rightarrow h_2$-transition can be seen with a notch in the line center as an effect of interference with the LO-phonon. Line shapes of this Fano-type are discussed for example in Ref.7. It is obvious, that only phonons with wavevector q $\approx$ 0 (phonons from allowed and intrinsic forbidden scattering) are able to decay resonantly into $h_0 \rightarrow h_2$-transitions and contribute to interference. So the intensity of the residual LO-phonon peak in the notch of the Fano-profile should originate primarily from phonons with finite wavevector q (phonons from extrinsic forbidden, impurity induced scattering). The identification of this phonon as a quantum well excitation is derived from its resonance behaviour.

The Fano-profile was fitted assuming a Gaussian-line shaped $h_0 \rightarrow h_2$-transition and using well documented relations[7]. The "asymmetry parameter" $\tilde{q}$ is determined by the ratio of the scattering amplitudes of the coupled excitations and was found to depend distinctly on the laser frequency ($-0.1 \geq \mathrm{Re}(\tilde{q}) \geq -1$). The simultaneous slight variation of the most of the other parameters can be assumed to be caused partly by sample inhomogeneities. Therefore the magnetic field dependence of the real and imaginary part of the phonon self-energy Δ_{ph} and Γ_{ph} and of the frequency difference between the $h_0 \rightarrow h_2$-transition and the LO-phonon must be considered for a fixed excitation energy (reasonably the nearly field independent frequency of resonance).

A softening and a reduction in lifetime of the phonon increasing with the field is observed, which seem not to saturate, when the $h_0 \rightarrow h_2$-transition and the LO-phonon coincide at about 7 T ($\Delta_{ph}(B = 7\,T) = -5\,cm^{-1}$, $\Gamma_{ph}(B = 7\,T) = 4.5\,cm^{-1}$). The further dependence beyond the point of resonant interaction must be a topic of future work.

REFERENCES

1) Mooradian, A. and Wright, G. B., Phys. Rev. Lett. 16, 999 (1966)
2) Cerdeira, F., Fjeldly, T.A. and Cardona, M., Phys. Rev. B8, 4734 (1973)
3) Olego, D. and Cardona, M., Phys. Rev. B23, 6592 (1981)
4) Heimann, D., Pinczuk, A., Gossard, A. C., Fasolino, A. and Altarelli, M., Proc. of the 18. Int. Conf. on the Phys. of Semicond., ed. by O. Engström, 617 (World Scientific, Singapore 1987)
5) Ando, T., J. Phys. Soc. Jap. 54, 1528 (1985)
6) Pinczuk, A., Worlock, J.M., Störmer, H. L., Dingle, R., Wiegmann, W. and Gossard, A. C., Sol. State Comm. 36, 43 (1980)
7) Bechstedt, F. and Peuker, K., phys. stat. sol. (b) 72, 743 (1975)

EFFECTS OF SUBSTRATE MISORIENTATION ON THE INTERFACE BROADENING OF GaAs/Al$_{0.3}$Ga$_{0.7}$Al SUPERLATTICES

Y.Chen[*,+], Y.Jin[+], X.Zhu[+], and S.L. Zhang[+]
[*]China Center of Advanced Science and Technology(World Laboratory)
and
[+]Department of Physics, Peking University, Beijing 100871, China

There is considerable current interest in the nature of the interfacial structure in GaAs/Al$_x$Ga$_{1-x}$As quantum wells and superlattices[1-8]. An example of this is the increased control of the interface quality during the growth process with molecular beam epitaxy(MBE). It has been shown that a growth interruption at GaAs-Al$_x$Ga$_{1-x}$As interfaces leads to a clear smoothing of the interface roughness[6], and that an appropriate setting of the growth temperature can minimize the interface broadening[4]. So far most of the studies have been performed on the structures grown on GaAs (001) oriented substrate. Here we report an experimental investigation of the effects of the substrate misorientation on the interface broadening of GaAs/Al$_{0.3}$Ga$_{0.7}$ superlattices. Raman scattering spectra are analyzed of the superlattices grown on GaAs (001) substrate exactly oriented or slightly misoriented. The results are discussed in terms of broadening at GaAs-Al$_x$Ga$_{1-x}$As interfaces.

The studied samples are two GaAs/Al$_{0.3}$Ga$_{0.7}$As superlattices grown by MBE on GaAs (001) substrate exactly oriented(Sample A) and 4° misoriented toward [110] direction(Sample B). The growth was monitored on Sample B by using real-time reflection high energy electron diffraction(RHEED)[8]. In brief, the two samples consist of 160 periods of alternating GaAs and Al$_x$Ga$_{1-x}$As layers. The nominal values are 0.3 for the Al content and 31Å for the thicknesses of GaAs and Al$_{0.3}$Ga$_{0.7}$As sublayers(d_1 and d_2). The effective period $d=d_1+d_2$ of each sample was determined with a computer controlled simple X-ray diffractometer.

On Fig.1 are shown the X-ray diffraction patterns of the superlattice ±1 satellites surrounding the (002) principal reflection(The doublets of the satellites were due to the spectral distribution of the X-ray source). The effective period measured is 62Å for Sample A, as expected from the growth condition recorded by RHEED and 67Å for Sample B, a value which is larger than that of Sample A. As for the X-ray diffraction line width, the satellites of Sample B are better resolved than that of Sample A,which indicates a better interface quality of the sample with the misoriented substrate.

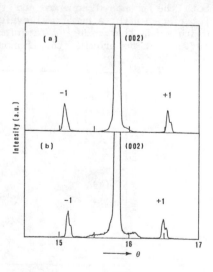

Fig.1 X-ray diffraction patterns of Sample A(a) and Sample B(b) in the vicinity of the (002) principal reflection.

The Raman scattering experiments were performed at liquid nitrogen temperature in backscattering configuration of $z(xy)\bar{z}$ on (001) faces. The incident light was 514.5 nm line of an Ar^+ laser with a typical power of 400mW. The scattered light was analyzed by a Spex double monochromator(1 m focal length) and detected by a conventional photon counting system.

Fig. 2 displays Raman spectra in the frequency region of optical phonons. Four peaks are identified for both samples. The peaks labeled LO_1, LO_3 and LO_5 are attributed to the optical phonons confined in GaAs layers. The peak labeled LO_A is the GaAs-like LO phonon arising from the $Al_xGa_{1-x}As$ barriers. Clearly the phonon features of Sample B have been shifted with respect to the corresponding ones of Sample A. This may be resulted from a derivation of the period thickness, of the force constant[10], and/or of the interface width[4]. Since the change of the crystal symmetry due to the 4° misorientation should not produce a noticeable phonon frequency shift, we ignore this effect in the following discussion. As obtained above for the X-ray diffraction data, the substrate misorientation has resulted in a thicker effective period, which may correspond to derivations of the well and/or barrier thicknesses. We found that within the linear chain model, the observed frequency shifts of the $LO_m(m=1,3,5)$ phonons could be reproduced by increasing the GaAs layer thickness by two monolayers.

Another plausible explanation of the observed frequency shifts is due to the interface broadening model. Previously Jusserand and his collaborators have introduced an interface broadening with an "erf profile" for the local aluminum concentration along the growth direction[4], in order to interpret the growth temperature dependence of the phonons frequencies and Raman intensities. They showed that both the phonon frequency and the Raman line width depend on the interface width. A numerical evaluation as a function of the interface width shows that reducing interface width by one or two monolayers is sufficient to provide the phonon frequency shifts induced by the

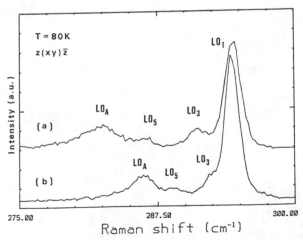

Fig.2. Raman spectra in the region of optical phonons of Sample A(a) and Sample B(b).

substrate misorientation. This indicates that Sample B has interfaces much sharper than that of Sample A. Such an attribution is supported by the fact that the Raman line widths of Sample B are all narrower than that of Sample A, in accordance with the X-ray diffraction and photoluminescence data[8,9]. Thus we conclude that a misorientation of the substrate by 4° toward to [110] direction has an effect of reducing interface broadening.

The reduction of the interface broadening of Sample B can be understood in terms of growth induced atomic-steps at $GaAs-Al_xGa_{1-x}As$ interfaces. It is known that the atomic-steps structure crucially depends on the growth condition, e.g. on the growth temperature and the growth interruptions. For an optimized grown on GaAs (001) substrate, the lateral size of the atomic steps structures is about 200Å at $Al_xGa_{1-x}As$-on-GaAs interfaces(rough), but is much smaller at $GaAs$-on-$Al_xGa_{1-x}As$ interfaces(pseudo-smooth). In the case of using misoriented substrate, a 4° misorientation toward [110] direction gives rise to regularly spaced steps, serving as nucleation centers during interface formations. Although the sticking coefficient and the thermodiffusion length of Ga and Al have not been changed, the misoriented induced steps might change the growth dynamics, in particularly, for the interface formations. The results suggest that in average large smooth areas can be formed at interfaces with the choice of 4° substrate misorientation toward [110] direction. As previously discussed by Massies et al, a detail analysis of the effects of the substrate misorientation on the growth dynamics of the interfaces should consider the anisotropy of Ga and Al surface diffusion coefficient, taking into account the respective role of arsenic dangling bond and steps direction at the surface.

In summary we have investigated the effects of the substrate misorientation on the structural properties of the $GaAs/Al_{0.3}Ga_{0.7}Al$ superlattice. The results of the Raman scattering characterization have shown that the interface quality of the superlattices could be improved by using the 4° misorientation of the GaAs (001) substrate toward [110] direction.

Acknowledgment: One of us (Y.C.) would like to acknowledge J.Massies for providing high quality samples and for useful discussions.

References:
1) C.Weishbuch, R.Dingle, A.C.Gossard and W.Weigmann, Solid State Commun. 38,709(1981)
2) G.Bastard, C.Delalande, M.H.Meynadier, P.M.Frijink and M.Voos, Phys.Rev. B29,7042(1984)
3) C.Delalande, M.H.Meynadier,and M.Voos, Phys.Rev. B30,42(1985)
4) B.Jusserand, F.Alexandre, D.paquet, and G.Le Roux, Appl.Phys.Lett. 47,310(1985)
5) P.Auvray, M.Bandet, and A. Regreny, J.Appl. Phys.62, 456(1987)
6) M.Tanaka,H.Icinose,T.Furuta,Y.Ishida, and H.Sakaki,in Proceedings of MSSIII, Montpellier(1987)
7) Y.Chen,R.Cingolani,J.Massies, G.Neu, F.Turco, and J.C.Garcia, Il Nuovo Cimento D10, (1988)
8) C.Neri,C.Deparis,G.Neu,J.Massies, Y.Chen,and B.Gil,in Proceedings of Euro MBE (1989)
9) Y.Chen, J.Massies et al, unpublished
10)B.Jusserand, D.Paquet, and A.Regreny, Phys. Rev.B30,6245(1984)

6

Electron-Phonon Interaction

ELECTRON-PHONON INTERACTION IN SEMICONDUCTOR HETEROSTRUCTURES

W. Dietsche

Max-Planck-Institut für Festkörperforschung
7000 Stuttgart 80, Federal Republic of Germany

It is well known that layers of electrons can form near semiconductor interfaces. The most important examples of such systems are the MOS devices (Metal-Oxide-Silicon) and the GaAs-AlGaAs-heterostructures. These layers behave in many respect like two dimensional electron gases (2DEG).[1] The interaction of the 2DEG with phonons is important because it is the path by which most of excess energy in the 2DEG relaxes. The absorption and emission of phonons is also of basic interest because of the different dimensionalities of the two types of excitations. In this contribution I review our work on the emission and absorption of acoustical phonons by 2DEG's.[2-4] Since most of our work used superconducting tunnel junctions, we did not study effects in magnetic field. Such experiments will be reviewed by L.J. Challis.[5]

The emission of ballistic phonons by GaAs-heterostructures was observed by Chin et al.[6] but no spectral information was obtained. This was possible in our work where we used superconducting tunnel junctions as detectors and generators of phonons. Employing phonon imaging we measured the angular dependence of the phonon absorption and determined the type of the relevant electron-phonon interaction.

PHONON EMISSION IN MOS STRUCTURES

We studied the phonon emission in MOS structures by passing a current through the 2DEG and observing the emitted phonons. The lowest subband in (100) Si is circular in k-space and is filled up to a Fermi vector $k_F=(2\pi N_s/g_V)^{1/2}$ where N_s is the electron density and $g_V=2$ is the valley degeneracy. The probability of phonon absorption or emission can be calculated using Fermi's golden rule:

$$\Gamma(\mathbf{q}) = \frac{2\pi}{\hbar} \sum_{\mathbf{k}_i} |M|^2 \delta(\epsilon_i + \hbar\omega - \epsilon_f) f(\mathbf{k}_i)[1 - f(\mathbf{k}_f)],$$

where f is the Fermi distribution function which includes a drift momentum $\mathbf{K}_d = m^* v_D/\hbar$ due to a transport current, $\mathbf{k}_i$ and $\mathbf{k}_f$ are the two-dimensional wave wave vectors of the initial and final electron states, respectively, and M is the matrix element. This matrix element is nonzero only if $\mathbf{k}_i - \mathbf{k}_f = \mathbf{q}_p$ where $\mathbf{q}_p$ is the component of the phonon wavevector in the electron plane. Since the electrons are localized in the perpendicular direction there is a large uncertainty of their perpendicular momentum leading to coupling to phonons with perpendicular components of less than 1/d. Here d is the "thickness" of the 2DEG. Quantitatively this is described by a form factor.

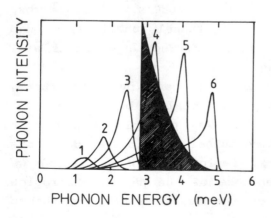

Fig. 1. Theoretical phonon-emission spectra for transverse acoustical (TA) phonons. The numbered traces correspond to k_F-values in units of 10^6 cm^{-1}. The spectra are tuned across the detector response function with increasing k_F.

An interesting feature at low temperatures is the existence of a maximum q_p value of the emitted phonons of about $2k_F$ originating from electron transitions along the diameter of the Fermi circle. Thus the spectrum of emitted phonons extends to maximum values of $q=2k_F/\sin\theta$ or of frequency $\Omega=v_s 2k_F/\sin\theta$ where v_s is the sound velocity. Calculated spectra of transverse phonons emitted in the [111]-direction are shown in Fig. 1 for different k_F values (traces marked 1 to 6). They show a distinct maximum just below $2k_F$ and then a rapid fall-off.

The experiments were performed with (100) Si samples of about 3mm thickness. On one side of the crystal a MOS structure was prepared.[2] Current pulses of about 100ns duration were passed through the 2DEG resulting in phonon pulses which were detected at the opposite surface. As detector a superconducting Pb tunnel junction was generally used. This detector has a cut-on frequency in the detection sensitivity which corresponds to the superconducting energy gap (650 GHz). At higher phonon frequencies there is an effective cut-off due to the isotope scattering in Si. The combined effects lead to a detector response function as indicated by the hatched area in Fig. 1. In the course of the experiment N_s was varied by changing the gate voltage attracting more or less electrons toward the Si-SiO2 interface. Consequently a sharp rise of the detector signal was expected when the phonon emission spectra were tuned across the detector response function.

Experimental results are shown in Fig. 2 (solid lines). Here the TA and LA signals are plotted vs. N_s. The emission direction was in both cases near [111]. Steep increases were indeed observed where expected (arrows). Since the LA phonon spectrum extends to higher frequencies (they have a higher sound velocity) the LA signal increase occurs already at smaller densities. Comparison with theory (dashed lines) shows good agreement for small N_s values but considerable deviations at larger N_s. Similar agreement with theory at small and disagreement at large N_s was observed if the emission angle was varied.[2]

We believe that the deviations are due to electron transitions within and between higher subbands which lie close to E_F at large N_s. Higher subbands were completely neglected in the theory. To test this assumption uniaxial stress was applied to the sample. This shifts the different subband with respect to each other and particularly the E_0' band which lies near the zone edge starts to be populated. The effect on the phonon emission is seen in Fig. 3: the anomalous signal fall-off occurs already at smaller electron densities and the phonon emission disappears altogether at higher stresses. In the region of the signal fall-off in Fig.'s 2 and 3 there is actually an excess of low-frequency phonons as is apparent if an Al-junction detector with 100 GHz cut-on is used .[7]

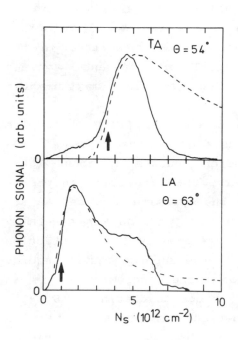

Fig. 2 Experimental phonon emission as function of electron density for two phonon modes.

This excess phonon intensity also shifts to smaller electron densities under uniaxial stress. Thus it seems as if the upper subbands lead to the opening of new emission channels of low-frequency phonons.

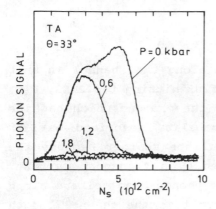

Fig. 3   Same as Fig. 2 for TA phonons emitted under a smaller angle. Stress effectively eliminates the emission of phonons with frequencies in the detection band.

PHONON ABSORPTION EXPERIMENTS IN GAAS-HETEROSTRUCTURES

The most straightforward way to measure phonon absorption is to monitor the phonon flux through a semiconductor interface and record the change of signal with and without a 2DEG. This was indeed done by Hensel et al.[8] However, interferences of the phonons reflected off the 2DEG and off the semiconductor interface obscured the absorption measurement. In order to measure the phonon absorption directly we utilised the phonon drag effect.[3,4] The experimental setup is shown in Fig. 4. This time the 2DEG device was prepared on a GaAs-AlGaAs heterostructure grown by molecular-beam epitaxy on GaAs wafers of 0.6 mm thickness and (100) orientation. The 2DEG had a charge density of 5.3×10^{11} cm^{-2}. A structure was etched out consisting of two contact areas (0.5×0.8 mm² each) connected by a 50×80 μm² bridge. A voltmeter was attached to the contacts. Monochromatic phonons were generated by a Pb tunnel junction (0.2×0.2 mm²) placed on the opposite surface of the sample. Phonons from the generator hit the 2DEG and, if absorbed they transfer their parallel momentum to the 2DEG. This additional momentum will rapidly be distributed over all electrons leading to a drift velocity of the electrons. Let the channel of the 2DEG structure point into the x-direction then

$$\dot{Q}_x = \int q_x [\Gamma(\mathbf{q})/v] F(\mathbf{q})\, d\mathbf{q}$$

is the total transfered momentum along the channel in unit time. Here $F(\mathbf{q})\,d\mathbf{q}$ is the phonon flux hitting the 2DEG, $\Gamma(\mathbf{q})$ is the absorption probability of the phonons which has the same form as the emission probability described earlier except that the $2k_F$ cut-off should be sharper because the electron temperature is not rised by a transport current. Under steady state conditions the phonon-drag voltage along the channel will be $V=hQ_x l/e$ where l is the channel length and e is the unit charge.

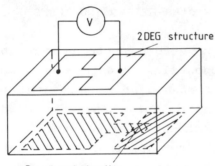

2DEG structure

Pb - tunnel junction

Fig. 4 Experimental set-up for a phonon-drag experiment with a tunnel junction generator.

Experimental results are shown in Fig. 5. Data obtained with the phonon generator being placed in three different directions from the channel are presented. The phonon-drag voltages are plotted vs. frequency $f=\Omega/2\pi$. Absolute intensity values can not be compared because of different generator efficiencies. In each trace, we find a steady rise of the signal until a cut-off is reached which we relate to the $2k_F$-condition. In one direction two cut-offs were observed, in this case two phonon modes contributed to the signal. The spectral width of the cut-offs is presently limited by the angular resolution of the experiment. If we read the cut-off frequencies from the experimental data (dashed lines) and use average focusing-corrected absorption angles we find a $k_F=(1.85\pm0.15)\times10^6$ cm^{-1} which is in excellent agreement with the 1.82×10^6 cm^{-1} value calculated from the electron density.

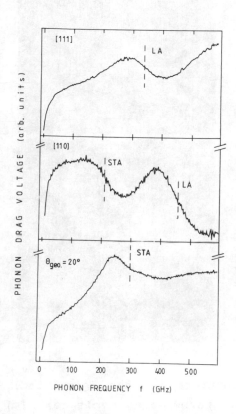

Fig. 5 Phonon-drag voltages as function of phonon frequency for three different directions.

The phonon-drag technique was also used by us to measure the angular dependence of the phonon absorption. For this end it was combined with the phonon-imaging technique.[9] The Pb-junction generator was replaced by an Al film onto which a laser (HeNe, 1mW) was focused. The laser was raster-scanned across the surface and phonons emanating from the heated spot hit the 2DEG channel from differnt directions. Simultaneously the phonon-drag voltage was recorded and displayed as grey tones on a TV monitor. The phonon flux in such an experiment is modulated by the phonon focusing. The focusing pattern expected in a (100) GaAs sample is shown in Fig. 6 (a). Bright areas correspond to high phonon fluxes.

The measured phonon-drag image is shown in Fig. 6 (b). Zero voltage corresponds to an average grey tone while positive and negative values are displayed brighter and

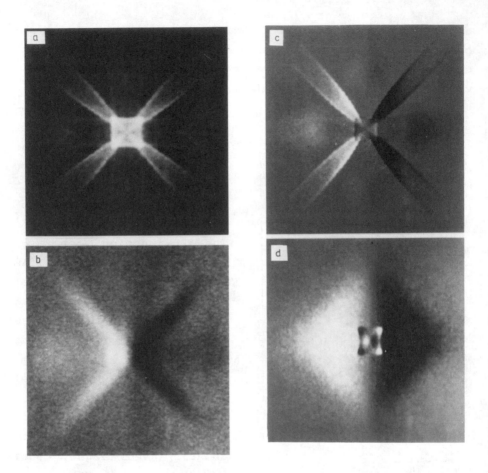

Fig. 6   (a) Phonon focusing pattern in GaAs (simulated); (b) measured image of the phonon-drag voltage; (c) calculated image of the phonon-drag pattern using piezoelectric interaction; (d) same as (c) but using deformation potential interaction.

darker, respectively. The maximum voltages were of the order of 0.5 μV. Sharp features are visible which coincide with some of those of the focusing pattern. As expected for the phonon-drag effect the voltages changed sign whenever the q_x-component of the incident phonons reversed its sign.

Most prominent are the FTA-ridges along the {100} planes.
On the other hand, the STA structures along the {110}
planes do not show up in this measurement. The two broad
structures which are centered at [111]-directions are LA
phonons as revealed by time-resolved measurements. The
sharpness of the structures demonstrates that only the
bridge in the 2DEG device contributed to the observed
pattern while phonon drag voltages in the relatively large
contact areas seemed to average out. A possible exception
is dicussed below.

The absence of the STA modes and the relatively strong
appearance of the LA phonons is indicative of a strong
phonon-polarisation dependence of the electron-phonon interaction. In GaAs there are two types of coupling
possible: deformation potential and piezoelectric. In the
first case the square of the matrix element M is
proportional to $(\Xi_d \, \mathbf{q}\mathbf{a})^2$ where Ξ_d is the deformation
potential and $\mathbf{a}$ is the unit polarisation vector of the
phonon with wavevector $\mathbf{q}$. For piezoelectric interaction, on
the other hand, the corresponding expression is
$e_{14}^2 \, (a_x q_y q_z + a_y q_x q_z + a_z q_x q_y)^2$.[10] where e14 is a piezoelectric constant In both cases screening is neglected. For a
comparison with experiment the phonon focusing calculations
were repeated but the resulting phonon fluxes were
multiplied by a number proportional to q_x times the phonon
absorption probability. In Fig. 6 (c) and (d) the resulting
theoretical phonon-drag images are shown if piezoelectric
and deformation-potential interaction is assumed,
respectively.

Obviously, the piezoelectric pattern agrees much
better with the experimental data than the deformation
potential one. Therefore one can conclude that the coupling
in GaAs heterostructures is of the piezoelectric type.
There is one area in the experimental image which does not
agree with the theoretical one: near the [100] direction we
observed a much higher signal than expected. This

phenomenon is not yet understood. In this region there are very high fluxes incident on the contact areas also and it is possible that the different electric field components do not cancel out. Unfortunately, this large signal obscured the drift of electrons towards the phonon generator that is expected due to the different signs of the x component of the group velocity and the q vector in this region. Only small dips were found in the observed signal which are almost invisible in Fig. 6 (b).

In conclusion, it was shown that using phonon spectroscopy information about the emission and absorption spectra of 2DEG's can be obtained. The most interesting feature is the $2k_F$ cut-off. The different dimensionalities of electron gases and phonons led to cut-off frequencies which were dependent on emission angle and phonon mode. Using phonon imaging technique the dependence of the phonon interaction on direction or, more precisely, on polarisation was measured. In GaAs heterostructures good agreement with piezoelectric interaction was found.

1) T.Ando, A.B. Fowler, and F. Stern, Rev. Mod. Phys **54**, 437 (1982).
2) M. Rothenfusser, L. Köster, and W. Dietsche, Phys. Rev. **B34**, 5518 (1986).
3) H. Karl, W. Dietsche, A. Fischer, and G. Ploog, Phys.Rev.Lett. **61**, 2360 (1988).
4) A. Lega, H. Karl, W. Dietsche, A. Fischer and G. Ploog, Surf. Science ,in print.
5) L.J. Challis, this volume.
6) M.A. Chin, V. Narayanamurti, H. L. Störmer, and J.C.M. Hwang, in **Phonon Scattering in Condensed Matter** ed. by W. EIsenmenger et al. (Springer 1984) p. 331.
7) W. Dietsche, M. Rothenfusser, and L. Köster, Physica Scripta **T25**, 194 (1989).
8) J.C. Hensel, R.C. Dynes, and D.C. Tsui, J. Phys. (Coll.) **42**, C6-308(1981).
9) G.A. Northrop and J.P. Wolfe in **Nonequilibrium Phonon Dynamics** ed. by W.E. Bron (Plenum)p.165.
10) B.K. Ridley, **Quantum Processes in Semiconductors** (CLarendon, Oxford, 1982)

PHONO-CONDUCTION SPECTROSCOPY OF SHALLOW STATES IN SEMICONDUCTORS

KURT LASSMANN, MARTIN GIENGER, PETER GROSS

1. Physikalisches Institut, Univ. Stuttgart
Pfaffenwaldring 57, D-7000 Stuttgart 80, FRG

1. INTRODUCTION

There is a long-standing interest in the problem of dynamics of carriers excited from and trapped by shallow states in semiconductors. High frequency phonons in many cases mediate the relaxation of the carriers from free to bound states. Specific experimental evidence of the phonons involved could give direct information on the processes effective in relaxation but as yet is missing.

The process inverse to relaxation, the excitation of carriers from bound to free states by nonthermal phonons has been developed in the last years as a new spectroscopic technique called PIC (phonon induced conductivity) [1]) or phonoconductivity: Superconducting tunnelling junctions are used as high frequency phonon generators with a bias tunable sharp spectral component. Sensitive detection of the excitation is possible by the change in conductivity. It has been shown by this technique that

* Al-junctions transmit quasimonochromatic phonons up to at least 14 meV into Ge, i.e. frequencies well beyond the Debye frequency of the lowest TA-branch in Ge;

* strong phonon coupling to D^-/A^+ states of overcharged donors/acceptors allows sensitive investigation of their properties;

* one-phonon excitation from the ground state (GS) to the conduction band (CB) is possible for donors in Ge since the many valley structure of the CB facilitates the conservation of the large phonon k-vector in the transition.

A problem connected with such a one-phonon transition is that the deformation potential coupling may be drastically reduced by the condition of momentum conservation because the wavelength of the corresponding phonons may be much smaller than the extent of the wave function of the shallow bound states. Therefore, to understand the experimentally observed rapid decay of nonequilibrium free carriers at low temperatures intermediate states have been invoked for relaxation such as the quasi-hydrogenic ladder of excited states of the trap ("Lax cascade", [2]) or the weak binding of an additional carrier by a neutral trap in analogy to H^-. [3]

2. EXPERIMENTAL

A typical experimental setup may be as follows [4]: An Al-junction as tunable phonon source is evaporated on one of the 15 mm x 5 mm faces of the 2 mm thick samples. The junction films are relatively thin (~15 nm) to prevent excessive down-conversion of the high frequency phonons by reabsorption within the films. 100 nm thick Al-contact films are evaporated on the side opposite to the junction for measurement of the conductance change. It is the bias U to the superconducting tunnelling junction (bath temperature typically 1 K) that determines the maximum energy ($eU-2\angle_{Al}$) of the phonons emitted by the tunnelling quasiparticles. This maximum phonon frequency is filtered by Lock-In technique if the junction bias is modulated (For a review see e.g. Eisenmenger [5]).

The side opposite to the junction is irradiated with visible light from an incandescent lamp on top of the cryostat via a glass fibre rod to produce free carriers (electron-hole pairs) for a finite resistance of the sample (typically 1 MΩ) to measure the phonon induced resistance changes. The carriers diffusing and drifting through the sample are partly trapped by the neutral donors/acceptors to form D^-/A^+-states with binding energies around 2 meV.

A property specific to phonoconductivity as opposed to FIR-photoconductivity is the strong gradient in phonon intensity away from the phonon generator not so much from geometrical spreading but more because of the small mean free path (mfp) of high frequency phonons. This allows distinguishing configurations for additional information as detailed in [6]: Measuring the resistance change between the contacts opposite to the generator means a long phonon path; measuring across the sample may mean a short phonon path if by appropriate polarity of the measuring bias the appropriate type of carriers for additional D^-/A^+ production is drawn underneath the phonon generator where the phonon intensity is high. Epitaxially doped layered samples additionally allow to locate the detection zone and thus e.g. to estimate the effective phonon mfp. A first example was given in [7]. A low or ohmic contact resistance is not essential in these measurements but the modulation frequency is limited to about 1 kHz because of the large electronic time constant for a high resistance. Higher modulation frequencies are possible by the installation of a transimpedance amplifier with small heat dissipation close to the sample within the cryostat rendering a better signal to noise ratio and, in addition, some information on the carrier dynamics in the sample. Similar devices have been applied by several authors in low temperature photoconductivity (see e.g. Haller [8]).

Stress may be applied by tearing a yoke against the sample with a wire strained from top of the cryostat by a screw. It is measured either with resistive strain gauges near the top of the cryostat or with a piezoelectric strain gauge adjacent to the sample. The analysis of our experiments (see below) indicates that "zero" stress may be indefinite up to 50 bar because of the needs of a firm positioning of the narrow samples. Some inhomogeneous residual stress may be due to the fact that the electrical connection to the junction and contact films is made by indium cones pressed against the sample. Near the surface there is another source of residual strains: the surface damage that may be produced by polishing with .25 um diamond grain which was the case for most of the Ge measurements described below. These strains will be more important for the highest phonon frequencies where the mfp is in the um range. Two types of processes determine the phonon mfp at high frequencies and low temperatures: elastic isotope scattering $-$ E_{ph}^{-4} and anharmonic decay $-$ E_{ph}^{-5}. Only rough estimates are possible in the high frequency range: With dominant elastic scattering the resulting diffusion length may range from several 1000 um at 2 meV ($-$ the threshold of the D^--states) to several um at 12 meV ($-$ the binding energy of the D^0.) Thus the 2 meV phonons may excite the D^- across the whole thickness of the sample increasing the carrier density everywhere across the sample whereas the 12 meV-phonons ionize the D^0 only in a thin layer beneath the junction generating a thin space charge layer which will mainly influence the contact resistance.

3.   RESULTS

As mentioned before a basic problem in the phonoionization of shallow impurities in semiconductors is the k-conservation: the large extent of the impurity wavefunction as compared to the wavelength of the corresponding high energy phonons leads to a strong reduction of the interaction. There are two classes of shallow states where the restriction is relieved:
1) The binding energy of D^-/A^+-states is only 1/20 of that of the corresponding neutral states within the effective mass approximation

(EMA) whereas the extent of the wavefunction is larger by only a factor of about 2.

2) In the case of a many valley band structure such as the CB of Ge the bound neutral states are made up of Bloch functions of different valleys which may facilitate a large k-vector transfer by intervalley scattering in the transition.

Both situations have been investigated by PIC measurements.

3.1. Phonoconductivity response of D^-/A^+-states in Si and Ge

In the case of D^-/A^+-states in Si and Ge the same conductivity thresholds are found as with FIR-conductivity showing that the excitation is in fact by a one-phonon process. Differences to FIR connected with the nontheless large phonon vector show up in the steepness and most spectacularly in the stress dependence of the phonon response 9) 10). By the inclusion of details of the valence band structure in the calculation Haug and Sigmund 11) obtained good qualitative agreement for the stress dependence of the phonoconductivity response in the case of Si:B$^+$ and Si:In$^+$. More detailed comparison of the phonoconductivity results with the so far restricted results of theory as well as FIR might be necessary for a complete understanding of the situation. Apart from the case of Si:B the only other acceptor where the stress dependence can be compared is the double acceptor Ge:Be 12) where quite similar characteristic differences are found as for Si:B as shown in Fig.1.

So far we have found no indication of the existence of excited A$^+$-states in Si as postulated from a model to explain the multiplet structure of acceptor bound excitons in Si 13). In contrast to the continuous chemical shift of the multiplet splitting seen in the exciton case for the acceptors B, Al, Ga, and In we find for the binding energy of the corresponding A$^+$-states only for In$^+$ a chemical shift (i. e. influence of the central cell potential) from the effective mass value of about 2 meV (as observed for B$^+$, Al$^+$, Ga$^+$) to

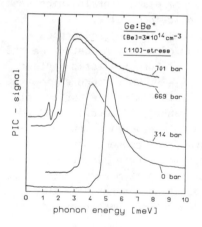

Fig. 1. Stress dependence of the phonoconductivity response of Ge:Be$^+$. The sharp peak appearing with increasing stress is similarly observed 14) in the case of Si:B$^+$, Si:Al$^+$, Si:Ga$^+$. It is, however, not found in the stress dependence of the FIR response for the measured cases of Si:B$^+$ and Ge:Be$^+$.

about 6 meV [4]) which may be regarded as another example for the so-called shallow-to-deep instability of the binding energy with increasing central cell attraction.

The phonon interaction with these states is rather strong. From the resistance under illumination it is estimated that the detection threshold is below 10^9 cm^{-3} D$^-$/A$^+$ centers for our standard experimental set up. At lower illumination intensities depopulation by the phonon irradiation is observed [14]).

3.2 Phonoconductivity response of D^0-states in Ge

We have found phonon induced conductivity thresholds corresponding to the respective D^0-binding energies for Ge-samples containing Sb, P, and As with dopant concentrations of $<10^{12}$ cm^{-3} to 6×10^{14} cm^{-3} [15]). This means that Al-junctions emit primary phonons as determined by the junction bias up to the Debye frequencies of the transverse acoustic phonon branches in Ge. Fig. 2 gives an overview of the phonoconductivity response in n-Ge: The threshold near 2 meV is due to the excitation of the D$^-$-states and the steps near 9.9 meV, 12.4 meV and 13.4 meV are ascribed to the one-phonon ionisation of Sb^0, P^0, and As^0, respectively. The high sensitivity for Sb is evident from the PIC signal of the As-doped sample where the Sb concentration should be below 10^{12} cm^{-3}, the detection limit in photoluminescence measurements of the sample (K. Thonke, Stuttgart, private communication). The Sb^0-signal of a sample doped with 3×10^{14} cm^{-3} Sb is shown in Fig. 3 in detail. The threshold at 9.9 meV (obtained by extrapolating both the

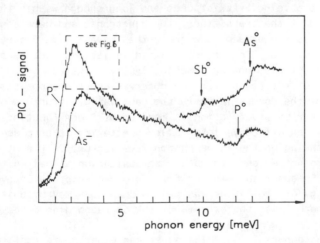

Fig. 2. PIC signal of Ge:P and Ge:P,As. The threshold at 2 meV belong to the D$^-$; the steps above about 10 meV to the D^0 as indicated. P = $2.4\cdot10^{14}$ cm^{-3}; As = $6.0\cdot10^{14}$ cm^{-3}; Sb $< 10^{12}$ cm^{-3}.

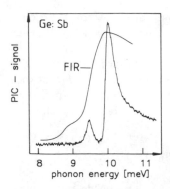

phonon energy [meV]

Fig. 3. PIC signal of
Ge:Sb. The precursor $2\Delta_{Al} =$
.6 meV front of the rise at
9.9 meV is a spectral
property of the Al-junction.
The threshold for PIC is
sharper than for FIR. A peak
in the phonon density of
states may be one reason.
Sb $= 3.0\cdot10^{14}$cm^{-3} (PIC).
Sb $= 1.5\cdot10^{14}$cm^{-3} (FIR) [16].

"base line" on the low energy side and the
turning point tangent of the threshold to
a common foot point) is much sharper than
any corresponding FIR-response which appears to be sensitive to a distribution of
binding energies of donor complexes somewhat below the binding energy of the isolated donor (see e.g. [16]). The optical
values for the binding energy are therefore obtained from the sharp transitions
to higher excited states extrapolating
then to the continuum with the EMA. The
optical value of 10.3 meV thus obtained
for Sb is distinctly above the PIC-thresh-
old. The feature in front of this threshold is a spectral precursor emitted by the
Al-phonon source at an energy $2\Delta = 0.6$ meV
before the main line. It can be distinguished only for sharp and prominent spectral structures on the detecting side. It
proves that the gap of the Al-junction has
not been reduced by the injection of quasiparticles and phonons. The sharpness and
intensity of the signal is specific to Sb^0
and may be due to the slow-TA peak in the
phonon density of states in Ge. This would then mean that phonons with
near zero group velocity exist deep and long enough within the substrate to induce the transitions. The electronic and defect state of
the contact and subsurface zone could then have a significant influence on the response. So far, correlated variations in this sense were
not observed but also not systematically investigated.

One important reason for the observation of the one-step phonoionization of the donors in Ge is the many valley structure of the CB.
Since the donor GS wavefunction is made up of Bloch functions of the 4
111 -valleys an effective intervalley scattering takes place in the
transition taking up the large phonon wave vector. On the other hand,
so far we did not succeed to obtain an ionization signal neither for
the 5.8 meV Sn-donor in GaAs nor for the Ga-acceptor in Ge even if the
binding energy of the latter was pulled down to 8 meV by uniaxial
stress. In both cases the extrema of the corresponding bands are at
k = 0.

For the donors in Ge uniaxial stress experiments confirm the
picture of intervalley scattering. Uniaxial stress shifts the valleys
of the CB and splits the degenerate bound states of the donors. As a
consequence there will be a downshift of the binding energy and level

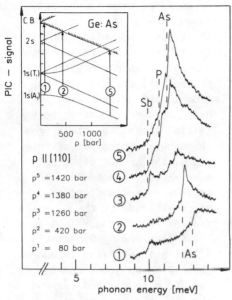

Fig. 4. Stress dependence of threshold height and position of the As-doped sample of Fig. 2 containing besides residual Sb also some P ($\cong 10^{13}$ cm^{-3}). The threshold of P is visible only in curve 4 when enhanced enough by level crossing.

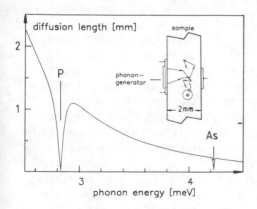

Fig. 5. Estimated diffusion length due to isotope scattering and singlet-triplet resonant scattering (As, P = $5 \cdot 10^{14}$ cm^{-3}, a_B = 3.65 nm, v_{TA} = 3.28 km/sec).

crossings with the lowest valley(s). The downshift is seen in the experiment as a shift of the threshold whereas the level crossings are manifest by maxima in the response.

This is shown in Fig. 4 for the As doped sample of Fig. 2 containing residual Sb ($< 10^{12}$ cm^{-3}) and P (-10^{13} cm^{-3}): The shift of the As threshold from 13.4 meV down to 11.3 meV as well as the variation of the threshold height according to the crossings for all three donors is clearly seen. The threshold of the P^0 is only visible for the crossing condition. In contrast to P and As the expected small shift of the threshold of Sb did not show up. This may be due to an effective pinning of the response to the steep peak of the phonon density of states.

Phonon transitions to the 1s triplet can also be observed at small stress in the case of P and As: For these donors the 1s-singlet <-> 1s-triplet GS splitting is in the energy regime of the D$^-$-response. If by the corresponding resonant scattering the phonon density becomes higher within the zone sensitive for D$^-$-detection, an increase of the signal is expected. An estimate for the transverse phonons shows (Fig. 5) that this extra scattering is strong at 2 meV (P) and less significant at 4 meV (As) when compared to the isotopic background. In the experiment we obtain comparable signal in-

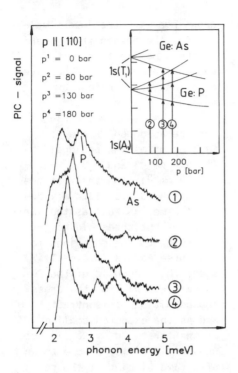

Fig. 6. Stress dependence of the D° singlet-triplet scattering superimposed on the D^{-}-signal.
$P = 2.4 \cdot 10^{14} cm^{-3}$,
$As \cong 10^{13} cm^{-3}$.

creases for both donors as shown in Fig. 6. At "zero" stress we find splittings of the 1s triplet differing from run to run which we attribute to residual uniaxial stress from mounting (typically 30 bar). Taking account of this initial stress we obtain from the stress dependence of this signal the ground state splitting in accordance with the optically determined values. (Phonon scattering by the GS splitting of the Sb donor in Ge was first observed by Dynes et al. [17]) by a 1 meV fixed frequency setup with Sn junctions as emitter and detector and variing the stress.)

From the threshold shift (with the initial stress accounted for) we obtain the stress dependence of the binding energy as shown in Fig. 7 for the case of P. The full lines are the expected energy differences taking (i) the optically determined binding energy at zero stress, (ii) a deformation potential constant of 16 eV for the GS splitting and the CB shifts. Our values are consistently smaller. The large discrepancy for the 100 direction is not understood. The correct orientation is ascertained directly in our experiments by the fact the GS-singlet-triplet scattering shows no splitting nor shift in this case. A slight misorientation, however, gives a fourfold structure in the level crossing expected for a nonsymmetry direction where all four valleys shift differently. Optical transitions from the GS to excited states do not show any shift nor splitting for 100 -stress. Extrapolating back to zero stress we obtain 9.9±.05 meV, 12.4±.05 meV, and

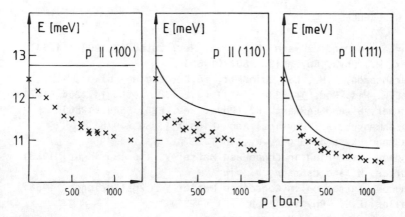

Fig. 7. Stress dependence of the P°-PIC threshold. Full lines are calculated with the optical values for the binding energies. The discrepancy for [100]-direction is discussed in the text.

13.4±.1 meV for Sb, P, and As, respectively. The difference to the optical values [18]) of 10.3 meV, 12.8 meV, and 14.2 meV, respectively, apparently increases with increasing depth of the donor. Values derived from the temperature dependence of Hall measurements (Lopez and Koenig [19])) are somewhat below the PIC thresholds.
By the combination of the DPC for the band shift, the value of the ground state splitting at zero stress and the stress and threshold energy for level crossing we obtain the DPCs for the ground states of Sb, P, and As as 16.5 ± .5 eV, 16.0 ± .5 eV, and 15.7 ± .5 eV, respectively, i.e. there seems to be a small chemical shift of these constants.

4.   SUMMARY

Phonoconduction spectroscopy of shallow states in semiconductors by help of superconduction Al-junctions as phonon sources with high spectral resolution is a sensitive new technique with phonon frequencies well up in the Debye range. The large k-vector of the phonons involved opens new possibilities for the investigation of electron dynamics and states in semiconductors.

Financial support by the Deutsche Forschungsgemeinschaft is gratefully acknowledged.

5.   REFERENCES

1) Burger, W. and Laßmann, K., Phys. Rev. Lett. 53, 2035 (1984)
2) Lax, M., Phys. Rev. 119, 1502 (1960)
3) Gershenzon, E.M., Ladyzhinskii, Yu.P., and Mel'nikov, A.P.,
   ZhETF. Pis. Red. 14, 380 (1971) (J.E.T.P. Lett. 14, 256 (1971))
4) Burger, W. and Laßmann, K., Phys. Rev. B33, 5868 (1986)
5) Eisenmenger, W., Physical Acoustics XII, 80 (1976)
6) Laßmann, K. and Burger, W., in
   Phonon Scattering in Condensed Matter V, 116 (Springer, 1986)
7) Burger, W. and Laßmann, K., in
   Phonon Scattering in Condensed Matter V, 126 (Springer, 1986)
8) Haller, E.E., Physica 146B, 201 (1987)
9) Groß, P., Gienger, M., and Laßmann, K.,
   Japan J. Appl. Phys., 26, 873 (1987)
10) Sugimoto, N., Narita, S., Taniguchi, M., and Kobayashi, M.,
    Solid St. Comm., 30, 395 (1979)
11) Haug, R. and Sigmund, E., Phys. Rev. B40, to be published (1989)
12) McMurray, R.E., Jr., Solid St. Comm., 53, 1127 (1985)
13) Vouk, M.A., and Lightowlers, E.C., J. Luminescence, 15, 357 (1977)
14) Groß, P., Gienger, M., and Laßmann, K., these Proceedings
15) Gienger, M., Groß, P., and Laßmann, K.,
    Proc. 3rd Int. Conf. on Shallow Impurities in Semiconductors,
    Inst. of Physics Conf. Series 95, 173 (1988)
16) Nagasaka, K. and Narita, S.,
    J. Phys. Soc. Jap. 35, 788 and 797 (1973)
17) Dynes, R.C., Narayanamurti, V., and Chin, M.,
    Phys. Rev. Lett. 26, 181 (1971)
18) Reuszer, J.H. and Fisher, P.,Phys. Rev. 140, A245 (1965)
19) Lopez, A.A. and Koenig, S.H.,
    Proc. IX Int. Conf. on the Physics of Semiconductors, Moscow,
    ed. S.M. Ryvkin (Leningrad: Nauka), 1061 (1968)

MICROSCOPIC THEORY OF INTERVALLEY SCATTERING IN GaAs: $\vec{k}$-DEPENDENCE OF INTERVALLEY DEFORMATION POTENTIALS

STEFAN ZOLLNER, SUDHA GOPALAN*, and MANUEL CARDONA

Max-Planck-Institut für Festkörperforschung,
Heisenbergstr. 1, D-7000 Stuttgart 80, FRG

*The University of Western Ontario, Department of Physics,
London, Ontario, CANADA N6A 3K7

1. INTRODUCTION

We have recently used the rigid-pseudoion method (RPIM) to calculate intervalley deformation potentials (IDPs) at high-symmetry points for III-V-semiconductors. [1] Our results agree with some recent experiments, [2] but for a full understanding of the measured scattering times it is necessary to calculate the IDPs as a function of the phonon wave vector and perform an integration over the Brillouin zone to include all energy-conserving intervalley transitions.

2. THE RIGID-PSEUDOION METHOD

In the framework of the rigid-pseudoion method, [3] the electron-phonon interaction Hamiltonian to first order in phonon displacement is given by the gradient of the crystal potential, which we calculate from an empirical local pseudopotential band structure. [4] The IDP is given by multiplying this gradient with the phonon polarization vector (obtained from a parametrized phonon shell model[5,6]) and summing over the basis.[1]

3. INTERVALLEY DEFORMATION POTENTIALS

We first discuss the IDPs for scattering between the Γ- and L-valleys in GaAs. An investigation of the selection rules connecting the Γ- and L-points in a zincblende crystal[1] shows that ΓL-scattering is allowed for the LA and LO phonons, but LO phonon scattering is forbidden in Ge (diamond symmetry) and therefore should be small in GaAs. The calculated IDPs, in Conwell's[7] notation, are $D(\Gamma, L, LA)=3$ eV/Å and $D(\Gamma, L, LO)=0.4$ eV/Å. The experimental result[2] is $D(\Gamma, L, LO)=3.5$ eV/Å and $D(\Gamma, L, LA)=$small.

The discrepancy between the experimental and theoretical result can be explained as follows: The calculated results are for energy non-conserving transitions connecting high symmetry points, whereas the experiment studies energy-conserving transitions from the Γ-valley to the L-valley. In order to investigate this difference we have calculated the $\vec{k}$-dependent IDPs $D(\vec{k}, L, j)$ for initial points $\vec{k}$ near Γ along the $\langle 100 \rangle$-direction, see Fig. 1(a). It can be seen that D decreases for the LA phonon and increases for the LO phonon. Above the energy threshold for ΓL-scattering (vertical line) the LO phonon gives the dominant intervalley scattering contribution. $D(\vec{k}, L, LA + LO)=3.5$ eV/Å is nearly independent of $\vec{k}$ and in very good agreement with the experiment. [2] The transverse phonons are either even or odd for this symmetry direction upon a reflection in a $(01\bar{1})$-plane. Both electronic states are even, therefore the odd phonons are not allowed to couple. The even phonons give a small contribution, shown in Fig. 1(a), and can be neglected in the experiments. [2] Similar calculations for initial $\vec{k}$-points along the $\langle 110 \rangle$- and $\langle 111 \rangle$-directions had the same result.

For scattering between the Γ-point and the X-point (X_1-symmetry, As at the origin), only the LO phonon is allowed in GaAs. We find $D(\Gamma, X_1, LO)=2.9\,\text{eV/Å}$ from our calculations, whereas Monte-Carlo simulations[8] usually assume a much larger IDP around $10\,\text{eV/Å}$. Again we study the $\vec{k}$-dependence of the IDP, but now for an initial point along the $\langle 111 \rangle$-direction. $D(\vec{k}, X_1, LO)$ decreases slightly from $3\,\text{eV/Å}$ down to $2\,\text{eV/Å}$, whereas $D(\vec{k}, X_1, LA)$ remains small. The interesting result of this calculation is the importance of the TA^+ and TO^+ phonons (same selection rules as before). Above the energy threshold of $480\,\text{meV}$ (vertical line), they give the dominant contribution to intervalley scattering with IDPs of about $D(\vec{k}, X_1, TA^+)=D(\vec{k}, X_1, TO^+)=2.5\,\text{eV/Å}$. Results for other symmetry directions give somewhat different numbers, but are qualitatively the same. As the TA^+ phonon has a very low energy, it may be responsible for the short ΓX-scattering times observed in GaAs. [8] Similar results for the $\vec{k}$-dependence of the IDPs have been obtained for GaP. [9]

4. SCATTERING TIMES

The calculation of the scattering time $\tau(\vec{k})$ for a given electron with wave vector $\vec{k}$ to a different valley is possible by summing over all phonon wave vectors $\vec{q}$ that lead to a final electron state with the correct energy. This summation is an integral over the Brillouin zone constrained by two delta functions, which can be calculated using the tetrahedron method. [10,11] The experimentally observed scattering times, however, are more difficult to calculate as $\tau(\vec{k})$ has to be weighted with the electron distribution function which is in general a not well known function of time, excitation density, etc. This makes a second integration over the Brillouin zone necessary.

In order to demonstrate the method, we calculate the return time for an electron at the L-point in GaAs to the Γ-valley. The dimensionless, temperature-independent electron-phonon spectral function (see Ref. 10 for a detailed description of the formalism), which contains all the

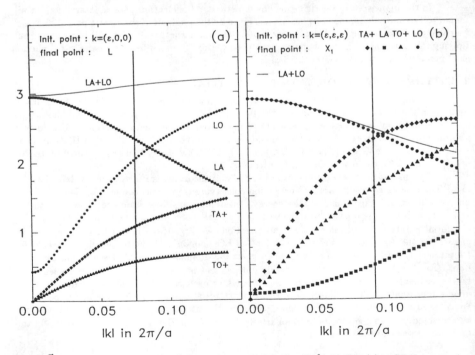

Fig. 1: $\vec{k}$-dependence of intervalley deformation potentials (in eV/Å) for ΓL- (a) and ΓX_1-scattering (b) in GaAs.

necessary information for the intervalley scattering time as a function of temperature, is shown in Fig. 2. An integration over the phonon occupation factor yields $L\Gamma$ return times of about 5 ps for 0 K and 1.8 ps for 300 K, in reasonable agreement with Monte-Carlos simulations for recent ultrafast laser experiments. [12,13]

5. SUMMARY

The $\vec{k}$-dependence of $\Gamma - X$ and $\Gamma - L$ intervalley deformation potentials was calculated with the rigid-pseudoion method. By integrating these IDPs with the tetrahedron method we find the return times of an electron from the L-point to the Γ-valley. The results are in reasonable agreement with recent experiments.

6. REFERENCES

[1]S. Zollner, Sudha Gopalan, and M. Cardona, Appl. Phys. Lett. 54, 614 (1989).

[2]J. A. Kash, R. G. Ulbrich, and J. C. Tsang, to appear in Solid State Electron.

[3]P. B. Allen and M. Cardona, Phys. Rev. B 27, 4760 (1983).

[4]J. P. Walter and M. L. Cohen, Phys. Rev. 183, 763 (1969).

[5]K. Kunc and H. Bilz, Solid State Commun. 19, 1027 (1976).

[6]K. Kunc and O. H. Nielsen, Computer Physics Commun. 17, 413 (1979).

[7]E. M. Conwell, *High field transport in semiconductors*, (Academic, New York, 1967).

[8]K. F. Brennan, D. H. Park, K. Hess, and M. A. Littlejohn, J. Appl. Phys. 63, 5004 (1988).

[9]S. Zollner, J. Kircher, M. Cardona, and Sudha Gopalan, to appear in Solid State Electron.

[10]P. Lautenschlager, P. B. Allen, and M. Cardona, Phys. Rev. B 33, 5501 (1986).

[11]Sudha Gopalan, P. Lautenschlager, and M. Cardona, Phys. Rev. B 35, 5577 (1987).

[12]A. Katz and R. R. Alfano, Appl. Phys. Lett. 53, 1065 (1988).

[13]J. Shah, B. Deveaud, T. C. Damen, W. T. Tsang, A. C. Gossard, and P. Lugli, Phys. Rev. Lett. 59, 2222 (1987).

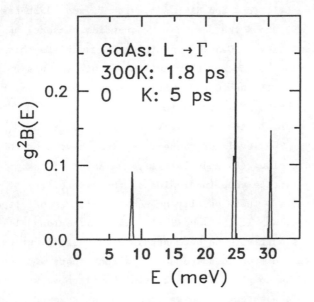

Fig. 2: Temperature-independent electron-phonon spectral function for the return of an electron from the L-point to the Γ-valley.

PHONON SCATTERING BY ORTHORHOMBICALLY DISTORTED JAHN-TELLER IONS IN
III-V SEMICONDUCTORS

L J Challis*, B Salce+, N Butler*, M Sahraoui-Tahar* and W Ulrici°

*Department of Physics, University of Nottingham, Nottingham NG7 2RD
+Centre d-Etudes Nucleaires, Service des Basses Temperatures, 85X,
38041 Grenoble Cedex, France
°Akademie der Wissenschaften der DDR, Berlin, Zentralinstitut fur
Elektronphysik, 1086, Berlin, DDR

Ions with T_1 orbital ground states in T_d sites (Cr^{3+}, V^{2+}, Ni^{2+})
exhibit unusual properties in GaAs, GaP and InP clearly seen in phonon
scattering and attributable to orthorhombic Jahn-Teller distortions.
The reduced thermal resistivity W/W_o (W_o = pure material[1]) of
GaAs:Ni[2], GaP:Ni[3], GaP:V[3], GaAs:Cr[5] and InP:Cr[6] with a
few ppm of dopant shows strong resonant scattering above 1K; fig 1, attributed to V^{2+}, Cr^{3+} (d^3) and Ni^{2+} (d^8) with T_1 ground states (the
upper scale in (d) shows the dominant frequencies present, $h\nu \sim 4kT$).
The lack of scattering in GaAs:V containing V^{2+} suggests this ion has
the low spin state, ^2E and not the usual high spin (Hund's rule) state
$^4T_1$. In GaAs:Ni, GaP:Ni and GaP:V, the scattering can be described by
$\tau^{-1} = C\omega^4/(\omega_o{}^2 - \omega^2)^2$, shown by solid lines, with $\omega_o/2\pi$ in GHz=300
(GaAs:Ni), 500 (GaP:Ni) and 380 (GaP:V) and in GaAs:Ni, $C\alpha[Ni^{2+}]$ determined optically. The levels of these T_1 systems have been shown to be
sensitive to both <100> (E) and <111> (T_2) uniaxial stress.[2,3,5,6]

This strong lattice coupling indicates the probability of strong
Jahn-Teller effects. These usually quench the effects of either <100>
or <111> stress: <100> if the Jahn-Teller wells lie along <111> and
vice-versa. So sensitivity to both stresses is unusual but explicable
if the Jahn-Teller distortions are so large that there are non-linear
terms in either the coupling or the lattice Hamiltonian when the <110>
saddle points on the potential energy surface can descend below those
along <100> and <111> and become minima. The Jahn-Teller wells are

then orthorhombic with the property that the levels are sensitive to both <100> and <111> stress. Tunnelling leads to an excited T_2 level above the ground T_1 and so to resonant scattering.

The stress dependence of these 6 systems is broadly consistent with this model as is the resonant scattering in the 3 Ni^{2+} and V^{2+} systems. The 3 Cr^{3+} systems are more complicated with scattering at several frequencies. Stress data shows these ions to be the most strongly coupled and from evidence of inhomogenous distribution, it seems possible that the scattering is due to ion clusters coupled through their strain fields[2]. Scattering at ~10GHz in GaAs:V, GaP:V and GaP:Ni is tentatively ascribed to T_1 ions in complex with a defect.

We are grateful to Mr D Arnaud, Dr B Cockayne, Mr J-A Favre, Mr J M Martinod, Mr W R MacEwan and Mr W B Roys for help with sample preparation and measurements; Prof C A Bates, Prof B Clerjaud, Dr J L Dunn and Dr M C M O'Brien for discussions and the Algerian Government, the British Council, the Royal Society and the Science and Engineering Research Council, for financial support.

REFERENCES

1. This was calculated for GaP (indirect gap) as W_o is very sensitive to traces of shallow impurities.

2. Challis, L. J., Salce, B., Butler, N., Sahraoui-Tahar, M. and Ulrici, W., J. Phys: Condensed Matter, in press

3. Sahraoui-Tahar, M., Salce, B., Challis, L. J., Butler, N., Ulrici, W. and Cockayne, B., submitted for publication.

4. Butler, N., Challis, L. J., Sahraoui-Tahar, M., Salce, B. and Ulrici, W., J. Phys: Condensed Matter, 1, 1191, (1989).

5. Challis, L. J., Locatelli, M., Ramdane, A. and Salce, B., J. Phys C: Solid State Phys, 15, 1419, (1982).

6. Butler, N., Jouglar, J., Salce, B., Challis, L. J. and Vuillermoz P. L., J. Phys C: Solid State Phys, 18, L725, (1985); Butler, N., Jouglar, J., Salce, B., Challis, L. J., Ramdane, A. and Vuillermoz, P. L., Proc 5th Int Conf of Phonon Scattering in Condensed Matter, Urbana, (1986), eds A. C. Anderson and J. P. Wolfe (Berlin, Springer), 123.

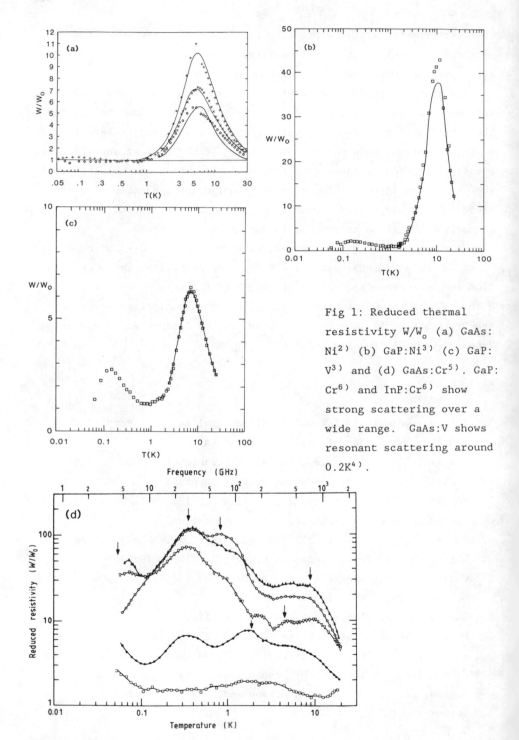

Fig 1: Reduced thermal resistivity W/W_o (a) GaAs: $Ni^{2)}$ (b) GaP:$Ni^{3)}$ (c) GaP: $V^{3)}$ and (d) GaAs:$Cr^{5)}$. GaP: $Cr^{6)}$ and InP:$Cr^{6)}$ show strong scattering over a wide range. GaAs:V shows resonant scattering around $0.2K^{4)}$.

INELASTIC LIGHT SCATTERING BY PHONON AND FREE ELECTRON EXCITATIONS IN SEMICONDUCTORS WITH NONPARABOLIC DISPERSION OF ENERGY BANDS

B.H.BAIRAMOV, V.A.VOITENKO, I.P.IPATOVA AND V.V.TOPOROV

A.F.Ioffe Physico-Technical Institute, Academy of
Sciences of the USSR, 194021, Leningrad, USSR

Inelastic light scattering by two-phonon difference excitations, involving longitudional and transverse optical phonons at the critical points near the Brillouin zone boundaries, superimposed on the quasi-elastic scattering from individual electron single-particle excitations is studied experimentally and theoretically in InP samples with nonparabolic dispersion of energy bands for electron consentration varying from 10^8 to 10^{19} cm^{-3}.

The unscreened proceses of quasi-elastic scattering are described in a number of review articles (see [1]). The line shapes of such a scattering received little attention, nevertheless they carry important information about the kinetics of fluctuations giving rise to light scattering processes. The detailed experimental investigation of the quasi-elastic line shapes was given for n-GaAs [2], n-Si and n-Ge [3]. Tsen and Bray [2] observed the transformation of Gaussian profile to Lorentzian profile in case of light scattering by spin density fluctuations. They notice narrowing of the spectrum with decreasing collision time. The Lorentzian profile (with half-width $\Gamma = q^2 D$, D-being diffusion coefficient) itself can become narrower with increasing impurity consentration due to decreasing D.Contreras et al.[3] have shown that in heavily doped n-Si and n-Ge single-particle spectra have Lorentzian profiles. These experiments are accounted for in [4], where it is shown that these profiles reflected diffusional motion of free carriers.

Our quasi-elastic Raman scattering measurements were performed using 1064.1 nm line of YAG:Nd laser with incident beam along [112] direction and scattered light collected at 90° along [111] direction of the crystal. Typical spectra for progressively increasing free carrier consentrations are given in Fig.1-4.

It is found that quasi-elastic light scattering spectra by single-particle excitations of electrons at very low consentration range (n $\approx$ $5\cdot10^{15}$ cm^{-3})

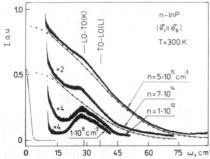 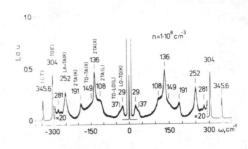

Fig.1 Raman spectra of n-type InP samples in low consentration range showing the dependence of the scattering
line shape on consentration. Solid curves obtained after
subtraction of the difference frequency combination two-
phonon contribution, represent spectra for single-particle
scattering from conduction electrons. The theoretical fir
with Gaussian approximation is shown by open circles with
halfwidth $\Gamma=35.0 cm^{-1}$ for sample with $n=5\cdot 10^{15}cm^{-3}$ and
$\Gamma=24.9 cm^{-1}$ for sample with $n=7\cdot 10^{14}cm^{-1}$. $e_i \parallel e_s$. T=300K.

Fig.2 First- and second-order Stokes and anti-Stokes
Raman scattering spectra of the semi-insulating n-type
InP sample with $n=1\times 10^8 cm^{-3}$. The scan is linear in wavelength and positions of the lines are indicated in wave-
numbers. $e_i \perp e_s$. T=300 K.

for $e_i \parallel e_s$ scattering configuration are accurately described by Gaussian predicted for Maxwell electron distribution due to electron density fluctuations. As we illustrate in Fig.2, there is considerable broadening of the
spectrum for samples with higher electron consentrations.
With increasing consentration the Gaussian line
shape of single particle spectra becomes Lorentzian.
For intermediate consentration range from $3\cdot 10^{16}$ to
$5\cdot 10^{17}cm^{-3}$ narrowing of the Lorentzian spectra (with
$\Gamma=q^2 D$) due to the spin-density mechanism is demonstrated
(Fig.3). With further increasing consentration from
$5\cdot 10^{17}$ to $1\cdot 10^{18}cm^{-3}$ broadening of Lorentzian spectra due
to the unscreened momentum density fluctuation mechanism
(with halfwidths $\Gamma=1/\tau_{p2}$, τ_{p2} - being the relaxiation
time of the second spherical harmonic) is obtained for
$e_i \perp e_s$ scattering configuration.

REFERENCES

1   Abstreiter, G., Pinczuk, A. and Cardona, M., In:
     "Light Scattering in Solids", ed. by Cardona M. and
     Güntherodt. Springer 1984, 15.

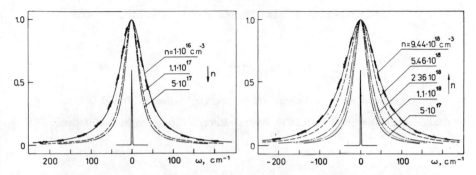

Fig.3 The theoretical spectra with Lorentzian line
shape and halfwidth $\Gamma = q^2 D$ for spin-density fluctuation
scattering mechanism. They fit reasonable well to the
quasi-elastic single-particle scattering spectra obtained
from experimental spectra of n-type InP samples, for the
electron consentration range varying from $n=3 \times 10^{16}$ to
$5 \times 10^{17} cm^{-3}$ for $e_i \perp e_s$ scattering configuration at T=300K,
after subtraction of the overton two-phonon contribution.
These spectra directly demonstrate the narrowing of the
single-particle scattering with increasing consentration
of conduction electrons.

Fig.4 The theoretical spectra with the Lorentzian
line shape and halfwidth $\Gamma = 1/p_2$ for momentum density
fluctuation mechanism. They very accurately fit to the
experimental spectra of n-type InP samples for the elec-
tron consentration range varying from 5×10^{17} to 9.44×10^{18}
cm^{-3} for $e_i \perp e_s$ scattering configuration at T=300 K,
after correction for the two-phonon scattering contribution. These spectra directly demonstrate the broadening
of single-particle scattering spectra with increasing
consentrationnin contrast to the previous case for intermediate consentration range, considered in Fig.4.

2   Tsen, K.T. and Bray, R., Solid State Commun. <u>45</u>,
    685 (1983).
3   Contreras, G., Sood, A.K. and Cardona, M., Phys. Rev.
    B<u>32</u>, 924 (1985), 930.
4   Ipatova, I.P., Subashiev, A.V., Voitenko, V.A., Solid
    State Commun. <u>37</u>, 893 (1981).
5   Bairamov, B.H., Voitenko, V.A., Ipatova, I.P., Suba-
    shiev, A.V., Toporov, V.V., Ihne, E., Sov. Phys. Sol.
    State <u>28</u>, 754 (1986).

EVIDENCE FOR STRONG ELECTRON-PHONON COUPLING IN GaP:V

J.L. DUNN[*], C.A. BATES[*] and W. ULRICI [+]

[*]Department of Physics, University of Nottingham, Nottingham NG7 2RD, U.K.

[+] Akademie der Wissenshaften der DDR, Zentralinstitüt für Electronenphysik, 1086 Berlin, DDR

INTRODUCTION

The shape and structure of the optical absorption or photoluminescence spectrum from magnetic impurity ions in semiconductors is known to depend upon both the nature and strength of the coupling between the electrons of the magnetic ions and the lattice phonons. The coupling between ions which form orbital T_1 triplet states and phonons of t_2-symmetry is particularly strong in many III-V semiconductors, which consequently undergo strong $T \otimes t_2$ Jahn-Teller (JT) effects. Until very recently, a theory of these JT systems had not been developed in sufficient detail to give consistent results in the simultaneous modelling of the zero phonon line (ZPL) and the band. It will be shown here how recently published calculations of the second-order JT reduction factors involved in spin-orbit coupling[1] can unify the models of the band and ZPL for the GaP:V^{3+} system. The GaAs:V^{3+} and InP:V^{3+} systems will also be discussed briefly.

THE GaP:V^{3+} SYSTEM

The absorption spectrum from vanadium-doped GaP samples consists of a triple-peaked absorption band with a maximum at ~ 9600 cm^{-1} and resonances at 8698.7, 8713.1 and 8762.7 cm^{-1} (Figure 1). The original interpretation of this spectrum associated all three resonances at $\sim$ 8700 cm^{-1} with the $^3A_2 \rightarrow {}^3T_1$(F) ZPL of $V^{3+2)}$. However, this is unsatisfactory for two reasons. Firstly, the resonance at 8762.7 cm^{-1} is much broader than the other two resonances, and well-separated from them. Secondly, it is impossible to fit the results using any simple JT model employing strong coupling. An alternative explanation of the results in which the two sharp resonances are associated with the ZPL and the third broad resonance with either a phonon replica or an excited vibronic state is given below.

As an A-state cannot be split, the structure of the ZPL must arise entirely from splittings within the $^3T_1$(F) states. These states can be modelled using isomorphic $l = 1$ states, for which the main perturbation is spin-orbit coupling. A suitable effective Hamiltonian is

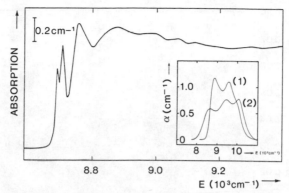

Figure 1: Part of the optical absorption spectrum of the internal $^3A_2 \to {}^3T_1$ (F) transition of V^{3+} ions in GaP measured at T = 5 K. The inset shows the entire broad-band absorption spectrum of this transition measured at T = 78 K (marked 1) and 300 K (marked 2).

$$H_{eff} = a\,l.S + b(l.S)^2 + c(E_\theta'E_\theta{}^S + E_\varepsilon'E_\varepsilon{}^S) \tag{1}$$

where $E_\theta'E_\theta{}^S$ and $E_\varepsilon'E_\varepsilon{}^S$ are orbital and spin operators. The most important contributions to the parameters a, b and c originate from the JT effect. If the JT effect is of a strong $T \otimes t_2$ type, the parameters are dominated by the second order reduction factor $f_b{}^t$ such that:

$$a = \tfrac{1}{2}b = -\tfrac{3}{4}c = -N_{Tt}{}^2(k_1{}^{T_1})^2\,\lambda^2\,f_b{}^t \tag{2}$$

where $(k_1{}^{T_1})$ (= - 3/2) is the isomorphic constant and N_{Tt} is the normalising factor for the $l=1$ states. (N.B. a is always negative in strong coupling.) The J = 2 (T_2) state then has energy $|a|$ relative to the J = 0 (A_1) state, and the J = 1 (T_1) and J = 2 (E) states both have energy $3|a|$. We propose that transitions to the J = 0 (A_1) and J = 2 (T_2) states cannot be resolved within the error of the experiment, resulting in two sharp resonances only. The separation of the resonances suggests that

$$a = -\tfrac{9}{4}N_{Tt}{}^2\,\lambda^2\,f_b{}^t \approx (-4.8 \pm 1.0)\ cm^{-1} \tag{3}$$

This value can be used to estimate the strength of the $T \otimes t_2$ coupling.

Figure 2 shows a plot of the factor $N_{Tt}{}^2\,\hbar\omega_T f_b{}^t$ as a function of the Huang-Rhys factor S_t (= $E_T/\hbar\omega_T$, where E_T is the JT energy and $\hbar\omega_T$ is the energy of the photon quantum). Taking $\lambda = \hbar\omega_T = 100\ cm^{-1}$ gives a value for S_t of 8.1. From measurements of the separation between the side and central peaks of the absorption band as a function of temperature, Ulrici et al[2] deduced a value for S_t of 7 ± 2. This is consistent with the value obtained from the ZPL.

THE GaAs:V^{3+} AND InP:V^{3+} SYSTEMS

Absorption measurements have also been carried out on the GaAs:V^{3+} and InP:V^{3+} systems[3,4]. Both show a triple-peaked band similar to that of GaP:V^{3+}, indicating a strong $T \otimes t_2$ JT effect (with $S_t \sim 10$ for GaAs), but with one ZPL component only. In addition, for the

798

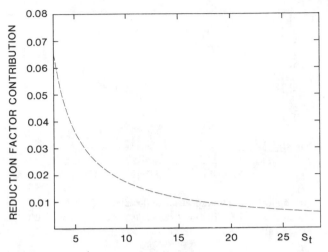

Figure 2: A plot of $N_{Tt}{}^2 \hbar \omega_T f_b{}^t$ as a function of S_t.

second-order spin-orbit coupling splittings in the $^3T_1$ (F) state to be less than the observed line width (~ 3 cm^{-1}), the relations (2) would require a value of ~ 40 for S_t. This is very unrealistic as it would imply a very large E_T such that the ZPL would have zero intensity.

We believe that, in these systems, there is also coupling to e-type phonon modes making the JT effect a mixture of $T \otimes t_2$ and $T \otimes (e + t_2)$. Unfortunately JT theory has not yet been extended to include mixed systems of this type so that a quantitative analysis cannot be undertaken at present. However, Dunn and Bates[5] have shown that, for a pure $T \otimes (e + t_2)$ JT system, the second-order reduction factors are significantly smaller than they are for a pure $T \otimes t_2$ JT system. We therefore expect the e-coupling to decrease the $T \otimes t_2$ second-order spin-orbit coupling splitting. This would allow the ZPL results to be described with a much smaller value of S_t.

CONCLUSIONS

In conclusion, we have provided an explanation of the optical data concerning the $^3A_2 \rightarrow$ $^3T_1$ (F) transition of V^{3+} in GaP in terms of a strong $T \otimes t_2$ JT effect and including second order JT-generated spin-orbit coupling terms. The corresponding data for V^{3+} ions in both GaAs and InP again suggest a $T \otimes t_2$ coupling but with a significant additional e-coupling.

REFERENCES

1)   Bates, C.A. and Dunn, J.L., J. Phys.: Condens. Matter 1, 2605-16, (1989).

2)   Ulrici, W., Eaves, L., Friedland, K. and Halliday, D.P., Phys. Stat. Sol.(b) 141, 191-202, (1987).

3)   Ulrici, W., Friedland, K., Eaves, L. and Halliday, D.P., Phys. Stat. Sol.(b) 131, 719-28, (1985).

4)   Clerjaud, B., Côte, D., Naud, C., Bremond, G., Guillot, G. and Nouailhat, A., J. Cryst. Growth 83, 194-197, (1987).

5)   Dunn, J.L. and Bates, C.A., J. Phys.: Condens. Matter 1, 2617-29, (1989).

ACOUSTIC PROPERTIES OF DISORDERED INTERACTING ELECTRON SYSTEM

T. SOTA and K. SUZUKI

Department of electrical Engineering, Waseda University, Shinjuku, Tokyo 169, Japan

1. INTRODUCTION

Impure semiconductors are suitable for studying the disordered electron system because the carrier concentration and the degree of disorder are easily varied by impurity doping. Many experimental and theoretical studies have been done. We calculated previously the acoustic properties of heavily doped many-valley semiconductors in the weak localization regime where both localization and interaction effects were taken into account on the same footing.[1] Sachdev pointed out[2] that we did not consider renormalization of interaction parameters and got on erroneous result. Although it is desirable to construct a theory with Finkelstein's renormalization scheme of mutual interaction[3] and a full account of the localization effect, it seems to be impossible at present.

In this paper we derive expressions for the attenuation coefficient α and the change Δv in the sound velocity taking account of the diffusion enhanced Coulomb interaction in heavily doped many-valley semiconductors at the metallic side far from the critical region for the metal-nonmetal transition, and examine in detail the behavior of α. Quantum corrections to α are different from the result by Sachdev.

2. RESULTS AND DISCUSSION

Calculations are performed under the following assumptions for our system. (i)Homogeneity of the system. (ii)ν equivalent valleys with an isotropic effective mass m_0. (iii)$k_B T \ll \varepsilon_F$(the Fermi energy). (iv)The deformation potential coupling for the electron-phonon interaction. (v)τ_2(the intervalley scattering time)$\gg \tau_0$(the total scattering time). (vi)Constant interaction parameters F_1 and F_2 between electrons. (vii)Only the intravalley processes for both the electron-phonon interaction and the interaction between electrons.

α and Δv are derived from the phonon self-energy Π using the relations $\alpha = -(\omega/v)\mathrm{Im}\Pi$ and $\Delta v = (v/2)\mathrm{Re}\Pi$. Π is determined from the density response function R defined by

$$R_{ij}(\mathbf{q},\omega) = \chi_{ij}(\mathbf{q},\omega) + V(q)R_{m,n}\chi_{im}(\mathbf{q},\omega)R_{nj}(\mathbf{q},\omega). \qquad (1)$$

Here $V(q) = 4\pi e^2/\varepsilon q^2$, $i,j,m,$ and n are the valley indices and χ is an irreducible part of R determined by the short range interaction alone. When we calculate χ using the diagrammatic method, we adopt the particle number conservation law as a postulate and get a benefit of Finkelstein's renormalization scheme. Final expressions of Δv and α, for $\omega\tau_2 \ll 1$, are as follows.

$$\Delta v = -(1/\rho v)\Sigma_i \mid C_{is}\mid^2 N_0 (1-2N_0 F_1 + N_0 F_2), \qquad (2)$$

$$\alpha = (1/\rho v^3)\Sigma_i \mid C_{is}\mid^2 N_0 (1-2N_0 F_1 + N_0 F_2)^2 \omega^2 \tau_2', \qquad (3)$$

$$\tau_2' = \nu A_4 (1-2A_3) \tau_2^{-1}. \qquad (4)$$

Here ρ is tha mass density, C_{is} is the deformation potential coupling constant for shear components, N_0 is the density-of-states per spin at ε_F in each valley. The quantities A_3 and A_4 represent the effect of the diffusion enhanced Coulomb interaction, and give quantum corrections to α. Quantum corrections do not occur in Δv

Equations (3) and (4) can be reduced to more simple expressions in some special case as follows.

$$\alpha = \alpha_0 \Delta(T), \qquad (5)$$

with

$$\alpha_0 = 2\Sigma_i \mid C_{is}\mid^2 N_0 (1-2N_0 F_1 + N_0 F_2)^2 \omega^2 \tau_2/\nu, \qquad (6)$$

$$\Delta(T) = 1 + \frac{3^{3/2} N_0 (F_1 - 2F_2)}{4(\varepsilon F \tau_0)^2} \delta(T). \qquad (7)$$

For $\omega_m \gg \nu/\tau_2$ $\delta(T)$ becomes

$$\delta(T) = -\frac{2\nu-4}{2\pi\nu} + \frac{2\nu-4}{4\nu}\phi^+, \qquad (8)$$

and for $\omega_m \ll \nu/\tau_2$

$$\delta(T) = -\frac{2\nu-4}{2\pi\nu}\frac{\tau_0}{\tau_2} + \frac{2\nu+5}{4\nu^{3/2}}(\frac{\tau_0}{\tau_2})^{1/2} - \frac{1}{4\nu}\phi^-,$$

$$(9)$$

where $\phi^{\pm} = 2a^{\pm} + (2\pi T\tau_0)^{1/2}\{(n_0^{\pm})^{-1/2} - 2(n_0^{\pm})^{1/2}$

$$+2(n_0{}^\pm-1/2)^{1/2}-2^{1/2}$$

$$+\sum_{m=1}^{m=n_0{}^\pm-1}[1/m^{1/2}-2((m+1/2)^{1/2}-(m-1/2)^{1/2})]\} \qquad (10)$$

with $a^+=1$, $a^-=(\tau_0/\tau_2)^{1/2}$, $n_0{}^+=[1/2\pi T\tau_0]$, and $n_0{}^-=[1/2\pi T\tau_2]$. Here ω_m is the characteristic frequency of the thermal fluctuation and [y] means to take the maximum integer which does not exceed y.

We have performed numerical calculations using Eqs. (5)-(10), though they are not shown here. The following has been found. For $\omega_m \gg \nu/\tau_2$, where the thermal fluctuation dominates the system, α increases as T decreases. For $\omega_m \ll \nu/\tau_2$ α has only a weak T-dependence, decreases with decreasing T, and the magnitude of α is much smaller than that for $\omega_m \gg \nu/\tau_2$. This implies that when the intervalley scattering dominates the system the quantum corrections are suppressed.

With respect to the aspect of α over a full temperature range for a given τ_2, we may expect that α reaches its maximum value at about T corresponding to $\omega_m \sim \nu/\tau_2$ and then decreases as decreasing T. Taking account of the condition $\varepsilon_F\tau_0 \gg 1$, which permits the perturbational calculation, $\Delta\alpha=\alpha-\alpha_0$ is estimated to be at most 10% of α_0. The temperature region where $\Delta\alpha$ becomes appreciable will be relatively narrow.

It is clear from the argument given above that the T-dependence of α is determined by the competition between the thermal fluctuation and the intervalley scattering. Taking account of the values of τ_2 deduced from experiments, it follws that $\Delta\alpha$ in Sb-doped Ge may be measured more easily than $\Delta\alpha$ in As-doped Ge for $\omega_m \gg \nu/\tau_2$, but it is not easy to measure $\Delta\alpha$ in both As- and Sb-doped Ge for $\omega_m \ll \nu/\tau_2$ and in Si for T<1 K.

Finally we compare our results with Sachdev's ones. He considered only the temperature region $\omega_m \gg \nu/\tau_2$. Therefore his theory can not be applied to Si where $\omega_m \ll \nu/\tau_2$ is always satisfied for T<1 K. He also concluded that α decreases with decreasing T. This result is different from ours.

References
1. Sota, T. and Suzuki, K., Phys. Rev. B33, 8458(1986).
2. Sachdev, S., Phys. Rev. Lett. 58, 2590(1987).
3. Finkelstein, A. M., Pis'ma Zh Eksp. Fiz. 37, 436(1983) [JETP Lett. 37, 517(1983)]; Zh. Eksp. Teor. Fiz. 84, 168(1983) [Sov. Phys. JETP 57, 97(1983)]; ibid. 86, 367(1984) [59, 212(1984)]: Z. Phys B 56, 189(1984).

POINT-CONTACT SPECTROSCOPY OF THE ELECTRON-PHONON INTERACTION IN $(La_{1-x}Pr_x)Ni_5$ COMPOUNDS

P.Samuely, M.Reiffers
Institute of Experimental Physics, Slovak Academy of Sciences,
043 53 Košice, Czechoslovakia

A.I.Akimenko, N.M.Ponomarenko, I.K.Yanson
Physico-Technical Institute of Low-Temperatures, Ukr.Academy of
Sciences, Kharkov, USSR

The electron-phonon interaction (EPI) in $(La_{1-x}Pr_x)Ni_5$, where
x=0, 0.1, 0.7 and 1 is studied by point contact (PC) spectroscopy. In $LaNi_5$ the PC spectrum corresponding to the EPI
function g (the phonon density of states modulated by the
matrix element of the EPI) is found. Increasing the Pr content in samples the energy position of the phonon peaks in
the spectra remains unchanged. But an additional structure
reflecting another electron scattering mechanism emerges.

PC spectroscopy showed to be a usefull method for the RNi_5
(R-rare earth) compounds. It enabled the direct study of an interaction between conduction electrons and local excitations - the crystal electric field (CEF) levels in $PrNi_5$[1,2,3] and the 4f levels in
the valence fluctuations system $CeNi_5$[4].

PC spectrum (the second derivative $d^2U/dI^2(U)$ of the voltage U
with respect to the current I) was measured with a conventional modulation technique. Point contacts were obtained by a touch of two
prisms in liquid helium. The prisms were prepared from the
$(La_{1-x}Pr_x)Ni_5$ specimens. Some of the specimens were characterised by
a low residual resistivity ratio, RRR = 5, 2.6, 2.1 and 23 for x=0,
0.1, 0.7 and 1, respectively.

Due to low RRR related to the short elastic mean free path of
electrons l_i the homocontacts $LaNi_5$-$LaNi_5$ mostly do not reveal

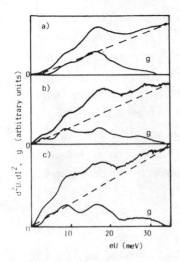

Fig.1: PC spectra of LaNi$_5$-LaNi$_5$-(a), LaNi$_5$-Cu(b) and LaNi$_5$-Ag(c). g- the LaNi$_5$-PC EPI function obtained from the spectrum. --- background signal.

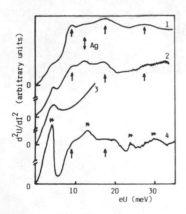

Fig.2: PC spectra of $(La_{1-x}Pr_x)Ni_5$-Ag. x=0, 0.1, 0.7 and 1 for curves 1-4, resp. Peaks related to: phonons ($\uparrow$), transitions between CEF levels ($*$)[1].

spectroscopic features. We improved the situation by the heterocontacts with Cu and Ag.

Heterocontact spectrum represents a sum of the PC EPI functions divided by the Fermi velocities g_1/v_{F1} + + g_2/v_{F2} [2] (provided the symmetrical heterocontact). In many heterocontacts the Cu(Ag) EPI peak was seen besides the LaNi$_5$ structure and always an increase of spectral intensity at the related energy (8÷15 meV for Ag and 14÷22 meV for Cu). Thus we conclude that g/v$_F$ value for LaNi$_5$ in this energy range is very similar to that of noble metals (g/v$_F$ = = 9.6.10^8 s/m for Cu[2]). We obtained the functions proportional to the LaNi$_5$ EPI function g by subtracting the Cu(Ag) part of spectrum together with the background signal. The Cu(Ag) part was taken as a half of spectral intensity at the Cu(Ag) peak energy. The shape of resulting functions is very similar. The functions reveal all the peaks as the LaNi$_5$ phonon density of states measured by neutron spectroscopy[5].

In particular the peaks at 8÷9 meV, 17÷18 meV, and 27÷28 meV and a shoulder at about 4.5 meV can be seen in Fig.1.

In Fig.2 the heterocontact spectra of diluted compounds $(La_{1-x}Pr_x)Ni_5$ with Ag as a counter electrode are presented. Substitution of La by Pr does not influence the observed phonon energies. For the Pr content x=0.7 and 1 a peak at 4.5 meV related to the transition of the Pr^{3+} ion in CEF from the ground state Γ_4 to the excited one Γ_{5A} dominates in the spectrum. Peak at 24 meV previously interpreted as phononic[2] we consider to reflect a transition between two excited levels $\Gamma_{5A} \rightarrow \Gamma_{5B}$. The transition requires just this energy and is sufficiently probable. The lifetime of Pr^{3+} in the Γ_{5A} state $(10^{-12}$ s as estimated from the intrinsic peak width 1 meV) is much longer than time during the electron passes through the point contact $(10^{-14}$ s).

REFERENCES

1. Akimenko, A.I., Ponomarenko, N.M., Yanson, I.K., Jánoš, Š. and Reiffers, M., Sov.Phys.Solid State <u>26</u>, 1374 (1984)
2. Akimenko, A.I., Ponomarenko, N.M. and Yanson, I.K., Sov.J.Low Temp.Phys. <u>12</u>, 342 (1986)
3. Reiffers, M., Naidyuk, Yu.G., Jansen, A.G.M., Wyder, P., Yanson, I.K., Gignoux, D. and Schmitt, D., Phys.Rev.Lett. <u>62</u>, 1560 (1989)
4. Akimenko, A.I., Ponomarenko, N.M. and Yanson, I.K., Sov.Phys. Solid State <u>28</u>, 615 (1986)
5. Bührer, W., Furrer, A., Hälg, W. and Schlapbach, L., J.Phys.<u>F9</u>, L141 (1979)

HIGH RESOLUTION STUDY OF HOT HOLE MAGNETOPHONON RESONANCE IN p-InSb IN HIGH MAGNETIC FIELDS UP TO 40T

NORIHIKO KAMATA[A)], KOJI YAMADA[A),B)] and NOBORU MIURA[B)]

A)Saitama University, Shimo-Ohkubo, Urawa, Saitama 338, Japan
B)Institute for Solid State Physics, University of Tokyo, Roppongi, Minato-ku, Tokyo 106, Japan

INTRODUCTION

In order to discuss the energy relaxation mechanism of hot holes in p-InSb[1)], magnetophonon resonance(MPR) was investigated in pulsed high magnetic fields up to 40T. An improved measuring system for a high resolution study has been developed, so that detailed discussion on the fine structure of the MPR spectra became possible in spite of the low hole mobility and the complexity of the valence band structure.

Experimental data of the longitudinal MPR along <100> and <111> axes under a temperature range between 4.2K and 30K are compared with the energy band model of Pidgeon and Brown[2)]. Besides the transition from Landau level to acceptor state and the inter-Landau level transition[1)], many additional peaks were resolved clearly and a possible mechanism was discussed.

EXPERIMENTAL PROCEDURES

The samples were Ge doped p-type InSb with $N_A-N_D=1.6 \times 10^{14} cm^{-3}$ and $\mu_p=57000 cm^2/Vs$ at T=25K, respectively.

Synchronized with the pulsed magnetic field with a 5ms half width, the electric fields between 10V/cm and 30V/cm were applied to the sample via the extra-low resistance($R_{on}<20m\Omega$) FET switch. The lattice temperature rise due to the current flow is estimated to be negligible in this experimental condition.

The MPR signal was differentiated by the RC circuit and was recorded in a transient recorder with a 16bit resolution for each $2\mu s$ interval. The exact d^2R/dB^2 spectra were calculated by a host computer taking into account the phase shift of the RC circuit and the discrimination of impulse noise etc.

RESULTS AND DISCUSSION

Energy band model

Pidgeon and Brown[2] introduced the interaction between 2 conduction and 6 valence bands within the framework of Kane[3] and Luttinger[4] models to explain the magneto-optical spectra of InSb. This model was used to discuss the light hole transition of hot holes for magnetic fields below 3T[5]. A calculation of Landau energy was done under the same model[2] with a slight modification of the matrix element[6].

The LO phonon energy around the Γ point of 24.3meV and the acceptor binding energy of 9.8meV[7] were used to obtain the resonant condition. The magnetic field dependence of the acceptor binding energy is small since heavy hole transition is concerned $(\gamma=(1/2\hbar\omega_c)/R^* < 0.25)$[8],[9].

B//⟨111⟩ direction

The heavy hole Landau level to acceptor transition with the index number[2] starting from N=2 and the heavy hole inter Landau level transition of N>4 were assigned as shown in Fig. 1. A reasonable agreement between experiment and the model was obtained. Doubly peaked structures corresponding to paired spin states were clearly seen in the inter-Landau level transition especially at N=6,8 and 10.

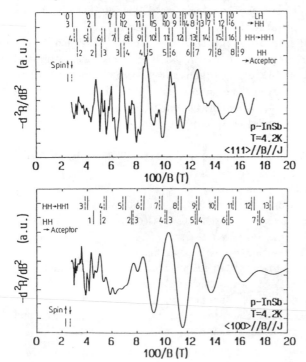

Fig.1 d^2R/dB^2 vs. 1/B for ⟨111⟩//B at T=4.2K. Major resonances were assigned to HH to acceptor and HH inter-Landau level transitions. A possible resonance of LH to HH transition is also shown.

Fig.2 d^2R/dB^2 vs. 1/B for ⟨100⟩//B at T=4.2K. The resonant position of HH to acceptor and inter-Landau level transitions were shown.

B//<100> direction

The assignment for <100> direction is shown in Fig.2. Again the dominant resonance came from the transition of heavy hole to the acceptor state starting from N=1. The peak shift due to the electric field was not clear within the electric field condition(15.4 - 20.4V/cm).

Additional transitions

Besides major resonances, many additional peaks were observed as shown in Figs. 1 and 2. Some of these in <111> case coincided with the resonance of the light hole to heavy hole transition(labeled as LH→HH in Fig. 1). Since a considerable mixing between light hole and heavy hole bands exists in InSb, this transition has a possibility to contribute to MPR. However the LH→HH resonance depends more strongly on the band parameters used in the model calculation than other major resonances. The dependence of the resonant field B for LH0→HH2 transition on each band parameter is estimated as follows:$dB/d\gamma_1=-2.2(T)$, $dB/d\gamma_2=-3.0(T)$, $dB/d\gamma_3=-3.4(T)$, $dB/d\Delta=-2.5(T/eV)$, $dB/dP^2=4.9\times10^{-4}(T/N^2s^2)$ and $dB/dT=3.5\times10^{-3}(T/K)$. A detailed calculation of the transition probability should be necessary.

CONCLUSIONS

A reasonable agreement between experiment and the energy band model was obtained for <111> and <100> directions. In addition to the major resonant series of Landau to acceptor levels and inter-Landau level transitions, many fine structures were clearly resolved in MPR spectra. A possible mechanism of LH→HH transition was discussed.

ACKNOWLEDGEMENTS

The authors wish to thank Sumitomo Electric Industries Ltd. for supplying the sample.

REFERENCES

1)YAMADA, K. and MIURA, N., Physica 134B, 179(1985).

2)PIDGEON, C. R. and BROWN, R. N., Phys. Rev. 146, 575(1966).

3)KANE, E. O., Semicond. and Semimetals, 1, 75(1966).

4)LUTTINGER, J. M., Phys. Rev., 102, 1030(1956).

5)SHIMOMAE, K., KASAI, K. and HAMAGUCHI, C., J. Phys. Soc. Jap. 49, 1060(1980).

6)KIM, R. S. and NARITA, S., Phys. Stat. Sol. (b) 73, 741(1976).

7)SHARAN, R. and HEASELL, E. L., J. Phys. Chem. Solids 31, 541(1970).

8)PUTLEY, E. H., Semicond. and Semimetals, 1, 289(1966).

9)MAKADO, P. C. and McGILL N. C., J. Phys. C 19, 873(1986).

MICROCONTACT SPECTRA OF THE ELECTRON-PHONON INTERACTION FOR METALS.

Zhernov A.P., Kulagina T.N.
I.V.Kurchatov Institute of Atomic Energy,
Moscow, USSR

In the microscopic approach we calculated the micro-contact (MC) spectra $S_{pc}(E)$ of the electron-phonon inter-action (EPI) for nontransition metals and 3d-metals and carried out the comparison with experimental data [1-4]. The basic results of the our work are discussed below briefly.

Firstly, let's consider the problem of the verification of a contact model. Metal Zn is a strong anisotropic one. In [1] the spectra $S_{pc}(E)$ were determined for some orientations of a microbridge axis relatively of crystallographic axes $\vec{n}$. There are spectra $S_{pc}^{1}(E)$ and $S_{pc}^{2}(E)$ on fig.1 for the orientation n along the hexagonal axes ($\vec{n} \parallel \vec{c}$) and for the perpendicular direction ($\vec{n} \perp \vec{c}$). The designing spectra are in a good agreement with the experimental data. This is testified for the successful choice of the microcontact model (see also[2]). As a result of the another test for a microbridge model let's consider the values $\lambda_{pc} = 2 \int (dE/E) S_{pc}(E)$, which are the integral ones over a MC spectrum. There are values of λ_{pc}, which were got for alkali metals Na, K and Li, hexagonal metals Mg, Zn and Be, and for Al, in table. These λ_{pc} correspond to the case of polycrystals and to the plane-hole-model. Values of λ_{pc} differ by 30% on the experimental

data. The making more precise of the microbridge models
and the experiment technique has to improve an agreement.
As an example, illustating some existing problems, let's
consider MC spectra for Al (fig.1). Using the pseudopotential formalism the experimental phonon density of states
and the Eliashberg spectrum $S_E(E)$ are described with a
high accuracy. As for the $S_{pc}(E)$, that it were described
with a less accuracy, especially for frequensies, correspond to the longitudinal modes [3].

Then, we would like to pay attention for the next
circumstance: one may attract the MC EPI spectra for the
analysis of cleanly dynamical problems. Firstly, these
spectra are integral characteristics over all phonon modes.
Secondly, the intensities of general peaks of $S_{pc}(E)$ depend on polarization vectors essentially. Besides, the do-
ubtless dignity of MC spectra is their high resolution, as
a rule. Data about MC spectra were used for definition of
force constants in [1], where Zn and Be were studied. Note,
that intensities of general peaks are sensible to peculiarities of the Fermi surface geometry. By another words,
the small groups of low velocities electrons may essential
influence on the structure of $S_{pc}(E)$ [4] .

On the basis of the model of Labbe-Friedel-Barisiĉ
we calculated MC spectra for five transitional metals: V,
W, Mo, Nb and Ta. The structure of these spectra is similar to phonon density states spectra (fig.1). Note, this
model allows to determine MC spectra relatively simply.
This model would be used, for example, in consideration
of different contact models and for estimates of a background in the MC-spectroscopy of 3d-metals.

The MC-method allows to get a direct detailing information about metal properties on the level of EPI spectra.
It is very perspective one. The using of more developing
models of contacts will allow to accurate the presentation
about metals parameters.

810

REFERENCES

1. Zhernov A.P. and Trenin A.E., Fiz.Met.Metalloved. <u>66</u>, 640 (1988).
2. Zhernov A.P. et al., Fiz.Nizk.Temp. <u>8</u>, 713 (1982).
3. Zhernov A.P. and Kulagina T.N., Phys.St.Sol.(b) <u>147</u>, 148 (1988).
4. Zhernov A.P., Kulagin V.D., Kulagina T.N., J.Phys.F <u>15</u>, 579 (1985).

Table. Values of theoretical and experimental parameters λ_{pc}^{th} and λ_{pc}^{exp} .

	Na	K	Li	Mg	Zn	Be	Al
λ_{pc}^{th}	0.055	0.057	0.6	0.13	0.16	0.42	0.42
λ_{pc}^{exp}	0.064	0.11	0.45	0.16	0.23	0.35	0.48

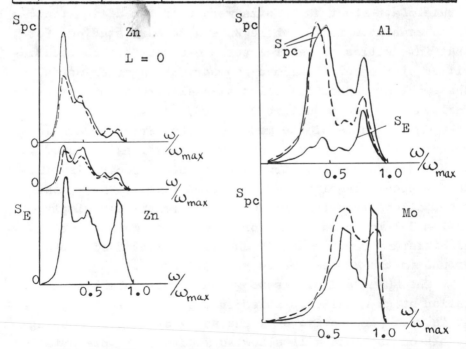

Fig.1   EPI spectra
(――― - theory,   ――― - experiment).

GENERATION OF TRANSVERSE PHONONS BY MODE CONVERSION IN NON-EQUILIBRIUM METAL FILMS

N. Perrin*, M.N. Wybourne** and J.K. Wigmore***

* Groupe de Physique des Solides de l'E.N.S., 24 rue Lhomond,
  75231 Paris Cedex 05, France;

** Department of Physics, University of Oregon, Eugene, OR 97403, USA;

*** Department of Physics, University of Lancaster, Lancaster LA1 4YW, UK.

For many years a description of the distribution of phonons injected into a substrate from a heated film has been an interesting and vexing question. The acoustic mismatch model[1] was the first attempt to describe the situation. Later a more realistic model was introduced[2] to account for the frequency dependence of the electron-phonon interaction in the metal heater. This model predicted a phonon distribution that differed significantly from a Bose-Einstein distribution.

Recent phonon pulse experiments[3], in which a large number of zone-boundary phonons were generated in the heater, showed significant departures from both models. In particular, it was found that at high excitation power densities the dominant phonon energy injected into the substrate was approximately half the zone-boundary energy of the heater material. Other experiments have also reported similar deficiencies in the high frequency components of a phonon pulse as discussed in reference 5.

In this paper we describe a phonon radiation model in which we have incorporated spontaneous phonon decay of a phonon of wavevector q within the heater film. We show that for an elastically isotropic metal film in the steady state, the spontaneous phonon decay modifies the longitudinal phonon spectrum, produces non-equilibrium transverse phonon spectra and reduces the temperature attained by the electrons upon heating by the applied electric field.

We proceed by describing the evolution of the phonon distributions by a series of rate equations that are coupled by the three-phonon processes, $L \rightarrow L + T_2$, $L \rightarrow T_1 + T_1$ and $L \rightarrow T_2 + T_2$. Each process is governed by the conservation of energy and momentum and has a characteristic relaxation rate, $\tau_1^{-1}(q)$, $\tau_2^{-1}(q)$ and $\tau_3^{-1}(q)$ respectively, that is proprotional to q^5 in the Debye model[4]. For simplicity we neglect umklapp processes and assume the Fermi surface of the metal to be spherical. Thus the electrons only couple to longitudinal phonons and the scattering is described by a rate $\tau_{ep}(q)^{-1}$ proportional to q. From the coupled rate equations the following steady state phonon distributions are obtained :

$$N_L(q) = \overline{N}_L(q,T_0) + M(q) \quad \text{for } q > q_{ZB}/2$$

$$N_L(q) = \bar{N}_L(q,T_0) + \tau_{ep}(q)/\{\tau_1(2q) \times (1+x(q))\} \times M(2q) + M(q) \qquad \text{for } q \leqslant q_{ZB}/2$$

$$N_{T_1}(q) = \bar{N}_{T_1}(q,T_0) + \left\{2\tau_b^{T_1}(q)/\tau_2(2q/c)\right\} \times M(2q/c)$$

$$N_{T_2}(q) = \bar{N}_{T_2}(q,T_0) + \tau_b^{T_2}(q) \times \left\{1/\tau_1(2q/c) + 2/\tau_3(2q/c)\right\} \times M(2q/c)$$

$$\text{where } M(q) = 1/(1+x(q)) \times \left\{\bar{N}_L(q,T_e) - \bar{N}_L(q,T_0)\right\}$$

$$\text{and} \quad x(q) = \tau_{ep}(q) \times \left\{1/\tau_b^L + 1/\tau_1(q) + 1/\tau_2(q) + 1/\tau_3(q)\right\} ;$$

q_{ZB} is the zone-boundary wavevector and c is the ratio of the longitudinal to transverse velocity of sound. The phonon escape time for each mode j is $\tau_b^j = 4\eta d/v^j$ where d is the film thickness, v^j the sound velocity and η is the frequency independent acoustic mismatch parameter. The steady state electron temperature T_e is determined from the energy balance between the electrons and longitudinal phonons.

Using this model we have calculated the phonon spectra generated by heating a 10nm thick gold film on an Al_2O_3 substrate at an ambient temperature of 7 K. The values of the scattering rates we have used are, $\tau_{ep}^{-1} = 2.1q$ s^{-1}, $(\tau_b^L)^{-1} = 0.68 \times 10^{10}$s^{-1}, $(\tau_b^{T1})^{-1} = (\tau_b^{T2})^{-1} = 0.26 \times 10^{10}$s^{-1}, $\tau_1^{-1} = 7 \times 10^{-41}q^5$ s^{-1}, $\tau_2^{-1} = 2 \times 10^{-40}q^5$ s^{-1} and $\tau_3^{-1} = 4 \times 10^{-40}q^5$ s^{-1}. Details of the calculations that lead to these values are given elsewhere[5]. For an applied electric field of 1.55×10^5 Vm^{-1}, the energy densities for the various modes are shown in figure 1. Also shown in the figure are the energy densities calculated without the down conversion process. It is seen that the effect of the three phonon processes is to reduce the maximum of the longitudinal phonon distribution to below half the zone-boundary energy and to introduce non-equilibrium transverse phonon distributions at $\omega > 3 \times 10^{12}$ s^{-1}. We have calculated the frequency of the maximum of the longitudinal phonon distribution as a function of applied electric field, figure 2. This frequency deviates significantly from the value without phonon down conversion and is seen to tend towards a limit close to half the zone-boundary of the heater material, as observed experimentally[3].

Finally, we have calculated the ratio of the electron temperatures with and without the three phonon processes, as a function of electric field, figure 3. We find that for fields greater than 10^3 Vm^{-1} the rise in the electron temperature is lower than expected. This may be attributed to more efficient dissipation of heat from the electron system that results from the spontaneous phonon decay. We note that the increase in the ratio of the electron temperatures at fields in excess of 10^6 Vm^{-1} arises from increased electron scattering that increases the resistivity of the film.

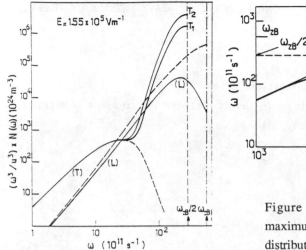

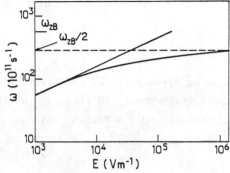

Figure 2 : frequency of the maximum of the longitunal phonon distribution

Figure 1 : energy densities for the various modes

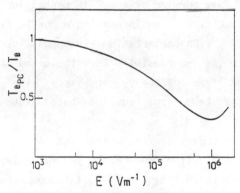

Figure 3 : ratio of the electron temperatures with (T_{epc}) and without (T_e) phonon down conversion.

1] Little, W.A., Can. J. Phys. 37, 334 (1959).

2] Perrin, N. and Budd, H., Phys. Rev. Lett. 28, 1701 (1972).

3] Wybourne, M.N., Wigmore, J.K. and Perrin, N., J. Phys : Condensed Matter (1989, in Press).

4] Berke, A., Mayer, A.P. and Wehner, R.K., J. Phys. C : Solid State Physics 21, 2305 (1988).

5] Perrin, N., Wybourne, M.N. and Wigmore, J.K. Phys. Rev. B (1989, in Press).

FIELD DEPENDENCE OF SOUND VELOCITY IN HEAVY FERMION SYSTEMS

R.J. Wojciechowski. L. Kowalewski
Solid State Theory Division, Institute of Physics, A. Mickiewicz University, 60-769 Poznań, Poland

1. Introduction

External magnetic field has a profound influence on the sound propagation in heavy fermion systems. In particular the oscillations of the sound velocity with an increasing magnetic field were observed experimentally [1,2]. Characteristic features of the electronic subsystem which is responsible for these oscillations can be interpreted in a form of heavy and very slow quasiparticles in the vicinity of the Fermi surface below the spin fluctuation temperature T^{*}.

The sound propagation effects are well defined in heavy fermion systems since the electronic Grüneisen parameter is extremely large. However, well separated singularities of the quasiparticle density of states are necessary that oscillatory dependence of the sound velocity occur.

The aim of the paper is to estimate the influence of magnetic field on the sound velocity and to discuss conditions for its oscillations with an increasing field.

2. The Model

To discuss the electronic subsystem we use the following conduction electron energy in a magnetic field

$$E_{1,k_z,\sigma} = \frac{k_z^2}{2m} + \omega_c \left(1 + \frac{1}{2} \right) + \frac{1}{2}\,\sigma g\mu_B H. \tag{1}$$

We neglect an influence of an external magnetic field on the bare 4f electron levels and on the hybridization parameter.

To describe the electronic subsystem we resort to the mean field version of the periodic Anderson Hamiltonian in the limit of infinite U [3] with an external magnetic field. After having performed a canonical transformation we get the following quasiparticle Hamiltonian:

$$H_{MF} = \sum_{1,k_y,k_z,\sigma} \tilde{E}_{1,k_z,\sigma}^{(\beta)}\, d_{1,k_y,k_z,\sigma}^{(\beta)+}\, d_{1,k_y,k_z,\sigma}^{(\beta)}\;, \tag{2}$$

where

$$\tilde{E}_{1,k_z,\sigma}^{(\beta)} = \frac{1}{2}\left\{ E_{1,k_z,\sigma} + E_{4f} \pm \sqrt{(E_{1,k_z,\sigma} - E_{4f})^2 + 4\tilde{V}^2} \right\}, \tag{3}$$

β is the band index $\pm$; E_{4f} is the renormalized 4f level and $\tilde{V}^2 = (1-n_{4f})V$ is the renormalized hybridization matrix element, n_{4f} is the mean value of the number of 4f electrons at each site.

To derive the sound velocity we assume that the dominant mechanism of the quasiparticle-phonon interaction is due to the strong volume dependence of the spin fluctuation temperature T^* leading to the large electronic Grüneisen parameter [4] $\eta = -\dfrac{v}{T^*}\dfrac{dT^*}{dv}$.

The quasipartricle-phonon interaction Hamiltonian in the basis of states in a magnetic field takes the form:

$$H_{q-ph} = \sum_{\substack{q,1,k_y,k_z \\ 1',\sigma,\beta}} A_{1,1'}(T^*,q_x)\, d_{1,k_y,k_z,\sigma}^{(\beta)+}\, d_{1',k_y+q_y,k_z+q_z,\sigma}^{(\beta)}\, Y_q\;, \tag{4}$$

where:

$$A_{1,1'}(T^*,q_x) = \frac{\eta a |q|}{\sqrt{2MN\omega_q}}\, T^*\, \frac{\theta(2\pi)^2}{L_y\,L_z} \int_{-\infty}^{\infty} h_1^*(\theta(x-x_k))h_{1'}(\theta(x-x_k)_{y'})e^{iq_x x}\, dx\;,$$

$\theta = \sqrt{m\omega_c}$, ω_c is the cyclotron frequency, h_1 are the Landau harmonic oscillator functions and $Y_q = (a_q + a_{-q}^+)$.

3. Calculation of the Acoustic Longitudinal Phonon Velocity

An influence of the magnetic field on the sound velocity can be estimated with the aid of the Dyson equations for the phonon Green's function $D(q,\omega) = \langle\langle Y_q | Y_{-q} \rangle\rangle_\omega$.

The renormalized phonon frequency is obtained from the poles of the Green function $D(q,\omega)$ which is calculated to the second order with respect to quasiparticle-phonon interaction strength.

The oscillating part of the sound velocity follows from the phonon self-energy $\Pi(q,\omega)$

$$\Pi(q,\omega) = \sum_{\substack{1,1' \\ k_z,\sigma}} \frac{n_{1',k_y-q_y,k_z-q_z,\sigma} - n_{1',k_y,k_z,\sigma}}{\omega - (\tilde{E}_{1,k_z,\sigma} - \tilde{E}_{1',k_z-q_z,\sigma})} A^2_{1,1'} \cdot (T^*,q_x) \quad (5)$$

where $n_{1',k_y,k_z,\sigma} = \langle d^+_{1,k_y,k_z,\sigma} d_{1,k_y,k_z,\sigma} \rangle$.

In the long wavelength limit we obtain the sound velocity, in the form

$$(v')^2 = v_0^2 \left(1 - \lim_{q \to 0} \frac{2}{\omega q} \Pi(q,0)\right), \quad (6)$$

where v_0 is the bare sound velocity.

On the basis of our analysis it seems reasonable to assume that the magnetic field dependence of the sound velocity in heavy fermion systems results from the field dependence of quasiparticles while the latter arises from the field dependence of the conduction electrons only.

Using the slave boson MFA Hamiltonian for the Anderson model we obtain magnetic oscillations of the longitudinal sound velocity in the limit of large magnetic field and at zero temperature. The resulting renormalized sound velocity yields precisely the same enhancement of the oscillation frequence as that obtained from the dHvA effect [5].

1. Thalmeier, P. et al, Europhys. Lett. 4, 1177 (1987).

2. Nikl, D. et al, Phys. Rev.B35, 6864 (1987).

3. Coleman, P., Phys. Rev. B29, 3035 (1984).

4. Fulde, P. J. Phys. F18, 601 (1988)

5. Rasul, J.W., Phys. Rev. B39, 663 (1989).

Ultrashort Solitons in Coupled Electron-Phonon Systems

K. Hasenburg, G. Mahler and E. Sigmund

Institut für Theoretische Physik, Universität Stuttgart,
Pfaffenwaldring 57, 7000 Stuttgart 80, FRG

Introduction: Without using the rotating wave approximation, we numerically show that the one-dimensional discrete model system of equidistant local electronic two-level centers coupled to acoustic phonons exhibits soliton solutions with half-widths down to 6 lattice constants. These solitons are dynamically stable solutions where phonon energy is absorbed and reemitted by the electronic systems. Despite the discreteness of the lattice the solitons are extremely stable[1].

The model Hamiltonian is

$$H = \sum_n \frac{1}{2M} p_n^2 \; + \; \sum_n \frac{1}{2} D(u_{n+1} - u_n)^2 \; + \; \hbar\Omega \sum_n N_{2_n} \qquad (1)$$
$$+ \; \sum_n (u_{n+1} - u_n)(\hbar\lambda\sigma_n^+ + \hbar\lambda^*\sigma_n^-).$$

Mass M and spring constant D are the parameters of the phonon system and u_n and p_n displacement and momentum operators of M at lattice site n. $\hbar\Omega$ is the energy splitting of the two-level systems, N_{2_n} the occupation number operator of the upper level of the nth center and σ_n^+ lifts one electron from the lower to the upper level at site n. The last term of eq. (1) is a deformation potential interaction with coupling constant λ. N_{2_n}, σ_n^+ and σ_n^- obey the Pauli spin algebra commutator relations, with σ_n^+ and σ_n^- corresponding to the spin-flip operators.

Equations of Motion: The time evolution of the model system is calculated within the Heisenberg picture. In order to solve the resulting set of equations numerically, we proceed from equations of operators to equations of expectation values. These c-numbers are the matrix elements of the operators taken over

818

coherent states. This restriction is justified because in our model we consider coherent excitations with non-zero mean values for u_n and p_n etc. in both the electronic and the vibrational systems. Furthermore, since we are interested in nonlinear effects, excitation energies are expected to be rather high. Thus the preconditions are satisfied to confine ourselves to expectation values in a good approximation.[2] We thus obtain

$$\ddot{u}_n = \frac{D}{M}(u_{n-1} - 2u_n + u_{n+1}) + \frac{2\hbar}{M}\,\mathrm{Re}(\lambda(\sigma_n^+ - \sigma_{n-1}^+)) \tag{3a}$$

$$\dot{\sigma}_n^+ = i\Omega\sigma_n^+ + i(1 - 2N_{2_n})\,\lambda^*(u_{n+1} - u_n) \tag{3b}$$

$$\dot{N}_{2_n} = 2(u_{n+1} - u_n)\,\mathrm{Im}(\lambda\sigma_n^+) \tag{3c}$$

The origin of the nonlinearity is the saturation term $(1 - 2N_{2_n})$ in equation (3b) which arises from the Fermi commutation relation of the electronic operators.

Results: The common analytic approach for the case of phonons, which are in resonance with the 2-level-centers, employs both the rotating wave approximation (RWA) and the continuum approximation (analogous to the case of self-induced transparency in nonlinear optics[3]). This approach yields the Sine-Gordon nonlinear differential equation. In this paper, however, we present a direct numerical solution of (3) for a system of 600 lattice sites (cyclic boundary conditions) **without** using RWA and continuum approximation.[1] This allows for both resonant high frequency phonons (i.e., wavelength $\lambda_{res} \sim$ lattice constant a_o) **and** the search for "ultrashort" solitons[4] (i.e., half-width in the same order of

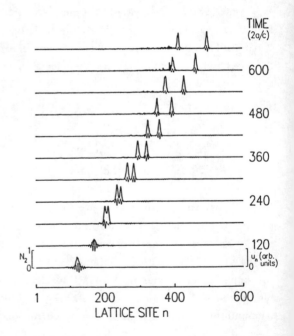

Fig. 1: Pair formation of "ultrashort" solitons. Displacement u_n (thin line) and occupation number N_{2_n} (heavy line) for successive time steps t. Initial condition is a resonant sech-enveloped phonon pulse, with $N_{2_n} = \sigma_n = 0$.

magnitude as λ_{res}). Fig. 1 shows that we have found such solitons with $\lambda_{res} = 6a_0$ and soliton half-width $\Delta \sim \lambda_{res}$. For times larger than in fig. 1 the solitons decay, not because of the lattice discreteness, but by interaction with those phonons which are not bound to the soliton.

The continuum approximation of the analytical approach replaces the difference terms in eq. (3) by the corresponding derivatives. Its range of validity has been tested by comparison with the direct numerical solution for different relations λ_{res}/a_0. It comes out that already for $\lambda_{res} \leq 12a_0$ the soliton total energy begins to deviate considerably! The RWA has been tested by comparison to the numerical results with decreasing relation Δ/λ_{res}. For small soliton velocities the RWA astonishingly reproduces both shape and total energy of the soliton down to $\Delta \sim 1.5\ \lambda_{res}$!

Nonlinear Heterostructures: The saturation term in eqn. (3) gives rise to soliton solutions. This effect allows to accomplish even functional elements by nonlinear heterostructures. These are obtained by a space-dependent electron- phonon coupling constant λ: An "acoustic shutter", e.g., is depicted in fig. 2: In region A, where the electron-phonon coupling λ is set to zero, a phonon mode is switched on adiabatically. When passing through the interface to region B - where λ has the finite value 0.2 - the running plane wave is transformed into a chain of solitons. The second interface B-A now transformes the solitons *completely* back into a *train of phonon pulses.*

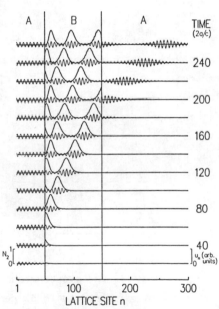

Fig. 2: Indications as in fig. 1. A single linear phonon mode (region A: $\lambda = 0$) is transformed into a chain of phonon pulses while passing the interface region B with $\lambda = 0.2$: "Acoustic shutter".

References:

[1]: Hasenburg, K., Sigmund, E. and Mahler, G., Phys. Scr. (1989), in print, and Phys. Rev. B, to be published (1989)

[2]: Haken, H. and Schenzle, A., Z. Phys. **258**, 231 (1973)

[3]: McCall, S.L. and Hahn, E.L., Phys. Rev. **183**, 457 (1969)

[4]: Kujawski, A., Z. Phys. B**66**, 271 (1987)

ULTRASONIC STUDY OF MAGNETIC FREEZE-OUT IN n-InSb AT LOW TEMPERATURES

J.H. Page[§] and F. Guillon[†]

§ Physics Department, University of Manitoba, Winnipeg, Manitoba, Canada
† Département de Physique, Université d'Ottawa, Ottawa, Ontario, Canada

Current studies of magnetotransport in semiconductors are focussing primarily on magnetic-field-induced localization in 3D narrow-gap materials and on the quantum Hall effect in 2D heterostructures. In piezoelectric semiconductors, these electronic properties can be investigated using ultrasonic techniques because the strong acoustoelectric coupling leads to changes in the acoustic velocity and attenuation. Hutson and White[1] showed that these changes are given by

$$\frac{\Delta v}{v_0} = \frac{\kappa^2}{2} \left[1 + \frac{\omega_C}{\omega_D} + \left(\frac{\omega}{\omega_D}\right)^2 \right] \Bigg/ \left[1 + \left(\frac{\omega_C}{\omega}\right)^2 \left(1 + \frac{\omega^2}{\omega_C \omega_D}\right)^2 \right] \qquad (1)$$

$$\alpha = \frac{4.34 \ \kappa^2}{v_0} \ \omega_C \Bigg/ \left[1 + \left(\frac{\omega_C}{\omega}\right)^2 \left(1 + \frac{\omega^2}{\omega_C \omega_D}\right)^2 \right] \qquad (2)$$

where v, α and ω are the ultrasonic velocity, attenuation and frequency, κ^2 is the electromechanical coupling constant, and the attenuation is expressed in dB per unit length. The dependence on electrical properties is determined by the dielectric relaxation frequency $\omega_C = \sigma/\varepsilon\varepsilon_0$, where σ is the electrical conductivity and ε the relative dielectric permittivity, and the diffusion frequency $\omega_D = v^2/fD$, where D is the diffusion constant and f accounts for the fraction of the acoustically produced space charge that is mobile. Recent attenuation measurements[2] as a function of magnetic field in n-InSb have shown relaxation peaks that suggest an unusually large contribution from the diffusion term in eq.(2). The purpose of the

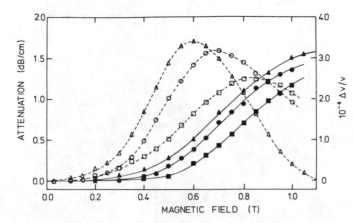

Fig. 1 Ultrasonic velocity (solid symbols) and attenuation (open symbols) at 68 MHz (□), 120 MHz (○), & 165 MHz (△). The temperature is 1.5 K.

present work is to use both ultrasonic velocity and attenuation measurements to investigate these apparent diffusion effects more fully, and hence to attempt a more quantitative investigation of magnetic freeze-out in n-InSb.

The present experiments were performed on a n-InSb sample with a low excess donor concentration N_D-N_A of $6.3 \times 10^{19}/m^3$. Longitudinal ultrasonic waves were used with wavevector $\underline{q} \parallel [111]$ and magnetic field $\underline{B} \perp \underline{q}$. To avoid complications from hot electron effects, the data were taken at low powers (typically $P_{ac} \sim 1-5$ μW) where the data were independent of the acoustic intensity.

Figure 1 shows the field dependence of $\Delta v/v$ and α at 1.5 K for several ultrasonic frequencies. These data cannot be interpreted using the Hutson-White model unless the terms in ω_D are included, since (i) the height of the attenuation peaks does not increase linearly with frequency and (ii) at high frequencies the maximum in α does not occur at the same field as the midpoint of the step ($\kappa^2/4$) in $\Delta v/v$ (cf. Ref. 1). The data were analysed quantitatively by solving eqs.(1) and (2) for ω_c and ω_D in terms of $\Delta v/v_0$ and α, and using measured values of $\kappa^2 = 6.4 \times 10^{-4}$ (Ref. 3) and $v_0 = 3740$ m/s to obtain the values of ω_c and ω_D shown in Fig. 2. From the relation $\omega_c = \sigma_{xx}/\varepsilon\varepsilon_0 = ne\mu/\varepsilon\varepsilon_0(1 + \mu^2 B^2)$, and the value of the mobility $\mu = 2.0$ m²/V·s deduced by comparing attenuation data in longitudinal and transverse fields, the field dependence of the conduction band concentration n was determined. This was used to evaluate the donor binding energy E_d from standard relations for $n \propto \exp(-E_d/k_B T)$. The results are shown

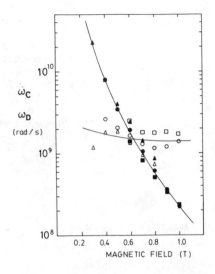

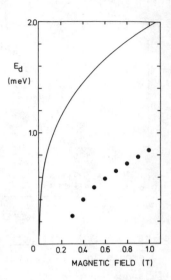

Fig. 2 Field dependence of ω_c (solid symbols) and ω_D (open symbols).

Fig. 3 Field dependence of the donor binding energy E_d.

in Fig. 3 where it can be seen that E_d increases with magnetic field causing the carriers to be "frozen out" of the conduction band. In this range of fields our values of E_d are substantially smaller than predicted by the YKA model[4] (solid curve). By contrast with ω_c, the parameter ω_D appears to be almost independent of magnetic field for the data analysed in Fig. 2. In a transverse field the diffusion frequency ω_D is $ev^2(1+\mu^2B^2)/f\mu k_BT$, giving values of order 10^{11} rad/s if f=1 or larger still if f<1. Since the values shown in Fig. 2 are about two orders of magnitude smaller than this, we conclude that for n-InSb, the parameter that plays the role of ω_D in eqs. (1) and (2) cannot be ascribed to diffusion effects but must characterize some other phenomenon. Further work is planned to investigate the origin of this new relaxation effect and to extend the measurements to higher fields and other doping concentrations.

References

1. Hutson, A.R. and White, D.L., J. Appl. Phys. 33, 40 (1962).
2. Guillon, F., Fernandez, B. and Cheeke, J.D.N., to be published.
3. Quirion, G., Poirier, M. and Cheeke, J., J. Phys. C 20, 917 (1987)
4. Yafet, Y., Keyes, R.W., and Adams, E.W., J. Phys. Chem. Solids 1, 137 (1956).

EFFECT OF OPTICAL PHONONS ON THE EXCITONIC SELF-TRAPPING RATE

Michael Schreiber and Hans–Jörg Kmiecik
Institut für Physik, Universität Dortmund, 4600 Dortmund 50, Federal Republic of Germany

We determine the adiabatic potential energy surface for excitons in a deformable lattice by variation with regard to the spatial extent of the excitonic wave function. To describe the autolocalization process we analyze the potential barrier separating free and self–trapped excitonic states and calculate the tunneling rate within the model of multi–phonon non–radiative transitions. Special emphasis is on the influence of optical phonons in alkali halides.

1. INTRODUCTION

According to the Franck–Condon principle the optical absorption process in insulator and semiconductor crystals yields excitonic Bloch waves in a rigid lattice. These so–called free excitons (FE) are mobile within the whole crystal. Due to the exciton–lattice interaction, however, the exciton can induce a local lattice distortion at which an excitonic state can be localized. If the interaction is sufficiently strong, this self–trapped exciton (STE) is energetically lower than the FE so that a strongly Stokes–shifted luminescence can be observed in addition to the FE–emission.[1]

In alkali halides this autolocalization process is known to be determined by the self–trapping (ST) of a hole within the halogen sublattice,[2] where the lattice distortion leads to the formation of a quasi–molecule of halogen ions along the $\langle 100 \rangle$–direction in simple cubic compounds like CsI, and along the $\langle 110 \rangle$–direction in fcc crystals like RbI. Subsequently at these V_k–centres an electron can be trapped; this process further reduces the halide distance and thus stabilizes the deformation.[2]

Taking only the coupling of the excitons to acoustic phonons into account, we have already described this ST process in rare gas solids[3] as well as in alkali halides.[3,4] In the present paper we quantitatively analyse the influence of optical phonons which can be expected to significantly change the previously obtained results.

2. THE MODEL

Our model Hamiltonian[3,5] comprises the kinetic energy of the FE, characterized by the half width B of the exciton band, the local harmonic lattice vibrations of suitably averaged phonon frequency ω, the coupling of the excitons to acoustical phonons of linear dispersion in a deformation potential ansatz and the Fröhlich–type interaction with dispersionless optical phonons in agreement with the ansatz of Nasu and Toyozawa.[6] In the adiabatic approximation the coupling leads to a static local distortion, characterized by an exponential decay constant α for the spatial extent and a configurational coordinate Q for the amplitude of the interaction mode.

The lowest adiabatic potential surface is obtained by minimizing the expectation

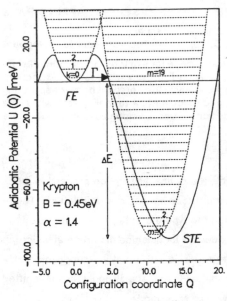

value of the Hamiltonian with respect to the localization length of the STE. For fixed α, a typical result is shown in Fig. 1. The FE and STE potential valleys are separated by a potential barrier (PB) which the system has to tunnel through before quickly relaxing to the STE state with an energy gain ΔE.

Neglecting the dependence of the exciton localization length on the configurational coordinate during the tunneling process, one can employ second order perturbation theory to calculate the tunneling rate (i.e., the ST rate) at vanishing temperature from Fermi's golden rule.[3),5),6)] This treatment is based on the theory of multi-phonon non-radiative transitions and effectively replaces the adiabatic potential by two independent harmonic oscillators as shown in Fig. 1.

Considering the spatial extent α^{-1} of the distortion as a variational parameter, too, we determine the maximal ST rate Γ_{STE} as well as the minimal height U_B of the PB which can be compared to experimental data.

3. RESULTS AND DISCUSSION

As the input parameters B und ΔE are not precisely known from experiment, results for various values of B are plotted in Fig. 2 in dependence on ΔE. Data obtained previously[4)] for the coupling to acoustic phonons only are shown by thick lines, here the coupling constant S_a is varied to give the required ΔE in each case. We have then fixed S_a at certain values corresponding to $\Delta E = -0.4 eV$, $-0.8 eV$. $-1.2 eV$, and $-1.6 eV$ and increased the coupling constant S_o which governs the Fröhlich interaction of the excitons and the optical phonons. As a consequence new values of ΔE result. Due to the delicate mixing[5),6)] of acoustic and optical modes entering the configuational coordinate Q, the energy gain ΔE first decreases and then increases with increasing S_o (see thin lines in Fig. 2).

An important parameter in this ansatz is the Wannier radius γ of the exciton, because the Fröhlich interaction crucially depends on the internal structure of the exciton. We have computed[5)] the influence of the optical phonons for various values of γ, for sc and fcc lattices, and for input parameters B, ΔE, and ω which are typical for CsI, RbI, and KI.[2)]

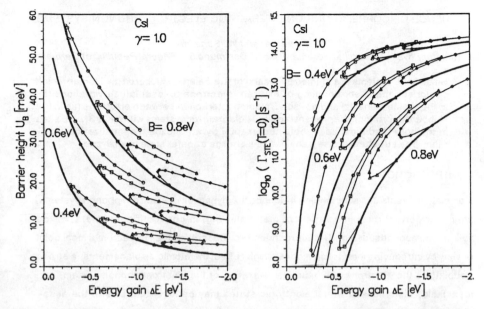

Fig.2: Minimal barrier heights U_B and maximal tunneling rates Γ_{STE} for $S_0=0$ (thick lines) and $S_a=const.$ (thin lines, symbols mark an increase of S_0 by 4).

A detailed analysis yields the following results:[5] For fixed ΔE, a more extended lattice distortion and, correspondingly, a less localized excitonic wave function has to be adopted by the system during the ST process if optical phonons are taken into account. As a consequence, for all B and γ, the maximal rates are always increased. This effect is more pronounced for Wannier excitons ($\gamma > 1$) than for Frenkel excitons ($\gamma < 1$). The exciton localization length which corresponds to the minimal barrier height is of the order of two lattice constants. As a consequence the coupling to optical phonons is very effective for Wannier excitons with $\gamma=2$ and yields a distinct reduction of the barrier height. On the other hand, for Frenkel excitons the barrier heights increase with increasing coupling S_0.

In conclusion, we have demonstrated how optical phonons alter the essential charac-teristics of the ST process. A comparison with available experimental data[3],[4],[5] shows that the inclusion of the optical phonons is necessary to achieve satisfactory agreement.

1 Nishimura,H., Ohhigashi,C., Tanaka,Y., Tomura,M., J. Phys. Soc. Jpn. 43, 157 (1977)
2 Lushchik, Ch. B., in: "Excitons", ed. Rashba, E. I., Sturge, M. D., (North–Holland, Amsterdam 1982), p.505; Iida, T. and Monnier, R., phys. stat. sol (b) 74, 91 (1976); Higashimura, T., Nakaoka, Y., and Iida, T., J. Phys. C17, 4127 (1984)
3 Kmiecik, H. J. and Schreiber, M., J. Lumin. 37, 191 (1987); Kmiecik, H. J., Schreiber, M., Kloiber, T., Kruse, M., and Zimmerer, G., J. Lumin. 38, 93 (1987)
4 Schreiber, M., Kmiecik, H. J., and Schmidt, W., J. Lumin. 40/41, 453 (1988)
5 Kmiecik, H. J., Doctor thesis, Universität Dortmund, 1989
6 Nasu, K. and Toyozawa, Y., J. Phys. Soc. Japan 50, 235 (1981)

INFLUENCE OF CORRELATIONS ON INTERACTING ELECTRON-PHONON SYSTEMS

M. Schreiber and W. Schmidt
Institut für Physik, Universität Dortmund, 4600 Dortmund 50, Federal Republic of Germany

Small finite systems are studied by means of the Peierls-Hubbard Hamiltonian which comprises the linear coupling of itinerant electrons to local lattice vibrations (as in the Holstein ansatz) and on-site Coulomb interaction between electrons (as in the Hubbard approach). We present the phase diagram for systems with $N=2, 3, \ldots, 8$ sites in the half-filled band case, obtained exactly by a Lanczos diagonalization algorithm. The structural phase transition to a charge transfer state is analyzed.

1. INTRODUCTION

Correlation effects can substantially influence the dynamics of electron-phonon systems, playing important roles in a variety of materials, e.g. in conducting polymers, mixed valence compounds, or high T_c superconductors. Describing the lattice from a local point of view by introducing interaction modes which reflect the atomic displacements, a strong coupling to the electrons can induce a relaxation of the electron-phonon system, i.e., local lattice distortions at which electronic states may be localized. Due to the associated energy gain this tendency can be interpreted as an effectively attractive electron-electron interaction, favouring charge-transfer states. In contrast, the on-site Coulomb interaction strongly disfavours such charge enhancements on certain sites and tends to uniformly distribute the electrons. In this paper we analyze this competition for small finite systems, for which we can obtain exact results from a direct diagonalization of our simple model Hamiltonian, described in the next chapter.

2. MODEL HAMILTONIAN

We consider a one-dimensioal model described by the Hamiltonian

$$H = -B/2 \sum_i \sum_s (c_{i,s}^+ c_{i+1,s} + c_{i+1,s}^+ c_{i,s}) + U \sum_i n_{i+} n_{i-} + \tfrac{1}{2} \sum_i Q_i^2 - S^{1/2} \sum_i n_i Q_i \qquad (1)$$

which was originally proposed by Toyozawa.[1] The sums over i include N sites of a finite chain with periodic boundary conditions and the spin index s takes the values + and −. The ratio of the basic interactions is determined by three parameters:

The bandwidth B characterizes the itinerant electrons, described by the hopping between nearest neighbours. The Coulomb energy U reflects the short-range electronic correlations as in the Hubbard model. The lattice vibrations are described simply by Einstein oscillators. In the adiabatic approximation only the harmonic potential term in (1) has to be taken into account, defining the configurational coordinates Q of the interaction modes appropriately. Finally, the lattice relaxation energy S is determined as the energy gain due to the linear coupling of the electron density n to the interaction modes.

We consider small finite systems of up to N=8 sites with an equal number of electrons, i.e., the half-filled band case. As the model Hamiltonian does not mix states with different total spin s, it is possible to restrict the solution of the secular equation to a Hilbert space with significantly reduced dimension (e.g. 1764 for s=0 instead of 12870 in the N=8 case). The respective projection operators[2] for the construction of the subspace can be straightforwardly defined and numerically employed. A further classification of the basis states by means of the irreducible representations of the 2N-dimensional Dieder group of the possible rotations and reflections of the N sites, however, allows a distribution of the eigenstates of the adiabatic Hamiltonian only in the case Q=0. In the non-adiabatic treatment this analysis is indispensible.[2]

Consequently, matrices of size 3, 8, 20, 75, 175, 784, and 1764 had to be solved for the N=2, 3,..., 8-systems, respectively, which can be effectively performed by a Lanczos diagonalization algorithm, because of the sparseness of the involved matrices. The resulting lowest adiabatic energy, however, depends on the N configurational coordinates Q_i (out of which the trivial fully symmetric linear combination can be separated) requiring an (N−1)-dimensional minimization to obtain the stable ground state.

As the results, which are described in the next chapter, depend only on the ratio of the input parameters, it is convenient to use a triangular representation of the parameter space, where the relative parameters of a given point are determined as usual[1] by its distance from the edges.

3. RESULTS

The obtained phase diagrams and the type of the resulting ground states as well as their degeneracy are displayed in fig. 1. The results corroborate earlier findings[3],[4] that

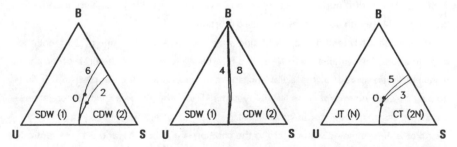

Fig. 1: Phase diagrams for the ground state of an N-site N-electron system. N is given in the figure. The ground state phases are characterized by the electronic configuration as spin-density (SDW) or charge-density waves (CDW), or by the lattice deformation as Jahn-Teller (JT) or charge-transfer (CT) distortion. The degeneracy is shown in parentheses. Thick lines mark first-order, thin lines second-order phase transitions.

systems with N=4n-2, N=4n and N=2n+1 behave differently (n being an integer). The discrepancies can be attributed, on one hand, to the different degeneracy of the states at the Fermi level in the band limit, and on the other hand, to the unpaired spin in systems with odd N.

In all systems two extreme situations can be distinguished defining the two possible system phases: If the electron-phonon interaction significantly dominates the correlation effects, the ground state suffers a relaxation from $Q_1=0$ to a distorted lattice configuration ($Q_1 \neq 0$). Simultaneously the charge density becomes inhomogeneous, namely enhanced at sites with $Q_1 > 0$ and weakened where $Q_1 < 0$. Thus in the even-numbered systems a charge-density wave is formed; in the dimer case, this is simply the ionic state. Strong correlation on the other hand leads to an uniform charge distribution, but spin ordering. For example, in systems with even N a spin-density wave is stabilized.

The transition between these two limiting cases depends on the strength of the transfer interaction. For small bandwidth B there is always a first-order phase transition between these two regimes. Larger B leads to a second-order transition except for the N=4n-systems. The corresponding phase boundary ends on the BS-edge of the triangle, reflecting a Peierls transition for these cases. However, with increasing system size the critical point O approaches the B-vertex, so that the phase diagrams assimilate to the N=4n-case. The ground state was always found to have the lowest possible total spin. As a consequence, in the even-numbered systems the charge transfer state is doubly degenerate, while the spin-density wave is non-degenerate.

In the odd-numbered systems with strong electron-phonon coupling, the unpaired spin leads to 2N degenerate charge transfer minima. In contrast to the above-mentioned spin-density waves in these systems the ground state is distorted even for small S. Due to the Jahn-Teller effect it suffers a spontaneous symmetry breaking resulting in an N-fold degenerate state with comparatively small distortion.

We have numerically tested that the ground state is always given by one of the mentioned configurations for systems up to N=8. It is interesting to note that the first-order phase transition is an artifact of the adiabatic approximation and will disappear in the non-adiabatic treatment except in the N=4n-systems.[2] Here the phase transition persists because the vibronic states for large U and large S are of different symmetry (namely symmetric or anti-symmetric with regard to the rotation i→i+1, i.e., A1 and B1, respectively) so that no hybridization occurs.

1 Toyozawa, Y., J. Phys. Soc. Jpn. 50, 1861 (1981)
2 Schmidt, W., Doctor thesis, Universität Dortmund 1989 and to be published.
3 Takimoto, J. and Toyozawa, Y., J. Phys. Soc. Jpn. 52, 4331 (1983)
4 Weber, S. and Büttner, H., Solid State Comm. 56, 395 (1984)

AVERAGE ELECTRON-PHONON INTERACTION, $<I^2>$,
FOR HCP METALS

C. -G. Jiang, G. Fletcher and J. L. Fry

Department of Physics
The University of Texas at Arlington
Arlington, Texas 76019 USA

1. INTRODUCTION

This paper reports calculations of the Fermi surface averaged electron-phonon interaction, $<I^2>$, for hcp metals using an orthogonal-tight-binding approximation (OTBA).[1] While monatomic crystal structures may be treated within the rigid-muffin-tin approximation (RMTA),[2] crystals with more than one atom per primitive cell require additional, sometimes doubtful approximations. The purpose of this paper is to obtain $<I^2>$ for the hcp elements for comparison with other estimated values, and to prepare for dealing with the more complex issue of the electron-phonon interaction in copper-oxide materials with many atoms per primitive cell.

2. METHOD

The OTBA is straightrorward, but can be tedious and computationally intensive. For the hcp structure with two identical atoms per unit cell the quantity to be computed is

$$<I^2> = \frac{FS \int d^2p/v_p \ FS' \int d^2p'/v_p' \ \Sigma(j) \ |\Sigma(\ell) <p|\vec{e}_j(q,\ell).\nabla U_\ell|p'>e^{i\vec{q}.\vec{r}_\ell}|^2}{(\ FS \int (d^2p/v_p))^2}$$

where $\vec{e}_j(q,\ell)$ is the polarization vector for phonon branch j with momentum $\vec{q} = \vec{p} - \vec{p}'$, U_ℓ is the potential of an ion at $\vec{r}_\ell$, and v_p is the electron velocity at momentum $\vec{p}$ on the Fermi surface (FS). Using Frolich's tight-binding approximation[3] this integral may be expressed in terms of gradients of a tight-binding parameterization of first principles band structure calculations. In this work two-centered, orthogonal parameters for the hcp elements were taken from the book by Papaconstantopoulos.[4] Gradients of the parameters were computed in spherical coordinates using the scaling laws of Harrison[5] to evaluate the radial part. The final expression for $<I^2>$ is too lengthy to reproduce here, but involves sums of products of various Fermi surface integrals similar to the formulas of Varma et al.[6] Surface integrals were evaluated using a version of the analytic tetrahedron method.[7]

3. RESULTS

Table I summarizes the results for 15 hcp elements. The Fermi energy, E_F, and density of states at E_F, $D(E_F)$, are given in columns 2 and 3, and $<I^2>$ values from this work (OTBA), empirical estimates,[8-9], and RMTA for cubic phases[10] are given in columns 4, 5 and 6 respectively.

4. CONCLUSION

The $<I^2>$ values of OTBA are sometimes sensitive to the input band structure (shape of FS etc.), but agreement between OTBA, RMTA and empirical estimates using very different techniques suggests that a basic understanding of the average-electron phonon interaction has been achieved. Neither RMTA nor OTBA approaches are expected to work well for free electron metals (e.g. Be in Table I). A direct comparison of RMTA and OTBA will be made for hcp elements in future work.

5. ACKNOWLEDGMENT

This research was supported by The Robert A. Welch Foundation.

6. REFERENCES

1.   Fletcher, G., et al. Phys. Rev. B <u>37</u>, 4944 (1988).
2.   Gaspari, G. D. and B. L. Gyorffy, Phys. Rev. Lett. <u>28</u>, 801 (1972).
3.   Froelich, H., in <u>Perspectives in Modern Physics</u>, edited by R. E. Marshak (Interscience, New York, 1966).
4.   Papaconstantopoulos, D. A., <u>Handbook of the Band Structures of Elemental Solids</u> (Plenum, New York, 1986).
5.   Harrison, W. A., <u>Electronic Structure and Properties of Solids</u> (Freeman, San Francisco, 1980).
6.   Varma, C. M., et al., Phys. Rev. B <u>19</u>, 6130 (1979).

7. Pattnaik, P. C. et al., International J. Quantum Chem. **15**, 499 (1981).

8. Hopfield, J. J., Phys. Rev. **186**, 443 (1969).

9. Butler, W. H., Phys. Rev. B **15**, 5267 (1977).

10. Papaconstantopoulos, D. A., et al., Phys. Rev. B **15**, 4221 (1977).

Table I. Fermi surface averaged electron-phonon interaction, $<I^2>$, in $(Ry/a.u.)^2$, E_F in Ry and $N(E_F)$ in States/Ry.cell.

Element	E_F	$N(E_F)$	$<I^2>$		
			OTBA	Empirical	RMTA
Be	0.766	1.1	0.1155		0.0262
Mg	0.403	9.8	0.0182		0.0008
Sc	0.429	59.8	0.0032		0.0042
Ti	0.590	29.0	0.0050	0.0077[a]	0.0080
Co	0.683	85.1	0.0026		0.0060
Co↑	0.695	9.4	0.0043		
Co↓	0.695	42.1	0.0052		
Zn	0.397	6.9	0.0058		0.0052
Y	0.395	53.3	0.0033	0.0008[b]	0.0044
Zr	0.544	28.6	0.0061	0.0082[b]	0.0112
Tc	0.700	47.3	0.0110	0.0244[b]	0.0212
Ru	0.763	25.4	0.0122	0.0167[b]	0.0181
Cd	0.215	8.0	0.0032		0.0030
Hf	0.572	26.2	0.0076	0.0119[a]	
Re	0.799	16.4	0.0183	0.0352[a]	
Os	0.849	18.9	0.0189		
Tl	0.343	15.4	0.0056		

[a]Ref. 8   [b]Ref. 9

ON PHONON SCATTERING IN NiPt ALLOYS

H. LITSCHEL, I. ISARIE,
Faculty of Mechanical Engineering, Sibiu, Romania

Measurements were taken of the temperature dependence of the electric resistivity (ϱ) [1], spontaneous intensity of magnetization (σ_S) [2] and Hall effect in $Ni_{1-x}Pt_x$ alloys (x=1,2,3,4,5,8,10,15,20,25,30 at%) from 77 K up to their ferromagnetic Curie temperatures. The samples of $Ni_{1-x}Pt_x$ were prepared by using high purity metals (99.99% Ni and 99.999% Pt). Melting was performed in an arc-melting furnace in an argon atmosphere. In order to eliminate any segregation and to obtain perfectly disordered solid solutions all samples were thermally treated in vacuum at 1173K for 24 hours and then cooled (150K/h). The electric resistivity and Hall effect measurements were taken using plates of under 0.2mm which had been laminated and polished to the desired thickness. After these operations, the samples were again thermally treated to eliminate texturing.

The phononic (ϱ_f), magnetic (ϱ_m), and residual (ϱ_o) parts of electric resistivity were separated and represented as temperature functions (ϱ_f and ϱ_m), respectively concentration function (ϱ_o). The numerical values of the residual resistivity of nickel-platinum alloys are the lowest in nickel plus 5d elements alloys which shows that the perturbation potential due to the impurity atoms is small. The same conclusion - small perturbation potential - can be drawn from the temperature dependence of the phonon part which is linear; and the curves corresponding to the various concentrations are very close.

The magnetic part of the resistivity grows relatively slightly with the increase of the platinum concentration, while the residual part of this resistivity decreases, which is completely in line with Mott's theoretical model.

In order to decipher the scattering mechanisms of the electrons of conductibility we analysed the dependences:

$$R_S = a\varrho + b\varrho^2; \quad R_S = \text{const } \varrho^n$$

$$R_S = \alpha + \beta\varrho_m; \quad R_S = A[\sigma_S^2(0) - \sigma_S^2(T)]$$

$$R_S = \text{const}[\sigma_S^2(0) - \sigma_S^2(T)]^m; \quad R_S = R_S(\varrho_f^2)$$

and we calculated the value of the coefficients a and b as well as the exponents n and m.

Some of the experimental data are given in figures 1 and 2.

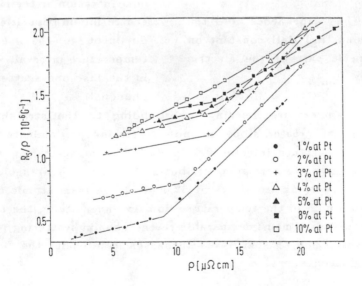

Fig. 1   The R_S/ϱ ratio dependence of electric resistivity (ϱ)

834

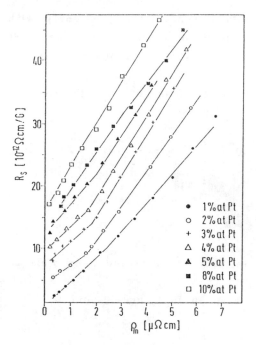

Fig. 2 The dependence of the spontaneous (R_S) Hall constant on the magnetic part of the electric resistivity (ϱ_m)

The analysis of the dependences yields the following main conclusions:

- the $R_S = R_S (\varrho_f^2)$ dependence is linear, the influence of the impurities upon the spontaneous Hall constant being stronger than their influence upon the residual resistivity;

- in the domain of temperature where the slope of the straight lines $R_S / \varrho = f(\varrho)$ is smaller, the essential contribution to R_S comes from the scattering of impurities and phonons (except for the samples of 1%, 5%, and 25% platinum atomic concentration, in the first two scattering on impurities being dominant in the mentioned temperature interval, whereas in the last one scattering on phonons);

- in the temperature domain corresponding to the straight lines $R_S / \varrho = f(\varrho)$ of greater slope, phonon scattering predominates;

- in the temperature interval $0.4\,T_C$ - $0.7\,T_C$ the scattering of spin waves makes an essential contribution to the spontaneous Hall constant. Exceptions are alloys of 5% and 8% platinum atomic concentration for which in the temperature domain under study, the scattering on impurities remains comparable to the phonon scattering process, whereas scattering on spin waves surpasses in intensity the first two scattering mechanisms.

References

1) H. Litschel and I. Pop., J. Phys. Chem. Solids, 46, 1421 (1985)
2) H. Litschel and I. Pop., Studia, Ser. Physica, 1, 3 (1988)

ELEKTRON-PHONON INTERACTION IN NiTi

M. KEIL, M. MÜLLNER, W. WEBER*

Institut für Kernphysik, J. W. Goethe Universität,
D-6000 Frankfurt 90, FRG
* Institut für Nukleare Festkörperphysik, Kernforschungs-
zentrum Karlsruhe, D-7500 Karlsruhe, FRG

In order to understand the atomistic mechanism of the shape memory effect and to interprete our experimental results on phonon dispersion /1/, we studied the lattice dynamics of equiatomic NiTi in the austenitic CsCl-phase taking into account the electron-phonon interaction by means of the model of Varma and Weber /2/. In this model the dynamical matrix is divided into two parts: a long range part, which is calculated from the band structure by perturbation theory. This part consists of the generalized susceptibility χ and the electron-phonon matrix elements; and a short range part, determined by a next nearest neighbour Born-von Kármán fit.

The basis of our calculations was a tight-binding fit to a selfconsistent tight binding band structure calculation. The results are very similar to

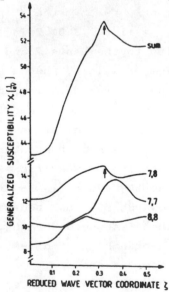

Figure 1: Generalized susceptibility χ in Σ-direction

those reported by Papaconstantopoulos et al. /3/,
with subtle differences
especially at the Fermi
surfaces.

In fig. 1 the generalized
susceptibility χ in Σ-
direction is shown toge-
ther with its components,
the respective band-band
transitions. Only bands
crossing the Fermi level
are considered. The Q-
position of the χ-maxi-
mum, that predominantly
consists of band 7 to 8
transitions, is in agree-

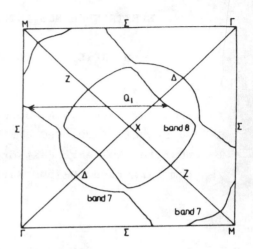

Figure 2: Fermi surfaces in
(001)-plane with nesting
vector Q_I

ment with the appropriate nesting wave vector
$Q_I = (1/3 \; 1/3 \; 0) \; 2\pi/a$ between the Fermi surfaces belonging to the bands 7 and 8 (fig. 2).

The resulting phonon dispersion curves are compared with
the neutron data in fig. 3. The main result is the prediction of the condensed T_2A soft mode in Σ-direction near Q_I
due to the electron-phonon interaction. Thus the phase
transition to the intermediate phase is caused by a strong
coupling of specific electrons near the Fermi energy.

The LA-branch in Σ-direction is also drastically lowered
towards the zone boundary where it becomes degenerate with
the T_2A-phonon branch. Another anomaly is present in the
LA-branch in Λ-direction, which is renormalized over a
wide Q-range.

The derived phonon density of states predicts a gap between optical and acoustical phonons as in our Born-von
Kármán calculations in contrast to the experimental results /4/ (cf. the previous contributions).

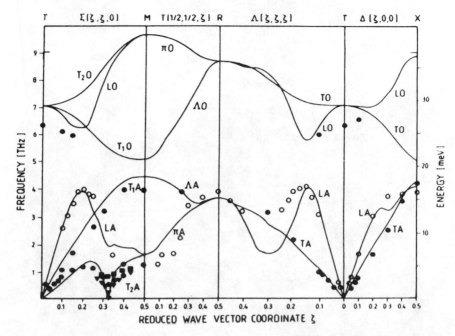

Figure 3: Calculated phonon dispersion. Experimental Data messured by Herget et al. /1/ (dots). ■ 500 K/ o ● 400 K/ ▼ 320 K

/1/ G. Herget, M. Müllner, G. Eckold,
Annual Report IKF-48 (1988) 73
Proc. Int. Symp. "The Martensitic Transfomation in Science and Technology", 9/10 March 1989, Bochum FRG, in print
/2/ C. M. Varma, W. Weber, Phys. Rev. B19 (1979) 6142
/3/ D. A. Papaconstantopoulos, G. N. Kamm, P. N. Poulopoulos, Solid State Commun. 41 (1982) 93
/4/ G. Herget, M. Müllner, J. B. Suck, R. Schmidt, H. Wipf Europhysics Lett., in print

This work was financially supported by the German Federal Minister of Research and Technology under Contract No. BMFT 03-MU1FRA.

7

Surface Phonons

SURFACE DYNAMICS AND SURFACE RECONSTRUCTION

F.W. de Wette* and A.D. Kulkarni
Department of Physics, University of Texas
Austin, Texas 78712, USA

U. Schröder and J. Prade
University of Regensburg, D-8400 Regensburg, FRG

W. Kress
Max-Planck-Institut für Festkörperforschung,
Heisenbergstrasse 1, D-7000 Stuttgart 80, FRG

ABSTRACT

Surface structure (relaxation, reconstruction) and surface dynamics are governed by the interaction potentials between the particles at and near the surface. Recent theoretical work concerning these topics is reviewed and compared to available experimental data. Emphasis is on crystals with closed-shell electronic configurations, since for these, the potentials are to a large extent independent of the spatial arrangement of the interacting ions, hence governing bulk as well as surface properties. In contrast, in covalent crystals and metals the interaction potentials are subject to the boundary conditions provided by the surface, thus requiring inclusion of the electronic system in structural calculations.

Examples of the interplay between structural and dynamical properties and the role of soft modes in surface reconstruction are discussed.

1.INTRODUCTION

The study of surface dynamics and surface structure and their interrelation has been greatly stimulated by the refinement of experimental techniques such as LEED, high-resolution EELS, He-diffraction and inelastic He-scattering. Both the structure and the dynamics of crystal surfaces are to a large degree determined by the character of the binding of the crystal; this determines how the crystal is terminated at the surface, and whether and how the forces at the surface are changed. The four main categories of crystal binding give rise to distinctly different surface structures and dynamics. For instance, the saturated electronic configurations in *molecular crystals* and *solid noble gases* lead to conventionally simple pair-potentials, which are hardly modified in the presence of a surface. As a result, quite detailed (albeit not first-principle) calculations of static and dynamic surface properties can be made using bulk pair interaction models. A similar situation exists for the *ionic crystals*, due to the near to closed-shell electronic configurations in those materials. Also here, the pair interactions are barely changed near the surface, but the presence of Coulomb interactions gives rise to distinct differences with the molecular crystals. The *perovskites* display a more interesting set of phenomena. On the one hand the large polarizabilities can give rise to a ferroelectric transition, while on the other hand the competition between short-range and Coulomb interactions can lead to a variety of other structural transitions. In general the effects of the surface on structure as well as on dynamics are enhanced, and the interrelation between surface structure and dynamics becomes more involved, especially in the neighborhood of bulk structural transitions. Nevertheless, away from such transitions, the surface structure is a simple *relaxation*, and the surface dynamics is mainly determined by the relaxation-modified forces at the surface. In these compounds, effects due to non-central forces which may be enhanced by the lower symmetry at the surface, are usually small.

In *covalent crystals* the surface termination of the crystal can create dangling bonds which will give rise to electronic rearrangements at the surface; this usually leads to *surface reconstruction* rather than relaxation. A much studied case is the Si(111) (2x1) surface, where the reconstruction, believed to be due to the formation of Π-bonded chains, gives rise to some interesting dynamical features.

Finally, in *metals* the surface structure is determined by a combination of short-range interactions and surface redistribution of the conduction electron density which smoothes the surface. The latter provides a mechanism (especially in d-band metals) which will limit surface relaxation of ionic planes, and may even drive a net inward displacement.

Because of the important role played by the valence and conduction electrons in both the structure and the dynamics of the surfaces of covalent crystals and metals, first-principle approaches to the surface structure and dynamics are in a sense a greater necessity for these compounds than for the ionic crystals and perovskites. For the latter a fair amount of underlying physics is already contained in interaction models, such as the shell model, the bond charge model, etc.

Because of space limitations of this presentation we can only discuss of few examples, chosen from the alkali halides, perovskites and covalent crystals. The situation for the metals will be reviewed in the contribution of A. Eguiluz.

For a more extended review of the subject (and extensive literature quotations) we refer to a recent conference presentation by Ludwig[1], and to a forthcoming review article[2]

2. CALCULATIONAL APPROACHES

The two common approaches used in surface dynamics calculations are the *slab method* and the *Green's function method*. The slab method gives direct dynamical information: i.e. bulk and surface phonon eigenvectors throughout the system, which, however, is limited to a finite number of layers. The Green's function method represents a true semi-infinite crystal and is particularly suitable for obtaining densities of states and surface resonances. The interaction models used in the calculations should represent the most important features of the particle interactions. Thus, for instance for alkali halides and perovskites, both Coulomb interactions and ionic polarizabilities are crucially important. These are taken into account, together with the short-range interactions, in *shell models* (SM's). In covalent crystals the main interactions are short-range and can be represented by force constants, possibly complemented with dipole interactions, or by, for instance, the *bond charge model*.

A usual starting point for a surface calculation is a bulk interaction model. However, such a model may contain effects of bulk symmetry (e.g. masking of non-central interactions) which can be undone of the surface. This can, in particular, be a problem with force constants models, for which it is not a-priori clear nor unique how the force constants should be modified at the surface. Of course, first principle (e.g. total energy) calculations should give the ultimate answers to these questions. But the surface calculations then have to deal with overwhelming computational and practical difficulties. For instance, except for the simplest symmetry-determined cases, the frozen surface phonons are basically unknown. And in order to give full dynamical data, even first principle calculations have to rely on model parametrizations.

3. SELECTED EXAMPLES

a. (001) Surfaces Of Alkali Halides[3,4]

Because of the closed-shell electronic configurations and simple structures of the alkali halides, the surface relaxations are small[3], as are the effects of the relaxation on the dynamics[4]. The outer surface usually relaxes inwards and rumples as a result of the difference in polarizability and/or ionic size; relaxations diminish rapidly away from the surface. The effects of the relaxation on the surface vibrations are in general quite small, but largest for optical modes vibrating perpendicular to the surface. All these effects are most pronounced for the heaviest compounds (RbBr, RbI). Soft mode behavior is not expected - and not found - because of the great structural stability of the alkali halides.

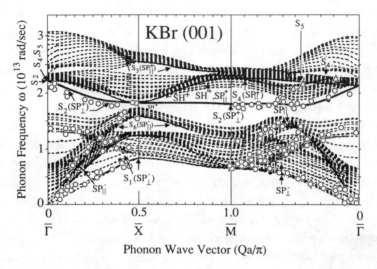

Fig. 1. Calculated and measured surface phonon dispersion curves of KBr(001) (After Refs. 4 and 5a, respectively).

In general, the agreement between measured and calculated surface phonons is quite good, even for unrelaxed surfaces. As an example we show in Fig. 1 results of recent He-scattering measurements of surface-phonon dispersion curves of KBr(001)[5a]. The agreement is such that there is no point in trying to improve upon it by modifying the bulk SM used in the calculations. On the other hand, a predicted surface mode *above* the bulk continuum in RbBr[6] has not been found experimentally[5b]; it may well be that appearance of this mode in the calculations is an artefact of the bulk SM's for RbBr and RbI.

b. Perovskites: (001) Surfaces Of KMnF$_3$[7,8]

Perovskites, while also ionic-type solids, form an interesting contrast to the alkali halides in that many of them exhibit one or more structural phase transitions which in many cases are associated with softening of certain bulk phonon modes. This raises the question about the role of soft surface modes in surface reconstruction,

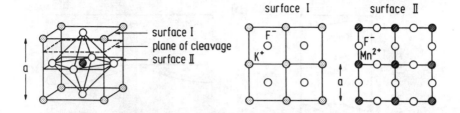

Fig. 2. Perovskite unit cell, (001)plane of cleavage ; geometries of surfaces I and II.

and about the relationship between surface reconstruction and bulk phase transition. In this connection the K-F(001) surface (surface I, Fig. 2) of KMnF$_3$ provides an interesting example.

Bulk KMnF$_3$ exhibits a displacive second order transition at $T_c^{bulk} = 186$ K which is associated with a soft phonon mode at the R point of the bulk BZ. Well above T_c^{bulk} the two KMnF$_3$(001) surfaces show relaxation - not reconstruction - similar to that of the alkali halides, but roughly twice as large; the relaxation patterns are shown in Fig. 4. The main characteristics of these z-relaxations are: "rumpling" (positive and negative ions relax by different amounts); reduction of the first interplanar layer; and rapid diminuation of relaxation away from the surface. Turning to the dynamics of these surfaces it is interesting to note that the *unrelaxed* surface I is dynamically unstable, due to an unstable surface phonon mode at the $\bar{M}$ point of the surface BZ (SBZ) which is peeled down from the lower edge of the soft bulk band. For the relaxed surface this mode is stable (cf. Fig. 4 (b)). In a quasi-harmonic treatment, the interaction constant A$_1$, which governs this soft mode behavior, becomes temp-

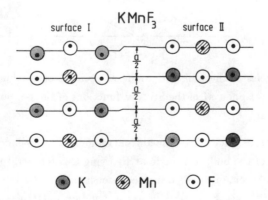

Fig. 3. Relaxation of surfaces I and II of KMnF$_3$, viewed along the [010] direction. Displacements in the z direction are enlarged by a factor of 3.

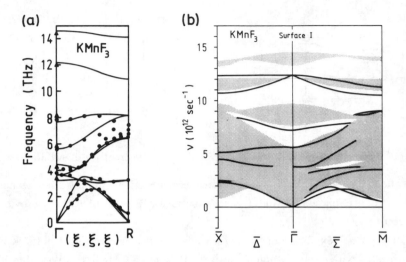

Fig. 4. (a) Bulk phonon dispersion curves along ΓR of the bulk BZ. Fulldrawn curves fitted to room temperature measurements of Ref. 9(a); dashed curves fitted to 200 K measurements of Ref. 9(b). (b) Surface phonon dispersion curves of the KF surface (surface I) of $KMnF_3(001)$ for wave vectors along $\bar{\Gamma}\bar{X}$ and $\bar{\Gamma}\bar{M}$ of the SBZ. The shaded areas represent the bulk bands.

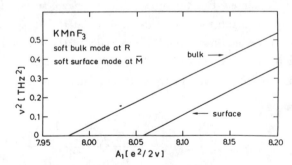

Fig. 5. Frequencies of the bulk soft mode at point R of the bulk BZ and of the surface soft mode at point $\bar{M}$ of the SBZ, as functions of the temperature dependent interaction constant A_1.

erature dependent, as can be determined from temperature dependent measurements of the frequency of the bulk soft mode at R. Using this temperature dependence of A_1 in a slab calculation, one finds a surface reconstruction at a temperature $T_c^{surface}$ slightly above T_c^{bulk} (see Fig. 5). The reconstruction pattern is shown in Fig. 6. Recent He-scattering experiments[10] have shown a c(2x2) superstructure of this sur-

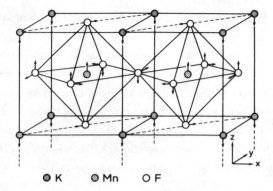

Fig. 6. Reconstruction pattern at the KF surface (surface I) of KMnF₃ resulting from the surface soft mode at $\bar{M}$ (After Refs. 7,8).

face. Below $T_c^{surface}$ the reconstructed surface is stable, i.e. it does not exhibit a soft mode. Questions as to what happen for temperatures between the surface reconstruction and bulk transition temperatures (and whether in some fashion the surface reconstruction triggers the bulk transition) are as yet unanswered.

c. Silicon(001) (2x1) Surface

Covalent crystals with the diamond structure exhibit a variety of surface reconstructions. The most frequently studied surface is Si(111), which exhibits (7x7) periodicity below 1150 K, but shows a (2x1) structure after cleavage; it is common-

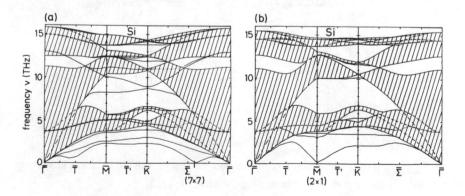

Fig. 7. (a) Surface phonons at a Si(111) surface with 7x7 soft mode, resulting from decreased surface force constants; (b) Surface phonons at a Si(111) surface with 2x1 soft mode, in the dipole model (After Refs. 1,11).

ly believed that this reconstruction is associated with the formation of Π-bonded chains.

Ludwig and co-workers[1,11] have extensively studied the structures and the dynamics of Si(111), in the context of force constant models, including dipolar interactions. In the pure force constant model, lowering of the surface force constants leads to the 7x7 reconstruction, while in the long-range dipole model, increasing the dipole parameters leads to the 2x1 reconstruction. Both reconstructions occur largely independently of the amount of parameter change, and always at the same $\vec{q}$-vectors. Figs. 6a and 6b show the soft modes, causing the 7x7 and 2x1 reconstructions, respectively.

An early EELS investigation[12] of Si(111) (2x1) revealed a 56 meV optical mode, while recent He-scattering measurements[13] detected, in addition to the Rayleigh mode, a dispersion-less mode at about 10 meV (Fig. 7). An interesting controversy surrounds the interpretation of these modes. While Alerhand et al.[14], on the basis of a semi-empirical tight-binding calculation, considers the chain geometry of Si(111) (2x1) essential for the interpretation of these modes, in the interpretation of both Ludwig et al.[1,10] and Miglio et al.[15] (bond charge calculation) these modes are folded versions of surface modes which are intrinsic to the unreconstructed Si(111) surface. We believe the latter interpretation to be the correct one.

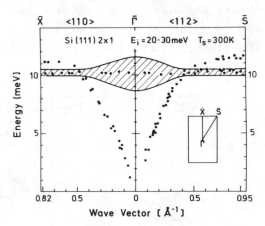

Fig. 8. Measured surface-phonon dispersion curves of the Si(111) (2x1) surface along the ΓX and ΓS direction in the SBZ (After Ref. 13).

REFERENCES

*Supported by grants from the National Science Foundation (DMR-8816301) and the Robert A. Welch Foundation (F-433).

1.) Ludwig, W., in Festkörperprobleme 29, Advances in Solid State Physics (Vieweg, Braunschweig 1989), p. 107.

2.) de Wette, F.W., in Dynamical Properties of Solids, G.K. Horton and A.A. Maradudin, eds. North-Holland (Amsterdam 1990) Vol. 6.

3.) de Wette, F.W., Kress, W. and Schröder, U., Phys. Rev. B32, 4143 (1985).

4.) Kress, W., de Wette, F.W. and Schröder, U., Phys. Rev. B35, 5783 (1987).

5.) a. Chern, G., Skofronick, J.G., Brug, W.P. and Safron, S.A., Phys. Rev. B39, 12828 (1989).
    b. ibid, B39, 12838 (1989).

6.) de Wette, F.W., Kulkarni, A.D., Schröder, U. and Kress, W., Phys. Rev. B35, 2476 (1987).

7.) Prade, J., Kulkarni, A.D., de Wette, F.W., Reiger, R., Schröder, U. and Kress, W., Surface Science 211/212, 329 (1989).

8.) Reiger, R., Prade, J., Schröder, U., de Wette, F.W., Kulkarni, A.D. and Kress, W., Phys. Rev. B39, 7938 (1989).

9.) (a) Lehner, N., Rau, H., Strobel, K., Geick, R., Heger, G., Bouillot, J., Renker, B., Rousseau, M., and Stirling, W. G., J. Phys. C 15, 6545 (1982)
    (b) Gesi, K., Axe, J. D., Shirane, G., and Linz, A., Phys. Rev. B5, 1933 (1972)

10.) Vollmer, R. and Toennies, J.P., Verhandl. DPG (VI) 24, 66 (1969).

11.) Goldammer, W., Ludwig, W., Zierau, W. and Falter, C., Surf. Sci. 141, 139 (1984).

12.) Ibach, H., Phys. Rev. Lett. 27, 253 (1971).

13.) Harten, U., Toennies, J.P. and Wöll, Ch., Phys. Rev. Lett. 57, 2947 (1986).

14.) Alerhand, O.L., Allan, D.C. and Mele, E.J., Phys. Rev. Lett. 55, 2700 (1985); Alerhand, O.L. and Mele, E.J., Phys. Rev. Lett. 59, 657 (1987); Phys. Rev. B37, 2536 (1988).

15.) Miglio, L., Santini, P., Ruggerone, P. and Benedek, G., Phys. Rev. Lett. 62, 3070 (1989).

MICROSCOPIC THEORY OF SURFACE PHONONS IN METALS

Adolfo G. Eguiluz
Department of Physics
Montana State University
Bozeman, MT 59717, USA

ABSTRACT

We present an outline of recent progress made in the microscopic evaluation of interatomic force constants and surface phonon dispersion relations in metal surfaces. Concepts introduced in phenomenological models of surface phonons, such as force-constant stiffening/softening and surface-stress effects, are discussed in the light of the microscopic approach for the case of aluminum. The pair potential at the surface is shown to be non-central. The calculated vibrational spectra for Al(111) reveals striking similarities with the main features of recently calculated spectra for Cu(111) and Ag(111). For Al(110) the spectrum shows a resonance that provides a mechanism for a previously unexplained feature of experiment.

1.   INTRODUCTION

In recent years the dispersion relations of surface phonons have been measured for sp-bonded-(Al[1]), noble-(Ag,[2] Cu,[3] Au,[4]) and transition-(Ni,[5] Pt,[6] Pd,[7]) metal surfaces by the complementary techniques of helium atom-surface scattering[2] and electron-energy loss spectroscopy.[5] Comparison of experimental and theoretical dispersion curves (and associated loss intensities[8,9]) has provided a good deal of information about the nature of the interatomic force constants (FC's) at the surface. Up to the present time this analysis has been mostly based on the use of Born-von Karman lattice dynamical models, which do not, in general, lend themselves to drawing unique conclusions about the surface FC's. Witness for example the long-lived controversy[10-13] stirred up by the original interpretation[10] of the surface phonon spectra of the (111) surface of the noble metals[14] in terms of a drastic FC-softening at the surface.

Because of the complexity of the surface-screening problem, it is only very recently that the theoretical work has begun to go beyond the simple-model stage.[11,13,15-18] The microscopic theory faces the challenge of providing a fundamental explanation for the various physical mechanisms that have been proposed on the basis of simple models. That challenge is now beginning to be met.

2. MICROSCOPIC THEORIES

2.1 Density-Response Method

The electrons and ions in a metal interact via a term in the Hamiltonian of the form

$$\hat{H}_{i,el} = \int d^3x \; \hat{n}(\vec{x}) \sum_l \mathrm{v}(\vec{x} - \vec{R}(l)) \quad , \tag{1}$$

where $\hat{n}(\vec{x})$ is the density operator for the electrons, and $\mathrm{v}(\vec{x} - \vec{R}(l))$ is the electron-ion potential energy. Setting $\vec{R}(l) = \vec{x}(l) + \vec{u}(l)$, where $\vec{u}(l)$ is the instantaneous displacement of the lth ion from its equilibrium position $\vec{x}(l)$, the Feynman-Hellmann theorem[19] leads us to the following result for the *electronic contribution* to the FC's coupling a pair of ions for $l' \neq l$:

$$\Phi_{\mu\nu}^{(el)}(l,l') = \int d^3x \int d^3x' \frac{\partial \mathrm{v}(\vec{x} - \vec{x}(l))}{\partial x_\mu(l)} \chi(\vec{x}\vec{x}') \frac{\partial \mathrm{v}(\vec{x}' - \vec{x}(l'))}{\partial x_\nu(l')}, \tag{2}$$

where μ, ν denote Cartesian directions, and where $\chi(\vec{x}\vec{x}')$ is the static *density response function* for interacting electrons.[19] For $l' = l$ we use the condition of infinitesimal translational invariance. The total FC's are given by the sum of Eq. (2) and the direct ion-ion FC's.

The above formalism is exact (within the harmonic and adiabatic approximations). The main difficulty associated with its implementation for a metal surface is an accurate evaluation of χ. The only available surface-phonon work based on Eq. (2) is for nearly-free electron metals,[15,17,18] for which χ is approximated by the response function for electrons *in a uniform background*, and v is modelled by a local pseudopotential. This procedure corresponds to second-order pseudopotential perturbation theory. The results given below for Al were obtained using the formalism developed by this author for the computation of χ for a slab.[20] As is now well-known,[20,21] a self-consistent evaluation of χ and the surface electronic structure is necessary. The entire program is carried out within the local-density approximation (LDA) to density functional theory. A preliminary account of the first non-perturbative evaluation of Eq. (2) for a metal surface is given elsewhere in these proceedings.[22]

2.2   The "Direct" Method

A straight implementation of the "frozen-phonon" method[23] for the surface problem is precluded by the lower symmetry of the surface environment. One simply cannot guess the dependence of the phonon eigenvectors on distance from the surface. Ho and Bohnen have successfully implemented a variation of the direct method for the case of free-electron metals,[16] and very recently also for Au(110).[24] Their procedure consists in computing selected elements of the dynamical matrix for high-symmetry points of the surface Brillouin zone (SBZ). Small distortions along three orthogonal directions are introduced in the surface layer, with a pattern of displacements that corresponds to a wave vector at the center, or at the boundary, of the SBZ. Total-energy calculations are performed, and interplanar forces are calculated using the Feynman-Hellmann theorem.

The power of the method is that it relies on state-of-the-art band-structure technology (with use of norm-conserving pseudopotentials). Its main limitation is, at the present time, that it does not provide entire *dispersion relations* from first-principles.

2.3   The Embedded-Atom Method

The embedded-atom method[25] is an elaboration of the quasiatom and effective-medium theories.[26] Viewing every atom in the solid as being embedded in the host of the remaining atoms, the total energy of the solid is written as[25]

$$E_{tot} = \sum_{l} F(\rho_l) + \frac{1}{2} \sum_{l,l'}' \phi(|\vec{R}(l) - \vec{R}(l')|) \quad, \tag{3}$$

where ϕ is a short-ranged pair potential, ρ_l is the electron density at the position of the lth atom due to all the other atoms in the solid, and $F(\rho)$ is the *embedding energy*.[25] If one approximates ρ_l by a linear superposition of the electron densities of all the other atoms, E_{tot} becomes an explicit function of the atomic positions; one can then determine the interatomic FC's $\Phi_{\mu\nu}(l,l')$ by differentiation.

This method, which has recently been applied for noble metal surfaces,[13] is phenomenological; the functions F and ϕ are determined by fitting to the known values of the lattice constant, elastic constants, vacancy-formation energy, and sublimation energy. Its main appeal is that it incorporates many-atom forces (through F), and that the environment dependence of the FC's includes some surface effects automatically.

3. THEORETICAL RESULTS AND THEIR COMPARISON WITH EXPERIMENT

We present selected results for the (100), (110), and (111) surfaces of Al, obtained by the density-response method. Contact is briefly made with results of the direct method for the first two surfaces. The spectral densities obtained for Al(111) prove to be quite similar to those obtained for Cu(111) and Ag(111) by the embedded-atom method. For Al(110) the spectrum shows a "longitudinal resonance" that accounts for a hitherto unexplained feature of the helium atom time-of-flight spectrum for this surface.

3.1 Al(100)

The full interatomic FC-matrix $\Phi_{\mu\nu}(l, l')$ was computed for a 17-layer slab from the solution of the screening problem in LDA. The slab was then stretched by adding an arbitrary number of inner layers, for which we used FC's calculated for bulk Al.[18] The dynamical matrix was obtained from the FC's by summing over shells of nearest neighbors (nn's) in real space, including up to twelve shells. The Heine-Abarenkov pseudopotential was used; its core radius and well depth were determined for an ion in the bulk in Ref. 17.

Detailed numerical results for the surface phonon dispersion curves are given in Ref. 18, from which we extract the following conclusions:

(i) The dispersion curve for the Rayleigh wave (RW) lies appreciably *higher* than the curve obtained from the use of bulk FC's. This behavior, which has also been observed for other FCC metals,[3,5] is not, for Al(100), due to a *stiffening* of the FC coupling nn's in the outmost two layers. This FC actually slightly *softens*, a reflection of the fact that the calculated relaxation corresponds to a small (~0.7%) *expansion* of the first interlayer spacing.

(ii) The upward shift of the RW frequency is mostly determined by the surface value of the *tangential* FC coupling first-nn's in the outmost layer for displacements orthogonal to the bond and to the surface.

855

(iii)   For a simple first-nn model,[3] the FC in (ii) is proportional to the radial derivative
         of the (central) pair potential at the equilibrium position, which is, in this model,
         directly related to surface stress.[3] However, the first-principles pair potential
         is *non-central* near the surface (see Fig. 1). Its derivative *normal* to the surface
         is what affects the RW frequency at the zone boundary, and not its *lateral*
         derivative, which is the one that contributes to surface stress. We then have that
         a finite surface stress[27] does not find its way into the RW dispersion curve for
         large wave vectors, as it does in the simple models.[3]

(iv)   The first-principles perturbative calculation yields a RW frequency at $\overline{X}$ that is
         ~9% higher than the experimental value of Mohamed and Kesmodel.[1] A 1%
         adjustment of the *electronic contribution* to the surface force field brings the
         theoretical value into agreement with experiment, and with the non-perturbative
         result of Ho and Bohnen.[16]

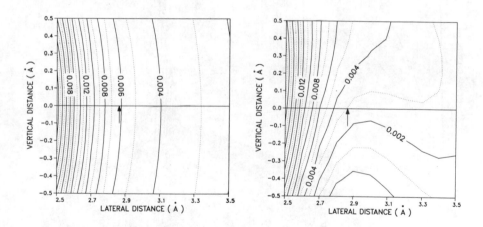

Fig. 1. Equipotential lines for the pair potential for Al(100) (in Ry) drawn on a
plane normal to the surface, in the neighborhood of the third (left panel) and first
(right panel) atomic planes, respectively. The arrow indicates the first nn
position. Note that the equipotential lines in the latter case are not circles, *i.e.*,
the pair potential is not central.

856

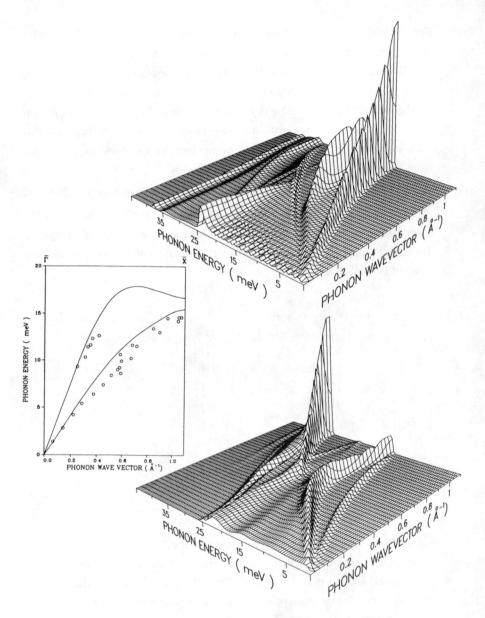

Fig. 2.   Al(110): Spectral densities for surface phonons polarized in the saggital plane, in the outmost layer, and dispersion curves for the RW and longitudinal resonance along $\overline{\Gamma}\,\overline{X}$.   Circles: helium atom-scattering data.[1]

3.2   Al(110)

The surface phonon calculation proceeds as outlined above for the (100) surface. The much larger (oscillatory) relaxation experienced by this open surface[28] has appreciable quantitative impact on the phonon spectrum. The discussion that follows refers to phonon propagation along $\overline{\Gamma}\,\overline{X}$ for the relaxed surface.[18]

This surface was investigated experimentally be means of helium atom-surface scattering by Toennies and Wöll,[1] in order to test a proposed explanation for the "anomalous" surface mode or longitudinal resonance found first in the (111) surface of the noble metals,[2] and subsequently in the (110) surface as well;[1,7] the resonance is attributed to a surface-induced reduction in sp-d hybridization.[10]

Figure 2 shows, for the outmost atomic layer, the spectral density function for phonon eigenmodes polarized in the saggital plane. The upper panel refers to polarization *normal* to the surface, and the lower one to *longitudinal* polarization. We draw attention to an important feature of the results: *a resonance embedded in the continuum of bulk modes is clearly visible.* For small and intermediate wave vectors both polarizations contribute to the resonance with significant spectral weight (competitive with that for the RW, for the case of normal polarization). For large wave vectors the mode emerges from the continuum, and becomes mostly polarized along the surface normal at $\overline{X}$.

Figure 2 also shows the dispersion curves for the RW[29] and the resonance just described, together with the experimental data of Toennies and Wöll.[1] Our theoretical results provide a mechanism for the observed "anomalous" loss, in the absence of the sp-d hybridization mechanism proposed for the noble metals.

The "kinematical" mechanism at play in the present case is basically the one proposed by Persson *et al.*[30] for the (110) surface of FCC metals, namely the avoided crossing of the "pseudoband gap" mode[30] clearly seen in the upper panel of Fig. 2 for small wave vectors and the resonance discussed above, which derives from the longitudinal mode in the gap at $\overline{X}$ (lower panel).

3.3   Al(111)

The discussion given in Section 3.1 for the RW dispersion curve for Al(100) applies for the present case as well. (The calculated relaxation for Al(111) also amounts to a small *expansion* of the outmost interlayer spacing.[18]) The frequency of the RW for large wave vectors is a very sensitive function of the calculated value of the tangential FC coupling first nn's for displacements normal to the surface. With the

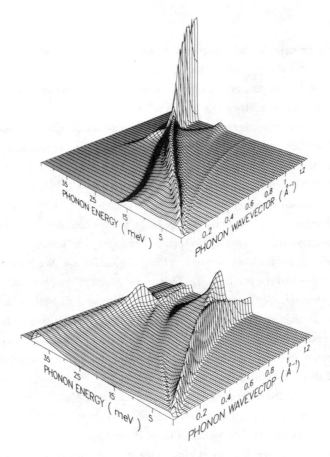

Fig. 3. Al(111): Phonon spectral densities along $\overline{\Gamma}\,\overline{M}$. Upper panel: longitudinal modes for the first atomic layer. Lower panel: modes polarized along the surface normal in the second layer.

adjustment indicated in Section 3.1 (iv) we obtain excellent agreement with the dispersion relations measured by Lock et al.[1] along *four* directions in the SBZ.

Now, the helium atom-surface scattering experiment performed by Lock et al.[1] does not reveal a longitudinal resonance for Al(111). In view of the current thrust in the field of surface lattice dynamics, we think it is worthwhile noting that the phonon spectral densities we have calculated for Al(111) along $\overline{\Gamma}\,\overline{M}$ are virtually identical to those reported very recently by Nelson et al.[13] for Cu(111) and Ag(111). These authors (who used the embedded-atom method[25]) attribute the anomalous surface mode observed for these surfaces to a hybridization between the longitudinal resonance present for small wave vectors in the *first-layer* spectrum (upper panel in Fig. 3) and a mode polarized along the surface normal, present for large wave vectors in the *second-layer* spectrum between the RW and the gap (lower panel in Fig. 3). We would like to propose that Al(111) be investigated experimentally by electron-energy loss spectroscopy.

Acknowledgements

The results reported in Section 3 were obtained in collaboration with Jorge A. Gaspar. This work was partially supported by NSF Grant No. DMR-8603820, by MONTS Grant No. 196504, and by the San Diego Supercomputer Center.

4.   REFERENCES

1.   Toennies, J.P., and Wöll, Ch., Phys. Rev. B 36, 4475 (1987); Lock, A., Toennies, J.P., Wöll, Ch., Bortolani, V., Franchini, A., and Santoro, G., Phys. Rev. B 37, 7087 (1988); Mohamed, M.H., and Kesmodel, L.L., Phys. Rev. B 37, 6519 (1988).

2.   Doak, R.B., Harten, U., and Toennies, J.P., Phys. Rev. Lett. 51, 578 (1983); Bracco, G., Tatarek, R., Tommasini, F., Linke, U., and Persson, M., Phys. Rev. B 36, 2928 (1987).

3.   Wuttig, M., Franchy, R., and Ibach, H., Z. Phys. B 65, 71 (1986); Hall, B.M., Mills, D.L., Mohamed, M.H., and Kesmodel, L.L., Phys. Rev. B 38, 5856 (1988); Harten, U., Toennies, J.P., and Wöll, Ch., Faraday, Discuss. Chem. Soc. 80, 1 (1985).

4.   Bortolani, V., Santoro, G., Harten, U., and Toennies, J.P., Surf. Sci., 148, 82 (1984); Stroscio, J.A., Persson, M., Bare, S.R., and Ho, W., Phys. Rev. Lett. 54, 1428 (1985).

5.   Lehwald, S., Szeftel, J.M., Ibach, H., Rahman, T.S., and Mills, D.L., Phys. Rev. Lett. 50, 518 (1983); Lehwald, S., *et al.*, Surf. Sci., 192 131 (1987).

6.   Harten, U., Toennies, J.P., Wöll, Ch., and Zhang, G., Phys. Rev. Lett. 55, 2308 (1985).

7.   Lahee, A.M., Toennies, J.P., and Wöll, Ch., Surf. Sci., 191, 529 (1987).

8.   Tong, S.Y., Li, C.H., and Mills, D.L., Phys. Rev. Lett. 44, 407 (1980).

9.   Celli, V., Benedek, G., Harten, U., Toennies, J.P., Doak, R.B., and Bortolani, V., Surf. Sci. 143, L376 (1984).

10.   Bortolani, V., Franchini, A., Nizzoli, F., and Santoro, G., Phys. Rev. Lett. 52, 429 (1983).

11.   Jayanthi, C.S., Bilz, H., Kress, W., and Benedek, G., Phys. Rev. Lett. 59, 795 (1987).

12. Hall, B.M., *et al.*, in Ref. 3.

13. Nelson, J.S., Daw, M.S., and Sowa, E.C., Phys. Rev. B 40, 1465 (1989); Phys. Rev. Lett. 61, 1977 (1988).

14. Doak, R.B., *et al.*, in Ref. 2.

15. Beatrice, C., and Calandra, C., Phys. Rev. B 10, 6130 (1983).

16. Ho, K.M., and Bohnen, K.P., Phys. Rev. Lett. 56, 934 (1986); Phys. Rev. B 38, 12897 (1988); Bohnen, K.P., and Ho, K.M., Surf. Sci. 207, 105 (1988).

17. Eguiluz, A.G., Maradudin, A.A., and Wallis, R.F., Phys. Rev. Lett. 60, 309 (1988).

18. Gaspar, J.A., and Eguiluz, A.G., submitted to Phys. Rev. B, and to be published.

19. Feynmann, R.P., Phys. Rev. 56, 340 (1939); Hellmann, H., *Einfuhrung in die Quantenchemie* (Deuticke, Leipzig, 1937).

20. Eguiluz, A.G., Phys. Rev. B 31, 3303 (1985); Phys. Scripta 36, 651 (1987).

21. Liebsch, A., Phys. Rev. B 32, 6255 (1985).

22. Quong, A., Maradudin, A.A., Wallis, R.F., Gaspar, J.A., and Eguiluz, A.G., in this volume.

23. Kunc, K., and Martin, R.M., Phys. Rev. B 24, 2311 (1981).

24. Mills, D.L., private communication.

25. Daw, M.S., and Baskes, M.I., Phys. Rev. B 29, 6443 (1984).

26. Stott, M.J., and Zaremba, E., Phys. Rev. B 22, 1564 (1980); Norskov, J.K., and Lang, N.D., Phys. Rev. B 21, 2131 (1980).

27. Needs, R.J., and Godfrey, M.J., Phys. Scripta T 19, 391 (1987).

28. Eguiluz, A.G., Phys. Rev. B 35, 5473 (1987).

29. The actual first-principles value of the RW frequency at $\overline{X}$ has been lowered by ~ 1 meV by the same 1% adjustment of the electronic contribution of the surface force field alluded to for Al(100). The first-principles value is slightly lower (~ 0.5 meV) than the value obtained by Ho and Bohnen (Ref. 16), which is 2.5 meV higher than experiment (Ref. 1).

30. Persson, M., Stroscio, J.A., and Ho, W., Phys. Scr. 36, 548 (1987).

INVESTIGATION OF SURFACE PHONONS BY HE ATOM SCATTERING

E. Hulpke

MPI für Strömungsforschung

Bunsenstr. 10, 3400 Göttingen,

1. Introduction:

Inelastic scattering of He atoms from surfaces has been applied to study surface phonon dispersion curves since 1980 [1]. The success of these early experiments and the instant availability of surface lattice dynamical calculations [2] stimulated an increased activity in this field. Today surface phonons on a large number of clean and adsorbate covered surfaces have been measured and most of this work is comprehensively discussed in recent review articles on the different aspects of phonon inelastic He scattering [3,4].

Table 1 shows a compilation of the systems investigated. Alkali halide surfaces were - mainly because of the simplicity of their preparation - the first ones to be studied. Strongly corrugated, they show pronounced diffraction and are particularly inert towards surface contamination. Therefore these surfaces still serve as prototype systems to very precisely test lattice dynamical calculations and scattering theory [5]. The surprising behaviour of metal surfaces, the lattice dynamics of which deviates significantly from that of the truncated bulk [6], was one of the motivations for investigating such a variety of different metals; additional phonon modes were observed and calculations based on force constant schemes which reproduce the bulk phonon dispersion could provide for these anomalies only, if the force constants at the surface were - sometimes drastically - changed [3]. Thus one of the goals of these experiments was to search for systematic patterns in these changes of the interatomic forces in the surface. Surface phonons are sensitive to the presence of adatoms because of mass loading effects [7] and, in the case of chemisorption, because of effective force constant changes. Quenching of electronic surface states on e.g. hydrogen adsorption leads in some cases to a restoration of bulk like force constants at saturation coverage [8,9].

Most recently interests have concentrated on three groups of problems.

Table 1: Systems for which surface phonon dispersion curves are measured using He atom scattering

Insulators
LiF(100),   NaF(100)
NaCl(100), KCl(100)
KBr(100),   RbBr(100)
RbCl(100), NaI(100)
MgO(100)

Semiconductors
GaAs(100)
Si(111) (2x1)

Layered material
Graphite(0001)
GaSe(0001)
$TaSe_2$-2H,-1T(001); TaS_2

Metals
Al(111) Al(110)
Ag(111) Ag(110)
Au(111) Au(110)
Cu(111) Cu(110) Cu(100)
Ni(100)
Pb(110)
Pd(110)
Pt(111)
W(001)
Mo(001)
Nb(001)

Perovskites
$KMnF_3$(001)

Adsorbate covered surfaces
O/Ag(110)   ; Kr,Xe/Ag(111)
Kr,Xe/Graphit
N/Mo(001)   ; H/Mo(001)
O/Ni(100)   ; CO/Ni(100)
H/Pt(111)   ; CO/Pt(111) ; Ar, Kr,Xe/Pt(111)
Cu/Si(111)   ; Pb/Cu(111);

All references can be found in the review article by J.P. Toennies [4]

Two dimensional structural phase transitions (SPT): Very strong anomalies in surface vibrations have been observed on the (001) surfaces of W and Mo which are connected with the well known structural instability of these surfaces [10,11]. Quite in analogy to a class of SPTs in bulk solids the softening of a surface phonon mode provides insight into the mechanism which drives the reconstruction of W(001) and Mo(001).

Thin films: Surface vibrations in layered crystals [12] and physisorbed thin films [13,14] have in some cases been studied before and display interesting deviations from the bulk behaviour. Most recently He scattering experiments have been performed on epitaxially grown metal films. The first surprising observation in the system Pb/-Cu(111) was the behaviour of the film growth itself: long range order appears to be dramatically influenced by an interplay of surface, film/substrate interface and the electronic band structure of the film [15]. Band structure effects also seem to bear upon the dispersion of surface phonons [16].

Stepped surfaces: Steps represent defects which are unique to surfaces. In vicinal surfaces these defects can be prepared to feature a long range order and thus resemble two dimensional superlattices. Stepped surfaces and in particular two dimensional phase transitions on these surfaces have been the topic of a series of He scattering experiments [17,18]. The very interesting question as to the lattice dynamical behaviour of such systems is, however, at its infancy [19]. Calculations [20] indicate that the spectrum of surface phonons contains a wealth of additional modes which will provide information on interatomic forces that bind step atoms.

All these three issues will be treated in separate contributions at this conference. The primary theme of this article will be to present a brief introduction to phonon inelastic He scattering, to its advantages and its shortcomings and to illustrate its power by discussing the results obtained on the clean W(001) surface.

2. Experimental aspects

Fig. 1 shows a schematic of the scattering experiment. An intense, monochromatic ($\Delta E/E \sim 2\%$) He atom beam impinges on the surface. He atom kinetic energies can be chosen between 7 and 70 meV. The beam is chopped into bursts of a few μsec duration to facilitate time of flight (TOF) analysis of the scattered He. Extensive differential pumping prevents surface contamination via the high vacuum stages and provides for low background signals in the detector. Scattered He atoms travel along a flight tube of typically 1-2 m length, are ionized and counted. The dynamical range of the He scattering method supercedes -due to the low background - that of all the other diffractive methods used in surface science by far. Intensities between 10 cps and 10^7 cps can be routinely measured. The angle between incoming and scattered beam (θ_{SD}) is usually kept fixed (e.g. 90°) and the amount of momentum transferred parallel to the surface Q is changed by a variation of the polar angle θ_i. Large dynamical range and energy resolution down to 200 μeV are characteristic for these machines.

The theory of inelastic He atom scattering from surfaces is well understood [21,23]. It provides valuable help to the experimentalist as to the effect that changes in beam energy or surface temperature have on the signal. Fig. 2 shows a plot of the differential reflection coefficient $d^2R/dE_f d\Omega$ for different amounts of parallel momentum transfer ΔK and different surface temperatures. Notice the logarithmic scale on the ordinate which, for larger Q, indicates a considerable loss in intensity and thus large measuring times which on surfaces are usually prohibited because of increasing surface contamination. The differential reflection coefficient also provides for a "selection rule" in that waves which are polarized in the surface perpendicular to ΔK (shear horizontal waves) cannot be detected. In fact, any information on the polarization of the wave the He atom has been scattered from is contained in this reflection coefficient. Thus any assignment of a polarization character to a measured phonon mode relies on a comparison with theory, i.e. on calculated relative scattered intensities. The onset of multiple phonon excitation can be conservatively estimated from the Weare criterion [23]. Single phonon excitation prevails for sufficiently small surface temperatures and beam energies. This criterion puts an upper limit to the usable beam energy and thus to the vibrational energies to be studied. Phonon energies and wavevectors are extracted from energy losses and scattering angles using conservation laws for energy and parallel momentum transfer to the surface.

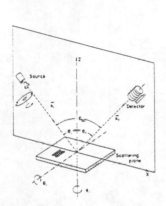

Fig.1. Schematic of the experiment

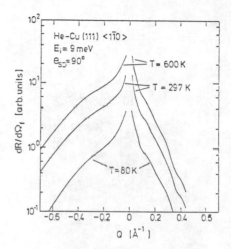

Fig.2. Calculated inelastic cross sections for different surface temperatures neglecting the Debye Waller Effect[4]

3. Two dimensional structural phase transitions and surface phonons.

Two dimensional SPTs are a well known phenomenon on solid surfaces, where they are usually called "temperature induced reconstruction" [24]. Very intensive studies have been devoted to the reconstruction of the clean W(001) surface which at temperatures above ~ 400 K displays a (1x1) diffraction pattern when irradiated with X-rays [25], electrons [26] or He atoms [10]. At temperatures below 220 K the pattern is changed and corresponds to a c(2x2) periodicity in the surface. Structural analysis of the low temperature phase yields a real space arrangement of surface atoms in which the atoms are displaced from their bulk terminal positions by ~ 0.2Å and form diagonal rows of zig-zag chains [27]. It should be noted that this structure resembles the displacement patterns of two of the three possible phonon modes at the $\bar{M}$ point of the surface Brillouin zone (SBZ).

Fig. 3 shows a series of phonon dispersion curves along the $\bar{\Gamma}$-$\bar{M}$

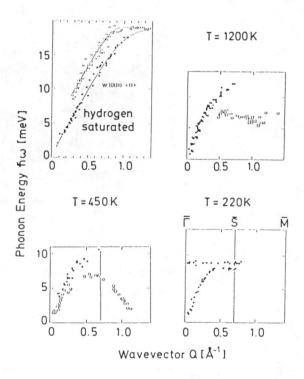

Fig.3. Phonon dispersion curves on "stabilized" (cf. text) and clean W(001) for different temperatures

direction which were measured with He atom scattering. The first panel displays such curves that were obtained at a saturation coverage of hydrogen atoms. In this case the restoration of bulk termination like conditions as discussed before has taken place and the phonons resemble a behaviour which would be observed if the clean surface were structurally stable. Calculations [9] indicate that the upper mode has longitudinal, the lower one transverse polarization (Rayleigh wave). Two phonon modes are also found on the clean surface at temperatures of ~ 1200 K. The lower mode, which in this case has longitudinal polarization (cf. ref. 28) exhibits a curious indentation at Q ~ 1.15 Å^{-1}. At lower temperatures the mode softens considerably. Below 220 K the surface is reconstructed. $\bar{S}$ represents the new SBZ boundary and the backfolding of the Rayleigh wave results in an optical mode and in the formation of a gap at $\bar{S}$. Reconstruction is obviously accompanied by the softening of a longitudinal phonon. The He scattering results therefore indicate that the (1x1) → c(2x2) SPT on W(001) is of the displacive type.

An interesting aspect of these experimental results is the fact that the observed phonon anomaly does not occur at the $\bar{M}$ point as one would expect from the LT phase periodicity but at an incommensurate (IC) Q_s. The presence of an IC superstructure on this surface is also documented in the He diffraction patterns along $\bar{\Gamma}$-$\bar{M}$ shown in Fig. 4.

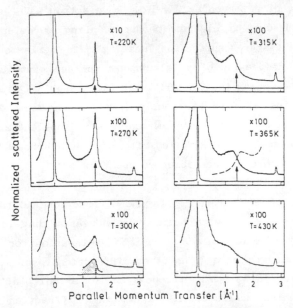

Fig.4. He atom diffraction pattern for different surface temperatures, cf. text.

The sharp 1/2 order beam at $\Delta K = 1.4\ \overset{\circ}{A}^{-1}$ found in the LT phase broadens and appears to move more and more towards IC positions in K space when the surface temperature is raised. Note that the specular ($\Delta K = 0$) and the first order reflection ($\Delta K = 2.8\ \overset{\circ}{A}^{-1}$) are not affected. Also note that the intensity scale in all panels has been blown up by the amount indicated and that the shaded peak represents the elastic contribution to the intensity which has been separated out using the TOF technique. This surprising behaviour has not been found in LEED, X-ray or Ne beam diffraction [29], it has, however, been reproduced by other groups [29,30].

Reasons for this interesting and still somewhat controversial discrepancy can be given in terms of one of the early models used to explain the reconstruction of W(001) [31]. The model suggests that the reconstruction is driven by electron-phonon-coupling leading to a giant Kohn anomaly which is accompanied by a charge density wave (CDW), a mechanism not uncommon in bulk crystal SPTs for systems with a quasi one dimensional metallic character [32]. The model predicts the softening of a phonon at $Q = 2k_F$ (k_F electron wave vector at the Fermi energy) due to electron-phonon-coupling, provided that the Fermi surface features sufficient "nesting". Photoemission measurements of the dispersion of an electronic surface state on W(001)[33] had led to abandoning the CDW model for this case [27]. The Fermi line offers a series of parallel sections, the connecting $Q = 2k_F$ is, however, always different from $\bar{\Gamma}$-$\bar{M}$ and thus in apparent contradiction to the commensurate c(2x2) periodicity of the LT phase that had at all lower temperatures been observed in LEED.

The discovery of IC features in the He diffraction and the behaviour of the surface phonons with temperature suggests that we revive the CDW model. It also provides a plausible explanation for the discrepancies between He atom and the other diffraction results. The IC CDW with the period $2\Pi/Q_s = \pi/k_F$ produces an IC periodic lattice distortion (PLD) which manifests itself in the diffraction pattern by satellite reflections on the sharp integer order beams. The satellites appear at $G \pm Q_S$ (G reciprocal lattice vector) and their intensities depend on $|G \pm Q_S|$ [34]. In the present case of an almost commensurate Q_S the beams $G_{00} + Q_S$ and $G_{11} - Q_S$ are close to one another and thus have similar intensities. LEED and X-ray diffraction will, because of the broadness of these satellite beams, be unable to resolve the "splitting" of the 1/2 order spot. Eikonal calculations for a hard wall quasi IC corrugated model potential (resembling the reconstructed surface) show [35] that neighboring satellite intensities in He scattering are quite different (30:1) indicating that in Fig. 4 all (11)-satellites are suppressed. The above calculation has also been performed for Ne using the experimental parameters of ref. 29. Due to the much smaller wavelength of the Ne the satellite intensities in this case resemble the situation discussed for LEED.

4. Conclusions

Inelastic He atom scattering has emerged as a powerful method to study surface vibrations. Phonons on a variety of different surfaces have been measured allowing for a better understanding of how interatomic forces in surfaces deviate from those in the bulk. This deviation is related to the reduction in coordination number and the resulting response of the electronic system. Therefore phonon studies on stepped surfaces will further improve our knowledge on this subject. They will also provide information on the lattice dynamics on such systems which is influenced by finite size effects due to finite terrace dimensions and point defects on the step edges. Finite size effects (in terms of film thickness) as well as the influence of the electronic band structure also play a role in phonons on the surfaces of metal films. The measurement of surface phonons in systems featuring SPTs can be used to find out about the nature and the driving forces of these transitions. The comparision of related systems of this kind, as W and Mo, very favourably enhances the understanding of these processes.

References:
1. Brusdeylins, G., Doak, R.B., and Toennies, J.P.,
   Phys. Rev. Lett. 44 (1980) 1417
2. Chen, T.S., de Wette, F.W., and Alldredge, G.P.,
   Phys. Rev. B15 (1977) 1167; Benedek, G., Surf. Sci. 61 (1976) 603
3. Toennies, J.P., Solvay Conf. on Surf. Sci.
   Springer Series in Surface Science, 14 (1988)
4. Toennies, J.P., in "Surface Phonons",
   Kress ed. to appear Springer (1989?)
5. Chern, G., Skofronik, J.G., Brug, W.P., and Safron, S.A., Phys.
   Rev. B39 (1989) 12828
6. Harten, U., Toennies, J.P., and Wöll, Ch.,
   Faraday Discuss. Chem. Soc. 80 (1985) 137
7. Alldredge, G.P., Allen, R.E., and de Wette, F.W.,
   Phys. Rev. B4 (1987) 1682
8. Harten, U., et al., Phys. Rev. B38 (1988 I) 3305
9. Ernst, H.-J., Hulpke, E., Toennies, J.P., and Wöll, Chr.,
   to be published Surface Sci. (1989)
10. Ernst, H.-J., et al., Phys. Rev. Lett. 58 (1987) 1941
11. Hulpke, E., and Smilgies, D.-M., Phys. Rev. B40 (1989 I) 1338,
    cf. also contribution of D.-M. Smilgies et al. to this volume
12. cf. Benedek, G., et al., Phys. Rev. Lett 60 (1988) 1037,
    cf. also contribution by Brusdeylins et al. in this volume
13. Toennies, J.P., and Vollmer, R., Phys. Rev. B40 (1989)...
14. Kern, K., et al., Phys. Rev. B35 (1987) 886
15. Hinch, B.J., Koziol, C., Toennies, J.P., and Zhang, G.,
    to appear in Europhysics Lett. (1989)
16. cf. contribution by Zhang, G., et al. to this volume
17. cf. Fabre, F., et al., Solid State Comm. 64 (1987) 1125
18. Conrad, E., et al., J. Chem. Phys. 84 (1986) 1015
19. Witte, G., Diplomthesis Göttingen 1989
20. cf. contribution by Lock, A., et al. to this volume
21. Bortolani, V., and Levi, A.C., La Rivisita del Nuovo Cimento 9
    No. 11 (1986)
22. Eichenauer, D., Harten, U., Toennies, J.P., and Celli, V.,
    J. Chem. Phys. 86 (1987) 3693
23. Weare, J.H., J. Chem. Phys. 61 (1974) 2900
24. Zangwill, A., Physics at Surfaces, Cambridge University Press,
    Cambridge 1988
25. Robinson, I.K. et al., Phys. Rev. Lett. 62 (1989) 1294
26. Felter, T.E., et al., Phys. Rev. Lett. 38 (1977) 1138
27. King, D.A., Physica Scripta T4, 34 (1983) 38
28. Wang, X.W., and Weber, W., Phys. Rev. Lett. 58 (1987) 1452
29. Schweizer, E.K., and Rettner, C.T.,
    Surface Sci. Lett. 208 (1989) L29
30. Salanon, B., and Lapujoulade, J.,
    Surface Sci. Lett 173 (1986) L6
31. Tosatti, E., Solid State Commun. 25 (1978) 637
32. Wilson, J.A., Di Salvo, F.J., and Mahajan, S., Advances in
    Physics XXIV, London 1975, p 117 ff.
33. Campuzano, J.C. et al., Phys. Rev. Lett. 45 (1980) 1649
34. Overhauser, A.W., Phys. Rev. B3 (1971) 3173
35. Ernst, H.-J., Hulpke, E., and Toennies, J.P., to be published

INELASTIC NEUTRON SCATTERING FROM ADSORBATES

H.J.Lauter

Institut Laue-Langevin, BP 156X, F-38042 Grenoble, France

Introduction

The phase diagrams of a monolayer of adsorbed gases or light molecules look similar as the phase diagrams in 3-dimensions (3-D) if the adatom-adatom interaction is governed by van-der Waals forces. This means that the usual coexistence regions, the triple point and the critical point are present as well in 2-dimensions (2-D) as in 3-D [1]. The phase diagrams can be studied by adsorption isotherm or heat capacity measurements, which reveal mainly the coexistence regions or the phase boundaries, respectively. However, the substrate can not always be regarded to be ideally flat. In many cases the adsorbate does see the adsorption sites of the substrate and locks into a commensurate phase (c-phase). In the case of graphite as substrate the ($\sqrt{3}$ x $\sqrt{3}$) R 30° overstructure is seen in many cases. This structure is shown in fig.1. The heavier rare gases and the light molecules (N_2, CD_4) exhibit the C-phase only if the lattice parameter of the dense plane in 3-dimensions is close to the nearest neighbor distance in the C-phase. This is the case because the nearest neighbor distance does nearly not change in the densest plane if the adjacent planes are taken off which means if the dimensionality changes from 3-D to 2-D. The quantum gases however show all a C-phase in their 2-dimensional (2-D) phase diagram despite a much denser nearest neighbor distance in 3-D. This is due to the zero-point motion which gives a repulsive contribution to the nearest neighbor interaction and consequently a high compressibility to the system. Thus if as before the dimensionality is changed and many of nearest neighbor atoms are missing the 2-D lattice expands. If in addition the corrugation of the adsorption potential is added, the quantum gases recognize the dilute density structure of the C-phase as the ground state.

The phase diagram of D_2 on graphite is shown in figure 2 as an example of an adsorbed quantum gas. The location of the phase boundaries has been determined by heat capacity measurements [2]. The definite

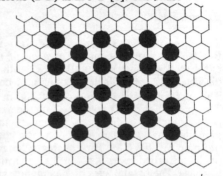

Fig.1: Model of the commensurate ($\sqrt{3}$ x $\sqrt{3}$) R 30° overstructure on graphite.

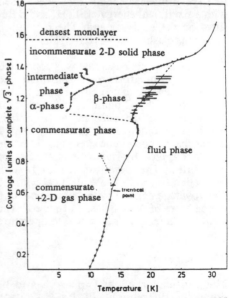

Fig.2: Phase-diagram of D_2 on graphite [2].

872

allocation of the different phases to structures is given by scattering techniques. In this case it has been done by neutron diffraction and LEED [3,4].

The inelastic neutron scattering gives additional information of the interaction between the adsorbed particles itself but also between the adsorbate and the substrate. In particular interesting is the search for the phonon gap at the zone center which characterizes the loss of the translational invariance of the adsorbed layer in the C-phase. The transition from the C-phase to the incommensurate higher density phase is characterized by different intermediate phases as seen in figure 1. Theories of the commensutrate-incommensurate transition predict domain walls in the transition region. In these domain wall phases the commensurate phase is still locally present. Thus the study of the phonon gap will reveal interesting features.

Experimental

The experimental set-up is described in Ref.5. The sample consists of a stack of exfoliated graphite sheets (Papyex) with a diameter of 2 cm and a height of about 7 cm. The surface area is in the order of 200 m^2. The graphite is a 2-D powder. Only the axis perpendicular to the basal planes of the graphite shows a certain order of a mosaicity of about 30° FWHM. The coherence length of the adsorbed layer is about 300 Å in the C-phase. The basal planes of the graphite are parallel to the scattering plane of the neutron spectrometer. For the inelastic studies the IN3 spectrometer of the ILL has been used with a fixed final energy of 1THz and a Be-filter on the analyser end. The resolution across the elastic line was 0.03 THz.

The density of the adsorbate was controlled by adsorption isotherms. But also the highest intensity of the Bragg-peak of the adsorbate in the C-phase as a function of coverage at constant temperature determines the best commensurate phase $\rho=1$. $\rho=1$ means that all adsorption sites in the C-phase are occupied by adsorbed atoms or molecules. The definition of $\rho=1$ with diffraction is within 2-3% identical with the $\rho=1$ coverage defined by the highest melting temperature of the C-phase in figure 2. The small difference may be a real temperature effect.

Measurements

The verification of the C-phase has to be done by diffraction. In fig.3 neutron diffraction patterns are shown at various

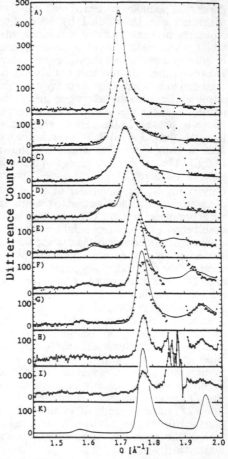

Fig.3: Diffraction pattern of D_2 on graphite in the C-phase and α-phase at various average densities of the adsorbate; T=2K. The solid lines are fits with the model in fig.7 [3].

average adsorbate densities [3]. The spectrum A at a slightly overfilled C-phase shows the maximum intensity at a momentum transfer $Q=1.702$ Å^{-1} which corresponds for a triangular lattice to 4.26 Å, the nearest neighbor distance in the C-phase (fig.1). All the spectra shown, the elastic and the inelastic ones, show difference counts. The signal from the empty cell without adsorbate has always been subtracted.

The inelastic neutron measurements have been taken at two different momentum transfers Q. The reason for that is depicted in fig.4 which shows the reciprocal space of the triangular lattice of the C-phase. The scan taken with a $Q=1.7$ Å^{-1} collects all excitation with wave vector q along the circle with radius Q with the help of the Bragg-points which are marked by the vector τ. These excitations with wave vector q have mainly transverse polarization and the highest intensity is expected from the zone boundary phonons because of the high density of states. If a phonon gap exists at the zone center a second high density of states is expected at this point. As the 2-D powder averaging crosses also the Γ-points a second peak is expected in energy in the scan with a $Q=1.7$ Å^{-1}.

The scan taken with $Q=0.85$ Å^{-1} collects the longitudinal zone boundary phonons due to the 2-D powder averaging (fig.4). So with the correct choice of the momentum transfer different modes can be separated even in a powder-like sample.

The scans with the different momentum transfers are shown in figure 5. The data points in fig.5a ($Q=1.7$ Å^{-1}) show clearly a double peak. The one at lower energies represents the phonon gap, whereas the one at higher energies shows the collected transverse zone boundary phonons. The scan with a $Q=0.85$ Å^{-1} (fig.5b) shows only one peak which results from collected longitudinal zone boundary phonons. The fit to the data is a two parameter fit. The adsorbate molecules are thought to be connected by each other by a net of springs

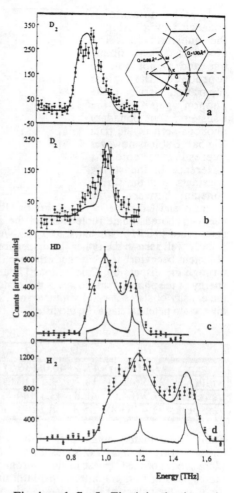

Fig.4 and fig.5: Fig.4 is the inset in fig.5a. Fig.4 shows the reciprocal space of a triangular lattice. The circles are shown on which phonons are collected due to the powder averaging for $Q=1.7$ Å^{-1} and $Q=0.85$ Å^{-1}. Q is the total momentum transfer; τ is a reciprocal lattice vector and q is the phonon wave vector [3]. Fig.5: Neutron inelastic data of the hydrogen isotopes adsorbed on graphite in the C-phase [8].

and each molecule is connected by a spring to the substrate [5]. The model is a very simple one and is not able to reproduce the clear separation between the phonon gap and the transverse zone boundary phonons seen in the data (fig.5a). But it is however interesting to note the difference in the spring constants. α is the spring constant between the adatoms and β the one

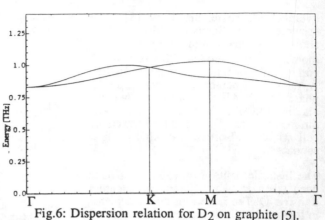

Fig.6: Dispersion relation for D_2 on graphite [5].

between the adsorbate molecule and the substrate at the adsorption site. α has been determined to 0.016 N/m and β to 0.182 N/m. The ratio is about 1/10. This small ratio is equally well seen in the rather flat dispersion in fig.6. Thus the molecule exhibits nearly "Einstein behavior" [6], but the influence of adatom-adatom interaction is still visible through the dispersion. Calculations [7] are in good agreement with the value of the energy of the phonon gap, however the adatom-adatom interaction visible by the width of the density of states is still by a factor of 1.5 too small (see table 1). Thus the tail of the interaction potential has to be modified.

	H_2	HD	D_2
z.c.gap	47.3 (46.6)	43.2	40.0 (36.9)
width	27.5 (42.1)	14.7	9.5 (14.8)
trans.peak	57.9 (64.9)	48.8	43.3*(44.2)
long.peak	71.4 (83.8)	55.8	48.1 (50.3)

Table 1: Parameters obtained from the fits characterizing the density of states for the in-plane modes of the C-phase of the hydrogen atoms adsorbed on graphite. In parenthesis are given the values of Ref.7.. Indicated are the following: z.c.gap is the zone center gap energy, width is the width of the density of states, trans.peak and long.peak are the peaks arising from the transverse and longitudinal phonons in the density of states, respectively. All values are in Kelvin (48 K = 1THz = 4.14 meV). (* The experimental data of fig.5a suggest a value of 44.4K for the transverse peak)

The effect of the isotope mass can be probed by using in addition HD and H_2 as adsorbate [8]. The spectra are shown in fig.5c and 5d. Both spectra make use of the incoherent scattering cross section of HD and H_2. The density of states is seen as a solid line in the lower part of the figures. The line through the points is the density of states folded with the resolution of the instrument. The characteristic values of the dispersion relations are summarized in table 1. The same model as described for D_2 has been applied to the other isotopes. The theory [7] describes well the isotope shift which is not only due to the different mass but also due to the anharmonicity of the potentials. The isotope effect is seen with decreasing mass as well in the shift of the phonon gap to higher energies as in an increasing width of the density of states. As with D_2 the width of the density of states is wider by the same factor 1.5 (table 1).

Once the inelastic signals are understood in the C-phase the adjacent phases can be investigated. This will be described first for the α-phase. In fig.3 several scans are shown taken at different coverages and constant temperature. They follow a path from the

C-phase into the α-phase (fig.2). The diffraction peak shifts with increasing density to higher Q-values according to the compression. At the same time satellites are moving outwards from the main peak. The feature around 1.88 Å^{-1} is the (002) graphite peak, which is due to an interference

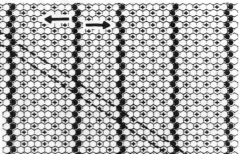

Fig.7: Striped superheavy domain wall model for D_2 on graphite [3].

phenomena. It is of no importance for this study and cuts unfortunately out a certain region in Q, where the signal of the adsorbate is not observable. The fit to the data has been done with a model of striped superheavy domain walls [3,4] which is depicted in fig.7 [3]. For a more detailed discussion the reader is referred to the Refs.3 and 4. It should only be mentioned that the crosses in fig.7 represent the molecules still in the commensurate position. The filled circles mark the molecules in the domain walls in an ideal position. However they are too close and a certain relaxation indicated by the arrows takes place. In the applied model the distance between the domain walls has a distribution which depends on the domain wall density. The width of the domain walls is fairly small. It extends only across 4 rows of molecules. This is again due to the high compressibilty of the 2-D quantum gases asmentioned in the introduction.

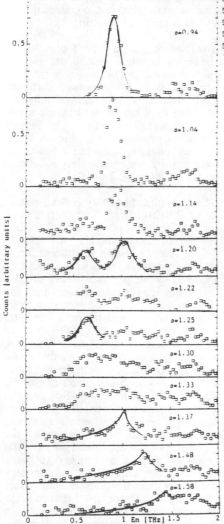

Fig.8: Neutron inelastic data of D_2 on graphite at T=4K for various coverages (Q=1.7 Å^{-1}) [9]. The lines are guides to the eye.

The inelastic study of the domain wall structure is shown in fig.8 [9]. The top two spectra are taken in the C-phase and show the phonon gap and the transverse zone boundary phonon in one peak due to the relaxed instrument resolution (0.6 THz) used for these scans. If the average density is further increased the α-phase is entered (see fig.2) and the signal around 1 THz starts to decrease. This is to understand because the amount of the molecules in the C-phase is decreasing

876

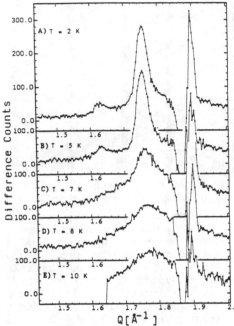

Fig.9: Diffraction pattern of D_2 on graphite at constant coverage $\rho=1.16$ (see fig.2) at various temperatures showing the α-β transition [3b].

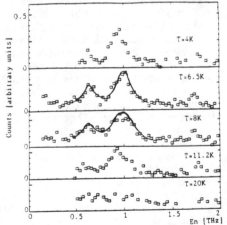

Fig.10: Neutron inelastic data of D_2 on graphite at constant coverage $\rho=1.16$ (see fig.2) at various temperatures across the α-β transition [9]. The lines are guides to the eye.

due to the model of the domain walls. At the same time a signal at about 0.6 THz appears which finally at $\rho=1.25$ is the only signal to be seen. This signal is to attribute to an excitation of the domain walls which is still to calculate. This signal is best at the densest α-phase ($\rho=1.25$, see fig.2). This is a first example of the usefulness of the inelastic studies to complete and to understand the events in the phase diagram.

The following scans in fig.8 show that in the region between $\rho=1.30$ and $\rho=1.33$ the signals can not be any more resolved probably due to too many different excitation in this phase which was modeled by a hexagonal heavy domain wall structure [3,4]. In the region beyond the density of $\rho=1.33$ the inelastic response changes again and indicates the pure transverse zone boundary phonon of an incommensurate 2-D solid.

The next object to study is the β-phase. In fig.9 diffraction patterns are shown taken at constant coverage $\rho=1.16$ as a function of temperature [3]. At T=2K and 5K the already known satellite structure of the α-phase is seen, which could be modeled by striped superheavy domain walls. For higher temperatures than the α-β transition (fig.2) the spectrum looks like a liquid structure factor in particular if the temperature is raised. The structure of this "reentrant liquid" was not known, but got new interest because this phase seems to be separated from the normal 2-D liquid by very broad peaks in specific heat (fig.2). The inelastic neutron measurements are shown in fig.10. They have also been taken with the relaxed resolution (fig.8). So again the excitations of the commensurate parts (phonon gap and transverse zone boundary) are seen and at lower energy the signal from the domain walls. Here no change in the spectrum is seen if the temperature crosses the α-β transition at 7.2 K. The consequence is that the domain walls do still exist in the β-phase because the excitation belonging to them are still visible and also the excitations

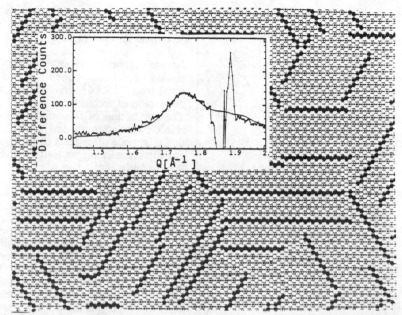

Fig.11: Model for the β-phase. It is made up of patches of the α–phase structure [3b]. The inset shows an overlay of a measured spectrum in the β-phase (spectrum c in fig.9) and calculated one using the structure from the same figure [3b].

from a commensurate phase. The solution for a structural picture is to introduce patches of domain walls as depicted in fig.11. This model still allows for the inelastic features but the structure factor seen as inset in fig.11 [3b] fits the data. This is a proof that the β–phase is a disordered domain wall phase. The evolution with temperature can be modeled by shorter and shorter domain walls until melting around 20K. Perhaps the unbinding of domain walls can be treated in the class of Kosterlitz-Thouless transitions [10].

The second selected example of excitations in adsorbates is the system of helium on graphite. This time not only a monolayer is looked at but a film. The film is composed of two solid layers adjacent to the substrate (below 2K) and subsequent liquid layers (see e.g.Ref.13). So the liquid ^{4}He film has two boundaries, the solid-liquid one and the liquid-gas one. Excitations can propagate along these interfaces, which are the freezing-melting wave [11] and the ripplon [12], respectively. Any excitation with a dispersion like a freezing wave could not yet be measured with neutron scattering. But in addition to the signals arising from the bulk ^{4}He some modes could be detected which have no dispersion (the energy does not change as a function with wave vector). These modes (at 0.4 meV, 0.6 meV....) are localized at the solid-liquid interface because they still exist if the sample cell is completely filled with helium (this means that the liquid-gas interface is suppressed) [13]. On the other hand these modes disappear between a coverage of 2.5 and 4 layers as shown in fig.12. This range has unfortunately not been investigated in more detail. But for coverages beyond 4 the intensity of this mode is no function of coverage in agreement with the explanation that it is bound at the solid-liquid interface. These modes can be taken to explain the high transmission of phonons through the interface between a soilid and liquid helium (Kapitza-resistance). In contrast the bulk signal increases linearly with coverage. It extrapolates to zero at a coverage of 5 layers.

878

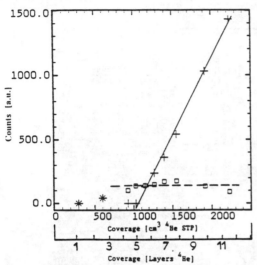

Fig.12: Intensities of the bulk signal (+) and the interface signal (□) as a function of coverage (^{4}He on graphite powder [11]). The signal at a coverage of a monolayer and at 2.6 layers is marked by (*) (it is a constant background !); T=0.8K.

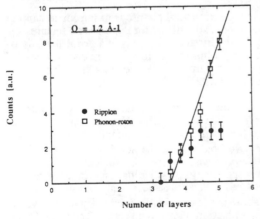

Fig.13: Intensities of the bulk signal (□) and the ripplon (●) as a function of coverage (^{4}He on Papyex-graphite); T=0.6K.

This indicates that in addition to the two solid layers 3 liquid layers may not contribute to superfluidity.

At the second interface a ripplon should be visible. Indeed it could be measured [13,14] and shows the dispersion expected from theory [12]. The attribution of this mode to be a ripplon was strengthened by the fact that this mode disappears if the liquid-gas interface is suppressed by filling the sample cell completely with helium. The fig.13 shows the evolution of the signal height of the ripplon with coverage. Again the saturation shows that this mode is bound to an interface. It becomes visible above 3 layers (two of them solid !). Like in fig.12 the evolution of the bulk signal is exhibited. The extrapolation is made only with coverages below 5 layers. This shows that the bulk signal disappears at about 3.5 layers. There is no discrepancy with the dependence in fig.12 because here in fig.13 the scale is much finer and in principle a no linear tail of the extrapolated bulk signal is seen. A rough extrapolation from still higher coverages extrapolates to about 4.5 layers in agreements with fig.12.

In conclusion it has been shown that even on powder like adsorbates a lot of information can be drawn from inelastic neutron studies from adsorbates. These studies give insight into the dynamical behavior of the adsorbates which is also very valuable for the modelling of 2-D phases in the demonstrated case of the commensurate-incommensurate transition. The second example concerned the excitations at the interfaces of bulk helium. Here long outstanding features could be verified.

[1] Thomy A., Duval X. and Regnier J., Surf.Sci.Rep.**1**, 1 (1981)

[2] Freimuth H. and Wiechert H., Surf.Sci. **178**, 716 (1986)

[3a] Schildberg H.P., Lauter H.J., Freimuth H., Wiechert H. and Haensel, Jap.J.Appl.Phys.**26**, 345 (1987),(Proc.18th Int.Conf.on Low Temperature Physics, Kyoto);

[3b] Schildberg H.P., Thesis, University of Kiel R.F.A. (1988)

[4] Cui.J, Fain S.C., Freimuth H., Wiechert H., Schidberg H.P. and Lauter H.J., Phys.Rev.Lett.**60**, 1848 (1987)

[5] Frank V.L.P., Lauter H.J. and Leiderer P., Phys.Rev.Lett.**61**, 436 (1988)

[6] Nielsen M., McTague J.P. and Passell L, Phase Transitions in Surface Films, Plenum Press (1980) p.127

[7] Novaco A.D., Phys.Rev.Lett.**60**, 2058 (1988)

[8] Lauter H.J., Frank V.L.P., Leiderer P. and Wiechert H., PhysicaB**156&157**, 280 (1989)

[9] Frank V.L.P, Lauter H.J. and Leiderer P., Jap.J.Appl.Phys.**26**, 347 (1987),(Proc.18th Int.Conf.on Low Temperature Physics, Kyoto)

[10] Kosterlitz M. and Thouless D.L., Prog in Low Temp.Phys. **VII B**, 371 (1987)

[11] Keshishev K.O., Parshin A.Ya. and Babkin A.B., Sov.Phys. JETP **53**, 362 (1981)

[12] Edwards D.O. and Saam W.F., Prog. in Low Temp.Phys. **VII A**, 283 (1978)

[13] Lauter H.J., Frank V.L.P., Godfrin H. and Leiderer P., Elementary Excitations in Quantum Fluids, Ohbayashi K and Watabe M. Eds.,Springer Series in Solid-State Sciences 79 (1989) p.99;

[14] Godfrin H., Frank V.L.P., Lauter H.J. and Leider P., this conference

Phonons of Thin Epitaxial Pb Films Grown on a Cu(111) Surface

C. Koziol, J.P. Toennies and G. Zhang

Max-Planck-Institut für Strömungsforschung

Bunsenstraße 10, 3400 Göttingen, FRG

During the past several years there have been numerous measurements of phonons on clean and with adsorbate covered single crystal surfaces using inelastic He scattering [1] and inelastic electron scattering [2]. It is well known that the surface modes are attributed to a "peeling off" from the bulk band edges produced by the surface boundary conditions and have an exponential decay into the bulk. Therefore it is of great interest to know how the phonons look like in very thin metallic films of only a few atomic monolayers (ML) and how they develop with thickness in comparison to the rare gas overlayers physisorbed on smooth surfaces [3,4], which have been studied quite intensively. However there have been only very few studies of metals deposited on metal surfaces [5]. The system Pb on Cu(111) is a good candidate for such a study since the energy ranges of the modes for the both materials are very different because of the very different masses and therefore the modes do not overlap much. The interaction between the adlayer and the substrate is metallic bonding instead of the weak physisorption for the rare gas adsorption. Moreover, Pb grows epitaxially ML by ML with two-dimensional islands, fcc(111) oriented, incommensurately on the Cu(111) substrate at the sample temperatures below 200 K [6,7]. The angular distributions from the He scattering show very sharp diffraction peaks with each completed layer which indicates that the films are well-ordered [7].

Our experiments were performed with a high resolution helium atom scattering apparatus which is much the same as that described earlier [8]. In addition a Knudsen cell was used to evaporate Pb perpendicular to the Cu(111) surface during the He scattering [9]. The determination of the thicknesses of the films was simply followed by observing the oscillations in the He specular scattered intensity during the deposition [9]. Fig 1 shows the phonon dispersion curves of a 0.5 ML (solid circles) and a 1 ML (open circles) Pb on a Cu(111) surface at a substrate temperature of 160 K. The solid lines indicate the Rayleigh waves in the same directions of the Cu(111) substrate in the continuum limit as calculated from the elastic constants [10]. In the middle of Fig.1 the reciprocal lattice with the first and the second Brillouin zone and also the irreducible part of the Brilliuon zone is sketched. The data points shown in Fig.1 were measured along the azimuth directions $\bar{\Sigma}(\bar{\Gamma}\bar{M})$, $\bar{T}(\bar{\Gamma}\bar{K})$ and $\bar{T}'(\bar{K}\bar{M})$. For both coverages we have observed two inelastic modes along the $\bar{\Sigma}$ and $\bar{T}$. The lower one is very sharp, as shown in Fig.2, and has a strong

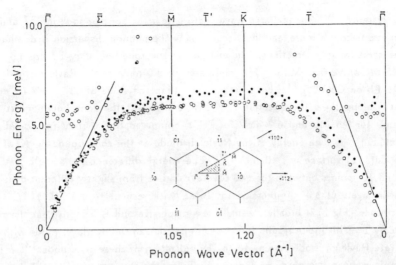

Fig.1: Dispersion relations of 0.5 ML and 1 ML Pb evaporated on a Cu(111) surface at the sample temperature of 160 K.

Pb / Cu (111) < 110 >

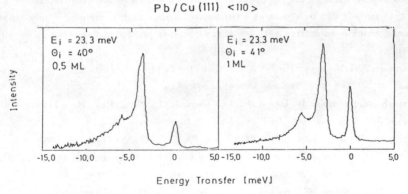

Energy Transfer [meV]

Fig.2: Energy loss spectra of 0.5 ML and 1 ML Pb on Cu(111) along $\bar{\Gamma}\bar{K}$ direction. The peaks at zero energy transfer are attributed to elastic scattering from defects.

dispersion up to the zone boundaries $\bar{M}$ and $\bar{K}$. It can be assumed that this mode is the vibrational mode of the adlayer atom against the substrate and has a vertical polarisation as in the case of rare gas overlayers[3,4,13]. For small wave vectors near the $\bar{\Gamma}$-point this mode has the characteristics of the Rayleigh mode of the substrate Cu(111) as can be seen from the good agreement to the acoustic velocities of sound (Fig.1). This behavior can be easily understood because of the long range interactions of the substrate mode due to the long wave length and has been observed for many systems [3,5]. For the large wave vectors near the zone boundaries $\bar{M}$ and $\bar{K}$ this mode becomes more lead like because now the phonon dispersion is dominated by the short range interactions. The energies at the zone boundary $\bar{M}$ are 6.5 meV for 0.5 ML and 6.0 meV for 1 ML, compared to 4.0 meV of the Rayleigh mode for 32 ML Pb on Cu(111) [11], which are smaller than the value of 13.3 meV for clean Cu(111) along the same direction [12]. The energies at the zone boundary $\bar{K}$ are 6.7 meV for 0.5 ML and 6.2 meV for 1 ML compared to 5.0 meV for 50 ML Pb on Cu(111)[11] . The energy of the Rayleigh mode at the zone boundary $\bar{K}$ of the clean Cu(111) surface is 14.0 meV[12]. The energy difference (0.5 meV) at the $\bar{M}$- and $\bar{K}$- points between the two coverages arises from slightly different densities of the adlayers (1-3% compression for 1 ML with respect to 0.5 ML)[6,7]. The upper mode (Fig.1) is broader, weaker, less dispersive and is located near the zone boundary $\bar{\Gamma}$. This mode results from the intersection of the overlayer mode with the substrate Rayleigh mode leading to the hybridization of these two modes[13] . This phenomenon was also observed experimentally for the rare gas overlayers[3], but was not seen for the metallic overlayers of Ag grown on Cu(100) and Ni(100) surfaces which is probably due to the low spectral density of this vertically polarized mode[5] . In fact, this mode can only exist as a surface resonance and, therefore, is very broad and weak which is consistent with the measurements presented here (Fig.2).

References

1) Toennies, J.P., J. Vac. Sci. Technol A5(4) (1987) 440

2) Ibach, H., J. Vac. Sci. Technol. A5(4) (1987) 419

3) Kern, K., Zeppenfeld, P., David, R., and Comsa, G., Phys. Rev. B 35 (1987) 886; Toennies, J.P., and Vollmer, R., Phys. Rev. B 40 (1989)

4) Gibson, K.D., and Sibener, S.J., J. Chem. Phys. 88 (1988) 7862 and 7893

5) Daum, W., Journal of Electron Spectrocopy and related Phenomena 44 (1987) 271

6) Meyer, G., Michailov, M., and Henzler, M., Surf. Sci. 202 (1988) 125

7) Hinch, B. J., Koziol, C., Toennies, J. P., and Zhang, G., in preparation

8) Bortolani, V., Franchini, A., Santoro, G., Toennies, J. P., Wöll, Ch., and Zhang, G., Phys. Rev. B. 40 (1989)

9) Hinch, B. J., Koziol, C., Toennies, J. P., and Zhang, G., Europhysics Letters, in print

10) Farnell, G. W., in "Physical Acoustics", Vol. 6, Mason, W. P., ed., Academic Press New York (1976), p. 109

11) Koziol, C., Toennies, J. P., and Zhang, G., in preparation

12) Harten, U., Toennies, J. P., and Wöll, Ch., Faraday Discuss. Chem. Soc. 80 (1985) 137

13) Hall, B., Mills, D. L., and Black, J. E., Phys. Rev. B 32 (1985) 4932

OXYGEN PHASES CHEMISORBED ON Ag(110) STUDIED BY INELASTIC He ATOM SCATTERING

G.Bracco, P.Cantini, R.Tatarek and G.Vandoni

Dipartimento di Fisica, via Dodecaneso 33, I-16146 Genova (Italy)

The oxygen-covered Ag(110) surface is characterized by a rich variety of different ordered structures which have been the subject of several experimental and theoretical studies because of the specific activity of this surface for the epoxidation of ethylene and its omologues[1]. In particular, if sample oxidation occurs at surface temperatures above $T_S \simeq 190$ K and with not too high pressure dosing, the p(n×1) LEED diffraction patterns, with $2 \leq n \leq 7$, are observed[2]. The general explanation of the LEED results assumes p(n×1) oxygen overlayers with the oxygen atoms chemisorbed in long bridge sites forming chains along the $< 001 >$ direction[3]. To check whether these structures are due only to the arrays of O atoms or also to the reconstructed outermost Ag layers, the surface dynamics of a few oxygen phases chemisorbed on Ag(110) was investigated by inelastic He atom scattering with time-of-flight (TOF) detection. In this paper preliminary results on the p(3×1)O-Ag(110) surface are reported and comparison with recent data on the bare substrate is made[4].

The Ag(110) surface, cleaned by ion bombardment and annealed at 750 K, is exposed to an oxygen pressure in the 10^{-6}mbar range while cooling to room temperature. The layer formation is controlled by continuously recording the scattered intensity at the angular position where the $(\bar{\frac{1}{3}},0)$ diffraction peak appears. At $T_S \simeq 340$ K, the oxygen flux is switched off and the sample annealed at about 370 K for ~40 minutes. This procedure gives rise to sharp and intense diffraction peaks, with low background in between, when angular distribution scans are measured. More than 70 TOF spectra were taken along the $< 001 >$ direction with the surface held at room temperature, beam energy $E_b = 17.5$ meV, angles of incidence Θ_i ranging from $42°$ to $70°$ and scattering angles $\Theta_f = 110.3° - \Theta_i$. A typical energy converted TOF spectrum is shown in fig.1 a. The incoherent elastic peak E and a few peaks associated with surface vibrational modes are observed both for positive and negative energy losses corresponding to phonon creation and annihilation processes respectively. The analysis of the

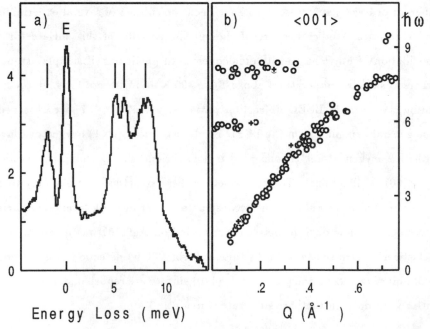

Fig.1, a) Energy loss spectrum measured along the $< 001 >$ direction of the p(n×1)O-Ag(110) surface with angle of incidence $\Theta_i = 45.2°$ (the scattered intensity I is given in arbitrary units). b) Energy versus parallel momentum transfer Q of the detected phonons reported in a reduced-zone plot (crosses correspond to the inelastic features shown in part a)).

spectra, carried out with the usual background subtraction procedure, yields the description of the surface phonon spectrum presented in fig.1 b.

The Rayleigh mode and two dispersionless surface resonances are well described. The comparable effectiveness in transferring energy to the He atoms can be easily estimated with the help of fig.1 a, which also shows the larger width of the resonance at ~ 8 meV. The most striking evidence of the oxygen presence comes out from the shifting of the substrate Rayleigh wave to a much higher frequency at the Brillouin-zone boundary [4] in a similar way as for the p(2×1) phase [5]. Moreover, because of the backfolding mechanism the Ag(110) phonon spectrum signs the present one via the 6 meV mode [6].

To understand the origin of the ~ 8 meV mode as well as the scattering of the data describing the Rayleigh wave in the range 0.35 Å^{-1} $\leq Q \leq 0.6$ Å^{-1} lattice dynamical calculations are now in progress.

REFERENCES

1. Segeth, W., Wijngaard, J. H. and Sawatzky, G. A., Surf. Sci. <u>194</u>, 615 (1988).

2. Campbell, C. T. and Paffet, M. T., Surf. Sci. <u>143</u>, 517 (1984).

3. Engelhardt, H. A. and Menzel, D., Surf. Sci. <u>57</u>, 591 (1976).

4. Tatarek, R., Bracco, G., Tommasini, F., Franchini, A., Bortolani, V., Santoro, G. and Wallis, R. F., Surf. Sci. <u>211/212</u>,314 (1989).

5. Yang, L., Rahman, T. S., Bracco, G. and Tatarek, R., unpublished.

6. Santoro, G., private communication.

INTERFACE OPTICAL PHONONS :
A NON-DESTRUCTIVE AND QUANTITATIVE STUDY BY
HIGH RESOLUTION ELECTRON ENERGY LOSS SPECTROSCOPY

P.A. Thiry, R. Sporken, J.J. Pireaux, R. Caudano

Ph. Lambin*, J.P. Vigneron and A.A. Lucas

Institute for Studies in Interface Sciences

Facultés Universitaires Notre-Dame de la Paix

Rue de Bruxelles 61, B-5000 Namur, Belgium

Long-wavelength interface optical modes have been theoretically predicted by the dielectric function theory and observed by High Resolution Electron Energy Loss Spectroscopy (HREELS) for layered polar materials. Al-As type phonons have been measured at the interfaces of a GaAs-AlGaAs superlattice and CaF_2 interface phonons in thin epitaxial layers on Si(111). The limitations of the technique are illustrated with the epitaxial system AlSb/Sb(111).

In a HREELS experiment, an incident beam of slow electrons (< 10 eV) is backscattered on a solid surface and its energy and momentum exchanges with the target are analyzed. Three interaction mechanisms have been identified and theoretically described[1] : the dipole, impact and resonance scatterings. The resonance scattering is a particular case of the impact scattering : both require a contact with the short-ranged potentials of the target atoms. This can result in large momentum transfers associated with microscopic excitations. On the contrary, small momentum exchanges are involved when the interaction proceeds by a long-ranged dipolar coupling between the electric field of the moving electron and the polarization field of a dielectric material surface. In this case, if the substrate is infrared active, a HREEL spectrum taken in the near specular geometry, will show characteristic energy losses and gains related to the excitation or annihilation of macroscopic surface optical phonons (Fuchs-Kliewer modes)[2]. The polarization field generated by these vibrations extends quite far in the vacuum and in the sample. Its range is, in fact, determined by the attenuation depth of the FK phonons and coincides with the average wavelength of these surface modes : it can be estimated at 100 Å. Because of the long range properties of the dipole scattering, this value of 100 Å can also be considered as a typical probing depth for a HREELS analysis performed in specular configuration on a polar material.

The theoretical formulation of the dipole interaction in HREELS (the so-called dielectric theory) starts from a macroscopic description which uses the long-wavelength limit of the infrared surface response of the target, expressed as a function of the frequency dependent dielectric function $\varepsilon(\omega)$[3]. The first step of the calculation of a HREEL spectrum consists in evaluating a classical energy loss probability $P_{cl}(\omega)$. For this purpose, the work done on the moving electron by the target polarization, is calculated for non-penetrating trajectories. However, slow electrons mean free paths are not easy to determine in ionic materials. According to the universal curve[4], they may be even fairly large. If this is the case, the theory predicts a possible excitation of longitudinal optical (LO) bulk phonons, which have a vanishing polarization field outside the sample[5]. Up to now, such LO modes have never been observed in the HREEL spectra of ionic insulators, even with the best demonstrated energy resolution of 2.5 meV[6]. These conclusions should warrant the application of the dielectric theory in its "non-penetrating" formulation, which moreover proved very successful in the interpretation of HREEL spectra of isotropic as well as of anisotropic polar materials[2,6].

Similar calculations were then performed for more sophisticated targets made of superimposed layers of different isotropic dielectric functions[3]. The above mentioned formalism is still valid, provided that, in the expression of the surface response of the target, $\varepsilon(\omega)$ is replaced, case by case, by an effective dielectric function $\xi(\omega)$. A simple example is an epitaxial film of CaF_2 on a thick Si(111) substrate, both materials being crystalline and isotropic. In this case, the "surface" response of the target exhibits two wavevector dependent poles. The corresponding contribution to the energy loss spectrum is made of two peaks. A high intensity peak related to the excitation of a surface FK mode and a lower intensity peak appearing at a slightly lower energy which is attributed to the excitation of an interface FK phonon. The intensity ratio of the two peaks increases with the thickness of the film. According to the predictions of the theory, both surface and interface phonons were observed[7] and their frequency dispersion was measured with respect to the film thickness . As a matter of fact, the interface mode could even be followed up to a CaF_2 layer thickness of several hundreds Å. These results unquestionably established the capacity of HREELS to investigate buried interfaces.

A compositional semiconductor superlattice presents another challenging system for HREELS, both from the experimental and theoretical points of view. Here again however, the dielectric theory could easily be reformulated with the help of the continued fraction formalism[3]. The resulting effective dielectric function $\xi(\omega)$ is characterized by multiple poles which give rise to a complicated structure of FK phonons branches in the (ω, k) plane. Some isolated modes can be unambiguously identified as surface modes.

The other modes which accumulate in continua, are interface modes. Some of these latter modes were observed in the HREEL spectra recorded on a GaAs/$Al_{0.3}Ga_{0.7}As$ (100 Å/100 Å) 25 periods superlattice[8]. Besides the GaAs surface phonon, Al-As type interface phonons were clearly resolved but the lack of resolution hindered the observation of the GaAs type interface vibrations continuum which was too close to the GaAs surface phonon.

From these two remarkable examples, it would be premature to conclude that HREELS is dedicated to the study of interfaces within a probing depth of some hundreds Å. There are some important shortcomings and they will show up in the last example. Thick epitaxial layers of AlSb have been grown on a hot Sb(111) crystal surface and analyzed by HREELS[9]. No interface phonon was visible in this case and, furthermore, no thickness-dependent frequency shift was observed for the surface phonon of AlSb. This illustrates the limitations of HREELS, which again were predicted by the dielectric theory. Indeed, the conducting substrate screens any optical contribution from its interface and the low oscillator strength associated to the AlSb phonon is not sufficient to produce an observable frequency shift. Nevertheless, the FK mode intensity could be closely related to the AlSb layer thickness but in this case, a direct observation of the interface was not possible by HREELS.

In conclusion, the interface capacity of HREELS has been clearly demonstrated, but only a limited number of layered systems is concerned which consist mainly of ionic insulators and undoped semiconductors. However, for these kinds of materials, HREELS offers an interesting non-destructive and non-invasive interface investigation tool.

REFERENCES

* Senior Research Assistant of the National Fund for Scientific Research (Belgium).
1. Thiry, P.A., Liehr, M., Pireaux, J.J. and Caudano, R., Physica Scripta 35, 368 (1987)
2. Thiry, P.A., Liehr, M., Pireaux, J.J. and Caudano, R., Phys. Rev. B29, 4824 (1984)
3. Lambin, Ph., Vigneron, J.P. and Lucas, A.A., Phys. Rev. B32, 8203 (1985)
4. Seah, M.P. and Dench, W.A., Surf. Interf. Anal. 1, 1 (1979)
5. Lucas, A.A., Vigneron, J.P., Lambin, Ph., Thiry, P.A., Liehr, M., Pireaux, J.J. and Caudano, R., Int. J. Quant. Chem., ACS 19, 687 (1986)
6. Liehr, M., Thiry, P.A., Pireaux, J.J. and Caudano, R., J. Vac. Sci. Technol. A2, 1079 (1984)
7. Liehr, M., Thiry, P.A., Pireaux, J.J. and Caudano, R., Phys. Rev. B34, 7471 (1986)
8. Lambin, Ph., Vigneron, J.P., Lucas, A.A., Thiry, P.A., Liehr, M., Pireaux, J.J. and Caudano, R., Phys. Rev. Lett. 56, 1842 (1986)
9. Sporken, R., Thiry, P.A., Xhonneux, P., Caudano, R. and Delrue, J.P., Appl. Surf. Sci. (in press)

ELASTIC AND INELASTIC HE ATOM SCATTERING ON NAI(001)<110>.*

J. G. SKOFRONICK, G. CHERN, W. P. BRUG, J. DUAN AND S. A. SAFRON, DEPARTMENTS OF PHYSICS AND CHEMISTRY, FLORIDA STATE UNIVERSITY, TALLAHASSEE, FL USA 32306 AND J. R. MANSON, DEPARTMENT OF PHYSICS AND ASTRONOMY, CLEMSON UNIVERSITY, CLEMSON, SC USA 29631.

A monoenergetic thermal energy beam has been scattered from the NaI(001)<110> surface and the elastic and inelastic scattering behavior has been measured.[1] Two series of time-of-flight measurements have been carried out: one study provided the surface dispersion curves for this direction, while the second was a measure of the elastic and inelastic scattering intensity in the specular direction as a function of target temperature where the Debye-Waller and multiphonon scattering effects become important.

The measured surface dispersion curves showed the Rayleigh mode, the acoustic longitudinal S_6 mode and the crossing S_8 mode. The first two were in quite good agreement to previous theoretical models, but the S_8 mode was not evident from the calculations.[2,3] However, in the unpublished work of Benedek, the S_8 mode showed up as a very broad peak in the surface projected density of states[4]. The existence of the S_8 mode now seems to be a general feature of the alkali halide surface vibrations[5] on occasion appearing as a sagittal resonance mechanism[6] or sometimes as a folding mechanism[7].

Measurements of the elastic and inelastic scattering intensity in the specular direction versus the target temperature have been done for NaI(001)<110> over the temperature range of 120K to 723K. As expected, the specular intensity decreases with target temperature in accordance with the well-known Debye-Waller effect. However, there is an inelastic "foot" immediately below the elastic peak which increases relative to the elastic peak with temperature due to multiphonon scattering. In addition to these two scattering contributions, there is also a background from incoherent diffuse scattering as well as from the residual gas in the detector. Several spectra exhibiting this behavior are shown in Fig. 1.

A considerable amount of information can be extracted from these measurements by using a model calculation based on small energy multiphonon exchange collisions.[8] The results show that the elastic peak intensity $I_G(T)$ as a function of temperature, T, is governed approximately by a Debye-Waller factor and

$$I_G(T) = I_G(0)\exp(-2W) \approx I_G(0)\exp(-\alpha T) \qquad (1)$$

where W is the Debye-Waller factor and the last exponential represents the harmonic approximation. For inelastic collisions with small energy losses close to the elastic peak, the inelastic scattering "foot" has a Debye-Waller factor very nearly the same as the elastic case and thus it can be shown that the inelastic intensity of the "foot", $I_{inel}(T)$, can be written

$$I_{inel}(T) \approx I_{inel}(0)\exp(-2W)F(T) \qquad (2)$$

where the function $F(T)$ provides information on multiphonon collisions. In the high temperature limit, for crystal temperatures greater than the Debye temperature, the experiment gives the left hand sides of Eqs. 1 and 2 and thus the functional form of $F(T)$ can be obtained. Its temperature dependence is nearly that of an increasing exponential in T, and can be regarded as a power series where each T^n term gives the relative intensity for the exchange of n phonons. For the NaI(001) case at T = 523K, the term with the largest contribution has $n \approx 9$ which implies that the dominant interaction involves approximately 9 phonons and that phonon exchanges from 18-20 may still be important.

We acknowledge discussions with G. Bishop and with G. Benedek.

* Supported by US DOE Grant DE-FG05-85-ER45208.

REFERENCES

1. Chern, G., Skofronick, J. G., Brug, W. P., and Safron, S. A. , Phys. Rev. B39, 12828 (1989).

2. Benedek, G., Surf. Sci. 61, 603 (1976).

3. Chen, T. S., deWette, F. W. and Alldridge, G. P., Phys. Rev. B15,

1167 (1977).

4.   Benedek, G. Private Communication.

5.   Safron, S. A., Chern, G., Brug, W. P., Skofronick, J. G. and Benedek, G., This conference.

6.   deWette, F. W., Kress, W. and Schroeder, U., Phys. Rev. B33, 2835 (1986).

7.   Benedek, G., Miglio, Brusdeylins, G., Skofronick, J. G., and Toennies, J. P., Phys. Rev. B35, 6593 (1987).

8.   Chern, G., Brug, W. P., Duan, J., Safron, S. A., Skofronick, J. G., and Manson, J. R., (In Preparation).

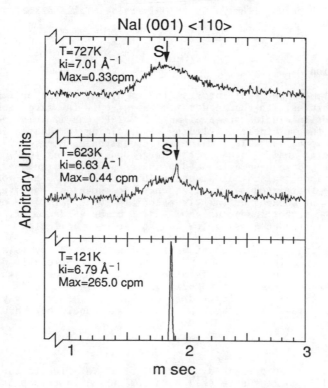

Figure 1.   Representative spectra of the scattered specular beam from the NaI(001) surface in the <110> direction as a function of target temperature.   The lower spectrum shows the large elastic signal which occurs for the cold target, while the middle one has only a small elastic peak (S) with the diffuse inelastic "foot" along with a uniform background.   The upper curve does not show an elastic peak.

CHARGE DENSITY WAVE COUPLED PHONONS ON A 1T-TAS (001) SINGLE
CRYSTAL SURFACE MEASURED BY HELIUM ATOM SCATTERING

G.Brusdeylins, F.Hofmann, P.Ruggerone, J.P.Toennies, R. Vollmer

Max-Planck-Institut fuer Stroemungsforschung, Bunsenstrasse 10,
Sonderforschungsbereich 126, Lotzestrasse 16-18, D-3400 Goet-
tingen, FRG,

G. Benedek

Dipartimento di Fisica dell'Universita', Via Celoria 16,
I-20133 Milano, Italy,

and J.G. Skofronick

Department of Physics, Florida State University, Tallahassee,
FL 32306, USA

1. Introduction

      In spite of the great interest in the dynamical properties of
layered structures with charge density waves, only few experimental
phonon dispersion studies are available [1-3]. Primarily this seems
to be due to the intrinsic difficulty in obtaining thick stacking-
fault-free samples suitable for inelastic neutron spectroscopy. In
transition metal chalcogenides with charge density waves (CDW) inelastic neutron data have evidenced peculiar properties such as Kohn
anomalies in the longitudinal acoustic (LA) branch [1-4], but most of
the data were restricted to low frequency and momentum. Moreover the
neutron study of phonons definitely associated with satellite peaks
appeared to be rather inconclusive [1] and no clear evidence of phonons coupled to (induced by) a CDW has so far been collected.

      Since surface properties of weakly bound layered crystals -apart
from a few remarkable exceptions [5-6] - should not appreciably differ
from the bulk ones, helium atom scattering (HAS) has been shown to be
a viable technique to extract information on static [7] and dynamical
properties [8] for very thin layered crystal samples. In this way, as
shown in Ref. 8, the optical branches of 2H-TaSe$_2$ could be measured for
the first time.

2. Experimental

      In this work we present a HAS study of 1T-TaS$_2$ in its low tem-
perature (120 K) commensurate ($\sqrt{13}x\sqrt{13}$) CDW phase. In this phase the
CDW hexagonal superlattice (g-vector = 0.595 Å^{-1}) is rotated 13.9°
with respect to the hexagonal crystal lattice (G-vector = 2.146 Å^{-1}).
The crystal [9] has been cleaved in situ. The apparatus, the sample preparation and measurement procedure are those described for previous
studies of layered crystals [5-6]. In Fig. 1 we show the diffraction
pattern at 120 K along the g direction (R13.9). The superstructure
peaks are fairly intense, indicating a CDW corrugation as large as

about ± 0.2 Å. An accurate study of the temperature dependence both on cooling and warming and of hysteresis effects in diffraction peak intensities will be presented elsewhere.

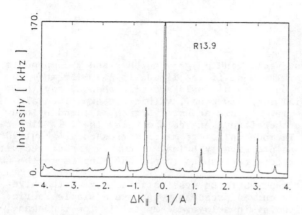

Fig.1 CDW diffraction pattern of the $\sqrt{13}\times\sqrt{13}$ commensurate phase at T= 120 K. Only CDW superlattice diffraction peaks are visible in this direction (rotated by 13.9° with respect to <100>), k_i=8.56 Å^{-1}.

Time-of-flight spectra show several well-resolved inelastic peaks in the acoustic region and weaker structures in the optical region of the crystal. Here we restrict our analysis to the acoustic region. The experimental points, folded back into the positive momentum and energy transfer quadrant are shown in Fig.2 for the two crystallographic directions of the normal lattice parallel to the layer (left) and for the g-direction of the CDW superlattice (right).

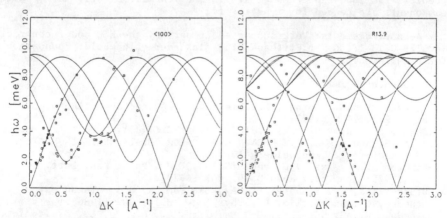

Fig. 2   Dispersion curves of the Rayleigh mode of 1T-TaS$_2$ in the CDW commensurate phase at 120 K along the normal lattice (left frame) and CDW superlattice (right frame) symmetry directions.

The points are disseminated in various regions of the $(\Delta K, \hbar\omega)$ plane, but they can be easily fitted by a single sinusoidal dispersion curve centered at the superstructure g-vectors contained in the irreducible part of the lattice Brillouin zone. This is expressed by

894

$$\omega^2 / A = \sin^2 \left(-\frac{K_x + g_x}{G'} \right) + \sin^2 \left(-\frac{K_x + g_x}{2G'} + 3\frac{K_y + g_y}{2G'} \right) +$$

$$\sin^2 \left(-\frac{K_x + g_x}{2G'} + 3\frac{K_y + g_y}{2G'} \right) , \qquad G' = \sqrt{3}\, G/2\pi$$

where K is the phonon wavevector and g_x, g_y are the x and y components of the CDW lattice peaks. The best-fit amplitude is $A = 6.50$ meV. The data can clearly be attributed to a single acoustic phonon, say the Rayleigh mode. The initial slope is equal, within the experimental accuracy, to that of the lowest bulk acoustic branch as found by neutron scattering [1] , and therefore these data display the full dispersion of the c-polarized TA mode up to the maximum frequency (about 10 meV). Many experimental points clearly belong to curves associated with nonzero g-values and therefore correspond to phonons coupled to the CDW.

The sinusoidal fit appears to be less satisfactory in the highest part of the dispersion curve, presumably because a single Fourier component for a dispersion law in a layered crystal is too crude an approximation. Note, however, that there is no anomaly of the kind reported for $TaSe_2$ near the normal-to-incommensurate CDW phase transition [6] ; here we are much below the corresponding transition temperature (543 K). On the other hand no LA mode (the one showing a deep anomaly in the bulk) is here observed. The few high frequency phonons seen close to either $K = 0$ or $K = g$ are spread between 8 and 10 meV. This agrees with the lowest Raman active frequencies of A symmetry (normal to the layers), 10.04 and 8.85 meV [2] . The sinusoidal fitting predicts two zone-center modes at 9.3 and 7.0 meV; clearly more Fourier components are needed for a better fit.

Acknowledgements: Work supported in part by the EEC under contract No ST2P-0013-1F(CD). P.R. thanks the Alexander v. Humboldt Foundation for a fellowship.

3. References

1) Ziebeck, K.R.A., Dorner, B., Sterling, W.G. and Schoellhorn, A., J. Phys. F 7, 1139 (1977).
2) Sugai, S., Phys. Stat. Solidi (b) 129, 13 (1985).
3) Bilz, H. and Kress, W., PHONON DISPERSION RELATIONS IN INSULATORS (Springer Verlag, Heidelberg 1979).
4) Moncton, D.E., Axe, J.D. and DiSalvo, F.J., Phys. Rev. Lett. 34, 734 (1975) and Phys. Rev. B 16, 801 (1977).
5) Brusdeylins, G., Rechsteiner, A., Skofronick, J.G., Toennies, J.P., Benedek, G., and Miglio, L., Phys. Rev. B 34, 902 (1986).
6) Benedek, G., Brusdeylins, G., Heimlich, C., Miglio, L., Skofronick, J.G.,Toennies, J.P. and Vollmer, R., Phys. Rev. Letters 60, 1307 (1988).
7) Cantini, P., in DYNAMICS OF GAS-SURFACE INTERACTIONS, ed. by Benedek, G. and Valbusa, U. (Springer Verlag, Heidelberg 1982) p. 84.
8) Benedek, G., Miglio, L., Brusdeylins, G., Heimlich, C., Skofronick, J.G., Toennies, J.P., Europhys. Letters 5, 253 (1988).
9) Provided by R. Claessen, Institut fuer Experimentalphysik, D-2300 Kiel 1, Olshausenstr. 40.

SURFACE PHONON DISPERSION CURVES OF IONIC INSULATORS MEASURED BY HIGH RESOLUTION ELECTRON ENERGY LOSS SPECTROSCOPY

J.L. Longueville, P.A. Thiry, J.J. Pireaux and R. Caudano

Laboratoire Interdisciplinaire de Spectroscopie Electronique, Institute for Research in Interface Sciences, Facultés Universitaires Notre-Dame de la Paix, Rue de Bruxelles 61, B-5000 Namur (Belgium)

High Resolution Electron Energy Loss Spectroscopy (HREELS) has been used to measure the surface phonon dispersion curves of three fluorides: NaF(001), LiF(001) and CaF_2(111). The HREEL spectra show characteristic peaks of the Rayleigh mode (S1), and of the Lucas mode (S4) along the $\overline{\Gamma}$-$\overline{M}$ and $\overline{\Gamma}$-$\overline{K}$ directions. A good agreement is found with He-beam scattering results and theoretical calculations.

1. INTRODUCTION

The measurement of dispersion curves by HREELS was first reported by Lehwald et al.[1] in 1983. In this case a metallic surface, Ni(100) was used and the Rayleigh mode (S1) could be observed along one direction of the surface Brillouin zone (SBZ) . On the other hand, since many years, acoustical surface phonons dispersion curves were routinely measured by He-beam scattering on ionic insulators[2] which provided the easiest surfaces to analyze by this technique. However, today, He-beam scattering experiments have been successfully carried out on metals and, at the expense of some energy resolution, extended also to the observation of optical surface phonons on ionic samples[3]. Such insulating targets could not until recently be analyzed by HREELS because of charging problems. Experimental work done by Liehr et al.[4] demonstrated that the "flood gun" technique could achieve a sufficient surface-potential stabilization in order to provide a stable capture of the reflected electron beam .

With the help of this new experimental setup, we were able to measure by HREELS the surface phonons of three fluorides : NaF(001), LiF(001) and CaF_2(111). The measurements performed at low energy in the specular configuration, where dipolar scattering dominates, gave information on the infrared dielectric response of the material and allowed to determine I.R. optical constants. An off-specular geometry and a higher electron energy were selected to record the dispersion curves along the most important symmetry directions of the SBZ. In this case, impact scattering was the dominant interaction mechanism[5].

2. DIPOLAR RESULTS

HREEL spectra were first measured in the specular geometry ($\theta = 45°$) and at low electron impact energy (5 to 10 eV). Charge neutralization of the fluoride samples was obtained by irradiation with the defocused beam of an auxiliary electron gun, the intensity of which was about 0.1 µA for an electron energy of 1 keV. The HREEL spectra recorded on the three fluorides were very similar. They all show the characteristic multiple energy losses and gains related to the excitation and annihilation of one surface phonon. This vibrational mode is analogous to those first discovered by Fuchs and Kliewer (F-K) for ionic slabs[6] and is traditionally named after them. The following energies were measured :

$$\omega_{FK} (NaF) = 355 \text{ cm}^{-1}; \quad \omega_{FK} (LiF) = 573 \text{ cm}^{-1}; \quad \omega_{FK} (CaF_2) = 412 \text{ cm}^{-1}.$$

The experimental conditions described above, favour the dipolar interaction mechanism between the moving electron and the polarization field associated with the vibration of the target. Consequently, the HREELS data could be interpreted on the basis of the dielectric function theory of energy loss proposed by Lucas and Sunjic[7] in 1972 . The recent formulation of this theory provides analytical expressions for the calculation of a theoretical spectrum depending only on the behaviour of the dielectric function in the infrared frequency range. These formulae were used to determine the infrared optical constants of NaF, LiF and CaF_2 (resonance frequency ω_{TO}, oscillator strength $\Delta\varepsilon$, and damping factor γ). The results will be extensively presented and discussed in another paper. Let us mention that the resulting values are in good agreement with data available from other techniques (I.R. reflectivity spectroscopy, principallly).

3. DISPERSION CURVES

The dispersion measurements were performed in a non-specular geometry and with higher electron energies (60 to 100 eV) in order to provide enough momentum transfer without favouring too much multiphonon and bulk phonon excitation. At these high energies, no charging effect was observed on our samples so that we could avoid the use of the neutralization gun. HREEL spectra were recorded for some principal directions of the two-dimensional SBZ : $\bar{\Gamma}-\bar{M}$ for NaF and LiF, $\bar{\Gamma}-\bar{M}$ and $\bar{\Gamma}-\bar{K}$ for CaF_2 . The observation of low energy surface phonons was, most of the time, difficult because of a relatively important incoherent elastic contribution accompanied by its F-K phonons satellites. Such a high incoherent elastic intensity was not found in the He-beam experiments. In our case, it is certainly related to a residual surface roughness due to imperfect sample preparation and could not be eliminated for our samples by modifying the cleaning procedure (heating in vacuo after cleaving).

It was necessary to apply a data treatment in order to extract precise energy values, especially for the low frequency peaks. The resulting dispersion curves are displayed in Figs. 1-3. They all show mainly three curves :
  - the Rayleigh mode (S1), situated below the projection of the acoustic bulk phonons;
  - an optical surface mode, identified as the micoscopic mode S4 (Lucas mode);
  - the Fuchs-Kliewer mode, showing no significant dispersion throughout the SBZ. This lack of dispersion was expected since the F-K mode intensity was always observed proportional to the incoherent elastic peak intensity, demonstrating that no actual momentum had been transferred to this mode.

In Figs. 1 and 2, the results for NaF and LiF are presented superimposed on theoretical calculations based on a Green's function formalism[8]. As can be seen, our data fit rather well the theoretical curves, especially for the S1 mode. As far as LiF is concerned, the most important result of this work is the confirmation of the behaviour of the Rayleigh mode near the zone boundary : our data indeed confirm the bending of the dispersion curve near the M point, which was first reported by Brusdeylins et al. from their measurements of inelastic He-atom scattering[2]. This effect can be accounted for by modifying the surface polarization and introducing the relaxation in the theoretical model.

Concerning CaF_2, no theoretical surface phonon dispersion curves are available so far for the (111) surface. However, dynamical calculations performed on CaF_2(110) slabs[9] reveal the existence of many new surface modes that cannot be easily compared to our experimental results.

4. ACKNOWLEDGEMENTS

This work is supported by the Belgian F.J.B.R. and S.P.P.S. One of us (J.L.L.) is grateful to the I.R.S.I.A. for financial support.

5. REFERENCES

1. Lehwald S., Szeftel J.M., Ibach H., Rahman T.S., Mills D.L., Phys. Rev. Lett. 50, 518 (1983)
2. Brusdeylins G., Doak R.B., and Toennies J.P., Phys. Rev. B 27, 3662 (1983)
3. Brusdeylins G., Rechsteiner R., Skofronick J.G., Toennies J.P., Benedek G. and Miglio L., Phys. Rev. Lett. 54, 466 (1985)
4. Liehr M., Thiry P.A., Pireaux J.J., and Caudano R., Phys. Rev. B 33, 5682 (1986)
5. Thiry P.A., Liehr M., Pireaux J.J., and Caudano R., Physica Scripta 35, 368 (1987)
6. Fuchs R. and Kliewer K.L., Phys. Rev. 140, A2076 (1965)
7. A.A. Lucas and M. Sunjic, Prog. Surf. Sci. 2, Pt. 2, 75 (1972)
8. Benedek G., Brivio G.P., Miglio L., and Velasco V.R., Phys. Rev. B 26, 497 (1982)
9. G. Lakshmi, F.W. de Wette, and R. Srinivasan, Surf. Sci. 94, 232(1980)

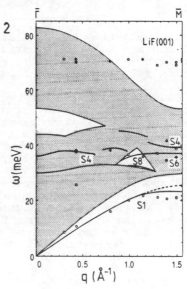

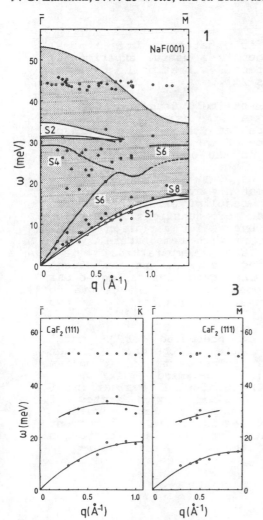

Figs. 1 and 2 : Surface phonons dispersion curves of NaF(001) and LiF(001) along the $\bar{\Gamma}$-$\bar{M}$ direction. HREELS results are reported on theroretical curves from ref. 8. Shaded regions correspond to the surface-projected bulk bands.

Fig. 3 : Surface phonons dispersion measured by HREELS on CaF$_2$(111). Full lines are simple polynomial interpolations between experimental points.

COHERENT PHONON REPRESENTATION FOR He METAL SURFACE
COLLISION SCATTERING

A. KHATER

Service de Physique des Atomes et des Surfaces,
Centre d'Etudes Nucléaires de Saclay,
91191 Gif-sur-Yvette Cedex, France.

The elastic and inelastic scattering of thermal helium atoms from solid surfaces is a useful experimental technique to study surface vibrational states, particularly in metal surfaces /1/.

Calculations have been extensively developed for the theory of elastic He scattering, using perturbation techniques, via a modulation of the static surface potential by the thermal motion of the crystal atoms /2/, and for inelastic scattering in the one-phonon distorted wave Born approximation /3,4/.

Reasonable agreement between a variety of theoretical calculations and experimental results, exists in several cases. Nevertheless a number of questions remain of interest. The stronger than Debye-Waller variation of the elastic scattering cross-section with temperature, which has been parametrized at high temperatures by a temperature dependent Debye frequency /2/, and the strong decrease of the one-phonon inelastic scattering cross section when the wave vector goes out from the centre to the edge of the surface Brillouin zone /4/, are two examples of such questions.

In this letter, a model is presented for He metal surface scattering, in the coherent phonon representation. This approach yields a unified formalism for elastic, and inelastic one- and multi-phonon scattering. It also permits the analytical evaluation of all contributions from the virtual phonon states of surfaces atoms, in the harmonic approximation. Theoretical results for soft potentials, are in agreement with experimental measurements for specular scattering from the Cu(100) surface up to the melting temperature.

The exact scattering wave function can be written asymptotically away from the surface as :

$$\psi = A_1 (\exp i\vec{k}_1 \cdot \vec{r}) + A_2 \exp (i\vec{k}_2 \cdot \vec{r}) \qquad (1)$$

$\vec{k}_1$ and $\vec{k}_2$ are the incident and scattered wave vectors, respectively, and A_1 and A_2, are amplitudes, of the incident and scattered states. The interaction between the incident He atom and the solid surface, is assumed to be the sum of pairwise potentials between the He and individual atoms. This is reasonable in general for metal surfaces that are not compact. To ensure that the surface boundary condition of vanishing wavefunction is satisfied, the rate of change of the momentum flux in this representation, yields consequently a scattering Hamiltonian between thermal helium and solid surface atoms, in the form :

$$H' = z^{-1} \sum_{\alpha \vec{\ell}} \hbar \, \Gamma(\vec{\ell}) \, (n_{1\alpha} \, k_1 - n_{2\alpha} \, k_2)$$
$$\times \frac{1}{2} \, (A_1 A_2{}^* + A_1{}^* A_2) \, u_\alpha (\vec{\ell}) \qquad (2)$$

where $\vec{\ell}$ identifies the position vector of a surface atom, z is the number of surface atoms contributing predominantly to the scattering ; $\Gamma(\vec{\ell})$ is the transition rate for scattering the helium atoms between the two states $|\vec{k}_1 >$ and $|\vec{k}_2 >$; $n_{1\alpha}$ and $n_{2\alpha}$ are the corresponding direction cosines in the cartesian component α ; $u_\alpha(\vec{\ell})$ denotes the displacement of surface atoms. The form of equation (2) for the scattering Hamiltonian is different from previous work /3/.

The Hamiltonian for the motion of the surface atoms contains consequently, a ground potential, and under the collision with the helium beam, an additional perturbation given by equation (2). It is possible to express the vibrational states, in the harmonic approximation, of both the ground and the driven Hamiltonians, in a coherent phonon representation /5/. This permits an analytical evaluation of the contributions of all possible transitions between virtual phonon states between the excited and the ground Hamiltonians. The novel results consequently go beyond previous work /2/, where only a limited number of contributions were evaluated. The scattering Hamiltonian of equation (2), leads to a scattering cross section in the form :

$$\frac{d\sigma}{d\Omega} \sim \frac{m}{M} \frac{1}{k_{1z}} \sum_{\vec{q}j} \frac{1}{\omega_{\vec{q}j}} (n (\omega_{\vec{q}j}) + \frac{1}{0}) \sum_{\alpha\beta}$$

$$(n_{1\alpha}k_1 - n_{2\alpha}k_2) (n_{1\beta}k_1 - n_{2\beta}k_2) e_\alpha (\vec{q}_j) e_\beta (\vec{q}_j)$$

$$|\gamma(\vec{q})|^2 \delta(\vec{k}_2-\vec{k}_1-\vec{q}) \sum_{m,n} |<a,n|b,m>|^2 e^{-\beta m \hbar \omega}$$

$$(1 - e^{-\beta \hbar \omega}) \delta[\hbar\omega_{ab} + (m-n) \hbar\omega - \hbar (E_2 - E_1)] \qquad (3)$$

$\vec{q}$ characterizes the annihilated or created phonon. The $\vec{q}$ dependent contribution is decoupled from the energy conservation condition were $|a,n>$ and $|b,m>$ denote coherent phonon states, and the summation is carried over all possible virtual states therein. $\hbar\omega_{ab}$ is the energy difference between the minima of the potential wells ; and $\gamma(\vec{q})$ is a structure factor defined by :

$$z^{-1} \sum_{\vec{\ell}} M_{\vec{\ell}}^{-\frac{1}{2}} \Gamma(\vec{\ell}) \exp (i\vec{q}.\vec{\ell})$$

where $M_{\vec{\ell}}$ is the $\vec{\ell}$ atomic mass.

To calculate $\Gamma(\vec{\ell})$ over incident and scattered states, I choose a soft potential of the form exp (-Kr). In the limit of low temperature T, equation (3) then yields that the intensity of the specular peak for which $k_1 = k_2$, varies as exp (- C T^4), where C is a product of constant terms. At higher temperatures the variation of this intensity can be expressed as exp(-C'(1+T/2T$_m$)^2 T), T_m being the melting temperature, which result is in agreement with experimental measurements for specular He scattering from the Cu(100) surface /6/.

Acknowledgements : the autor acknowledges useful discussions with G. Armand, J. Lapujoulade, R.J. Manson and B. Salanon, and acknowledges the support of C. Boiziau.

References

1. Toennies, J.P., J. Vac. Sci. Technol. A2, 1055 (1984).
2. Armand G., Gorse, D., Lapujoulade, J. and Mason, J.R., Europhysics Lett. 3, 1113 (1987), and references therein.
3. Bortolani, V., Franchini, A., Nizzoli, F. and Santoro, G., Phys. Rev. Lett. 52, 429 (1984).
4. Eichenauer, D., Harten, U. and Toennies, J.P. and Celli, V., J. Chem. Phys. 86, 3693 (1987).
5. Haken, H., "Quantum Field Theory of Solids", North-Holland Pub. Co. (Amsterdam, 1976).
6. Gorse, D. and Lapujoulade, J., Surface Sci. 162, 847 (1985).

LATTICE DYNAMICAL CALCULATIONS FOR STEPPED SURFACES

A.Lock, J.P.Toennies, G.Witte

Max-Planck-Institut für Strömungsforschung
Bunsenstraße 10, 3400 Göttingen, FRG

Introduction

A great deal of theoretical and experimental work has been published dealing with surface vibrations [1,2]. Thorough studies of vibrational surface modes in realistic crystal models have been carried out for some low Miller indexed surfaces for fcc and bcc metals [3,4]. There are, however, only few studies of high Miller indexed surfaces of fcc crystals [5,6]. The most complete report, for a variety of stepped surfaces, was presented by Black and Bopp [6].

The present work is aimed at making available the full dispersion curves as well as a detailed analysis of surface modes of vibration for high Miller indexed surfaces. We have developed a computer program that is capable of dealing with simple surfaces of both fcc and bcc crystal lattices of arbitrary Miller indices. The extension to crystals with a lattice basis is in progress. When handling the different structures appropriate interaction potentials can be chosen. In this study we restrict ourselves to one of the simplest cases: calculations for a fcc crystal with (221) surfaces using a force constant scheme with only next nearest neighbour forces.

Method of Computation

The (221) surface of a fcc crystal consists of (111) indexed terraces four atoms wide (Fig 1b) which are tilted 15.81° with respect to the macroscopic surface. The step edges of this surface run along the [$\bar{1}$10] direction, while the unit vector perpendicular to the step that lies in the macroscopic surface points into the [$\bar{1}\bar{1}$4] direction (Fig. 1a). There exist two types of calculations to compute surface phonon dispersion curves: the slab method introduced by Allen, Alldredge and de Wette [1] and the Greens function method. We have chosen the slab technique for all our calculations.

The slab method requires the construction of the dynamical matrix for a N-layer slab with 3N times 3N terms in it. Diagonalization of the dynamical matrix leads to 3N modes of vibration of the slab each with its own eigenvector. Those modes in which the eigenvectors show up large components at the surfaces and small ones in the interior of the slab are called "surface modes".

To show the essential vibrational features of stepped surfaces we set up the dynamical matrix for a 144-layer slab using the nickel force constant of Black and Bopp [6]. The elements of the dynamical matrix are calculated analytically by means of a computer algebra program which drastically reduces the time needed and ensures that the matrix elements are correct. This program will be described elsewhere [7]. The matrix is then diagonalized using a standard diagonalizer for Hermitian matrices. Subsequently the eigenvectors are analyzed.

Results

The dispersion relations of the 144-layer slab for a path around the irreducible element of the surface Brillouin zone (Fig. 1c) are shown in Fig. 2. The results can be summarized as follows : on the fcc (221) surface there are two **new** surface modes due to the presence of steps. They are labeled O_1 and O_2. Inspection of the polarization vectors reveals that they are optical surface modes (Fig. 3). Along the $\bar{\Gamma}\bar{X}$ direction all other modes, except the Rayleigh mode (S_1), change their polarization at points of avoided crossing. At the $\bar{X}$-point two modes become degenerate. The O_2 mode can be considered as a step localized mode because at the zone boundary atom 1 has the largest amplitude of vibration some 4 times larger than that of atom 3, while atoms 2 and 3 are nearly at rest. Along $\bar{S}\bar{Y}$ the two lower modes again show an avoided crossing. One of the optical modes mixes with the bulk modes and can be followed only with difficulty. In the $\bar{\Gamma}\bar{Y}$ region the modes can formally be seen as a result of backfolding of the (111) surface mode in the $\bar{\Gamma}\bar{M}$ direction. An energy gap appears to exist at the $\bar{Y}$-point. Results for other surfaces along with an extended discussion will be given in a subsequent paper.

References

[1]   R.E. Allen, G.P. Alldredge, F.W. de Wette, Phys. Rev. B $\underline{4}$, (1971) 1648

[2]   A.A. Maradudin, E.M. Montroll, G.H. Weiss, I.P. Ipatova, in: Solid State Physics, Suppl 3 (Academic Press, New York, 1971)

[3]   J.E. Black, D.A. Campbell, R.F. Wallis, Surf. Sci. $\underline{115}$, (1982) 161

[4]   J.E. Black, F.C. Shanes, R.F. Wallis, Surf. Sci. $\underline{133}$, (1983) 199

[5]   G. Armand, P. Masri, Surf. Sci $\underline{130}$, (1983) 89

[6]   J.E. Black, P. Bopp, Surf. Sci. $\underline{140}$, (1984) 275

[7]   R. Berndt et al. in preparation

Figures

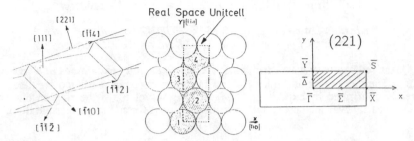

Fig. 1 Geometry of the fcc (221) surface with real space unit cell and surface Brillouin zone.

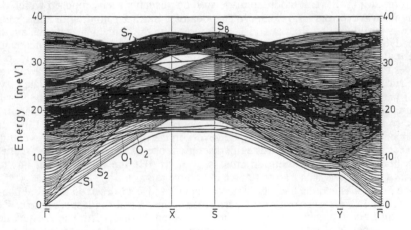

Fig. 2 Dispersion relations of a 144 layer slab for the path given in Fig. 1c.

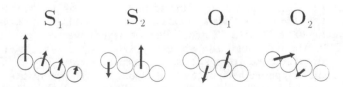

Fig. 3 Displacement patterns of surface modes near the $\bar{X}$-point.

DIRECT OBSERVATION OF RIPPLONS IN ^{4}He FILMS BY NEUTRON SCATTERING

H.Godfrin[1], V.L.P.Frank[1,2], H.J.Lauter[1] and P.Leiderer[2]

1 Institut Laue-Langevin, BP 156X, F-38042 Grenoble, France
2 University of Konstanz, D-7750 Konstanz, West Germany

Quantized capillary waves (ripplons) are the elementary excitations of a free liquid surface. Their existence in bulk ^{4}He and in films has been predicted by theory and indirectly confirmed by experiment[1,2]. At long wavelengths the ripplon dispersion relation is easily evaluated using hydrodynamic relations for an incompressible fluid:

$$\omega^2 = (\alpha_0/\rho_0) k^3 \qquad \{1\}$$

where α_0 is the zero temperature surface tension, ρ_0 the ^{4}He density at zero pressure and k the wavevector. The temperature dependence of the surface tension ($\alpha(T)$) at very low temperatures can be deduced from the ripplon dispersion relation. Detailed measurements[3] of $\alpha(T)$ revealed a much larger temperature dependence than expected from formula {1}. Several modified dispersion curves have been proposed which differ mainly for wavevectors above 0.5Å^{-1}. The idea of a ´surface roton´, with a minimum at ~2K, was introduced by Reut and Fisher[4] improving the agreement with the available thermodynamic data. Edwards et al.[1,3], taking into account the curvature dependence of α, were able to fit the experimental data on the excess surface entropy. Their model involves two parameters: a length δ=d(ln α_0)/dK where K=$(r_1{}^{-1} + r_2{}^{-1})$ is the curvature of the surface, and an area a = dδ/dK. Within the precision of the entropy data, several sets of parameters have been used (a=+1.5Å^2, δ=0 [3] and a=+1.0Å^2, δ=-0.336Å [1]), the latter giving a somewhat better agreement. Such a large variation in the parameters corresponds to very different ripplon dispersion curves at wavevectors ~1Å^{-1}. Little direct experimental evidence is available[5], however, on the ripplon dispersion curve at these wavevectors. Such a study requires a microscopic probe like inelastic neutron scattering (INS), but due to the low neutron cross section of ^{4}He the measurement has to be performed on samples with a large surface to volume ratio.

We have measured the inelastic structure factor of ^{4}He adsorbed on the basal plane of graphite. The INS measurements were performed at the time of fligth spectrometer IN6 at the Institut Laue-Langevin´s reactor using a wavelength of 5.12Å. The elastic energy resolution depends sligthly on the momentum transfer Q due to sample size effects, increasing from 80 to 110μeV with scattering angle.

The sample consisted of 31.70g of Papyex[6] sheets oriented with their c–axis normal to the scattering plane. The temperature of the sample was kept at 0.65K for all the measurements. An adequate annealing of the adsorbed films was performed after each change in coverage. The total surface area was determined by adsorption isotherms and neutron diffraction to be 730m^2 ± 2%. The ^{4}He monolayer coverage (0.112 atoms/Å^2) was 304cc STP. The data obtained before any ^{4}He was adsorbed were used as background and substracted from subsequent measurements. The result of two measurements will be reported here: a) at a total coverage of 0.448atoms/Å^2, equivalent roughly to 5 atomic layers and b) a scan with the cell filled with bulk superfluid ^{4}He.

The first and second layers are solid at these coverages and their density is well known[7,8] (1st layer=0.115at/Å^2, 2nd layer=0.094at/Å^2). Thus, 0.209at/Å^2 correspond to the solid and the remaining amount of ^{4}He to the liquid layers. We use a mean density of 0.078at/Å^2 for a liquid layer to evaluate the thickness of the film. This value has to be taken with care, since microscopic calculations showed that the density of the bulk liquid-vacuum interface decreases slowly within a distance of ~5Å[9].

The result of the measurement for 5 layers is depicted in figure 1 as a contour plot of $S(Q,\omega)$ in order to give a general overview of all channels and detectors. One can easily recognize two excitation branches. The higher energy one agrees well with the bulk phonon-roton dispersion relation (solid line in fig.1). The lower branch, located at about half the energy of the previous one, is the main object of this paper. Evidence of the existence of this branch has been found previously on measurements[5] done on a different substrate (Vulcan III graphite powder), together with the observation of dispersionless modes. In the present experiment no dispersionless modes were observed for $Q < 1.5\text{Å}^{-1}$; this may be due to the larger coherence length or to the preferential orientation of our Papyex sample. The absence of these flat modes enables us to determine the lower branch dispersion relation unambiguously up to $Q \sim 1.5\text{Å}^{-1}$.

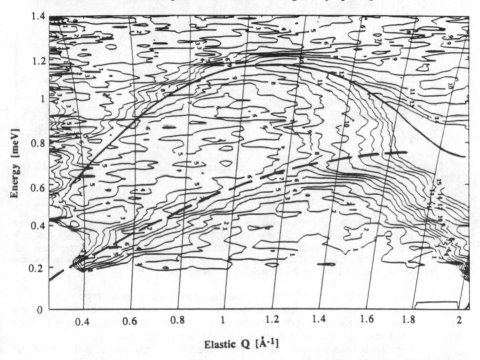

Figure 1: Contour plot of $S(Q,\omega)$ (arbitrary units) for a coverage of $0.448\text{at}/\text{Å}^2$. The contour lines 1 to 9 are separated by 10 units, lines 10 to 15 by 30 units. The thick solid line is the bulk ^{4}He phonon-roton dispersion relation. The elastic Q values are given in abscissa, constant Q lines are indicated by thin solid lines in the graph. The dashed line corresponds to the ripplon dispersion curve determined by Edwards and Saam[1], using $a = + 1.0\text{Å}^2$, $\delta = - 0.336\text{Å}$.

To determine the origin of the lower excitation branch we performed a measurement with the cell filled with bulk liquid. This procedure suppresses the liquid-vapour interface, but does not affect the adsorbed solid-liquid interface. As seen in figure 2, the phonon-roton part is strongly enhanced while the peak at 0.47meV is suppressed. This is observed for all the measured spectra. Therefore, we conclude that the low energy excitation branch belongs to the liquid free surface and is identified as a ripplon.

The experimental ripplon dispersion curve displays a strong downward curvature and seems to merge with the bulk roton minimum for $Q\sim2\text{Å}^{-1}$. Due to the high intensity of the roton signal it is difficult to follow the ripplon peak at these wavevectors.

Below $Q=1.5\text{Å}^{-1}$ our result agrees well with the calculation of Edwards and Saam[1] for $a=+1.0\text{Å}^2$, $\delta=-0.336\text{Å}$ and does not agree with the dispersion curves given in references 3 and 9.

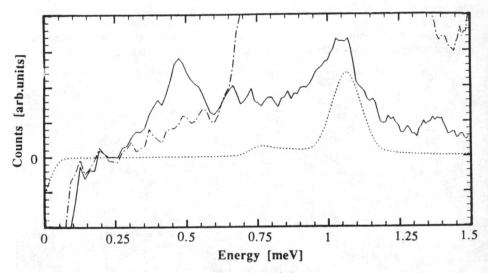

Figure 2: Inelastic spectrum (arbitrary units) for $Q=0.8\text{Å}^{-1}$ as a function of energy. The solid line corresponds to the measurement with coverage 0.448at/Å^2 (5 atomic layers). The peak at $\sim0.47\text{meV}$ corresponds to the ripplon and the one at $\sim1.05\text{meV}$ to the bulk phonon. Dashed line: same spectrum with the cell filled with bulk liquid; note that the ripplon peak has now disappeared. Dotted line: bulk spectrum divided by 100; the peak at 0.75meV is due to multiple scattering.

In conclusion, we have directly determined the ripplon spectrum at large wavevectors for the first time. With the cell filled with ^{4}He, the surface mode is suppressed, showing clearly that this excitation corresponds to the liquid-gas interface. Measurements of the temperature and coverage dependence and at higher Q are in progress.

This work has been partially supported by the West Germany Federal Ministry of Research and Technique (BMFT).

[1]   Edwards D.O. and Saam W.F., Prog. in Low Temp.Phys., Ed. D.F.Brewer, Vol. VII A, 283 (1978), North Holland
[2]   Vinen W.F., Springer Series in Sol.St.Sci., Ed. K.Ohbayashi & M.Watabe, Vol.79, 189 (1989), Springer Verlag.
[3]   Edwards D.O., Eckardt J.R. and Gasparini F.M., Phys.Rev. A 9, 2070 (1974)
[4]   Reut S and Fisher I.Z., Sov.Phys. JETP 33, 981 (1971)
[5]   Lauter H.J., Frank V.L.P., Godfrin H. and Leiderer P., Springer Series in Sol.St.Sci., Ed. K.Ohbayashi & M.Watabe, Vol.79, 99 (1989), Springer Verlag.
[6]   Papyex is produced by Carbone Lorraine, 45 Rue des Acacias, F-75821 Paris Cedex 17
[7]   Lauter H.J., Schildberg H.P., Godfrin H., Wiechert H. and Haensel R., Can.J.Phys. 65, 1435 (1987) and references therein.
[8]   Schildberg H.P., PhD Thesis, Kiel University, West Germany, 1988
[9]   Krotscheck E., Stringari S. and Treiner J., Phys.Rev. B 35, 4754 (1987)

LATTICE DYNAMICS CALCULATIONS ON ADSORBED MOLECULAR LAYERS WITH LARGE AMPLITUDE MOTIONS

T.H.M. van den Berg

Institute of Theoretical Chemistry, University of Nijmegen,
Toernooiveld, 6525 ED Nijmegen The Netherlands

Adsorbed monolayers of N_2 molecules on graphite occur in a rich variety of interesting quasi-twodimensional phases. Neutron and low-energy electron diffraction and heat-capacity measurements have shown that the low temperature ground state of a N_2 adlayer on graphite is an ordered phase with a commensurate $(\sqrt{3} \times \sqrt{3})R30°$ center of mass structure. Adlayer lattice vibration frequencies have not been measured, yet, but recent developments on adsorbed rare gas layers suggest that helium scattering is a promising measuring technique. In this contribution we present quantummechanical lattice dynamics calculations for the low temperature structure, which may be valuable for future experiments.

In order to describe the interaction between the adsorbed molecules we use an *ab initio* N_2-N_2 potential, which is available either as an atom-atom model or as a spherical expansion in terms of rigid rotor functions. From dynamics calculations on bulk nitrogen, which were in good agreement with experiment, we have concluded that the spherical expansion provides the best description of the potential anisotropy. Adopting the spherical expansion model, the intermolecular potential is further expanded into center of mass displacements up to quartic terms inclusive.

A similar expansion is used for the adlayer-substrate interaction which has been obtained from a semi-empirical Lennard Jones 12-6 atom-atom potential model. Because of the two-dimensional translation symmetry parallel to the substrate surface the molecule-substrate expansion coefficients are written as a two-dimensional Fourier series, using new analytical formulas. From a detailed numerical analysis, we have concluded that the combined spherical, displacement and Fourier expansion of the N_2-graphite potential is rapidly convergent; higher order Fourier terms are only needed for the substrate toplayer. Furthermore, it was shown that the molecule-substrate interaction strongly determines the out-of-plane potential anisotropy, whereas the in-plane anisotropy is dominated by the admolecule-admolecule interaction.

We have first performed self-consistent Mean Field calculations in a finite basis of spherical harmonics for the orientational motions and three-dimensional harmonic oscillator functions for the translational vibrations. The full potential anisotropy is retained and displacement anharmonicities up to fourth order are included. This procedure is particularly suitable for large amplitude orientational motions because we start with a free rotor basis. In order to regain the coupling between the molecular motions, which has been neglected at the MF level, we have applied the Time Dependent Hartree method. This is a linear response model relating the coupled and uncoupled response functions, or generalized susceptibilities, by way of

$$\chi^{-1}(q, z) = (\chi^0(z))^{-1} + W(q). \tag{1}$$

Here, $\chi^0(z)$ represents the single particle MF susceptibility matrix in a basis of dynamical variables, which are either free rotor functions or powers of molecular displacement coordinates. The dispersion is introduced by the coupling matrix $W(q)$, which comprises Fourier transformed potential expansion coefficients. The poles of $\chi(q, z)$ provide the collective excitation frequencies. Using a finite basis of MF states, Eq. (1) can be transformed into

$$\chi(q, z) = Q^T [z - M(q)]^{-1} PQ \tag{2}$$

with TDH matrix

$$M(q) = \varepsilon - PQW(q)Q^T. \tag{3}$$

The diagonal matrices ε and P contain MF excitation energies and occupation numbers and Q is the transformation matrix from the dynamical variables to MF excitation and de-excitation operators. Without further approximations, we have diagonalized the TDH matrix $M(q)$ of Eq. (3), which yields the poles of $\chi(q, z)$, i.e. the phonon frequencies, as eigenvalues. Apart from (small) finite basis errors, this procedure is exact at the TDH level.

As an example we show some calculated TDH phonon dispersion curves in Fig. 1. Because of the two-dimensional spacegroup symmetry $(p2gg)$, the phonon frequencies, apart from crossings, are not degenerate in the interior of the Brillouin zone (ΓX and $Y\Gamma$), whereas time-reversal causes a two-fold degeneracy at the boundary (XSY). From the calculations we conclude that the out-of-plane and in-plane excitations are largely decoupled both at the MF and TDH level. In addition, an extra (not symmetry determined) rotation-translation decoupling occurs at the center Γ of the Brillouin zone. The in-plane dynamics appears to be dominated by the *ab initio* N_2-N_2 potential, whereas the N_2-graphite interaction only influences the two (in-plane) acoustic modes around Γ. On the contrary, the out-of-plane excitation frequencies consist of a large q-independent contribution of the admolecule-substrate interaction and a relatively small q-dependent part caused by the admolecule-admolecule interaction.

Although we have started the calculations with a free rotor basis, the rotational MF wave functions are strongly localized and the corresponding set of energies resembles the spectrum of a two-dimensional harmonic oscillator. The two lowest lying excited states possess a nodal plane, either perpendicular or parallel to the substrate surface, resembling the first excited states of an in-plane

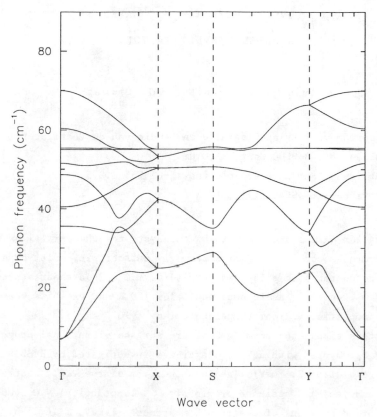

Fig. 1. Calculated TDH phonon dispersion curves.

and an out-of-plane (quasi) harmonic oscillator. Analysis of the higher excited states confirm this simple model. As far as the in-plane motions are concerned, the translational MF spectrum can also be interpreted with a harmonic model. By contrast, for the out-of-plane displacements, anharmonic potential terms appear to be important. The inclusion of only the cubic anharmonicities, however, causes the artifact that the molecules can easily escape from the surface. So quartic anharmonic terms are essential for the out-of-plane translational dynamics.

In addition, we have also performed harmonic lattice dynamics calculations, which, apart from the out-of-plane frequencies, are in good agreement with the TDH results, at least when the same spherically expanded potential is used. So we conclude that the dynamics of the low-temperature commensurate N_2 adlayer on graphite is mainly harmonic, i.e. the molecules are strongly localised and orientationally locked.

V_2O_5 COVERED WITH A V_6O_{13} LAYER :
A HREELS INVESTIGATION.

H. Poelman [1], J. Vennik [1], G. Dalmai [2].

1 Laboratorium voor Kristallografie en Studie van de Vaste Stof,
   Krijgslaan 281, B-9000 Gent, BELGIUM
2 Laboratoire de Surfaces et Interfaces -UA 253 I.S.E.N.-,
   41 Bld Vauban, F-59046 Lille, FRANCE

Vanadium oxides are known to be catalysts for the partial oxidation of hydrocarbons [1]. V_2O_5 is a strongly anisotropic, multi-phonon mode semiconductor with a layered structure. V_6O_{13} is a shear structure of V_2O_5, which is semiconducting below 150 K and exhibits weakly metallic properties at room temperature (n_e : 10^{20} cm^{-3}).

In this paper, the room temperature surface vibrational properties of V_2O_5 covered with a V_6O_{13} layer are investigated by HREELS (optimal resolution 2.5 meV). The loss spectrum of the V_2O_5 substrate has been analysed and assigned previously [2]. Topotactical V_6O_{13} upon V_2O_5 is formed by means of electron bombardment (1 keV, 1 µA, 5 h) and subsequent heating (550 °C, 30 min). The transition is monitored by digital LEED. The LEED spots of the resulting V_6O_{13} pattern are very large, denoting a surface with a high concentration of defects.

Upon transition, the V_2O_5 loss spectrum quickly changes into a spectrum with a loss continuum and two broad loss structures (56 and 123 meV) (fig.1: upper curve), and the FWHM progressively increases from 2.9 meV (V_2O_5) to 16.6 meV (V_6O_{13} layer). The origin of this large FWHM is twofold: the presence of, on the one hand, the metallic overlayer [3] and of the surface defects on the other. While in general the FWHM for metal layers deposited upon semiconductor substrates tends to decrease once a certain thickness is reached [3], for V_6O_{13} on V_2O_5, the resolution remained about 16 meV for increasing V_6O_{13} thicknesses, reflecting the strong influence of the surface defects.

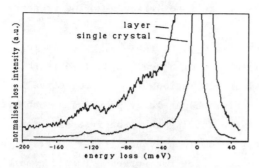

Fig.1: experimental HREEL spectra
upper: V_6O_{13} upon V_2O_5 (FWHM=16.6 meV)
lower: single crystal V_6O_{13} (FWHM=4.7 meV)

Table 1: losses (meV) for single crystal V_6O_{13} and for V_6O_{13} upon V_2O_5

single crystal	layer
11.0	
20.9	
31.6	
49.1	56
70.1	
115.7	
128.5	123

The loss structure can arise from the underlying V_2O_5 substrate phonons. In order to check the validity of this explanation, several single loss simulations were performed based on the single scattering theory of Demuth and Persson for isotropic materials [3] and adapted to anisotropic substrates. V_2O_5 is described by a dielectric tensor with 15 IR active modes. Metallic V_6O_{13} here is assumed to be isotropic and characterised by a mean plasma frequency (1.87 eV) and relaxation time ($7.7 \cdot 10^{-12}$ s) [4]. The simulations are convoluted with a gaussian of 16.6 meV.

Figure 2 shows the normalised simulations of V_6O_{13} layers with thickness 2, 20 and 200 Å as well as the experimental spectrum. Clearly, any distinct loss structure disappears as the layer grows thicker, leaving only a broadened elastic peak. From comparison with the experimental spectrum, the layer thickness is estimated to be close to 200Å.

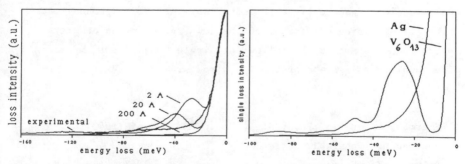

Fig.2: experimental layer spectrum and single loss simulations of V_6O_{13} upon V_2O_5 convoluted with a 16.6 meV gaussian.

Fig.3: single loss simulation of 2 Å Ag and V_6O_{13} upon V_2O_5 convoluted with a 5 meV gaussian.

Hence, one must conclude that there's no evidence for the presence of substrate phonon losses.

In view of the weakly metallic character of V_6O_{13}, it is interesting to compare its effect with that of a real metal layer such as Ag. Figure 3 shows two single loss simulations for a hypothetical 2 Å Ag and V_6O_{13} layer on V_2O_5. Clearly, the Ag layer provides a much better screening than the V_6O_{13} layer, showing that the free electron concentration of the latter is too low to prevent dipolar interaction. Thus, for a thicker layer, this weak electron screening could possibly leave the strongest V_6O_{13} metal phonons detectable, so that dielectric properties of a "weak" metal such as V_6O_{13} should be described by both its plasma frequency and its vibrational modes. From this point of view, the broad loss structures can be regarded as due to intense V_6O_{13} lattice vibrations whose dipole field is not efficiently screened.

In order to develop this hypothesis, measurements were performed on single crystal V_6O_{13} at room temperature. The lower curve in figure 1 shows the typical loss spectrum with a resolution of 4.73 meV. In the metallic phase, V_6O_{13} has 27 IR active modes of which only 7 are present in the experimental spectrum. The two highest energy losses can be assigned to V-O stretching modes. In view of the very large difference in resolution, the broad loss structures of the layer can be regarded as due to non-resolved V_6O_{13} phonons (table 1).

In conclusion, the surface transition of V_2O_5 into V_6O_{13} has been well established by HREELS. Upon comparison with the single crystal, the layer loss spectrum is assumed to represent V_6O_{13} phonon modes which are not screened because of the low free electron density. Future IR measurements will allow for single loss simulations and a more complete assignment of the V_6O_{13} modes.

References

1 Fiermans L. et al., Phys. Stat. Sol. (a) 59, 485 (1980)

2 Poelman H., Vennik J., Dalmai G.,J. Electr. Spectr. 44, 251 (1987)

3 Demuth J.E., Persson B.N.J.,Appl. Surf. Sci. 22/23, 415 (1985)

4 Van Hove W., Clauws P., Vennik J., Sol. Stat. Comm. 33, 11 (1980)

PHONON GAP ENERGY AS A FUNCTION OF TEMPERATURE FOR COMMENSURATE MONOLAYERS ON GRAPHITE

V.L.P.Frank[1,2], H.J.Lauter[1] and P.Leiderer[2]

1 Institut Laue-Langevin, BP 156X, F-38042 Grenoble, France
2 University of Konstanz, D-7750 Konstanz, West Germany

Gases physisorbed on adequate substrates present a large variety of phases and phase transitions. Graphite has been one of the most frequently studied substrates, since well characterized samples with a large surface to mass ratio can easily be produced. The observed phase transitions are the result of a delicate balance between the interaction of the adsorbed gas and the substrate, and between the adsorbed molecules themselves. Unfortunately, the knowledge of the details of these interaction potentials is sparse since not many experimental techniques are available. The adsorption potential itself has been determined mainly with molecular beam scattering, but the magnitude of the in-plane corrugation of the adsorption potential is very difficult to obtain in this way. It is this corrugation that forces some physisorbed substances into a commensurate phase and produces, due to the lost translational invariance, an energy gap at the zone centre in the acoustic branches of the phonon spectrum. Thus, the determination of this gap is a direct measure of the adsorption potential corrugation. The temperature renormalization of the phonon spectrum gives insight into the anharmonic terms of the adsorption and intermolecular potentials. Recently, several measurements of the gap became available and allowed a quantitative comparison with theoretical models of the adsorption potentials[1].

We present here the results of inelastic neutron scattering experiments that determined the phonon gap (Δ), and its dependence on temperature, for various substances, together with a model that reproduces the observed features. The studied adsorbates present all a registered $\sqrt{3} * \sqrt{3}$ R30 phase ($a_{nn} = 4.26$Å) and can be grouped mainly into two classes according to their quantum character. The hydrogen isotopes (H_2, HD, D_2) and ^3He are typical quantum gases: their interaction potential is weak, they exhibit a large zero point motion and a very large compressibility. Nitrogen (N_2) and deuterated-methane (CD_4), on the other hand, are much heavier molecules, with consequently a smaller zero point motion forming less compressible monolayers. These differences are evidenced in the 2-dimensional phonon spectra as shown in Table I. The

	H_2	HD	D_2	^3He	N_2	CD_4
Phonon Gap	47.3	43.2	40.0	13.5	19.3	15.9
DOS width	27.5	14.7	9.5	-	64.2	76.7
DOS centre	61.0	50.6	44.2	-	51.4	55.3
W/C ratio	0.45	0.29	0.21	-	1.25	1.39
T_m	20.5	19.4	18.5	3.05	~50*	~55*

Table I: Some characteristic values of the 2-dimensional phonon density of states (DOS) and melting temperature (T_m) of the commensurate phase. (Values are given in Kelvin. Row 4 gives the ratio of the values in rows 2 and 3).
*The phase diagrams for N_2 and CD_4 present a commensurate region that extends to higher temperatures when the coverage is slightly higher than the commensurate one. Details of the respective phase diagram can be found in ref. [2] and [3].

quantum gases present very little dispersion and resemble, to a first approximation, Einstein oscillators. The reason for this behaviour is the large zero point motion which favours the commensurate phase, even if the nearest neighbour distance is ~20% larger than that of the 3D solid. Due to the strong interaction between the adsorbed molecules, the dispersion curves of N_2 and CD_4 cover a wider energy range. In this case the 3D lattice parameter matches within some percent the one of the 2D commensurate phase.

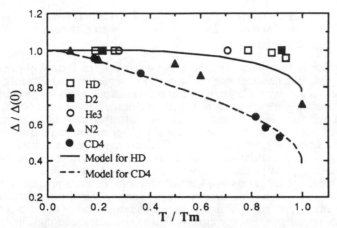

Figure.1: Normalized zone centre phonon gap vs. reduced temperature for several gases. The lines are the result of the model calculation. (Δ stands for the zone center phonon gap)

Figure 1 shows the temperature renormalization of the phonon gap due to the strong anharmonicity of the adsorbate potential. The $q=0$ mode depends only on the curvature of the adsorption potential and not on the intermolecular forces. The effective curvature is a weighted average over the root mean square amplitude of vibration, which is in turn determined by the effective curvature of the potential. This quasi-harmonic approximation must thus be solved in a self-consistent way. On the other hand, the intermolecular potential is also anharmonic and can be expanded in terms of the two-particle correlation function. This additional anharmonicity renormalizes the whole frequency spectrum of the phonons.

More insight into the problem can be gained using the following model hamiltonian to obtain the corresponding dispersion curves [4,5]:

$$ H = \frac{1}{2}\sum_{l\alpha} m\, \dot{u}_{l\alpha}^2 + \frac{1}{2}\sum_{\substack{l\alpha \\ l'\beta}} \phi_{ll'\alpha\beta}\, u_{l\alpha} u_{l'\beta} + \frac{1}{2}\sum_{\substack{l\alpha \\ \beta}} \theta_{l\alpha\beta}\, u_{l\alpha} u_{l\beta} $$

where $\alpha, \beta = 1,2$ are the coordinate components, and l, l' designate the molecular sites,

$$ \phi_{ll'\alpha\beta} = \frac{\partial^2 U}{\partial u_{l\alpha}\partial u_{r\beta}} + \frac{1}{2}\frac{\partial^4 U}{\partial^2 u_{l\alpha}\partial^2 u_{r\beta}}\left\langle \left| \bar{u}_l - \bar{u}_{l'} \right|^2 \right\rangle , \quad \theta_{l\alpha\beta} = \frac{\partial^2 V}{\partial u_{l\alpha}\partial u_{l\beta}} + \frac{1}{2}\frac{\partial^4 V}{\partial^2 u_{l\alpha}\partial^2 u_{l\beta}}\left\langle \left| \bar{u}_l \right|^2 \right\rangle $$

with $U(r_l - r_{l'})$ the inter-molecular potential and $V(r_l)$ the substrate-adsorbate interaction. The anharmonic part of the potential is partially taken into account by the bi-quadratic term. All the derivatives are evaluated at the equilibrium position. A similar treatment for several gases can be found in ref. [4] The two thermal averages are expressed as:

$$\left\langle \left| \overline{u}_1 - \overline{u}_{1'} \right|^2 \right\rangle = \frac{\hbar}{Nm} \sum_j \sum_{BZ} \coth\left(\frac{\beta\hbar\omega_j(\overline{q})}{2}\right) \frac{1}{\omega_j(\overline{q})} \left\{ 1 - \cos\left(\overline{q} \cdot (\overline{R}_1 - \overline{R}_{1'})\right) \right\}$$

$$\left\langle \left| \overline{u}_1 \right|^2 \right\rangle = \frac{\hbar}{2Nm} \sum_j \sum_{BZ} \coth\left(\frac{\beta\hbar\omega_j(\overline{q})}{2}\right) \frac{1}{\omega_j(\overline{q})}$$

where $\beta = 1/k_B T$.

The dynamical matrix was evaluated using this ansatz for the potentials and the structure of the $\sqrt{3}*\sqrt{3}$ R30 hexagonal lattice. The corresponding expressions were iterated until convergence was obtained for each temperature. The resulting temperature renormalization of the phonon gap obtained with this model for HD and CD_4 are also indicated in fig.1. The melting of this structure was defined as the temperature at which $<u^2>$ diverged. For HD we used the Silvera-Goldman[6] intermolecular potential and for the CD_4 molecule a 6-12 Lennard-Jones potential ($\epsilon = 137K$, $\sigma = 3.68\text{Å}$). The adsorbate-substrate potential was modeled after Steele[7]. A calculation done for D_2 using the same microscopic model as in ref. 8 indicates that the gap is less affected by temperature[9], more in agreement with the experimental data.

The agreement between the data and the model is reasonable taking into account the incomplete knowledge of the potential parameters involved. In this treatment a mean field approach has been used which should break down near the melting transition. Nevertheless the qualitative features of the temperature dependence of the phonon gap are reproduced: for the quantum gases, the rms vibrational amplitude is determined by their zero point motion and is little affected by thermal population of phonons resulting in a very weak variation of the gap with temperature. For the heavier gases, on the other hand, $k_B T_m$ at the melting temperature lies well above the cutoff energy of the phonon DOS and the melting transition is driven by thermally excited phonons.

We would like to thank D.Strauch for very helpful discussions and A.Novaco for sharing his results before publication. This work has been partially supported by the Federal Ministry of Research and Technique (BMFT) of the Federal Republic of Germany.

[1] Frank V.L.P., Lauter H.J. and Leiderer P., Phys.Rev.Lett., 61,436 (1988); Lauter H.J., Frank V.L.P., Leiderer P. and Wiechert H., Physica B. 156&157, 280 (1989); Hansen F.Y., Frank V.L.P., Taub H., Bruch L.W., Lauter H.J. and Dennison J.R., submitted to Phys.Rev.Lett. ; see also papers presented at this conference.
[2] Chan M.H.W., Migone A.D., Miner K.D. and Li Z.R., Phys.Rev.B, 30, 2681 (1984)
[3] Kim H.K., Zhang Q.M., Chan M.H.W., Phys.Rev.B, 34, 4699 (1986)
[4] Hakim T.M., Glyde H.R. and Chui S.T., Phys.Rev.B, 37, 974 (1988)
[5] Bilz. H, Strauch D. and Wehner R.K., Handbuch der Physik, Vol. XXV/2d, 1984
[6] Silvera I.F. and Goldman V., J.Chem.Phys., 69, 4209 (1978)
[7] Steele W. A., Surf.Sci., 36, 317 (1973)
[8] Novaco A.D., Phys.Rev.Lett..61, 436 (1988)
[9] Novaco A.D., private communication

ROTATIONAL EXCITATION AND OUT-OF-PLANE VIBRATION OF para-H$_2$ AND HD ADSORBED ON GRAPHITE

J.L.Armony[1], V.L.P.Frank[1,2], H.J.Lauter[1] and P.Leiderer[2]

1 Institut Laue-Langevin, BP 156X, F-38042 Grenoble, France
2 University of Konstanz, D-7750 Konstanz, West Germany

Recently detailed inelastic neutron measurements[1] of the in-plane excitations of hydrogen isotopes (H$_2$, HD and D$_2$) adsorbed on graphite have been reported and good agreement with theoretical calculations[2] has been found. A basic assumption in this calculations is that the molecules are in the $J=0$ rotational ground state. To investigate the validity of this assumption a more detailed analysis of the quantum mechanical state of this system has been performed[3]. It showed that the lowest energy state has a 0.999 $J=0$ component indeed , confirming the previous hypothesis. As an additional feature, the calculations predict a splitting of the first excited state into a doublet formed by the $M_J=0$ and $M_J=\pm 1$ states. This splitting is due to the anisotropy of the adsorption potential and thus the molecule feels a different interaction when the axis of rotation, classically speaking, is perpendicular or parallel to the adsorption plane. However, the magnitude of this effect as measured by NMR measurements[4] is an order of magnitude smaller than the theoretical prediction. This suggests that the treatment of the coupling between rotational and translational modes must still be refined.

We have performed neutron inelastic scattering measurements in order to observe the rotational transition of adsorbed H$_2$ and HD and also the out of plane vibration of these molecules. We used two types of substrate: a) a graphite powder[5] and b) oriented graphite foils[6]. The coherence length of these two substrates is 40Å and 300Å, respectively. The spectra taken before any gas was adsorbed were used as background and subtracted from the following runs. The measurements were carried out at the three axis spectrometer IN3 at the Institute Laue-Langevin. We used a fixed final energy of 0.98THz, both a Cu111 and a PG002 monochromator, a PG002 analyser and a Be filter positioned after the sample. The resolution obtained from the elastic line using a vanadium sample was 0.033THz (FWHM) for the PG002 monochromator. The FWHM calculated energy resolution at 3.5THZ (the energy of the $J=0\rightarrow 1$ transition for H$_2$) was 0.21THz and 0.12THz for the PG002 and the Cu111 monochromators, respectively. (1THz = 4.14meV = 48.28K)

Graphite Powder Substrate

The spectra taken with an adsorbed commensurate monolayer of H$_2$ present two peaks between 3 and 5.5THz (fig.1). The first one at an energy of 3.53THz is identified as the $J=0\rightarrow 1$ rotational transition. Its value coincides very accurately with the one quoted in the literature for a free H$_2$ molecule[7] showing little influence of the presence of the substrate. The expected splitting of 0.64THz[3] is not detected, in agreement with the NMR data[4]. A spectrum taken at 0.5K did not show any difference to the one at 3.7K. The broader peak at 4.7THz can be assigned to the out-of-plane mode of the molecule vibrating against the substrate. The large width of this mode is due to its coupling to the phonons of the graphite substrate[8]. This mode disappears when the H$_2$ layer melts at 18K. The rotational transition peak, on the other hand, only loses intensity due to the increased Debye-Waller factor.

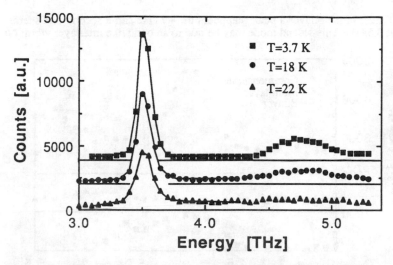

Figure 1: Spectra taken in the commensurate phase of H_2 adsorbed on graphite powder as a function of temperature ($Q=2.30\text{Å}^{-1}$). The successive runs are displaced by 2000 counts.

Oriented graphite substrate

To be able to study the polarization of the above mentioned modes in more detail we used an oriented graphite[6] substrate. The graphite platelets were mounted with their c-axis in the scattering plane, thus allowing one to choose between Q transfer parallel and perpendicular to the adsorption plane (in the following called in-plane (IP) and out-of-plane (OP) polarizations, respectively).

Rotational transition

In the commensurate and full monolayers of H_2, the rotational transition peak shows no difference at all between both polarizations and with respect to the powder substrate. A side peak appears at 3.41THz (FWHM = 0.09THz) when increasing the coverage to a bilayer in the OP configuration. No change is seen in the IP spectra. The origin of this second peak is not clear: it could be due either to a vibrational excitation of the double layer or to a splitting of the rotational transition. The second possibility seems unlikely, since this side peak shows up as an extra intensity when comparing the two polarizations, but a more detailed study should be performed.

The rotational transition for HD is found at 2.68THz, very close to the free molecule value[7]. No splitting could be observed within the experimental resolution (FWHM = 0.21THz, predicted splitting 0.64THz). Furthermore, the transition energy does not depend on coverage up to the full monolayer.

Out-of-plane vibration

Similar to the graphite powder substrate, the OP vibration shows up for both gases, H_2 and HD, as a broad feature, due to the coupling to the graphite surface modes. For H_2, the OP mode again appears at 4.7THz, whereas for HD it is found at 3.6THz due to the higher molecular mass. In contrast to expectation little change in intensity is observed between IP and OP configurations, indicating that this mode does not have a well defined polarization. Increasing the coverage to a full monolayer affects this mode strongly : the define peak disappears, and only a diffuse intensity distribution remains (fig.2). In the bilayer a peak reappears at 4.7THz and a second one seems to show up at 5.5THz. This second mode may be due to an optic like interlayer vibration.

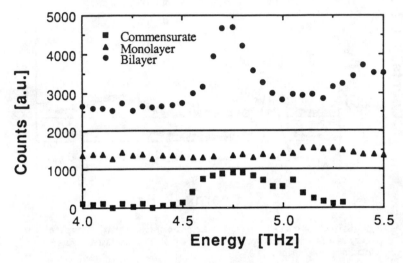

Figure2: Spectra of H_2 adsorbed on Papyex as a function of coverage (T=4K, Q=2.30Å^{-1}). The successive runs are displaced by 1000 counts.

In summary, we have investigated two high-frequency excitations in H_2 and HD adsorbed on graphite: i) The $J=0 \rightarrow 1$ rotational transition, for which no splitting could be observed within the experimental resolution, indicating that the effect discussed in ref. 3 must be at least a factor of five smaller than predicted; ii) The out of plane mode, whose frequency at 4.7THz and 3.6THz for H_2 and HD, respectively, agrees well with the one obtained using Steele's adsorption potential.

We would like to thank A. Novaco for very helpful discussions. This work has been partially supported by the Federal Ministry of Research and Technique (BMFT) of the Federal Republic of Germany.

[1]   Frank V.L.P., Lauter H.J and Leiderer P., Phys.Rev.Lett., 61, 436 (1988)
       Lauter H.J., Frank V.L.P., Leiderer P. and Wiechert H., Physica B, 156&157, 280 (1989)
[2]   Novaco A.D., Phys.Rev.Lett., 60, 2058 (1988)
[3]   Novaco A.D.and Wroblewski J.P., Phys. Rev. B, 39, 11364 (1989)
[4]   Kubik P.R., Hardy W.N. and Glattli H., Can. J. Phys. 63, 605 (1985)
[5]   Vulcan III is produced by the National Physical Laboratory, UK.
[6]   Papyex is produced by Carbon Lorraine, 45 Rue des Acacias, F-75821 Paris Cedex 17
[7]   Silvera I.F., Rev.Mod.Phys. 52, 393 (1980)
[8]   de Rouffignac E., Alldredge G.P. and de Wette F.W., Phys.Rev.B, 24, 6050 (1981)
[9]   Steele W.A., Surf.Sci. 36, 317 (1973)

EXCITATIONS IN THE COMMENSURATE PHASE OF CD$_4$ ADSORBED ON GRAPHITE

T. Moeller[1,2], H.J. Lauter[1], V.L.P. Frank[1,2] and P. Leiderer[2]

1 *Institut Laue -Langevin, BP 156X, F-38042 Grenoble, France*
2 *University of Konstanz, D-7750 Konstanz, West Germany*

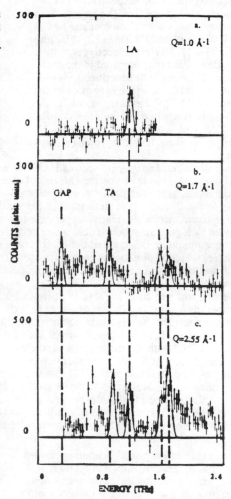

The phase diagram of methane on the basal planes of graphite is still in progress to be refined [1]. One of the very interesting features is the 'lock in' of the methane molecules into the commensurate over-structure and the phase transitions connected to it. In the present paper we report about the excitations in the commensurate phase (completely filled), in particular the phonon gap, its dependence on temperature and the onset of rotational and translational diffusion.

The experimental cell is the same as used in ref. 2 (Papyex graphite, a quasi 2-dimensional powder). The energy resolution of the IN3 spectrometer at ILL was 0.033 THz across the elastic line [3]. In fig. 1 the spectra are taken with the momentum transfer parallel to the basal planes of the graphite (the scattered intensity from the unloaded sample is always subtracted). All spectra are taken at 10 K. The scan at monentum transfer Q=1.0 Å^{-1} focusses on the longitudinal phonon at the zone boundary [3], which is seen at 1.3 THz (fig. 1a).

The scan at Q=1.7 Å^{-1} (the value of the (10) Bragg peak of the commensurate structure) picks up intensity from the phonon gap and the transverse phonons at the zone boundary, which are identified with energies of 0.3 THz and 1.0 THz, respectively (fig.1b). The phonon gap energy is determined by the curvature of the corrugation of the adatom-substrate potential.

In Ref. 4 a gap of 0.8 THz was calculated for CH$_4$ .Taking into account the isotope mass difference this gives 0.7 THz for CD$_4$. This value seems to be high.

Fig. 1: Spectra at 10 K. Q-vector is always parallel to the surface. Energy gap (GAP), transverse (TA) and longitudinal (LA) acoustic phonon are depicted.

920

The dispersion has been recalculated in harmonic approximation (Lenard-Jones parameters $\varepsilon=137K$ and $\sigma=3.68Å$ [5]) using Ref. 6. This gives 0.35 THz in good agreement with the data.

Using the same calculation one obtains 0.85 THz and 1.38 THz for the transverse and the longitudinal zone boundary phonon, respectively. These modes are given by the interaction between the adsorbed molecules. Ref. 4 (and 7) calculates 1.2 THz (1.1 THz) and 1.73 THz (1.7 THz), respectively, which are somewhat high compared to the data. The discrepancy will be reduced, if cubic terms are included in the expansion of the potential [4,7]. The assignment of the modes is supported by the scan done at $Q=2.55$ Å^{-1}, which is shown in fig. 1c. Here both zone boundary modes show up.

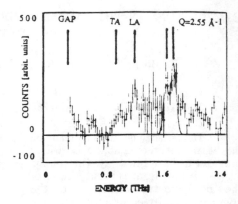

Fig. 2: Out-of-plane spectrum at 10 K.
(Q-vector perpendicular to the surface)

Other dominant features are seen at 1.75 THz and 1.85 THz. Here the out-of plane mode and librational mode do merge. Our calculation gives a value of 2.0 THz for the out-of-plane mode, also intensity from a librational mode is predicted around 1.7 THz [7]. The scan in fig. 2 with momentum transfer perpendicular to the graphite surface shows that the in-plane modes are nearly suppressed and replaced by a broad feature. This may be caused by a graphite mode visible by increased error bars.

If the temperature is increased there appears between 30 K and 40 K a broad quasielastic feature, which is due to rotational diffusion (fig. 4). This transition could not be found with specific heat measurements [8] but calculations [9] agree with the detected transition temperature.

The phonon gap at 0.26 THz is still visible and can be followed as a function of temperature (fig.3). The phonon gap shifts to lower energies due to the strong anharmonicity of the adsorbate potential and the increased mean square displacement of the molecules. The model used is explained in ref. 10. It is interesting to note that a good fit is obtained with the melting temperature of 55 K [8] instead with the transition of the expanded solid at 48 K. The expanded phase is thought to be a solid phase due to the shape of the diffraction peaks [11], but with a larger lattice parameter with respect to the commensurate phase. At 55 K the expanded phase melts into a liquid. This result leaves the possibility open that the expanded phase is a domain-wall-phase despite theoretical predictions [12]. We have already indications (fig. 3) that the phonon gap exists at 50 K. This would be similar to D_2 on graphite where the phonon gap exists in a domain-wall-phase [13].

Beyond the melting at 55 K another contribution adds to the rotational diffusion, which is the translational diffusion in a liquid (fig. 4).

In conclusion the existence of acoustic modes could be seen at low temperatures. The phonon gap proves the commensurate phase. The temperature dependence of the phonon gap shows the anharmonicity of the adsorption potential and gives some new ideas about the expanded phase . The diffusive component reveals the onset of rotational and subsequent translational motion.

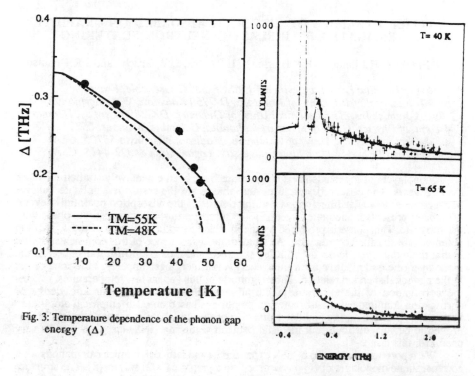

Fig. 3: Temperature dependence of the phonon gap
energy (Δ)

Fig. 4: Onset of rotational and translational
diffusion. Both spectra are taken with
momentum transfer Q=1.7 Å^{-1}parallel
to the surface. At 40 K the phonon gap
is visible.

References

[1] Vora P., Sinha S.K. and Crawford R.K., Phys. Rev. Lett. 43, 704 (1979)
[2] Frank V.L.P., Lauter H.J., Hansen F.Y., Taub H., Bruch L.W. and Dennison J.R.
"Corrugation in the N_2-potential ...", this conference
[3] Frank V.L.P., Lauter H.J. and Leiderer P., Phys. Rev. Lett. 61, 436 (1988)
[4] Hakim T.M., Glyde H.R. and Chui S.T., Phys. Rev. B 37, 974 (1988)
[5] Phillips J.M. and Hruska C.D., Phys. Rev. B 39, 5425 (1989)
[6] Steele W.A., Surf. Sci. 36, 317 (1973)
[7] Maki K. anf Klein M.L., J. Chem. Phys. 74, 1488 (1981)
[8] Kim H.K., Zhang Q.M. and Chan M.H.W., Phys. Rev. B 34, 4699 (1986)
[9] Bruno J. and Giri M.R., Phys. Rev. B 29, 5190 (1984)
[10] Frank V.L.P., Lauter H.J. and Leiderer P."Phonon gap energy ...", this conference
[11] Beaume R., Suzanne J., Coulomb J.P. and Glanchant A., Surf. Sci. 137, L1117
(1984)
[12] Phillips J.M., Phys. Rev. B 29, 4821 (1984)
[13] Frank V.L.P., Lauter H.J. and Leiderer P., Jap. J. Appl. Phys., Vol. 26, 347
(1987)

LATTICE DYNAMICS OF NITROGEN ADSORBED ON GRAPHITE INVESTIGATED BY INELASTIC NEUTRON SCATERING

V.L.P.Frank[1,2], H.J.Lauter[1], F.Y.Hansen[3] , H.Taub[4] , L.W.Bruch[5] and J.R.Dennison[6]

1 Institut Laue-Langevin, BP 156X, F-38042 Grenoble, France
2 Fac. of Physics, Univ. of Konstanz, D-7750 Konstanz, West Germany
3 Fysisk-Kemisk Inst.,The Technical Univ. of Denmark, DK-2800 Lyngby, Denmark
4 Dept. of Physics, Univ. of Missouri-Columbia, Columbia, Missouri 65211, USA
5 Dept. of Physics, Univ. of Wisconsin, Madison, Wisconsin 53706, USA
6 Dept. of Physics, Utah State University, Logan, Utah 84322-4415, USA

Monolayers of gases adsorbed on graphite have been extensively studied for more than 15 years. The observed structures and phases are the result of a delicate balance between intermolecular interactions and the details of the adsorption potential. Several adsorbates present, at adequate coverages and temperatures, a commensurate phase with the substrate. This phase is produced by the in-plane periodic variation of the adsorption potential, the so called corrugation. An immediate consequence of this commensurability is that the monolayer 'locks in´ with the substrate and thus a uniform translation of the monolayer (the q=0 mode) costs a finite energy. An energy gap appears at the zone center in the phonon dispersion relation. The magnitude of this gap, at low temperatures, reflects the corrugation of the substrate at the adsorption site. Most of our knowledge of corrugation amplitudes for adsorbates on graphite is based on semi-empirical atom-atom potentials[1]. The experimental determination of the adsorbate-substrate corrugation has proved to be very difficult, but inelastic neutron scattering (INS) has shown to be very useful for this task[2-4].

We report here the results of INS experiments and lattice dynamics calculations for a commensurate monolayer of N_2 physisorbed on a graphite (002) surface[5]. The structure and phase transitions of this system have been extensively studied theoretically[6-8] and experimentally[9;10]. However, despite detailed lattice dynamics calculations[11], there have been no previous measurements of the phonon spectra of this system.

The INS experiments were taken with the three-axis spectrometer IN3 using the same technique as for the D_2 experiments[2,3]. The spectrometer was operated with a fixed final energy (E_F=5.04 meV), a Be filter after the sample and a horizontally focusing analyzer giving an elastic resolution of 0.07THz FWHM and increasing to 0.12THz FWHM at an energy transfer of 2THz. The sample cell was filled with Papyex[12] sheets which could be mounted so that the majority of the graphite c-axes were either parallel or perpendicular to the wavevector transfer Q (denoted parallel and perpendicular configurations, respectively). The scattering from the sample cell without adsorbed N_2 was substracted as the background. The commensurate monolayer corresponded to a coverage of 202cc. STP of N_2.

The unit cell of the commensurate √3x3 herringbone monolayer phase is shown in Fig. 1(a). The modelling of the film was done using the intermolecular interaction given by the X1 model of Murthy et.al.[13] supplemented by the McLachlan substrate mediated interaction[14] and Steele´s model of the N_2-graphite interaction[15], minimizing the energy including the zero-point energy. The calculated structure reproduces the observed commensurate phase.

Ignoring the stretching of the intramolecular bond, which lies at a much higher energy than the lattice vibrations, each molecule has five degrees of freedom and there are two molecules per unit cell. The ten branches of the phonon dispersion relations were calculated performing harmonic lattice dynamics on the obtained structure, and are depicted in Fig. 1(b) for selected symmetry directions in the Brillouin zone. These calculations omitted the McLachlan substrate mediated interaction.

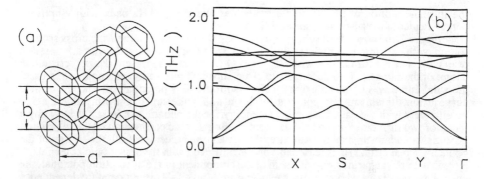

Figure 1: a) Projection of the commensurate N_2 monolayer unit cell on the graphite surface. The cell dimensions are $a=7.38\text{Å}^{-1}$ ($=3\,a_g$) and $b=4.26\text{Å}^{-1}$ ($=\sqrt{3}\,a_g$) where $a_g=2.46\text{Å}^{-1}$ is the graphite lattice constant. The N-N bond of the molecules is parallel to the surface. b) Phonon dispersion relations of the N_2 monolayer using the isotropic adatom-substrate potential. (Figures taken from Ref.5)

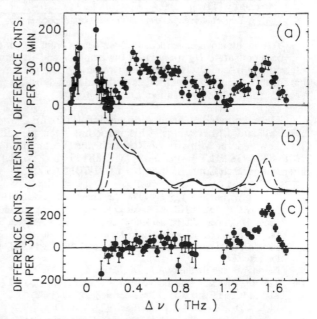

Figure 2: a) Neutron energy-loss spectrum from the N_2 monolayer taken at 4K with $Q=1.70\text{Å}^{-1}$ in the parallel configuration. b) Calculated one-phonon inelastic coherent cross sections after spherical averaging and folding with the instrumental resolution function. Two different N_2–graphite potentials were assumed: isotropic atom-atom (C-N) potential of Ref.15 (solid curve); and anisotropic atom-atom potential of Ref.17 (dashed curve). c) Neutron spectra from the N_2 monolayer taken at 4K with Q in the perpendicular configuration: $Q=1.70\text{Å}^{-1}$ ($\Delta v < 1$THz) and $Q=2.25\text{Å}^{-1}$ ($\Delta v > 1$THz). (Figures taken from Ref.5)

Figure 2(a) shows the difference INS spectrum taken at $Q=1.70\text{Å}^{-1}$ in the parallel configuration. This Q value gives a good sampling of the region of reciprocal space near the center of the Brillouin zone[5, 16]. Two peaks can be recognized at ~0.4THz and 1.5THz. No other features are discernible up to 2.5THz. Due to the large intensity of the elastic peak, whose width agrees with the experimental resolution, no peaks can be determined below 0.2THz. Scans taken in the perpendicular configuration are depicted in Fig. 2(c), the peak at 0.4THz is very difficult to resolve and only the peak at 1.5THz can be recognized. These features are consistent with assigning the lower-frequency mode to

924

the in plane motion of the adsorbate molecules and the 1.5THz peak with vibrations perpendicular to the adsorption surface[5].

Figure 2(b) shows the result[5] of evaluating the one-phonon inelastic cross section, folded with the instrumental resolution and spherically averaged over all possible orientations of the film lattice with respect to Q (in order to represent the polycrystalline nature of the sample). Two different adsorbate-substrate interaction potentials were used: the solid curve was obtained using Steele´s isotropic C-N potential[15] and the dashed curve using the anisotropic potential of Joshi and Tildesley[17]. Both models agree qualitatively with the data, but predict an energy for the phonon gap which is lower by a factor of two than the experimental value, indicating that the corrugation of the interaction potential is underestimated in both cases. However, the agreement with the energy of the bouncing mode at 1.5THz is reasonable and is improved with the anisotropic potential.

Due to the anharmonicity of the in plane component of the adsorption potential, the energy gap decreases with increasing temperature (Table I). This temperature dependence can be calculated using a self consistent phonon theory and qualitative agreement is found[18].

Temperature [K]	4.5	10.0	25.0	30.0	50.0
Gap energy [THz]	0.45	0.43	0.42	0.39	0.32

Table I: Temperature dependence of the zone center phonon gap.

To conclude, we have measured the zone-center gap in the lowest phonon branches of a commensurate N_2 monolayer on graphite and its temperature dependence. The inferred gap frequency is a factor of two larger than present models predict, suggesting that for this system current models of the adatom-substrate potential give an inadequate representation of the corrugation.

We acknowledge helpful discussion with S.F.O'Shea, G.Cardini, J.Z.Larese, L.Passell and J.M.Hastings. This work has been partially supported by the West Germany Federal Ministry of Research and Technique (V.L.P.F), NSF Grants No. DMR-8704938 (H.T.) and No. DMR-8817761 (L.W.B.) and The Danish Natural Science Research Council Grant No. M 11-7015 (F.Y.H.).

[1]   Bruch L.W., Surf.Sci. 125, 194 (1983), and references contained therein.
[2]   Frank V.L.P., Lauter H.J. and Leiderer P., Phys.Rev.Lett. 61, 436 (1988)
[3]   Lauter H.J., Frank V.L.P., Leiderer P. and Wiechert H., Physica B 156&157, 280 (1989)
[4]   See also other contributions presented at this conference.
[5]   Hansen F.Y., Frank V.L.P., Taub H., Bruch L.W., Lauter H.J. and Dennison J.R., submitted to Phys.Rev.Lett.
[6]   Harris A.B. and Berlinsky A.J., Can.J.Phys. 57, 1852 (1979)
[7]   Talbot J., Tildesley D.J. and Steele W.A., Molec.Phys 51, 1331 (1984)
[8]   Kuchta B. and Etters R.D., J.Phys.Chem. 88, 2793 (1988)
[9]   Diehl R.D. and Fain S.C.,Jr., Surf.Sci. 125, 116 (1983)
[10]   Wang R., Wang S.-K., Taub H., Newton J.C and Schechter H., Phys.Rev.B 35, 5841 (1987)
[11]   Cardini G. and O´Shea S.F., Surf.Sci. 154, 231 (1985)
[12]   Papyex is produced by Carbon Lorraine, 45 rue des Acacias, 75821 Paris 17, France.
[13]   Murthy C.S., Singer K. and McDonald I.R., Mol.Phys. 44,1 35 (1981)
[14]   Bruch L.W., J.Chem.Phys. 79, 3148 (1983)
[15]   Steele W.A., J.Phys.(Paris) 38, C4-61 (1978)
[16]   Taub H., Carneiro K., Kjems J.K., Passell L. and McTague J.P., Phys.Rev.B 16, 4551 (1977)
[17]   Joshi Y.P. and Tildesley D.J., Mol.Phys. 55, 999 (1985)
[18]   Frank V.L.P., Lauter H.J. and Leiderer P., presented at this conference.

REFLECTION OF TERAHERTZ-PHONONS AT CRYSTAL SURFACES

U.Happek, R.O.Pohl[1] and K.F.Renk

Institut für Angewandte Physik, Universität Regensburg,

8400 Regensburg, Fed. Rep. of Germany

([1] permanent address: LASSP, Cornell University, Ithaca, N.Y., U.S.A.)

We have studied the reflection of acoustic THz-phonons at the interface between fluorid crystals and $HeII$ in a frequency range between 0.2 THz and 1.1 THz. For "real" surfaces, i.e. being not ideal planes compared to the phonon wavelength, we obtain a constant phonon reflectivity R=0.8 for $\nu \geq 0.4$ THz which is not noticeably influenced by surface treatments. The phonon reflectivity seems to be determined by a damaged layer in a region near the crystal surface.

The lifetime of nonequilibrium phonons in crystals immersed in liquid helium is limited by the spontaneous anharmonic decay[1] which can be described in terms of a decay rate $\tau_a^{-1} = \lambda_a \cdot \nu^5$, where λ_a is an anharmonicity parameter, and by phonon losses due to spatial escape from the crystal volume. The latter process is influenced by the phonon reflectivity of the crystal surface and elastic phonon scattering at isotopes and impurities which leads to diffusive propagation of the phonons, characterized by a diffusion constant $D(\nu)$ that shows a ν^{-4} dependence. The rate equation describing the phonon distribution when the decay of high frequency phonons can be neglected reads $\partial f(\nu)/\partial t = \tau_a^{-1} f(\nu) + D(\nu)\nabla^2 f(\nu)$. Expanding an initially homogeneous distribution into spatial modes, higher modes decay very fast and only the fundamental mode has to be considered. The solution of the diffusive part of the rate equation is $f(\vec{r}) = f_o \cdot \cos(k_x x) \cdot \cos(k_y y) \cdot \cos(k_z z) \cdot \exp(-t/\tau_{esc})$, with $\tau_{esc}^{-1} = (k_x^2 + k_y^2 + k_z^2) \cdot D$. The k_i are connected with the reflectivity R of the crystal surface and the diffusion constant by the equation $k_i \cdot \tan(k_i \frac{a_i}{2}) = \frac{1}{4} \cdot c \cdot (1 - R)/D$, $a_x \cdot a_y \cdot a_z$ being the volume of the crystal and c an average velocity of sound. From the spatial phonon distribution (i.e. k_i) and the phonon decay time $\tau = (1/\tau_a + 1/\tau_{esc})^{-1}$ both D and R can be determined.

Phonons were generated homogeneously in the crystal volume by multiphonon absorption of pulsed radiation of a CO_2 laser with a frequency of 28 THz, leading primarily to the generation of optical phonons which decay very fast into high frequency acoustic phonons[2]. For phonon detection we used the vibronic sideband spectrometer, for this purpose the crystals are weakly doped with Eu^{2+}ions[3]. Phonon-induced fluorescence was

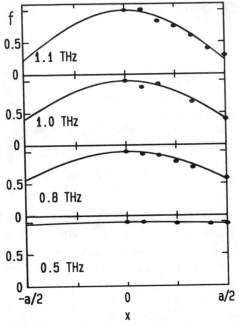

Fig. 1:   Spatial phonon distributions for different detection frequencies.

observed from a small volume within the crystal, by shifting this volume, the spatial distribution of the phonons could be investigated. The samples were immersed in superfluid helium. Some time after the phonon excitation the decrease of the phonon-induced signals at different points in the crystal can be described by the same time constant, i.e. the shape of the spatial distribution remains constant. Fig. 1 shows the spatial distributions at different detection frequencies. At 1.1 THz, the distribution decreases towards the crystal surface which is due to strong elastic scattering. At the surface, there is still a large number of phonons, indicating a high reflectivity. At 1.0 THz and 0.8 THz, the mean free path of the phonons has increased, leading to a more flat distribution. At a detection frequency of 0.5 THz, the distribution is almost constant and the phonon losses are given by $\tau_{esc}^{-1} = \frac{c}{4} \cdot \frac{A}{V} \cdot (1 - R)$, where A and V are the surface area and the volume of the crystal, respectively. With $\tau_{esc} = 6.7$ μs and a crystal size of $8 \cdot 8 \cdot 10$ mm^3 we obtain a reflectivity of the surface R(0.5 THz)=0.79. At higher frequencies, the analysis of the decay time and the spatial distribution gives again a phonon reflectivity around R=0.79, as an average value we obtain R=0.79 $\pm$ 0.02. This value includes both specular and diffusive reflection of TA and LA phonons. We have measured the phonon reflectivity in crystals of different surface quality, varying from good optical quality to sandblasted and find for all samples a phonon reflection of R=0.80 $\pm$ 0.02 for $\nu \geq 0.4$ THz. We obtain the same value from measurements on SrF_2 crystals and from the analysis of data for the phonon decay time in diamond crystals[2].

We covered 85% of a sample with vacuum grease or a 4000 Å layer of lead and found no change in the reflectivity (fig.2). For frequencies below 0.4 THz, the reflectivity

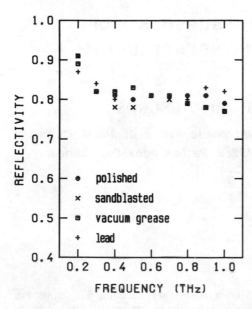

Fig. 2: Spectral phonon reflectivity for different surface treatments.

increases, indicating the onset of the regime where the acoustic mismatch theory can be applied. Thus, not the direct contact between the crystal and $HeII$ determines the phonon reflectivity but a subsurface region, the liquid helium serves as a heat sink only. For high frequency phonons even highly polished surfaces are damaged regions, a further degradation of this region or a covering of the surface has no effect. Phonons are scattered at boundaries within the crystal bulk, which explains that covering of the surface does not influence the reflectivity and that surfaces of different solids show the same reflectivity.

The results can be described by the model of diffusive mismatch[4]. Assuming that acoustic correlations are completely destroyed by diffusive scattering, this model gives a frequency independent reflection coefficient R=0.5 for a boundary within the subsurface crystal bulk. The measured value can be understood in the framework of this model when the subsurface region consists of several boundaries and parts of the transmitted phonons reenter the crystal bulk.

To summarize, we have measured the absolute reflectivity of the interface between mechanically polished fluorid crystals and $HeII$. We find a constant reflection coefficient of R=0.8 for frequencies between 0.4 and 1.1 THz, which cannot be influenced by surface treatments. We attribute the reflection to boundaries within the crystal material near the surface, caused by damages in a subsurface region.

One of us (R.O.P.) thanks the A.v.Humboldt Foundation for financial support.

1) Baumgartner,R., Engelhardt,M. and Renk,K.F., Phys Rev.Lett., 47, 1403 (1985).

2) Happek,U., Ayant,Y., Buisson,R. and Renk,K.F., Europhys. Lett., 3, 1001 (1987).

3) Bron,W.E. and Grill,W., Phys.Rev B 21,5303, 5313 (1977).

4) Swartz,E.T. and Pohl,R.O., Rev. Mod. Phys. 61 (1989) 605.

EXPERIMENTAL MEASUREMENTS OF
4He SOLID LIQUID INTERFACE INERTIA

J. POITRENAUD and P. LEGROS

Laboratoire d'Ultrasons (*), Université Pierre et Marie Curie,
Tour 13, 4 place Jussieu, 75252 Paris Cedex 05, France.

1. INTRODUCTION

The solid liquid interface of 4He is known to present roughening transitions. For temperatures above the transition, melting is a very fast process. Previous experiments [1, 2, 3] show that sound transmission is a very good tool to study 4He crystal growth. It was found that the transmission rapidly decreases as the temperature is lowered although the acoustic mismatch theory predicts temperature independence. However at low temperatures in the case of very high frequency sound wave, the transmission is expected to be partially restored if the work of tension forces at the interface is involved. This theory [4] predicts a frequency dependence of sound transmission and a T^{-5} variation of the Kapitza resistance R_K as experimentally observed. But the order of magnitude of the calculated value of R_K is too high. Thus Puech and Castaing (PC) [5] have introduced the notion of growth-inertia. This new concept reconciles the experience and the theory. But the calculation incorporates the effect of a wide band of phonons. This fact triggered our interest to investigate the transmission of a high frequency monochromatic ultrasound wave through a rough 4He interface.

2. THEORY

The modified version of fast melting theory shows that the phonon transmission τ is enhanced at low tempertaures for high enough frequencies in place of rapidly decreasing with T. The amplitude τ of the transmitted energy can be written [6] :

$$\tau = 4 \; z_L \; z_C / \left| z_L + z_C + z_L \; z_C / \zeta \right|^2 \tag{1}$$

z_L and z_C are the acoustic impedances of the liquid and the crystal and ζ is defined as a

---

(*) Associated with the Centre National de la Recherche Scientifique

complex interface impedance which is the sum of an inertial impedance ζ_σ and a friction one ζ_K :

$$\zeta = \left(\frac{\rho_C \rho_L}{\rho_C - \rho_L}\right)^2 \left(K^{-1} + \frac{i\omega\sigma}{\rho_C \rho_L}\right) = \zeta_K + i\zeta_\sigma \qquad (2)$$

ρ_C and ρ_L are respectively the crystal and liquid densities, ω the ultrasound wave pulsation and K the isothermal growth coefficient. The mobility of the rough interface is limited by the bulk thermal excitations [1-3, 7, 8], that is, by the rotons of the liquid and the phonons of the liquid and the solid. K^{-1} consists in three terms :

$$\rho_C \, K^{-1} = A + BT^4 + C \exp\left(-\Delta/T\right) \qquad (3)$$

A is a residual damping, Δ the minimum roton energy. It can be calculated that ζ_σ becomes larger than ζ_K at $T \leq .5$ K if the frequency $\nu = 30$ MHz. Consequently it is expected that at sufficiently high frequency the ultrasound transmission coefficient instead of decreasing continuously when T is lowered tends towards a constant value.

3. EXPERIENCE AND RESULTS

The experiments have been performed at three frequencies 15, 45 and 75 MHz and between .4 and 1.1 K. The thickness of the crystal and the velocity of sound C_C, from which we deduced the crystallographic orientation θ of the sample, were determined from the analysis of successive transmitted pulses. Figure 1 shows an example of the temperature dependence of τ for one sample at three frequencies after correction for attenuation of the bulk liquid and solid. It appears clearly that τ is function of frequency. Taking $\Delta = 7.2$ K, A, B, C and σ were deduced from (1) (2) in two steps. For high temperatures $T > .8$ K, the rotons monitor the interface mobility and C is deduced while, for low temperatures, $T < .8$ K, the acoustic impedances z_L and z_C become negligible with regard to ζ_K and ζ_σ, then A, B and σ can be determined. The subsequent fit for sample C between experimental points and calculated curves is illustrated by figure 1. Table I gives A B, C and σ values for three different orientated samples. σ values are to be compared with P estimation of 2.4×10^{-10} g/cm^2 deduced from R_K experimental values analysis.

sample	C_C (m/s)	frequency (MHz)	A (cm/s)	B (cm/s/K^4)	C x 10^{-5}	σ x 10^{10} (g/cm^2)
a ($\theta = 8°$)	538	15	.04 ± .04	5.4 ± .6	.72 ± .13	3.1 ± .2
		45	.65 ± .2	.02 ± .01	.72 ± .13	3.1 ± .2
b ($\theta = 18°$)	520	45	.6 ± .2	8.5 ± .8	2.9 ± .3	7.1 ± .4
		75	1.34 ± .2	18 ± 2	2.5 ± .3	7.1 ± .4
c ($\theta < 3°$)	540	15	.04 ± .2	1.95 ± .2	.29 ± .10	1.6 ± .2
		45	0.0 ± .01	3.3 ± .1	.29 ± .10	1.5 ± .1
		75	.06 ± .02	3.1 ± .2	.29 ± .10	1.7 ± .1

930

We wish to emphasize that for the same sample the same σ value permits to take into account the variation of τ at all the frequencies. Figure 2 shows that the extrapolated τ values at T = 0, at 15, 45 and 75 MHz vary quadratically with ν as is expected. Furthermore the anisotropy of σ has been put in evidence.

In conclusion, we have presented an extension of previous measurements of ultrasound transmission through a rough ^{4}He interface at frequencies which allow to get a clear evidence of liquid-solid interface mass inertia σ and its anisotropy. Additionally the values of the mobility coefficient K determined from our data agree with those of previous studies.

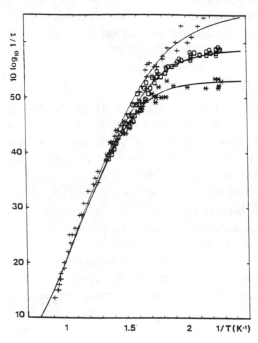

[1] CASTAING, B., BALIBAR, S. and LAROCHE, C., J. Physique **41** (1980) 897.

[2] CASTAING, B., J. Physique Lett. **45** (1984) L-233.

[3] MOELTER, M.J., MANNING, M.B., ELBAUM, C., Phys. Rev. **B 34** (1986) 4924.

[4] MARCHENKO, V.I., PARSHIN, A., J.E.T.P. Lett. **31** (1980) 724.

[5] PUECH, L., CASTAING, B., J. Physique Lett. **43** (1982) L-601.

[6] UWAHA, M. and NOZIERES, P., J. Physique **46** (1985) 109.

[7] KESHISHEV, K.O., PARSHIN, A. Ya. and BABKIN, A.B., Sov. Phys. JETP **53** (1981) 362.

[8] BODENSHOHN, J., LEIDERER, P., Proc. of the 4th International Conference on Phonon Scattering in Condensed Matter (Springer Verlag, 1984), p. 266.

Figure 1. - Comparison between the temperature dependence of the sound transmission data for sample c and the fit which is described in the text. $\Delta = 7.2$ K .

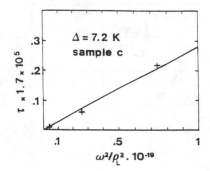

Figure 2. - T = 0 extrapolated values of τ vs. ω^2/ρ_L^2 for sample c. The slope of the obtained straight line is σ^2.

Surface Phonons from Ab-Initio Molecular Dynamics: Si(111)2x1

F. Ancilotto, W. Andreoni

IBM Research Division, Zurich Research Laboratory,
8803 Rüschlikon, Switzerland

A. Selloni, R. Car and M. Parrinello(*)

International School for Advanced Studies,
Strada costiera 11, 34100 Trieste, Italy

The current understanding of the observed 2×1 reconstruction of the Si (111) surface at low temperature is based on Pandey's π-bonded chain model[1] in which the bulk-terminated surface reconstructs by forming tilted chains of surface atoms along the $(\bar{1}10)$ direction, where neighboring dangling bonds form extended quasi-one-dimensional states. The quasi-one-dimensional nature of the surface chains is expected to give rise to unique vibrational properties. Electron energy loss experiments have established the existence of a strongly dipole-active surface resonance at $\sim$ 56 meV.[2] Subsequent experiments,[3] while confirming this result, have provided grounds for a conflicting interpretation of the character of this mode. Theory has yet not been able to clarify this issue, since semi-empirical calculations attribute this mode either to a polymer-like, longitudinal optical vibration along the tópmost chains[4] or to a more isotropic, subsurface vibration polarized normal to them.[5] At lower frequencies He-scattering data[6] reveal, in addition to the Rayleigh mode, a dispersionless excitation at $\sim$ 10 meV. This mode has been associated[4] with atomic motions where the topmost chain as a whole oscillates normal to the surface.

It is known that both the pseudopotential and the Local Density Functional formalisms can be used to obtain accurate information about lattice dynamical properties of semiconductor crystals. We present here results of both equilibrium structure and vibrational properties of the Si(111) 2×1 surface, which we have obtained from a computer simulation based on the first-principle Molecular-Dynamics (MD) method introduced by two of us.[7] A detailed account of our calculations will be given elsewhere.[8] A periodically repeated slab of 64 atoms with (2×4) periodicity in the surface plane is used to model an isolated surface. Neighboring slabs are separated by a vacuum region corresponding to $\sim$ 8 bulk layers. An additional layer of hydrogen atoms is added to saturate the dangling bonds on one surface. The Kohn-Sham orbitals at the Γ-point have been expanded in $\sim$ 3000 plane waves, corresponding to an energy cutoff of 8 Ry. Three **k**-points of the surface BZ are folded at Γ

due to our (2×4) supercell, i.e. $\bar{\Gamma}$, $\bar{J}$ and the point half-way between them (hereafter $\bar{J}/2$).

An equilibrium surface geometry has been obtained by relaxing the original Pandey's chain structure.[1] The optimized structure is found to be $\sim$ 0.42 eV/(surface atom) lower than the unreconstructed surface. In Table I our structural data are compared with experiment[9] (LEED) and with the results of previous pseudopotential total energy calculations.[10] In Table II the positions of the electronic surface states at $\bar{\Gamma}$, $\bar{J}/2$ and $\bar{J}$ are also compared with experiment[11] and with the results of Ref. [10].

Table I. Buckling of different layers (in Å). The inset shows the equilibrium structure.

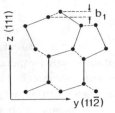

Layer	This work	LEED (Ref. [9])	Ref. [10]
1	0.49	$b_1 = 0.38 \pm 0.08$	0.1
2	0.04	0.09	0.05
3	0.02	0.07	0.08
4	0.27	0.20	0.30
5	0.16	0.13	0.20
6	0.01	0.03	−

Table II. Surface states energies (eV).

	This work	Expt. (Ref. [11])	Ref. [10]
$E_s(\bar{\Gamma}) - E_s(\bar{J})$	− 0.51	− 0.65	− 0.55
$E_s(\bar{\Gamma}) - E_s(\bar{J}/2)$	0.22	0.18	0.22

The phonon density of states (PDOS) has been obtained from the ionic trajectories $\{R_i(t)\}$ generated in a microcanonical MD run at T $\sim$ 120 K, by taking the Fourier transform of the velocity auto-correlation function $< \Sigma_{i\alpha} v_{i\alpha}(t)v_{i\alpha}(0) >$, $v_{i\alpha}(t)$ being the α-th component of the i-th mobile ion. Similarly, the k-resolved phonon spectral density has been calculated using the correlation function $< v_\alpha(k,t)v_\alpha(-k,0) >$, where $v_\alpha(k,t) = \Sigma_i e^{ik \cdot R_i(t)} v_{i\alpha}(t)$. The main components of the phonon eigenvectors for selected modes were identified using the correlation function $< v_Q(t)v_Q(0) >$, $v_Q = \Sigma_i \mathbf{v}_i \cdot \hat{\mathbf{Q}}_i$ being the projection of the ion velocities onto a guessed displacement field $\hat{\mathbf{Q}}_i$. The ionic trajectories have been followed for a time $t \sim$ 1 ps, i.e. twice the period of the lowest energy ($\sim$ 10 meV) mode detected in our simulations.

By comparing the layer-resolved PDOS and the results of a 48 atoms-bulk simulation we found that the surface features of our 2×1 structure are localized in the two topmost layers and the behavior of the III and IV layer is almost "bulk"-like. The rms atom displacement from equilibrium is $\sim$ 0.2 a.u. at the surface and tends to $\sim$ 0.1 a.u. in the innermost layers. On the first layer the rms displacement normal to the surface is twice as large as that parallel to it, where in the innermost layers the expected isotropy is recovered. Figure 1 shows the calculated phonon dispersion in the direction $\overline{\Gamma}\,\overline{J}$ of the surface BZ. To each mode a symbol is associated which indicates where the mode is mainly localized. Below the bulk LTO edge, in the region where the 56 meV loss peak is observed, two peaks in the PDOS are found. One (D2) is localized mainly on the second layer and is polarized along the chain direction. The other (IC) corresponds to intrachain vibration in the first and second layer and is polarized in a direction normal to the chains. A definite assignment to the experimental peak requires the calculation of the surface phonon dipole activity. This calculation is in progress. At a lower energy we also identified the polymer-like mode of the first layer (D) to which in Ref. [4] the 56 meV loss peak is ascribed. In agreement with He-scattering measurements,[6] we found a flat mode at ~ 9 meV, polarized in the x-z plane. The position of the Rayleigh wave (R) at $\overline{J}$ is ~ 6 meV higher than experimentally observed.

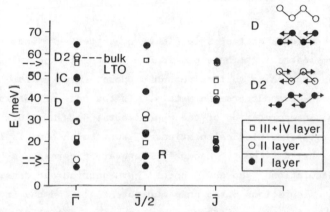

Fig. 1. Phonon dispersion along the $\overline{\Gamma}\,\overline{J}$ direction. The arrows indicate the experimental data.[2,6] The inset shows the atom displacements for the D and D2 modes.

REFERENCES

(*) Present address: IBM Research Division, Zurich Research Laboratory, 8803 Rüschlikon, Switzerland
1. Pandey K.C., Phys. Rev. Lett. **47**, 1913 (1981); **49**, 233 (1982).
2. Ibach H., Phys. Rev. Lett. **27**, 253 (1971).
3. Di Nardo N.J., Thompson W.A., Schell-Sorokin A.J. and Demuth J.E., Phys. Rev. **B32**, 3007 (1986); Del Pennino U., Betti M.G., Mariani C., Nannarone S., Bertoni C.M., Abbati I., Braicovich L. and Rizzi A., Phys. Rev. **B39**, 10380 (1989).
4. Alerhand O.L. and Mele E.J., Phys. Rev. **B37**, 2536 (1988).
5. Miglio L., Santini P., Ruggerone P. and Benedek G., Phys. Rev. Lett. **62**, 3070 (1989).
6. Harten U., Toennies J.P. and Woll C., Phys. Rev. Lett. **57**, 2947 (1986).
7. Car R. and Parrinello M., Phys. Rev. Lett. **55**, 2471 (1985).
8. Ancilotto F., Andreoni W., Selloni A., Car R. and Parrinello M., to be published.
9. Himpsel F.J., Marcus P.M., Tromp R., Batra I.P., Cook M.R., Jona F. and Liu H., Phys. Rev. **B30**, 2257 (1984).
10. Northrup J.E. and Cohen M.L., Physica **117B**, 774 (1983).
11. Uhrberg R.I., Hannson G.V., Nicholls J.M. and Flodstrom S.A., Phys. Rev. Lett. **48**, 1032 (1982).

FIRST-PRINCIPLES SCREENING CALCULATION OF THE SURFACE PHONON
DISPERSION CURVES AT THE (100) SURFACE OF SODIUM

A. A. QUONG, A. A. MARADUDIN, AND R. F. WALLIS
Department of Physics, University of California
Irvine, California 92717, USA

J. A. GASPAR AND A. G. EGUILUZ
Department of Physics, Montana State University
Bozeman, Montana 59717, USA

INTRODUCTION

The techniques of inelastic electron scattering[1] and inelastic helium atom scattering[2] have provided a wealth of experimental information concerning surface phonon dispersion curves for insulators, semiconductors and metals. In the case of metals, the theoretical interpretation of the experimental data has suffered, because first-principles calculations of surface phonon dispersion curves are very difficult. Recently, however, progress has been made on such calculations. Ho and Bohnen[3] have carried out density functional calculations based on pseudopotentials and the frozen phonon method for zone boundary surface phonons on Al(110). Eguiluz, Maradudin and Wallis[4] have combined density functional theory with linear response theory and an expansion of the electronic ground state energy to second order in the pseudopotential to calculate full surface phonon dispersion curves for Al(110). A somewhat similar, but nonself-consistent, approach has been used by Beatrice and Calandra[5] to investigate surface phonons on Na(100).

THEORETICAL DEVELOPMENT

We report calculations for unrelaxed Na(100) carried out using density functional and linear response theory with the pseudopotential interaction treated nonperturbatively. This procedure includes three-body and higher-body interactions that are neglected in Refs. 4 and 5. A local Heine-Abarenkov pseudopotential was used.

We first calculated self-consistently the one-electron solutions to the Kohn-Sham equations including the electron-ion pseudopotential interaction and exchange and correlation in the LDA. The one-electron functions have the form of two-dimensional Bloch states. These functions were then used to calculate the response function for non-interacting electrons $\chi^{(0)}$. The response function χ for inter-acting electrons was calculated by solving the integral equation $\chi = \chi^{(0)} + \chi^{(0)}V\chi$ where V contains both Coulomb and exchange-correlation contributions. The nondiagonal elements of the dynamical matrix $D_{\alpha\beta}(\ell\kappa,\ell'\kappa')$ can be expressed in terms of χ as

$$D_{\alpha\beta}(\ell\kappa,\ell'\kappa') = \frac{1}{M}\left\{\iint d^3r\,d^3r'\,\nabla_\alpha V_b(\vec{r}-\vec{R}(\ell\kappa))\chi(\vec{r},\vec{r}')\right.$$

$$\left. \times\,\nabla_\beta V_b(\vec{r}'-\vec{R}(\ell'\kappa'))\right\} + D_{\alpha\beta}^{ion}(\ell\kappa,\ell'\kappa') \qquad (1)$$

where $V_b(\vec{r}-\vec{R}(\ell\kappa))$ is the interaction potential of an electron with an ion $\ell\kappa$ at $\vec{R}(\ell\kappa)$, M is the ionic mass, and $D_{\alpha\beta}^{ion}(\ell\kappa,\ell'\kappa')$ is the direct ion-ion contribution. The diagonal elements of the dynamical matrix were calculated using infinitesimal translational invariance.

NUMERICAL RESULTS

The dynamical matrix was diagonalized for a 13-layer slab of sodium bounded by (100) planes. The results for the phonon energies are shown in Fig. 1 for the principal directions in the surface Brillouin zone. The Rayleigh mode lying below the bulk continuum is clearly evident. Between the $\overline{X}$ and $\overline{M}$ points there are two gaps in the bulk spectrum. In the lower frequency gap are two surface modes,

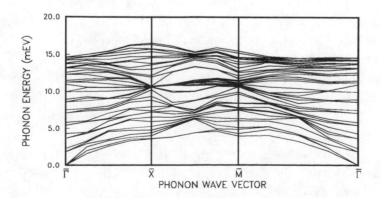

Fig. 1. Nonperturbative dispersion curves for a Na(100) slab.

each split by the finite thickness of the slab, and in the higher frequency gap another surface mode is clearly visible.

For comparison, we have carried out calculations for Na(100) in which the electron-ion interaction is treated perturbatively to second order. The results shown in Fig. 2 for a 51-layer slab are quite similar to those in Fig. 1 with the exception of an additional surface mode very near the top of the upper gap. We do not obtain a local minimum[5] of the Rayleigh mode between the $\bar{X}$ and $\bar{M}$ points.

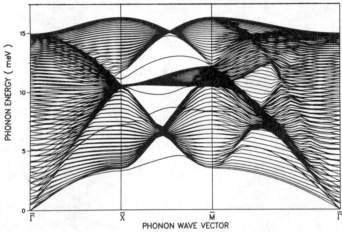

Fig. 2. Perturbative dispersion curves for a Na(100) slab

ACKNOWLEDGMENTS

The work of three of the authors (A.A.Q., A.A.M., and R.F.W.) was supported in part by NSF Grant No. DMR-8815866 and the work of the other author (A.G.E.) was supported in part by MONTS Grant No. 196504 and by NSF Grant No. DMR-8603820. All authors acknowledge the award of CRAY computer time by the San Diego Supercomputer Center.

REFERENCES
1. Lehwald, S., Szeftel, J. M., Ibach, H., Rahman, T. S., and Mills, D. L., Phys. Rev. Lett. 50, 518 (1983).
2. Doak, R. B., Harten, U., and Toennies, J. P., Phys. Rev. Lett. 51, 578 (1983).
3. Ho, K. M. and Bohnen, K. P., Phys. Rev. Lett. 56, 934 (1986).
4. Eguiluz, A. G., Maradudin, A. A., and Wallis, R. F., Phys. Rev. Lett. 60, 309 (1988).
5. Beatrice, C. and Calandra, C., Phys. Rev. B28, 6130 (1983).

SURFACE PHONONS IN GaAs (110) AND Ge (111):2x1 : A BOND CHARGE MODEL CALCULATION

P.Santini, *L.Miglio*, and *G.Benedek*

Dipartimento di Fisica dell'Universita'
via Celoria 16, I-20133 Milano, Italy

and

P.Ruggerone

Max-Planck-Institut für Strömungsforschung
Bunsenstraße 10, D-3400 Göttingen, FRG

Introduction

The Bond Charge Model has proved to be a reliable tool for the investigation of semiconductor surface dynamics even in the case of extensive reconstruction. In particular a recent work of ours[1] investigated the surface dynamics of the Si (111):2x1 and explained all the existing experimental features. In the present work we extend our approach to the polar semiconductor GaAs (110) and to Ge (111):2x1, showing that some surface dispersion curves are exclusively determined by the structural configuration of the topmost atoms, independent of the substrate orientation. A comparison between these new calculations for GaAs (110), Ge (111):2x1 and the previous one for Si (111):2x1 demonstrates that some relevant features of the surface crystallography can be investigated by dynamical studies.

Both GaAs (110) and Ge (111):2x1 are characterized by zig-zag chain arrangements of the topmost atoms, which are slightly tilted in order to minimize the energy [2],[3] (as in the case of Si(111):2x1[1]). This geometrical similarity and the nearly equal masses of Ga, As and Ge make the comparison between the two systems worthwhile and interesting. Our slab calculations are performed with no disposable parameters except for the changes required by the bond angles rearrangements and the static equilibrium condition. This approach, in turn, demonstrates that the Bond Charge Model validity extends beyond the tetrahedral bulk configuration.

Details of the model

As in the case of Si (111):2x1 the atomic configurations as estimated by total energy and experimental techniques[2],[3] are taken as input data for our calculations. In particular the difference in vertical coordinates of the topmost atoms constituting the surface chains (tilt parameter) is assumed to be 0.68Å for GaAs (110) and 0.256Å for Ge (111):2x1. The bond charges are always located in the bond midpoints in Ge (even along the tilted chains), while their positions in the GaAs case share the bond lengths with a 3:5 ratio. According to *ab-initio* calculations[5] of the charge density maps the As cores bear one dangling bond charge representing a saturated orbital. Static equilibrium conditions are imposed on cores and bond charges in order to confirm the bond charge positions and to obtain the new values of the first derivatives for the core-core and core-bond charge potentials. Further details about our modelling procedure will be found in Ref.1 for the (111) surface and in Ref.6 for the GaAs case. All the calculations are performed in the slab configuration with 24 atomic layers for Ge and 23 for GaAs. The electronic degrees of freedom (bond charges) are eliminated only after the perturbation of the surface has been introduced (adiabatic approximation), so that the main features of the electronic rearrangement in the topmost layer are taken into account.

Results and discussion

In Fig.1 and in Fig.2 we display the surface modes (solid lines) and the surface projection of the bulk bands (shaded areas) for the GaAs (110) and for the Ge (111):2x1 respectively. For what concerns Ge no comparison with experimental data is possible, since neither He-scattering nor EELS measurments are available, while a fairly good agreement is obtained for the low energy dispersion curves of GaAs (110)[6] to the Harten and Toennies[7] and to the Doak and Nguyen[8] time-of-flight data.

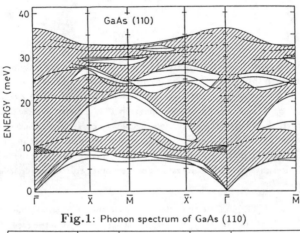

Fig.1: Phonon spectrum of GaAs (110)

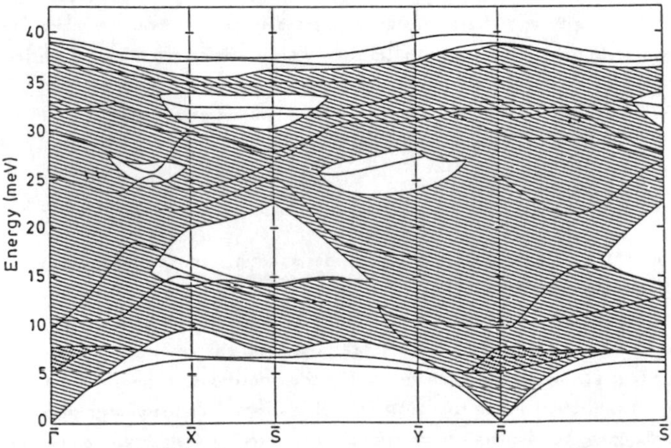

Fig.2: Ge (111):2x1 phonon spectrum

The bulk bands span the same energy range in both cases due to the similarity of masses and force constants, even if the different substrate orientations give a dissimilar distribution of the gaps. In the high energy part of the spectrum two dispersion curves are found on top of the bulk band for the Ge case, which are produced (as in the case of silicon[1]) by the fivefold ring atomic structure just beneath the topmost layer. These tightly bounded structures take their origin from the Pandey-2x1-reconstruction and are obviously not present in the GaAs case. However, since the topmost atomic layer is arranged in a chain structure both in GaAs and Ge, the dynamical fingerprint of this configuration can be found in the peculiar dispersion of the sharply localized vibration starting at 25 meV in GaAs and at 31.7 meV in Ge. The displacement pattern of the latter optical mode is always parallel to the chain direction and it is present also in the Si (111):2x1 case (at 52.5 meV in $\overline{\Gamma}$ [1]). Actually, except for a rescaling of the frequencies, the surface dispersion relations of Ge are nearly identical to the silicon case, as could be expected by the fact that the structure and electronic features[2],[9] are equal as well. It is worthwhile to note, however, that in both these covalent systems the gaps at $\overline{X}$ between the two fivefold rings originated branches and between the Rayleigh wave and the upper lying flat branch are a measure of the structural tilt. In GaAs (110), on the other hand, the dynamical inequivalence of the two atoms constituting the surface chains is augmented by their intrinsic difference (one is Ga and the other is As). Moreover a shear polarized dispersion branch is present between the Rayleigh wave and the 10 meV branch. Actually the relevant differencies between the GaAs (110) dispersion curves and the Ge (111):2x1 ones can be found in those branches which display a larger penetration into the bulk structures, owing to the different orientation of the latters.

A final comment should be added for the presence of two modes in the gap between $\overline{X}$ and $\overline{S}$, both in Ge and in Si. The latter are localized at the interface between the reconstructed overlayers and the substrate, showing that, for the case of strongly reconstructed surfaces, the surface projection of interface originated modes may play an important role[1].

Acknowledgments

The present calculations, performed on a Cray XMP computer, was financially supported by the Consiglio Nazionale delle Ricerche della Repubblica Italiana. One of us (P.R.) thanks the Alexander-von-Humboldt Foundation for a fellowship.

References

1) Miglio, L., Santini, P., Ruggerone, P., and Benedek, G., Phys.Rev.Lett. 62, 3070 (1989).

2) Northrup, J.E. and Cohen, M.L., Phys.Rev. B27, 6553 (1983).

3) Qian, G., Martin, R.M. and Chadi, D.J., Phys.Rev B37, 1303 (1988).

4) Pandey, K.C., Phys.Rev.Lett. 49, 223 (1982).

5) Chelikowski, J.R. and Cohen, M.L., Phys.Rev. B20 , 4150 (1979).

6) Santini, P., Miglio, L. and Benedek, G., Harten, U., Ruggerone, P. and Toennies, J.P., to be published.

7) Harten, U. and Toennies, J.P., Europhys.Lett. 4, 833 (1987).

8) Doak, R.B. and Nguyen, D.B., J.Electron Spectrosc.Relat.Phenom. 44, 281 (1987).

9) Northrup, J.E. and Cohen, M.L., Phys.Rev.Lett. 49, 1349 (1982).

STABILITY OF THE BCC(001) SURFACES : A COMPARISON
BETWEEN Mo(001) AND Nb(001)

M. Hüppauff, E. Hulpke, D.-M. Smilgies

Max-Planck-Institut für Strömungsforschung, D-3400 Göttingen, FRG

The bcc(001) surfaces of W and Mo are known to undergo a two dimensional structural phase transition when cooled below room temperature [1,2,3]. These reconstructions can be described in terms of a periodic lattice distortion in the $\langle 110 \rangle$ direction which is commensurate with respect to the underlying lattice in the case of W(001), but incommensurate in the case of Mo(001). The periodic lattice distortions give rise to extra spots in the diffraction patterns as observed by low energy electron diffraction, a c(2x2) structure for W(001) [1,2,3] and a quartet of spots centered at the $(\frac{1}{2} \; \frac{1}{2})$ position and split in the $\langle 110 \rangle$ directions for Mo(001) [3]. Recently we have been able to observe these split spots also in helium atom beam diffraction. However, the bcc(001) surfaces of Ta and Nb are found to be stable in a temperature range from 15 K to 1000 K [4,5], i.e. no extra diffraction spots could be seen.

We have performed Helium atom scattering experiments on the clean Mo(001) and Nb(001) surfaces in a search for the lattice dynamical origin of this behavior. For Mo(001) in addition to the Rayleigh mode an anomalous, low energy phonon mode was observed in the $\langle 110 \rangle$ azimuth which softens from about 8 meV at a surface temperature of 1000 K to about 2 meV close to room temperature [6]. In the $\langle 100 \rangle$ direction also a second mode, lower in energy than the Rayleigh mode was found which, however, did not change when approaching the phase transition temperature of about 250 K [6]. This result is very similar to previous findings for W(001) [7]. In contrast the Nb(001) surface shows a regular behavior of the two measured phonon modes without any pronounced temperature dependence. The two modes are probably the Rayleigh mode and a longitudinal resonance, as known from many close-packed metal surfaces. These phonon dispersion curves resemble those observed on the hydrogen saturated p(1x1)-2H/Mo(001) surface. This adsorbate structure features of two hydrogen atoms per surface unit cell occupying the bridge positions between neighboring molybdenum atoms [8,9]. Hydrogen is known to quench occupied electronic surface states [10,11] and strong occupied surface states close to the Fermi level have indeed been found for the clean Mo(001) surface [12] which give rise to a strong peak in the local density of states close to the Fermi energy [13]. On Nb(001) there is also an electronic surface resonance [14], it does, however, not contribute much to the local density of states [15]. Therefore the similiarity between the clean Nb surface and the hydrogen saturated Mo surface on one hand and the anomalous dynamics of the clean Mo surface on the other hand can be regarded as evidence for an electronic origin of the reconstruction of the clean Mo(001) surface.

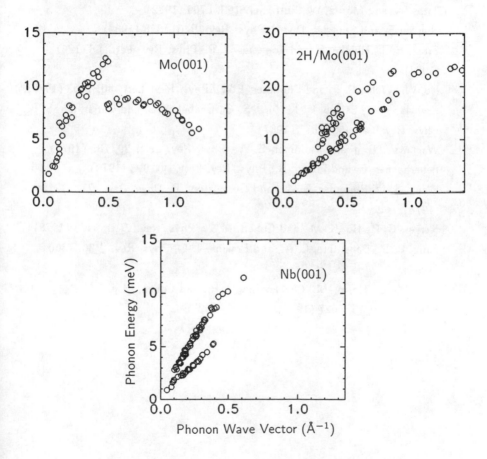

Fig.1 Surface phonon dispersion in the ⟨110⟩ direction
for Mo(001), 2H/Mo(001), and Nb(001).

References :

1)   Yonehara, K. and Schmidt, L.D., Surf. Sci. <u>25</u>, 238 (1971).

2)   Debe, M. K. and King, D. A., J. Phys. <u>C'10</u>, L303 (1977).

3)   Felter, T. E., Barker, R. A. and Estrup, P. J., Phys. Rev. Lett. <u>38</u> ,1138 (1977).

4)   Melmed, A. J., Ceyer, S. T., Tung, R. T., and Graham, W. R., Surf. Sci. <u>111</u>, L701 (1981).

5)   Titov, A. and Moritz, W., Surf. Sci. <u>123</u>, L709 (1982).

6)   Hulpke, E. and Smilgies, D.-M., Phys. Rev. <u>B40</u>, 1338 (1989).

7)   Ernst, H.-J., Hulpke, E. and Toennies, J. P., Phys. Rev. Lett. <u>58</u>, 1941 (1987).

8)   Ho, W., Willis, R. F. and Plummer, E.W., Phys. Rev. Lett. <u>40</u>, 1463 (1978).

9)   Prybyla, J. A., Estrup, P. J., Ying, S. C., Chabal, Y. J. and Christman, S. B., Phys. Rev. Lett. <u>58</u>, 1877 (1987).

10)   Waclawski, B. J. and Plummer, E. W., Phys. Rev. Lett. <u>29</u>, 783 (1972).

11)   Feuerbacher, B. and Fitton, B., Phys. Rev. Lett. <u>30</u>, 923 (1973).

12)   Weng, S.-L., Plummer, E. W., and Gustafsson, T., Phys. Rev. <u>B18</u>, 1718 (1978).

13)   Kerker, G. P., Ho, K.-M., and Cohen, M. L., Phys. Rev. <u>B18</u>, 5473 (1978).

14)   Fang, B. S., Ballentine, C. A., and Erskine, J. L., Phys. Rev. <u>B36</u>, 7360 (1987).

15)   Louie, G. L., Ho, K.-M., Chelikowsky, J. R., and Cohen, M. L., Phys. Rev. <u>B15</u>, 5627 (1977).

LATTICE DYNAMICS AND MEAN–SQUARE DISPLACEMENTS
OF THE RECONSTRUCTED Si (001)–(2x1) SURFACE

A. Mazur und J. Pollmann

Institut für Theoretische Physik II, Universität Münster, D–4400 Münster, FRG

We report lattice–dynamical calculations of surface phonons for semi-infinite Si (001)–(2x1) which are based on a total energy scheme. The only input is an empirical tight–binding electronic bulk bandstructure, the lattice constant and the Raman frequency. Complete information on the vibronic properties including displacement correlation functions is obtained from the surface phonon Greenfunction.

1.   INTRODUCTION

Surface phonons at Si (001) have been studied previously by various methods. Ludwig and collaborators[1] used a semiinfinite geometry together with phenomenological force constant models and studied the vibronic properties of the ideal surface. Tiersten et al.[2] treated semiinfinite systems employing the Keating model. Allan and Mele[3] analyzed vibronic properties of Si (001)–(2x1) using a slab geometry together with force constants calculated from a total energy ansatz. In this paper, we report a study of surface vibrational properties of Si (001)–(2x1) which fullfils simultaneously the following three requirements: (a) the optimal surface–atomic configuration is calculated at the very beginning by total energy minimization, (b) lattice–dynamical calculations are carried out for a truely semiinfinite system and (c) the atomic force constants of this system are calculated.

2.   THEORY

The calculations are based on the total energy ansatz for covalent semiconductors as suggested by Chadi[4]. The electronic properties, entering the total energy of the semiinfinite system, are treated by scattering theory[5] using electronic Greenfunctions. For bulk Si, the formalism yields a very realistic description of the phonon dispersions, as we have shown recently[6]. It can be applied to semiinfinite

944

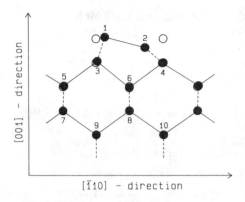

Fig.1 *Side view of the dimer geometry*

systems, as well. The theory proceeds in four distinct steps. First, the forces on all atoms are calculated and the energy is minimized by changing the atomic configuration until all forces vanish. Second, the atomic force con stants are calculated for the ener gy—optimized semiinfinite structure and the dynamical matrix is set up. It is $\infty \times \infty$ in size due to the broken sym metry perpendicular to the surface and can not be diagonalized directly. Therefore, we apply scattering theory, once more, and calculate in a third step the surface phonon Greenfunction. In the fourth step we determine surface bound states, resonances and thermodynamic properties like displacement correlation functions from the surface phonon Greenfunction.

3.   RESULTS

Our energy minimization yields an asymmetric dimer geometry in agreement with the results of previous theoretical und experimental surface structure determinations. A side view is shown in Fig. 1 where atomic distances are drawn to scale.

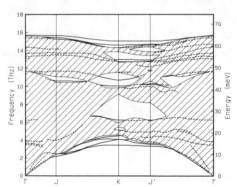

Fig.2 *Phonon dispersions at Si(001)—(2x1)*

In Fig.2 we present the surface vibronic spectrum of the reconstructed surface. The projected bulk phonon branches are indicated by hatchings. We see, that the reconstructed Si (001) surface gives rise to a number of bound surface phonon states (full lines) and pronounced resonances (dashed lines). The gross—features in Fig. 2 are similar to the slab results of Allan und Mele[3], although there occur significant differences due to the fact that we use a semiinfinite geometry. A detailed discussion of these differences and their origins will be given elsewhere.

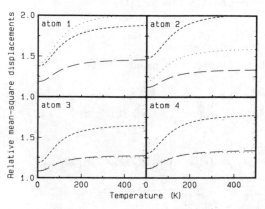

Thermodynamic properties can be calculated using quasiboson statistics. Mean–square displacements of the atoms near the surface are of particular importance. At the surface they are larger than in the bulk and show pronounced orientational dependencies. To highlight these effects, we have plotted in Fig. 3 as function of temperature re-

Fig.3 Relative mean–square displacements lative mean–square displacements which are given by the ratios of the anisotropic surface values to the isotropic bulk value. We see very pronounced enhancement factors between 1.5 and 2 for the surface atoms (see Fig. 1 for reference). In particular, for dimer atom 1, the surface–parallel mean–square displacement in the dimer direction (dotted line) is even larger than the surface–perpendicular value (short dashed line). We find, that the enhancement factors approach the value of 1 only very slowly as function of layer number. A detailed discussion of these results and their meaning for the interpretation of surface scattering experiments will be given in a forthcoming publication[7].

We gratefully acknowledge financial support of this work by the Deutsche Forschungsgemeinschaft under Contract No. Po–215/4–1.

4.   REFERENCES

[1]   see, e.g., Ludwig, W. in Festkörperprobleme/Advances in Solid State Physics, Vol. 29, ed. by U. Rössler (Vieweg, Braunschweig 1989), p 107

[2]   Tiersten S., Ying S.C. and Reineke T.L., Phys. Rev. **B33**, 4062 (1986)

[3]   Allan D.C. and Mele E.J., Phys. Rev. **B35**, 5533 (1987)

[4]   Chadi D.J., Phys. Rev. Lett. **41**, 1062 (1978)

[5]   Pollmann J., Kalla R., Krüger P., Mazur A. and Wolfgarten G., Appl. Phys. **A41**, 21 (1980)

[6]   Mazur A. and Pollmann J., Phys. Rev. **B39**, 5261 (1989)

[7]   Mazur A. and Pollmann J., submitted to Surf. Sci.

SURFACE DYNAMICS OF SrTiO$_3$

J. Prade and U. Schröder
Universität Regensburg, D–8400 Regensburg, FRG

W. Kress
Max-Planck-Institut für Festkörperforschung, D–7000 Stuttgart 80, FRG

A.D. Kulkarni and F.W. de Wette [*]
Department of Physics, University of Texas, Austin, Texas 78712, USA

Abstract

We present results for the surface relaxation of (001) surfaces of the oxidic perovskite SrTiO$_3$. Our calculations show an excellent agreement with recent measurements of the Sr-O-surface of this perovskite. In addition, we show our results of the surface phonon calculations of the relaxed Sr-O-surface.

Model

The recent developments in growing thin films of high-T$_c$-superconducting compounds have focused new interest on the oxidic perovskite SrTiO$_3$, which is widely used as a substrate for these films. In addition, new measurements[1] of the (001) surface structure of SrTiO$_3$ have been reported. In this paper we present theoretical results of the relaxation and the dynamics of the (001) surfaces of this material.

At temperatures down to 110 K SrTiO$_3$ has the simple cubic perovskite structure. At a temperature around 105 K it undergoes an antidistortive structural phase transition, which leads to a tetragonal crystal structure. This phase transition results from a soft mode at the R-point (R'$_{15}$-mode, also called R$_{25}$), which drives the structural phase change. As we have pointed out earlier[2] and is evident from the structure, perovskites should have two different types of (001) surfaces. In the case of SrTiO$_3$ these are a Sr-O-surface and a Ti-O$_2$-surface, surfaces I and II, respectively. This has been confirmed experimentally.[1]

Our investigations are based on shell model calculations, in which we take into account-long range Coulomb interactions, short-range overlap interactions of neighbouring ions, which we assume to be of the Born-Mayer-type, and polarizabilities of the ions. For these surface calculations we carry over the interactions from a bulk model, which has been developed to account for the elastic, optical, and phonon data of $SrTiO_3$.[3] No change of this model nor refitting of its parameters has been done in transferring the bulk interactions to our slab calculations. In this model the polarizability of the O^{2-}-ions is assumed to be not isotropic. Possibly this has to be changed at surface II, since this surface cuts through the O^{2-}-oktahedron. Therefore, in this paper we will discuss surface I only.

Results

Surface I

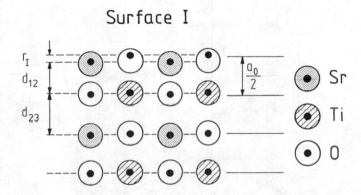

Fig. 1: Relaxation pattern of surface I of $SrTiO_3$. Displacements are enhanced by a factor of 2.

Fig. 1 shows the result of the z-relaxation of surfaces I of $SrTiO_3$ and in Table I we compare our results, calculated with room temperature data, with the experimental values[1] of the z-relaxation at 120 K.

Table I: Comparison of calculated and measured values of the changes of interlayer distances $\Delta d_{ij}/d_0$ and surface rumpling r. d_0 represents the bulk distance, $\Delta d_{ij} = d_{ij} - d_0$, $r = z(O^{2-}) - z(Sr^{2+})$.

	Surface I	
	calc.	exp.
$r[\text{Å}]$	.179	.16($\pm$.08)
$\Delta d_{12}/d_0$ [%]	-9.5	$-10(\pm 2)$
$\Delta d_{23}/d_0$ [%]	2.9	4(± 2)

It is evident, that there is excellent agreement between measured and calculated values of the z-relaxation for the Sr-O-surface.

In Fig. 2 we present the surface phonon dispersion of surface I. No experimental data on surface phonons of $SrTiO_3$ are currently available for comparison. We draw attention to the lowest surface phonon at the $\overline{M}$-point, which is the surface counterpart to the soft R_{25} bulk mode. This feature is similar to what we found earlier for $KMnF_3$.[2]

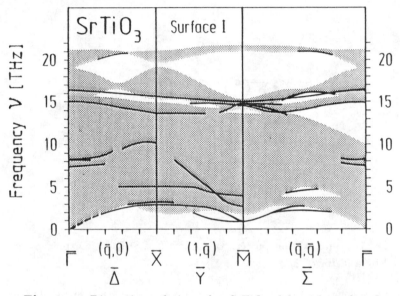

Fig. 2: Dispersion relation of a $SrTiO_3$-slab with surface I terminations. Thick lines represent the identified surface phonons, whereas the shaded area indicates the surface projected bulk bands.

We are presently investigating the relaxation and the dynamics of surface II and the temperature dependence of the lowest surface mode at the $\overline{M}$-point.

References

*) Supported by the NSF and the Robert A. Welch Foundation.
1) Bickel, N., Schmidt, G., Heinz, K., and Müller, K., Phys. Rev. Lett. **62**, 2009 (1989).
2) Prade, J., Kulkarni, A.D., de Wette, F.W., Reiger, R., and Schröder, U., Surface Science **211/212**, 329 (1989).
3) Becher, R., PhD dissertation, Regensburg 1989 (unpublished).

INTERPRETATION OF FEATURES IN THE SURFACE VIBRATIONAL MODES OF KBR AND RBBR.*

S. A. SAFRON, G. CHERN, W. P. BRUG AND J. G. SKOFRONICK, DEPARTMENTS OF CHEMISTRY AND PHYSICS, FLORIDA STATE UNIVERSITY, TALLAHASSEE, FLORIDA 32306 AND G. BENEDEK, DIPARTIMENTO DI FISICA DELL'UNIVERSITA, VIA CELORIA 16, I-20133 MILANO, ITALIA.

High-resolution He atom-surface scattering experiments were carried out first on the LiF(001) crystal[1] and since then on seven other alkali halides,[1-6] a sufficient number now that the general features of the scattering common to all these materials can be discerned. In all of these crystals the Rayleigh wave has been observed and, for most, mapped out across the entire surface Brillouin zone (SBZ). In many, longitudinal and optical modes have also been seen across the SBZ.

In NaF($\overline{\Gamma X}$)[2] and KCl($\overline{\Gamma M}$)[1] crossing modes were found. These modes appear as optical modes in the acoustic bands in that they have non-zero frequency at $\overline{\Gamma}$ and become very close in energy to the Rayleigh modes at $\overline{M}$ or $\overline{X}$. An explanation to account for them was offered by Benedek et al.[2] for NaF, and indirectly also for KCl. Their argument follows from that of Segall and Foldy[7] who suggested that isobaric crystals (alkali and halide of nearly equal mass) in the bulk should exhibit "extended symmetry"; namely, the vibrational modes should resemble those of the quasi-monatomic crystal with half the lattice spacing. Then, some of the optical branches of the isobaric alkali halide should arise as a "folding-back" of the purely acoustic branches in the extended monatomic Brillouin zone (BZ) into the true fcc BZ.

This explanation, however, was countered by de Wette et al.[8] who found evidence for crossing modes in slab dynamics calculations

for alkali halides clearly not isobaric, such as KBr. They argue
that these modes, "sagittal resonances", arise because of the
hybridization between the bulk TA_1 and TO_1 in the [110] direction.
They point out that crossing modes might be found whenever the bulk
dispersion curves have avoided crossings in this region.

In this laboratory we have carried out high-resolution He
scattering experiments on RbBr and KBr[5]. For both we find crossing
modes in both high-symmetry SBZ directions. On the basis of the
above arguments we thought that the mechanism responsible in KBr was
that of the sagittal resonance given by de Wette and co-workers and
that in RbBr, an isobaric alkali halide, it was the folding described
by Benedek et al.

Since these experiments, we have looked at NaI[6] which is
neither isobaric nor does it have an avoided crossing in the right
region, but which does appear to have a crossing mode in $\overline{\Gamma X}$ and
possibly also in $\overline{\Gamma M}$!

The observation that so many of these materials have crossing
modes suggests that it may be a more general feature of the alkali
halide lattice vibrations than explained by either the folding or
sagittal resonance arguments. In going back to the bulk dispersion
curves[9] one notices a curious fact; namely, that at the $\overline{\Gamma}$ point the
frequency of the measured crossing mode is very nearly the same as
the bulk LA frequency at X (in the [100] direction) which is the same
as the extension of the TA_2 mode in the [110] direction at (1,1,0),
an X point of the next BZ. The polarization of this mode is [001],
and on the surface it should project to a sagittal plane mode.
Further, in the calculations of the surface projected density of
states for KBr by Benedek[10] the crossing modes do not show up as
true surface resonances, but rather as distributions which peak near
the frequencies of the TA_2 mode at the corresponding points of the
surface zone!

Thus, we suggest that the origin of the crossing modes in $\overline{\Gamma X}$ is
the TA_2 mode which has the proper polarization to project as a
sagittal plane mode. Moreover, in the bulk this mode extends over
the region from Γ to X (i.e., (0,0,0) to (1,1,0)), while $\overline{\Gamma X}$ is
exactly half of this (from (0,0) to (½,½)). That is, it appears that

the crossing mode arises from the folded portion of the extension of this mode into the next BZ. It is reminiscent of, but really quite different from the earlier ideas of Foldy and co-workers.

For the other high symmetry direction $\overline{\Gamma M}$, continuing this reasoning, the crossing mode would arise from the projection of the transverse bulk branch from X to W in the first half of this region (from $(0,0)$ to $(\frac{1}{2},0)$) which also has a frequency at X $(\overline{\Gamma})$ of the LA mode in the [100] direction. For the remainder of $\overline{\Gamma M}$, the other transverse bulk mode from W to X with the frequency of the TA mode in the [100] direction at X $(\overline{M})$ should also project onto the SBZ with the proper polarization for a sagittal plane mode.

We conclude that a zeroth order explanation for the appearance of peaks in the inelastic spectra termed as crossing modes in the alkali halides, except possibly for LiF, is the result of the projection of the bulk modes existing for the most part at the edge of the BZ.

*Supported by U. S. D. O. E. grant number DE-FG05-85ER45208.

REFERENCES

1. Brusdeylins, G., Doak, R. B., and Toennies, J. P., Phys. Rev. B27, 3662 (1983).

2. Benedek, G., Brusdeylins, G., Miglio, L., Rechsteiner, R., Skofronick, J. G., and Toennies, J. P., Phys. Rev. B28, 2104 (1983).

3. Benedek, G., Brusdeylins, G., Doak, R. B., Skofronick, J. G., and Toennies, J. P., Phys. Rev. B28, 2104 (1983).

4. Chern, G., Brug, W. P., Safron, S. A., and Skofronick, J. G., J. Vac. Sci. Technol.A 7, 2094 (1989).

5. Chern, G., Skofronick, J. G., Brug, W. P., and Safron, S. A., Phys. Rev. B39, 12838 (1989); Phys. Rev. B39, 12828 (1989).

6. Chern, G., Brug, W. P., Duan, J., Safron, S. A., and Skofronick, J. G. (unpublished).

7. Segall, B., and Foldy, L. L., Sol. State Comm. 47, 593 (1983).

8. De Wette, F. W., Kress, W., and Schroeder, U., Phys. Rev. B33, 2835 (1986).

9. Bilz, H., and Kress, W., in Phonon Dispersion Relations in Insulators, Springer Press, New York (1979).

10. Benedek, G., (unpublished).

VIBRATIONAL PROPERTIES OF THE NON-IDEAL
AlAs/GaAs (001) INTERFACE

D Kechrakos and J C Inkson

University of Exeter, Department of Physics, Stocker Road, Exeter EX4 4QL, UK.

Abstract. We use a 2-parameter Valence Field Force model and Green's Function (GF) techniques to study the vibrational density of states (VDOS) at the AlAs/GaAs(001) interface containing steps and islands. The parallel confinement of the optical phonons is shown and new asymmetric As sites are compared.

A number of experiments have shown that the interfaces (IF) of GaAs quantum-well structures grown by state-of-art Molecular Beam Epitaxy have the asperities of one atomic layer (~2.8 Å)[1-4]. The long-range interface roughness has been quite successfully analysed in terms of linear chain models for phonons[4] and electrons[1]. However, the study of short-range roughness requires a knowledge of the parallel dispersion of the excitations. To this end, we present here a full 3D microscopic calculation of the local vibrational spectrum near atomic scale imperfections at a single GaAs-on-AlAs(001) IF, for which much evidence of short-range roughness exhists[1-4]. The study of simple isolated defect structures at a single interface serves as a preliminary stage to the study of a distribution of defects at a multi-interface system, ie a quantum-well or a superlattice.

We use a 2-parameter VFF model[7] that includes 1nn bond-stretching and 2nn bond-bending forces. We thus bypass the long-range forces, as they are of secondary importance for off-centre optical phonons; the latter are particularly sensitive to the short-range interface roughness[4]. The parameters of the model are chosen to fit the maximum and the X-point TA frequencies of each crystal. The bond-bending parameter at the interface region is taken as the average of the two crystal values. We model the islands at the GaAs-on-AlAs interface by small clusters of Al atoms that substitute isotopically the Ga ones in the bottom Ga plane. This model corresponds to an AlAs monolayer asperity right at the interface. Furthermore, we make these islands of infinite length in one direction and keep them finite in the second one, by introducing infinite succesive lines (rows) of Al atoms. This structure corresponds to an interface step of finite width. The symbol $<l_1 l_2 l_3>_n$ is used to denote the crystallographic orientation of the step and the number, n, of Al rows it consists of.

The VDOS for the bulk crystals projected on the two atoms in the unit cell is shown in Fig.1. The substantial mass difference between As and Al opens a band gap in the AlAs spectrum in the region of the GaAs optical phonons. This fact determines the vanishingly small penetration of the GaAs optical phonons in the first Al plane when the interface is formed (Fig.2), ie the GaAs optical phonon confinement. The confinement of the AlAs optical phonons (~12THz) is enen stronger as they lie above the maximum frequency of the GaAs crystal. The creation of the interface gives rise to an asymmetric arsenic site (As*), which exhibits a two-mode behaviour as shown in Fig2, through the extra peak at ~8THz. The local VDOS at the Al sites for different defect structures is shown in Fig.3. The δ-peak arising from the single Al defect in bulk GaAs gradually develops to a broad band around 11THz as the number of the directly bonded Al sites increases. Note that the middle atom in a 3-row step (~5Å) exhibits a behaviour very similar to an Al in an infinitely wide "step" (Fig.2). No pronounced extra peaks with respect to the clean IF are seen, which implies the absence of any resonances or local states associated with these interface defects. The parallel confinement of the GaAs optical phonons is obvious from Fig.3 and of the AlAs ones from Fig.4, where the VDOS at Ga sites next to an Al step is shown. The formation of IF steps creates new As* sites with all possible local environments. In Fig.5, the variation of the heights of the two peaks due to the GaAs (~8THz) and AlAs(~12THz) optical phonons can be seen. A similar behaviour should show in the polarised Raman spectrum, which is determined by the vibrations of asymmetric sites [6].

In conclusion, we have done a microscopic calculation of the vibrational spectrum near atomic scale imperfections at the AlAs/GaAs(001) interface. We did not find any pronounced resonances or local states due to these imperfections. The parallel confinement of the optical phonons of both materials was shown and the two-mode behaviour of certain new asymmetric As sites was compared.

References

[1] Tanaka, M., Sakaki, H.and Yoshino, J., J. Appl. Phys. 25, L155 (1986)

[2] Tanaka, M., Ichinose, H., Furuta, T., Ishida, Y. and Sakaki H., J. de Phys. C5,101 (1987)

[3] Jusserand, B., Alexandre, F., Paquet, D. and Le Roux, G., Appl. Phys. Lett. 47, 301 (1985)

[4] Fasol, G., Tanaka, M., Sakaki, H. and Horikoshi, Y., Phys. Rev. B38, 6056 (1988)

[5] Deans, M. and Inkson, J. C., Semic. Sci. Tech. 4, 138 (1989)

[6] Wang, C. and Barrio, R. A., Phys. Rev. Lett. 61, 191 (1989).

954

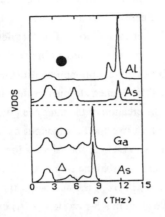

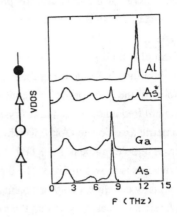

Fig.1
Bulk crystals.

Fig.2
Perfect interface.

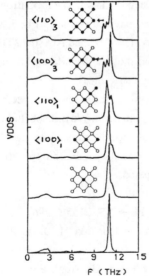

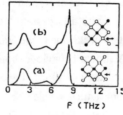

Fig.4
Ga sites next to
interface steps
(a) next to <100>$_1$,
(b) next to <110>$_1$.

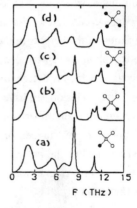

Fig.3
Al defect sites. The
bottom Ga-plane
structure is indicated.

Fig.5
Asymmetric As sites
(a) above <100>$_1$,
(b) above <110>$_3$,
(c) at perfect IF,
(d) below <100>$_1$.

ATTENUATION OF SURFACE ACOUSTIC WAVES DUE TO CUBIC ANHARMONICITY

A. P. Mayer, Xuemei Huang,
A. A. Maradudin, and R. F. Wallis

Department of Physics and Institute for
Surface and Interface Science
University of California Irvine, CA 92717, USA

ABSTRACT. The intrinsic attenuation of Rayleigh waves is calculated in the hydrodynamic regime at temperatures where normal processes dominate. The phonon viscosities are obtained from an approximate solution of the phonon Boltzmann equation on the basis of nonlinear elasticity theory. An amplitude-dependent attenuation for Love waves due to coupling to a leaky second harmonic is also discussed.

1.   INTRODUCTION

Previous calculations of the attenuation of surface acoustic waves in the hydrodynamic regime have been of a largely phenomenological character[1-5]. In this investigation, we attempt to relate the attenuation coefficient of Rayleigh waves to the microscopic couplings of the underlying material, also in the hydrodynamic regime. In a first study, we avoid an extensive lattice dynamical calculation by working at temperatures where Umklapp processes are negligible and the thermal phonons can be described within continuum theory.

2.   PHONON VISCOSITIES AND THERMAL CONDUCTIVITY

In the regime under consideration, the propagation of acoustic waves is governed by the following coupled equations for the displacement field $\mathbf{u}$, the local temperature T and the drift velocity $\mathbf{V}$[6]:

$$P_{kl}\, \dot{V}_l - \nu_{klmn}\, V_{m|ln} + S\, T_{|k} = \mu_{mnkl}\, \dot{u}_{m|ln} \tag{1}$$

$$C_V\, \dot{T} - \chi_{kl}\, T_{|kl} + T_o\, S\, V_{k|k} = -T_o\, C_{klmn}\, \alpha_{kl}\, \dot{u}_{m|n} \tag{2}$$

$$\rho\, \ddot{u}_k = C_{klmn}\, u_{m|ln} + \eta_{klmn}\dot{u}_{m|ln} - C_{klmn}\alpha_{mn} T_{|l} + \mu_{klmn}\, V_{m|ln}\ . \tag{3}$$

where C_V and S are the specific heat and entropy densities, respectively. The symbol ,m denotes the derivative with respect to the mth spatial coordinate. As boundary conditions we require that the surface be stress-free and that the normal component of the heat flux and the normal and parallel components of the drift velocity have to vanish at the surface. We impose the latter two conditions on the argument, that at the surface, due to the presence of the surface and due to surface roughness, the distribution of thermal phonons relaxes to a local rather than a drifting local equilibrium. The coefficients in Eqs. (1-3) are calculated for an isotropic elastic medium in terms of the two linear and three nonlinear Lamé constants, and the mass density (ρ). The viscosity tensors η, μ, and ν and the hydrodynamic thermal conductivity tensor χ can be expressed in terms of the effective inverse of the Peierls collision operator in the phonon Boltzmann equation[6]. We have evaluated these expressions numerically with a collision operator in which the matrix elements and phonon frequencies are calculated from the Lamé constants using nonlinear elasticity theory. The collinear processes of the longitudinal acoustic phonons present in the dispersionless theory have been included. For an isotropic elastic continuum, there are only four independent viscosities and one independent element of χ. The results for these five quantities at $T_o = 30K$ are listed in Table 1 together with the Lamé constants used in the calculation, which approximate the elastic moduli of BaF_2.

Table 1.

Lamé constants $[10^{10}\ N/m^2]$					Viscosities at $T_o = 30K\ [10^{-5}\ Ns/m^2]$			
λ	μ	α	β	γ	η_{11}	η_{44}	μ_{44}	ν_{44}
4.10	2.53	-20.6	-12.1	-2.7	5.7	0.9	1.6	21.8

$\rho = 4890.0\ kg/m^3$ $\qquad\qquad$ $\chi = 2.3\ N/(sK)$

3. ATTENUATION OF RAYLEIGH WAVES

The damping constant Γ of Rayleigh waves is calculated from Equs. (1-3) in a perturbative way. In a first step, the fluctuations of T and V due to the presence of the Rayleigh wave as external perturbation are calculated from Equs. (1,2). The boundary conditions on V and the heat flux are satisfied with a solution of (1,2) with the r.h.s. equated to zero. The result for T and V is then inserted into Equ. (3) and the attenuation coefficient is obtained by projecting (3) on the Rayleigh wave displacement field in a similar way as in Ref. 3). Preliminary results are summarised in Table 2. They indicate that it is crucial to satisfy the boundary conditions correctly. Γ_η represents the contribution of the viscosity tensor η to the attenuation coefficient.

Table 2. Attenuation coefficient
of Rayleigh waves

$T[K]$	$\nu[MHz]$	$\Gamma[s^{-1}]$	$\Gamma_\eta\,[s^{-1}]$
30	1.0	$2.9 \cdot 10^1$	$7.8 \cdot 10^{-3}$
30	5.5	$1.5 \cdot 10^2$	$2.4 \cdot 10^{-1}$
30	10.0	$2.7 \cdot 10^2$	$7.8 \cdot 10^{-1}$
20	1.0	5.4	$1.2 \cdot 10^{-2}$

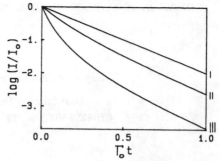

Fig. 1. Dependence of the Love wave intensity on time t for $\gamma I_o / \Gamma_o$ ≪ 1 (I), =1 (II), =5 (III).

4. INTENSITY-DEPENDENT ATTENUATION OF LOVE WAVES

A different attenuation mechanism for surface acoustic waves, which is also due to cubic anharmonicity, is coupling to a leaky second harmonic. This phenomenon can occur for Love waves of the lowest branch propagating along the (100) direction of a (001) surface, if the elastic moduli of the substrate satisfy the inequality $c_{11}(c_{11}-c_{44})>(c_{44}+c_{12})^2$. The attenuation coefficient Γ of the Love wave then depends on the intensity I as $\Gamma=\Gamma_o+\gamma I_o$. This is expected to influence the temporal behaviour of the intensity I in the way illustrated in Fig. 1 (I_o is the initial intensity).

ACKNOWLEDGMENT

One of us (APM) would like to thank Prof. R. K. Wehner for stimulating discussions and the Deutsche Forschungsgemeinschaft for financial support. The work of XH, AAM and RFW was supported in part by NSF Grant No. 88-15866.

REFERENCES

1) Maris, H. J., Phys. Rev. 188, 1308 (1969).
2) King, P. J. and Sheard, F. W., J. Appl. Phys. 40, 5189 (1969).
3) Sinha, B. K. and Tiersten, H. F., J. Appl. Phys. 52, 7196 (1981).
4) Lardner, R. W., J. Appl. Phys. 55, 3251 (1984).
5) Jurczyk, G. K. and Klemens, P. G., J. Appl. Phys. 58, 2593 (1985).
6) Gurevich, V. L., "Transport in Phonon Systems",
   (North-Holland, Amsterdam, 1986).

CORRELATED DETECTION OF α-PARTICLES ON Si-SURFACES

C. Hagen, W. Rothmund, A. Zehnder

Paul Scherrer Institute
CH-5232 Villigen PSI, Switzerland

Superconducting tunneling junctions (STJ) are well suited for the detection of phonons, owing to the fast pair breaking time of phonons of energy greater than 2Δ [1]. Two $Sn/SnO_x/Sn$ STJ were evaporated 100 μm apart onto the (100) surface of a silicon wafer, polished on one side (thickness 330 μm). The junctions were aligned with one of the main crystal axes. The whole substrate (10×13 mm^2) was irradiated by a 2μCi ^{241}Am α-source with an average penetration depth of the α-particles of 12 μm . Three different substrate purities (20 Ωcm, 500 Ωcm and 5000 Ωcm) were investigated and the substrates were irradiated from the top or the back side. The signals of each junction were measured independently with two charge-sensitive preamplifiers. Pulse height and rise time of each signal were measured as well as the time difference between the two correlated pulses. Calibration was done by applying a test pulse to each preamplifier, which could be varied in amplitude and rise time. The operating temperature was between 0.4 K and 1.0 K.

A typical energy correlation spectrum is shown in fig. 1. The energy was calculated conservatively, assuming the minimum pair-breaking energy of 1.2 meV instead of the observed 5 meV [2]. For each energy deposited in STJ1, there are up to four distinct energy releases detected in STJ2, which reflects anisotropic propagation of the phonons in the silicon crystals (see below). The width of the distribution of the measured time differences was typically 300 ns FWHM, which gives a lower limit for the phonon propagation velocity of about 700 m/s.

The energy threshold of the signals was above the electronic threshold. This indicates that phonons are thermalized below 2Δ after covering a well defined distance. For some substrates a strong temperature dependence of the threshold energy was observed as shown in fig. 2 for a temperature of 0.9 K.

Fig. 3 and fig. 4 show the difference between front- and back-side irradiation. When irradiating the back side, the high energy events in the scatter plots disappear, but four traces in the plot are still visible, however the two inner traces are closer together. From the coincidence rate and the activity of the α-source we obtain a sensitive area of 1.2 mm^2. The single rate of a STJ was about 1.2 times the measured coincidence rate of 400 s^{-1} and 300 s^{-1} for front- and back-side irradiation, respectively.

The measurements of fig. 1 to fig. 4 were carried out in vacuum. When covering the substrate with a superfluid ^{4}He film ($\approx$ 10 Å thickness), the signal amplitude as well as the rate decreased drastically ($\approx$ 10 s^{-1}). The effect of the ^{4}He film on the substrate of fig. 3 and 4 is shown in fig. 5 and 6. The widening of the pattern reflects an increased energy loss between junction 1 and 2, a fact that indicates surface rather than bulk effects.

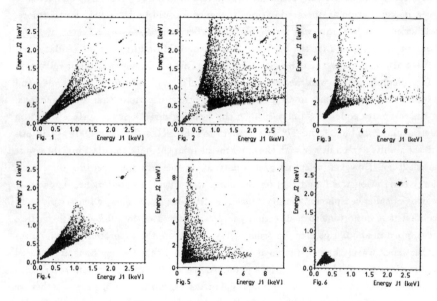

Energy scatter plots of correlated pulses for front-side irradiation at 0.6 K (fig. 1) and 0.9 K (fig. 2). The difference between front- and back-side irradiation is seen in fig. 3 and fig. 4, respectively. The effect of a ^{4}He-film on the substrate for front-side irradiation is shown in fig. 5 and for back-side irradiation in fig. 6.

In order to further estimate the role of surface effects, we evaporated 3 STJ. Between STJ1 and STJ2 was a scratch 50 μm wide and 5 μm deep, that should suppress bulk phonon effects for front irradiated substrates. The measurements obtained with the pair STJ1 and STJ2 were compared to those of STJ2 and STJ3. We found no marked differences between the patterns of the energy scatter plots and the event rate was the same in both cases (350 s^{-1}).

In order to evaluate the influence of multiple reflections of the phonons at the substrate surfaces, we clamped it to the heat sink by only small area contacts or glued it to the cold finger. No differences in the behavior was found, suggesting the absence of multiple phonon reflection.

We performed 24 runs with 13 different STJ pairs. In all cases we obtained the four trace patterns more or less pronounced, regardless of the substrate purity.

As already mentioned the energy correlation shown in fig. 1 could be explained with the anisotropic elastic properties of silicon leading to phonon focusing. The focusing directions were calculated numerically following the method described by Northrop and Wolfe[3]. However, no correlation as observed in our experiments could be obtained. Furthermore, the calculated detection rate was lower than the experimental rate by a factor of 10 for the front and a factor of 2 for the back irradiated sample. In addition, the calculated energies were always smaller than the measured ones. This result suggests the involvement of surface phonons which were neglected in the simulations. As shown by Camley and Maradudin[4] surface phonons in Ge are focused in other directions than bulk phonons. Using the computer codes of Camley et al.[4], the focusing directions for surface waves on Si were calculated to be $12°$ and $27°$ with respect to the $<010>$- and the $<001>$-axis. Although the results agreed better with the experiment than the bulk calculations, they were not satisfactory. However, selecting events only along the principal focusing directions, an energy correlation was found that agreed fairly well with the experiment. Apparently the regions of enhanced phonon density cause the structure in the energy scatter plots. In this context it is important to realize that the power generated by a 5.5 MeV α-particle that is stopped in about 1 ps is of the order of 1 W. Therefore the generated phonon pulses are rather shock waves than normal heat pulses. Assuming that the phonons are focused in the directions as calculated for surface phonons, one could imagine that in the regions of strong enhancement solitons can propagate. Further calculations will show whether we deal with power fluxes of about 300 W/mm^2, which are needed for soliton generation [5].

We are indebted to Prof. A.A. Maradudin, Prof. R.E. Camley and Prof. J.P. Wolfe for helpful discussions and for making their computer programs available to us, and to P. Jokinen for developing the electronic circuitry.

1) S. B. Kaplan, C. C. Chi, D. L. Langenberg, J. J. Chang, S. Jafarey and D. J. Scalapino, Phys. Rev. B14, 4854 (1976)

2) D. Twerenbold and A. Zehnder, J. Appl. Phys. 61, 1 (1987)

3) G.A. Northrop and J.P. Wolfe, Phys. Rev. B22, 6196 (1980).

4) R.E. Camley and A.A. Maradudin, Phys. Rev. B27, 1959 (1983).

5) V. I. Nayanov, JETP Lett. 44, 314 (1987)

STUDY OF SURFACE AND INTERFACE PHONONS IN GaAs/AlAs SUPERLATTICES BY RAMAN AND FAR INFRARED SPECTROSCOPY

T Dumelow[1,2], A Hamilton[1], T J Parker[1], B Samson[2], S R P Smith[2], D R Tilley[2], D Hilton[3], K Moore[3], and C T B Foxon[3].

1)   Department of Physics, Royal Holloway and Bedford New College, Egham, Surrey TW20 0EX, UK.

2)   Department of Physics, University of Essex, Wivenhoe Park, Colchester, Essex CO4 3SQ, UK.

3)   Philips Research Laboratories, Cross Oak Lane, Redhill, Surrey RH1 5HA, UK.

INTRODUCTION

The interaction of light with phonons may give rise to a range of polariton modes depending on the experiment employed. Here we investigate modes propagating parallel to the interfaces of a short period superlattice (SL). In general, these modes are localised at some sort of boundary, with an exponentially decaying amplitude to either side. We report observations of two such polariton types: (i) interface (IF) phonons, observed by resonant Raman scattering, in which the modes are pinned to the individual SL layer boundaries; and (ii) surface polaritons (SP), observed by attenuated total reflection (ATR) far infrared spectroscopy, in which the modes are pinned to the vacuum/SL boundary at the surface of the sample, but extend over many layers. Guided wave modes, confined within the SL, are also observed in ATR on samples containing an SL of finite thickness.

The value of the in-plane wavevector k_x is of crucial importance in observing the above modes. In the resonant Raman case, defect scattering may induce k_x values of the order of $k_x d > 1$, where d is a single layer thickness, this being the k_x region in which IF modes are observed[1]. There is a range of IF modes throughout the reststrahlen region of each constituent layer, but they approach a limiting value ω_∞ at large k_x when $\varepsilon_1(\omega) = -\varepsilon_2$, as shown in Fig. 1(a). In the ATR case, $k_x d \ll 1$, and the dielectric properties reduce to those of a single uniaxial medium. The dielectric tensor may then be described in terms of the bulk properties of the SL's two component media[2], corrected, if necessary, to take account of confinement[3]. We have used this model, with damping ignored, to calculate the SP dispersion curves shown in Fig. 1(b).

EXPERIMENTAL MEASUREMENTS

The sample consisted of 1 μm of GaAs/AlAs SL, with 0·1 μm GaAs cladding layers, grown by molecular beam epitaxy on a GaAs substrate. Each SL layer was 4 monolayers thick. All measurements were carried out at room temperature. For the Raman experiments the top cladding layer was etched from the sample; the measurements were performed at resonance in the region of the main SL photoluminescence peak (~2·08 eV). For the ATR experiments the sample was positioned ~10 μm from the base of a 20° silicon prism placed in the output stage of a Michelson interferometer[5].

Fig. 2(a) shows typical Raman spectra in the AlAs optic phonon region, both away from resonance and at resonance. A prominent IF mode centred close to ω_- (see Fig. 1(a)) appears in the resonant spectrum.

ATR spectra are shown in Fig. 2(b). SP or guided wave dips occur where their dispersion curves intersect with the scan line. The SP dispersion curve shown in Fig. 1(b) does not give exact agreement when applied to SL's of finite thickness, but we have used the same dielectric function, with the overall sample structure and phonon damping taken into account, to calculate the theoretical spectra included in Fig. 2(b). These show excellent agreement with experiment.

ACKNOWLEDGEMENTS

The authors wish to thank the UK Science and Engineering Research Council for supporting this work and research studentships for AH and BS.

REFERENCES

1) Sood A.K., Menendez J., Cardona M. and Ploog K., Phys. Rev. Letters 19, 2115 (1985).

2) Raj N. and Tilley D.R., Solid State Communications 55, 373 (1985).

3) Chu H. and Chang Y-C, Phys. Rev. B 38, 12369 (1988).

4) Dumelow T., El Gohary A.R., Hamilton A., Maslin K.A., Parker T.J., Raj N., Samson B., Smith S.R.P., Tilley D.R.T., Dobson P.J., Foxon C.T.B., Hilton D., and Moore K., Materials Science and Engineering B, in press.

5) Dumelow T. and Parker T.J., Int. J. of IR and MM Waves, to be published.

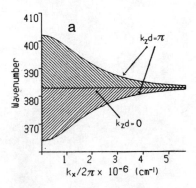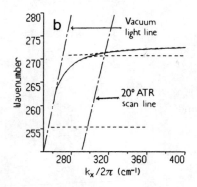

Fig. 1: *Calculated dispersion curves for a GaAs/AlAs SL whose layers are 4 monolayers thick. (a) Interface modes in the AlAs optic phonon region of an infinite SL, constrained within the limits $k_z d = 0, \pi$ where k_z is the wavevector of the collective excitation in the direction normal to the interfaces. (b) Surface polaritons in the GaAs optic phonon region of a semi-infinite SL - the solid curve is calculated considering only first order confinement, whereas the dashed curves take account of higher order terms. Confined phonon frequencies are taken from the experimental values reported in reference 4.*

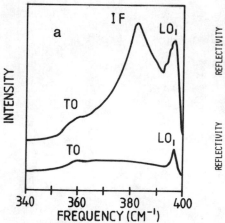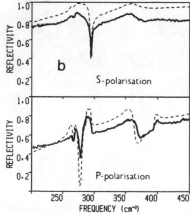

Fig. 2: *Room temperature spectra for a GaAs/AlAs superlattice sample whose layers are 4 monolayers thick: (a) Raman results in the AlAs optic phonon region - the top scan is resonant with the GaAs well transition $(\omega_L = 2 \cdot 08 eV)$, and the bottom scan is non-resonant $(\omega_L = 2 \cdot 41 eV)$. (b) Experimental (solid curves) and theoretical (dashed curves) ATR spectra. The s-polarised spectra show a prominent guided wave feature, whilst those in p-polarisation show a mixture of surface polariton and guided wave modes.*

8

Low Dimensional Systems

PHONON EMISSION AND SCATTERING IN A TWO-DIMENSIONAL ELECTRON GAS IN QUANTIZING MAGNETIC FIELDS.

L J Challis, A J Kent and V W Rampton

Department of Physics, University of Nottingham, Nottingham, NG7 2RD, UK

1. INTRODUCTION

Magnetic quantization leads to changes in both the phonon emission that occurs when a current is passed through the 2DEG and the absorption and re-emission from an incident phonon beam.[1] We review here some of the work carried out on (100) Si MOSFETS and GaAs/(Al-Ga)As heterostructures for which ν_c = 147B GHz and 466B GHz respectively.

2. PHONON EMISSION

2.1 Low Electron Temperatures $(kT_e \ll \hbar\omega_c \ (< \hbar\omega_{LO}))$

When the Fermi level E_F lies between 2 Landau levels, the large difference in Hall voltage across the 3D contacts and across the 2DEG causes the source-drain current to enter and leave the 2DEG at two diagonally opposite corners. As a result, when I -> 0, the source-drain resistance R_{SD} becomes equal to the quantized Hall resistance $R_H = h/ie^2$ (i is the number of filled Landau levels) even though the longitudinal resistivity is vanishingly small. Since $\underline{E}.\underline{J} = 0$, the current flows along the equipotentials without dissipation over most of the sample and the resistance R_{SD} and dissipation I^2R_{SD} are believed to be located in the corners where the current has to cross equipotentials to reach the contacts.

The power input for $kT_e \ll \hbar\omega_c$ is too small for pulse techniques so CW studies have been made on a Si MOSFET by measuring the temperatures opposite the middle and corners of the 2DEG[2] as the sheet density n_s is increased with a gate while $I^2 R_{SD}$ is kept constant by computer control. Momentum conservation and phonon focussing cause phonon emission to be largely in directions close to the normal to the 2DEG so that the surface temperature opposite a small area on the 2DEG is a measure of the phonon intensity emitted. Fig 1 shows the temperatures opposite (a) the middle, T_M, and (b) a 2DEG corner, T_c, as a function of n_s. When the system is in the Quantum Hall regime, the middle cools and two of the corners warm up demonstrating the movement of the dissipation to the corners. Reversing the field causes these corners to cool and oscillate in phase with T_M as the dissipation moves to the other pair. Reversing the current produces no detectable change showing that the dissipation is equally shared between the 2 corners.

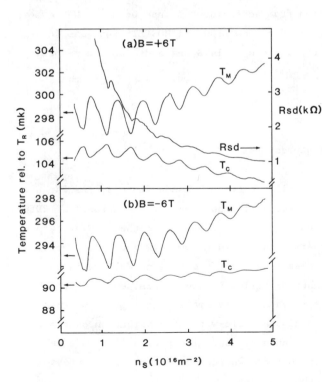

Fig 1: Temperatures T_M and T_c opposite the middle and one corner of a 2 DEG for $P=360\mu W$[2] .
(a) B=+6T, (b) B=-6T.

2.2 Medium Electron Temperatures ($kT_e \sim \hbar\omega_c$)

The emission should now be a mixture of cyclotron phonons[3] and lower frequencies from intra-Landau level transitions and has been studied using pulse techniques both with discrete bolometers[4] and an imaging system[5]. The 5mm Si substrate is used as a low-pass filter since isotope scattering drops the ballistic transmission from 1 at low frequencies to $>\frac{1}{2}$ by 800GHz ($>\frac{1}{2}$ because of phonon focussing[6]). Fig 2 shows the signal for $P = 1.5Wmm^{-2}$ drops by 15% at 5.3T ($\nu_c = $ 800GHz) suggesting cyclotron phonons form ~30% of the total. However, ~30% is presumably also emitted upwards and thermalised so the total cyclotron emission may approach 60% with ~40% at lower frequencies. This ratio should increase with mobility.

The imaging system developed to study magnetic field changes in angular distribution, has recently been extended to a 2D array; fig 3[7], where parallel 100μm strips of Cu and CdS on the lower substrate surface form electrodes and bolometers respectively. CdS is semi-insulating in the dark but illumination for a short time with a 100μm laser beam produces persistent donors so that the illuminated area acts as a bolometer and can be raster-scanned to image the phonon intensity.

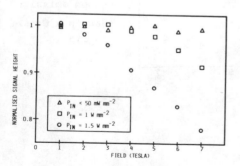

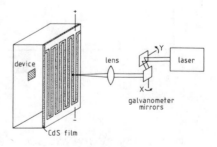

Fig 2: Phonon signal detected by a bolometer as a function of field[4]

Fig 3: Phonon imaging system[7]

2.3 High Electron Temperatures ($kT_e \sim \hbar\omega_{LO}$ ($> \hbar\omega_c$))

At these temperatures, the phonon emission should be a mixture of cyclotron phonons, optical phonons and some lower frequency phonons from both intra-Landau level transitions and two-phonon processes such as $n\omega_c = \omega_{LO} + \omega_{TA}$. Resonant generation of optical phonons can occur when $n\omega_c = \omega_{LO}$ (magneto-phonon resonance) leading to an increase in optical phonons and a corresponding decrease in cyclotron phonons. Fig 4 shows that this results in a substantial increase in the number of phonons arriving at or near the TA phonon ballistic transit time suggesting that the production and decay of LO phonons is faster than that of the cyclotron phonons. This is consistent with theoretical calculations showing rather long production times of cyclotron phonons [3], [9] which are predominantly LA in GaAs.

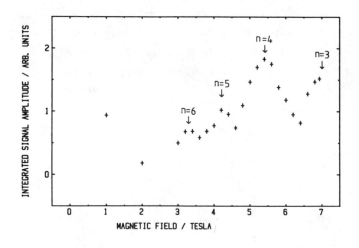

Fig 4
Magneto-phonon
resonances in
phonon emission
seen in a GaAs
2DEG[8]

3. PHONON ABSORPTION AND SCATTERING

The field dependence of the transmission has been studied in both Si MOSFETS and GaAs/(AlGa)As heterostructures and fig 5 shows data for GaAs/(AlGa)As[10].

For B >2T, ν_c lies above the isotope limit and the transmission loss is dominated by intra-Landau level transitions leading to signal minima when E_F lies within a Landau level. The phase change for B<2T is attributed to a switch to predominantly cyclotron phonon scattering. Similar results have been seen in Si MOSFETS[4] and measurements as a function of the sheet density n_s allowed the magnitude of the loss to be measured. AT B=0 this was ~6% for n_s ~ 5 x $10^{16}m^{-2}$ so if the loss is largely due to reflection, this suggests a 40% difference in the velocity of sound in the 2DEG from that in Si. The loss falls with magnetic fields presumably because the 2DEG becomes transparent at frequencies between ν_c and Γ (the Landau level width).

The absorption from a phonon beam incident on a 2DEG can be measured from its rise in temperature determined from the source-drain resistance $R_{SD}(T_e)$. The first experiment [11], on a GaAs/(AlGa)As heterostructure, was carried out at 0.2K ($\hbar\omega_c \gg kT_e$) so the absorption was due to intra-Landau level transitions with $\nu < \Gamma$ ~ 20GHz. The heater power was constant and the phonon intensity at low frequencies was varied by passing the pulse through ^{4}He which down-converts to < 20GHz a proportion which decreases with increasing pressure. Fig 6 suggests that the absorption increases with Landau level index N.

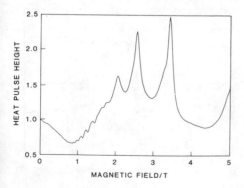

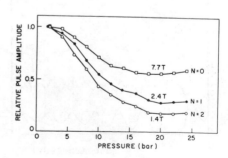

Fig 5: LA phonon transmission through a 2DEG in GaAs[10]

Fig 6: Low frequency absorption by a 2DEG in GaAs[11]

Absorption at higher frequencies was investigated using a Si MOSFET and results[12] are shown in fig 7 as a function of n_s for B = 7T. At the lower phonon powers, the rise in electron temperature is greatest when E_F lies within a Landau level implying that intra-Landau level transitions are largely responsible. However, at the higher powers, additional peaks occur when E_F is mid-gap suggesting that strong absorption is also occurring at the cyclotron frequency and this was confirmed at lower fields by showing that the rise in T_e was proportional to the intensity at ν_c. (This was increased by increasing the power to the heater). So this system provides a frequency-selective phonon detector. The band-width is large in Si, ~ 300GHz, because of the modest mobility but band-widths of a few GHz should be possible with GaAs/(AlGa)As heterostructures.

Field dependent phonon scattering from a 2DEG has also been seen in the thermal conductivity below 1K of a 50μm GaAs wafer with multilayer heterostructures on one face. The surface scattering is largely specular and so sensitive to diffuse scattering (from absorption and re-emission) that is greatest when E_F lies in a Landau level as seen in fig 8[13].

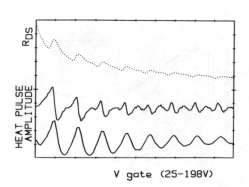

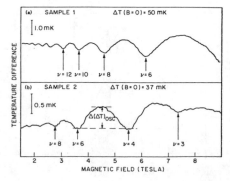

Figure 7: Phonon absorption by a Si MOSFET[12]. Middle curve = 5Wmm^{-2} lower curve = 0.5Wmm^{-2}

Fig 8: Oscillations in ΔT caused by phonon scattering by a 2DEG in GaAs[13]

The transmission loss in a 2DEG in a GaAs/(AlGa)As heterostructure from longitudinal ultrasonics at 9.36GHz[14] has been detected with a CdS bolometer; fig 9. The ultrasound modulates m* and so $\hbar\omega_c$ and it is proposed that the attenuation at normal incidence and the background at oblique incidence, is due to relaxation. Its strength suggests it may be near the peak at $\omega\tau \sim 1$ i.e. $\tau \sim 2 \times 10^{-11}$s $\sim\tau$(electron-electron). At oblique incidence, Shubnikov-de Haas oscillations for B > 1T are attributed to Joule heat loss due to piezo-electric fields and the signal minimum at 0.36T to the first observation in 2D of magnetoacoustic geometric resonance. The ultrasound produces modulation in the 2DEG plane, with wavefront separation $\lambda_{11} = \lambda/\sin\theta = 0.70$ μm. Geometric resonances should occur when $n\lambda_{11} \sim$ the electron orbit diameter, $2\hbar k_F/Be \sim 0.55\mu$m so lower field resonances (n=2,3...) may be observable at higher mobilities. These effects may be compared with magnetoresistance oscillations seen recently using static periodic modulation[15].

Shubnikov-de Haas effects can also be seen using surface acoustic waves (SAW)[16], [14] and provided the first observations of ultrasonic interaction with a 2DEG. The SAW are generated in a range from 70 - 600MHz using interdigital transducers. Examples of data are in fig 10.

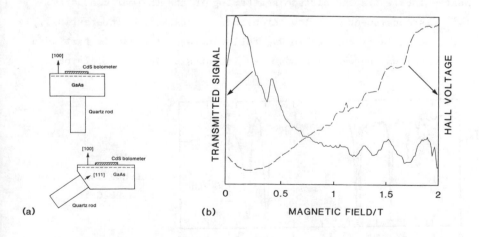

Fig 9: Ultrasonic transmission by a 2DEG in GaAs (a) experimental arrangement (b) results[14]

4. DISCUSSION

For low intensity pulses incident on a 2DEG, absorption should occur in a time $\tau_p \sim 10^{-9}$ s, the average for phonon absorption/emission by an electron. Thermal equilibrium should be rapid through electron-electron scattering with $\tau_e \sim 10^{-11}$ s and R_{SD} should change at the same rate since it depends on T_e through screening. So there should be no further net absorption for pulses longer than τ_p (absorption = emission) but loss continues since half the emission is away from the detector. This should be largely spontaneous so only a small proportion of the reflection from the incident beam should be specular.

This behaviour should change for high intensity pulses. The re-emission should now be largely stimulated with phonons emitted with $\theta_f = \pm\theta_i$ corresponding to a high proportion of specular reflection (q_{11} and q are both conserved so that $q_\perp$ is fixed in magnitude but not in sign). The average phonon energy absorbed $>>kT_e$ so thermalisation is slower and some of the re-emission may occur before thermalisation: resonance fluorescence. (A description of the reflection using the density response function[17] would seem equivalent to assuming stimulated emission.) The observation of strong specular reflection from a 2DEG in Si[18] suggests the high intensity picture is appropriate for heat pulses while the diffuse scattering in the thermal conductivity[13] is consistent with low intensities as expected. Interestingly, the time scale of changes in R_{SD} following a heat pulse is fast when E_F lies midway between 2 Landau levels but slow in zero field or when

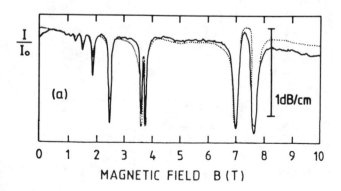

Fig 10:
Attenuation of SAW by a 2DEG in GaAs[16]

E_F lies inside a level[12] and we suggest this is due to differences in the nature of the resistance change. In the first case, an increase in scattering occurs when electrons are excited to higher Landau levels but, in the second two cases, through temperature dependent changes in screening which continue until equilibrium.

5. ACKNOWLEDGEMENTS

We are very grateful to our collaborators: P J A Carter, J Cooper, A G Every, G A Hardy, P Hawker, N P Hewett, D C Hurley, K B McEnaney, T Miyasato, D Neilson, M I Newton, F F Ouali, P A Russell, F W Sheard, G A Toombs and Y B Wahab, for invaluable support from Nottingham and Southampton colleagues for samples and to GEC, NATO and SERC for financial support.

6. REFERENCES

1. Earlier discussions of some of this work have been given by Challis, L. J., Kent, A. J. and Rampton, V. W., Application of High Magnetic Fields in Semiconductor Physics ed G Landwehr (Springer, Berlin) 1989 in press and by Challis, L. J., Toombs, G. A. and Sheard, F. W., Physics of Phonons, T Paskiewicz, Lect Notes in Physics, <u>285</u>, (Springer, Berlin) p 348, 1987.
2. Ouali, F. F., Hewett, N. P., Russell, P. A. and Challis, L. J., these proceedings.
3. For theoretical analysis see: Toombs, G. A., Sheard, F. W., Neilson, D. and Challis, L. J., Solid State Commun. <u>64</u>, 577 (1987)
4. Kent, A. J., Rampton, V. W., Newton, M. I., Carter, P. J. A., Hardy, G. A., Russell, P. A. and Challis, L. J.: Surf Sci, <u>196</u>, 410 (1988).
5. Hurley, D. C., Hardy, G. A., Kent, A. J., Rampton, V. W. and Challis, L. J., Application of High Magnetic Fields in Semiconductor Physics, ed G Landwehr (Springer, Berlin) 1989 in press.
6. Fieseler, M., Wenderoth, M. and Ulbrich, R. G., Proc 19th Conf on Physics of Semiconductors, Warsaw, ed W Zawadski, (Polish Academy of Sciences, Warsaw), 1477, (1988).

7. Kent, A. J., Hardy, G. A., Hawker, P. and Hurley, D. C. these proceedings.

8. Hawker, P., Kent, A. J., Challis, L. J., Henini, M. and Hughes, O.H., J Phys:Condensed Matter $\underline{1}$, 1153, (1989).

9. We are very grateful to Prof Y Levinson for correspondence on this point.

10. Newton, M.I., Carter, P. J. A., Rampton, V. W., Henini, M. and Hughes, O.H., Proc 19th Conf on Physics of Semiconductors, ed W Zawadski (Polish Academy of Sciences, Warsaw), 335, (1988).

11. Eisenstein, J.P., Narayanamurti, V., Stormer, H.L., Cho, A. Y. and Hwang, J. C. M., Proc 5th Int Conf on Phonon Scattering in Condensed Matter, Urbana, 1986, ed A C Anderson and J P Wolfe, (Springer, Berlin), Solid St Science Series, $\underline{68}$, 401, (1986).

12. Kent, A. J., Hardy, G. A., Hawker, P., Rampton, V. W., Newton, M. I., Russell, P. A. and Challis, L. J., Phys Rev Letts $\underline{61}$, 180, (1988).

13. Eisenstein, J. P., Gossard, A. C. and Narayanamurti, V., Phys Rev Letts, $\underline{59}$, 1341, (1987); Eisenstein, J. P., Gossard, A. C. and Narayanamurti, V., Surf Sci, $\underline{196}$, 445, (1988).

14. Rampton, V. W., Newton, M. I., Carter, P. J. A., Henini, M., Hughes, O. H., Heath, M., Davies, M., Challis, L. J. and Kent, A. J., Act Phys Slovaka, (1989), in press.

15. Weiss, D., Klitzing, K. v, Ploog, K. and Weimann, G., The Application of High Magnetic Fields in Semiconductor Physics, ed G Landwehr (Springer, Berlin), 1989, in press; Winkler, R. W., Kotthaus, J. P. and Ploog, K., Phys Rev Letts $\underline{62}$, 1177, (1989).

16. Wixforth, A,. Kotthaus, J. P. and Weimann, G., Phys Rev Letts $\underline{56}$, 2104, (1986); Wixforth, A. and Kotthaus, J. P., Application of High Magnetic Fields in Semiconductor Physics, ed G Landwehr (Springer, Berlin) 1989, in press

17. Hensel, J. C., Halperin, B. I. and Dynes, R. C., Phys Rev Letts, $\underline{51}$, 2302, (1983)

18. Rampton, V. W., Wahab, Y. bin., Newton, M. I., Carter, P. J. A., McEnaney, K., Henini, M. and Hughes, O. H. these proceedings.

COMPUTER SIMULATION OF INTRINSIC LOCALIZED VIBRATIONS IN ANHARMONIC CHAIN

V.M. Burlakov, S.A. Kiselev, V.N. Pyrkov*

Institute of Spectroscopy, USSR Ac. of Sciences, Troitsk, Moscow region, USSR, 142092
*Lebedev Physical Institute, USSR Ac. of Sciences, Moscow, USSR

1. INTRODUCTION

It is well understood now that the appearance of localized vibrational (LV) excitations in the harmonic lattice is caused by mass or force constant defects (see, for example, /1/). It was theoretically predicted recently that the localization may also occur in ideal anharmonic lattices /2/. Here we present the results of investigations of intrinsic LV (ILV) properties in a linear anharmonic chain by molecular dynamics.

2. COMPUTER SIMULATION TECHNIQUE

The investigation procedure was that of excitation of ILV in a linear chain of 128 particles with the potential function

$$V=\sum_n [\frac{K_2}{2}(x_{n+1}-x_n)^2 + \frac{K_3}{3}(x_{n+1}-x_n)^3 + \frac{K_4}{4}(x_{n+1}-x_n)^4] \qquad (1)$$

where K_2, K_3 and K_4 are constants.

The ILV was excited by making the initial displacements x_o of opposite sign for two adjacent particles. The remaining chain stayed at rest with zero displacements (classical zero temperature limit). For the calculation procedure circular boundary conditions were used. As the main characteristic of the ILV motion a

Fourier-component of the displacement $A(\omega,n)$ of any particle computed through the time interval 128 ω_m was used (ω_m is determined below).

3. RESULTS AND DISCUSSION

An example of spatial distribution of $A(\omega,n)$ is represented in Fig.1. For $a_0=1.0$ the splitting of ILV from delocalized excitations band (DEB) is almost invisible. But this splitting increases with a_0.

The cutoff frequency of delocalized excitations spectrum depends on the amplitudes of particle vibrations a_{hf} in the high-frequency mode:

$$\omega_m = 2\sqrt{(K_2+K_4/2\ a_{hf}^z)} \tag{2}$$

In all our computer experiments the second term under square root was too small compared to the first one thus $\omega_m \cong 2\sqrt{K_2}$.

For higher values of a_0 the splitting of ILV from DEB is more obvious and the frequency of ILV can be approximated by

$$\omega_{eff} = \sqrt{3K_2+9K_4\ \bar{a}^2} \tag{3}$$

This value corresponds to the quasi-harmonic frequency of a symmetric vibration of an anharmonic oscillator consisting of two vibrating atoms and with two edjacent atoms remaining at rest.

The main difference of ILV excitation in a two-atomic chain from that in a single-atomic chain is the decrease of localization threshold. This fact corresponds to the localization threshold in the harmonic lattice when it is determined by the strength of the defect and the width of DEB. It is important to discuss the stability of the results obtained above under conditions of nonzero temperature. The temperature in a linear chain may be simulated by chaotic initial displacements of particles. These chaotic displacements were set up using a random numerical generator.

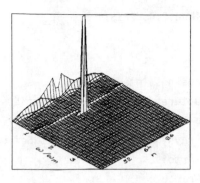

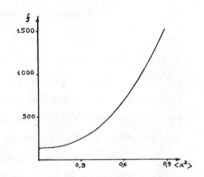

Fig.1   Fig.2

The T-dependence of the localization threshold thus determined is presented in Fig.2. Since the thermodynamics of ILV excitations is analogous to that of vacancies the number of ILV at temperature T is determined by

$$n= \frac{N}{N_l} \exp[-f/k_B T],$$ (4)

where N is the total number of particles, N_l is the number of particles effectively involved into ILV (as discussed above $N_l \cong 2$), k_B is Boltzman constant, f is ILV energy equal to $3K_2 a_0^2 + 9/2 \cdot a_0^4$. The temperature T can be estimated by the classical oscillator formula of the mean-square displacement $T = m\omega^2/2 \cdot \langle a^2 \rangle$. As shown in Fig.2, the T-dependence of f/T ratio is not monotonous but has a minimum at $\langle a^2 \rangle \cong 0.3$. It means that there is a maximum in the excitation probability of ILV P_{ILV} at some value of T. But it is worth noting that this maximum may significantly change or even disappear in quantum consideration.

4. ACKNOWLEDGEMENTS

The authors wish to thank professor V.M.Agranovich and his colleagues for useful discussions.

5. REFERENCES

1. A.S.Barker, A.J.Sievers, Rev. Mod. Phys., 47 Suppl. N2, (1975).

2. A.J.Sievers, S. Takeno, Phys. Rev. Lett., 1988, v.61

SANDWICH PHONONS IN B.C.C. CRYSTALS

A. JBARA, A. AKJOUJ, B. SYLLA, L. DOBRZYNSKI
Equipe Internationale de Dynamique des Interfaces
Laboratoire de Dynamique et Structure des Matériaux, Associé
(n°801) au Centre National de la Recherche Scientifique, Unité
de Formation et de Recherche en Physique, Université des Sciences
et Techniques de Lille-Flandres-Artois,
59655 Villeneuve d'Ascq Cedex, France.

Acoustical and optical localized phonons within a cristalline slab sandwiched between two others crystals were predicted before[1,2] with the help of models using monoatomic simple cubic crystals.

The present paper presents such sandwich phonons for body centered cubic crystals.

We start from an infinite body centered cubic crystal of atoms of mass m_i ; the index i refers to three different crystals used to build the composite studied here. The distance between second nearest neighbour atoms is called a and is supposed to be the same in the three crystals out of which are build the crystal sandwiches studied here. The interatomic interactions within each crystal i are β_i between first-nearest neighbour atoms only and such that there is no interaction between the three space components of the displacement vector $\vec{u}$. Such a Montroll-Potts model[3] of b.c.c. crystals describes the transverse polarized phonons and does not address to the phonons polarized within the saggital plane, which is the plane containing the propagation vector $\vec{k}_{//}$ parallel to the planar interfaces.

The composites under study here are formed by a finite slab i=2 of 2L(001) atomic planes of a given b.c.c. crystal embedded in an another (i=1) b.c.c. crystal. The interface interatomic interactions were taken to be $\beta_I=(\beta_1+\beta_2)/2$. So the only parameters appearing in this study are the first nearest neighbour interactions (β_1 and β_2) and the masses (m_1 and m_2) caracterizing the two b.c.c crystals out

of which the crystalline sandwich studied here is build. With the help
of the interface response theory[4] an explicit expression[5] giving a
relation between the frequencies ω of the sandwich phonons and the
propagation vector $\vec{k}_{//}$ was obtained. Due to lack of space, we present
here only a few illustrative curves. On these curves, $\Omega=\omega/\omega_M$, where
ω_M is the maximum frequency of the host b.c.c crystal, k_1 is the component of $\vec{k}_{//}$ along the x_1 direction of space. The shaded areas give the
bulk phonons of the host crystal and the other dispersion curves are
those of the sandwich phonons for 2L=6 (001) atomic planes of another
b.c.c crystal. Let us stress that these sandwich phonons are truely
localized or confined phonons within the sandwich.

REFERENCES

1. Akjouj, A., Sylla, B., Zielinski, P. and Dobrzynski, L., J.Phys.
   C20, 6137 (1987).
2. Sylla, B. More, M. and Dobrzynski, L., Surface Sci. 213, 588 (1988).
3. Montroll, E.N. and Potts, R.B., Phys.Rev. 102, 72 (1956).
4. Dobrzynski, L., Surface Sci. Rept. 6, 119 (1986).
5. Jbara, A., Thèse de l'Université de Lille I, 1990.

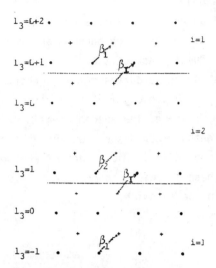

$l_3=L+2$

$l_3=L+1$ $i=L$

$l_3=L$

          $i=2$

$l_3=1$

$l_3=0$

$l_3=-1$ $i=1$

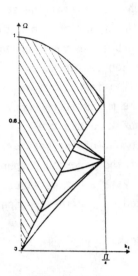

Figure 1. Geometry of the sandwich
. Atom of type 1
+ Atom of type 2

Figure 2. Dispersion curves
for a sandwich of six atomic
planes : $m_2/m_1=4, \beta_2/\beta_1=1.5$

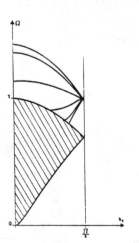

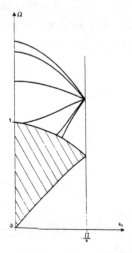

Figure 3. Dispersion curves for a
sandwich of six atomic planes :
$m_2/m_1=1.5, \beta_2/\beta_1=3$

Figure 4. Dispersion curves
for a sandwich of six atomic
planes : $m_2/m_1=0.5, \beta_2/\beta_1=1.5$

FRENKEL-KONTOROVA MODEL WITH LONG-RANGE INTERPARTICLE INTERACTIONS

O.M.Braun, Yu.S.Kivshar[+] and I.I.Zelenskaya

Institute for Physics, 46 Science Avenue, Kiev, USSR
[+]Institute for Low Temperature Physics and Engineering
47 Lenin Avenue, Kharkov, USSR

The Frenkel-Kontorova (FK) model [1] is successfully used to explain the dynamics of a number of physical objects such as dislocations in solids, domain walls in magnetic crystals, adsorbed layers on surfaces, and so on. The discrete Hamiltonian of the FK model,

$$H = \sum_n \frac{1}{2} \dot{u}_n^2 + (1 - \cos u_n) + \frac{1}{2} \sum_{k \neq 0} V(u_{n+k} - u_n), \qquad (1)$$

describes a chain of unit-mass atoms in a periodic potential with the interparticle interaction $V(x)$. In the usual FK model [1] it is assumed that nearest neighbouring atoms interact only, $k = \pm 1$, and $V(x)$ is harmonic, i.e. may be presented as $V(x) = \frac{1}{2}(\ell/\pi)^2 (x-a)^2$. Here a is the equilibrium distance between atoms and the parameter ℓ characterizes the intensity of the interaction. In case $a = 2\pi q$, q being an integer, the ground state of the system corresponds to a periodic structure of atoms with the period a which is commensurate with the period $a_s = 2\pi$ of the external potential in (1). In the continuum limit, i.e. for $V(a) \gg 1$, simple transformations of (1) yield the well-known Hamiltonian of the sine-Gordon (SG) model. The SG equation, which follows from the SG Hamiltonian, has a topological solution, the kink (or anti-kink): $u(x) = 4 \tan^{-1}\left\{ \exp\left[-\sigma(x-X)/d\right]\right\}$. The kink with $\sigma = +1$

(or anti-kink with $\sigma = -1$) corresponds to the minimally possible contraction (extention) of the commensurate structure of atoms, $d = 2\ell q$ being the kink size.

The report aims to present an extended version of the FK model which describes a chain of atoms in a periodic potential with long-range interparticle interactions. A physically important example of such an interaction is the dipole-dipole repulsion of atoms chemisorbed on a metal surface /2/. The interaction may be described by the power potential for $|x| \gg 1$,

$$V(x) \simeq V_0 \, (2\pi/|x|)^n , \quad n \geqslant 1 . \tag{2}$$

In the continuum limit the Hamiltonian (1),(2) leads to an integro-differential version of the SG equation. As a result, the kink solution has power-like "tails",

$$u(x) \approx u(\, |x| = \infty \,) + 2\pi(a/d)^{n-1}/(n+1) \, |x|^{n+1}, \quad |x| \gg 1.$$

The latter stipulates the power law of interaction between kinks in the system,

$$V_{int}(x) = V_{int}^{(0)}(x) + V_{int}^{(1)}(x) ,$$

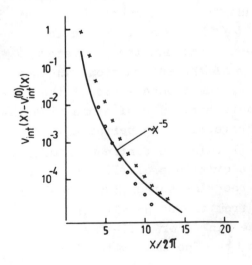

Fig.1

Interaction energy of kinks in the case of dipole interatomic interaction (n = 3). Results of numerical calculations are indicated by crosses for kinks, and circles for anti-kinks.

$$V_{int}^{(0)}(x) \approx \sigma_1 \sigma_2 V_0 / \ q^{2n+1} (\ |x|/a)^n, \ V_{int}^{(1)}(x) \sim |x|^{-(n+2)},$$

σ_1 and σ_2 being kinks' polarities. We have compared analytical results with the computer calculations of the energy $V_{int}^{(1)}(x)$ which we have done for a chain of more than 20 particles with dipole interparticle interactions $(n = 3 \ )$ (Fig.1).

The extended FK model with a long-range interaction of kinks leads to a number of new effects. For example, for rational concentrations, $\theta = p/2\pi q$ (p atoms and q wells), the ground state of the system is connected with the corresponding commensurate structure. When θ is increased, these structures are changing from one to another. As a result, the amplitude of the Peierls energy (an effective periodic potential for kink motion) $\epsilon_p(\theta)$ takes the form of the Devil's staircase and for each rational θ undergoes a step (see Fig.2).

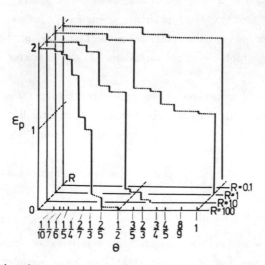

Fig.2

The Peierls energy as a function of the coverage parameter θ for the dipole interaction between atoms (n=3) at various values of the parameter

$R = V(2\pi)/2$.

It is important that the similar dependences may be observed experimentally in quasi-1D-systems/3, 2 /.

1. Frenkel,J. and Kontorova,T.,Phys.Z.Sow.13,1(1938).
2. Bolshov, L.A., et al., Usp.Fiz.Nauk 122,125 (1977).
3. Braun, O.M., et al., J.Phys. C 21, 3881 (1988).

THE EFFECT OF PERIODIC SURFACE ROUGHNESS ON PHONON DISPERSION IN NARROW WIRES

J. Seyler and M. N. Wybourne

Department of Physics
University of Oregon
Eugene, OR 97403 USA

In the elastic continuum limit, wire structures with free-boundaries have been shown to have many dispersion branches close to the zone-center.[1] These branches arise from spatial quantization of the phonon spectrum and depend on the geometry of the wire. The fact that the free-surface determines the nature of the dispersion branches suggests that surface irregularities will modify the phonon spectrum. This is particularly important for wires of nanometer scale because the methods used to produce the wires inevitably lead to variations of the diameter by as much as 5%.[2]

In this paper we introduce a method to determine the modification of the phonon spectrum in wires with surface roughness. In particular we demonstrate the effect of a periodic surface structure on the lowest phonon branch.

The displacement vector $\vec{u}$ for wave propagation in isotropic elastic media satisfies the following wave equation:

$$\frac{\delta^2 \vec{u}}{\delta t^2} = c_t^2 \nabla^2 \vec{u} + \left(c_l^2 - c_t^2 \right) \text{grad div } \vec{u} \tag{1}$$

where c_t and c_l are the transverse and longitudinal velocities of sound. We have introduced surface roughness of a cylindrical wire by a modulation of the radius, $R' = R(1 + \chi f(z))$, where R is the wire radius, χ is the amplitude of roughness and $f(z) = \sin \gamma z$ with γ being the wave vector of the roughness. We solved equation (1) in cylindrical coordinates by looking for solutions of the form,

$$\vec{u}\,(r,\ \theta,\ z) = \vec{\mathcal{R}}\,(r,\ z)\ e^{iqz}\ e^{im\theta}\ e^{i\omega t} \tag{2}$$

where the radial part of the wave function, $\vec{\mathcal{R}}(r,\ z)$, incorporates the effect of the surface roughness. The coupling between the r and z components makes it impossible to solve equation (1) using a separation of variables technique.

To overcome this problem we have adopted the following procedure. The derivatives with respect to z were eliminated for each vector component of $\vec{\mathcal{R}}$ by introducing a coordinate transformation $r' = r(1 + \chi\ f(z))$ which yields the following relation,

$$\frac{\delta\vec{\mathcal{R}}}{\delta z} = r\ \frac{\chi\ \dfrac{\delta f(z)}{\delta z}}{1+\chi f(z)}\ \frac{\delta\vec{\mathcal{R}}}{\delta r} \tag{3}$$

The wave functions were then determined as power series by de-coupling the longitudinal and transverse components into $\vec{u} = \vec{u}_1 + \vec{u}_t^{\ 3)}$ and using equation (3) to solve equation (1).

The solutions to the equations of motion have to satisfy the boundary conditions that express the condition that the surface of the wire be stress-free. This requires that the normal components of the stress tensor T_{ij} on the surface vanish, that is $T_{r'r'} = T_{r'\theta'} = T_{r'z'} = 0$ at $r' = R'$. The transformed stress tensor components $T_{r'j'}$ are given by,

$$T_{r'j'} = \sum_j \frac{\delta r}{\delta r'}\ \frac{\delta j}{\delta j'}\ T_{rj}$$

with

$$T_{rj} = \left[B \sum_{\nu=0}^{\infty} b_\nu^j (\beta r)^{m+\nu} \sin m\theta + A \sum_\nu a_\nu^j (\alpha r)^{m+\nu} \sin m\theta + C \sum_\nu c_\nu^j (\alpha r)^{m+\nu} \cos m\theta \right] e^{iqz};$$

where $j = r,\ \theta,\ z;$ $\quad \alpha^2 = \dfrac{\omega^2}{c_t^2} - q^2$ and $\beta^2 = \dfrac{\omega^2}{c_1^2} - q^2.$

The coefficients b_ν^j, a_ν^j, c_ν^j are functions of r and z only and B, A and C are constants. T_{rj} are evaluated at n points, each at position z_n, along the axis of the rod. The number of slices taken cover one period of the surface profile function $f(z)$. The dispersion relation follows from the condition that Det $| P^{\nu,n} | = 0$, where the elements $P^{\nu,n}$ are given by the matrices,

$$P^{\nu,n} = \begin{bmatrix} b_\nu^r(z_n) & b_\nu^\theta(z_n) & b_\nu^z(z_n) \\ a_\nu^r(z_n) & a_\nu^\theta(z_n) & a_\nu^z(z_n) \\ c_\nu^r(z_n) & c_\nu^\theta(z_n) & c_\nu^z(z_n) \end{bmatrix}.$$

988

We have applied the model to wires made from a material that has a velocity of sound ratio $c_l/c_t = 2$. Further, we have only considered the case of axial symmetry, that is m=0. In the absence of surface roughness, the predicted dispersion data are shown in figure 1. In this figure ρ and E are density and Young's modulus respectively. This data agrees with previously published phonon dispersion in similar wire structures.[1] The effect of introducing a sinusoidal surface roughness with an amplitude of up to 5% of the radius, is demonstrated in figure 2. As expected, band-gaps are formed in the phonon spectrum. For the lowest phonon branch, the band-gap has a linear dependence on the amplitude of the roughness and a quadratic dependence on the period. From the numerical results the band-gap can be described by, $\omega_{gap} = \chi \, \gamma^2 c_t R$. We note that the dependence of the band-gap on the amplitude and period of the roughness is the same as that found by Glass et al.[4] for Rayleigh surface waves propagating across a grating structure.

The work herein was supported by the National Science Foundation under Grant DMR-8713884.

1) Grigoryan, V. G. and Sedrakyan, D. G., Sov. Phys. Acoust. <u>29</u> 281 (1983).
2) Lee, K. L., Ahmed, M., Kelly, M. J. and Wybourne, M. N., Electronics Letters <u>20</u> 2289 (1984).
3) Landau, L. D. and Lifshitz, E. M., Theory of Elasticity. Addison-Wesley, (1959).
4) Glass, N. E., Loudon, R. and Maradudin, A. A., Phys. Rev. B<u>24</u> 6843 (1981).

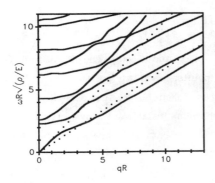

Figure 1. Phonon dispersion without surface roughness. Dotted lines show the bulk dispersion.

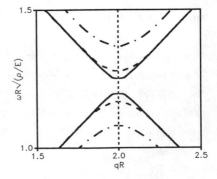

Figure 2. Band-gap in the lowest phonon branch. $\gamma = 2/R$, $\chi = $ (—) 1%, (— — —) 2% and (• — • —) 5%.

ON TEMPERATURE BEHAVIOUR OF ELASTIK
PROPERTIES OF LAYERED CRYSTALS

N.I.Lebedev and A.S.Sigov

Institute of Radioengineering, Electronics and Automation,
117454 Moscow, USSR

We consider the nonlinear interaction between the Lifshitz mode and other acoustic modes of layered crystals. There are obtained two interesting results being accurate in all orders of the interaction (which is expected to be the greatest numerically for crystals of the graphite type): i) the longitudinal sound mode softens sufficiently far away from the melting temperature and becomes of the Lifshitz type itself; ii) the negative thermal expansion coefficient due to the Lifshitz mode does not depend on temperature within the whole classical region.

The acoustic spectrum of layered crystals, such as graphite, BN, GaS, GaSe, has the well known form[1,2)]

$$\zeta \omega_1^2 = (\lambda + \mu) K_\perp^2 + \alpha_1 K_z^2 ,$$
$$\zeta \omega_2^2 = 2\mu K_\perp^2 + \alpha_1 K_z^2 ,$$
$$\zeta \omega_3^2 = \alpha_1 K_\perp^2 + \alpha_2 K_\perp^2 + \beta K_\perp^4 .$$

(1)

Here the z-axis is perpendicular to the plane of layers, $K_\perp^2 = K_x^2 + K_y^2$, $\lambda = \lambda_{xxxx} - \lambda_{xyxy}/2$, $\mu = \lambda_{xyxy}/2$, $\alpha_1 = \lambda_{xzxz}$, $\alpha_2 = \lambda_{zzzz}$. Due to the weakness of the interlayer interaction the values of λ and μ are much greater than those of α_1 and α_2. The mode ω_3 is named the Lifshitz mode[2)]. It is just the mode determining some specific properties of the layered crystals: the deviation from T^3 law for the low-temperature specific heat, the negative thermal expansion coefficient along the layers, etc[1,2].

The Lifshitz mode is "soft" for all the directions of the wave vector $\vec{K}$. Therefore the fluctuations for this mode are very intensive and their anharmonic interactions must be taken into account far away from the melting transition point T_{mel} . The strong temperature dependence of the elastic properties of layered crystals may be an evidence of such interactions.

We begin with neglecting the weak interlayer interactions. Then the system is described by the Hamiltonian of a single elastic layer[3]

$$H = \int d^2r \left\{ \frac{\lambda}{2} u_{\ell\ell}^2 + \mu \left(u_{ik}^2 - \frac{1}{2} \delta_{ik} u_{\ell\ell}^2 \right) + \frac{\beta}{2} \left(\frac{\partial^2 u_z}{\partial x^2} + \frac{\partial^2 u_z}{\partial y^2} \right)^2 \right\} , \qquad (2)$$

$$u_{ik} = \frac{1}{2} \left(\frac{\partial u_i}{\partial x_k} + \frac{\partial u_k}{\partial x_i} + \frac{\partial u_z}{\partial x_i} \frac{\partial u_z}{\partial x_k} \right), \quad i, k = x, y . \qquad (3)$$

The nonlinearity arises due to the last term in Eq.(3). The Hamiltonian (2) is fairly complicated but it can be simplified for $\mu/\lambda \ll 1$. Indeed, such a simplification can't be performed for any layered crystal, but in reality the corresponding values of μ/λ are: 0.2 for graphite, 0.27 for GaS, and 0.3 for GaSe[2] . For $\mu=0$ one obtains for the Green function of ω_1 mode in the first order in the interaction

$$G_1(\vec{K}) = \frac{1}{\lambda K^2} + \frac{T}{2\pi \beta^2} \frac{1}{K^4} \ln \frac{K_0}{K} . \qquad (4)$$

The value of K_0 is defined by the interlayer interaction: $K_0 = (K_m \alpha_2/\beta)^{1/2}$. We have shown that all the higher-order corrections to Eq.(4) are identically zero. Thus it is valid also in the case when the second term is much greater than the first one. The vanishing of higher-order

term for the Hamiltonian (2) with $\mu=0$ has been already
demonstrated (in a different way)[4] , but Eq.(4) has not
been displayed in papers available. With account of the
weak interlayer interaction Eq.(4) is valid for $K>K_o$. The
thermal expansion coefficient is another quantity that
has no corrections due to higher-order terms. That is why
it does not depend on temperature in the classical region.
The details of calculations and discussion will be published later[5].

It should be mentioned in conclusion that for lay-
ered crystals with small μ/λ the ω_1 mode softens due to
the interaction with the Lifshitz mode and becomes of the
Lifshitz type itself if the temperature is high enough.
This temperature can be estimated[5] as $T\sim0.1T_{mel}$. Such
calculation results may be verified experimentally for
graphite or BN. Another result is the absence of temperature dependence of the negative thermal expansion coefficient due to the Lifshitz mode. Let us note that in BN
this coefficient is really independent of temperature within the range of about 500K[6].

REFERENCES

1. Lifshitz, I.M., Zh.Eksp.Teor.Fiz. 22, 475(1952).
2. Belenkii, G.L., Salaev, A.J. and Suleimanov, R.A., Usp.
   Fiz.Nauk 55, 87(1988).
3. Landau, L.D. and Lifshitz, E.M., "The Theory of Elasticity", Nauk. Moscow (1987).
4. Nelson, D.R. and Peliti, L., J.Phys. (Paris) 48, 1085
   (1987).
5. Lebedev, N.I. and Sigov, A.S., Fiz.Tverd.Tela, in press.
6. Yates, B., Overy, H.J. and Pirgon, O., Phil. Mag. 32,
   847(1975).

THERMALIZATION OF 2DEG IN QUANTUM WELL DUE TO SPONTANEOUS EMISSION OF ACOUSTIC PHONONS

F W Sheard

Department of Physics, University of Nottingham, Nottingham NG7 2RD, England

Electrons may be injected into a quantum well by suitably biasing a resonant tunnelling double-barrier structure (DBS). If the electron kinetic energy is too small for optic phonon emission then, at low temperatures, energy relaxation occurs via spontaneous emission of acoustic phonons. To observe the consequent thermalization of the two-dimensional electron gas (2DEG) in the well, the electron storage time τ_2 must be greater than the energy relaxation time τ_{ph}.

The conduction-band profile in a DBS under bias is shown in the Figure. In a lightly doped emitter a 2DEG is formed in the accumulation layer and is degenerate at low temperatures. Resonant tunnelling occurs when the voltage drop V_1 across the emitter barrier brings the

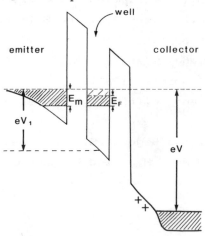

bound state in the well into coincidence with the bound state in the emitter. In the sequential theory of resonant tunnelling[1], electrons tunnel from the emitter into the well (transition rate W_1) and occupy states of transverse motion up to a maximum energy E_m, which is the same as the Fermi energy of the emitter 2DEG. This follows because transverse momentum is conserved in tunnelling. The electrons in the well then tunnel out into unoccupied states in the collector (transition rate $W_2 = 1/\tau_2$).

The occupation number of a state of transverse kinetic energy E in the well, is determined dynamically by the rate equation[2]

$$\dot{f} = (1 - f)W_1 - fW_2 - (f - f^\circ)W_{ph}, \tag{1}$$

where the latter term takes account of energy relaxation and $W_{ph} = 1/\tau_{ph}$. Here f° is an equilibrium Fermi-Dirac distribution with Fermi energy E_F. At liquid He temperatures we can take $f^\circ = 1$ $(E < E_F)$ and $f^\circ = 0$ $(E > E_F)$. The steady-state solution $(\dot{f} = 0)$ is given by

$$f = \begin{cases} (W_1 + W_{ph})/W_{tot} & (0 < E < E_F) \\ W_1/W_{tot} & (E_F < E < E_m), \end{cases} \tag{2}$$

where $W_{tot} = W_1 + W_2 + W_{ph}$ and $f = 0$ $(E > E_m)$. The position of the relaxed Fermi level E_F is determined by the condition that the energy relaxation term conserves the number of electrons. If W_{ph} is taken to be constant this gives

$$E_F = E_m W_1/(W_1 + W_2),$$

which is independent of W_{ph} in this simple model. The distribution function (2) gives an areal density of electrons in the well $n = DE_F$, where $D = m^*/\pi\hbar^2$ is the 2D density of states. This is the same as if one assumed unit occupancy $f = 1$ for $E < E_F$. The rate W_1 is a resonant function of the voltage drop V_1. At the threshold for resonant tunnelling $W_1 \sim 0$ so $E_F \sim 0$ and rises to a maximum value $E_F \simeq E_m$ at the resonant peak where $W_1 \gg W_2$. This change requires a small change in V_1 but occurs over an extended range of applied bias V owing to the screening effect of the charge buildup in the well[2].

This thermalization effect has been observed experimentally[3] in a GaAs/(AlGa)As DBS with well width w = 5.8 nm. The relaxed Fermi energy E_F was determined from the periodicity of magneto-oscillations in the tunnel current due to Landau levels passing through the quasi-Fermi level. In order to have an appreciable discontinuity in f at $E = E_F$, we require $W_{ph} \gg W_2$ ie $\tau_2 \gg \tau_{ph}$. We have calculated the rate of spontaneous emission of longitudinal acoustic phonons for an electron of wave vector k in the lowest subband of a GaAs quantum well. Previous calculations[4] refer only to elevated temperatures ($k_B T \gg$ average emitted phonon energy). The result is

$$W_{ph} = \frac{3\Xi^2 m^* k}{\pi \rho s \hbar^2 w} ,$$

where Ξ, ρ, s and m^* are deformation potential, density, sound velocity and effective mass respectively. For a typical electron energy E = 5 meV, $W_{ph} = 3 \times 10^9$ s^{-1} and hence $\tau_{ph} \sim 0.3$ ns. The tunnelling escape rate $W_2 = 1/\tau_2 = \nu T_2$ can be obtained from the collector-barrier transmission coefficient T_2 and attempt rate ν of an electron in the quantum well. For a 11 nm thick $Al_{0.4}Ga_{0.6}As$ barrier[3] this gives $\tau_2 \sim 1$ μs which is indeed $\gg \tau_{ph}$. These figures confirm that thermalization of a 2DEG in a quantum well via spontaneous acoustic phonon emission can be achieved in a suitably tailored DBS.

REFERENCES

1. Luryi, S., Appl. Phys. Lett. 47, 490 (1985).

2. Sheard, F.W. and Toombs, G.A., Appl. Phys. Lett. 52, 1228 (1988).

3. Leadbeater, M.L., Alves, E.S., Sheard, F.W., Eaves, L., Henini, M., Hughes, O.H. and Toombs, G.A., submitted to Phys. Rev. Lett.

4. Hess, K., Appl. Phys. Lett. 35, 484 (1979).

AN INVESTIGATION OF THE PHONONS EMITTED BY A HOT TWO DIMENSIONAL ELECTRON GAS (2DEG) IN A GALLIUM ARSENIDE/ALUMINIUM GALLIUM ARSENIDE HETEROJUNCTION

A.J. KENT, P. HAWKER, M. HENINI, O.H. HUGHES AND L.J. CHALLIS

Department of Physics, University of Nottingham, University Park, Nottingham NG7 2RD. UK.

The primary energy relaxation process for a hot 2DEG in the GaAs/AlGaAs heterojunction is through the emission of phonons. This has been observed experimentally using heat pulse techniques[1]. At low electron temperatures (T_e) only piezoelectrically coupled transverse (TA) modes were detected in contradiction to theoretical calculations[2] which indicate that at low T_e the electrons are coupled more strongly to longitudinal (LA) modes via the deformation potential. At higher T_e evidence of optic phonon emission is seen. Such studies can give more direct information about the electron phonon interaction than may be obtained, for example, from the dependence of the mobility on T_e.

We are using heat pulse techniques to investigate the phonon emission by a hot 2DEG in the GaAs/AlGaAs heterojunction, measurements of its power, angular and magnetic field dependence have been made.

The (100) GaAs/AlGaAs heterojunction was MBE grown on a 5mm thick SI GaAs wafer. The 2DEG carrier concentration was $4.7 \times 10^{15} m^{-2}$ and its 4.2K mobility was $25 m^2 V^{-1} s^{-1}$. An active area (0.3x0.3mm) was defined by etching the layers and electrical contacts made by diffusing indium. A spatially resolving CdS bolometer (described elsewhere in these proceedings) was fabricated on the sample. Electrical pulses of 100ns duration with power densities between 0.1 and 10 Wmm^{-2} were applied to the device. A reflectometer was used to measure the power input and device resistance during the pulse which was compared with measurements of resistance as a function of temperature to obtain a value for T_e.

Figure 1 shows the heat pulse detected using the above arrangement at 0.1 Wmm^{-2} input ($T_e \approx 40K$) the fast rise and the long decaying tail are typical of all traces obtained in this experiment. From the time of flight we conclude that we are observing only TA mode phonons. In GaAs only LA modes couple to electrons via the deformation potential. The emission of all modes is possible via the piezoelectric interaction. However, because the deformation potential coupling to LA modes is much stronger, if the signal we observe were due to piezo- coupled TA modes then we should expect to see a strong LA mode also. Furthermore average energy loss rates due to acoustic phonon emission, of the order 10^{-13} W/electron, are unable to account for the minimum electrical power input of 2.5×10^{-11} W/electron.

Figure 2 shows the emission anisotropy at $T_e = 125K$. The peak around $\theta = 0^0$ is due to the phonon focusing of TA modes in GaAs. However, compared with the pattern obtained with a 3D metal film heater the peak is rounded and less pronounced.

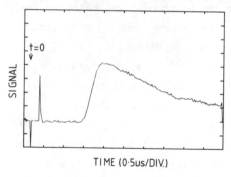

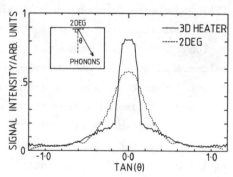

Fig.1 Heat pulse signal from an electrically heated 2DEG in a GaAs/AlGaAs heterojunction.

Fig.2 Anisotropy of phonon intensity in the (110) plane, θ is the angle between the emission direction and the source normal.

We attribute the TA signal to the decay products from optic phonon emission. After emission the optic modes decay to zone boundary LA modes and then to TA modes, this process takes place in less than 100ns [3]. The TA modes cannot decay further and reach to the detector. The shape of the heat pulse and the rounded phonon focusing pattern are characteristic of the propagation of dispersive TA modes in GaAs [4].

Upon application of a quantising magnetic field cyclotron phonon emission is possible[5]. However high frequency cut—off's due to the restrictions of momentum conservation perpendicular to the plane of the 2DEG during the emission process make this even less efficient than acoustic phonon emission in zero field. The emission of optical phonons is still therefore the primary energy loss process in our power range. Figure 3 shows the intensity of the emitted pulse as a function of magnetic field. The peaks observed are attributed to magnetophonon resonant emission of LO phonons at $n\hbar\omega_c = \hbar\omega_{LO}$[6].

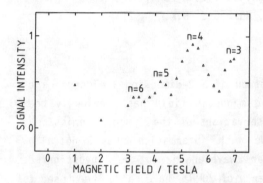

Fig.3 Phonon intensity as a function of magnetic field applied normal to the 2DEG. Magnetophonon resonances of order n are indicated.

This work is being supported by a grant from the Science and Engineering Research Council of the UK.

REFERENCES

1) Chin M A, Narayanamurti V, Stormer H L and Hwang J C M (1984) Phonon Scattering in Condensed Matter vol. IV, ed. W Eisenmenger, K Lassman and S Dottinger (Berlin:Springer) p328

2) Vass E, Solid State Commun. 61, 127 (1987).

3) Ulbrich R G (1985) Nonequilibrium Phonon Dynamics, ed. W E Bron (New York:Plenum) p101.

4) Wolfe J P and Northrop G A, (as ref. 1)

5) Kent A J, Rampton V W, Newton M I, Carter P J, Hardy G A, Hawker P, Russell P A and Challis L J, Surf. Sci. 196, 410 (1988).

6) Hawker P, Kent A J, Challis L J, Henini M and Hughes O H, J. Phys.: Condens. Matter 1, 1153 (1989).

THE INTERACTION OF 9.3GHz ULTRASONIC WAVES WITH THE 2-DEG
OF A GaAs HETEROJUNCTION

P.J.A.Carter, V.W.Rampton, M.I.Newton*, K.McEnaney,
M.Henini and O.H.Hughes
Department of Physics,
University of Nottingham, University Park,
Nottingham, NG7 2RD, U.K.
* Department of Physical Sciences,
Trent Polytechnic, Clifton Lane,
Nottingham, NG11 8NS, U.K.

1.INTRODUCTION

The electron-phonon interaction of a 2-dimensional electron gas in the presence of a quantising magnetic field has previously been studied using heat pulses and observations of the phonon emission[1]. Although these techniques provide much information they do not allow effects specific to a particular wavelength to be investigated. Surface acoustic waves at a frequency of 70MHz have also been used to probe a 2-DEG[2]. At this frequency it is appropriate to describe the electron-phonon coupling in terms of macroscopic variables. We have conducted experiments using 9.3GHz bulk longitudinal ultrasonic waves. Propagation down both the [100] and [111] directions has been studied in magnetic fields up to 2 Tesla and temperatures down to 2K.

2.EXPERIMENTS

Heterojunctions were grown by MBE on the (100) face of a 5mm thick semi-insulating GaAs wafer. Hall measurements made at 4.2K indicate a mobility of $20m^2/Vs$ and carrier density of $4.8 \times 10^{15}m^{-2}$. For propagation in the [100] direction a 3mm

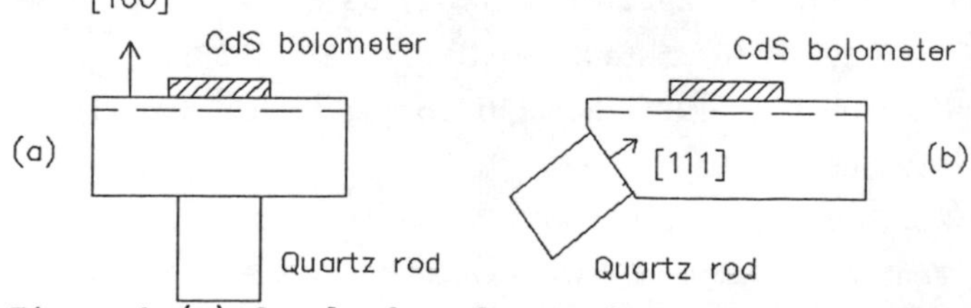

Figure 1 (a) Sample for ultrasonic propagation along [100] (b) Sample for ultrasonic propagation along [111].

diameter quartz rod was cemented with epoxy to the polished back face of the sample. The ultrasonic waves had normal incidence to the 2-DEG, figure 1(a). In the case of propagation down the [111] direction the ultrasonic waves are incident at 35.3° to the 2-DEG, figure 1(b). The other end of the quartz rod was inserted into a resonant microwave cavity that could be supplied with 500nsec microwave pulses. The ultrasound was detected by a cadmium sulphide bolometer[3] fabricated on the MBE grown face of the sample.

3.RESULTS

The height of the first ultrasonic echo detected by the bolometer for propagation in the [100] direction as a function of magnetic field can be seen in figure 2. This indicates an increase in the overall attenuation that reaches 18% by 2T. Figure 3 shows a result for an ultrasonic wave propagated down the [111] direction, also shown are Hall measurements made on the same

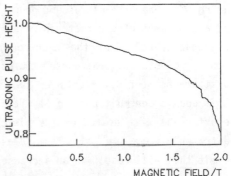

Figure 2 Detected ultrasonic pulse height for [100] propagation

sample. In the magnetic field region between 0 and 1T there is an overall increase in the attenuation of about 20%. Above 1T oscillations are observed as a function of magnetic field. These reach a magnitude of 6% by 2T and are superimposed on a relatively flat background. The echo height maxima of these oscillations correspond to the plateaus in the Hall measurements. In addition to these features there is a pronounced echo height minimum at $0.33\pm0.02T$, this has a magnitude of 5%. Rotation of the magnetic field indicated that all observed features depended on the

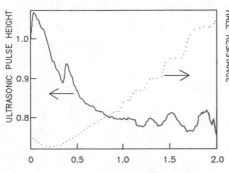

Figure 3 Detected ultrasonic pulse height for [111] propagation

component of magnetic field normal to the plane of the 2-DEG.

4.DISCUSSION

Ultrasonic waves propagating down the [100] direction produce only compressional strains relative to the 2-DEG. The electron energy is a function of strain and we might expect an attenuation if the period of the ultrasound is comparable to the relaxation time to restore thermal equilibrium, this will increase with the magnetic field. However it seems likely that at 2K the thermal relaxation time in zero field is already longer than the period, so we infer that the electron temperature is raised. When the ultrasound propagates down the [111] direction there is a component of shear strain relative to the 2-DEG. This can couple piezoelectrically to the electrons, so the attenuation follows the density of states. This causes the oscillations observed above 1T. The distance between wavefronts in the plane of the 2-DEG is 700nm. When this equals odd multiples of half the diameter of the electron orbit the electron is accelerated by the piezoelectric field associated with the ultrasonic wave. In the semi-classical, high electron screening limit this will result in an increase in the 2-DEG conductivity, which corresponds to a decrease in attenuation[4]. Conversely even multiples of half the diameter of electron orbit correspond to attenuation maxima, the effect is known as the acoustic geometric resonance. For the carrier density of our samples the principle attenuation maximum is expected at about 0.32T, this accounts for the feature observed at 0.33T. In 3D it is necessary to add a phase factor to the effective wavelength to predict the exact magnetic field position of the attenuation maxima. It is a requirement for this effect that the electron completes most of an orbit before being scattered, this can be achieved in a high mobility heterojunction.

5.ACKNOWLEDGEMENTS

This work was supported by the SERC of the U.K.

6.REFERENCES

1)Challis L.J., Kent A.J and Rampton V.W., Int. conf. The Application of High Magnetic Fields in Semiconductor Physics, Wurzburg (1988), in press.
2)Wixforth A. and Kotthaus J.P., Phys.Rev.Letts.56,2104 (1986)
3)Rampton V.W. and Newton M.I., J.Phys.D21,1572 (1988)
4)Cohen M.H., Harrison M.J. and Harrison W.A., Phys.Rev.117,937 (1960)

ZONE CENTER PHONON GAP IN THE COMMENSURATE PHASE OF ^{3}He ADSORBED ON GRAPHITE

V.L.P.Frank[1,2], H.J.Lauter[1], H.Godfrin[1] and P.Leiderer[2]

1 Institut Laue-Langevin, BP 156X, F-38042 Grenoble, France
2 University of Konstanz, D-7750 Konstanz, West Germany

Rare gases adsorbed on the basal plane of graphite are well defined systems which allow a detailed study of two-dimensional (2D) matter. Recently, measurements of the lateral variation of the adsorption potential have been performed on hydrogen isotopes, methane and nitrogen in the commensurate phase[1- 4], providing quantitative values against which the theoretical models of the adsorbate-substrate interaction can be tested. The lack of translational invariance in the commensurate phase produces a gap at the zone center in the acoustic branch of the phonon dispersion relation. The magnitude of this gap at low temperatures is a measure of the corrugation of the adsorption potential.

The case of ^{3}He is of particular interest due to its high quantum character. ^{3}He displays a commensurate ($\sqrt{3} \times \sqrt{3}$)R30 phase at adequate temperatures and coverages[5]. The theory of the stability of the commensurate phase against an intrinsic incommensurate one has been studied by Ni and Bruch[6]. The main difficulties are the limited knowledge of the corrugation and the large zero point motion of these light atoms. Clearly, the determination of the phonon gap would be useful for further development of the theories. Unfortunately, this quantity cannot be obtained by heat capacity measurements (see the discussion in ref. 5) and it is experimentally very difficult to perform neutron inelastic measurements on this system due to the large neutron absorption cross section of ^{3}He.

The measurements were performed on the three axis spectrometer IN14. The instrument was used in the fixed final wavevector (K_F) mode, with $K_F = 1.56$Å^{-1}, a Be filter after the sample and a horizontally curved analyzer. The energy resolution at zero energy transfer was 0.055THz (Vanadium standard). (1THz = 48.28K).

The sample consisted of 31.70g of Papyex[7] sheets oriented with their c–axis normal to the scattering plane. The total surface area was determined by adsorption isotherms and neutron diffraction to be 730m^2 ± 2%. The amount of ^{3}He corresponding to the commensurate coverage was 172cc STP, which defines a relative coverage $\rho = 1$.

Usually difference spectra are presented, where the scan taken with the graphite sample before adsorbing the gas is subtracted as background. With ^{3}He this is not longer possible, since its large neutron absorption modifies drastically the signal coming from the graphite substrate. An alternative solution is to use the data obtained at the same coverage, but higher temperatures where the layer melts (above 3K for ^{3}He[5]). This, however, produces a broadening of the elastic peak due to the higher diffusion coefficient of the fluid (quasielastic scattering). At still higher temperatures desorption occurs and again the signal from the substrate is modified. Thus, we present here data were no background has been subtracted.

The graphite sample, and hence the adsorbed ^{3}He layer, are essentially a 2D powder, since the in-plane crystallographic directions are not oriented. Thus, the measured signal consists of a directional average over all phonons for which scattering is allowed. The scan taken at a momentum transfer Q = 1.70Å^{-1} focuses on the phonons at the zone center and the transverse phonons at the zone boundary, and the one at Q = 0.85Å^{-1} focuses on the longitudinal zone boundary phonon[1, 2, 8].

Figure 1 depicts two spectra of the ^{3}He layer for $\rho = 0.9$, one at 0.85K and one at 14K (for Q = 1.70Å^{-1}). At low temperatures an asymmetric peak appears, well separated from the elastic line, with a low energy edge at ~0.27THz. No other features are discernible up to 2THz. The shape of this peak is similar to the one observed for D_2[1,2], but due to the low signal produced by the ^{3}He layer it is very difficult to resolve the gap and the transverse phonon. The bare graphite background does not show any peaks in our measuring range. To confirm that this peak is due to the commensurate solid the same scan was performed at 14K where the ^{3}He is a 2D fluid and only a small amount of gas is desorbed. As expected, this scan does not show any discernible peak (fig.1).

According to theoretical calculations[6] the gap should be ~0.7THz. However, as claimed by the authors, the neglect of relaxation and correlations in this model may overestimate its magnitude. It is not expected that these corrections will reduce the calculated gap below 0.1THz, our experimental lower energy limit due to the high intensity of the elastic line. Thus, we identify the lower edge of the measured peak with the zone center phonon gap.

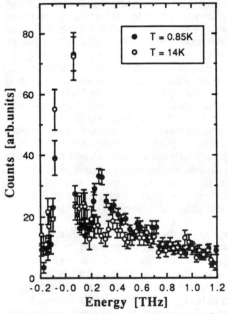

Figure 1: Neutron inelastic spectra for Q=1.70Å-1 at two temperatures and a coverage ρ = 0.9. The gap is found at 0.27 THz and disappears when the layer is fluid at 14K.

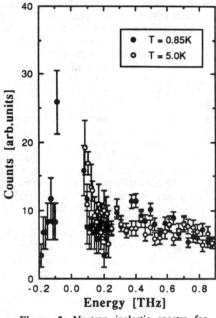

Figure 2: Neutron inelastic spectra for Q=0.85Å-1 at two temperatures and coverage ρ = 0.9. This scan focusses on the longitudinal zone boundary phonon.

The spectra taken at Q=0.85Å-1 are shown in figure 2. Some intensity is seen at ~0.38THz in the scan at low temperature which migth be due to the longitudinal zone boundary phonon. No trace of the peak at low energy can be seen, confirming its identification as the phonon gap. At 5K the adsorbed layer is melted and the intensity attributed to the longitudinal peak has disappeared (fig.2).

Using an Einstein excitation energy of ~0.3THz as an approximation to the real density of states, one can evaluate the root-mean square displacement of a ^{3}He atom to be 0.18 of the nearest neighbour distance (L = 4.26Å). This can be compared with the value of 0.16 calculated by Ni and Bruch[6].

To investigate the melting of the commensurate phase one should in principle take inelastic neutron spectra as a function of temperature. This is not possible in practice due to the very long counting times needed to obtain an acceptable statistics. Therefore, we performed measurements at two defined energies: 0.17 and. 0.26THz, the minimum and maximum of the peak of figure 1. Their difference gives an estimation of the gap´s magnitude, provided that the gap´s energy varies little with temperature. This has been checked with two scans taken at 0.85 and 2.50K and is also predicted theoretically. The results are shown in figure 3 and one observes clearly that at 3K the gap can no longer be resolved. This agrees with the melting temperature determined by specific heat measurements (the ^{3}He layer melts at 3K for $\rho=1$ and at about 2.7K for $\rho=0.9$ [5,9]).

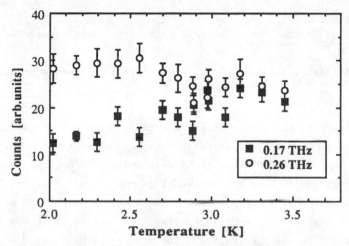

Figure 3: Neutron intensity vs. temperature for $Q = 1.70\text{Å}^{-1}$ at two energies. The energies are choosen to represent the temperature dependence of the phonon gap of figure 1. The difference vanishes and melting occurs at 3K. (Coverage $\rho = 1$).

In conclusion, we have determined the phonon gap of the commensurate phase of ^{3}He adsorbed on graphite ($\Delta = 0.27$THz) and the width of the phonon density of states is roughly estimated to be 0.11THz. These values allow a quantitative test of present theoretical models and particularly quantum Monte Carlo simulations of the ^{3}He on graphite system which are currently developed to understand 2D nuclear magnetic behavior at low temperatures[10].

This work has been partially supported by the West Germany Federal Ministry of Research and Technique (BMFT).

[1]   Frank V.L.P., Lauter H.J. and Leiderer P., Phys.Rev.Lett. 61,436 (1988)
[2]   Lauter H.J., Frank V.L.P., Leiderer P. and Wiechert H., Physica B. 156&157, 280 (1989)
[3]   Hansen F.Y., Frank V.L.P., Taub H., Bruch L.W., Lauter H.J. and Dennison J.R., submitted to Phys.Rev.Lett.
[4]   See also papers presented at this conference.
[5]   Bretz M., Dash J.G., Hickernell D.C., McLean E.O. and Vilches O.E., Phys.Rev. A 8, 1589 (1973)
[6]   Ni X.-Z. and Bruch L.W., Phys.Rev. B 33, 4584 (1986)
[7]   Papyex is produced by Carbone Lorraine, 45 Rue des Acacias, F-75821 Paris Cedex 17
[8]   Taub H., Carneiro K., Kjems J.K. and Passell L., Phys.Rev B 16, 4551 (1977)
[9]   Hering S., PhD. thesis, University of Washington, 1974
[10] Godfrin H., Ruel R.R. and Osheroff D.D., Phys.Rev.Lett. 60, 305 (1988)

PHONON EMISSION IN THE QUANTUM HALL REGIME

F F Ouali, N P Hewett*, P A Russell and L J Challis

Department of Physics, University of Nottingham, Nottingham, NG7 2RS

* Now at British Telecom Research Laboratories, Martlesham Heath,
  Ipswich

In a field B, the states of a two-dimensional electron gas (2DEG) form broadened Landau levels separated by $\hbar\omega_c$ (ω_c = Be/m*; ν_c =147B GHz for Si). When E_F lies between two levels and $kT_e \ll \hbar\omega_c$, ρ_{xx} ->0 and the Hall resistivity R_H = h/ie^2 where i is the number of filled Landau bands. The large difference in Hall voltage across the 3D contacts and the 2DEG causes the source-drain current to enter and leave the 2DEG at two diagonally opposite corners and the source-drain resistance $R_{SD}=R_H$[1]. The situation, assuming bulk flow is shown schematically in Fig 1. Since $\underline{E}.\underline{J}$ = 0, the current flows along the equipotentials over most of the sample and the resistance R_{SD} and dissipation I^2R_{SD} are likely to be in the corners where the current has to move between equipotentials to reach the contacts. The power necessary to approximate to these conditions is too small for pulses and studies have been made using a CW technique; fig 2. The 2DEG measuring 2 x 3mm^2 is within a Si sample measuring 0.4 x 5 x 20mm^3 connected in vacuo through a copper link to a 1K helium bath. The temperature is measured at 3 points on the surface facing the 2DEG by 4 thermometers (2 matched pairs) mounted in cages 15cm above the sample and connected by Cu wires to two 0.25mm^2 squares and a strip of silver foil attached to the Si with GE varnish. The contacts are (a) opposite the middle of the 2DEG, M, (b) opposite one corner, C, and (c) close to the bath contact, R. T_M and T_c measure the emission from the middle and corner of the 2DEG, which occurs in directions close to the normal because of momentum conservation and phonon focussing.

Fig 3(a) shows these temperatures (relative to T_R) as a function of n_s for B = +6T and $P = I^2 R_{SD} = 360\mu W$. Whenever the system is in the Quantum Hall regime, the middle cools and the corner warms up demonstrating the movement of the dissipation. T_M is unaffected by field reversal, fig 3(b), but T_c falls and now oscillates in phase with T_M confirming that the current entry and exit points switch to the other two corners. No detectable change in oscillation amplitude occurs to ($\lesssim 3\%$) when the current direction is reversed suggesting that the dissipation is equally divided between the entry and exit corners. Any decrease in the total phonon current obtained from T_R when the system enters the Quantum Hall regime is $<5 \times 10^{-4} P$ and this places an upper limit to the increase in far infra-red emission from the corners. This FIR has been seen in GaAs[2] but not yet in Si.

The dissipation mechanism is still unknown but it may occur by the QUILLS[3] process : as a result of the strong E-fields , electrons can tunnel into empty Landau states separated by $\Delta y = \{(2i + 1)^{\frac{1}{2}} + (2i' +1)^{\frac{1}{2}}\} l_B$ where $l_B = (\hbar/eB)^{\frac{1}{2}}$ · The two states differ by $\Delta k_x = \Delta y / l_B^2$ and tunnelling is accompanied by phonon emission with $q_{11} = \Delta k_x$ to conserve in-plane momentum. The hot electrons then relax by emitting cyclotron phonons and photons. For the conditions used, the two types of phonons have comparable energy . This mechanism could not however account for the dissipation at low currents, $eV_H < \hbar\omega_c$.

We are very grateful to Mr J Cooper for his generous help with the experiments, Dr C M K Starbuck and his colleagues, Dept of Electronics and Information Engineering, Southampton University for samples and SERC and the GEC Hirst Research Laboratory for financial support.

REFERENCES

1. Fang F F and Stiles P J., Phys Rev, B27, 6487, (1983).
2. von Klitzing K, Ebert G, Kleinmichel N, Obloh H, Dorda G and Weimann G, Proc 17th Int Conf on Physics of Semiconductors, San Francisco, eds J D Chadi and W A Harrison (Springer, Berlin), 471, (1984).
3. Eaves L and Sheard F W., Semicond. Sci Technolog., 1, 346, (1986).

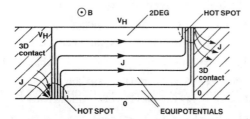

Fig 1: Equipotentials and current flow in the Quantum Hall regime.

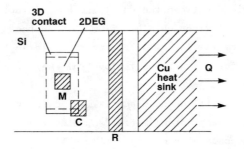

Fig 2: Thermometer contacts M, C and R on opposite face to 2DEG. The Si wafer is connected to a He bath by a Cu link.

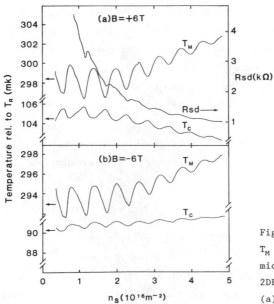

Fig 3: R_{SD} and temperatures T_M and T_C opposite the middle and one corner of a 2DEG, $P=360\mu W$, $T_R=1.57K$. (a) B +6T (b) B -6T.

REFLECTION OF PHONONS BY THE 2-DEG OF A SI-MOSFET AND A
GaAs HETEROJUNCTION

V.W.Rampton, Y.bin Wahab[*], M.I.Newton[#], P.J.A.Carter,
K.McEnaney, M.Henini and O.H.Hughes
Department of Physics,
University of Nottingham, University Park, Nottingham, NG7 2RD, U.K.
* Department of Physics, Faculty of Science,
University of Technology of Malaysia,
Karung Berkunci 791, 80990 Johor Bahru,Malaysia.
Department of Physical Sciences, Trent Polytechnic, Clifton Lane,
Nottingham, NG11 8NS, U.K.

1. INTRODUCTION

We have made experiments to study the reflection of ballistic heat pulses by a 2-DEG; we used an aluminium superconducting tunnel-junction as detector and our heater was excited by a current pulse. We found that the reflected signal increased with electron density in a Si-MOSFET but decreased in a GaAs/AlGaAs heterojunction. A similar experiment has been reported by Hensel et al.[1]. They used a Si-MOSFET and a constantan film heated by a laser pulse as the generator and an aluminium superconducting bolometer as the detector. They found a 2% decrease in detected signal when the electron sheet density was increased. The full interpretation of their results[2] took into account reflection from the 2-DEG, reflection from the Si/SiO_2 interface, phonon mode conversion and interference between the various reflected beams.

2. EXPERIMENTS

Two samples were used, one was a Si-MOSFET and the other a

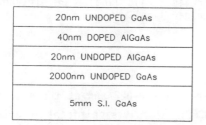
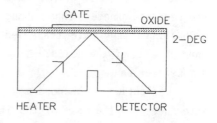

Figure 1(a) GaAs layer structure **(b)** Heater and detector geometry.

GaAs/AlGaAs heterostructure. The samples were grown on wafers which were 5mm. thick. The Si-MOSFET had an oxide thickness of 800nm which gave a charging rate of $2.67 \times 10^{14} m^{-2} V^{-1}$ and a threshold voltage of -5 V. The GaAs/AlGaAs layer structure is shown in figure 1(a), layers of SiO and gold were evaporated on top to form a gate. The electron sheet density could be increased by a small amount by the gate voltage from a value of about $9 \times 10^{15} m^{-2}$. The arrangement of heater and detector for both specimens is shown in figure 1(b). Experiments were made at 1.06K using 500nsec current pulses through the heater.

3. RESULTS

The detected signal in the case of the Si-MOSFET was very small when the gate was biassed to threshold but when a positive gate voltage was applied the detected signal increased rapidly as shown in figure 2. In the case of the GaAs/AlGaAs heterostructure the detected signal at all gate voltages was much larger than that from the Si-MOSFET and decreased as the gate voltage was increased. The results are shown in figure 3.

4. DISCUSSION

4.1 Detection Spectrum.

The aluminium tunnel junction detects phonons at frequencies above about 70GHz. Isotope scattering of phonons in both Si and GaAs[3] removes phonons from the ballistic pulse which have frequencies greater than about 600GHz when they have travelled through 17mm of the solid.

4.2 Reflection

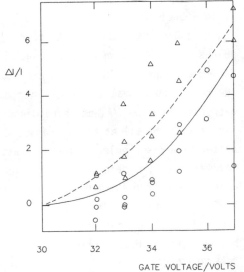

Figure 2 Relative change in detected ballistic phonon pulse in Si-MOSFET. Circles (solid line) - LA. Triangles (dashed line) - TA.

Reflection of phonons may occur at the 2-DEG and at the material interfaces. In the case of the Si-MOSFET the only significant interface is the Si/SiO_2 interface. Any phonons entering the oxide, which is amorphous, will be scattered and will not return in the reflected ballistic pulse. Acoustic mismatch at a perfect Si/SiO_2 interface implies a reflection of a few percent.[2] However the interface is probably rough on the scale of phonon wavelengths at frequencies above 70GHz. Thus we do not expect to observe appreciable specular phonon reflection from the Si/SiO_2 interface in the ballistic pulse. Hence in the case of the Si-MOSFET we observe a ballistic pulse reflected by the 2-DEG; the intensity increases with the electron sheet density. Hensel and Dynes[1] detector was capable of detecting low frequency phonons specularly reflected by the Si/SiO_2 interface as well as by the 2-DEG. The GaAs/AlGaAs heterostructure on

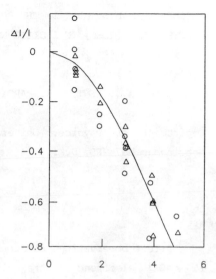

GATE VOLTAGE/VOLTS

Figure 3 Relative change in ballistic phonon pulse in GaAs heterojunction. Circles - LA. Triangles - TA.

the other hand has MBE grown interfaces which are of high quality. The acoustic mismatch between GaAs and AlGaAs with 30% Al is very small but the surface in contact with SiO should give a very large reflection as the mismatch is large[4]; at an angle of incidence of 54° the TA mode is totally reflected while the LA mode is near to total reflection. Our GaAs/AlGaAs heterojunction experiment gave a large intensity reflection from the GaAs/SiO interface; this reflection is reduced by phonon scattering by the 2-DEG as the electron sheet density increases.

5. ACKNOWLEDGEMENTS

We thank Prof L.J.Challis, Dr.A.J.Kent and Dr.F.W.Sheard for discussions, Dr.C.M.K.Starbuck(Southampton) for making the MOSFET and S.E.R.C. for financial support.

6. REFERENCES

1) Hensel J.C., Dynes R.C. and Tsui D.C., Phys.Rev.B28,124 (1983)
2) Hensel J.C. Halperin B.I.and Dynes R.C., Phys.Rev.Lett.51,2302 (1983)
3) Tamura S., Phys.Rev. B30,849 (1984)
4) Fröhlich H. and Stab H., Acta Phys.Slov.37,223 (1987)

THE SPATIAL ANISOTROPY OF THE PHONON EMISSION BY A HOT TWO DIMENSIONAL
ELECTRON GAS (2DEG) IN A SILICON DEVICE IN A QUANTISING MAGNETIC FIELD

A.J. KENT, G.A. HARDY, P. HAWKER AND D.C. HURLEY

Department of Physics, University of Nottingham, University Park,
Nottingham NG7 2RD, U.K.

Direct observation of acoustic phonons emitted by moderately hot
($T_e \lesssim 40K$) electrons in the silicon inversion layer can give more
detailed information regarding the electron-phonon interaction than
could hitherto be obtained from electron transport measurements. The
phonon emission is expected to display spatial anisotropies which can
be seen to arise from the application of energy and momentum
conservation rules to the emission process: The maximum component of
the phonon wavevector parallel to the 2DEG ($q_{//}$) is $2k_F$, where k_F is
the radius of the Fermi circle in k-space. This restricts the emitted
phonons to within a cone of semi-angle $\sin^{-1}(2k_F v_s/\omega_q)$ to the normal to
the 2DEG, where v_s is the phonon velocity and ω_q its frequency.
Evidence of this $2k_F$ cut-off has been observed experimentally[1] using
frequency selective phonon detectors. In the presence of a quantising
magnetic field the emission of cyclotron phonons associated with
inter-Landau level electronic transitions is possible, evidence of this
has been obtained by using the isotope scattering in Si as a low pass
phonon filter[2]. Localisation of the electrons on the scale of the
magnetic length is expected to further restrict $q_{//}$ with the effect of
concentrating the emission to still smaller angles[3].

We have used phonon imaging techniques to produce linescans of the
emitted phonon intensity which we compare with monte-carlo simulations
of the phonon focusing patterns obtained when the phonon emission is
assumed to have an angular dependence as calculated theoretically for a
hot 2DEG in Si.

The (100) Si MOSFET had a gate area of 1mm x 1mm and a mobility of $4500 \text{cm}^2 \text{V}^{-1} \text{s}^{-1}$. On the opposite surface of the 5mm thick $1000 \, \Omega \text{cm}$ Si wafer we fabricated a spatially resolving CdS bolometer (described elsewhere in these proceedings). The 2DEG was heated by applying 100ns duration electrical pulses to the drain source electrodes.

Figure 1 shows the detected intensity for transverse (TA) mode phonons as a function of position for a line corresponding to phonon propagation in the (110) plane, θ is the phonon propagation direction relative to the normal to the 2DEG, <001>. Results are shown for a MOSFET in zero field and at 4 Tesla, also for comparison we show the anisotropy using a 3D metal film heater in place of the 2DEG. All results were obtained for a constant electrical input of 40mW.

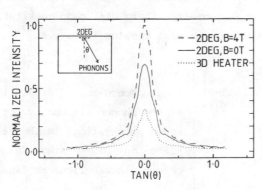

Fig.1 Linescans of the TA phonon intensity for emission by a hot 2DEG, $N_S = 7 \times 10^{15}$ electrons/m^2, and a 3D heater. (Sample temperature=2K).

The dominant feature in the anisotropy, a large peak around $\theta = 0^{\circ}$, is due to phonon focusing of TA modes in the Si crystal. However it is clear that the peak is significantly more intense in the case of the MOSFET and even more so in an applied magnetic field. As the power input is constant this indicates that the emission is being concentrated to smaller angles, probably at the expense of directions outside the range of our scan ($\theta = 0-45^{\circ}$).

Numerical calculations of the angular dependence of the phonon emission, in the absence of screening, have been performed following the method of Toombs et. al.[3]. We have taken account of the finite mobility in our device by approximating the density of states in the Landau levels with a Gaussian function. Figure 2 shows the results of this calculation for B=0 and 4T, normalised for constant power. The concentration of the emission to directions $\theta \leqslant 40^{\circ}$ is clear.

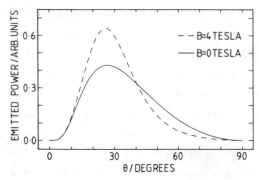

Fig.2. Theoretical calculations of phonon emission as a function of angle θ to the normal to the 2DEG. We have assumed an electron temperature, $T_e = 15K$.

These results have been used as the initial distribution of phonons as a function of q-vector direction in a monte-carlo phonon focusing simulation program [4]. Figure 3, shows the resulting predicted intensities for our experimental geometry. There is good qualitative agreement between these and the experimental results in fig. 1.

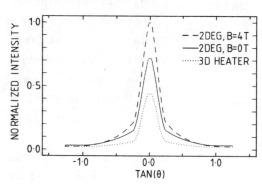

Fig.3 Monte-carlo simulations of the phonon intensity for TA modes propagating in the (110) plane in Si. Using initial phonon distributions as fig.2.

This work is being supported by a grant from the Science and Engineering Research Council of the UK.

REFERENCES

1) Rothenfusser M, Koster L and Dietche W, Phys. Rev. B34, 5518 (1986).
2) Kent A J, Rampton V W, Newton M I, Carter P J A, Hardy G A, Hawker P Russell P A and Challis L J, Surf. Sci. 196, 410 (1988).
3) Toombs G A, Sheard F W, Nielson D and Challis L J, Solid State Commun. 64, 577 (1987).
4) Northrop G A, Comp. Phys. Commum. 28, 103 (1982).

PHONON SCATTERING BY 2D ELECTRONS IN GaAs/AlGaAs
HETEROSTRUCTURES AT HIGH MAGNETIC FIELDS

S. Tamura and H. Kitagawa

Department of Engineering Science, Hokkaido University, Sapporo 060, Japan

We study theoretically the magneto-oscillations of thermal conductance observed recently by Eisenstein, et al.[1] in GaAs/AlGaAs heterostructures at low temperatures. To do this we calculate the thermal-phonon lifetime which is limited by the scattering with Landau-quantized electrons [2] at high magnetic fields. Although their direct contribution is very small (typically one part in 10^5 of the phonon contribution at 0.5 K),[1] the 2D electrons in the heterolayers can influence the heat conduction through the electron-phonon interaction. Let K and δK be the lattice thermal conductance of a GaAs/AlGaAs multilayer system and its modulation due to electron-phonon coupling, respectively. Thus, we have

$$\frac{\delta(\Delta T)}{\Delta T} = -\frac{\delta K}{K} ,$$

(1)

where $\delta(\Delta T)$ and ΔT are the temperature differences induced by the electron-phonon coupling in the heterolayers and other scattering mechanisms present in the system, respectively. The experimental result suggests that the magnitude of Eq.(1) is of the order of 10^{-2} or less.[1] The expression for the thermal conductance δK per unit volume is derived from an argument based on the Boltzman equation as

$$\delta K = -\frac{1}{V}\sum_{\lambda}\left[\tau_{\lambda}^{e\text{-}p}\right]^{-1}\tau_{\lambda}^2 C_{ph} v_{\lambda}^2 \cos^2\psi_{\lambda} ,$$

(2)

where V is the normalization volume, $\lambda \equiv (q,j)$ represents both wave vector q and mode j of phonons, $\tau^{e\text{-}p}$ and τ are the lifetimes of phonons limited by the electron-phonon interaction and other scattering mechanisms (τ is determined by boundary scattering at low temperatures), v is the sound velocity, ψ is the angle between the temperature gradient and sound-velocity vector, and C_{ph} is the heat capacity per normal mode of phonons.

In a magnetic field B applied normal to the interfaces of heterolayers (along z-axis), the transverse 2D motion (in x-y plane) of electrons is quantized and their energy spectrum is split into discrete Landau levels with energies E_n, where n is the Landau index. Taking these considerations into account, we find

$$\left[\tau_\lambda^{e\text{-}p}\right]^{-1} = 2\pi\omega_\lambda \sum_{\kappa\kappa'} \left|M_{\kappa\kappa'}^\lambda\right|^2 D_\kappa(E_F) D_{\kappa'}(E_F - \hbar\omega_\lambda) \quad , \qquad (3)$$

where M is the matrix element for the electron-phonon interaction (the sum of the deformation and piezoelectric potentials), D_κ denotes the DOS of electrons in the states specified by $\kappa \equiv (n,\sigma,k_y)$, σ discriminates the orientation of electron spin and k_y is the wave number in the y-direction.

In the experimental temperature range 0.1 to 1K, almost no phonons which can contribute to inter-Landau level transitions are excited since the typical energies of phonons and inter-Landau levels are 0.1meV (at T=1K) and 2meV (at B=1T), respectively. Hence, only intra-Landau-level transitions are responsible for the phonon scattering by 2D electrons. These transitions are indeed possible in the presence of scattering centers owing to the broadening of Landau levels, otherwise the sharp Landau levels prohibit the possibility of such transitions. In the presence of strong magnetic fields the DOS is a sum of Gaussian functions with a level width $\Gamma = \gamma B^{1/2}$, where γ is a coefficient independent of magnetic field.[2,3]

The screening of the electron-phonon interaction is another important effect to be taken into account because the wavelength of the relevant phonons is very long. An important aspect of the screening is the fact that the screening length is determined by the DOS at the Fermi level and hence it oscillates with magnetic field.[4] Thus the screening is very effective if a Landau level is nearly half-filled and it breaks down if the Landau level is either completely filled or empty. We shall find that this property of the screening of electron-phonon interaction is crucial in understanding experimental findings.

Figure 1 displays the calculated magnetic-field dependence of $\delta(\Delta T)/\Delta T$ at 0.35K. The numerical calculation was performed with the same parameters given by Eisenstein et al.[1] for their sample 2, e.g., the number of periods is fifty, and the 2D carrier density is $N_S = 5.3 \times 10^{11} \text{cm}^{-2}$. In our calculation the unknown prefactor of the Landau-level width was varied as a parameter and spin gaps were neglected. Figure 1(a) is a result obtained without screening effects, whereas Fig. 1(b) for three different values of the coefficient of the level width includes screening effects. Comparing with the experimental data, the amplitude of oscillation in Fig.1(a) is about thirty-five times larger than the experiment [from experimental data the amplitude of oscillation of $\delta(\Delta T)/\Delta T$ is 0.015 at B=4.5T],[1] while in the results including the screening effects the magnitudes are in accord with the experiment within a factor of two.

Figure 2 displays the temperature dependence of the oscillatory amplitude $\delta(\Delta T)$ caused by the electron-phonon interaction. The plotted $\delta(\Delta T)$ corresponds to the difference between the local maximum of the temperature difference at B=4.5T and its neighboring local minima. Here, we have chosen $\gamma = 1\text{meV T}^{-1/2}$ because the sample 2 used in the thermal-conductance measurement is practically identical to a sample used in the magnetization measurement,[5] and the latter result was fitted favorably with the same γ. The solid line at the top is the result obtained without screening, which exhibits the temperature dependence close to T^{-4} power law. This temperature dependence is much stronger than T^{-3}-behavior to which the experimental $\delta(\Delta T)$ approximately obeys from 0.2 to 0.8 K. More importantly, this calculation overestimates the strength of $\delta(\Delta T)$ by more than one order of magnitude at temperatures below 1K. In these respects, it is interesting to see whether inclusion of the screening effects lead to the correct temperature dependence of δ(ΔT) in addition to the overall magnitude. Indeed, the bold-solid line shows the result including screening effects, which describes semiquantitatively the expected temperature dependence as well as the magnitude.

To conclude, we have calculated the temperature difference in GaAs/AlGaAs heterostructures caused by the scattering of thermal phonons by 2D electrons at high magnetic fields. Screening effects are essential to reproduce quantitatively both the magnetic field and temperature dependence of the thermal conductance below 1K.

One of the authors (S.T) acknowledges the Foundation for C&C Promotion for financial support.

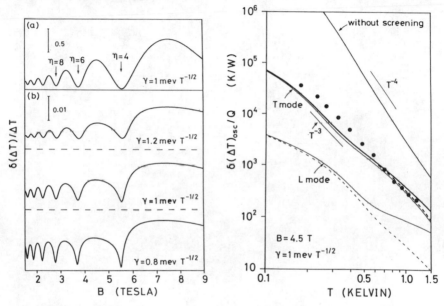

Fig.1 Magnetic-field dependence of temperature difference caused by electron-phonon interaction, (a) without screening effect and (b) with screening effect. Arrows indicate the magnetic fields at which the filling factor takes integer values.

Fig.2 Calculated temperature dependence of oscillatory amplitude δ(ΔT)osc at B=4.5T (bold line) together with experimental data (dots) from ref. 1. (Q is the applied heat flux.) The Landau-level width Γ= (1meV/T^{1/2})B^{1/2} = 2.12 meV is assumed. Dashed lines are the results obtained without deformation potential coupling.

1   Eisenstein, J. P. , Gossard, A. C. , and Narayanamurti,V. , Phys. Rev. Lett. 59,1341 (1987).
2   See, for example, Ando, T., Fowler, A. B., and Stern, F. , Rev. Mod. Phys. 54, 437 (1982).
3   Ando, T. , J. Phys. Soc. Jpn. 37, 622 (1974).
4   Ando, T. and Murayama, Y., J. Phys. Soc. Jpn. 54, 1519 (1985).
5   Eisenstein, J. P. , Stormer, H. L. , Narayanamurti, V. , Cho, A. Y. , Gossard, A. C. , and Tu, C. W. , Phys. Rev. Lett. 55, 875 (1985).

9

Excitations in Quantum Matter

PHONONS, ROTONS AND QUANTUM EVAPORATION IN LIQUID ^{4}HE

A F G Wyatt

Department of Physics
University of Exeter
Exeter
EX4 4QL
UK

1. INTRODUCTION

Quantum evaporation [1-4] by phonons and rotons in liquid ^{4}He has improved our knowledge of these excitations very considerably. It gives information on the lifetimes and scattering of the high energy excitations which has been impossible to obtain in other ways. For example we now know that the lifetimes of phonons and rotons can be some five orders of magnitude longer than the previously measured upper bound [5], and 4 phonon scattering has been observed [6].

The phenomenon of quantum evaporation is the annihilation of a single excitation at the surface of the liquid and the consequent liberation of a single atom into the space above the liquid. The energy of the excitation goes in overcoming the binding energy (7.16K) of the atom to the liquid and in giving the atom kinetic energy. The boundary conditions for this process have now been investigated in some detail and the basic picture of conservation of energy and parallel momentum is clearly established.

For rotons the process of quantum evaporation is particularly important as it enables rotons to be detected in the presence of much larger phonon fluxes which would dominate the signal from a detector in the liquid. Rotons are transformed into ballistic atoms at the free liquid surface which are readily detected, whereas rotons have a very low probability of transmitting their energy into a solid in the liquid. Properties of phonons and rotons are deduced from time of flight and angular measurements.

It is the possibility of propogating beams of phonons and rotons over many millimeters that direectly shows that these excitations have long lifetimes. The concept of elementary excitations or quasiparticles depends on their lifetime τ, or equivalently their uncertainty in energy $\delta\omega = \tau^{-1}$, where hω is the energy of the excitation. If $\omega\tau >> 1$, the excitation is well defined. It is therefore important to know τ for all the excitations under different conditions of pressure and temperature. Quantum evaporation is useful in the long lifetime limit.

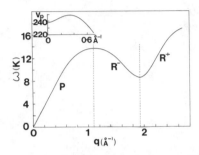

Fig. 1. The dispersion curve for liquid ^{4}He at Obar [7] showing the phonons (P) and rotons (R^- and R^+ which have negative and positive group velocities respectively). Inset is shown the phase velocity as a function of wave vector [8] which shows the upward dispersion.

The lifetimes of phonons and rotons might be limited by both spontaneous decay and scattering. Phonons with energies $\omega \leq 10K$, in ^{4}He at zero pressure, do spontaneously decay [9] due to the upward curvature of the dispersion curve (see fig 1). The lifetime varies as $\sim \omega^{-5}$ and is thought to be of order 10^{-10} s at $\omega \sim 9K$ [10,11]. For $\omega > 10K$ the lifetimes can be $> 10^{-6}$ s when the thermal phonon density is low. Quantum evaporation indicates that the majority of rotons are stable against spontaneous decay, but it may be that a narrow band of rotons also spontaneously decay to a roton and a phonon, which is only allowed if the roton group velocity exceeds the ultrasonic phonon velocity [12].

However it is clear from transport measurements such as viscosity and attenuation of second sound and from spectroscopy that there can be interactions between the elementary excitations. These scatterings can reduce the lifetimes to $< 10^{-11}$ s [5] when the thermal population of rotons is high. Neutron and Raman scattering can create pairs of excitations, [13, 14,15] albeit with low probability, associated with the high densities of states at the maxon and roton turning points. The Raman lifetime, which is the pair lifetime, is the same as the single roton lifetime for temperatures $T > 1.1K$. It has been suggested that the interactions lead to weakly bound pairs [16,17] but this seems to be an open question [18].

Phonons can interact with other phonons and with rotons. The interactions with phonons can be $P_1 + P_2 \rightarrow P_3$, i.e. the reverse of 3 phonon decay (3PP) and $P_1 + P_2 \rightarrow P_3 + P_4$ the 4 phonon process (4PP). Phonons above the energy ω_c can only interact by the relative weak 4PP [19]. Higher order processes are presumed to be even weaker. Phonons can be scattered by rotons, i.e. $P_1 + R_1 \rightarrow P_2 + R_2$

and $P_1 + R_1 \to R_2 + R_3$. If the liquid He is at a finite temperature then the scattering will be with the ambient phonons and rotons. For $T < 0.6K$ there are essentially no rotons and for $T > 1.2K$ rotons predominate, between these temperatures phonons will be scattered by both phonons and rotons.

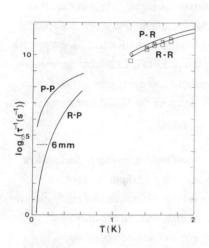

Fig. 2. The inverse lifetimes of high energy phonons and rotons due to scattering from thermal excitations as a function of temperature. The lines are theoretical; the first and second letters indicate the excitation being scattered and causing the scattering respectively. P - P, P - R and R - R from ref 19 and R - P ref 20. The squares and triangles are measured phonon ($q = 0.6 \text{Å}^{-1}$) and roton lifetimes respectively [5]. The dashed line labeled 6 mm indicates the inverse lifetime corresponding to 6 mm mean free path and velocity 240 ms^{-1}.

The scattering rate for a high energy phonon, of wavevector q in a thermal population of low energy phonons by 4PP is $\tau_{PP}^{-1} = 2.12 \ 10^{10} \ q^4 \ T^3 \ s^{-1}$ [19] (q is in Å^{-1} throughout) and the scattering angle is small. For the scattering of phonons by rotons ($P_1 + R_1 \to P_2 + R_2$), the scattering rate is proportional to the thermal roton density so $\tau_{PR}^{-1} = 1.34 \ 10^{14} \ q^4 \ T^{12} \ e^{-\Delta/T} \ s^{-1}$ [19]. These phonon scattering rates are indicated in fig 2. The attenuation of high energy phonons, $\omega > 10K$, is due to an interplay between 3PP and 4PP effects [21].

Rotons are scattered by phonons and rotons, i.e. $R_1 + P_1 \to R_2 + P_2$, $R_1 + P_1 \to R_2 + R_3$, and $R_1 + R_2 \to R_3 + R_4$ and $R_1 + R_2 \to R_3 + P_1$. If rotons are injected into liquid He at $T < 0.6K$ the roton-phonon scattering rate is $\tau_{RP}^{-1} = 1.9 \ 10^9 \ T^7 \ s^{-1}$ [21] and for $T > 1.2K$ the scattering by rotons is proportional to the thermal roton density and so $\tau_{RR}^{-1} = 9.81 \ 10^{12} \ T^{1/2} \ e^{-\Delta/T} \ s^{-1}$ [19] for $R_1 + R_2 \to R_3 + R_4$. These are shown in fig 2.

The ability to create and detect ballistic beams of excitations means we are free from just considering scattering amongst thermal populations of excitations with the averages over angles and wave vectors that obscures the details of the interaction. We can imagine the scattering of a beam of ballistic excitations by another beam. The first intersecting beams experiments have recently been made and will be discussed further in § 3.

2.   QUANTUM EVAPORATION

Consider an excitation incident on the free surface of liquid ^{4}He ,there are many possibilities and some are shown in figs 3 a and b. The one to one processes are the excitation i, speculary reflects ii, mode changes to another excitation and iii, evaporates an atom into the free space above the liquid. This last process is known as quantum evaporation. It is conceivable that besides evaporation a low energy excitation is also created. This might be a phonon in the bulk or a ripplon on the liquid surface as shown in fig 3b.

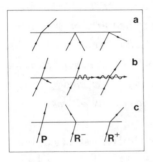

Fig. 3. Some processes due to an incident excitation at the free liquid surface.
a. evaporation, reflection and mode change;
b. evaporation and the creation of a phonon, ripplon and two ripplons;
c. evaporation by a phonon P, R^- and R^+ rotons, showing the atom directions.

The boundary conditions for all these possibilities is conservation of energy and component of momentum parallel to the surface. This latter condition arises from the translational symmetry of the surface. As the momenta of phonons and R^- and R^+ rotons (see fig 1) are very different even though their energies are similar, the directions of evaporated atoms are quite different for the same angle of incidence. As the R^- roton has its momentum oppositely directed to its group velocity, the atom is evaporated into the negative quadrant. This is shown in fig 3c.

An extensive range of measurements have established that quantum evaporation does indeed occur in a one to one process [4, 22]. The two conservation laws lead to 3 features, the total time of flight for the excitation and atom, the angle of refraction and the angular dispersion. This last property is basically the dependence of the angle of refraction on the excitation energy which is apparent as the measurements are made with a continuous spectrum of excitations.

The angles of refraction for different energy excitations, for an angle of incidence of 15°, is shown in fig 4 a. It can be seen that the atoms evaporated by phonons have almost no angular dispersion and are refracted a little towards the normal. In contrast both R⁻ and R⁺ rotons evaporate atoms at large angles to the normal and different energy rotons evaporate at quite different angles. (Note that the modulus of the refracted angle is used in fig 4 and that R⁻ roton-atom evaporation is at negative φ, as shown in the inset of fig 4a.)

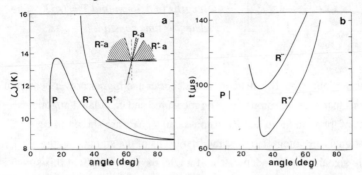

Fig. 4. a. The energy of the incident excitation is shown as a function of the modulus of the angle of the quantum evaporated atom, for an angle of incidence of 15º. The inset indicates the range of angles for a broad incident spectrum of excitations. b. The total time for the excitation and atom for equal path lengths of 6.5 mm is shown as a function of the evaporation angle.

The corresponding total times of flight are shown in figure 4b. The times are of course independent of angle and only depend on the excitation energy and the path lengths. However for a finite angle of incidence, energy is related to angle of refraction and so time can be shown against angle. Fig 4b shows that there is relatively little time of flight dispersion for phonon-atom signals. This arises because the dispersion of the phonons as a function of energy is opposite to that of the atoms, so there is a considerable degree of cancellation. In contrast the rotons disperse in the same way as the atoms and so the total time is a strong function of the excitation energy. The R⁺ rotons can have higher energies than phonons or R⁻ rotons and the group velocity is high at these high energies, so they produce the signals with the shortest times. The R⁻ rotons have similar energies to phonons but their group velocities are mainly slower, so R⁻ rotons produce the slowest signal.

The measured signals show the behaviour described above. For angles of incidence and refraction of 15º and 12º respectively, we expect a phonon-atom signal with little time dispersion and this can be seen in fig 5. For angles 15º and 40º we expect only a R⁺ roton-atom signal and this should be faster and more dispersed in time than the phonon-atom signal. This too can be seen in fig 5.

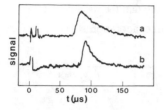

Fig. 5. a. R$^+$ roton-atom signal, -17 dB (ref 0.5W mm^{-2}) 10 μs; b. phonon-atom signal -14 dB, 0.05 μs; both for equal path lengths of 6.5 mm and T = 0.1K. The detector time constant is ~ 5 μs.

It has been found that it is better to use a shorter and higher power heater pulse to inject phonons than that for rotons [22]. The measured and calculated angular dependences of phonon-atom and R$^+$ roton-atom signals are shown in fig 6. The R$^+$ roton-atom signals are integrated over the first 90 μs of the signal. As the phonon-atom signal shape is independent of angle, the peak height of this signal is used to show the angular variation.

The peaks of the angular distributions occur at the calculated angles to within the experimental accuracy. For phonons it is possible to change the angle of incidence very considerably and in fig 6a it can be seen that the angle of refraction changes in a corresponding manner. Rotons cannot have a large angle of incidence and still evaportate atoms as the atoms cannot carry away the large component of momentum parallel to the surface. The R$^+$ roton-atom signals show a greater angular dispersion than the phonon-atom ones, as expected. The measured angular width of the phonon-atom distribution is essentially due entirely to the collimation. However the R$^+$ roton-atom distribution is much broader than the collimation width. These results and others are fully discussed elsewhere [22].

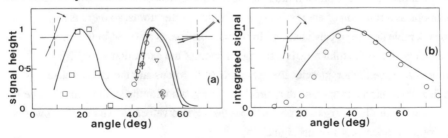

Fig. 6. a. Normalised signal height of phonon-atom signals as a function of the evaporation angle, for angles of incidence 25° (squares), 75° (triangles), 83° (circles). The solid lines are calculated. b. Integrated R$^+$ roton-atom signal as a function of evaporation angle for an angle of incidence of 14°.

The reader may have noticed that no examples of R- roton-atom evaporation have been given. This is because a heater in the liquid does not inject R- rotons into the ^{4}He and so R- roton-atom signals cannot be seen with the same experimental arrangements that shows phonon and R+ roton evaporation. R- rotons can be created by inelastic neutron scattering so they might be produced by condensing a beam of ^{4}He atoms. Quantum evaporation must be used to detect ballistic rotons, so the beam of R- rotons has to be reflected back to the liquid surface. The measured signals agree well with those calculated [23] which indicates that the boundery conditions apply equally well for R- rotons.

We conclude from this variety of measurements that quantum evaporation is well established although a few details still need clarifying [22]. There is a major unsolved question; why R- rotons are not produced by a heater in the liquid.

3. DISCUSSION

3.1 Roton Lifetimes

R+ roton-atom signals have a minimum total time which corresponds to R+ rotons of energy of ~ 14.0K for equal roton and atom path lengths. This energy is greater than the energy of the fastest rotons, ω ~ 12.3K. If the group velocity of these rotons is greater than the ultrasonic velocity then such a roton might spontaneously decay to a roton and a phonon [12]. If a substantial number of decays occurred over a distance of several millimeters then a hole would develop in the otherwise continuous spectrum of rotons. As roton number is conserved, the roton spectrum would increase at lower energies.

A hole in the roton spectrum will affect the received signal shape and will reduce the signal just after its start. Many roton-atom signals have been examined and there is no sign that this is happening which puts an upper limit on the hole of 2.21 - 2.27 Å^{-1} for roton decays in a liquid path of 6.5 mm [22]. If roton decay process exists, it is either very weak or only affects a very small range of roton wavevectors. The neutron measured $\omega(q)$ is not sufficiently good to say whether the maximum roton group velocity does exceed the ultrasonic velocity. Moreover the neutron line widths are too broad from instrument effects to show up decays which are other than strong.

Rotons can be scattered by thermal phonons and due to the small energy and momentum of phonons at T < 0.3K the roton normally only suffers a small change in energy and momentum at each collision. Such scatterings will broaden the energy spectrum of rotons, but it needs many scatterings to change the direction enough to reduce the chance of a roton finding a path to the detector. Some rotons show appreciable attenuation at T ~ 0.2K [24] which indicates that τ_{RP}^{-1} is much larger than one model suggests [20].

3.2 Phonon Lifetimes

Phonons with $\omega < \omega_c$ (10K) have very short mean free paths and do not contribute to the quantum evaporation signal when the liquid path is of order millimeters. Phonons with $\omega > \omega_c$ have very long mean free paths, > 10 mm, at T = 0. At finite ambient temperatures these phonons are scattered in the 4 phonon process which changes their energy and direction by small increments. Both of these affects reduce the signal, as the random walk in energy can take ω below ω_c where it decays strongly by 3PP,and the cumalitive change in direction either prevents the phonon getting through the collimation or evaporates an atom in the wrong direction to be detected. The theoretical 4PP mean free path is ~ 70 μm for $\omega = 11$K phonons at 0.1K. The indications are that the mean free paths are longer than this theory suggests [19], as there is little attenuation at this temperature [25].

3.3 Surface Effects

The surface of liquid ^{4}He must be structureless and at low temperatures, quite flat, as few ripplons are present. It is therefore not surprising that the one excitation to one atom process is dominant. However it is conceivable that besides the annihilation of the high energy excitation and the creation of a free atom, a low energy phonon or ripplon is created. This would have two effects. The energy of the atom would be reduced and so the signal would be slower, and the angle of the evaporated atom would change from the one to one value.

The angular width of the phonon-atom signal is essentially independent of the spectrum of phonon energies, as the atom direction is very close to the direction of the phonon. So any increase in width due to ripplon or phonon production should show here. In fact the measured width is somewhat less than we calculate, rather than wider. An order of magnitude upper limit on the wave vector of any created ripplon or phonon is ~ 0.1Å^{-1} [22].

4. SCATTERING EXPERIMENTS

One of the promising prospects opened up by quantum evaporation is the possibility of doing controlled scattering measurements. We can create beams of ballistic phonons and rotons and use dispersion to separate out different energies (although dispersion tends to cancel for phonon and atoms when the path lengths are equal, there is useful dispersion if the liquid path dominates). These beams can be made to intersect in the liquid and scattering effects measured.

The first such experiment has just been done in which a probe beam of phonons is propagated through a beam of rotons travelling predominantly at right angles to the phonon beam. The attenuation of the probe beam under different conditions enables the scattering cross section for phonon-roton collisions to be estimated. These results are presented in some detail in another paper to this conference [26].

5.   CONCLUSIONS

Quantum evaporation is now well established and accounts for a wide body of measurements of signal shape, arrival times and angular dependences. It has led to a better understanding of the injection processes for phonons and rotons [22].

As quantum evaporation is only sensitive to excitations with $\omega > 7.16K$ (the atom binding energy) it enables experiments to be made on these high energy excitations that are usually masked by the higher numbers of low energy phonons. Most importantly it allows ballistic rotons to be detected in the presence of large phonon fluxes.

The long excitation path lengths in the liquid ^{4}He enable the excitation lifetimes to be estimated. They are very long, at least 10^{-6} s at $T < 0.05K$. This emphasises how well the excitation picture describes ^{4}He.

Finally quantum evaporation is enabling scattering experiments to be done on ballistic excitations which should lead to a better understanding of the nature of the excitations, especially rotons.

6. ACKNOWLEDGEMENTS

Stimulating discussions with M Brown, A C Forbes, P Mulheran, M A H Tucker and G M Wyborn are warmly acknowledged.

7. REFERENCES

1. Baird, M J, Hope, F R and Wyatt, A F G, Nature $\underline{304}$, 325-6 (1983)
2. Wyatt, A F G, Baird, M J and Hope, F R, 75th Jubilee Conf on Helium-4 ed JGM Armitage, World Scientific Pub Co, Singapore, 117-22 (1983)
3. Hope, F R, Baird, M J and Wyatt, A F G, Phys Rev Letts $\underline{52}$, 1528-31 (1984)
4. Wyatt, A F G, Physica $\underline{126B}$, 392-9 (1984)
5. Mezei, F and Stirling, W G, Procs of the 75th Jubilee Conf on Liquid He, ed JGM Armitage, World Scientific Pub Co, Singapore, 111 (1983)
6. Baird, M J, Richards, B and Wyatt, AFG, Jap J App Phys $\underline{26-3}$, 387-8 (1987)
7. Cowley, R A and Woods, A D B, Can J Phys $\underline{49}$, 177-200 (1971)
8. Stirling, W, 75th Julbilee Conf on Helium-4, ed JGM Armitage, World Scientific Pub Co, Singapore, 109-10 (1983)
9. Maris, H J and Massey, W E, Phys Rev Letts $\underline{25}$, 220 (1970)
10. Slukin, T J and Bowley, R M, J Phys C: Solid State Phys $\underline{7}$, 1779-85 (1974)
11. Maris, H J, Rev Mod Phys $\underline{49}$, 341 (1977)
12. Pitaevskii, L T, Sov Phys JETP $\underline{9}$, 830-7 (1959)
13. Greytak, T J and Yan, J, Phys Rev Letts $\underline{22}$, 987 (1969)
14. Stirling, W G, Procs 2nd Int Conf on Phonon Physics, eds Kollar, J, Kroo, N, Menyhard, N and Siklos, T, World Scientific, 829 (1985)
15. Ohbyashi, K, Elementary excitations in quantum fluids, eds Ohbyashi, K and Watabe, M, Solid-State Sciences 79, Springer, 32-52 (1989)
16. Ruvalds, J and Zawadowski, A, Phys Rev Letts $\underline{25}$, 333 (1970)
17. Iwamoto, F, Prog Theor Phys (Japan) $\underline{44}$, 1135 (1970)
18. Iwamoto, F, Elementary excitations in quantum fluids, eds Ohbyashi, K and Watabe, M, Solid-State Sciences 79 Springer, 117-131 (1989)
19. Khalatnikov, K M, An introduction to the theory of superfluidity, W A Benjamin Inc NY, Amsterdam (1965)
20. Hyman, D S, Scully, M O, and Widom, A, Phys Rev $\underline{186}$, 231-8 (1969)
21. Wyatt, A F G, Jap J App Phys $\underline{26-3}$, 7-8 (1987)
22. Brown, M and Wyatt, A F G, to be published
23. Wyborn, G M and Wyatt, A F G, to be published
24. Wyatt, A F G and Brown, M, to be published
25. Tucker, M A H and Wyatt, A F G, Phonons 89 Conf Procs
26. Forbes, A C and Wyatt, A F G, Phonons 89 Conf Procs

TEMPERATURE DEPENDENCE OF THE PHONON-ROTON
EXCITATIONS IN LIQUID HELIUM-4

H. R. Glyde

Department of Physics, Univ. of Delaware, Newark, DE 19716,
U.S.A.

and

W. G. Stirling

Department of Physics, Keele University, Keele ST5 5BG, U.K.

and

S.E.R.C. Daresbury Laboratory, Warrington WA4 4AD, U.K.

Abstract

A new interpretation of the phonon-roton excitations in liquid
^{4}He is proposed. This is based on new detailed neutron scattering
measurements in both the superfluid and normal phases and on
comparison with other fluids.

****

Landau[1] and Feynman[2] proposed that superfluid ^{4}He supports
collective excitations. At low wave vectors Q these are phonons and
at higher Q, (Q = 1.925 Å^{-1}), rotons. They form part of the same
dispersion curve[1]. Our data suggests that the temperature dependence
of these two excitations is so different that the phonon and roton may
have a different origin. An interpretation of this difference can be
found in the dielectric formulation[3-6] of Bose liquids suggesting
that the excitations are collective phonons at low Q, but- single
quasiparticle excitations at higher Q (Q≥ 1.1 Å^{-1}).

Neutron scattering intensity from superfluid ^{4}He contains a sharp
peak superimposed on a broad background. The sharp component is
usually identified with collective excitation; the broad background
with excitation of pairs of weakly interacting particles . In Fig. 1
we show new data on the scattering intensity (the dynamic form factors

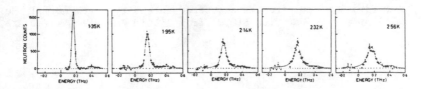

Fig. 1 Scattering intensity from liquid ^{4}He at Q = 0.4 Å^{-1}
(the phonon region). At T ≤ T_λ = 2.17 K, superfluid
phase, T > T_λ, normal phase.

$S(Q,\omega)$) of liquid ^{4}He at $Q = 0.4$ Å^{-1}. At low temperature ($T = 1.35$ K) the intensity is dominated by a single sharp peak with a very small, broad background at higher energy (ω). As T is increased, the sharp peak broadens. However, the basic shape of $S(Q,\omega)$ remains the same in the normal phase ($T > 2.17$ K); the intensity still lies predominantly within the same single peak. This suggests that $S(Q,\omega)$ is dominated by scattering from a single collective phonon in both the superfluid and normal phases.

In contrast, Woods and Svensson[7] and Talbot et al[8] presented data at $Q \geq 1.1$ Å^{-1} which suggested that the sharp component in $S(Q,\omega)$ disappeared at T_λ. That is, the observed intensity at low T in the superfluid phase contained a sharp peak superimposed on a significant broad background. As T was increased the intensity in the sharp peak decreased until it disappeared from $S(Q,\omega)$ at $T = T_\lambda$. In the normal phase ($T > T_\lambda$) $S(Q,\omega)$ contained only the broad component which was largely independent of temperature. The apparent disappearance of the sharp component from $S(Q,\omega)$ at T_λ, and the independence of $S(Q,\omega)$ with T for $T > T_\lambda$, is shown for the roton in our new data in Fig. 2.

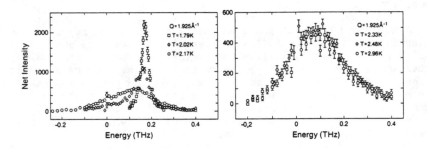

Fig. 2   Net scattering intensity from liquid ^{4}He at
        $Q = 1.925$ Å^{-1} (the roton region): Right,
        $T \leq T_\lambda$, the superfluid phase; Left, $T > T_\lambda$,
        the normal phase.

In Fig. 3 we compare the scattered intensity[8] from normal ^{4}He ($T \geq T_\lambda$) at $p = 20$ bar with $S(Q,\omega)$ from liquid ^{3}He at[9] $Q \approx 2.0$ Å^{-1}.

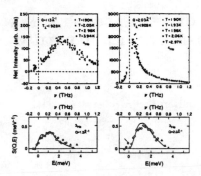

Fig. 3   Scattering intensity from normal ^{4}He at p = 20 bar at
         Q = 1.13 Å^{-1} and Q = 2.03 Å^{-1} (upper half); S(Q,E) in
         liquid ^{3}He at T = 40 mK (solid line) and T = 1.2 K
         (circles) (lower half).   Data from Refs. 8 and 9.

In both cases S(Q,ω) is broad and largely independent of T.   At these
Q values, ^{3}He does not support a collective excitation.   Rather S(Q,ω)
represents scattering from weakly interacting particle-hole
excitations.   The scattering at Q ≈ 1.2 Å^{-1} is broader in ^{4}He than in
^{3}He.   This suggests that normal ^{4}He also does not support collective
excitations at Q ≥ 1.1 Å^{-1}.   Most liquids do not.   The roton case is
not so clear, but the ratio of the FWHM to peak energy is large for
the roton.

    A possible interpretation is that liquid ^{4}He supports collective
phonons at low Q in both the normal and superfluid phases.   The dynamical response, χ, is taken up almost entirely by this single collective excitation.   At Q ≥ 1.1 Å^{-1} however, liquid ^{4}He does not support
a collective excitation.   In the normal phase of ^{4}He we observe scattering from pairs of particles and a broad S(Q,ω).   In the superfluid
phase at Q ≥ 1.1 Å^{-1}, we observe scattering from pairs of particles
(broad scattering) plus scattering from a single quasiparticle (sharp
component).   Because the superfluid has a condensate, the quasiparticle excitation[3-6] makes a contribution to χ and can be observed in
S(Q,ω).   The weight of the quasiparticle excitation in S(Q,ω) is proportional to the condensate fraction, n_0(T).   The sharp component in
S(Q,ω) due to scattering from single quasiparticles disappears from
S(Q,ω) at T = $T_λ$ since n_0(T) → 0 at T = $T_λ$.   This interpretation will
be discussed further in a forthcoming publication.

REFERENCES
1. L.D. Landau, J. Phys. U.S.S.R., **5**, 71 (1941); **11**, 91 (1947)
2. R.P. Feynman, Phys. Rev. **91**, 1301 (1953); **94**, 262 (1954).
3. N. Hugenholtz and D. Pines, Phys. Rev. **116**, 489 (1959).
4. J. Gavoret and P. Nozières, Ann. Phys. NY **28**, 349 (1964).
5. P.C. Hohenberg and P.C. Martin, Ann. Phys. NY **34**, 291 (1965).
6. A. Griffin and T.H. Cheung, Phys. Rev. **A7**, 2086 (1973).
7. A.D.B. Woods and E.C. Svensson, Phys. Rev. Lett.**41**, 974 (1977).
8. E.F. Talbot, H.R. Glyde, W.G. Stirling and E.C. Svensson, Phys.
   Rev. B **38**, 11 229 (1988).
9. K. Sköld and C. A. Pelizzari, Phil. Trans. R. Soc. Lond.
   B290 605 (1980).

THEORY OF ANOMALOUS PHONON
DISPERSION IN LIQUID HELIUM

H. J. MARIS

Department of Physics, Brown University, Providence, RI 02912 , USA

For phonons in crystalline solids the velocity usually decreases as the phonon wave-number increases. For phonons in superfluid helium-4 it was proposed by Maris and Massey[1], based on an analysis of ultrasonic experiments, that the velocity increases with increasing phonon momentum p, and this is referred to as anomalous phonon disper-sion. The existence of anomalous dispersion was subsequently confirmed in a variety of experiments[2]. The strength of the anomalous dispersion decreases as the pressure increases, and above about 20 bars the dispersion becomes normal.

In this paper we give a simple explanation of anomalous dispersion, and explain why the dispersion becomes normal at high pressures. Consider first how the excitations in ^{4}He vary with pressure, including the domain of negative pressure (tension). On general grounds we expect that the energy per atom must vary with molar volume qualitatively as shown in Fig. 1. It follows that the pressure P ($P = -dE/dV$) has a maximum negative value at the inflection point V_c. If we assume that E is an analytic function of V, then near V_c the bulk modulus $B \propto (V_c-V)$. Then for P close to P_c the sound velocity c varies as

$$c \propto B^{1/2} \propto (V_c - V)^{1/2} \propto (P - P_c)^{1/4} \qquad (1)$$

Allowing for this asymptotic behavior, we have used the experimental data[3] for c(P) with P > 0 to estimate P_c and c(P) for negative pressures. The result[4] is shown in Fig. 2. The value of P_c is found to be -8.9 ± 1 bars. Assuming that ^{4}He becomes unstable first at long wavelengths, the pressure dependence of the dispersion curve must be as shown in Fig. 3. Let us now write the dispersion curve for small p in the form

$$\epsilon = cp(1 + \alpha_2 p^2 + ...) \qquad (2)$$

It is clear that since c→0 as $P \to P_c$ α_2 must $\to \infty$. Thus, Fig. 3 provides an immediate qualitative explanation of the experimental observation that α_2 decreases as pressure increases (i.e., the dispersion becomes less anomalous and eventually normal).

One can make this discussion more quantitative by considering a phenomenological model for the energy density H in the liquid which is valid for disturbances varying fairly slowly in space and time. We write

$$H = U(\rho) + \frac{1}{2}\rho v^2 + \lambda \mid \nabla\rho \mid^2 + \frac{1}{2}\beta\dot{\rho}^2 \tag{3}$$

where $U(\rho)$ is the internal energy per unit volume of a static homogenous system, v is the fluid velocity, and λ and β are coefficients. This adds to the usual energy density in fluid mechanics terms in space and time-derivatives of the lowest order possible. It is straightforward to determine λ by the requirement that the energy of the free surface of the liquid have the correct value. This gives[4] $\lambda = 9.1 \times 10^{-7}$cgs. The term $\frac{1}{2}\beta\dot{\rho}^2$ is an extra internal energy that the liquid has when its density is changing. One can understand this on general quantum mechanical grounds as follows[4]. Let us look at a single atom in the liquid which moves in the potential V due to interactions with its neighbors. When the density is static the potential V will be independent of time and the atom will be in the ground state ψ_0. For a slow variation of the density, V will also change slowly and the wave function ψ of the atom will constantly adjust so that it is always fairly close to the ground state wave function for the instantaneous form of the potential V. The difference between ψ and ψ_0 is proportional to $\dot{V}$ (and hence to $\dot{\rho}$) and it is straightforward to show that the excess energy δE of the atom, relative to the energy in the instantaneous ground state, is $\hbar^2\Sigma_n \mid \dot{V}_{no} \mid^2 /E_{no}^3$. Thus, the coefficient β is given by

$$\beta = \frac{2\hbar^2\rho}{M} \sum_n \frac{1}{E_{no}^3} \mid \frac{d\dot{V}_{no}}{d\rho} \mid^2 \tag{4}$$

since the number of atoms in the liquid per unit volume is ρ/M.

Based on the energy density as given by Eq. (3) one can calculate the dispersion relation for long wavelength phonons. The result is of the form of Eq. (2) with

$$\alpha_2 = \rho\left(\frac{\lambda}{c^2} - \frac{\beta}{2}\right) \tag{5}$$

Accurate measurements of α_2 at 8 different densities have been made by Rugar and Foster[5]. Since, in principle c, λ, and β are all functions of ρ, it appears at first sight difficult to use these measurements of $\alpha_2(\rho)$ to test Eq. 5. However, we expect that the density dependence of c will be significantly larger than that of λ and β, since c varies rapidly with ρ and vanishes at the critical density ρ_c corresponding to the pressure P_c. If we take λ and β to be constants, then a plot of α_2/ρ as a function of $1/c^2$ should yield a straight line of slope λ and intercept $-\beta/2$. An excellent fit to the data is obtained in this way (Fig. 4), and the parameters obtained are $\lambda = 10.5 \times 10^7$ and $\beta = 1.55 \times 10^{-15}$

cgs. The value of λ is in very good agreement with the value estimated from the surface energy and the value of β is of the order of magnitude[4] predicted from Eq. (4).

In summary, there are two contributions to phonons dispersion in helium. There is a "spatial" dispersion (coming from the $\lambda \mid \nabla\rho \mid^2$ term) which is anomalous, and a "temporal" dispersion (from $\frac{1}{2}\beta\dot{\rho}^2$) which is "normal". The relative strengths of these contributions change with changing pressure largely as a result of the variation of the sound velocity.

This work was supported by the NSF under grant DMR-8719893.

1. Maris, H. J. and Massey, W. E., Phys. Rev. Lett. 25, 220 (1970).
2. For a review, see Maris, H. J., Rev. Mod. Phys. 49, 341 (1977).
3. Abraham, B. M., Eckstein, Y., Ketterson, J. B., Kuchnir, M., and Roach, P. R., Phys. Rev. A1, 250 (1970).
4. Xiong, Q., and Maris, H. J., J. Low Temp. Phys., to be published.
5. Rugar, D., and Foster, J. S., Phys. Rev. B30, 2595 (1984) .

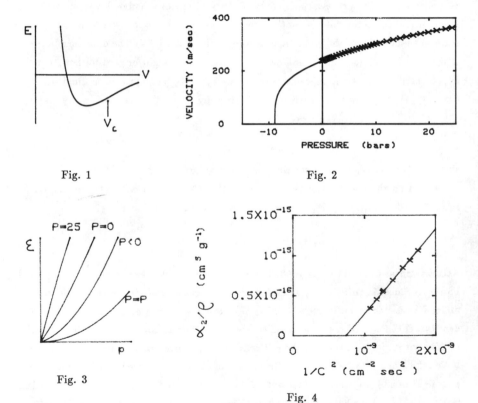

Fig. 1

Fig. 2

Fig. 3

Fig. 4

FINITE AMPLITUDE ACOUSTIC WAVES IN LIQUID HE-3 AND HE-4

M. Chapellier, J. Joffrin, M. Meisel*, A. Schuhl
Laboratoire de Physique des Solides, Bât. 510, Université Paris-Sud
F91405 Orsay Cédex (France)

The propagation of a finite amplitude wave in a nonlinear media can be characterized by two related effects : (1) distortion of the incident wave which, in its ultimate state, transforms into a saw-tooth profile with varying amplitude, and (2) the generation of higher harmonics of increasing order. In the low power regime, the amplitude of the nth harmonic grows like P^n_1 where P_1 is the power flowing into the fundamental fre- quency ; the amplitude of the fundamental wave is almost not affected by the nonlinear effects. In the high power regime, the conversion to higher frequencies is so efficient that saturation of all the waves takes place, whereby all of them approach a limited amplitude when measured at a given distance from the source. The physicists preparing a low temperature acoustic experiment in liquid ^{3}He or ^{4}He, face first with the open question of whether the linear and nonlinear regimes might exist in the acoustic cell [1].

EQUATION OF PROPAGATION

We consider the propagation of a distortion in a direction x of a liquid ; ρ is the specific density, v_0 the velocity of sound, a the local amplitude of vibration, p the local pressure, u the local internal energy per unit of mass, δp the pressure fluctuations and $\delta\rho=\rho-\rho_0$ the density fluctuations. Writing $v^2_0=\rho^2_0 \, \partial^2 u/\partial\rho^2$ the most general differential equation linking a and $\delta\rho$ is

$$[\partial^2/\partial t^2 + v^2_0 \, \partial^2/\partial x^2] \, \delta\rho = - \, \partial^2/\partial x^2 \, (\rho a^2 + v_0 \, \partial v_0/\partial\rho|_{\rho 0}\delta\rho^2) \qquad (1)$$

The second term contains all the non linear components ; in zero order $a=+v_0/\rho_0$, $\delta p= -v_0 \, \partial a/\partial x$; finally we get :

$$[-\partial^2/\partial t^2 + v^2_0 \, \partial^2/\partial x^2] \, \partial a/\partial x = + \gamma v^2_0 \, \partial^2/\partial x^2 \, [(\, \partial a/\partial x \,)] \qquad (2)$$

where $\gamma = (1 + \rho_0/v_0 \, \partial v_0/\partial\rho|_{\rho 0})$ is the effective nonlinear coefficient which may be measured directly in our experiment.

An analytical solutions for (2) has been established ; writing $v=a/a_0$ where a_0 is the velocity of the displacement at the origin of the propagation, $1/L=2\gamma \, a_0\omega/v^2_0$, and changing the variables (x,t) into (σ,y) by $\sigma=x/L$ and $y=\omega(t-x/v_0)$ replace (2) by : $\partial v/\partial\sigma=v\partial v/\partial y$. An implicit solution of this last equation is $v(y,\sigma)=\sin(y+\sigma v)$ and it is very convenient, especially if at x=0 the excitation of the wave is sinusoidal. Because we are interested in the amplitude of the different harmonics of the fundamental frequency, thus we perform a Fourier analysis for $v(\sigma,y)$ by writing

$$v = \sum_{n=1}^{\infty} B_n(\sigma) \sin ny$$

The B_n terms are the quantities measured when one changes the parameter σ. A numerical solution may be computed from the algebraic solutions [2].

EXPERIMENTAL CONFIGURATION AND PROCEDURE

The experimental acoustic path is filled with liquid ^{3}He or ^{4}He and is defined by the large faces of two pieces of quartz having the shape of a nail. The large circular faces are 10mm in diameter, are polished, optically flat, and are separated from each other by three small spacers 0.5mm thick ; thus, the propa- gation distance of the acoustic waves is 0.5mm or a multiple of it. The two heads of quartz are held parallel after alignment to better than 1 minute of arc at room temperature. The narrow sides of the quartz pieces have a diameter of 6.0mm. The overall length of one piece is 15mm while the other is 18mm. A LiNbO$_3$ transducer, 3mm in diameter, has been glued to the center of the small circular face of each quartz piece whose fundamental frequency, after gluing, is 83MHz. Each LiNbO$_3$ crystal terminates a coaxial line which provides electrical excitation This assembly can work in transmission at the fundamental frequency or at any odd multiple (3x83MHz=249MHz, 5x83MHz=415MHz) with an almost equally good efficiency. One LiNbO$_3$ is the emitter, generally operated at the fundamental frequency, while the other is the receiver, detecting at 83, 249 or 415MHz; it measures the transmitted acoustic power after one, three or five passages through the liquid. The acoustic assembly iis placed in a cell and attached to a home-made dilution refrigerator which is capable of cooling the liquid down to 17mK.

After selecting the emitting and receiving frequencies and fixing the temperature, a typical run involves the measurement of the transmitted power as a function of the emitted power for various echoes in the received pattern. For ^{4}He, the first few detected echoes are all distinct if the initial pulse length is shorter than 0.4μs; we have measured the amplitudes of the 111, 131 and 151 echoes. In the case of ^{3}He, we have restricted our analysis to the 111 detected signal.

EXPERIMENTAL RESULTS AND CONCLUSION

The figure shows an archetype example of the data obtained when the exciting frequency f=83 MHz and the resultant signal (the 111 echo in this case) is detected at some fixed distance from the origin and at the f, 3f and 5f harmonics of the incident wave. Since the temperature is very low (20mK), the attenuation associated with the liquid sample is zero. This curve for the signal detected at f should be aligned on a straight line of slope unity in the absence of anharmonicities. In fact the signal is clearly depressed on the high power side. The observed depression is about 6±1dB at the maximum power level. In addition, the data detected at 3f and 5f are plotted. These signals are easy to observe and appear, as one can imagine, when the incident wave starts to show depression. All of the experimental data

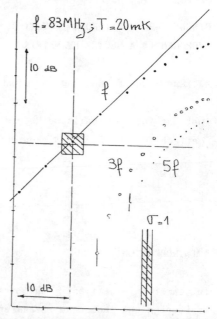

points may be superposed on the theoretical predictions. The two plots coincide nicely, and so, without any other interpretation or calculation, we may identify (to within ±1dB) the incident power corresponding to $\sigma=1$, the special value of the renormalized path length. It leads to a value of $\gamma = 3.7$.

We conclude by asking if ^{3}He and ^{4}He are as anharmonic as other systems? The answer is not clear : many substances have similar values for γ, including most solid. Nevertheless, we have verified that we cannot detect any harmonics at 3f and 5f in our quartz pieces at our maximum power.The absence of any observable effects is related to the fact that v_0 in quartz is much larger than in helium and that the strength of the nonlinearities are proportional to v_0^{-2}. Thus, by considering only the anharmonic constant γ, liquid helium cannot be distinguished from other condensed matter systems. On the other hand, it is very easy to reach acoustic power levels where anharmonic effects are important, and an experimentalist should always be careful in the evaluation of the acoustic flux if operation in the linear regime ($\sigma < 1$) is desired.

REFERENCES

1 - A. Hikata, H. Kwun, C. Elbaum, Phys. Rev. B 21, 3952 (1980)

2 - D. Blackstock, J. Acoust. Soc. Am. 39, 1019 (1966)

SINGLE- AND MULTIPHONON SCATTERING FROM A LOCAL MODE OF MOLECULAR HYDROGEN IN A DEUTERIUM MATRIX

W. Langel and E. Knözinger

Institut für Physikalische Chemie der Universität Siegen, Postfach 101240, D-5900 Siegen, Fed. Rep. of Germany

1 INTRODUCTION

In earlier experiments the transition from single phonon inelastic scattering at momentum transfers below 20 nm^{-1} via multiphonon scattering to single particle recoil at the hydrogen molecules above 50 nm^{-1} of momentum transfer was followed in pure hydrogen [1]. At the Argonne spallation source the recoil scattering was measured at up to 300 nm^{-1} and the momentum distribution of the hydrogen molecules was evaluated from the data [2]. The recoil scattering model may also be applied to inhomogeneous systems, e.g. to hydrogen on MoS_2 and WS_2. By fitting a recoil scattering function to the data, it was shown that hydrogen is adsorbed on these catalysts in a molecular form at higher pressures [3]. On the other hand, H_2 isolated in argon is rotating freely in its site, but no recoil scattering at higher momentum transfers was found [4]. From the well known results on pure hydrogen and deuterium it may be expected that a solution of H_2 in D_2 has some very interesting properties. In spite of the different mass of the two molecules and their equal interaction potentials, the phonon energies of both solids are nearly equal [5,6]. This is a quantum effect: The zero point energy of hydrogen is higher than that of deuterium. This leads to an expansion of the hydrogen crystal and higher next neighbour distances (by about 5%) than in deuterium under the same conditions. The effective potential between the molecules is reduced by this effect. In a classical force constant model [5,6], the force constants for deuterium are in fact higher by nearly a factor of two than for hydrogen.

In spite of their physical interest to the best of our knowledge no neutron studies on the mixtures have been done. During the experiment reported here the inelastic scattering of hydrogen deuterium mixtures was studied. Thereby previous work on several matrixisolated systems as well as on molecular hydrogen was continued.

2 RESULTS AND DISCUSSION

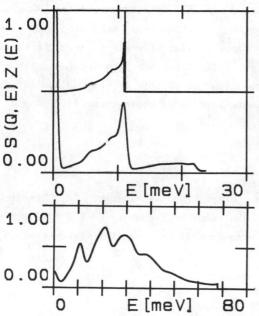

Fig.1: Local density of states of hydrogen in deuterium with a local mode at 11 meV and a strong excitation of resonant modes below that energy (a). Corresponding scattering functions in the single phonon ((b), Q=23 nm^{-1}) and impulse approximation regime ((c), Q=55 nm^{-1}).

The central question of the experiment presented here was, under which conditions diluted particles show recoil scattering. Older ideas considering binding forces and bound-free transitions [7] may not be applied reasonably to thermal and epithermal neutron scattering data. In the present work a local density of states for a diluted particle was calculated from the density of states of the environment and the mass and force constant defect of the particle itself [8]. The local density of states consists of the local mode itself and of a spectrum of modes resonant with the phonons of the matrix. The relative intensity of the resonant modes depends on the position of the local mode relative to the phonons of the matrix. From the local density of states a scattering function may be calculated. Fig.1 shows the result for different scattering conditions in the system H2 in D2. The mass defect of the hydrogen molecule is partly compensated by the fact that the force constant of the hydrogen-deuterium interaction is slightly lower than that of the deuterium-deuterium interaction. Therefore the local mode is very close to the phonon band and the contribution of the resonant modes is high. A program was developed during this work which allows to sum up multiphonon contributions [9] in a closed form to infinite order. The resulting scattering function is broadened in the single phonon regime (Fig.1b) and may be interpreted by the impulse approximation for particle with a mass of 2 at higher momentum transfers (Fig.1c). As will be shown later in a more explicit paper, this is in agreement with the experiment. By the analogous approach for hydrogen in argon it turns out, that there the contribution of local modes to the local density of states is low. As the calculation shows this is why the scattering function cannot be described by the impulse approximation [4].

3 ACKNOWLEDGEMENTS

This work has been funded by the German Federal Minister for Research and Technology (BMFT) under the contract number 03 KN1 SIE2. The neutron scattering experiments were done at the Institut Laue-Langevin, Grenoble.

4 REFERENCES

[1] W.Langel, J.Mol.Struct.,**143**(1986)1
[2] W.Langel, D.L.Price, R.O.Simmons, and P.Sokol,
     Phys.Rev.**B38**,11275-11283(1988)
[3] P.N.Jones, E.Knözinger, W.Langel, R.B.Moyes, and J.Tomkinson,
     Surface Science **207**,159-176(1988)
[4] W.Langel, Rev.Phys.Appl.**19**,755-758(1984)
[5] M.Nielsen, Phys.Rev.,**B7**(1973)1626
[6] M.Nielsen, H.Bjerrum Möller, Phys.Rev.,**B3**(1971)4383
[7] P.Eisenberger and P.M.Platzmann, Phys.Rev.**A2**,415(1972)
[8] P.D.Mannheim, Phys.Rev.**B5**,745(1972) and cit.there
[9] A.Sjölander, Ark.Fys.,**14**(1958)315

EXTENDED BOGOLIUBOV THEORY FOR A NON-IDEAL BOSE GAS

P. Mulheran

Department of Physics, The University, Stocker Road
Exeter, EX4 4QL,
U.K.

A non-linear transformation is introduced to diagonalise Bogoliubov's Hamiltonian to the order $\sqrt{n_o}$ in the interaction.

In 1947 Bogoliubov[1] introduced the first microscopic theory for a non-ideal Bose gas. His idea was that since the zero-momentum state was macroscopically occupied, the most significant interactions between the bosons were scatterings involving two or more condensate particles. Hence the Hamiltonian of the system was greatly simplified, and he reduced the dominant terms to a diagonal form using a canonical transformation. A few years later Breuckner and Sawada[2] also produced the same solution. They argued that in a system such as liquid helium, the inter-atomic potential would be dominated by hard core repulsion, so the appropriate matrix elemant of s-wave scattering to be used in the formula for the spectrum was a simple sinc-function. Indeed, a very good fit to the superfluid helium spectrum can be obtained in such a way. However, it was also realized that liquid helium is very far from being an ideal gas, and that the strong iteractions in the system greatly reduce the occupation of the condensed mode. Thus the higher order terms ignored by Bogoliubov are not negligible if one is to use this as a model for the superfluid.

In this paper a simple extension to Bogoliubov's theory is introduced in order to take account of some of these higher order interaction terms. Bogoliubov's Hamiltonian is written as :

$$H = E_o + \sum_k E_B(k)\alpha_k^+\alpha_k + H_1 + \text{terms } O(n_o)$$

where

$$H_1 = \sqrt{n_o} \sum_{kl} V(k) a^+_{l-k} (a_{-k} + a^+_k) a_l$$

The volume of the system is taken as unity, and n_o is the number of condensate bosons. $V(k)$ is the above mentioned scattering matrix element. The quasiparticle operators (α^+_k) are transforms of the free particle operators (a^+_k)

$$\alpha^+_k = \mu_k a^+_k + v_k a_{-k} \qquad \text{with} \qquad \mu_k^2 - v_k^2 = 1$$

$$\mu_k = \frac{\lambda_k + 1}{2\sqrt{\lambda_k}} \qquad\qquad \lambda_k = \sqrt{\frac{(\hbar k)^2 / 2m}{(\hbar k)^2/2m + 2n_o V(k)}}$$

The excitation spectrum is given by the formula

$$E_B(k) = \frac{(\hbar k)^2}{2m} \frac{1}{\lambda_k}$$

A new diagonal Hamiltonian ($\mathcal{H}$) is proposed to take account of the $\sqrt{n_o}$ terms in H :

$$\mathcal{H} = E_o + \sum_k E_B(k) \beta^+_k \beta_k$$

$E_B(k)$ is the Bogoliubov energy spectrum. The β^+_k operators are then written as a non-linear sum of α^+_k operators.

$$\beta^+_k = \alpha^+_k + \sum_l C(k,l) \alpha^+_{k+l} \alpha_l + \sum_l D(k,l) \alpha^+_{-l} \alpha^+_{k+l}$$

$$+ \sum_l F(k,l) \alpha_l \alpha_{-k-l} + \sum_l G(k,l) \alpha^+_{-l} \alpha_{-k-l}$$

Finally the coefficients C, D, F, G, which are $O(1/\sqrt{n_o})$, are to be expressed in terms of the Bogoliubov parameters. They are determined by the conditions that $\mathcal{H} = H$ up to the terms in H_1 and also that

$$[\beta_k, \beta^+_l] = \delta_{k,l} \qquad \text{and} \qquad [\beta^+_k, \beta^+_l] = 0 \quad \text{to } O(1/\sqrt{n_o}).$$

The result is

$$C(k-l,l) = -D(k,-l) \qquad \text{and} \qquad G(l,-k) = -D(k,-l)$$

$$D(k,l) = \frac{\sqrt{n_0}\,[V(k)\mu_{k+l}v_l(v_k-\mu_k)+V(k+l)v_k v_l(\mu_{k+l}-v_{k+l})+V(l)\mu_{k+l}\mu_k(\mu_l-v_l)]}{E_B(k)-E_B(k+l)-E_B(l)}$$

$$F(k,l) = \frac{\sqrt{n_0}\,[\,\text{NUM}\,]}{3[E_B(k)+E_B(k+l)+E_B(l)]}$$

$$\text{NUM} = 3V(k)\mu_{k+l}v_l(v_k-\mu_k)+V(k+l)(v_{k+l}-\mu_{k+l})(2\mu_k v_l+\mu_l v_k)$$
$$+V(l)(v_l-\mu_l)(2\mu_{k+l}v_k+\mu_k v_{k+l})$$

The β_k^+ states involve coupling between the Bogoliubov states. Miller, Pines and Nozières[3] proposed that such mixing should go some way to including the back-flow used by Feynman in his improved wave-function[4]. To see this the momentum distribution function of the bosons is calculated to the same order for the β_k^+ state :

$$n_k(q) = n_k(0) + \mu_q^2\,\delta_{k,q} + v_q^2\,\delta_{k,-q}$$

$$+\,[1-\delta_{k,q}]\{\,\mu_k(\,D(q,-k)+D(q,k-q))-v_k(\,D(q,-k)+D(q,k-q))\,\}^2$$

$$+\,[1-\delta_{k,-q}]\{\,\mu_k(\,F(k,q)+F(k,-k-q))-v_k(\,D(-k,-q)+D(-k,k+q))\,\}^2$$

$n_k(0)$ is the ground-state distibrution. The last two terms are the q-dependent fluid flow above the currents in the bare α_k^+ excitation.

In conclusion, it has been shown that Bogoliubov's work can be extended for more strongly interacting systems. In particular, both the simple form of the Hamiltonian and the excitation spectrum are retained.

[1] Bogoliubov,N.N., J. Phys. U.S.S.R. 11 , 23 (1947)
[2] Breuckner,K.A. and Sawada, K., Phys. Rev. 106 ,1117 (1957)
[3] Miller,A., Pines,D. and Nozières,P., Phys. Rev. 127 ,1452 (1962)
[4] Feynman,R.P., Phys. Rev. 94 ,262 (1954)

FOUR PHONON SCATTERING IN LIQUID ^{4}HE

M A H Tucker and A F G Wyatt

Department of Physics
University of Exeter
Exeter
EX4 4QL
UK

The four phonon scattering process (4PP) in liquid ^{4}He has received little experimental attention because most scattering of phonons is dominated by the much stronger three phonon process (3PP). In liquid ^{4}He at 0 bar, phonons with energy $\omega < 10$ K (strictly $\hbar\omega = kT$) undergo spontaneous decay into two or more lower energy phonons. However, phonons with $\omega > 10$ K can only interact by the 4PP (or higher order processes which are presumably weaker). A group of phonons propagating through the liquid therefore rapidly evolves into two populations, one of low energy phonons travelling at or near the ultrasonic velocity, and another of slower phonons with $\omega > 10$ K [1]. Due to the minimum energy of 7.16 K required to desorb an atom from the liquid, quantum evaporation provides a unique way of selectively studying phonons with $\omega > 10$ K. Rotons can also evaporate atoms, but this contribution to the measured signals can be minimised and distinguished from the evaporation by phonons.

In this experiment, a superconducting transition-edge bolometer is positioned vertically above, and facing, a thin metal film heater immersed in liquid ^{4}He. The system is uncollimated, and the height of the liquid surface above the heater can be varied. The temperature of the liquid is measured directly using a calibrated germanium resistance thermometer placed in the liquid. High energy phonons are produced from the heater with a 0.3 µs pulse at a power of -17 dB (ref 0.5 Wmm^{-2}). Other experiments show that a pulse of this duration and power produces a large phonon-atom signal with only a small contribution from evaporation by rotons [2]. The bolometer is naturally covered with a liquid ^{4}He film which absorbs $\sim 99\%$ of the atoms incident upon it.

At low temperatures, phonons and atoms travel ballistically over the distances involved in this experiment. As the temperature is raised, the number density of thermal phonons increases, and interactions between these and the high energy phonons will occur. Each scattering will cause a small change in the energy and the direction of the high energy phonon, both effects attenuating the evaporated atom signal. Should the phonon energy drop below 10 K, it rapidly decays to an energy insufficient to evaporate an atom. The cumulative effect of angular shifts can cause the evaporated atom to miss the bolometer.

Measured evaporation signals for different temperatures at one liquid depth are shown in figure 1a. Above 200 mK, the phonons are totally attenuated and the broad signal is due to R^+ roton-atom evaporation. At each temperature, the R^+ roton-atom contribution is subtracted off, and the phonon-atom signal integrated over time. This data is normalised to the low temperature value and is shown as a function of temperature in figure 1b for different liquid depths. It is clear that for a given temperature, the attenuation increases with liquid path length, as expected.

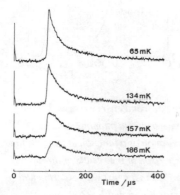
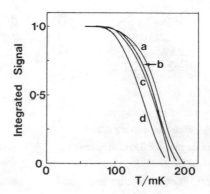

Figure 1a. Evaporation signals for a liquid depth of 5.9 mm corresponding to curve c in figure 1b. Liquid surface is 7.2 mm below the bolometer.

Figure 1b. Temperature dependence of integrated signal, normalised to lowest temperature value at four liquid depths; a) 3.4 mm, b) 4.65 mm, c) 5.9 mm, d) 8.1 mm.

The injected phonon spectrum for $\omega > 10$ K is assumed to be proportional to $\omega^2 \exp(-\omega/T_{eff})$, with $T_{eff} = 0.9$ K [1], and the theoretical 4PP mean free path to vary as $T^{-3} q^{-4}$ [3], where T is the temperature of the liquid and q is the wavevector of the injected phonon. The effect of the energy shifts after many collisions is modelled by a random walk of equal energy steps, $\delta\omega = \pm 1.1$ T, with complete loss if ω drops below 10 K [4]. The temperature dependent energy spectra of phonons arriving at the surface for a liquid depth of 6 mm are shown in figure 2. The areas under the curves are proportional to the total energy arriving at the surface of the liquid. The model is quite insensitive to reasonable values of T_{eff}. The angular scattering is simulated by a Monte Carlo program which calculates the probability of a strike on the bolometer. At each scattering, the allowed directions of the outgoing high energy phonon describe a cone of half-angle θ centred on the direction of the incoming high energy phonon. θ is calculated from an average over momenta and directions of the thermal phonons and is proportional to the temperature of the liquid. The results are compared in figure 3 with the experimental data.

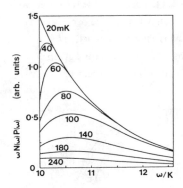

Figure 2. Energy spectra of high energy phonons arriving at the liquid surface for liquid temperatures of 20 mK to 240 mK. $N(\omega)$ is the injected phonon spectrum and $P(\omega)$ is the probability that a phonon of energy ω reaches the surface.

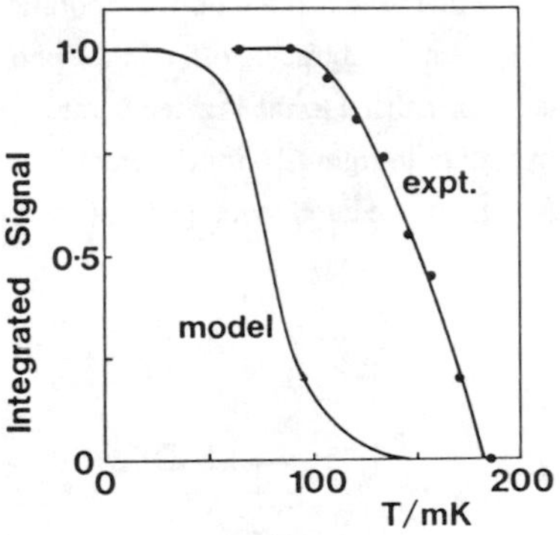

Figure 3. Total energy received at the bolometer due to phonon-atom evaporation as a function of temperature. Comparison of the results of the model described in the text with experimental data for liquid depth of 5.9 mm.

The model predicts that both the attenuation due solely to phonon loss by decay, and the attenuation due to angular scattering alone, are each greater than the measured attenuation. One possibility is that the theoretical value of the mean free path for the 4PP is too short.

REFERENCES

1   A F G Wyatt, N A Lockerbie and R A Sherlock, J Phys: Condensed Matter 1, 3507-22 (1989)

2   M Brown and A F G Wyatt, Japan J App Phys 26-3, 385-6 (1987)

3   I M Khalatnikov; An Introduction to the Theory of Superfluidity (Benjamin, New York, 1965)

4   A F G Wyatt, Japan J App Phys 26-3, 7-8 (1987)

PHONON-ROTON SCATTERING IN LIQUID HELIUM

A C Forbes and A F G Wyatt

Department of Physics
University of Exeter
Stocker Road
Exeter
EX4 4QL
UK

The excitation picture of superfluid helium is well established; the dispersion relation has been mapped out by inelastic neutron scattering [1]. The lifetime of these excitations is limited by their stability against spontaneous decay or interactions with other excitations. Current methods to determine the scattering involve thermal distributions of excitations with the consequent averages over angle and energy [2,3]. More information on the scattering process would be obtained if two excitations of known energy could be collided at a known angle. We report here an experiment which demonstrates this possibility.

In practice a probe beam of high energy phonons is passed through a beam of rotons. The phonons are scattered by the rotons and so the phonon beam is attenuated. The attenuation can be measured as a function of the roton number density.

Phonons and rotons can be injected into liquid He by pulse-heating a thin gold film. Most of the energy is radiated as low energy phonons but a small fraction goes into high energy phonons and rotons. Under the right conditions these excitations have long lifetimes and so can be collimated into ballistic beams. The phonons and rotons can be detected separately by quantum evaporation at the free liquid surface [4]. In this experiment phonons incident at 11° to the normal are detected by the atoms evaporated at 10° (see fig 1). Only phonons with $\omega > 10$ K reach the surface with enough energy to liberate atoms.

The flight time of the phonons through the roton cloud is short compared to the time development of the roton beam which is extended due to dispersion, so that at a given delay between h_1 and h_2 pulses the probe phonons effectively see an instantaneous roton density. The model we use to analyse our experiment considers a flux φ of probe phonons which are scattered by rotons of density n_2. The rate of scattering out of the phonon beam is $-d\varphi = \varphi \, n_2 \, \sigma \, dx$ hence we find

$$-\ln\,(\varphi/\varphi_o) = \int_0^L \sigma\, n_2\, dx\ =\sigma\,\bar{n}\,L \qquad (1,2)$$

Where φ_o is the unscattered phonon flux, L is the length of the ineraction volume and n_2 is the average roton density in the interaction volume.

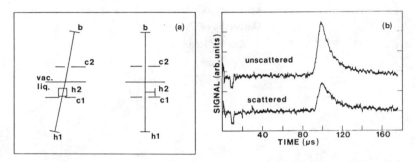

Fig 1 (a) The experimental arrangement is shown schematically. h_1 generates the phonon probe pulse and h_2 generates the scattering roton pulse. Phonon and atom path lengths in the liquid and vacuum are 6.5 mm. Collimation is by 0.8 mm diameter holes in the mica screens c_1 and c_2. (b) Shows a typical un-scattered and scattered probe signal. The phonon heat pulse (h_1) is 0.5 μs and -17 dB and the roton heat pulse (h_2) is 10 μs and -25 dB, fired 30 μs before the probe.

We expect n_2 to be proportional to the roton input power and so $-\ln\,(\varphi/\varphi_o)$ should vary linearly with this power. From the data in fig 2 we see this is so which confirms the validity of (1).

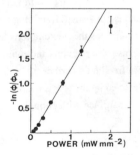

Fig 2 shows the linear variation of the attenuation of the phonon probe pulse with the scattering pulse power.

n_2 varies with the delay between h_1 and h_2 pulse due to the roton dispersion. We calculate n_2 (r,t) assuming that the rotons have an effective temperature (T_{eff} = 1.5 K) and are isotropically injected [5]. The absolute value of n_2 is uncertain because the fraction of heater energy going into roton production is not well known.

Measured attenuation and calculated n_2 are plotted against delay in fig 3. The good agreement shows that at the roton heater power used (-25 dB) the injected rotons behave ballistically down to velocities of 10 ms^{-1} (q = 1.935 Å^{-1}).

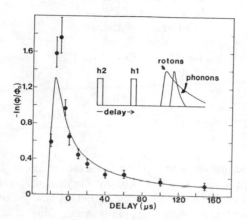

Fig 3 shows the dependance of the attenuation of the phonon probe pulse on the delay after the injection of the roton pulse by h_2 before the probe pulse is injected. The solid line shows the calculated roton number density n_2.

We estimate the magnitude of the phonon-roton cross-section: using roton production efficiency of 1% which is the order of magnitude estimated form the size of roton-atom signals [5], we find σ_{PR} = 7.0 x 10^{-17} m^2 for phonons with q = 0.7 Å^{-1}. This is much larger than the value calculated from the neutron linewidth data of Mezei and Stirling [2], 1.2 x 10^{-18} m^2 for q = 0.7 Å^{-1} phonons. The difference may be due to our experiment being more sensitive to small angle scattering than neutron spectoscopy.

We have directly observed the scattering of high energy phonons by rotons in cold liquid ^{4}He. The cross section for the scattering appears to be much higher than the values obtained from measurements on thermal population.

REFERENCES

1   Cowley R A and Woods A D B, Canadian Journal of Physics 49, 177 (1971)

2   Mezei F and Stirling W G, 75th Jubilee Conference on Helium, p 111

3   Mezei F, Physical Review Letters 44, no 24, p 1601 (1980)

4   Hope F R, Baird M T and Wyatt A F G, Physical Review Letters 52, no 17, p 1528 (1984)

5   Brown M and Wyatt A F G, This Conference

The Creation of R^+ rotons By Condensing ^{4}He Atoms

M Brown and A F G Wyatt
Department of Physics
University of Exeter
Stocker Road
Exeter, Devon EX4 4QL, UK

A beam of ^{4}He atoms incident on the free surface of liquid ^{4}He will condense with a probability approaching unity[1]. Their energy must end up as excitations in the bulk liquid, and in this paper we discuss whether this process is the reverse of quantum evaporation[2]. In particular, can a condensing atom create a single R^+ roton which inherits the energy of the atom and its component of momentum parallel to the surface? The energy of the roton created will be the sum of the kinetic energy of the atom and the condensation energy per atom of 7.16 K.

The condensation may, however, take place by the creation of multiple bulk excitations and intermediate excitations of the surface, instead of the direct production of a single phonon or roton. The angular distribution of the energy flux in the liquid created by a well collimated beam of atoms should give a good indication of which processes are occurring at the surface. The first measurements[3] of this energy flux showed a broad angular distribution which suggested that R^+ rotons and phonons were being created by the condensing atoms. However, the measured angular distribution was broader that expected and was shifted a little from the expected position. This effect was accounted for by assuming that some low energy ripplons were also being produced by the condensing atoms.

More recently it has been shown that condensing atoms can also produce ballistic R^- rotons which can be reflected back to the surface and made to re-evaporate atoms[4]. The measured angles of refraction at the liquid surface, the signal time of flight, and the detection by quantum evaporation, positively identifies the creation of these rotons. We describe here the results of looking directly in the liquid for R^+ rotons created by condensing atoms.

To measure the angular distribution of the energy flux in the liquid we use a detector and collimator which can be rotated about a point in the liquid's surface. A heater and collimator, placed above the surface, are used to generate a beam of atoms, directed at the same point on the surface, at an angle of incidence of 45°. Both path lengths are ~ 6.5 mm. This arrangement is shown schematically in the inset to figure 1. The thin film heater is naturally covered with a layer of liquid ^{4}He , which is partially evaporated when a 10 μs pulse is applied to the heater. The detector is a thin Zn film which is held at a fixed point on its normal–superconducting transition edge by a fast feedback circuit[5]. The detected signals are weak, and $\sim 10^5$ repetitions need to be averaged.

The detected signals are narrow and show no diffusive character, as can be seen in figure 1a. To obtain the angular distribution of the received energy these signals are integrated over time, as is shown in figure 1b for several different heater powers. At high powers there is a broad angular distribution which is similar to that found in the previous measurements[3]. However, when the heater power is reduced the distribution narrows considerably to a limiting distribution with a maximum at $\sim 10°$ to the normal. When the angle of incidence is reduced to $20°$, the position of the maximum detected signal shifts to $\sim 5°$.

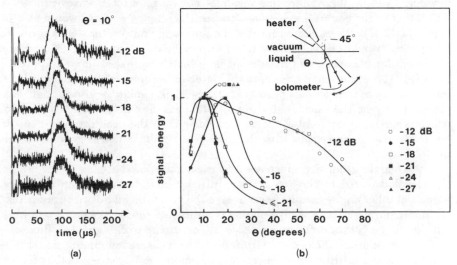

FIGURE 1 *(a) Signals measured in the liquid at $\theta = 10°$ due to condensing atoms, shown for different heater input powers. All powers are referred to $0\,\mathrm{dB} = 0.5\,\mathrm{Wmm}^{-2}$ at the heater. The detector time constant is approximately $5\,\mu s$. (b) Normalised integrated energy in the detected signals (integrated up to $140\,\mu s$ after the input pulse). The inset shows the experimental arrangement.*

In order to observe the quantum *evaporation* of atoms by R^+ rotons, the energy fluxes must be small[6], and from the results of figure 1 this also seems necessary for the corresponding condensation process. The measured angular position and angular width of the signals agree with those we calculate for the atom–R^+ roton process to within the experimental error. The minimum signal time of $\sim 70\,\mu s$ agrees well with the calculated time of $68\,\mu s$ for this process. It does therefore appear that ballistic R^+ rotons can be created at low atom fluxes, similar to the fluxes used to create R^- rotons[4].

We have estimated the atom flux on the surface of the liquid assuming isotropic emission of atoms from the heater, and the constant temperature bolometer system allows us to estimate the absolute detected energy. Hence we find that the product of the probabilities of an atom producing an R^+roton, and of the roton then giving its energy to the Zn detector film is $\sim 10^{-4}$. Most of this factor we believe to be associated with the second process. The low transmission of roton energy into the detector would explain why ballistic roton signals are not seen in the presence of large phonon fluxes, since the phonons have much higher transmission coefficients into the detector[7].

Due to the very small energy transmission coefficient for rotons we are not able to rule out other explanations for the detected signal. It may be that some rotons created at the surface spontaneously emit a small phonon[8]. However, only rotons which have a group velocity greater than the ultrasonic phonon velocity can do this, and it is not clear from the measured dispersion curve[9] how many rotons, if any, have this velocity. The range of roton wavevectors over which this might be possible is small, and so the phonon energies would also be small ($\hbar\omega \lesssim 1$ K). We estimate that the phonon energy flux on the detector due to this process would be 10^{-2} of the roton energy flux, but since the transmission coefficient for phonons into the detector is ~ 100 times that estimated above for rotons, the resulting energy flux into the detector is comparable. The calculated minimum signal time for this process is also $\sim 68\,\mu$s, similarly close to that measured. However, the calculated angle for the peak flux in the liquid is $\sim 17°$, which is further away from the measured angle than the peak roton flux would be, so the overall agreement with the measured results is not as good with this model.

When the atom flux is increased, by raising the power applied to the heater, the signal detected at the roton angle initially increases, but then saturates. Signals at other angles then increase to give the broad angular distribution. Also, the minimum signal time at high powers is noticeably shorter than at low powers. All this suggests that different processes are occurring with high atom fluxes.

In conclusion, the evidence from both the measured minimum signal time and the angular distribution of energy in the liquid strongly suggests that for low power atom beams condensing onto liquid ^{4}He the single atom–single R^+roton condensation process does occur.

References
[1] Nayak V U, Edwards D O, and Masuhara N 1983 *Phys. Rev. Lett.* **50** 990
[2] Wyatt A F G 1984 *Physica* **126B** 392
[3] Edwards D O, Ihas G G, and Tam C P 1977 *Phys. Rev. B* **16** 3122
[4] Wyborn G M and Wyatt A F G 1988 *JJAP* **supplement 26–3** 2095
[5] Wyatt A F G, Sherlock R A, and Allum D R 1982 *J. Phys. C: Solid State Phys.* **15** 1897
[6] Brown M and Wyatt A F G, to be published.
[7] Bradshaw T W and Wyatt A F G 1983 *J. Phys. C: Solid State Phys.* **16** 651
[8] Pitaevskii L P 1959 *Sov. Phys. JETP* **36(9)** 830
[9] Cowley R A and Woods A D B 1971 *Can. J. Phys* **49** 177

R$^+$roton–atom Evaporation Near The Critical Angle

M Brown and A F G Wyatt
Department of Physics
University of Exeter
Stocker Road
Exeter, Devon EX4 4QL, UK

An important test of the one–to–one quantum evaporation process for R$^+$rotons is to examine the behaviour near the critical angles of incidence at the surface, $\theta_c(q)$[1]. The critical angles arise because the momentum carried by an R$^+$roton is considerably greater than that of the atom which it desorbs: the need to conserve the component of momentum parallel to the plane of the liquid surface then prohibits the quantum evaporation process for any large angles of incidence. The range of values of $\theta_c(q)$ for R$^+$rotons is from $\theta_c \simeq 15°$ for rotons near the roton minimum in $\omega(q)$ ($q_{min} = 1.92\,\text{Å}^{-1}$), up to $\theta_c \simeq 30°$ for rotons with wavevectors $\simeq 2.6\,\text{Å}^{-1}$.

In another paper[2] we describe how a pulse of ballistic R$^+$rotons can be injected into liquid ^{4}He, and how the evaporated atoms can be detected. The shapes of the detected atom signals can be related to the spectrum of the rotons in the injected beam, and for suffciently low input pulse powers this spectrum can be characterised by a single effective temperature. Here we use this type of pulse shape analysis to analyse the wavevector dependence of the evaporation process close to the critical angles described above, a situation in which the shape $S(t)$ of the received signal is expected to change considerably as the angle of incidence of the rotons is varied.

The experimental arrangement used is shown schematically in figure 1a and is described in detail elsewhere[2]. The bolometer responds to the energy given up by the atoms when they condense onto the thin superfluid film covering its surface. If the bolometer were to touch the bulk liquid surface, capillary action would greatly increase the film thickness, and so nearly all of the condensation energy would be lost to the bulk directly, and not be coupled into the detector. We therefore place the bolometer at the detection angle $\phi = 75°$, where it is clear of the liquid surface, and then vary the angle of incidence of the excitations by moving the heater.

The expected behaviour as θ is varied can be inferred from figure 1b. This graph shows the relationship between the wavevector q of a roton and the angle of incidence θ at the surface, such that the evaporated atom leaves at an angle

$\phi = 75°$. For small values of θ, evaporation into this direction is by rotons with wavevectors close to the roton minimum at $q_{min} = 1.92\,\text{Å}^{-1}$. As θ is increased, evaporation by rotons with larger wavevectors can then be seen by the detector.

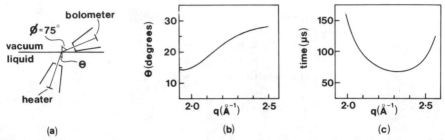

FIGURE 1 *(a) Experimental arrangement. (b) The calculated variation of the angle of incidence of an excitation with wavevector q, such that the evaporated atom is desorbed at an angle $\phi = 75°$. (c) The calculated total signal time for the quantum evaporation process as a function of wavevector q (see text).*

The total time of flight for a given R^+ roton–atom evaporation process can be expressed entirely in terms of the wavevector of the roton (this follows from the known dispersion of the rotons and free atoms, and the quantum evaporation conditions at the surface). For the roton and atom path lengths used in this experiment (both are $\sim 6.5\,\text{mm}$) these calculated total times of flight, as a function of the wavevector of the roton, are shown in figure 1c. Changing the angle of incidence from $\theta \simeq 15°$ to $\theta \simeq 30°$ can be seen to produce a change in the arrival time of the atoms detected at the bolometer. For small θ the total time is long, but reaches a minimum value of $68\,\mu s$ for the atoms evaporated by rotons with $q \sim 2.3\,\text{Å}^{-1}$, when θ is increased to $\sim 25°$.

In the actual experiment the collimation cannot be too restrictive, since there must be a sufficient atom flux at the bolometer to be detectable. Consequently, for any given heater and detector positions there is a finite range of angles of incidence and evaporation (extremal angles are $10°$ on each side of the nominal direction θ), and therefore also a range of roton wavevectors which can contribute to the signal. However, the overall differences between the behaviour at small and large angles of incidence should still be as described above.

Figure 2a shows the measured evaporation signal shapes obtained for several different heater positions. The input pulse used is $10\,\mu s$ long at a power of $2\,\text{mW}$ ($\equiv -24\,\text{dB}$ referred to $0.5\,\text{W}\,\text{mm}^{-2}$, as used in reference 2). As discussed in reference 2, this input pulse is of sufficently low power for the injected rotons to travel ballistically in the liquid, and the spectrum can be characterised by an effective temperature $T_{\text{eff}} \simeq 1.5\,\text{K}$. The expected behaviour is clearly being observed. The smaller values of θ produce a slow signal dominated by evaporation by rotons with wavevectors close to that at the roton minimum. Moving the heater to larger values of θ brings in processes involving larger wavevector rotons, which produce a narrower signal with a well defined minimum flight time of $\sim 70\,\mu s$, in agreement with the calculated value (shown by the dashed line in figure 2a).

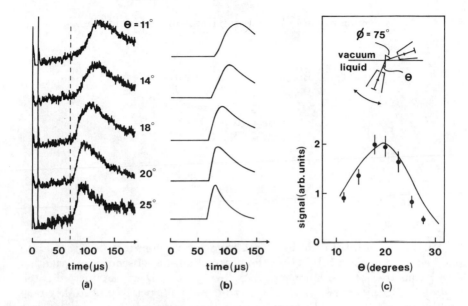

FIGURE 2 *(a) Measured evaporation signals for different values of θ. The detector time constant is approximately 5 μs. (b) Modelled signal shapes for the same angles θ as in (a). (c) The variation in the detected signal energy as a function of the heater position θ, integrated up to 150 μs after the start of the input pulse. The solid line is calculated from the modelled results shown in (b).*

To compare the results more fully with the assumption of quantum evaporation at the surface the Monte Carlo simulation described in reference 2 is used, which takes full account of the collimation geometry. Using the value $T_{\text{eff}} = 1.5\,\text{K}$ to characterise the injected roton spectrum the results of figure 2b are obtained. The overall agreement is very good, and is confirmed by comparing the way in which the total detected energy in the signals depends on θ with that predicted from the model, as shown in figure 2c.

In summary, it is clear that the desorption of atoms by R^+ rotons is via the one–to–one quantum evaporation process even when the angles of refraction are large. This type of refraction effect separates out different roton wavevectors, and should therefore be useful for making measurements of properties which depend directly on the excitation wavevector; for example, scattering effects in the bulk liquid during the propagation of the excitations to the surface.

References

[1] Wyatt A F G 1984 *Physica* **126B** 392
[2] Brown M and Wyatt A F G — 'The Injection of Large Wavevector Rotons into Liquid ^{4}He ' — this conference.

The Injection of Large Wavevector Rotons into Liquid ^{4}He

M Brown and A F G Wyatt
Department of Physics
University of Exeter
Stocker Road
Exeter, Devon EX4 4QL, UK

In this paper we describe how large wavevector rotons may be injected into liquid helium at ~ 100 mK by a thin film heater, and the subsequent propagation of the rotons over several millimetres. At low heater powers the injected roton density is sufficiently small for the rotons to travel ballistically, and so they may be collimated into a beam. However, at high powers, the injected rotons interact with each other, causing the propagation to be more diffusive.

Rotons couple very weakly into solid surfaces[1], so signals due to rotons injected by a broad band thermal source are very difficult to detect, since they are always swamped by a large flux of phonons whose detection efficiency is much higher. We therefore use the quantum evaporation process at the free surface to convert the energy of the rotons into the kinetic energy of free atoms. Quantum evaporation is a single excitation–single atom desorption process which conserves energy and the component of momentum in the plane of the surface[2]. The atoms evaporated in this way can then be detected with high efficiency, thus making it possible to deduce the behaviour of the rotons themselves. The rotons considered here are those with momenta greater than that at the roton minimum ($q_{min} = 1.92\,\text{Å}^{-1}$), and up to $q \sim 2.5\,\text{Å}^{-1}$.

The experimental arrangement used is shown schematically in figure 1, and is described in detail elsewhere[3]. A $10\,\mu$s heating pulse is applied to a thin metal film heater H, of area $1\,\text{mm} \times 1\,\text{mm}$. The two collimators C1 and C2 are made from mica sheet, $\sim 50\,\mu$m thick, in each of which is punched a rectangular slot $1\,\text{mm} \times 0.4\,\text{mm}$, with the narrow dimension in the plane of the figure. The path lengths in the liquid and in the vacuum are both ~ 6.5 mm. Evaporated atoms are detected by the energy they give up on condensation onto the thin superfluid film covering a zinc–transition edge bolometer B. The angles $\theta = 14°$ and $\phi = 40°$ are chosen to select out only evaporation by large wavevector rotons[3].

Figure 1a shows the effect of varying the power of the input pulse over the range -30 dB up to -12 dB, referred to $0.5\,\text{Wmm}^{-2}$ at the heater (i.e. from $0.5\,\text{mW}$ to $31.5\,\text{mW}$). For low powers $P \lesssim -24$ dB the detected pulse shapes can be understood using a model based on ballistic roton propagation in the liquid and quantum evaporation at the surface[3]. The roton spectrum injected by the heater is assumed to be characterised by a single effective temperature T_{eff} and based on the bulk liquid density of states, giving a distribution

$$n(q)\, dq \propto \frac{q^2}{\exp(\hbar\omega(q)/k_B T_{eff}) - 1}\, dq,$$

where $\hbar\omega(q)$ is the energy of a roton with wavevector q. As the input power is increased from $-30\,\mathrm{dB}$ to $-24\,\mathrm{dB}$ the bulk of the measured pulse shape moves to an earlier time. The model calculations (figure 1b) show that this behaviour is consistent with an increase in the effective temperature describing the roton spectrum. This increase in the value of T_{eff} changes the distribution of rotons so that more high energy rotons from well above the roton minimum are included. These are the ones which give rise to the fastest part of the evaporation signal.

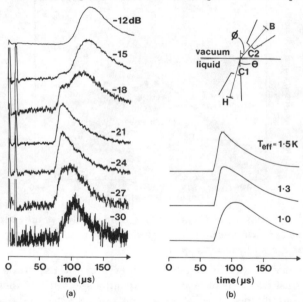

FIGURE 1   (a) Measured roton–atom pulse shapes for different heater powers (the signals are shown scaled to approximately the same peak height). Time is recorded from the start of the input pulse. The detector time constant is approximately 5 μs.   (b) Calculated pulse shapes for quantum evaporation by ballistic rotons (see text).

The model calculation allows an estimate to be made of the probability that a roton will reach the detector, taking into account the geometry of the experiment. Comparison with the measured low power results then gives an estimate of the fraction of the input pulse energy which is converted into ballistic rotons. Assuming that emission into the helium is isotropic (over the half space in front of the heater) and that both the evaporation process and the detection process at the bolometer are 100% efficient, we find that at $\sim -27\,\mathrm{dB}$ approximately 0.5% of the input energy is in the form of these large wavevector rotons. This is clearly a very rough estimate, but it does indicate that roton production is rather inefficient at the heater surface. Most of the energy emitted by the heater is in the form of low energy phonons with a characteristic temperature of $\sim 1\,\mathrm{K}^{(4)}$.

As the input power is increased above $\sim -21\,\mathrm{dB}$ the signal due to evaporation by *ballistic* rotons becomes *smaller*, and a second, slower component develops. This reduction in the amplitude of the fast ballistic signal must be due to scattering between the injected excitations. For $P \gtrsim -18\,\mathrm{dB}$ the signal becomes more diffusive and the total detected energy then increases rapidly with input power (figure 2). Roton production seems to become much more efficient at these higher powers.

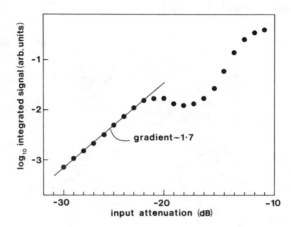

FIGURE 2   *The variation of the integrated evaporation signal energy as a function of input power (attenuation referred to $0.5\,\mathrm{W}\ \mathrm{mm}^{-2}$). Integration is up to $160\,\mu\mathrm{s}$ after the start of the input pulse.*

In summary, it is clear that large wavevector rotons can be injected into liquid helium with this type of heater. The energy spectrum of these excitations can be approximately described at low input powers ($P \lesssim -21\,\mathrm{dB}$) by a single thermal distribution. If the input power is not too high then the rotons propagate ballistically, although only a small fraction of the heater pulse energy is in this form. This fraction increases rapidly with heater power (once above $\sim -15\,\mathrm{dB}$), but scattering between the injected excitations then becomes important.

References
[1] Brown M and Wyatt A F G 'The Creation of $\mathrm{R^+}$rotons By Condensing ^{4}He Atoms' — this conference.
[2] Wyatt A F G 1984 *Physica* **126B** 392
[3] Brown M and Wyatt A F G — to be published.
[4] Wyatt A F G, Sherlock R A, and Allum D R 1982 *J. Phys. C: Solid State Phys.* **15** 1897

10

Nonlinear Phenomena

DYNAMICAL PHONON MEMORY EFFECTS AND PHONON SUPER AND SUBPROCESSES

Uno Kopvillem

Quantum Oceanology Laboratory;Pacific Oceanology
Institute,Radio St.7,Vladivostok 690032   USSR

1. Introduction

In the beginning of the 60-th it was realised that in
physics and generally in natural sciences there are global rules that make it possible to forecast the way of
evolution of puysics anf its practical applications [1] .
After the triumph of quantum electronics in 1959 the appearence of a new direction in acoustics-quantum acoustics
was predicted [2] .It is remarcable that so it really hap-
pened and now the definition"quantum acoustics"is a key-
word in physics classification scheme PACS.Then quantum
optics 1 and nonlinear and coherent infraacoustics(seis-
moacoustics) [3] and several other scientific directions
were successfully predicted 4,5[.With this progress a change in science accomplishment also occured.A scientist in
the field of quantum acoustics inevitabely also becomes a
specialist in NMR,EPR,quanyum electronics,quantum optics,
quantum biology et cetera.In one hands and mind occured
many of the ideas and technical performances of many sci-
ences.The structure of planning an experiment became very
complicated with unbounded possibilities with limits only
in economy.In quantum acoustics one starts from quantum
rules and atomic resonances.At date the most perspective
way in quantum acoustics is to exploit methods of phonon
eco- and coherent phonon avalanche spectroscopy.They have

in common global analitical applications and are suited
to construct new types of powerful sound generators and
supersensitive sound and single phonon detectors.The problem is to find monocrystals of high acoustical quality
with strong elastic multipole-phonon interactions as in
$MgO:Fe^{+2}$ and $RbMnF_3$.A very promising application of spin-
phonon interactions is the performance of gravitation wave detectors.It is possible to do quantum nondemolish-
ment experiments far beyond the quantum limits caused by
zero-point fluctuations of the elastic vacuum.Acoustical
coherence manifests itself also in biology and helps to
investigate life processes and to understand the cause of
spontaneous motions in living beings.Experiments show
that acoustical coherence displays unusual properties in
living systems.

In the field of infraacoustics many new dinamical effects were discovered with the help of laser strainmeters
that are not limited by bandwith problems.At low frequences the crust of the Earth is acoustically exstremely nonlinear what is caused by molecular and mechanical proper-
ties.The corresponding parameters exceed by 10^4 times the
known values for crystals at high frequences.As was expected from general analogy considerations in the realm
of infraacoustics it is possible to repeat all the effects
discovered in quantum acoustics as phase conjugation,ava-
lanche and echo phenomena.The change of frequency is not
a formality in nature.Super-and subprocesses will help to
predict the times and modes of earthquakes and volcanic
eruptions.

The general idea of this work is that zero-point oscillations are the foundation to unite super-and subprocesses in nature.By manipulating the phases of the elastic multipoles in matter it is possible to change the time duration and intensity of processes by a factor of
$[N(\lambda^2/4\pi a_0)]^2$ going from subprocesses to superpro-
cesses,where N, and a_0-correspondingly the number of

elastic multipoles,radiation wavelength and the square
section of the radiator.These enormous numbers illustrate the possibilities of sub-and superradiational quantum
acoustics.

2. Theoretical Considerations

We are interested in the cybernetics of the dynamical
acousticalholography on the microscopical level,how to
"teach" a huge system of particles $j(1,...,N)$ the ability
of selfmanipulation in phonon fields.This means the rea-
lisation of one type of bozon holography.The other is connected with photons and particles.In general a particle
may simultaneously possess many different physical multi-
poles(electromagnetic,elastic,mass,...)and interact with
the corresponding physical fields and can detect and generate these fields.The cross-effects are the most interesting ones.A system of elastic multipoles can interact
with a phonon field and absorb energy from this field.If
the elastic multipoles are connected with electromagnetic
(mass,...)multipoles the system of particles caused by
zero-point electromagnetic(gravitational,...)oscillations
will spontaneously and coherently generate electromagnetic
(gravitation,...) waves.So the vacuum helps to transform
the energy of one field into the energy of other fields
because all possible multipoles are connected with zero-
point oscillations.Only selforganised zero-point motion
can manipulate 10^{22} interacting particles.We can imagine
thys process as follows.We prepare a very complicated po-
limodal resonator for multifield resonances of phonons,
photons,...with a supply of potential energy and waveguides
for the new fields,generated in the resonator.Then we form
a spesial state for the beginning and launch the system.If
all was calculated correctly we get the expected behavior
of the system.From the point of view of mathematics all
the calculus is done according to the equation:

$$\langle Q(t) \rangle = Sp \int_0 \mathcal{L}(t)^{-1} Q(0) \mathcal{L}(t), \quad \mathcal{L}(t) = \overline{exp} \left[\frac{i}{\hbar} \int_0^t \mathcal{H}(t') dt', \right] \quad (1)$$

where ς -the dencity matrix at the starting point,$\langle Q(t)\rangle$ - the observed quantity at the time"t", $\mathcal{L}(t)$ _the evolution operator,$\mathcal{H}(t)$ -the hamiltonian of the system in the interaction frame,$\overline{exp}[...]$ -an ordered exponential operator.The dynamical properties of $\langle Q(t)\rangle$ depend from the Lie algebra generated by (1) and we shall call thys algebra the dinamical algebra of the physical situation.In many cases much may be said about the physical effect if the corresponding dynamical algebra is known.The detection of an echo means that the corresponding physical system has the dynamics SU(2) or SU(1,1).The intensity of the echo is:

for SU(2) $\quad \vec{K}_e = 2\vec{K}_2 - \vec{K}_1, \quad I = D_1 N^2 \sin^2\theta_1 \sin^4(\theta_2/2) \quad (2)$,

for SU(1,1) $\quad \vec{K}_e = 2\vec{K}_2 - \vec{K}_1, \quad I = D_2 N^2 \theta_1^2 sh^2\theta_2 \quad (3)$,

where $\vec{K}_1, \vec{K}_2$ and $\vec{K}_e$ -correspondingly the wave vectors of the first and second pulses and the echo; θ_1 and θ_2 -parameters of the pulses intensities,D_1 and D_2 -constants.Phonon echoes and avalanches were firs discussed in 6].The intensity of the avalanche depends of the coherent relaxation time τ_R and time of retardation t_d :

$$I(t) = \frac{E\tau}{2}\left(\frac{1}{T_2} - \frac{1}{\tau_R}\right)^2 sech^2\left[\frac{1}{\tau_R}(t-t_d)\right], \tau_R = \frac{\tau}{N}, \quad (4)$$

$$t_d = \left(\frac{1}{\tau_R} - \frac{1}{T_2}\right)^{-1} \ln\left[\frac{T_2 - \tau_R}{T_2}\sqrt{N}\right],$$

where T_2 is the phase relaxation time and τ is the spontaneous relaxation time of an isolated particle. E is the energy of the phonon.The launching of a phonon avalanche is caused by a very low intensity perturbation. The avalanche gives the possibility to generate very short coherent sound pulses.

Terasound amplification may be observed on 4-level systems if the upper two levels are lifetime broadened and the transition from them ends in the same continuum.Then there will be asymmetry between absorbtion and radiation of phonons,caused by interference of Fano type.An example may be corundum with Cr^{+3} in the exited state.

Phonon subprocesses occur when a system is in a correlated Dicke state (R,M), where M and R —correspondingly the quantum number and the macroscopical spin 4] .In these stqtes the elastic multipoles may be exited and may not radiate.The intensity is:

$$I = \gamma \, (R+M)(R-M+1)(R+M-1)(R-M+2) \qquad (5)$$

and we see that in many cases we get zero intensity where

$$\gamma = \left[4S(2S-1) \right]^{-1} I_0, \quad R = NS, \quad -R \le M \le R \; , \; \text{S-the micro-}$$

spin. These states may be important in biology to conserve energy for longe times.Space coherence may be used to trap coherent radiation by forming a situation where the space coherence is not appropriate for coherent padiation oe the trapped field. It is possible to construct a non-radiative hamiltonian from spin creation and destruction operators: S_+ and S_- :

$$\mathcal{H} = \sum_{j \mp k} \left(\mathcal{P}_{jk} \, S_+^j S_-^k + \mathcal{P}_{jk}^* \, S_+^k S_-^j \right), \qquad (6)$$

where $\mathcal{P}_{jk}$ is a tensor of flip-flop processes:if an elastic multipole (jk) radiates a phonon the elastic multipole(kj) absorbs thys phonon and energy is conserved.This is possible in crystals with axial symmetry and ctrong asymmetry in the polar direction.

New problems arise when we begin to investigate phonon super-and subprocesses in living beings where biological computers are active.From the point of view ofcybernetics coherence is a state with good information transport. The factor N^2 in the intensity of coherent radiation counts for the number of links between pairs of radiators.So it is possible to have N^4 links between radiators if we cluster all the radiators by pairs.This may occur if there is very strong interaction between the particles or if the biological computers are very active.Such an effect may be termed supercoherence.Our aim was to search for acoustical and electrical supercoherence in biology.Insight may be got by measuring the amplification(if present) of dielectric oscillations in living systems by enormous electric dipoles in membranes.The hamiltonian of the cell contains the system of gigantic electric dipoles and the interactions between the dipoles themselves and with the thermostat, the interaction between the dipoles and the metabolic pump, the operators of energy of the thermostat and pump. The interaction energy between the dipoles controlled by the biological computers is of the form 1-9]

$$\Delta E = A R^{-3} - B R^{-6}; \quad A = |\hbar e^2 z \gamma| M^{-1} (8 \bar{\omega})^{-1/2};$$

$$\gamma = \left\{ \left[\epsilon(\omega_+) \right]^{-1} - \left[\epsilon(\omega_-) \right]^{-1} \right\}; \quad \bar{\omega} = \frac{\omega_1 + \omega_2}{2};$$

$$\omega_{\pm} = \left\{ \frac{1}{2}(\omega_1^2 + \omega_2^2) \pm \left\{ \frac{1}{4}(\omega_1^2 - \omega_2^2)^2 + \frac{\beta_0^4}{(\Im_\pm)^2} \right\}^{1/2} \right\}^{1/2},$$

$$\beta_0 = \gamma^* e^2 \sqrt{z_1 z_2} M R^3 \tag{7}$$

where R is the distance between the dipoles, M-the molecular mass, e-the electronic charge, Z-number of charges, $\Im_+$ the dielectric constant at frequency ω_+, B-a constant, -a shape dependent constant, numbers 1 and 2 correspond to molecules 1 and 2; ω_1 and ω_2 -the frequencies of oscillation of notinteracting molecules. The dynamics of the cell is hardly nonlinear and may described by superalgebra osp(2|2) that contains SU(2) and SU(1,1). It follows that in a cell echo-and avalanche processes may occur what is possible to detect by means of the time dependence of the dielectric constant 9] :

$$\epsilon(\omega_0 \pm \delta\omega) = 1 + \frac{2\pi}{k_B T} \int \mu_0^2 \frac{1 \mp \omega_0 \Delta \delta\omega}{1 - \Delta(\delta\omega)^2};$$

$$\Delta \equiv \pi^2 \hbar^4 N / |\chi|^4 \Gamma^2(\omega_0);$$

$$\epsilon = v_m \epsilon_m + v_b \epsilon_b + (1 - v_m - v_b) \epsilon_w; \quad \epsilon_w = \epsilon_\infty + \frac{\epsilon_s - \epsilon_\infty}{1 + \omega^2 \tau^2},$$

where χ-an dipole-dipole interaction constant, $\int$ -density of the material, μ_0- mean value of the dipole moment, $\Gamma(\omega_0)$ -form-factor depending in complicated manner from the systems frequences, v_m and v_b -the volume fractions of dipolar dielectric constant and bound water constant , ϵ_w- free water constant, $\epsilon(0) = \epsilon_s$, $\epsilon(\infty) = \epsilon_\infty$, τ- dielectric relaxation time. Our guess is that $\epsilon(\omega_0 \pm \delta\omega)$ is proportional to the elastic multipole moment of the gigantic dipoles.

3. Experimental Realisations Of Dynamical Physical Effects In Nonlinear And Coherent Acoustics Contained In OSP(2|2)

All what follows has been predicted theoretically according to the line of thought that abstract algebraic structures may be relevant to predict new natural effects on the base of pieces of practical knoledge.Fig.1.Electron spin phonon echoes in $MgO:Fe^{+2}$,the vertical scale is in units of magnetoelastic stress $11]$ predicted in $6]$ .Fig.2. Spin echo generation with magnetic and sound pulses $12]$, predicted in $14]$.After the A-pulse we see acoustic superradiation and avalanche behavior,predicted in$1]$,$9]$.Fig.3.Nuclear spin phonon echo by sound pulses $1]$ and $2]$.Predicted in $1]$,$6]$ .Fig.4 and 5.Generation in the Earth crust with acoustical pulses 1 and 2 infrasound video-echo $\ni$,the vertical scale is in arbitrary displacement units,recorded by a laser interferometer in a continent-ocean transition region. On Fig.5 the vertical unit is 8 $\cdot$ 10^{-8}m.The pulses contain high frequency components,the video-echo generates no sound.Fig.$6]$. $7]$ The experimental discovery of two-level systems in the amorphous state of matter.a)The dependence of the intesity J of two-level system electromagnetic echo from the static magnetic field H in K gauss at 4'2 K at frequency $10^{10}s^{-1}$in clean quartz glass; σ) The same in quartz glass with $CeO_2+T_iO_2$.A phonon echo end electric dipole echo were predicted in glasses and discovered by other investigators.Fig.7.Discovery of acousto-electrical supercoherence in living beings $15]$ .A two-arm X-band interferometer was used to measure the dynamical dielectric constant;"a" and "b" - signals from isolated being A and isolated being B in the same conditions. "c" -signal from the living system of beings A and B.The effect of amplification and synchronisation is remarcable.Not living matter gives signals without any spades that are seen on all recordings a-c on Fig7.

References

1.Kopvillem U.Zh.ETF.42,1333(1962)

2.Korepanov V.and Kopvillem U.Zh.ETF.41,211(1961)

3.Kopvillem U.and Shubina R.Isv.Vus.Fisika,5,6(1963)

4.Kopvillem U. and Saburova R.Paraelectric Resonance.M. "Nauka",224 p.(1982)

5.Kopvillem U. and Prants S.Polarisation Echo.M."Nauka" 192 p.(1986)

6.Kopvillem U.and Nagibarov V.Zh.ETF.52,936(1967)

7.Kopvillem U.et al.UFN.105,767(1971)

8.Asadullin Ja. and Kopvillem U. FTT.9,2737(1967)

9.Kopvillem U.and Nagibarov V.Zh.ETF.54,312(1968)

10.Paul R.et al.Phys.Rev.A30,2676(1984)

11.Taylor D.R.and Bartlet I.G. Phys.Rev.Lett.30,2676(1973)

12.Golenishev-Kutuzov et al.Pisma Zh.ETF,25,296(1977)

13.Golenishev-Kutuzov et al.Zh.ETF.71,1070(1976)

14.Kopvillem U.and Koloskova N.Fizika Metallov i Metallo-vvedenie.10,818(1960)

15.Kopvillem U.and Sharipov R.Preprint,Laboratory of Qu-antum Oceanology,Academy of Sciences of the USSR,Vladivos-tok,USSR.

APPENDIX

In this appendix we give the parameters of a gigan-tic seismo-acoustical soliton generated by seismic avalan-che processes.The time is the summer time of Vladivostok.

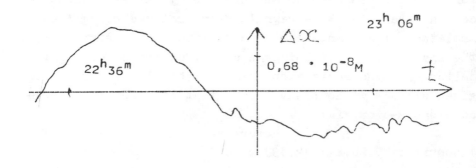

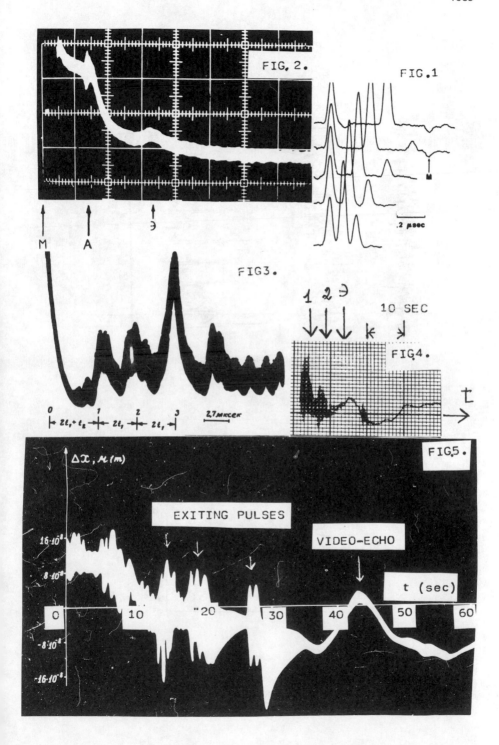

FIG. 2.

FIG.1

.2 μsec

M   A   Э

FIG3.

1 2 Э   10 SEC

FIG4.

t

0   1   2   3   2,7мксек

Δx, μ (m)

FIG.5.

EXITING PULSES

VIDEO-ECHO

t (sec)

16·10⁻⁸

8·10⁻⁸

0   10   20   30   40   50   60

-8·10⁻⁸

-16·10⁻⁸

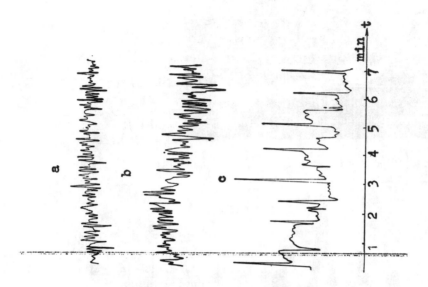

Fig.7.

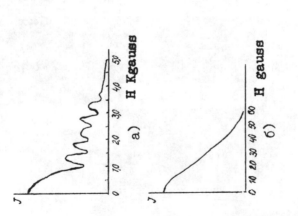

Fig.6.

PROPAGATION AND DEPHASING OF PICOSECOND PHONON-POLARITON PULSES IN AN UNIAXIAL CRYSTAL: LiIO3

F.VALLEE, G.M.GALE and C.FLYTZANIS
Laboratoire d'Optique Quantique du CNRS, Ecole Polytechnique
91128 Palaiseau Cedex - France

Ultrashort pulses of collective propagating excitations in crystals are very sensitive probes of dynamical and structural crystalline properties, yielding novel information on crystal behavior via their temporal and spatial evolution. In the optical phonons domain, the recent demonstration of a non-local nonlinear picosecond technique, namely the time and space resolved CARS, has allowed the creation and detection of Raman active phonon-polariton pulses in noncentrosymmetric cubic crystals [1]. In particular, using this technique in ammonium chloride (NH4Cl), the polariton characteristics were investigated in the vicinity of the crystal phase transition, pointing out the strong influence of crystal disorder on polariton dynamics [1], and new results on polariton interaction with a polar two phonon band (polariton Fermi resonance) were obtained [2].

We have generalized this technique to non-cubic crystals where the birefringence allows to circumvent the wave-vector restrictions associated to Raman techniques in cubic crystals; in uniaxial crystal, using adapted crystal orientation and beams polarization, the entire polariton dispersion curve can be observed [3]. We have performed time and space resolved CARS experiments in the uniaxial LiIO3 crystal, where we have investigated the upper and lower ordinary E_1 polariton, on both sides of the highest frequency Reststrahlen band (ω_{TO}=768cm^{-1}, ω_{LO}=843cm^{-1}).

In time and space resolved CARS experiments, a polariton wave packet is created at a certain time and point inside the crystal by coherent Raman scattering of two synchronized picosecond pulses. Its subsequent temporal and spatial evolution is directly monitored by coherent anti-Stokes Raman scattering of a probe pulse, which is delayed in time and displaced in space to follow the polariton inside the crystal, giving a direct access to polariton group velocity and dephasing time. Our group velocity measurements of ordinary polariton in LiIO3 are in very good agreement with the indirect value obtained from the derivative of the polariton dispersion curve. The frequency dependence of the measured polariton dephasing rate is depicted in Fig.1, for a fixed crystal temperature of 80K, showing a minimum of the dephasing rate in vicinity of the Reststrahlen band.

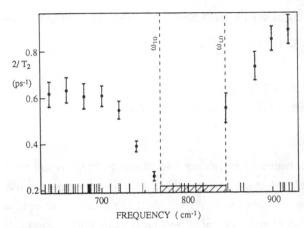

Fig.1: Frequency dependence of the measured polariton dephasing rate at T=80K. The full and dashed lines symbolize respectively the possible down- and up-conversion relaxation processes.

In ordered crystals, polariton dephasing is mainly due to anharmonic interaction with many phonons states, and hence the frequency dependence of the polariton dephasing rate mainly reflects the density of states of the accessible many phonons bands. At the lowest order, two processes intervene: down-conversion ($\omega_\pi \rightarrow \omega + \omega'$) into two lower energy phonons and up-conversion ($\omega_\pi + \omega \rightarrow \omega'$) due to scattering by thermal phonons. We have estimated the number of channels for these two mechanisms in LiIO$_3$ using the zero wave-vector phonons frequency [4], energy conservation and symmetry requirements. The results are plotted at the bottom of Fig.1, and show a good correlation between the number of accessible states and the measured polariton damping rate. Below the Reststrahlen band down and up-conversion processes are possible. The former correspond to polariton splitting into two phonons of comparable energy, and the latter to polariton conversion into ω_{TO}-ω_{LO} frequency region with absorption of a thermal lattice phonon ($\sim$100cm^{-1}). As the Reststrahlen band under investigation corresponds to the highest one phonon frequency region in LiIO$_3$, up-conversion is impossible for polaritons above the longitudinal mode, so that their dominant relaxation channel is down conversion into the ω_{TO}-ω_{LO} region with emission of a lattice phonon.

This interpretation of polariton relaxation receives additional support from our temperature dependent measurements as exemplified in Fig.2-3 for the 680cm^{-1} and 880cm^{-1} polaritons, respectively below and above the Reststrahlen band. For these particular polariton frequencies the dephasing rates can be written:

$$\Gamma(680) = \gamma^-(680)[1+2n(340)] + \gamma^+(680)[n(100)-n(780)]$$

$$\Gamma(880) = \gamma^-(880)[1+n(100)+n(780)]$$

where γ^- and γ^+ are frequency dependent average anharmonic coupling factors for respectively the down and up conversion processes, and n(ω) the Bose factor for the phonon of frequency ω. At low temperature, only the down-conversion process intervenes so that γ is fixed by low temperature measurements. Consequently, $\gamma^+(680)$ is used as a fitting paramater for the 680cm^{-1} polariton, whereas there is no real fitting parameter for the 880cm^{-1} polariton. Very good agreement between predicted (full lines of Fig.2-3) and measured temperature dependent dephasing rates is obtained for these two polariton frequencies. Similar agreement is obtained over the whole frequency range studied, which comfirms the previous interpretation of polariton relaxation.

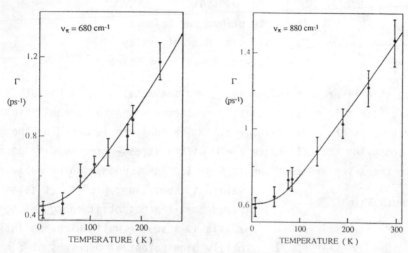

Fig.2-3: Temperature dependence of the polariton dephasing rate below (680cm^{-1}) and above (880cm^{-1}) the Reststrahlen band.

In conclusion,using a generalization of the time and space resolved CARS technique to uniaxial crystal, we have investigated the ordinary polariton dephasing rate in LiIO$_3$, on both sides of a Reststrahlen band. The frequency and temperature dependent measurements demonstrate the important role plays by the ω_{TO}-ω_{LO} frequency region in the polariton relaxation. These new results together with the previous ones clearly show that the time and space resolved CARS technique constitutes a powerful tool for the continuous probing of crystal anharmonicity as a function of frequency, which, together with temperature dependent measurements, opens up new possibilities for the investigation of anharmonic processes and phonon relaxation in solids.

References
1 G.M.Gale, F.Vallée and C.Flytzanis, Phys.Rev.Lett. 57, 1867 (1986)
2 F.Vallée, G.M.Gale and C.Flytzanis, Phys.Rev.Lett. 61, 2102 (1988)
3 L.A.Kulevsky, Yu.N.Polivanov and S.N.Poluektov, J.Ram.Spectr. 3, 239 (1975)
4 F.E.A.Melo and F.Cerdeira, Phys.Rev. B26, 720 (1982), and references therein

AMPLIFICATION OF RAYLEIGH SOUND WAVES IN
THIN BISMUTH FILMS AT LOW TEMPERATURES

C. C. Wu
Institute of Electronics
National Chiao Tung University
J. Tsai
Institute of Nuclear Science
National Tsing Hua University
Hsinchu, Taiwan, Republic of China

A thin layer with thickness d of a semimetal such as bismuth is grown epitaxially on an insulating substrate with the same elastic properties as the semimetal layer.[1] The analysis is made for the medium occupying the half-space $z \geq 0$ with a stress-free boundary parallel to the x-y plane as shown in Fig. 1. The propagation of Rayleigh wave $\vec{q}$ is kept along [110]. It is assumed that the potential energy along the z axis is a square well which has infinitely high potential barriers at $z = 0$ and $z = d$. The thickness of depletion layers becomes narrower by increasing the concentration of donor impurities or by making an epitaxial film with layered structure in which impurities are more highly doped in the layer near $z = 0$ and

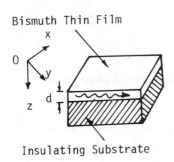

Bismuth Thin Film

Insulating Substrate

Fig. 1

$z = d$. The field operator $\psi(\vec{r})$ of electrons in the second quantization can be taken the form [2]

$$\psi(\vec{r}) = (2/V)^{\frac{1}{2}} \sum_{n=1}^{\infty} \sum_{\vec{k}} b_{\vec{k}n} \exp(i\vec{k}\cdot\vec{x}) \sin(\frac{n\pi z}{d}) , \qquad (1)$$

where $\vec{r} = (\vec{x},z) = (x,y,z)$, $\vec{k} = (k_x,k_y)$, V is the volume of the film, and $b_{\vec{k}n}$ is the operator of electrons satisfying the commutation relation of Fermi type. The energies of electrons $E_{\vec{k}n}$ in bismuth for the modified nonellipsoidal-nonparabolic (MNENP) model are given by the relation [3,4)]

$$E_{\vec{k}n}\left(1 + \frac{E_{\vec{k}n}}{E_g}\right) = \frac{\hbar^2 k_x^2}{2m_1} + \frac{\hbar^2 k_y^2}{2m_2}\left(1 - \frac{\hbar^2 \pi^2 n^2}{2m_3 E_g d^2}\right) + \frac{\hbar^4 k_y^4}{4m_2^2 E_g} - \frac{\hbar^4 k_x^2 k_y^2}{4m_1 m_2 E_g} + \frac{\hbar^2 \pi^2 n^2}{2m_3 d^2}, \quad (2)$$

where E_g is the energy gap between the conduction and valence bands, m_1, m_2, and m_3 are the effective masses of electrons along x-, y-, and z-axis, respectively.

The surface-phonon field operator can be expressed as [2)]

$$\vec{u}(\vec{r},t) = \sum_{\vec{q}} \left(\frac{\hbar}{2\rho\omega_{\vec{q}}S}\right)^{\frac{1}{2}}[a_{\vec{q}}\vec{u}_{\vec{q}}(z)\exp(i\vec{q}\cdot\vec{x} - i\omega_{\vec{q}}t) + (\text{Hermitian conjugate})],$$

$$(3)$$

where ρ is the mass density of the medium, $\omega_{\vec{q}} = c_R|\vec{q}|$ is the angular frequency of Rayleigh waves with velocity c_R, S is the surface area of the film, $a_{\vec{q}}$ is the operator of surface phonons obeying the commutation relation of Bose type, and $\vec{u}_{\vec{q}}(z)$ is the wave function of Rayleigh waves. Using the Green's function method with the Born approximation, the amplification coefficient of the surface phonon for the MNENP model can be obtained as

$$\alpha = \frac{8}{\pi^3}\left(\frac{x}{\hbar^2}\right)\frac{(m_1^2 m_2 E_g)^{\frac{1}{2}}}{k_B T}\left(\frac{q^2}{\rho c_R J}\right)\left(\frac{c_R}{c_\ell}\right)^4 \frac{A^2 C^2}{\left(1 + \frac{6\pi e^2 N^2 c_\ell^2}{\varepsilon_0 q^2 E_F c_R^2}\right)^2} \sum_{n=1}^{\infty}(-1)^{n+1}\exp\left[\frac{n(\frac{1}{2}E_g + E_F)}{k_B T}\right]$$

$$\times n \sum_{i,j} a_i^{\frac{1}{2}} a_j \left[\frac{S(\vec{q};i,j)}{T(\vec{q};i,j)}\right]^{\frac{1}{2}} K_{\frac{1}{4}}\left[\left(\frac{nE_g}{2k_B T}\right)\frac{S^2(\vec{q};i,j)}{T(\vec{q};i,j)}\right]\exp\{-\left(\frac{nE_g}{2k_B T}\right)[R(\vec{q};i,j)$$

$$- \frac{S^2(\vec{q};i,j)}{T(\vec{q};i,j)}]\}\frac{i^2 j^2[1 - (-1)^{i+j}\exp(-\pi A)]^2}{[A^2 + (i-j)^2]^2[A^2 + (i+j)^2]^2}, \quad (4)$$

where $x = v/c_R - 1$ is the drift parameter with the drift velocity of electrons v, C is the deformation potential, ε_0 is the dielectric constant, N is the electron concentration, E_F is the Fermi energy, c_ℓ is the velocity of the longitudinal sound wave, J and A are parameters defined from c_R, c_ℓ, and c_t (velocity of the transverse sound wave), and $K_p(z)$ is the modified Bessel function of order p. R, S, and T are functions of q, a_i, and a_j with

$$a_n = (1 + 2\pi^2\hbar^2 n^2/m_3 E_g d^2)^{\frac{3}{2}}. \quad (5)$$

The relevant values of physical parameters for bismuth are [5,6] m_1 = $m_0/172$, m_2 = $m_0/0.8$, m_3 = $m_0/88.5$ (m_0 is the free electron mass), ρ = 9.8 gm/cm³, N = 2.75 x 10^{17} cm⁻³, ε_0 = 10, c_R = 2.9 x 10^5 cm/sec, c_ℓ = 4.9 x 10^5 cm/sec, c_t = 3.8 x 10^5 cm/sec, E_g = 0.0153 eV, E_F = 0.0276 eV, C = 10 eV, d = 1 μm, and T = 4.2 K. The frequency dependence of the amplification coefficient at x = 100 (the applied electric field E = 960 mV/cm) is shown in Fig. 2. It shows that the amplification coefficient increases with the frequency rapidly, and after passing a maximum point the amplification coefficient decreases with the frequency. We also present some numerical results for the ellipsoidal-parabolic (EP) model, the ellipsoidal-nonparabolic (ENP) model, and the nonellipsoidal-nonparabolic (NENP) model. Results show that the amplification coefficients for EP, ENP, and MNENP models are much closer. The amplification coefficient versus the drift parameter x or the applied electric field E at ν = 3 GHz is shown in Fig. 3. We see that the NENP model in bismuth could not be used quite well in low temperatures.

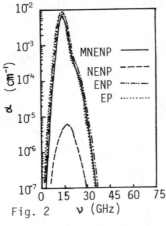

Fig. 2 ν (GHz)

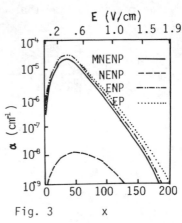

Fig. 3   x

1) Maissel, L. I. and Glang, L. *Handbook of Thin Film Technology*, McGraw-Hill Inc., New York, 1970.

2) Tamura, S and Sakuma, T., Phys. Rev. B15, 4948 (1977).

3) McClure, J. W. and Choi, K. H., Solid State Commun. 21, 1015 (1977).

4) Wu, C. C., Chen, M. H., and Lin, C. J., J. Phys. C:Solid State Phys. 16, L317 (1983).

5) Harrison, M. J. Phys. Rev. 119, 1260 (1960).

6) Fal'kovskiĭ, L. A., Usp. Fiz. Nauk 94, 3 (1968)[Sov. Phys.-Usp. 11, 1 (1969)].

FORCED STANDING SOLITONS AND CHAOS

Rongjue Wei

Institute of Acoustics, Nanjing University, Nanjing, China

The renewed interests of Faraday's oscillating water tank experiment in 1831 have led several recent important works[1-3] concerning either one of the two dominant phenomena in contemporary nonlinear physics--soliton and chaos. This paper will focus on some of the interesting problems left over by Ref.1 and presenting some experimental evidence of the connection between these two nonlinear phenomena out of this simple experiment.

The experimental set-up for this discovery[1] was simple--a long rectangular trough of water (which we adopted:$l_x \times l_y \times h=38 \times 2.7 \times 2$ cm^3, with h as the depth of water). It was driven vertically and pure-simple harmonically with a loudspeaker at a frequency of 2ν of about 10Hz. Above a certain threshold there was a paramatrically excited surface wave at half of the drive frequency. Considering the long trough as the two dimensional wave guide with the normal mode (p,q) governed by the equetion

$$\nu_{p,q} = c/2[(p/l_x)^2 + (q/l_y)^2]^{\frac{1}{2}} \qquad (1)$$

here c or the phase velocity mainly controlled by the wave number k and depth of water h, these authors observed at (0,1) mode some highly localized pulses or nonpropagating solitons in the x-direction and sloshing in the y-direction. We are going to supply some results that were not fully explored: (1) the effect of the trough dimensions, driving frequency and the amplitude on the formation etc. of the solitons, (2) the observation of multisoliton collisions and particularly, (3) the condition and behavior for this solitary system that exhibits chaos via bifurcations.

These standing solitons can be made moving in various ways. For a pair of same polarity they may either combine into one standing flat soliton (Fig.3b) or ready break up into two (Fig.3c) by "passing over" each other and repeat the same process indefinitely. The collision dynamics is one deserving further investigations, with some shown in Fig.3d. Its difference from Kdv solitons will be illustrated.

There have been several theoretical paper discussing the transition of solitary system to chaotic, e.g., by adding Kdv equation with some perturbed terms[5], but we will demonstrate it by increasing the vibrating amplitude. This is done by dipping a pair of electrodes into the trough. The peak to peak amplitude A(measured by a B/K accelerometer) of the vibrator vs time and corresponding FFT spectra are shown in Fig.4–Fig.6. It is rather interesting and intriguing to observe these phenomena, e.g., by comparing the period T (amplitude vs time) of Fig.4(about 1sec) with Fig.6a (T varies from 6 to 8sec. vs time) the latter shows the dominancy of subharmonics as time goes on. This furnishes another support for it enroutes to chaos via bifurcations.

Under simplified assumptions, both Larraza-Putterman[5] and Miles[6] arrived at nonlinear Schrodinger's equation for envelope soliton, under the assumption of $l_x = \infty$. We have worked out an equation of this type for finite l_x that obeys time reversal invariant and FPU recurrence so that the conservation of fluid mass is conserved. The calculated period of a successive collision is around 10 sec. showing agreement with the experiment[1]. Further theoretical as well as experimental investigations are in progress.

This paper will be presented with a video tape demonstration..

References:
1a. Wu,J., Keolian,R. and Rudnick,I.,Phys. Rev. Lett. 52, 1421(1984);
 b. Wu,J., Ph. D thesis, Phys. Dept., UCLA(1985).
 2. Keolian, R., Turkevich, L. A., Putterman, S. J. and Rudnick,I., Phys. Rev. Lett. 47, 1133(1981).
 3. Ciliberto, S. and Gollub, J. P., Phys. Rev. Lett. 52, 922(1984).
 4. Birner, B., Physica 19D, 238(1986).
 5. Larraza, A. and Putterman, S., Phys. Lett. 103A, 15(1984); J. of Fluid Mechanics, 148, 443(1984).
 6. Miles, J. J. Fluid Mech. 148, 450(19840.
 7. Xu,Y. G. Ph.D Thesis, Acoustics Inst., Nanjing Univ., Feb.(1986).

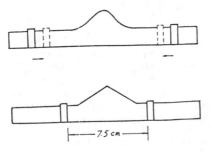

Fig.1 Effect of l_x on the generation and shape of a soliton.

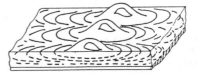

Fig.2 Array of soliton of $l_x = 54mm$, $l_y = 18cm$, exciting frequency 10 Hz for (0,6) mode.

Fig.3a Two solitons of same polarity.

Fig.3b The two solitons of same polarity then combine into one flat soliton.

Fig.3c The flat soliton start to break up.

Fig.3d Mode after collisions, for long trough, $l_z = 26cm$.

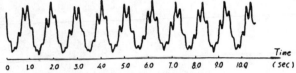

Fig.4 Electrods response at Stokes wave trough for $A = 0.76mm$, wave

amplitudes (in arbitrary linear unites) *vs* time in seconds.

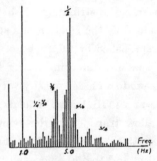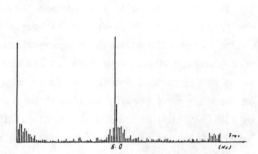

Fig.5a FFT spectrum analysis for Fig.4. Frequency resolution   Fig.5b FFT spectrum for soliton formed after outside

$\Delta\nu = 0.156 Hz$, amplitude in arbitrary linear unites.   disturbance. $\Delta\nu = 0.078 Hz$, $A = 0.8mm$.

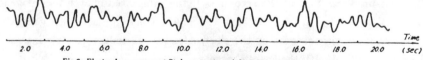

Fig.6a Electrodes response at Stokes wave trough for $0.76mm < A < 1.30mm$,

wave amplitude (in arbitrary linear unites) vs time in seconds.

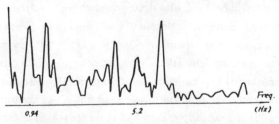

Fig.6b FFT continuous spectrum for $0.76mm < A < 1.30mm$, $\Delta\nu = 0.078 Hz$,

amplitude in arbitrary linear unites.

PHONON ENHANCED CHARGE TRANSFER IN II-IV COMPOUNDS

I.TERRY*,N.BENSAID*,D.J.MEREDITH*,J.K.WIGMORE* & B.LUNN+

* Physics Department, Lancaster University, Lancaster, U.K.
+ Department of Engineering Design & Manufacture, Hull University, Hull, U.K.

In a recent paper[1], it was shown that the energy distribution of shallow energy states in the $Bi_{12}GeO_{20}$ (BGO) can be obtained by the creation of a space charge hologram. A static charge distribution is established by the superposition of a travelling strain wave $S_1 \cos(\omega t - kx)$ and a spatially uniform electric field $E_2 \cos \omega t$. The actual transfer mechanism which leads to a charge density varying on the scale of an acoustic wavelength ($\simeq 0.5$ μm at 10 GHz) may be generated by either electric-field induced ionisation of shallow traps or carrier hopping between them. At 4K and in the absence of illumination, relaxation of the charge distribution via thermal ionisation may be very slow and the inhomogeneity will persist for long periods. Thus, following the application of a sequence of pulse pairs (the first produces S_1 and the second E_2), the integrated charge inhomogeneity can be observed by applying a single low power (read) pulse which generates a backward wave phonon echo by the electrostrictive or acoustoelectric interaction of its electric field with the space charge.

However, an electric field will also destroy the charge grating by homogeneous ionisation of trapped carriers. Because the tunnelling probability for field ionisation has an exponential dependence on the electric field, a microwave pulse can give rise to homogeneous ionisation through the appropriate term in its Fourier expansion. Under normal experimental conditions, a fully integrated hologram is addressed, non-destructively, by a low power read pulse. As the pulse amplitude is increased, the grating is gradually erased but at a rate which is determined by the energy distribution of the traps. This results from the sharp edge-like dependence of the tunnelling probability on trap depth for a given amplitude of read pulse.

Thus by ramping up the electric field of the read pulse, it is possible to achieve a pseudo-spectroscopic scan of the trap distribution. Figure 1 shows the result of applying the technique to BGO doped with Zn(0.9% PFU).

The condition of a long lived charge inhomogeneity imposes limitations on the type of materials that can be investigated. Whether or not storage is observed is a balance between a number of different relaxation processes. In materials which are resistive at low temperature, the number of free carriers is small prior to illumination with band-gap radiation and the electron traps (and also deeper recombination centres) are empty. If electrons are to be trapped in shallow states following illumination, a mechanism must be present which inhibits the recombination of electron and holes, but a detailed knowledge of the mechanism is not necessary for the method to work. The dynamical behaviour of the system during and after the application of microwave pulses can then be described by simple rate equations involving the trapped and space charge distributions[2] with the grating life time being determined by thermal ionisation of the traps and by the presence of free carriers.

Phonon echo measurements could not be made in all of the CdS and CdTe samples which were investigated. The fact that the size of the storage echoes is a function of the method of preparation and the number and type of impurity included in the melt is indicative of the extrinsic nature of the generation process. All samples of CdTe were prepared in the Department of Design and Manufacture at Hull University. Nominally pure CdTe, grown by the Bridgman method gave echoes which were too small for quantitative measurement but samples containing 100 ppm of indium and 0.01% manganese and the alloy $Cd_{0.96}Zn_{0.04}Te$ showed storage effects with echo amplitudes of a comparable size. Anion compensated CdS provided by IBM was known to give long lived storage echoes and was used for reference purposes. The distribution of states in CdS is shown in figure 2. It should be noted that it was possible to switch the sample from a state with a relatively short grating life-time, τ_1, (a few seconds) to one with τ_1 in excess of many hours by applying very large microwave pulses following irradiation at the band-gap frequency. This suggests that the barrier to recombination can be lowered by an electric field and free carriers, which would otherwise smear out the grating, are removed by recombination at a deep centre. Measurements of the microwave conductivity confirm this hypothesis. Storage echoes in all CdTe samples have short ($\simeq 100$ μs) liefetimes at 1K. Again, it is possible to switch their conducting state following band gap irradiation but a combination of microwave pulses and infra-red radiation was required to restore the dark microwave conductivity. The barrier to recombination is probably large in CdTe and the process only occurs if holes are released from deep traps by the i.r. However, the grating life-time is unaffected by the number of free carriers present. The origin of the short τ_1, is not clear. The variation of the storage echo amplitude with S_1 and E_2, as illustrated by the power dependences in figure 3, cannot be explained by the field ionisation model[2], and the charge grating could be the result of field assisted hopping. The decay of the grating (figure 4) and the twin peaks in the S_1 dependence of the echo amplitude suggest that there are two contributions to the grating.

1) Terry, I., Meredith, D.J. and Wigmore, J.K., Semiconductor Science and Technology (1989), to be published.

2) Shiren, N.S. and Melcher, R.L. Proc. IEEE Symp. Sonics and Ultrasonics (New York), 558, (1974).

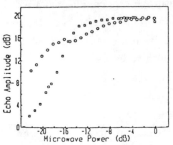

Fig 1. Normalised energy dependence of the density of filled states as a function of trap depth ϵ in BGO:Zn

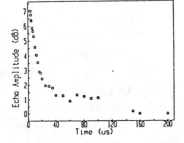

Fig 2. Normalised energy dependence of the density of filled states as a function of trap depth ϵ in CdS.

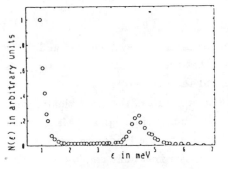

Fig 3. Storage echo amplitude as a function of microwave power P_1 (o) and P_2 (□) in CdTe:In.

Fig 4. Decay of the storage echo in CdTe:In showing evidence for two time constants.

ACOUSTIC SPECTRAL HOLE-BURNING IN RESONANT OSCILLATORS

F. Tsuruoka

Department of Physics, Kurume University

Kurume, Fukuoka 830

JAPAN

INTRODUCTION

In the energy absorption spectrum traces can be left through transforming the characteristics of some absorbers in narrow frequency region. The phenomenon is called spectral hole-burning. Photochemically induced hole-burning has been well studied and its application, for example, to memory system considered.[1]

Piezoelectric powder under resonant condition that is determined with applied frequency, its dimention, crystallographic angle and elastic constants resonantly absorbs alternating electric field energy. When shapes of particles are irregular and crystallographic angles are randomly distributed, the absorption frequencies are widely spread. The transformation of their acoustic properties may be introduced by application of high amplitude electric field to lead to acoustic spectral hole-burning in resonant oscillators.

EXPERIMENTAL METHOD

Sample powders were prepared by grinding commercial grade grains $KBrO_3$ and sieving them with two standard screens of 125 - 149 µm. Their electromechanical mode structures are extremely complex. Powders were set in a capacitor constructed of two copper plates (8 x 9 cm^2) separated by 1 mm. The number of contained particles was about 10^6. Rectangular electric field pulses to burn holes in powders were derived from a gated modulator (MATEC 5100 + 515A) and a cw generator (R&S SMX). A block diagram of the apparatus for spectrum measurement is shown in

1084

Fig. 1. Spectrum is expressed as electric energy dissipation in the condenser containing powder particles. A condenser with one piezoelectric particle is described by a electrical equivalent circuit composed of parallely connected a condenser and a series-resonant circuit with a resistance R. Electric energy dissipation in the equivalent circuit or the resistance R under constant voltage is expressed with the real part of its admittance (inverse of the impedance). The resistance R is proportional to the damping coefficient of particle oscillation.[2] A condenser with particle aggregate is represented by a circuit containing a number of parallel branches of series-resonant circuits. By the measurement apparatus with a hybrid the signal that is proportional to the real part of its admittance or the spectrum was obtained.

EXPERIMENTAL RESULTS

In Fig. 2 examples are shown. Curves (I) and (II) show the spectrum before and after field application, respectively. Apllied field strength E is 100 V/mm, frequency f 12.50 MHz, pulse width Δ 50 μs and repetition rate 20 Hz for 30 s at room temperature under 10^{-2} mmHg. Traces of applied pulses could be left and acoustic spectral hole-burning in piezoelectric powders has been achieved. The shape and half width of traces depended upon not only the Fourier component of rectangular pulse shape but also applied field strength. Increase of the signal or the real part of the admittance shows decrease of the resistance R in the equivalent circuit and of damping coefficient of particle oscillation. The latter was confirmed by measurement based

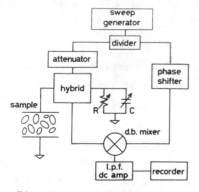

Fig. 1. A block diagram for spectrum measurement.

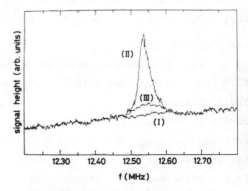

Fig. 2. Examples of the obtained spectrum.

on dynamic echo mechanism.[3] Curve (III) is obtained as follows. Hole burned particles were removed from the capacitor, well stirred and replaced. The 1st stirring treatment reduced height of the hole by about 85 %, but following repetitive treatment did not further change in its height. Height reduction of the hole is understood by consideration that piezoactive axes were diverted from electric field direction by stirring treatment. Curve (III) represents that intrinsic properties for oscillation of particle were transformed. Plastic deformation is thought to be introduced by large amplitude strain exerted. Strain amplitude S under resonant condition is estimated as $S = \varepsilon E \cdot 2\pi f \Delta$, where ε is the piezoelectric coefficient. With the typical value $\varepsilon \sim 3.3 \times 10^{-12}$ C/N for $KBrO_3$, strain S reaches higher than 10^{-3}. When a particle is broken, the real part of admittance of a condenser with the particle at the former resonant frequency decreases. But the experimental result shown in Fig. 2 does not support this estimation. For decrease of damping coefficient, we propose a role of dislocations introduced and multiplicated by large amplitude strain. The particles obtained by grinding had originally high dislocation density. Oscillation damping due to dislocations in crystal is well described by the Granato-Lücke model in which oscillation drives dislocations in a resistive atmosphere to dissipate oscillation energy.[4] Multiplicated and tangled dislocations do not effectively vibrate or contribute to energy dissipation.

In summary, traces of high amplitude electric field pulses could be left on the energy absorption spectrum of piezoelectric resonators and were found to endure despite stirring treatment. Experimental results were attempted to be explained by dislocations introduced by crushing on sample preparation and multiplied by large amplitude strain.

REFERENCES

1) Woerner, W.E., "Persistent Spectral Hole-Burning: Science and Applications", (Springer Verlag, Berlin 1988).
2) Marutake, M., Proc. IRE 49, 967 (1961).
3) Fossheim, K., Kajimura, K., Kazyaka, T.G., Melcher, R.L., and Shiren N.S., Phys. Rev. B17, 964 (1978).
4) Granato, A.V. and Lücke, L., J. Appl. Phys. 27, 583, 789 (1956).

NONLINEAR WAVEPACKET SCATTERING IN DISORDERED SYSTEMS

Yuri S. Kivshar

Institute for Low Temperature Physics and Engineering, 47 Lenin Avenue, Kharkov 310164, USSR

Wave propagation in nonlinear disordered media has
become an extensively studied subject recently (see, e.g.,
/1-3/). The combined effects of disorder and nonlinearity
introduce many novel and complex properties in the system
under consideration. Disorder in a linear chain will
generally lead to localization (see, e.g., /4/), such a
phenomenon first introduced by Anderson to describe the
exponentially decay of the electron wave function in
a disordered system. The Anderson localization can be
easily generalized to other waves like phonons. The presence of the nonlinearity may drastically change the behaviour of the system, and lead to a number of new effects.
In particular, nonlinearity may affect against disorder
and break localization effects.

To study the effect of nonlinearity on properties of
disordered systems one cann't use techniques developed
for the linear systems because the superposition principle
of waves does not hold in the nonlinear case. But, on the
other hand, strongly nonlinear systems have excitations in
the form of solitons, i.e. nonlinear wavepackets which can
propagate in pure systems without changing their shape or
velocity. This report aims to consider the problem of
coexistence of nonlinearity and disorder from the viewpoint
of strong nonlinearity but weak disorder intensity. Using
the perturbation theory for solitons the nonlinear behaviour can be accounted by a simple independent scattering

in the framework of the nonlinear Schrödinger (NLS) equation.

We start from the dimensionless NLS equation for the wave variable $u(x,t)$,

$$iu_t + u_{xx} + 2|u|^2 u = u \sum_n \mathcal{E}_n \delta(x - x_n) . \qquad (1)$$

The term in the r.h.s. of (1) describes random impurities, $\mathcal{E}_n$ being their intensities. In the simplest case $\mathcal{E}_n = \mathcal{E}$. Equation (1) is used to describe the dynamics of an envelope of phonon excitations under the action of an intensive pulse applied to the crystal, see /5/.

When the concentration p of the impurities is low, the scattering of a NLS soliton by many impurities can be approximately treated independently. The NLS soliton is characterized by two parameters, amplitude A and velocity V, or two physical values, energy E and "number of particles" N . The latter allows us to treat the soliton as a bound state of quasi-particles (phonons).

In the limit of independent scattering we may put $E_{i+1} = E_i \, T_E(E_i, N_i)$, $N_{i+1} = N_i \, T_N(E_i, N_i)$, where T_E and T_N are the NLS soliton transmission coefficients /6/, and i is number of an impurity. Using the results of /6/ for $\mathcal{E} \ll 1$ after simple calculations we may obtain the system of equations for the soliton parameters N=4A and V :

$$\frac{dN}{dz} = - (4V)^{-1} \int_0^\infty dy \, F(y, \alpha) , \qquad (2a)$$

$$\frac{dV}{dz} = -(8N)^{-1} \int_0^\infty dy (y^2 - 1) F(y, \alpha) - \frac{N}{128 \, V^2} \int_0^\infty dy F(y, \alpha), (2b)$$

where

$$F(y, \alpha) = \frac{[(y+1)^2 + \alpha^2]^2}{\cosh^2 [(\pi/4\alpha)(y^2 + \alpha^2 - 1)]} . \qquad (3)$$

and $\alpha \equiv N/4V$, $z = x/x_0$, $x_0 \equiv 16/\pi p \mathcal{E}^2$. In the linear limit, $N \ll 1$, simple analysis yields the following results, $V \approx V(0) = $ const, $N = N(0) \exp(-x/\ell)$, $E = E(0) \exp(-x/\ell)$, where $\ell = [p \mathcal{E}^2 / V^2(0)]^{-1}$ has the sense of localization length. In the nonlinear case, $N \gtrsim 1$, Eqs. (2),(3) may be investigated numerically [1]. As a result, the evolution of the system can be described as follows: in the case $\alpha < \alpha_c$, $\alpha_c \approx 1.285$, N tends asymptotically to zero while V goes to a constant, nonzero value, what implies $\alpha(\infty) = 0$. For the other case, $\alpha > \alpha_c$, the evolution drastically changes: both N and V get practically constant. The curves for some values of $N(0)$ are presented in Fig.1.

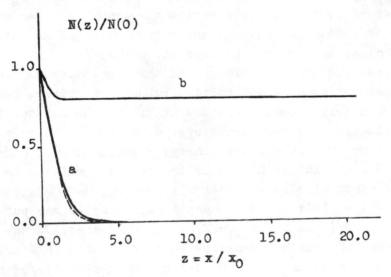

Fig.1. The number of particles N vs. the distance z. Solid lines are numerical curves at $V(0)=1$, and $N(0)=0.1$ (a), $N(0)=10$ (b). Dashed line is the analytical curve obtained for $N(0) \ll 1$.

So, we observe a decay of localization effects due to nonlinearity. The process is threshold in amplitude.

1. Devillard P. and Souillard B., J.Stat.Phys.43,423(1986).
2. Doucot B. and Raman R. J de Phys. 48, 509 (1987).
3. Li Q. et al. Phys. Rev.B, in press (1989).
4. Lifshitz I.M.,Gredeskul S.A., and Pastur L.A. Introduction into Theory of Disordered Systems,Wiley,N.Y.(1988).
5. Tappert F. and Varma C.,Phys.Rev.Lett. 25,1108(1970).
6. Kivshar Yu.S. et al., Phys.Lett. A125, 35 (1987).

[1] The calculations were performed by L.Vazquez and A.Sanchez (University Complutense,Madrid). A joint paper is in preparation.

PHONON BACKSCATTERING BY NORMAL PROCESSES[*]

Gerold Müller and Olaf Weis

Abteilung Festkörperphysik, Universität Ulm

Oberer Eselsberg, D-7900   Ulm, FRG

During our detailed quantitative study [1] of thermal phonon pulses in reflection geometry, we observed in all investigated crystal
plates (a-cut sapphire, [111]-cut silicon, X-cut and Z-cut quartz) that
phonon-backscattering processes due to phonon-phonon interactions occur
in the bulk, especially in the neighbourhood of the constantan radiator
where phonon density is highest.

Fig. 1 illustrates processes which can contribute to the backscattering of the emitted phonons. Scattering at static defects is proportional to the number of emitted phonons of the same kind and so rises
the detector signal linearly. Contrary, phonon-phonon interactions
between emitted phonons produce a detector signal increasing nonlinear

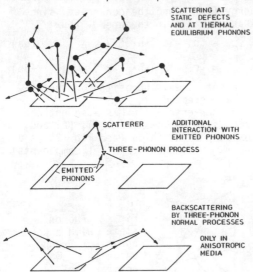

with the number of emitted phonons. At sufficient low radiation temperature, Umklapp-pro-
cesses are negligible and only
normal processes remain.

We demonstrate that within
anisotropic continuum acoustics
two emitted phonons can produce
in a three-phonon N-process one
phonon which has a group velocity pointing back to the radia-
tor/detector face, i.e. show
the same phonon-backscattering
as is thus far only attributed
to U-processes, where the energy flux is reversed to some extend.

Fig. 1: Different contributions
to phonon backscattering of
emitted phonons in a reflection
experiment

[*]   Supported by Deutsche Forschungsgemeinschaft

The phonon-energy flux is conserved under N-processes for a Debye solid [2], but this statement does not hold for a real isotropic or anisotropic solid, as will be shown. For a single three-phonon N-process between phonons with wave vector $\vec{q}$, $\vec{q}'$, $\vec{q}''$ and polarization σ, σ' and σ'', the conservation of energy requires for the phonon frequencies

$$\omega_{\vec{q}\sigma} = \omega_{\vec{q}'\sigma'} + \omega_{\vec{q}''\sigma''} \tag{1}$$

and the selection rule (= conservation of crystal momentum) demands

$$\vec{q} = \vec{q}' + \vec{q}''. \tag{2}$$

The energy flux is given by

$$\vec{S}_{\vec{q}\sigma} = \hbar \cdot \omega_{\vec{q}\sigma} \cdot \vec{w}_{\vec{q}\sigma}/V \qquad \text{and} \tag{3a}$$

$$\vec{S}_{\vec{q}'\sigma'} + \vec{S}_{\vec{q}''\sigma''} = \hbar \cdot \{\omega_{\vec{q}'\sigma'} \cdot \vec{w}_{\vec{q}'\sigma'} + \omega_{\vec{q}''\sigma''} \cdot \vec{w}_{\vec{q}''\sigma''}\}/V , \tag{3b}$$

where $\vec{w}_{\vec{q}\sigma} = \text{grad}_{\vec{q}} \, \omega_\sigma(\vec{q})$ denotes the group velocity and V the crystal volume.

In a <u>Debye solid</u> exists only one sort of phonons with the mean sound velocity $\bar{c}$ which is also the group velocity. As a consequence of (1) and (2) only collinear processes are allowed. Furthermore, (3a) and (3b) are equal and hence, the energy flux is conserved in three-phonon N-processes.

In an <u>isotropic solid</u>, the only allowed noncollinear processes are [3] in accordance with (1) and (2)

$$T + T \gtreqless L \qquad \text{and} \qquad T + L \gtreqless L, \tag{4}$$

where T denotes transverse and L longitudinal phonons. They have a group velocity equal to the phase velocity $\vec{c}_{\vec{q}\sigma} = c_\sigma \cdot \vec{e}_n$ which always points in direction of the wavenormal $\vec{e}_n$:

$$\vec{w}_{\vec{q}L} = \vec{c}_{\vec{q}L} = c_L \cdot \vec{e}_n \qquad \text{and} \qquad \vec{w}_{\vec{q}T} = \vec{c}_{\vec{q}T} = c_T \cdot \vec{e}_n. \tag{5}$$

According to (3) and (5), the energy flux cannot be conserved, since the two-phonon annihilation demands a change in the energy flux

$$\delta \vec{S}_{T+L \to L} \equiv \vec{S}_{\vec{q}L} - (\vec{S}_{\vec{q}''L} + \vec{S}_{\vec{q}'T}) = (c_L^2 - c_T^2) \cdot \hbar \cdot \vec{q}'/V , \tag{6a}$$

$$\delta \vec{S}_{T+T \to L} \equiv \vec{S}_{\vec{q}L} - (\vec{S}_{\vec{q}''T} + \vec{S}_{\vec{q}'T}) = (c_L^2 - c_T^2) \cdot \hbar \cdot \vec{q}/V . \tag{6b}$$

Of course, the sign in (6) at the right side is opposite for the reverse process in which two phonons are created.

In an <u>anisotropic solid</u>, the phonon-energy flux is also not conserved during a three-phonon N-process as follows from (1), (2) and (3). But in addition, there may exist phonon states (see Fig. 2) at a given frequency surface with $\vec{q}$ and $\vec{w}_{\vec{q}\sigma}$ having antiparallel components

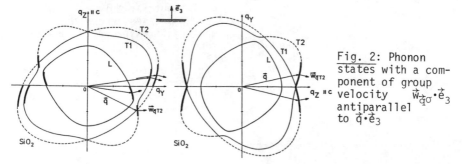

Fig. 2: Phonon states with a component of group velocity $\vec{w}_{\vec{q}\sigma} \cdot \vec{e}_3$ antiparallel to $\vec{q} \cdot \vec{e}_3$

perpendicular to the radiator/detector face. If such phonons are emitted, the $\vec{q}$-vector points backwards and therefore, the selection rule (2) can allow backscattering. On the other hand, if such phonons are produced in the interaction process, they can have a $\vec{q}$-vector pointing in forward direction, but propagate backward.

Hence, it is not allowed for a real solid to conclude from the conservation of quasimomentum that the energy flux is also conserved as is usually done in treating thermal conductivity. In principle, there is no difference between U- and N-processes, since they can be transfered into one another by redefining the basis cell in $\vec{q}$-space. But near the zone boundaries, inside the first Brillouin zone or outside, there exist much more phonon states which allow three-phonon backscattering than in the longwave acoustic branches.

In a typical steady state thermal conduction experiment at sufficient low temperatures, the phonon propagation is dominated by boundary scattering. Casimir [4] introduced the very doubtful assumption that incident phonons are absorbed at the boundary and thereafter diffusely reemitted with a distribution according to the 'local' temperature in order to get a thermal resistance. We demonstrate in Fig. 3 that even in the case of a perfect lateral crystal face where the parallel $\vec{q}$-component is conserved, the energy flux parallel to the crystal face is not conserved. Again phonon backscattering can occur in this two-phonon scattering process.

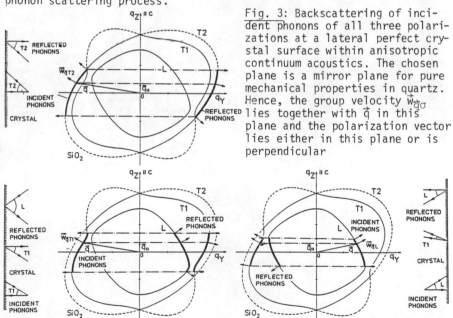

Fig. 3: Backscattering of incident phonons of all three polarizations at a lateral perfect crystal surface within anisotropic continuum acoustics. The chosen plane is a mirror plane for pure mechanical properties in quartz. Hence, the group velocity $\vec{w}_{\vec{q}\sigma}$ lies together with $\vec{q}$ in this plane and the polarization vector lies either in this plane or is perpendicular

1. G. Müller and O. Weis: paper submitted to Z.Phys.B-Condensed Matter
2. R. Peierls: Ann. Physik 3, 1055 (1929)
3. A. Herpin: Ann. de Phys. 7, 91 (1952)
4. H.B.G. Casimir: Physica 5, 495 (1938)

11

Phase Transitions

PHONONS AT MARTENSITIC PHASE TRANSITIONS

W. Petry[+], A. Heiming[+*], J. Trampenau[+$], M. Alba[+],

C. Herzig[$], H.R.Schober[&]

+ Institut Laue-Langevin, 156X, F-38042 Grenoble, France
* Freie Universität Berlin, Fachbereich Physik, D-1000 Berlin, FRG
* Inst. für Metallforschung, Univ. Münster, D-4400 Münster, FRG
& Inst. für Festkörperforschung der KFA-Jülich, D-5700 Jülich, FRG

MARTENSITIC PHASE TRANSITIONS

There exists no clear cut definition of what physicists understand under a "martensitic" phase transition. Originally used to describe the hard microconstituents of quenched steel (to honour the German scientist Martens) physicists call a large variety of transitions in metallic and other systems as martensitic, the characteristics of which can be summarized as follows [1] : The amount of transformation is i) virtually independent of time and ii) is characteristic of temperature. The transition is iii) very reversible, i.e. the same high temperature single crystal is obtained after a transformation cycle. iv) No change in the chemical composition and almost no change in volume is observed after the transition. There exists a v) definite relation between the orientation of the original high temperature structure and the new phase at low temperature or at elevated pressure. Therefore martensitic transformations are per definition of 1st order. They are displacive transitions which distinguish them from diffusive transitions which are often continuous and of 2nd order.

PHONON ANOMALIES IN ALLOYS

Following rigorously these definitions it is not expected to see any precursors of the approaching transition in the dynamical behaviour, i.e. in the phonon dispersion of the high symmetry phase. Table 1 shows a number of metallic alloys which deviate from this rule. Within the high symmetry phase dynamical as well as static precursors of the low symmetry phase are observed. Their temperature dependence

Table 1
Metallic alloys with displacive (martensitic) phase transitions
exhibiting phonon anomalies

Ni$_{1-x}$Ti$_x$	$x \simeq 0.5$	CsCl	$\rightarrow$ monoclinic	[2–6]
Ni$_{1-x}$Al$_x$	$x \sim 0.35$–0.5	CsCl	$\rightarrow$ 7R	[7–9]
Zr$_{1-x}$Nb$_x$	$x = 0.08$–0.24	bcc	$\nearrow \omega$ $\rightarrow$ hcp	[10–12]
Ti$_{1-x}$Cr$_x$	$x = 0.1$	bcc	$\nearrow \omega$ $\rightarrow$ hcp	[12]
AuCuZn$_2$		bcc	$\rightarrow$ fcc	[13]
CuZnAl		fcc	$\nearrow$ 2H $\rightarrow$ 95	[14–15]
CuAl$_3$(Ni)				[16]

and location in q-space reveals detailed information about the micro-scopic mechanism of the martensitic transformation. Two well studied examples, Ni$_{1-x}$Al$_x$ (x = 0.35–0.5)[7–9] and Ti$_{1-x}$Al$_x$ ($x \sim$ 0.5) [2–6] may serve as an illustration.

The high temperature phase of Ni$_{62.5}$Al$_{37.5}$ has the CsCl structure and transforms martensitically at the transition temperature $T_0 \sim$ 80 K to the modulated 7R structure. The well-known shape memory alloy Ni$_{50.5}$Ti$_{49.5}$ transforms from the high temperature ordered CsCl struc-ture to an intermediate or pre-martensitic R-phase at $T_0 \sim$ 320 K. In both cases T_0 is very sensitive to compositional variations and to the thermal pretreatment. Fig. 1 shows the transverse $T_1A[\xi\xi 0]$ phonon branch with [110] polarization measured at temperatures close to T_0 for both alloys. Both alloys show a remarkable softening of the $T_1A[\xi\xi 0]$ phonon branch at $\xi \sim$ 1/6 and ξ = 1/3, respectively, when approaching T_0. The soft and overdamped phonons freeze to marked elastic super-lattice reflections, the intensity of which increases critically upon approaching T_0. The phonons involved are those which achieve the displacements necessary to reach the new phase. It is important to note that the actual phase transitions occur at a finite phonon frequency, which distinguishes them definitely from soft mode transitions like in SrTiO$_3$ or KH$_2$PO$_4$ (KDP).

All these observations are in accordance with a series of theoreti-cal investigations over the last four years [17–22]. Precursor effects in terms of low frequency phonons which further decrease when the phase transition is approached have been predicted. Furthermore, they indicate a freezing of these phonons which manifests itself in an elastic diffuse scattering with a drastically increasing intensity close to the transition temperature T_0.

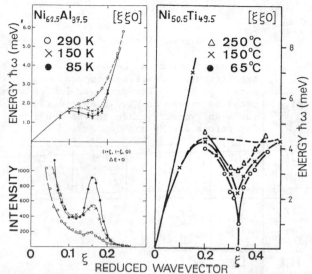

Fig. 1 Temperature dependence of the dispersion of the $T_{[110]}[\xi\xi0]$ mode for $Ni_{62.5}Al_{37.5}$ [7-9] and $Ni_{50.5}Ti_{19.5}$ [2-6] with transition temperatures $T_O = 80$ K and 320 K, respectively. For NiAl the arise of a superstructure peak is shown, too.

PHONON ANOMALIES IN PURE METALS

In contrast to the alloys, neutron scattering studies of precursor effects in <u>pure metals</u> yielded no or only a weak softening of particular phonons close to the martensitic phase transition. Table 2 presents metallic systems with displacive phase transitions and refers to some of the neutron scattering work. For instance in Li and Na phonons of small q of the relevant phonon branch become even harder when T_O is approached. <u>No elastic precursors</u> or charge-density-wave satellites have been found in Na and K. Li and Co show strong satellites above T_O. However, these have been explained by the coherent coexistence of the low temperature phase.

So, as a kind of conclusion of this difference between experimental observations on metallic alloys and pure metals one might ask to what extent phonon softening and central peaks are microscopic properties inherent to the martensitic phase transitions ? Or in other words : are the observed central peaks in alloys precursor of the new phase or simply due to the coherent existence of two phases ? To what extent is the condensation of soft and damped phonons to a central peak driven by point defects ?

Table 2

<table>
<tr><td>

Li
bcc-9R
74K
[23-25]

</td><td></td><td colspan="2">

Pure metals exhibiting displacive phase transitions

</td></tr>
</table>

Li
bcc-9R
74K
[23-25]

References refer to phonon work related to the phase transitions. Transition temperatures T_o are given for normal pressure. For pressure induced transitions the pressure refers to RT. Transitions are indicated from the high symmetry phase to the low symmetry phase. Exceptions are Mg and Fe.

Na bcc-hcp 51K [26]	**Mg** hcp-bcc 50kbar [30]				
K ? < 4K [27-28]	**Ca** bcc-fcc 721K [31-32]	**Sc** bcc-hcp 1610K [35]	**Ti** bcc $<^{\omega}_{hcp}$ 40kbar 1156K [39-40]	**Fe** bcc-fcc- 1667/ bcc 1183K [44]	**Co** fcc-hcp 673K [45]
	Sr bcc-fcc 878K [33]	**Y** bcc-hcp 1763K	**Zr** bcc $<^{\omega}_{hcp}$ 40kbar 1138K [41-42]		
Cs bcc-fcc 34/42kbar [29]	**Ba** bcc-hcp 62kbar [34]	**La** bcc-hcp-dhcp 1138/609K [36-38]	**Hf** bcc $<^{\omega}_{hcp}$ 60kbar 2015K [43]		

Tl
bcc-hcp
507
[46]

PHONON DISPERSION IN β–Ti, β–Zr AND β–Hf

With this in mind we performed extensive studies of inelastic and elastic neutron scattering of the bcc phase of the group 4 metals β–Ti, β–Zr, and β–Hf. By lowering the temperature the bcc lattice transforms to hcp and under pressure the trigonal ω structure (typical pressures at RT of 40 kbar) is obtained. These systems are particularly well suited to study the dynamics of phase transitions because : i) the parent bcc phase is of extreme simplicity and transforms martensitically into either of two new phases, ii) T_o occurs at high temperature : consequently, thermal equilibrium is always achieved during the measurements, iii) the defect concentration can be minimized and iv) theory [17-22] explicitly refers to these materials as model cases for bcc-hcp and bcc-ω transitions. However, these elements are extremely difficult to obtain as bcc single crystals, they have to be grown and oriented in situ on the spectrometer [47].

Figs. 2 to 4 show the phonon dispersion in β–Ti, β–Zr and β–Hf as measured by inelastic neutron scattering and fitted with a Born von Karman model taking into account force constants up to the fifth neighbour shell. The dispersion of all three elements resemble very much to each other, i.e. scale with the square root of the mass and the

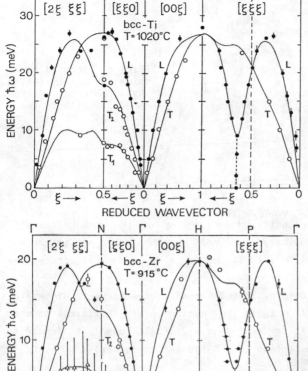

Fig. 2 Phonon dispersion of bcc-Ti fitted by a Born von Karman model with force constants to the 5th neighbour shell (full line) [40].

Fig. 3 Phonon dispersion of bcc-Zr fitted by a Born von Karman model with force constants to the 5th neighbour shell (full line). Bars for the $T_1[\xi\xi\,2\xi]$ and $T_1[\xi\xi0]$ phonons indicate the width of the phonons as fitted by a damped harmonic oscillator [41].

lattice constant. Most evidently the dispersion curves are dominated by two unusual properties : i) At $\vec{q} = 2/3(111)$ the $L[\xi\xi\xi]$ phonon branch shows a pronounced dip. ii) The whole $T_1[\xi\xi0]$ phonon branch with $[1\bar{1}0]$ polarization is of low frequency when compared to the other transverse phonons. iii) The same holds true for the $T_1[\xi\xi\,2\xi]$ phonon branch.

THE L2/3(111) PHONON

As indicated by the broken line in Figs. 2 and 3 phonon intensity reaches down to zero energy transfer for β-Ti and β-Zr. Energy scans at constant $\vec{q} = 2/3(111)$ as shown in Fig. 5 show the large distribution of inelastic intensity which can be best described in the classical picture of an (over)damped harmonic oscillation. The crucial point to note

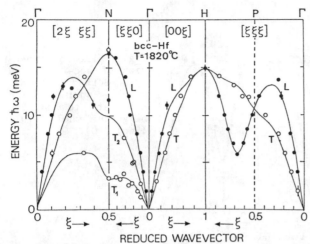

Fig. 4 Phonon disper-
sion of bcc-Hf fitted
by a Born von Karman
model with force
constants to the 5th
neighbour shell [43].

is that the elastic intensity shown on top of the broad inelastic
distribution can be entirely explained by the isotropic incoherent
scattering present at all q-values. Thus <u>no</u> elastic superstructure peak
exists at $\vec{q}$ = 2/3(111).

Similar scans have been performed for all three elements at several
temperatures extending from T_0 to the melting point. Different to what
one expects this scattering around $\vec{q}$ = 2/3(111) does not depend on
temperature (with the exception of a Debye-Waller factor).

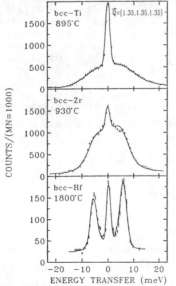

Fig. 5 Energy scans at constant
$\vec{q}$ = 2/3(111) for bcc-Ti, bcc-Zr
and bcc-Hf. All spectra are fitted
with a damped oscillator with
centre energies of 11(1)/7.5(5)/
5.9(1) meV and widths of 15(8)/
10(1)/1.6(2) meV for Ti, Zr and
Hf, respectively.

The L2/3(111) phonon corresponds to displacements of two neighbouring {111} planes towards each other, whereas every third {111} plane

stays at rest. When two neighbouring planes penetrate each other the ω-structure is formed with the crystallographic relation

$$(111)_\beta || (00.1)_\omega \text{ and } [\bar{1}01]_\beta || [01.0]_\omega.$$

Thus strongly damped fluctuations into the ω-phase are observed. Within the time window of the instrument ($\leq 10^{-10}$s) no stable ω-embryos exist. As indicated by the fwhm = 0.1 reciprocal lattice units of the phonon groups, this weakness of the bcc structure is well defined in q-space. In energy the phonons are (over)damped and lifetime of 0.8×10^{-13} s, 1.4×10^{-13} s and 8×10^{-13} s for β-Ti, β-Zr and β-Hf, respectively, are calculated from the phonon width. Because these properties do not change with temperature the weakness towards ω-fluctuations is an intrinsic bcc property.

For this mode the atomic motion can be viewed as a shearing of neighbouring [111] chains relative to each other, while each chain itself remains rigid [48-50]. The phonon frequency is determined by the weak restoring forces between the chains and therefore is low. All other modes will change the distances of the atoms along the chains and thus giving rise to extra restoring forces leading to higher phonon energies. It is the particular electronic screening which determines whether this general weakness of the bcc lattice towards the ω-structure is enhanced or not. Frozen phonon calculations of Ho et al [49-50] which include the particular bandstructure reveal that in group 4 metals the charge density is concentrated in d-bonds running in chains along [111] direction with very little interaction between neighbouring chains but strong interaction along a chain.

Filling up the d-band, the situation is inverted. In group 6 metals with 4 d-electrons it was found that d-bonds entangle the [111] chains and give rise to forces which oppose the shearing motion between [111] chains [49]. As a consequence the geometrical weakness is compensated

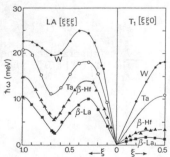

Fig. 6 Comparison of the L[ξξξ] and T_1[ξξ0] branch for the bcc phase of La, Hf, Ta and W. The lower the phonon energies the lower the d-electron density.

and no softening of the L2/3(111) phonon is observed. This picture of the d-electron density as the relevant control parameter is illustrated by Fig. 6 where the degree of softening of the L2/3(111) phonon nicely scales with the decreasing d-electron density.

THE $T_{[110]}1/2(110)$ PHONON

Following the crystallographic relation of the bcc-hcp (β–α) transition

$$(110)_\beta \parallel (00.1)_\alpha \quad \text{and} \quad [\bar{1}11]_\beta \parallel [\bar{2}1.0]_\alpha$$

it is just the $T_1[\xi\xi0]$ N-point phonon with $\xi = 1/2$ which displaces the neighbouring (110) planes in such a way that the hcp stacking sequence is achieved. Additionally two equivalent long wavelength shear modes – for instance $(1\bar{1}2)[\bar{1}11]$ and $(\bar{1}12)[1\bar{1}1]$ – are needed to squeeze the lattice into the final hexagonal basal plane. As with the ω-point the N-point phonon is also considerably damped. Furthermore on approaching T_0 the N-point phonon frequency decreases considerably (see Fig. 7) but the actual phase transition clearly occurs at finite frequency.

These observations may be compared to the results of frozen phonon calcultions of the N-point phonon in Zr [17] which have shown that the bcc phase is only stabilized by anharmonic contributions. For β-Zr at T = 1123°C these calculations yielded a phonon energy at the N-point of $\hbar\omega_N$ = 4.14(12) meV in good agreement with our experimental result $\hbar\omega_N$ = 5.4(1) for T = 1150°C. The temperature dependence of $\hbar\omega_N$ as calculated [17], $d\hbar\omega_N/dT = 8 \times 10^{-3}$ meV/K, and our measurements (Fig. 7) $d\hbar\omega_N/dT = 3.4(4) \times 10^{-3}$ meV/K do not agree so well. However, considering the problems of such calculations we regard this discrepancy as not so severe. A recent thermodynamic approach to the problem, expansion of the free energy in terms of the dynamical displacements, has independently suggested that a further small softening of the relevant

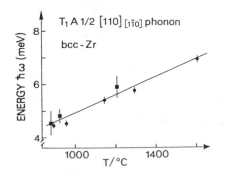

Fig.7 Temperature dependence of the $T_1 1/2[\xi\xi0]$ phonon in bcc-Zr.

low-energy phonon is sufficient to produce a lower minimum of the free energy for the low temperature phase [19,21].

As in the case of the ω-phonon no condensation of the damped N-point phonon to an elastic superstructure peak around $\vec{q} = 1/2(110)$ has been found.

RELATION BETWEEN THE L2/3(111) AND THE $T_1 1/2(110)$ PHONON ?

It has been shown that soft and damped phonons in two distinct regions of the q-space are characteristic features of the group 4 metals. The soft and temperature dependent $T_1[\xi\xi 0]$ phonon branch is the precursor of the martensitic $\beta \to \alpha$ transition. The soft L2/3(111) phonon reflects the intrinsic bcc property of a weak restoring force for displacements towards the ω-phase. It is appealing to ask whether the two soft phonons are related to each other.

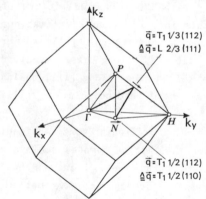

$\vec{q} = T_1 1/3(112)$
$\Delta\vec{q} = L\ 2/3\ (111)$

$\vec{q} = T_1 1/2(112)$
$\Delta\vec{q} = T_1 1/2(110)$

Fig. 8 Brillouin zone for bcc. Beside the usual main symmetry phonon branches in [$\xi\xi 0$], [00ξ] and [$\xi\xi\xi$] the off-symmetry branch [$\xi\xi 2\xi$] (thick line) is indicated. From $\vec{q} = 1/3(112)$ to $\vec{q} = 1/2(112)$ phonons along the Brillouin zone boundary are measured. The arrows indicate the change in polarization.

The answer is best visualized in reciprocal space. Fig. 8 shows the Brillouin zone of the bcc lattice. Both phonons, $T_1 1/2(110)$ and L2/3(111), lie on the zone boundary with polarizations as indicated by the arrows. Geometrical considerations show that the $T_1[1\bar{1}0]1/2(110)$ phonon is identical to the $T_1 1/2(112)$ phonon and the L2/3(111) phonon is identical to the $T_1 1/3(112)$ phonon. In other words, both phonons lie in the same off-symmetry $T_1[\xi\xi 2\xi]$ branch. Interesting enough the initial slope of this phonon branch is given by the long wavelength shear of (112) planes in the [11$\bar{1}$] direction, i.e. the soft shear needed for the $\beta \to \alpha$ transition. Fig. 7 shows the propagation vector [$\xi\xi 2\xi$] for $\xi = 0 \to 0.5$ reduced to the first Brillouin zone.

From Fig. 8 it is evident that <u>all</u> phonons needed for the phase transitions in group 4 metals lie on the same off-symmetry branch and it is anticipated that they are connected via a <u>valley of soft and (over)damped phonons on the Brillouin zone boundary along [112] direction</u>. That nature behaves like that, is best shown in the case of β-Zr in Fig. 3. Phonons along $T_1[\xi\xi 2\xi]$ are of low energy and as indicated by the bars a valley of overdamped low frequency phonons connects the $T_1 1/3(112)(= \omega$-point) phonon with the $T_1 1/2(112)$ (= N-point) phonon.

A PICTURE OF THE LATTICE VIBRATIONS IN REAL SPACE

Summarizing where soft modes are observed in q-space we state : along $[\xi\xi\xi]$ direction soft modes are observed in a narrow range around $\xi = 2/3$, i.e. indicating a correlation length in real space over roughly 10 lattice parameters along [111]. Perpendicular to $[\xi\xi\xi]$, namely in $[\xi\bar{\xi}0]$ and $[\xi\xi \bar{2}\xi]$ phonons propagate almost without dispersion, i.e. are <u>localized</u>. So we end up with the following picture of the lattice vibrations of group 4 metals in real space : excitations propagate along [111] chains. Excitations perpendicular to [111] do not propagate. If a given [111] chain vibrates along its direction neighbouring [111] chains do not follow this motion. Along [111] chains one has strong restoring forces whereas those perpendicular to [111] are weak.

The soft phonons are located on the Brillouin zone surface. Different to long wavelength shear modes these large q-modes achieve a maximum shift of neighbouring planes to each other, i.e. are best suited to achieve the displacement necessary for displacive or martensitic phase transitions.

DEFECT DRIVEN CONDENSATION OF SOFT AND OVERDAMPED MODES

With regard to the elastic precursors or central peaks all theoretical calculations insist on the freezing of the amplitudes of the low energy phonons which are related to the phase transition which then produce a kind of (quasi)elastic scattering. Our elastic measurements on <u>high purity</u> β-Ti and β-Zr in various Brillouin zones and at various temperatures show, however, to a high degree of precision no intensity which could be interpreted in terms of elastic precursors of the transition. The precursor effects which we find, namely the pronounced softening of overdamped phonons in the $T_1[\xi\xi 0]$ branch, are of purely dynamical nature.

It is obvious to suggest that the elastic precursors observed in metallic alloys are defect driven. To prove that for alloys on the basis of group 4 metals, we alloyed Ti and Zr with impurities as different as oxygen, nitrogen, Co, Nb or Cr.

Upon alloying in the order of 1 at% oxygen or nitrogen into β-Ti or β-Zr we found extremely temperature and sample dependent elastic peaks in all Brillouin zones. The width of these peaks never exceeded the instrumental resolution and their intensity increased exponentially upon approaching T_0. According to the phase diagram, an oxygen concentration of the order of 1 at% leads to the coexistence of α- and β-phases at temperatures just above T_0. Some additional tests unambiguously proved that these satellite peaks are only due to the coherent coexistence of α and β phase, caused by the alloying of interstitial impurities.

As stated before the diffuse scattering in pure samples around the ω-point at $\vec{q} = 2/3(111)$ is of purely inelastic nature (due to over-damped phonons). Careful analysis of this diffuse scattering in alloyed samples revealed an increase of this diffuse intensity, the intensity of which depends on the impurity and its concentration. Fig. 9 shows this increase of diffuse elastic intensity with respect to the neighbouring inelastic phonon intensity for different β-Zr alloys. In all cases this purely elastic intensity did not alter with temperature (with the exception of a Debye-Waller factor), i.e. turned out to be an intrinsic property of the bcc phase.

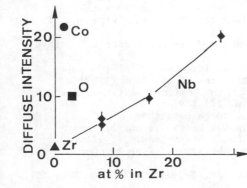

Fig. 9 Diffuse elastic intensity at $\vec{q} = 2/3(111)$ in units of the neighbouring L2/3(111) phonon intensity for different solutes in β-Zr.

So, two effects occur upon alloying : i) Mainly interstitial impurities shift the sample at a given temperature in a two phase region of coherent coexistence of β and α phase. This has nothing to do with an elastic precursor of the phase transition. ii) The overdamped and soft ω-phonon freezes upon alloying substituional and interstitial impurities to a static (within the time resolution of the method) displacement giving rise to an ω-like diffuse scattering. So, elastic precursors of the low symmetry phase can indeed be observed by introducing defects.

SUMMARY

The phonon dispersion of the bcc phase of the pure group 4 metals Ti, Zr and Hf is dominated by a valley of anomalous low frequency and strongly damped phonons along $[\xi\xi\,2\xi]$ propagation, which is related to the β-ω and β-α martensitic phase transitions. The low lying L2/3(111) phonon does not change with temperature and is due to a bcc inherent weakness towards ω-displacements. The $T_1$1/2(110) phonon decreases drastically upon approach T_0 and is a dynamical precursor of the β-α transition. Different to alloys the overdamped phonons, which achieve the displacements for the martensitic transition do not condensate to static displacements.

REFERENCES

[1]   Christian, Theory of Transformations in Metals and Alloys, Oxford, Pergamon Press (1975)
[2]   Satija S.K., Shapiro S.M., Salamon M.B., Wayman C.M., Phys. Rev. B 29, 6031 (1984)
[3]   Tietze H., Müllner M., Renker B., J. Phys. C 17, L529 (1984)
[4]   Müllner M., Tietze H., Eckold G., Assmus W., Proc. Int. Conf. on Martensitic Tranformations, ICOMAT 86, Nara (Japan), ed. by I. Tamura (The Japan Inst. of Metals) p. 159 (1987).
[5]   Herget G., Müllner M., Eckold G., Jex H., This conference.
[6]   Herget G., Müllner M., Suck J.-B., Schmidt R., Wipf H., This conference.
[7]   Shapiro, S.M., Yang B.X., Shirane G., Noda Y., Tanner L.E., Phys. Rev. Lett. 62, 1298 (1989)
[8]   Hallman E.D., Svensson E.C., This conference.
[9]   Gooding R.J., Krumhansl J.A., Phys. Rev. B 39, 1535 (1989)
[10] Moss S.C., Keating D.T., Axe J.D., in "Phase Transitions", ed. by Cross L.E. (Pergamon, New York) p. 179 (1973)
[11] Noda Y., Yamada Y., Shapiro S.M., to be published.
[12] Trampenau, J., Heiming A., Herzig C., Petry W., unpublished.
[13] Mori M., Yamada Y., Shirane G., Solid State Comm. 17, 127 (1975)
[14] Guénin G., Rios Jara D., Morin M., Delaey L., Pynn R., Godin P.F., J. Physique C4, 597 (1982)
[15] Zolliker M., Bührer W., Gotthardt R., Helvetica Physica Acta 61, 198 (1988)
[16] Tanahaski H., Morii Y., Iizumi M., Suzuki T., Otsuka K., ICOMAT 86, 163 (1987).
[17] Ye Y.-Y., Chen Y., Ho K.-M., Harmon B.N., Lindgard P.-A., Phys. Rev. Lett. 58, 1769 (1987)

[18] Chen Y., Ho K.-M., Harmon B.N., Phys. Rev. B $\underline{37}$, 283 (1988)
[19] Lindgard P.-A., Mouritsen O.G., Phys. Rev. Lett. $\underline{57}$, 2458 (1986).
[20] Dmitriev V.P., Rochal S.B., Gufa Yu.M., Toledano P., Phys. Rev. Lett. $\underline{60}$, 1958 (1988).
[21] Krumhansl J.A. and Gooding R.J., Phys. Rev. B $\underline{39}$, 3047 (1989).
[22] Kerr W.C. and Bishop A.R., Phys. Rev. B $\underline{34}$, 6295 (1986).
[23] Ernst G., Artner C., Blaschko O., Krexner G., Phys. Rev. B $\underline{33}$, 6465 (1986)
[24] Smith H.G., Phys. Rev. Lett. $\underline{58}$, 1228 (1987)
[25] Gooding R.J., Krumhansl J.A., Phys. Rev. B $\underline{38}$, 1695 (1989)
[26] Blaschko O., Krexner G., Phys. Rev. B $\underline{30}$, 1667 (1984)
[27] Giebultowicz T.M., Overhauser A.W., Werner S.A., Phys. Rev. Lett $\underline{56}$, 1485 (1986)
[28] Pintschovius L., Blaschko O., Krexner G., de Podesta M., Currat R., Phys. Rev. B $\underline{35}$, 9330 (1987)
[29] Mizuki J., Stassis C., Phys. Rev. B $\underline{34}$, 5890 (1986)
[30] Wentzcovitch R.M., Cohen M.L., Phys. Rev. B $\underline{37}$, 5571 (1988)
[31] Stassis C., Zaretsky J., Misemer D.K., Striver H.L., Harmon B.N., Nicklow R.M., Phys. Rev B $\underline{27}$, 3303 (1983)
[32] Heiroth M., Buchenau U., Schober H.R., Evers J., Phys. Rev. B $\underline{34}$, 6681 (1986)
[33] Mizuki J., Stassis C., Phys. Rev. B $\underline{32}$, 8372 (1985)
[34] Mizuki J., Chen Y., Ho K.-M., Stassis C., Phys. Rev. B $\underline{32}$, 666 (1985)
[35] Petry W., Heiming A., Trampenau J., Alba M., Herzig C., unpubl.
[36] Stassis C., Zarestky J., Solid State Comm. $\underline{52}$, 9 (1984)
[37] Stassis C., in Axe J.D., Nicklow R.M., Physics Today, Jan. 1985, p. 27
[38] Wang X.W., Harmon B.N., Chen Y., Ho K.-M., Stassis C., Phys. Rev. B $\underline{33}$, 3851 (1986)
[39] Petry W., Flottmann T., Heiming A., Trampenau J., Alba M., Vogl G., Phys. Rev. Lett. $\underline{61}$, 722 (1988)
[40] Petry W., Heiming A., Trampenau J., Alba M., Herzig C., Vogl G., Schober H.R., to be published.
[41] Heiming A., Petry W., Trampenau J., Alba M., Herzig C., Schober H.R., Vogl G., to be published.
[42] Stassis C., Zarestky J., Wakabayashi N., Phys. Rev. Lett. $\underline{41}$, 1726 (1978)
[43] Trampenau J., Petry W., Heiming A., Alba M., Herzig C., Miekelei W., Schober H.R., to be published.
[44] Zarestky J., Stassis C., Phys. Rev. B $\underline{35}$, 4500 (1987)
[45] Blaschko O., Krexner G., Pleschiutschnig J., Ernst G., Hitzenberger C., Karnthaler H.P., Korner A., Phys. Rev. Lett. $\underline{60}$, 2800 (1988)
[46] Iizumi M., J. Phys. Soc. of Japan $\underline{52}$, 549 (1983).
[47] Flottmann Th., Petry W., Serve R., Vogl G., Nuclear Instrum. & Methods A $\underline{260}$, 165 (1987).
[48] Falter C., Ludwig W., Selmke M., Zierau W., Phys. Lett. $\underline{90A}$, 250 (1982)
[49] Ho K.-M., Fu C.-L., Harmon B.N., Phys. Rev. B $\underline{28}$, 6687 (1983)
[50] Ho K.-M., Fu C.-L., Harmon B.N., Phys. Rev. B $\underline{29}$, 1575 (1984).

PHONON DISPERSION IN β-PHASE NICKEL ALUMINUM ALLOYS

E.D. HALLMAN
Department of Physics
Laurentian University
Sudbury, Ontario, Canada, P3E 2C6

E.C. SVENSSON
Atomic Energy of Canada Limited
Chalk River, Ontario, Canada, K0J 1J0

1. INTRODUCTION

$Ni_x Al_{1-x}$ alloys have the ordered CsCl (β) structure at room temperature for $0.45 < x < 0.63$. For $x > 0.50$, excess nickel ions are randomly distributed on aluminum sites[1]. A martensitic transformation occurs at low temperatures ($0 < T_m < 80$ K) for alloys with $0.58 < x < 0.63$. The β structure then transforms to a monoclinic 7R structure based on $Ni_2 Al$[2]. Shapiro et al.[3] have studied an anomaly in the $[\zeta\zeta 0]T_1 A$ phonon branch and the closely related satellite elastic peaks that appear at temperatures well above T_m as a result of the growth of distorted embryos of the low temperature phase within the β structure. Gooding and Krumhansl[4] have recently interpreted these interesting precursors of the martensitic transformation as resulting from phonon-strain interactions. To help advance the detailed understanding of Ni-Al alloys, we have, in neutron scattering measurements at Chalk River Nuclear Laboratories, determined the complete dispersion relations for $x = 0.50$ and $x = 0.58$, investigated phonon anomalies in two branches, and made a preliminary study of elastic satellite peaks for $x = 0.58$.

2. RESULTS AND DISCUSSION

The phonon dispersion relations are shown in Fig. 1. Solid and dashed lines represent 5th neighbour Born-von Karman force constant fits to our results for $x = 0.50$ and $x = 0.58$ (not shown). Note the marked dependence on composition especially for certain optic branches

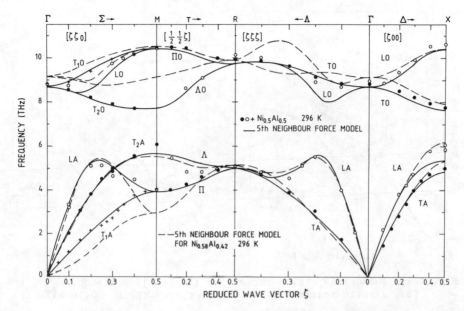

Fig. 1. Phonon dispersion relations for Ni-Al alloys.

and the $[\zeta\zeta 0]T_1$ acoustic branch. The force system is dominated by strong 1st and 2nd neighbour interactions. The 1st and 2nd neighbour force constants are given in Table 1. Note that, between x = 0.50 and x = 0.58, there are changes of 29-36% in the three largest force constants. The large changes observed for 2nd neighbour force constants support the assumptions made by Clapp et al[5] in modelling a martensitic transformation using molecular dynamics techniques.

Anomalies in the $[\zeta\zeta 0]T_1A$ and $[\zeta\zeta\zeta]TA$ branches were investigated in detail. Results for the $[\zeta\zeta 0]T_1A$ branch are shown in Fig. 2(a). The positions (indicated by arrows) of maximum deviation from a simple sine wave shape depend on composition and temperature. Elastic satellite peaks, observed in $[\zeta\zeta 0]$ directions near several reciprocal lattice points, were studied for $15 \leq T \leq 294$ K for the x = 0.58 alloy. Results obtained near (020) are shown in Fig. 2(b). Both

Table 1. Atomic Force Constants (dyn/cm) for the alloys studied.

AFC	x = 0.50	x = 0.58	AFC	x = 0.50	x = 0.58
$1XX^{NA}$	10122	13010	$1XY^{NA}$	10723	14585
$2XX^{NN}$	-1367	3831	$2YY^{NN}$	1154	-1976
$2XX^{AA}$	19756	12594	$2YY^{AA}$	-611	1414

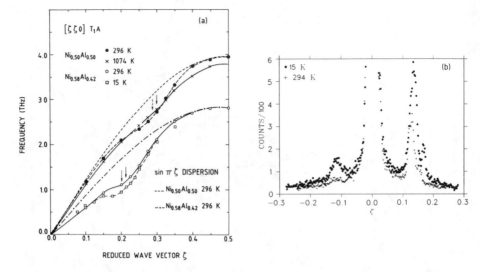

Fig. 2(a). Phonon dispersion curves for the $[\zeta\zeta 0]T_1 A$ branch.
(b). Elastic scattering at $(0,2+\zeta,-\zeta)$ for the x = 0.58 alloy.

the positions and intensities of the satellites vary with temperature. The large difference in satellite intensity on opposite sides of (020) is possibly attributable to macroscopic strains in our unannealed specimen. Note that there are two peaks on the "strong" side, at ζ = 0.134 and ζ = 0.167 at 15 K. As expected from theory[4], the $[\zeta\zeta 0]T_1 A$ phonon anomaly and the satellites occur at different ζ values.

The present study, in which we have obtained complete phonon dispersion relations and force constants for two Ni-Al alloys as well as new information on phonon anomalies and satellite peaks, represents a major broadening of our knowledge on this intriguing system.

3. REFERENCES

1. Enami, K., Nenno, S. and Shimizu, K., Trans. Jpn. Inst. Met. 14, 161 (1973).
2. Reynaud, F., Scripta Metall. 11, 765 (1977).
3. Shapiro, S.M., Yang, B.X., Shirane, G., Noda, Y. and Tanner, L.E., Phys. Rev. Lett. 62, 1298 (1989) and references therein.
4. Gooding, R.J. and Krumhansl, J.A., Phys. Rev. B 39, 1535 (1989).
5. Clapp, P.C., Rifkin, J., Kenyon, J. and Tanner, L.E., Met. Trans. A. 19A, 783 (1988).

ELASTIC ANOMALIES IN Fe₃Pt INVAR ALLOYS

U.Kawald*, J.Pelzl*, G.A.Saunders‡, P.Ngoepe‡ and H.Radhi‡

*Institut für Experimentalphysik VI, Ruhr–Universität Bochum, 4630 Bochum, FRG

‡School of Physics, University of Bath, Claverton Down, Bath BA2 7AY, UK

Abstract

Measurements of ultrasonic wave velocities under pressure in $Fe_{72}Pt_{28}$ show that $\partial C_{11}/\partial p$ and $\partial B^s/\partial p$ have large negative values in the ferromagnetic state but become positive in the paramagnetic state, evidencing a strong volume dependent magnetoelastic interaction. Negative values of the Grüneisen parameters of the longitudinal modes can account for the negative thermal expansion observed in the ferromagnetic state.

Iron platinum alloys with compositions near to Fe_3Pt show pronounced invar behaviour in their physical properties[1]. To establish the source of its negative thermal expansion, measurements have been made of the pressure dependencies of the elastic stiffness tensor components from room temperature up through T_C in monocrystalline $Fe_{72}Pt_{28}$. Such measurements determine the vibrational anharmonicities of the long wavelength acoustic phonons which contribute substantially to thermal expansion. The velocities of ultrasonic modes propagated along [001]– and [110]–directions have been measured between 298K and 388K and their pressure dependencies at selected temperatures in this range. The effects at room temperature of hydrostatic pressure on ultrasonic wave velocity are shown

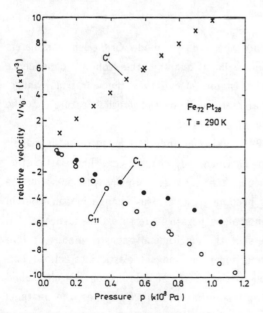

Figure 1: Hydrostatic pressure dependence of the natural velocity of ultrasonic waves propagated in $Fe_{72}Pt_{28}$.

in figure 1. An interesting observation is that application of pressure reduces the longitudinal velocities; the shear modes stiffen. The elastic stiffness constants at room temperature are : $C_{11} = 144$ GPa, $C_{12} = 109$ GPa, $C_{44} = 86$ GPa, C' $(= (C_{11} - C_{12})/2) = 17.6$ GPa, B^s $(= (C_{11} + 2C_{12})/3) = 120$ GPa.

The pressure derivatives of these elastic stiffnesses (table 1) show the extremely unusual feature that $\partial C_{11}/\partial p$ and $\partial B^s/\partial p$ are large negative quantities. At room temperature this material becomes easier to compress as pressure is applied!

Table 1: $\partial C_{ij}/\partial p$ for $Fe_{72}Pt_{28}$ $(T_c\sim 371K)$.

	Temperature		
	291 K	363 K	
		$p \to 0$	$p > 0.7 \times 10^8$ Pa
$\partial C_{11}/\partial p$	-23.9	+9.6	+27.4
$\partial C_{12}/\partial p$	-31	+1.2	+19.0
$\partial C_{44}/\partial p$	+4.7	+1.7	-8.5
$\partial C'/\partial p$	+3.7	+4.2	+4.2
$\partial B^s/\partial p$	-28.8	+4.0	+21.8

The data have been used to compute the acoustic mode Grüneisen parameters $\gamma(p,\underline{N})$ in the long wavelength limit which quantify the volume dependence $(-\partial \ln\omega_i/\partial \ln V)$ of the mode frequencies (figure 2) and thus the vibrational anharmonicity. It can be seen that this invar material shows the unusual feature that the $\gamma(p,\underline{N})$ for all the longitudinal acoustic modes are negative.

As the temperature is raised above 291K $\partial C_{11}/\partial p$ and $\partial B^s/\partial p$ decrease in magnitude until just below the Curie point T_c when they change sign. The elastic stiffness tensor components and bulk modulus then increase normally with pressure: the acoustic mode Grüneisen parameters become positive. This change in behaviour in the vicinity of T_c shows that the anomalous negative pressure dependence of C_{11} and B^s and the Grüneisen parameters of the longitudinal acoustic modes at zero wavevector are due to a strong volume dependent magnetoelastic interaction. This is established further by consideration of the effects of pressure on the ultrasound velocity at 363K; application of a pressure of 0.8×10^8 Pa takes the material through T_c and causes the magnetoelastic contribution to the elastic stiffness to become zero. The linear coefficient α of thermal expansion of disordered $Fe_{72}Pt_{28}$ is negative between about 270K and 400K[2]. In addition to the nonlinear acoustic

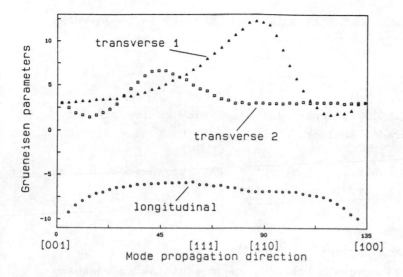

Figure 2: Orientation dependence of the acoustic mode Grüneisen parameters in the long wavelength limit for $Fe_{72}Pt_{28}$.

behaviour of a material under a finite strain, the vibrational anharmonicity determines its thermal expansion α. Negative values of the thermal Grüneisen parameters γ_{th} (given by $3\alpha VB^s/C_p$) are a feature of this invar material. γ_{th} stems from summation ($\Sigma C_i \gamma_i / \Sigma C_i$) over all the excited phonon modes. In the temperature range over which γ_{th} is negative, the effects of those modes which stiffen with decrease of volume (i.e. have positive γ_i) are more than counterbalanced by those which soften (have negative γ_i). Although ultrasonic measurements are confined to the long wavelength limit, it is clear that phonons in the longitudinal branch near $k=0$ where anharmonicities are dominated by the magnetoelastic interaction provide an important contribution to the invar behaviour.

1   E.F. Wassermann, Adv. Sol. Stat. Phys., 27, 85 (1987).

2   K. Sumiyama, M. Shiga, M. Morioka, Y. Nakamura, J. Phys. F: Metal Phys., 9, 1665 (1979).

COUPLED MODE SCATTERING PROCESSES AT 1q TO 2q PHASE
TRANSITIONS WITHIN INCOMMENSURATE PHASES

J. F. SCOTT, Department of Physics, University of
Colorado, Boulder, CO 80309-0390 USA

W. F. OLIVER, Department of Physics, Arizona State
University, Tempe, AZ 85287   USA

1. INTRODUCTION

It has been a matter of some dispute as to whether
phason-like excitations in the incommensurate phases of
insulators have a measurable cross section in light
scattering experiments.[1] It is experimentally and theoretically viewed[2] that these modes are overdamped at long
wavelength. Hence, the question is whether they may be
studied by dynamic central mode light scattering.[3] In this
short note we point out that very strong dynamic central
mode light scattering has been measured in incommensurate
barium sodium niobate;[3] that the observed linewidth follows
the Dq^2 dependence required for diffusion; that the inferred
diffusivity is about 20x greater than that for ordinary
thermal diffusion of entropy fluctuations at the same
temperature;[4,5] and that strong coupling between LA phonons
and the central mode creates a one-order of magnitude
enhancement in the Raman cross section. The central mode
peaks in scattering intensity not at T_I = 582K, nor at the
lock-in transition T_L = 543K, but at 565K, which is the point
within the IC phase where the modulation jumps[6] discontinuous-
ly from 1q (orthorhombic) to 2q (tetragonal). This transition is analogous to the APB wall roughening transition
originally predicted[7] in metallic CDW systems.

2. EXPERIMENT

Fig. 1 below shows the spectra of LA and TA phonons and of the dynamic central mode in the Brilluoin data of barium sodium niobate at 25, 265.5, 198, and 273°C. In longer papers elsewhere[8,9] we have shown that the width of the central mode varies exactly as q^2, that its intensity peaks at 565K, where the 1q to 2q transition occurs, and that it produces strong lineshape anomalies via its linear coupling with the LA phonons. The coupled modes have been deconvoluted via a formalism that is now standard.[10] It this analysis we find that nearly all (89%) of the central mode cross section is "borrowed" from the LA phonon; and nearly all of the 4% decrease in LA sound velocity[11] near T_L is due to phonon self energy renormalization due to relaxation into the central mode.

Independent measurements of thermal diffusion in this material in the same temperature range[4,5] show that it is 20x too slow to account for the observed spectra. Hence, at least two different diffusion mechanisms are inferred, as in lead germanate and other systems.[12]

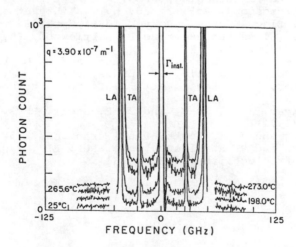

Fig. 1. Brillouin spectra of $Ba_2NaNb_5O_{15}$ at various T.

We suggest that the observed scattering process may be due to an intra-wall roughening of the anti-phase boundaries. Such a transformation was predicted by Rice et al.[7] Phason theories treat the APBs as smooth, structureless walls on an atomic scale. However, if we naively insert our experimental results into the phason theory of Levanyuk et al., we find from the ratio of intensities at different scattering angles

$$I(q) = (d^2 q^2 k_B T)(g + d^2 Dq^2)^{-1} \qquad (1.)$$

that the gap energy g for phasons is small but finite; and that the magnitude d of incommensurate modulation is 10-15% of the atomic spacing. This value may be reasonable, but our data need an extended theory of phasons that incorporates wall roughening.

3. REFERENCES

1.) N. I. Lebedev, A. P. Levanyuk, and A. S. Sigov, Laser Optics of Condensed Matter, ed. J. L. Birman et al., (Plenum, New York, 1988), p.273.
2.) R. Zeyher and W. Finger, Phys. Rev. Lett.49, 1833 (1981).
3.) W. F. Oliver and J. F. Scott, Ref. 1, p.263.
4.) G. H. Burkhart and R. R. Rice, J. Appl. Phys. 48, 1785 (1980).
5.) J. F. Scott, S.-J. Sheih et al., Phys. Rev. B (in press).
6.) Feng Duan and Pan Xiaoqing, Proc. XIth Int. Cong. Electron Microscopy, Vol. 2, ed. by T. Imura et al., (Jpn. Soc. Elec. Micros., Tokyo, 1986), p.1239.
7.) T. M. Rice, S. Whitehouse, and P. Littlewood, Phys. Rev. B24, 2751 (1981).
8.) W. F. Oliver, J. F. Scott, S. A. Lee, and S. M. Lindsay, Materials Sci. & Eng. A107 (in press, 1989).
9.) W. F. Oliver, J. F. Scott, et al., Ferroelectrics (in press, 1989).
10.) R. S. Katiyar, J. F. Ryan, and J. F. Scott, Phys. Rev. B4, 2635 (1971).
11.) P. W. Young and J. F. Scott, Phase Trans. 6, 175 (1986).
12.) K. B. Lyons and P. A. Fleury, Phys. Rev. B17, 2403 (1978); Ferroelectrics 35, 37 (1981).

ISOTOPE EFFECTS ON INCOMMENSURATE PHASE TRANSITIONS IN Rb_2ZnCl_4 and K_2SeO_4

T. Hidaka
Electrotechnical Laboratory,
1-1-4, Umezono, Tsukuba, Japan

INTRODUCTION

The phenomenon of the incommensurate phase transitions in ionic crystals such as Rb_2ZnCl_4, K_2SeO_4 or $NaNO_2$ is one of the very interesting phenomena in solid-state physics.   In K_2SeO_4, the incommensurate soft phonon has been observed with a neutron inelastic scattering experiment[1].   Therefore, the mechanism of their phase transitions are supposed to be *displacive*. However, most of all theories for the incommensurate phase transitions have been discussed to be due to the *order-disorder* type interactions between the nearest neighboring and the second- or the third-nearest neighboring pseudo spins[2,3]. Thus there is a discrepancy in the experimental result and the theory. By the way, recently, the Ti and Ba isotope effects on the ferroelectric phase transition in $BaTiO_3$ at $120\,^{\circ}C$ has been reported[4].   The remarkable fact was that with introducing light isotopes of Ti or Ba, the Curie temperature, T_c, rose promptly, and vice versa for heavy Ti isotope.   This isotope effect in $BaTiO_3$ differs completely from the H-D isotope effect in KDP type crystals. According the phase transition (low-temperature phase) in KDP, H or D atoms will be ordered to have the long-range coherence. The isotope effect on a crystal will give us a clear criterion to decide of which phase transition is the displacive or the order-disorder. In displacive-type phase transitions, their T_c rise with the light isotope introduction, whereas in order-disorder type transitions, T_c *may be reduced* with light-isotope introduction (this has not been confirmed except H-D replacing experiments). In this report we give firstly the isotope effects on the incommensurate-phase-transition onset temperature, T_i, of Rb_2ZnCl_4 and K_2SeO_4.

EXPERIMENTAL

[I] Rb and Zn isotope effects in Rb_2ZnCl_4

First, we report the Rb and Zn isotope effects on T_i in Rb_2ZnCl_4. Table 1 shows the isotope contents of samples A1, A2, B1 and B2, respectively. ^{85}Rb enriched RbCl and ^{87}Rb enriched one, and ^{64}Zn enriched ZnO ^{68}Zn enriched one were supplied by the Oak-Ridge National Laboratory. 0.01 N hydrochloric acid were added to those isotope enriched ZnO (and natural ZnO) to get $ZnCl_2$, and also add stoichiometric isotope enriched (or natural) RbCl. The samples A1 and A2 are the Rb-isotope-enriched materials (Zn is natural), and the samples B1 and B2 are the Zn-isotope enriched materials (Rb is natural). All crystals were grown by aqueous-solution method. T_i were determined using DSC method (with Rigaku, model DSC 8240 Differential Scanning Calorimeter). In Table 1, we see the small shifts in T_i (rising with light-isotope introduction) were observed. This means that the phase transition is *displacive* but the weights of Rb and Zn in the soft (incommensurate) mode is very low. Only Cl atoms move in it.

Table 1. Isotope contents and the incommensurate phase onset temperature, T_i. * means the values the same as the upper line.

Sample	Isotope contents (%)							$T_i(^{\circ}C)$
	^{85}Rb	^{87}Rb	^{64}Zn	^{66}Zn	^{67}Zn	^{68}Zn	^{69}Zn	
A1	99.78	0.22	48.89	27.81	4.11	18.56	0.62	29.4
A2	2.00	98.00	*	*	*	*	*	29.1
B1	72.15	27.85	99.85	0.14	0.01	0.01	0.01	28.2
B2	*	*	0.41	0.30	0.20	99.09	0.02	28.1

[2] Oxygen isotope effect in K_2SeO_4

Commercially available K_2SeO_4 was 3-times re-crystallized with aqueous-solution method. 100 mg of this purified K_2SeO_4 was resolved into 95 % ^{18}O enriched water, supplied by Shohkoh-Tsushoh Co.LTD., The chemical bond between Se and O is not so strong; ^{16}O in K_2SeO_4 is replaced somewhat by ^{18}O in the solvent $H_2^{18}O$ water. The re-crystallized K_2SeO_4 including ^{18}O was measured using DSC method. The result is shown in Fig.1. A small isotope shift (about 0.7K lowering in ^{18}O enriched material) was observed. Figure 2 shows the infrared absorption spectra of the ^{18}O enriched material and natural one. From Fig.2, we estimated that about 15-20 % of ^{16}O was replaced by ^{18}O in the material. Thus the oxygen isotope shift in K_2SeO_4 is concluded to be remarkably large to compare with Rb_2ZnCl_4. We may conclude that the incommensurate phase transition in K_2SeO_4 is surely displacive, and in it oxygen atoms mainly move; to be reasonably in agreement with the neutron-scattering experiments.

[1] Iizumi, M., et.al., Phys.Rev. B15, 4392 (1977).
[2] Bak, P. and von Boehm, J., Phys.Rev. B21 5297 (1980).
[3] Ishibashi, Y., J.Phys.Soc.Jpn, 51, 1220 (1982).
[4] Hidaka, T. and Oka, K., Phys.Rev. B35, 8502 (1987).

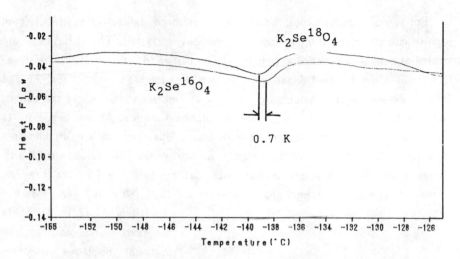

Fig.1. DSC data of natural K_2SeO_4 and ^{18}O enriched one. About 0.7 K isotope shift was observed.

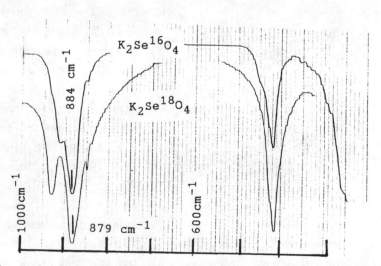

Fig.2. Infrared absorption data of natural K_2SeO_4 and ^{18}O enriched one. From this we estimated that 15-20% of ^{16}O was replaced by ^{18}O.

ISOTOPE INDUCED PHASE TRANSITIONS IN $(ND_4)_2TeCl_6$

S. Müller, U. Kawald, J.Pelzl

Institut für Experimentalphysik VI, Ruhr–Universität Bochum, 4630 Bochum, FRG

For several years the cubic ammonium hexachlorometallates $(NH_4)_2MeCl_6$ have been of interest in the study of molecular motions in crystals. The $(NH_4)^+$ ions are located at sites with tetrahedral symmetry. The barrier to rotation about the three–fold axis is a few hundred K and increases in the series Me = Pd, Pt, Sn, Pb and Te /1/. Because of the low activation energy, at ambient temperatures the ammo–nium tetrahedra undergo classical hindered rotational motion. At low temperatures (T < 50 K) the classical motion changes into a quantum mechanical tunneling. Recently it has been proved by chlorine nuclear quadrupole resonance (NQR), Raman and Brillouin scattering experiments that at least the Pb– and the Te–compound undergo a structural phase transition at $T_{c1} = 80$ K and $T_{c1} = 85$ K, respectively /2,3/. These transitions stand out from the rotational displacive transfor–mations in the other hexahalo–metallates because below T_{c1} the ^{35}Cl – NQR pattern pre–serves its single line spectrum indicating equivalent chlorine sites in the low symmetry phase. On the basis of the NQR and Raman results the symmetry of the low tempera–ture phase in $(NH_4)_2PbCl_6$ and $(NH_4)_2TeCl_6$ has been identi–fied as C_{3i}^2 /2,3/. Very re–cently for $(NH_4)_2PbCl_6$ this assignment has been confirmed by measurements of elastic neu–tron scattering /4/.

A similar behaviour has been

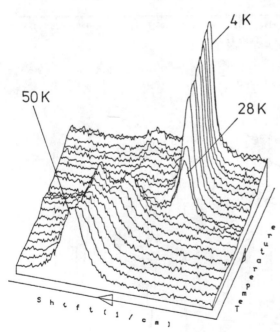

Figure 1: Temperature dependence of the totally symmetric ND_4 internal mode of $(ND_4)_2TeCl_6$ between 90 K and 4.2 K.

found for $(ND_4)_2TeCl_6$ /3/. The deuterated compound undergoes a transition at the same temperature $T_{c1} = 85$ K which preserves the Cl–NQR single line pattern, pointing towards a transition that is associated with a small angle rotation of the $TeCl_6$ octahedra about the threefold axis. However, whereas $(NH_4)_2TeCl_6$ shows no indication for a further structural change below T_{c1} new lines had appeared in the Raman spectrum of $(ND_4)_2TeCl_6$ at about 20 K /3/. In order to elucidate the origin of the different behaviour of the deuterated compound we have undertaken a detailed study of Raman and Brillouin scattering from $(ND_4)_2TeCl_6$ in the temperature range 4.2 K to 90 K.

Single crystals of $(ND_4)_2TeCl_6$ have been grown from solutions in heavy water. The degree of deuteration was nearly 100%. Raman and Brillouin scattering have been observed in backscattering from (100) and (111) crystal surfaces. Powder x–ray spectra have been recorded from grinded samples at three distinct temperatures.

The Raman spectrum can be divided into three regions due to the external lattice modes (0 to 100 cm^{-1}), the internal modes of the $TeCl_6$ octahedron (100 cm^{-1} to 400 cm^{-1}) and the internal modes of the ND_4 tetrahedron (1000 cm^{-1} to 3000 cm^{-1}). As the temperature is lowered from 85 K (T_{c1}) in all three regions

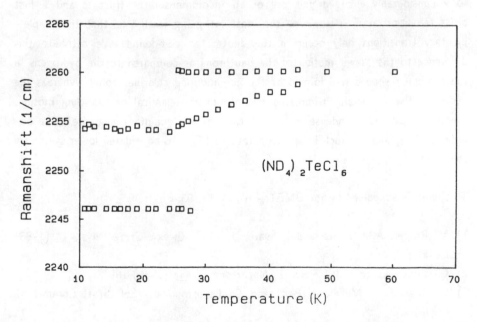

Figure 2: Temperature dependence of the frequency shift of the totally symmetric ND_4 internal mode.

dramatic changes of the line pattern are observed. Figure 1 shows the temperature evolution of the totally symmetric vibration of the ammonium ion at about 2250 cm^{-1} which transforms as A_{1g} in the high temperature cubic phase. The temperature dependence of its Raman shift is displayed in the graph of Figure 2. On decreasing the temperature, at about 50 K the line starts to split into at least two components. At 28 K the multi-line pattern changes again when one line disappears but another line emerges at lower energies. We identify T_{c2} = 50 K and T_{c3} = 28 K as transition temperatures because other lines also show distinct changes at these two temperatures. In the TeCl$_6$-internal mode region the formerly E_g and T_{2g} modes experience additional splitting at 50 K and 28 K. Most remarkably the E_g vibration has split into five components below T_{c3}. The external mode region is distinguished by the appearance of new lines at both transition temperatures and a soft mode in the intermediate phase between T_{c2} and T_{c3}. This soft mode seems to condense in the vicinity of T_{c2} at 50 K. Whereas at T_{c2} the new pattern gradually appears. At T_{c3} the spectra of all three regions change abruptly.

The present Raman observations together with the results of preliminary x-ray diffraction studies point towards a second order type transition at T_{c2} which leads to a considerably enlarged unit cell or an incommensurate structure and a first order transition at T_{c3}. However, the most intriguing feature is, that the supplementary transitions only occur in the deuterated compound. We attribute this difference to the slower motion of the deuterons as compared to the hydrogens in the triangular shaped well formed by the neighbouring chlorine atoms. Whereas the hydrogen always remains trifurcated because of the classical or tunneling motion, the deuterium may condense in one of the shallow minima giving rise to a disordered (T_{c2}) and at much lower temperatures (T_{c3}) to an ordered lower symmetry structure.

This work is supported by the BMFT, Project Nr. 03-PE2BOC

1) M. Prager, A.M. Raaen and I. Svare, J. Phys. C: Sol. State Phys., 16 (1983) L181
2) C. Dimitropoulos and J. Pelzl, Z. Naturforsch., 44a (1989) 109
3) U. Kawald, S. Müller, J. Pelzl and C. Dimitropoulos, Sol. Stat. Comm., 67 (1988) 239
4) R.L. Armstrong, B. Lo and B.M. Powell, to be published

DIELECTRIC RELEXATION IN Sr$_x$Ba$_{1-y}$Nb$_2$O$_6$ (SBN)

GÜNTHER SCHMIDT AND HOLM ARNDT
Sektion Physik
Martin-luther-Universität Halle-Wittenberg
4020 Halle, German Democratic Republic

Three features can be seen in the thermal expansion of SBN 75 (75 = 100x) (Figure 1): near $T_t \approx -40°C$ (new superlattice reflections were observed below $-75°C$ in SBN 50[1], near $T_m \approx 50°C$ where a diffuse phase transition (PT) is usually concluded from the dielectric-constant maximum[2], and above $T_i \approx 400°C$ (an ic modulation [3] was found to disappear in SBN 70 above $500°C$[4]. No PTs are indicated by the birefringence. The dielectric dispersion near T_t and T_m [5] indicate freezing processes rather than PTs. No dielectric data near T_i where the conductivity is high, have been reported.

Dielectric hysteresis below T_m suggests a field induced ferroelectric PT. When examined under the polarizing microscope lath shaped regions are seen to appear when a field is applied[6]. An orthorhombic distortion was detected in SBN 70 in this case[4]. The reverse transformation upon heating takes place below T_m already. The pattern disappears even upon cooling below $-100°C$ without loss of polar properties and reappears near T_t. (A gradual transition to a tetragonal re-entrant phase was observed in Ba$_2$NaNb$_5$O$_{15}$ (BNN) below $-170°C$.

We think that the properties of TTB-type crystals may be explained in the same way as proposed for BaTiO$_3$[8] and disordered (cubically stabilized) perovskites[9] because of structural similarity[2]. The Nb ions may be shifted from the oxygen-octahedron centers, and the PTs are accomplished by ordering of these dipoles. This, however, is impeded or

1124

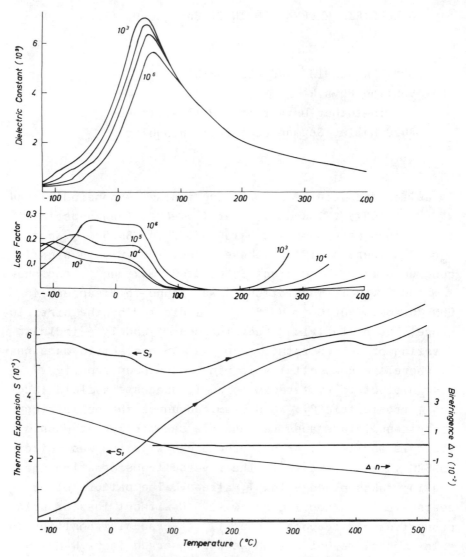

Figure 1   Temperature dependence of the dielectric
constant $_c$, loss factor, thermal expansion in a- (S_1)
and c-direction (S_3), and of the birefringence / n

even suppressed if the parent phase is stabilized as in
SBN because of the statistical and incomplete occupation
of cation sites[2]. One part of the NbO_6 dipoles will
freeze-in near T_i as at the diffuse ferroelastic PT in

BNN. Freezing-in with respect to the z-components with ferrodistortive short-range order leads to dielectric properties (Figure 1, 2) as known from PMN or PLZT[9].Ferroelectric regions can be induced by an electric field below T_m, which,however should be associated with reorientations of dipole components that froze-in below T_i already so that four kinds of orthorhombic[4] lath shaped ferroelectric regions are formed. The contrast seen in the polarizing microscope, however,tells more in favour of a monoclinic distortion. The contrast disappears below T_t probably because the remaining degrees of freedom of the NbO_6 dipoles freeze-in in such way that no direction perpendicular to the c-axis is preferred any longer.

REFERENCES

1. Bursill, L. A. and Peng Ju Lin, Acta Cryst. B43, 49 (1987).
2. Lines, M. E. and Glass, A. M., "Principles and Application of Ferroelectrics and Related Materials", Clarendon Press, Oxford (1977), pp. 280-284.
3. Schneck, J., Toledano, J. C., Whatmore, R. and Ainger, F. W., Ferroelectrics 36, 327 (1981).
4. Balagurov, A. M., Prokert, F. and Savenko, B. N., Phys. Stat. Sol. (a)103,131 (1987).
5. Kersten, O., Rost, A. and Schmidt, G., Phys. Stat. Sol. (a)75, 495 (1983).
6. Arndt, H. and Schmidt, G., Ferroelectrics, in press.
7. Schneck, J., Primot, J., Von der Mühll, R. and Ravez, J., Solid State Commun. 21, 57 (1977).
8. Comès, R., Lambert, M. and Guinier, A.J., Phys. Soc. Jpn. 28, Suppl., 195 (1970).
9. Schmidt, G., Ferroelectrics 78, 199 (1988). Schmidt, G., Phase Transitions, in press.

ELASTIC INVESTIGATIONS OF $K_2Ba(NO_2)_4$ NEAR 200 K

E. SCHUMANN (a), U. STRAUBE (a), G. SORGE (a)
AND L. A. SHUVALOV (b)
a) Sektion Physik der Martin-Luther-Universitaet,
   Halle/S., GDR
b) Institute of Crystallography, Academy of Sciences,
   Moscow, USSR

Abstract The temperature dependencies of some elastic properties of $K_2Ba(NO_2)_4$ were measured near the transition at 200 K. Up to now no significant dispersion effects of these properties could be found.

INTRODUCTION

Kaban (Potassium Barium Nitrite) - $K_2Ba(NO_2)_4$ is a substance with pure ferroelastic behaviour. It exhibits two phase transitions at 420 K and 200 K. Ivanov et al.[1] pointed out that the transition at $T_{\alpha\beta} = 420$ K is pure ferroelastic and of improper type. Unruh and Brueckner[2] observed a third transition at 460 K. Elastic properties at 420 K were already measured by Sorge et al.[3] and Luspin and Hauret[4]. But there are still very few informations on the low temperature transition at $T_{\beta\gamma} = 200$ K. A first indication of an elastic anomaly at 200 K was given by Arndt[5] with dilatational measurements. Therefore we tried to get more results from the $\beta\gamma$- transition.

EXPERIMENTAL

The ultrasonic measurements were performed using the Papadakis pulse overlap method and for lower frequencies the composite resonator technique. Unfortunately the samples were twinned perpendicular to the z-direction, giving rise to cracks induced by the ferroelastic state, perhaps because of very high internal stresses. It was impossible to force big samples into a ferroelastic monodomain state. The cleavage planes lie perpendicular to the z direction. Therefore only longitudinal sound waves in z direction could be used, because the cleavage planes disturb the sound wave in other cases.

RESULTS AND DISCUSSION

The results of the investigation of the elastic properties are shown in Figs. 1 to 4.

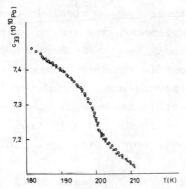

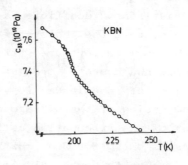

FIGURE 1: Temperature dependence of the elastic stiffness coefficient c_{33} at 2.5 MHz (ooo) and at 12.5 MHz (xxx)

FIGURE 2: Temperature dependence of the elastic stiffness coefficient c_{33} at 20 MHz

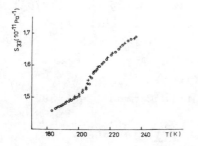

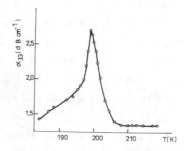

FIGURE 3: Temperature dependence of the elastic compliance s_{33} on cooling (xxx) and on heating (ooo)

FIGURE 4: Temperature dependence of the ultrasonic attenuation α_{33} at 2.5 MHz

The elastic stiffness coefficient c_{33} exhibits an anomalous smooth jump to higher values on cooling. The elastic properties should be frequency dependent similar to the dielectric properties, if we take in account a glass like process proposed by Unruh[6].

This was the reason to continue the measurement of ultrasound velocities at different frequencies (2.5 MHz, 12.5 MHz and 20 MHz). The elastic compliance s_{33} was determined to get information at still lower frequencies. To compare stiffness coefficients and compliances one has to take in mind that the whole stiffness matrix has to be converted. With room temperature stiffness coefficients determined by Luspin and Hauret[3] we get a good accordance. The determination of the losses is very uncertain because of the bad echo series of the ultrasound impulses. We can only find a peak of the attenuation at the transition temperature. This peak is shown for 2.5 MHz in Fig. 4.

Up to now no significant dispersion effects could be found in the inspected frequency region.

A first step to understand the temperature dependence of the elastic properties could be a thermodynamic description based on the Landau theory[7].

REFERENCES

1) Ivanov, N. R. et al., Kristallografiya, _23_, 443 (1978).
2) Unruh, H. G. and Brueckner, H. J., Jap. Journ. Appl. Phys., Suppl. _24_/2, 370 (1985).
3) Sorge, G., Straube, U., Shuvalov, L. A., phys. stat. sol. (a), _59_, 183 (1980).
4) Luspin, Y., Hauret, G., Solid State Commun., _46_, 397 (1983).
5) Arndt, H., Wiss. Z. Univ. Halle, _H.1_, 18 (1986).
6) Unruh, H.-G., phys. stat. sol. (b), _126_, 115 (1984).
7) Schumann, E., Straube, U., Sorge, G. and Shuvalov, L.A., to be published in the Proc. of the IMF-7, Saarbruecken

THERMAL EVOLUTION OF PHONONS IN Cu_3Au

E.C. SVENSSON
Atomic Energy of Canada Limited
Chalk River, Ontario, Canada, KOJ 1JO

E.D. HALLMAN
Department of Physics, Laurentian University
Sudbury, Ontario, Canada, P3E 2C6

B.D. GAULIN
Department of Physics, McMaster University
Hamilton, Ontario, Canada, L8S 4M1

1. INTRODUCTION

The order-disorder phase transition which occurs at temperature T_C in Cu_3Au is a prototype for discontinuous or first-order phase transitions. Much of the recent[1,2] and some of the earlier[3] x-ray and neutron scattering work on Cu_3Au has been motivated by the tantalizing possibility of observing effects related to a spinodal transition[4] expected to occur near T_c. In the hope of casting further light on this intriguing problem, we have carried out neutron scattering measurements, at temperatures from well below to somewhat above T_c, on an initially ordered sample of Cu_3Au, with particular emphasis on phonons at high symmetry points in the Brillouin zone.

2. RESULTS AND DISCUSSION

The measurements were made at the NRU reactor of Chalk River Nuclear Laboratories. Elastic and inelastic scans were carried out at seven temperatures from 280 to 692 K. The elastic intensity at the $(0,0,3)$ superlattice point gives a T_c of 657 ± 3 K for our sample. Consistent with previous measurements[2], marked changes in the elastic scattering only occur within ~ 4 K of T_c. The wave-vector transfers Q chosen for the inelastic scans were Γ and X points which are, respectively, zone centres (i.e. the ordering wave vector) and zone boundaries in the ordered phase. Results for the imaginary part of the dynamic susceptibility, $Im\chi(Q,\omega)$, at the Γ point $(2,2,1)$ are shown in Fig. 1. We see that the two well defined phonon peaks observed at low

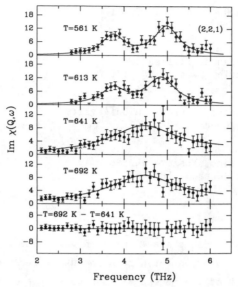

Fig.1. Imχ(Q,ω) for the Γ point (2,2,1). Curves are guides to the eye.

temperature broaden and merge as the temperature is raised, with most of the change occurring between 613 K (T_c-44 K) and 641 K (T_c-16 K). Between 641 K and 692 K, there is little further change as shown by the lowest panel of Fig. 1.

To summarize our results for (2,2,1), we have plotted, in the lower part of Fig. 2, Imχ(Q,ω) for approximately the frequencies of the two phonon peaks at 561 K and the minimum between them. The values shown are sums at four or five frequencies around the listed ν values ($\nu = \frac{\omega}{2\pi}$) with the values for 3.7 and 5.05 THz also added together since they exhibit the same qualitative behaviour. Here we see that the changes in the phonon modes for (2,2,1) are essentially complete at a temperature well below the region in which the sudden drop in the order parameter (top panel) occurs. The point of inflection in the curves for Imχ(Q,ω) is not very accurately determined by our measurements but there clearly is consistency between the temperature at which the changes in the phonon spectra are complete and the lower-spinodal temperature (arrow at $T_s = T_c$-24 K in Fig. 1) inferred earlier[3] from diffuse x-ray scattering measurements.

We observe somewhat more complex behaviour at the X point. Here one phonon peak has changed dramatically by 641 K while the other

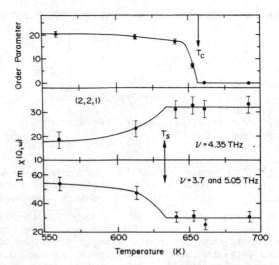

Fig.2. Top Panel: Values of the order parameter deduced from the intensity of the (0,0,3) superlattice peak. Bottom Panels: $\text{Im}\chi(Q,\omega)$ for (2,2,1) at selected frequencies. Curves are guides to the eye. Arrows indicate the order-disorder and lower-spinodal temperatures, T_c and $T_s = T_c - 24$ K (see text).

has changed relatively little but then changes markedly on passing through T_c. As one might have anticipated, the spinodal effects are thus more marked at Γ (the ordering wave vector) than at X. While our results give considerable support to the basic concept of spinodal ordering near a discontinuous phase transition, they also show that a more sophisticated theoretical approach is needed to provide an adequate quantitative understanding of all of our data.

3. REFERENCES

1. Noda, Y., Nishihara S. and Yamada, Y., J. Phys. Soc. Jap. 53, 4241 (1984). Nagler, S.E., Shannon, R.F., Harkless, C.R., Singh, M.A. and Nicklow, R.M., Phys. Rev. Lett. 61, 718 (1988). Ludwig Jr., K.F., Stephenson, G.B., Jordan-Sweet, J.L., Mainville, J., Yang, Y.S. and Sutton, M., Phys. Rev. Lett. 61, 1859 (1988).

2. Lander, G.H. and Brown, P.J., J. Phys. C, 18, 2017 (1985).

3. Chen, H., Cohen, J.B. and Ghosh, R., J. Phys. Chem. Sol., 38, 855 (1977).

4. Cook, H.E., J. Appl. Crys. 8, 132 (1975). Cowley, R.A., Adv. Phys. 29, 1 (1980).

THE POLARIZABILITY MODEL OF DISPLACIVE TYPE FERROELECTRIC SYSTEMS

A. BUSSMANN-HOLDER
Max-Planck-Institut für Festkörperforschung,
D-7000 Stuttgart 80, F.R.G.
Physikalisches Institut, Universität Bayreuth,
D-8580 Bayreuth, F.R.G.

Abstract   The phase transition mechanism of displacive type ferroelectric systems can be described in terms of a local on-site ϕ_4-potential which combines attractive harmonic electron-phonon interactions with fourth-order repulsive electron-two phonon terms. The various temperature regimes of the soft optic mode are quantitatively described within the model using the self-consistent phonon approximation. Going beyond this approximation interesting new solutions, such as periodic nonlinear waves, kinks, are discovered which describe the statics and dynamics of ferroelectric domain walls and pulse solutions.

Displacive type ferroelectric systems are in general oxides. Their phase transition is accompanied by a soft optic mode in the Brillouin zone center which determines the low temperature distorted structure.[1] The appearance of a q = 0 soft mode together with the instability of the oxygen ion suggests that a local phenomenon triggers the phase transition.[2] In a shell model description this has been taken into account by attributing anisotropic core-shell force constants to the oxygen ion where it is assumed that the coupling in the ferroelectric direction consists of a harmonic attractive electron-phonon interaction which is stabilized by a repulsive fourth-order electron-two phonon term.[2,3] The strong anisotropy of the typical ferroelectric properties allows the reduction of the three dimensional model to a pseudo-one dimensional one where only short-range interactions are considered.[4] The model Hamiltonian for such a diatomic system reads:

$$H = T + V + V_2$$

$$T = \sum_{i,n} \tfrac{1}{2} \{ m_i \dot{u}_{in}^2 + m_{ei} \dot{v}_{in}^2 \}$$

$$V = \sum_n \tfrac{1}{2} \{ f'(u_{1n+1} - u_{1n})^2 + f(u_{2n} - v_{1n})^2$$

$$+ f(u_{2n+1} - v_{1n})^2 \}$$

$$V_2 = \sum_n \{ g_2(u_{1n} - v_{1n})^2 + \tfrac{1}{2} g_4 (u_{1n} - v_{1n})^4 \} \ ,$$

where m_i, m_{ei}, u_{in} $(i = 1,2)$, v_{1n} refer to the ionic and shell masses and their respective displacements in the n'th cell. f', f are harmonic intersite phonon-phonon and electron-phonon couplings. g_2 and g_4 are the local harmonic attractive and nonlinear repulsive electron-phonon interactions. Treating V_2 in the self-consistent phonon approximation (SPA) defines an effective harmonic but temperature dependent local coupling constant $\rho(T)$ which now admits the calculation of the dispersion relations and its temperature dependence. A typical example of this procedure is shown in Fig. 1 for $KTaO_3$. Similar quantitatively good results are obtained for a variety of structurally different compounds.[5]

Going beyond the SPA leads to a double-well-problem which is closely related to ϕ_4-models. Solutions on the lattice are obtained in terms of periodic but nonlinear waves[8], periodons which indicate a tripling of the equivalent untit cell.[6] Incommensurate transitions with a mode softening at $qa \approx \frac{2\pi}{3}$ can be described with these solutions coupled to SPA phonons.[7,9]

REFERENCES

1. See e.g. M.E. Lines and A.M. Gloss, Principles and Applications of Ferroelectrics and Related Materials (Clarendon Press, Oxford, 1977)
2. R. Migoni, H. Bilz, and D. Bäuerle, Phys. Rev. Lett., 37, 1155 (1976)
3. R. Migoni, Thesis, Stuttgart (1976);
   C.H. Perry, R. Currat, H. Buhay, R.M. Migoni, W.G. Stirling, and G.D. Axe, Phys. Rev., B 39, 8666 (1989)

1134

4. H. Bilz, G. Benedek, and A. Bussmann-Holder, <u>Phys. Rev.</u>, B 35, 4840 (1987)
5. A. Bussmann-Holder, H. Bilz, and G. Benedek, <u>Phys Rev.</u>, B <u>39</u>, 9214 (1989)
6. H. Büttner and H. Bilz, <u>J. Phys.</u>, C <u>6</u>, 111 (1981); H. Büttner and H. Bilz, Recent Developments in Condensed Matter Physics, ed. G.T. Devreese (Plenum Press, New York, 1981) Vol. 1
7. H. Bilz, H. Büttner, A. Bussmann-Holder, W. Kress, and U. Schröder, <u>Phys. Rev. Lett.</u>, <u>48</u>, 264 (1982)
8. G. Benedek, A. Bussmann-Holder, and H. Bilz, <u>Phys. Rev.</u>, B <u>36</u>, 630 (1987)
9. A. Dobry, A. Greco. R. Migoni, and O. Zandron, <u>Phys. Rev.</u>, B <u>39</u>, 12182 (1989)

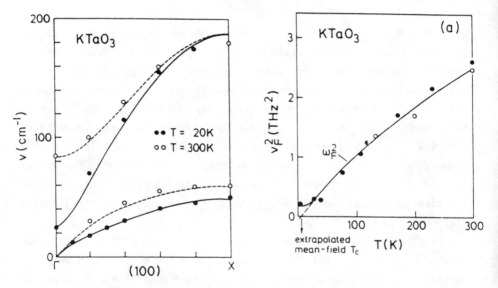

Fig.1) Comparison of theoretical (lines) and experimental (dots) dispersions and temperature dependence of the q=0 soft mode

PHONON CALCULATIONS IN THE CUBIC AND TETRAGONAL PHASES OF PbTiO$_3$

M.D. FONTANA, H. IDRISSI and C. CARABATOS-NEDELEC
Centre Lorrain d'Optique et Electronique des Solides
University of Metz and Supelec, 2 rue E. Belin,
57078 Metz Cedex 3, France

1. INTRODUCTION

Lead titanate PbTiO$_3$ is an important crystal in the oxydic perovskite family since its cubic-tetragonal phase transition (T_c = 496°C) is considered as a textbook example of the displacive mechanism[1]. A large frequency shift of the underdamped lowest vibrational mode was indeed detected by Raman scattering[2].

The paper concerns phonon calculations in cubic and tetragonal phases of PbTiO$_3$ in order to clarify the physical origin of this phonon softening. Very recently, Freire and Katiyar[3] have reported an interesting lattice dynamical investigation in tetragonal PbTiO$_3$ but the high optical dielectric constant[4] implying a large polarizability of the ions seems to be unconsistent with the rigid-ion approximation used in this study.

2. MODEL

In the cubic PbTiO$_3$ we apply the same shell model as previously used by Cowley[5] and Stirling[6] in SrTiO$_3$. Two parameters are needed for each short-range interaction between two ions κ and κ': $A_{\kappa\kappa'}$ and $B_{\kappa\kappa'}$. 14 parameters are employed within this model, 6 for short-range interactions, 2 for ionic charges, 3 for shell charges and 3 for core-shell couplings. In PbTiO$_3$ we do not assume an anisotropy in the oxygen polarizability, contrary to other oxydic perovskites such as SrTiO$_3$, KTaO$_3$[7], KTa$_{1-x}$Nb$_x$O$_3$[8] and BaTiO$_3$[9] because our Raman data display a weak Raman second-order scattering in the cubic phase[10].

To include all structural and symmetry changes related to the cubic-tetragonal phase transition without using supplementary model parameters, repulsive and electrostatic forces in the tetragonal phase are calculated from their values in the cubic phase.

Thus, we determine the short-range forces in the tetragonal phase as function of the cell parameters, the ionic displacements δ_κ and the parameters $A_{\kappa\kappa}$, and $B_{\kappa\kappa'}$, which are given with respect of the ionic cubic positions. By this way we try to describe the lattice dynamics of PbTiO$_3$ in both cubic and tetragonal phases with the same series of adjustable parameters, and possibly the same parameter values.

3. RESULTS

In the cubic phase, the dispersion curves are adjusted to the branches measured by inelastic neutron scattering at 510°C[11], whereas in the tetragonal phase, the phonons are fitted to the neutron data[11] together with Raman results obtained at room temperature [2,10].

The series of parameter values, reported in table, can provide an acceptable adjustment of the experimental data in all directions for the cubic and tetragonal phases. The figure displays the low-frequency phonon branches in the cubic and tetragonal phases, as well as the good agreement which is achieved between calculations and experimental data. Only the parameter A_1 is found to change its value from 22.85 at 510°C to 27.2 at 20°C. The phonon dispersion curves of PbTiO$_3$ can be thus described in the cubic and tetragonal phases within the framework of the same shell model. Results emphasize the special and important role of the axially short-range interaction between lead and oxygen ions in the phonon softening.

REFERENCES
1. Lines M.E. and Glass A.M., "Principles and Applications of Ferro-electrics and Related Materials "(Oxford Univ. Press) p248 (1977)
2. Burns G. and Scott B.A., Phys. Rev. B7, 3088 (1973)
3. Freire J.D. and Katiyar R.S. Phys. Rev. B37, 2074 (1988)
4. Kleemann W., Schäfer F.J. and Rytz D., Phys. Rev. B34, 7873 (1986)
5. Cowley R.A., Phys. Rev. 134, A981 (1964)
6. Stirling W., J. Phys. C. 5, 2711 (1972)
7. Migoni R., Bilz H. and Bäuerle D., Phys. Rev. Lett. 37, 1155(1976)
8. Kugel G.E., Fontana M.D. and Kress W., Phys. Rev. B35, 813 (1987)
9. Khatib D., Kugel G.E. and Godefroy L., Phase Transitions 9, 125 (1987) and to be published.
10. Fontana M.D., Idrissi H., Kugel G.E. and Wojcik K.,to be published
11. Shirane G., Axe J.D. and Harada J., Phys. Rev. B2, 155 (1970)

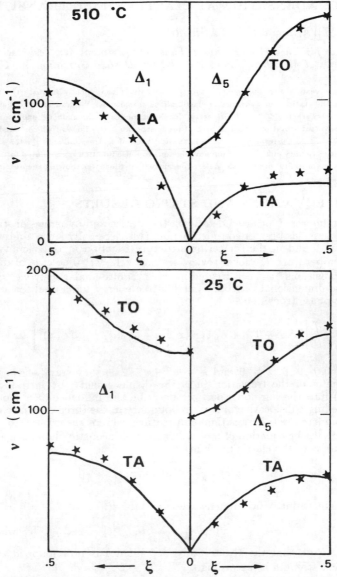

FIGURE   Fitted dispersion curves in cubic and tetragonal PbTiO$_3$
as function of reduced wave vector

TABLE   Model parameter values obtained in PbTiO$_3$ at 510°C. Only the
parameter A$_1$ is found to be dependent on the temperature. The   charges
Z and Y are given in units of the elementary charge e and the   short-
range forces in units of e^2/2v.

A$_1$(Pb-O)	B$_1$(Pb-O)	A$_2$(Ti-O)	B$_2$(Ti-O)	A$_3$(O-O)	B$_3$(O-O)	
22.85	-5.1	232	-1.8	1.4	-3.7	
Z(Pb)	Z(Ti)	Y(Pb)	Y(Ti)	Y(O)	K(Pb)	K(Ti)	K(O)
1.75	3.1	-5.4	-0.1	-1.8	1600	430	350

LOCAL CONDENSATION AT ELASTIC PHASE TRANSITIONS

F. SCHWABL and U.C. TÄUBER

Institut für Theoretische Physik, Physik-Department der Technischen Universität München, James-Franck-Str., D-8046 Garching, W. Germany

Abstract: We investigate the influence of short-range defects that locally increase the transition temperature on the statics and dynamics of elastic phase transitions. We find local condensation of the order parameter when $T \leq T_c^{loc}$. There are no localized modes in the acoustic phonon spectrum, but there is a resonant vibrational part in each of the scattering states instead which "condenses" at the defect for $T \rightarrow T_c^{loc}$. Correspondingly, the relaxation time of a stress-induced cluster diverges as $(T - T_c^{loc})^{-1}$ in the vicinity of T_c^{loc}. The phonon-phonon response function is calculated for a one- and a multi-defect system to first order in the defect concentration, the latter showing the development of a central peak.

1. MODEL EQUATIONS AND STATIC RESULTS

The influence of localized defects on the statics and dynamics of structural phase transitions has been of considerable interest, especially since a narrow central peak was found in the scattering structure function in the vicinity of the transition temperature T_c^0 (for a review see e.g. ref. [1]). The case of distortive structural transitions has been treated in ref. [2]; quite analagously we study the following one-dimensional Ginzburg-Landau free energy functional for a second-order elastic phase transition

$$\mathcal{F}[\epsilon] = \int \left[[a'(T - T_c^0) + U_0 \,\delta(x)] \,\epsilon(x)^2 + \frac{b}{2} \epsilon(x)^4 + d \,\epsilon'(x)^2 \right] dx \quad , \qquad (1)$$

where $a' > 0$, $b > 0$, $d > 0$ and $\epsilon = u' + u'^2/2$ is the strain serving as the order parameter of the transition (u is the displacement); by introducing the term $U_0 \delta(x)$ into the harmonic part we describe the influence of a short-range defect (modelling a dislocation or a grain boundary in the three-dimensional case) which locally increases the transition temperature (softens the crystal) for $U_0 < 0$. Following ref. [3], the equation of motion for the soft acoustic phonons is (ρ being the mass density of the elastic medium)

$$\rho \, \ddot{u}(x,t) = -\delta \mathcal{F}[u]/\delta u(x,t) \quad . \tag{2}$$

It is convenient to introduce dimensionless variables [2] $z = \sqrt{\frac{|a|}{d}}\, x$, $\tilde{t} = \sqrt{\frac{2a^2}{\rho d}}\, t$. $u(z) = \sqrt{\frac{b}{d}}\, u(x)$, $e(z) = \sqrt{\frac{b}{|a|}}\, \epsilon(x)$, $v_0 = \frac{1}{\sqrt{d|a|}} U_0$, $f[e] = \frac{b}{\sqrt{d|a|^3}} \mathcal{F}[\epsilon]$, where $a = a'(T - T_c^0)$. Note that the temperature dependence is now mapped onto the effective defect strength v_0 according to $T = T_c^0 + U_0^2/v_0^2 a' d$.

Within the framework of Ginzburg-Landau theory the equilibrium states $\bar{e}(z)$ are found from the stationarity condition $\frac{\delta f[e]}{\delta e(z)}\Big|_{e=\bar{e}} = 0$ with the restriction $f[\bar{e}] < \infty$, leading to the following nonlinear differential equation for $T > T_c^0$: $\bar{e}''(z) = [1 + v(z)]\,\bar{e}(z) + \bar{e}(z)^3$. The homogeneous solution $e_0 = 0$ is stable only when $v_0 > -2$, i.e. $T > T_c^{loc} = U_0^2/4a'd$. Below T_c^{loc} a cluster configuration

$$\bar{e}_C(z) = \pm \sqrt{2}\, [\sinh(|z| + \rho)]^{-1} \quad , \qquad \rho = \text{arcoth}\,(-v_0/2) \tag{3}$$

forms around the defect, its width diverging $\propto (T - T_c^0)^{-\frac{1}{2}}$, thereby continuously approaching the low-temperature phase, a phenomenon also termed *local condensation* [1),2)].

2. PHONON DYNAMICS FOR $T > T_c^{loc}$

Linearizing the fourth order differential equation (2) around the static solution $e_0 = 0$ we obtain an eigenvalue problem for the scattering states

$$\omega_k^2 u_k(z) = u_k''''(z) - u_k''(z) - v_0 u_k'(0) \delta'(z) \quad , \tag{4}$$

which is solved by $\omega_k^2 = k^2 (1 + k^2)$ and

$$u_k^+(z) = \frac{1}{\sqrt{\pi}} \cos kz \tag{5a}$$

$$u_k^-(z) = \frac{\text{sgn}(z)}{\sqrt{\pi}} \left[\cos \varphi_R(k) e^{-\sqrt{1+k^2}|z|} - \cos\left(k \mid z \mid + \varphi_R(k) \right) \right] \quad . \tag{5b}$$

Here $\tan \varphi_R(k) = \text{Im } R(k)/\text{Re } R(k)$ and

$$R(k) = \frac{ik \frac{v_0}{2}}{\left(\sqrt{1 + k^2} + ik \right) \left(\sqrt{1 + k^2} - ik + \frac{v_0}{2} \right)} \tag{6}$$

is the reflection coefficient (see Fig.1). The solutions (5) can be shown to form a complete set of orthonormalized eigenfunctions to eq.(4) [4)], which demonstrates that there is no localized mode in the spectrum in contrast to the distortive case [2)]. Note, however, that each of the antisymmetric scattering states (5b) contains a vibrational part $u_k^{loc}(z,t) = \text{sgn}(z) R(k) e^{-\sqrt{1+k^2}(|z|+ikt)}$ localized at the defect. Both the absolute value and the phase of $R(k)$ reach an extremum at a certain common wavenumber k_0 which is given by $k_0 \approx \sqrt{a'(T - T_c^{loc})/3d}$ in the vicinity of T_c^{loc} (in terms of the original variables), i.e. at T_c^{loc} the transmission amplitudes $1 - R(k)$ of the long-wavelength phonons vanish and the vibrational part condenses at the defect.

To illustrate this fact we study the relaxation of a stress-induced cluster $u^0(z) = \text{sgn}(z) e^{-\kappa|z|}$ $(\kappa > 0)$ by expanding the corresponding wavepacket in terms of the functions (5b). Its center of mass motion can then be found by the method of stationary phase yielding $z_m(\tilde{t}) = \pm(\tilde{t} - \tilde{t}_0)$ for $z \gtrless 0$, where near T_c^{loc} the relaxation time t_0 is given by

$$t_0 \approx \sqrt{2\rho d/a'^2} \, (T - T_c^{loc})^{-1} \tag{7}$$

(in the original variables) reflecting the fact that the corresponding displacement becomes stable at the local transition point.

3. RESPONSE FUNCTION AND CENTRAL PEAK

Introducing a heuristic damping coefficient $\gamma_k = D k^2 (1 + k^2)$ we proceed to discuss the phonon-phonon response function

$$\chi(z, z'; \omega) = \frac{1}{2\pi} \int_{-\infty}^{+\infty} \frac{u_k^+(z) u_k^+(z') + u_k^-(z) u_k^-(z')}{\omega_k^2 - \omega^2 - i \gamma_k \omega} \, dk \tag{8}$$

The integration can be performed analytically in the complex k-plane [4], the result being of the form $\chi(z, z'; \omega) = \chi_0(z - z'; \omega) + v_0 \chi_D(z, z'; \omega)$. As has been shown in ref. [5] by the means of a cumulant expansion we can calculate the corresponding response function for a *random multi-defect system* to first order in the defect concentration n using the formula

$$\chi(z - z'; \omega) = \chi_0(z - z'; \omega) + n\, v_0 \int_{-\infty}^{+\infty} \chi_D(z - z_D, z' - z_D; \omega)\, dz_D \quad . \tag{9}$$

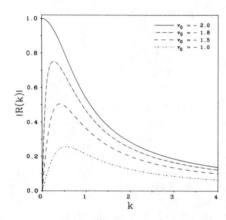

Fig.1: Reflection coefficient for different values of v_0

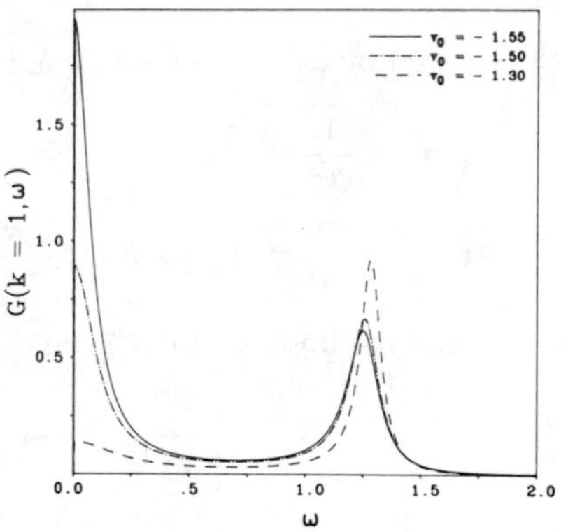

Fig.2: Correlation function $G(1, \omega)$ near $T_c^{loc}(n)$ $(n{=}0.1, D{=}0.01)$

Using the abbreviation $2\nu^2 = \sqrt{1 + 4\omega^2/(1 - iD\omega)} - 1$ we finally obtain the Fourier-transformed response function

$$\chi(k, \omega) = \frac{1}{2\pi\left((1 - iD\omega)\,k^2\left[(1 + k^2) + n\,v_0\,\dfrac{\sqrt{1+\nu^2} - i\nu}{\sqrt{1+\nu^2} - i\nu + \frac{v_0}{2}}\right] - \omega^2\right)} \quad , \tag{10}$$

where the term $\propto nv_0$ has been written in the form of a self-energy correction. The corresponding correlation function $G(k, \omega) = \operatorname{Im}\chi(k, \omega)/\omega$ is shown in fig.2 for $k = 1$, $n = 0.1$ and $D = 0.01$. There is a marked development of a *central peak* when the temperature approaches the shifted local transition temperature $T_c^{loc}(n) = T_c^{loc}(0) + n(1 + n)\,U_0^2/a'd$. The width of the central peak turns out to be independent of the phonon-damping D and behaves as $T - T_c^{loc}(n)$. Note that in fig.2 the soft phonon peak position is fixed due to the use of scaled variables.

We remark that the results of sections 2. and 3. also apply to the case of a first-order transition (e.g. described by a ϕ^6-model [6]) because the nonlinear terms disappear when eq.(2) is linearized around $e_0 = 0$.

References:

1) Bruce, A.D. and Cowley, R.A., Adv. Phys. **29**, 1 (1980)

2) Schmidt, H. and Schwabl, F., Z. Phys. B **30**, 197 (1978)

3) Landau, L.D. and Lifschitz, E.M., vol. VII, Theory of Elasticity, Pergamon Press (^{2}1969)

4) Schwabl, F. and Täuber, U.C., to be published

5) Reisinger, H. and Schwabl, F., Z. Phys. B **52**, 151 (1983)

6) Falk, F., Acta Met. **28**, 1773 (1980)

LATTICE DYNAMICS AND PHASE TRANSITIONS IN Pb_2CoWO_6

W. Bührer[*], M. Rüdlinger[*], P. Brüesch[#], H. Schmid[+]

[*]Labor für Neutronenstreuung ETHZ, CH-5303 Würenlingen

[#]Asea Brown Boveri Corporate Research, CH-5400 Baden

[+]Dept. de Chimie Physique, Universite de Geneve, CH-1211 Geneve

The elpasolite Pb_2CoWO_6 (PCW) shows, as a function of temperature, two structural phase transitions :

$$\text{Cubic} \xrightarrow[295 \text{ K}]{} \text{incomm.} \xrightarrow[235 \text{ K}]{} \text{orthorho./monocl.}[1]$$

In the cubic structure (space group O_h^5) the oxygen ions are shared between the Co and the Wo ions, and CoO_6 and WO_6 octahedra are formed. The elpasolite structure is closely related to the simple perovskite and to the anti-fluorite structure. In all three families displacive phase transitions driven by soft phonons are reported [2]. However, PCW shows very unusual behaviour: (i) both transitions are of first-order, and (ii) phases II and III can coexist in a rather broad temperature region. A theoretical model, based on the Landau theory of phase transitions, has been used for a tentative interpretation [3].

In the present work the lattice dynamics of the cubic phase has been investigated. The lowest energy phonon modes, measured by inelastic coherent neutron scattering (INS), are shown in Fig. 1. Temperature dependent, strongly overdamped phonons were observed for wavevectors along directions Δ, Z and Σ for all energies below ≈ 5 meV, and the plotted data are the results of damped oscillator fits to the neutron intensities. Experiments at different points in reciprocal space showed that rotational and translational modes contribute to the diffuse inelastic scattering. In order to get better insight into the higher frequency modes of PCW, infrared reflectivity measurements have been performed. All four IR-active modes (TO and LO) have been clearly observed, the lowest TO mode again being considerably broadened as in the INS spectra.

A lattice dynamical calculation with a rigid-ion model was initiated. The short-range part is described by a valence-force field with 3 stretching constants (Co-O, W-O, Pb-O) and one bending constant (Co-O-W); in addition there are 3 ionic charges as model parameters. The values, as given in Table 1, have been determined by a least-squares fit to the experimental frequencies (INS and IR). Experimental data and model calculation are displayed in Fig. 1 (along direction Δ, Z and Σ) and Table 2 (zone center Γ) for the low and high energy parts, respectively. In view of the simplicity of the model, the agreement is surprisingly good.

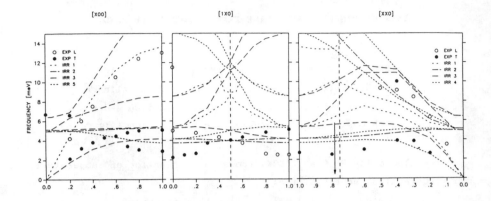

Fig. 1: Phonon dispersion of PCW along direction Δ, Z, and Σ at 420 K;
lines: best rigid-ion model calculation;
arrow: q of incommensurate phase II.

Table 1: Best-fit rigid-ion model parameters (charges in units of /e/,
force constants in dyn/cm).

ionic charges			stretch		
	W	0.91	stretch	C-O	12390
	Co	0.86		W-O	19280
	Pb	1.72		Pb-O	1690
	O	-0.87	bend	Co-O-W	10

Table 2: IR-active zone-center frequencies of PCW (in meV);

TO	experimental	calculation
	~ 2.5	6.4
	24.8	27.3
	32.2	32.7
	70.7	72.3
LO	14.3	16.1
	29.1	29.5
	51.5	44.3
	89.3	76.8

Phase II (incommensurate) is characterized by satellite peaks
appearing near the X points ((001), cubic notation) with wave-vectors
$q_1 = (\sigma,\sigma,1)$ $\sigma \approx 0.21$, and $q_2 = (\epsilon,1+2\epsilon,1-2\epsilon)$, $\epsilon \approx 0.07$; σ and ϵ both depend
on temperature. There is no lock-in transition to phase III with
satellite intensities at the X points and at $(0,m/2,n/2)$ (m,n odd).

The lattice dynamical calculation in the cubic phase shows that the eigenvectors of the modes Δ_3, Σ_2 and Σ_4 have mainly Pb displacements in the (x-y)-plane or parallel to z, respectively; the mode Δ_2 is the pure O-octahedron rotation. A condensation of one of the 'Pb-modes', together with a subsequent relaxation of O and W ions, could create atomic positions as obtained from the powder diagram analysis with space group Pmm2[4].

In a 'ideal' PCW crystal, the structural instabilities are given by the delicate balance of short- and long-range forces, and a slight anharmonicity of the interatomic potential can drive the transitions. This behaviour can be simulated within our simple model: a very small change of one of the parameters results in the condensation of one or several phonon modes (translational or rotational). In the 'real' crystal, instabilities can be influenced by local imperfections and disorder. Their importance is manifested by the coexistence of phases II and III below 235 K, and, furthermore, the existence of an orthorhombic and a monoclinic modification in phase III[4] (both factors probably depend on cooling rate as well as on sample preparation). Therefore, to conclude, the phase transitions in PCW are far from being understood.

[1] Brixel W. et al., Jap. J. Appl. Phys. Suppl. 24, 242 (1985)

[2] Aleksandrow K. and Misyul S., Kristallographiya 26, 1074 (1981)

[3] Maaroufi F., Thesis, Universite de Picardie (1988)

[4] Rüdlinger M., Diplomarbeit ETHZ, LNS Report 145 (1988/89)

A QUANTUM FIELD THEORY OF PHONONS AND MELTING

Toyoyuki Kitamura

Nagasahi Institite of Applied Science

Nagasaki 851-01, Japan

1. THE WARD-TAKAHASHI RELATIONS (WTR)

WTR associated with the spatially translational and rotational invariance are derived using the generator of the spatial translation and adding an infinitesimal symmetry breaking term $H_\varepsilon = -\varepsilon \int d^3x v(x) \psi^\dagger(x) \psi(x)$ to the invariant Hamiltonian H[1-5]:

$$0 = \sum <T_\tau \Psi_1(x_1) - -\nabla \Psi_i(x_i) - -\Psi_n(x_n)>_\varepsilon + \varepsilon \int d^4x <T_\tau \Psi_1(x_1) - - -\Psi_n(x_n) \nabla n(x)>_\varepsilon v(x), \quad (1)$$

where $<--->_\varepsilon$ means a statistical average, $n(x) = \Psi^\dagger(x)\Psi(x)$, $\Psi_i(x)$ stands for $\Psi(x), \Psi^\dagger(x)$ and $v(x)$ is the number density of the ground state atoms.

2. CRYSTALS

We specify the model[2-5]: an atom is assumed to be in a harmonic potential made up by the surrounding atoms. We consider two levels: the ground state $w_0(x)$ and the first excited states $w_i(x)$ ($i=x,y,z$). We take the Wannier functions: $\Psi(x) = \sum \phi_{\mu k}(x) a_{\mu k}, \phi_{\mu k} = \frac{1}{\sqrt{N}} \exp(ik \cdot R_n) w_\mu(x-R_n)$, where $\mu = 0,x,y,z$. In the two band model, we have the relations:

$$w_i(x) = -\zeta \nabla_i w_0(x), \qquad w_0(x) = \zeta \nabla_i w_i(x), \qquad \zeta = \sqrt{\frac{2\hbar}{m\omega}}, \quad (2)$$

where ω is the eigenfrequency of the harmonic oscillator. The Hamiltonian is given by

$$H = \sum \hbar \varepsilon_{\mu k}^0 a_{\mu k}^\dagger a_{\mu k} + \frac{\lambda}{2} \sum \Phi_{\tilde\mu\tilde\nu}(q) \rho_{\tilde\mu q}^\dagger \rho_{\tilde\nu q}, \quad (3), \qquad \Phi_{\tilde\mu\tilde\nu}(q) = \frac{1}{N^2} \sum_{m \neq n}$$

$$\exp[-iq \cdot (R_m - R_n)] \int d^3x d^3y w_\mu(x-R_m) w_{\mu'}(x-R_m) w_\nu(y-R_n) w_{\nu'}(y-R_n) V(x-y), (4)$$

$$\rho_{\mu q}^\dagger = \sum (a_{\mu k}^\dagger a_{\mu' k-q} + a_{\mu' k}^\dagger a_{\mu k-q}, \quad \tilde\mu = (\mu,\mu'), \tilde 0 = (0,0), \tilde i = (i,0), \quad (5)$$

Putting $v(x) = \sum w_0^2(x-R_n)$, $\tilde 0 = 0$, $\tilde i = i$, we obtain WTR for atomic and phonon Green's function, $G(k) = G(k,i\omega_n)$ and $D(q) = D(q,i\nu_n)$:

$$\Lambda_i \delta_{ij} = \frac{\varepsilon}{\hbar} (\sqrt{\pi}\zeta)^{-3} D_{ij}(0), \qquad \Lambda_i = \pm \frac{2}{\beta\hbar} \sum [G_0(k) - G_i(k)], \quad (6)$$

$$G_0(k) - G_i(k) = \frac{1}{\hbar} (\sqrt{\pi}\zeta)^{-3} G_0(k) G_i(k) [1 + \frac{1}{\hbar} \Gamma_{01i}(k.0.k) D_{1i}(0)], \quad (7)$$

In the RPA, the vertex part $\Gamma_{0ji} = \Phi_{ij}$, the self-energy for G is given by

$$\Sigma_0(k) = \quad 0\overset{0\bigcirc 0}{\vdots}0 \quad + \quad 0\overset{i\bigcirc i}{\vdots}0 \quad + \quad 0\overset{i}{\lessgtr}i\overset{}{\gtrless}0, \tag{8}$$

where $G(k): \longrightarrow$, $D_{ij}(q): \wedge\wedge\wedge$, $\phi_{ij}(q): \text{-----}, -\varepsilon v:\times$. The first ans second terms corresponds to the scattering of the periodic potential like
magnetizm and the last term to the mixing of the states like superconductivity. Using the relations of the potential functions, we can put

$$\left.\begin{array}{c}\varepsilon_{ik}\\ \varepsilon_{0k}\end{array}\right\} = \frac{\varepsilon_{0k}+\varepsilon_{1k}}{2} \pm \frac{\Delta}{2} \pm \sqrt{(\frac{\varepsilon_{0k}-\varepsilon_{1k}}{2})^2+\Delta^2}, \qquad \Delta = \frac{1}{\hbar}\Phi_{ii}(0)\lambda_i, \tag{9}$$

The gap equation is given by

$$\delta_{ij} = \frac{1}{\hbar}\Phi_{ij}(0)Q(0), \qquad Q(0) = \sum [f(\varepsilon_{1k})-f(\varepsilon_{0k})]/[\varepsilon_{1k}-\varepsilon_{0k}], \tag{10}$$

where $Q(0)$ is a bubble diagram and f is boson or fermion distribution
function. Defining $D_{ij}(q,\tau)=<T_\tau \rho_{iq}(\tau)\rho_{jq}^\dagger(0)>$, we obtain

$$D_{ij}(q) = Q(q)\delta_{1j} + \frac{1}{\hbar}Q(q)\Phi_{i1}(q)D_{1j}(q), \tag{11}$$

At low temperatures, $Q(q)=2N\omega/(i\nu)^2-\omega^2$, we obtain the dynamical equation

$$[(i\nu_n)^2\delta_{ij}-M_{i1}(q)]D_{1j}(q)=-2N\delta_{ij}, \quad M_{ij}(q)=-\frac{2N\omega}{\hbar}[\Phi_{ij}(q)-\Phi_{ij}(0)], \tag{12}$$

3. LIQUIDS

Atoms distribute randomly, but the radial distribution function
$g(x)=<\sum w_0^2(x-R_n)>$ breaks the spatially translational invariance in the
radial direction with an atom fixed at the original point[6], where <--->
is the ensemble average. Regarding $v(x)$ as $g(x)$ and using the cell
model, we obtain the WTR for a liquid. Despite of the existence of
$g(x)$, even within a few atomic distances, the rotational invariance is
satisfied. This fact relates to the importance of a longitudinal mode
in a liquid. The dominant peak near an atomic distance relates to the
existence of the energy gap on the spherical Brillouin zone with a radius of a half reciprocal atomic distance. The energy gap makes phonons
stable. The fact that the main contribution of the integration over
the whole region comes from the region without a few atomic distance
makes the calculation simple; the value of the integration is determined by the outer region and the integration over the whole region reduces the rank of the many body distribution function. Thus the equation
for a liquid obtained by replacing Φ_{ij} by $<\Phi_{ij}>$ in those for a crystal.
A longitudinal phonon spectrum has a roton type minimum near a recipro-

cal atomic distance. Transeverse modes are rather flat[6,7].

4. THERMODYNAMICS

The free energy is given by $\partial F/\partial\lambda = <H_{int}>$. The difference of the free energy between the crystalline and gaseous states is given by

$$\Delta F_s = \frac{N^2}{2}\Phi_{00} - \frac{N^2}{2}\Phi_{ii}(0) + \Delta F_{ss} + \Delta F_{sp}, \qquad (13)$$

Near the melting temperature, $\omega\beta\hbar$, $\varepsilon_\mu\beta\hbar < 1$, the gap equation leads to

$$\frac{1}{\lambda} = \beta\omega\hbar\alpha[1-(\frac{3\beta\hbar\Delta}{2})^2 + \frac{2}{15}(\frac{3\beta\hbar\Delta}{2})^4 + ----], \qquad \omega = -\frac{2N}{\hbar}\Phi_{ii}(0), \qquad (14)$$

where α is determined by the number of the nearest neighbours of the reciprocal lattice vectors and the volume of the integral region. The temperature T_s at the phonon instability in a crystal is determined by putting $\Delta=0$, $\lambda=1$. Eq.(14) with T_s and $\lambda=1$ is the G–L equation without the fluctuation parts. Using these equations, we obtain

$$\Delta F_{ss} = -\sigma\int_0^1 d\frac{1}{\lambda}\frac{\hbar^2}{8}\frac{\Delta^2}{\Phi_{ii}(0)} = -\frac{1}{18}\sigma\alpha Nk_B T_s(1-\frac{T}{T_s})^2, \qquad k_B T_s = \alpha\hbar\omega, \qquad (15)$$

Transforming the summation ν_n to the contour integration, we have

$$\Delta F_{sp} = -\frac{1}{2\beta\hbar}\sum\int d\lambda Tr\Phi(\mathbf{q})D(\mathbf{q}) = \frac{1}{\beta}\sum[ln2sinh\frac{\beta\omega_{\alpha q}}{2} - ln2sinh\frac{\beta\omega}{2}], \qquad (16)$$

we obtain the same equation for a liquid by replacing Φ_{ij} by $<\Phi_{ij}>$. The temperature T_1 at the phonon instability in a liquid is obtained by $k_B T_1 = \alpha h\omega_1$, $\omega_1 = -\frac{2N}{\hbar}<\Phi_{ii}(0)>$. Assuming the static elastic energies for both the states are equal, neglecting the fluctuation parts and noting the dominant mode for a liquid is a longitudinal, we can determine the melting temperature T_m by putting $\Delta F_{ss} = \Delta F_{1s}$:

$$T_m = \frac{\sqrt{\sigma\eta}-\eta}{\sqrt{\sigma\eta}-1}T_s, \qquad \frac{T_1}{T_s} = \frac{T_1}{T_s}, \qquad (17)$$

where T_s and T_1 are determined by the experimental phonon spectra.

REFERENCES

1. Kitamura, T. and Umezawa, H., Physica A121,67(1983).

2. Kitamura, T., Physica A128,427(1984).

3. Kitamura, T., Physica A135,21(1986).

4. Kitamura, T.,Phys. Lett. A117,81(1986), A118,341(1986).

5. Kitamura, T., Physica A144,29(1987).

6. Kitamura, T., Physica A151,303(1988).

7. Takeno, S. and Goda, M., Prog. Theor. Phys. 45,331(1971)

DISPLACIVE - ORDER-DISORDER CROSSOVER OF THE STRUCTURAL PHASE TRANSITION IN A-TCNB UNDER HYDROSTATIC PRESSURE

Ph. BOURGES*, C. ECOLIVET*, A. MIERZEJEWSKI**, Y. DELUGEARD*
and A. GIRARD*

* : Groupe de Physique Cristalline, URA au C.N.R.S. 040804
Universite de Rennes I, 35042 Rennes cedex, France.
** : Institute of Organic and Physical Chemistry, Technical
University of Wroclaw, 50-370 Wroclaw Poland.

Anthracene-tetracyanobenzene (A-TCNB) is a weak charge-transfer molecular crystal built of mixed stacks of donors (A) and acceptors (TCNB) molecules. These molecules are stacking along the $\vec{c}$ axis of the monoclinic lattice and the long axes of both molecules tend to be aligned along the $(2,0,\bar{1})$ direction. Around 200 K, a second order antiferrodistorsive phase transition occurs and the low temperature phase is identified by the reorientation of the A molecules in the $(\vec{a},\vec{b})$ plane. At atmospheric pressure, this SPT was thought to be order-disorder, but recently a soft mode was observed by Raman scattering for a protonated crystal [1] and by Raman and inelastic neutron scattering for a deuterated sample [2]. These experiments suggest that this SPT is mainly displacive in spite of so different soft mode characteristic parameters between both phases. It is also obvious that this SPT presents some order-disorder features. Indeed, two kinds of interactions could be responsible of this SPT : on the one hand, the steric hindrance between A molecules in $(\vec{a},\vec{b})$ plane yields to strong orientational coupling for the libration of A molecules around the axis perpendicular of their molecular plane, on the other hand the steric hindrance of the nitrogen atoms of TCNB with the hydrogen atoms of the A molecules in the $(1,0,2)$ plane gives a double well potential for the A molecules which could induce an order-disorder dynamics. From mean-field theory approach[3], we know that at atmospheric pressure the first one is preponderant and leads to the condensation of a soft mode.

As in molecular crystals, we can expect in A-TCNB some peculiar features when applying moderate hydrostatic pressures. So, we have performed Raman scattering under hydrostatic pressure (P) up to 4 kbar between 130 K and 293 K. Indeed, A-TCNB is very sensitive to slight changes of pressure.

In the Raman spectra, we have only recorded the low-frequency part because the other high-frequency modes are not especially affected by the transition. In the LT phase, we observe together the three components lying in this frequency region (2 Ag + 1 Bg) due to some depolarization by the pressure cell. According to harmonic lattice dynamics calculations [4], the pattern of the lowest frequency mode, which softens at atmospheric pressure, is mainly the libration of the A molecules around their plane normal. The frequency of the Bg mode, which pattern corresponds to the out-of-phase libration of the soft mode, follows in all the pressure range an evolution characteristic of a biquadratic coupling with the soft mode. Meanwhile, its linewidth increases with temperature more than usual. This behaviour, which also occurs at atmospheric pressure, is rather typical of the existence of static disorder in the crystal.

Therefore, the occurence of the frequency anomaly allows a rather accurate determination of the phase diagram and, apparently this SPT remains continuous in the whole pressure range. Pressure favours the low temperature phase since the phase diagram presents a large posi- tive slope dT_c/dP ($\approx$ 36 K/kbar near atmospheric pressure). This sign could be interpreted as an enhancement of the repulsion along the long molecular axis of both molecules though the compressibility is small along this $(2,0,\bar{1})$ direction. We can relate this fact to the isotopic effect : when replacing hydrogen by deuterium atomes, the transition temperature is lowered by about 15 K because C-D bonding is slightly shorter than C-H bonding.

Beyond the "soft mode", the other Ag mode does not shift appreciably and remains around 18 cm^{-1} . These two Ag modes couple linearly and give rise to a complex dynamic response with particular shapes. At .6 kbar, the mode softens and suggests a saturation frequency less than 10 cm^{-1} with a rather large damping. This unusual behaviour is quite similar to the soft mode behaviour at atmospheric pressure. But at higher pressures the frequency of the "soft mode" does not shift vs temperature and has always a large increasing damping as shown by the evolution of Raman spectra at 1.2 kbar(cf figure). So, the softening of the lowest lying mode progressively dissapears. Let us remark that the saturation frequency of this mode increases almost linearly with increasing P at a rate of 3.5 cm^{-1} /kbar. At 2.4 kbar, the "soft mode" becomes no longer overdamped. Then it loses its critical behaviour. Although it is very difficult to extract some quantitative information about the very low-frequency part of Raman spectra due to a large Rayleigh line, it seems that some additional intensity signs the existence of a quasi-elastic response.

All these observations indicate an order-disorder regime at high pressures. Such behaviour demonstrates a crossover of a rather displacive regime at atmospheric pressure to a typical order-disorder regime at higher pressures, and always in the interesting case of an continuous SPT.

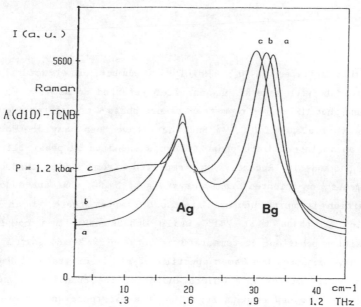

Figure : Evolution of Raman spectra at P = 1.2 kbar (T$_c$ = 242 K)
a) : T = 177 K ; b) : T = 196 K ; c) : T = 215 K

References
[1] Mierzejewski A., Ecolivet C., Pawley G.S., Luty T., Girard A. and Lemee M.H. : Sol. State Comm. 65, 431, (1988)
[2] Bourges Ph., Lemee-Cailleau M.H., Launois P., Moussa F., Cailleau H., Ecolivet C. and Mierzejewski A. : IMF 7th Int. Meeting on ferroelectricity Saarbrücken (1989)
[3] Luty T. and Kuchta B. : J. Chem. Phys., 85 (7), 4032, (1986)
[4] Brose K.H., Luty T. and Eckhardt C.J. : Chem., Phys., Lett., 137 (1987), 17

RAMAN STUDY OF THE STRUCTURAL PHASE TRANSITIONS IN Na_2SO_4

BYOUNG-KOO CHOI[*] and DAVID J. LOCKWOOD

Physics Division, National Research Council, Ottawa K1A 0R6, Canada

[*]Department of Science Education, Dankook University, Seoul 140-714, Korea

Sodium sulfate (Na_2SO_4) exhibits a number of structural phase transitions between several phases often referred to as I to V. It has been found that the stable room-temperature phase V transforms eventually to phase I at around 240°C.[1,2] But there have been many controversial reports on an intermediate phase (often assumed to be phase III or IV) between V and I. Recently we reported from ionic conductivity measurements[2] on oriented single crystals of Na_2SO_4 that the major V to I transition is quite unusual: in some samples there is an abrupt transition to phase I at 241°C while others exhibit a complicated pretransition behaviour at temperatures between 226°C and 241°C.

We have measured the Raman spectrum of single crystals of Na_2SO_4 to obtain a better microscopic understanding of the transitions and have obtained entirely new results regarding the existence and nature of an intermediate phase between V and I. In a majority of the samples the Raman spectrum indicated a direct first-order transition between phases V and I at 242°C without occurrence of an intermediate phase, but for other samples multiple transitions were observed. The sequence of transitions shown in Figure 1 was obtained at the first heating of a fresh Na_2SO_4(V) crystal, and both phases II and III appear before the final transition to phase I. Except for the first heating cycle, the subsequent sequences of transitions are III→I on heating and I→II→III on cooling in agreement with other techniques.[1,2] Hence the assignments in Figure 1 for phases II and III are adopted from the spectra measured at

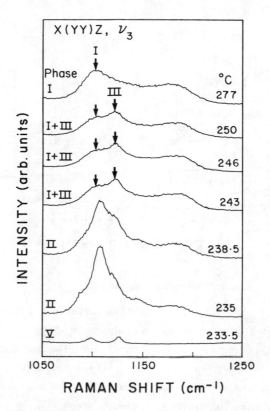

Figure 1.   The X(YY)Z Raman spectrum in the ν_3 region of Na_2SO_4 at various temperatures (X,Y,Z denote the phase V a,b,c axes).

the first cooling process (a full account of these results will be published elsewhere). For this sample, we found that phase V persists clearly to 233.5°C, in contrast to the many previous observations which give a V→III transition temperature at anywhere from 180 to 240°C. At 234°C, there is a sharp transition to phase II and this phase persists to around 239°C. After this, a mixed phase of III and I appears over a quite large temperature range. At 277°C, the pure spectrum of phase I is obtained, as shown in Figures 1 and 2. Thus only the V→II transition is well defined, and phase III does not appear on its own in the first heating process. Thus the unusual behaviour in the conductivity at the first heating process and the much mentioned 'sluggishness of transition'[1,2] probably arise from the mixed nature of phases III and I and the lack of a distinct transition temperature to phase I.

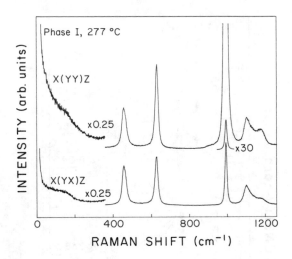

Figure 2.   Raman spectrum of $Na_2SO_4(I)$.

After the transition to phase I, no distinct external (lattice) vibrational modes are observed, as shown in Figure 2.   Instead, a Rayleigh wing-type scattering from the laser line and a broad shoulder at ~150 cm^{-1} are observed. This is a characteristic feature in the vibrational spectrum of fast ionic conductors such as α-AgI and it is thought to be due to disorder-induced breakdown of symmetry-based selection rules.[3] The Raman line widths of internal vibrations of sulfate ions in phase I are noticeably larger than in the other phases. The internal mode spectrum of phase I resembles much more the molten salt spectrum[4] than the ordered phase V or III spectrum, proving that orientational disorder of the sulfate ions is present even at temperatures around 242°C (at the first and subsequent cooling processes, phase I persists down to 234°C), which is far below the melting temperature of 884°C.

1.   See Saito, Y., Kobayashi, K. and Maruyama, T., Thermochimica Acta. 53, 289 (1982) and references therein.
2.   Choi, B. K. and Lockwood, D. J., Phys. Rev. B40, xxxx (1989).
3.   Hansen, R. C., Fjeldly, T. A. and Hochheimer, H. D., Phys. Stat. Sol. b70, 567 (1975).
4.   Walrafen, G. E., J. Chem. Phys. 43, 479 (1965).

PHONONS AND PHASE TRANSITIONS IN KDP-TYPE CRYSTALS

P. SIMON and F. GERVAIS

Centre de Recherches sur la Physique des Hautes Températures, C.N.R.S.

45071 ORLEANS Cedex 2, FRANCE

1. INTRODUCTION

The KDP-type compounds (MH_2XO_4, with $M = K,Rb,NH_4$; $X = P,As$) constitute a unique family where a parent paraelectric phase becomes ferroelectric or antiferroelectric upon cooling, depending of chemical composition. The phase transitions originate from ordering of the asymetric hydrogen bonds O-H- -O, but the triggering mechanism is a subject of controversy : is it a pure order-disorder one, as reviewed by Tokunaga and Tatsuzaki[1] or with displacive features for heavy ions ?[2,3] A comparative study by spectroscopic methods in the whole family can therefore shed new light on the physics of the phase transitions involved, and particularly on the mechanims of occurence of either ferroelectricity or antiferroelectricity. Another interest for a deeper understanding of the dynamic properties of this family is the study of the mixed system $Rb_{(1-x)}(NH_4)_xH_2PO_4$, which exhibits a proton glass behavior for chemical compositions situated in the middle of the phase diagram.[4]

This paper is devoted to the phonon spectra of KH_2PO_4, RbH_2PO_4, KH_2AsO_4 and $NH_4H_2PO_4$, hereafter denoted respectively KDP,RDP,KDA and ADP, in their high- and low-temperature phases. The former three undergo para-ferroelectric phase transitions while the latter becomes antiferroelectric at low temperature, due to the peculiar orientation of ammonium ions. The phonon responses are obtained by infrared reflectivity spectroscopy followed by simulation of the experimental spectra with the factorized form of the dielectric function.[5]

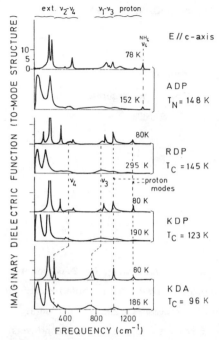

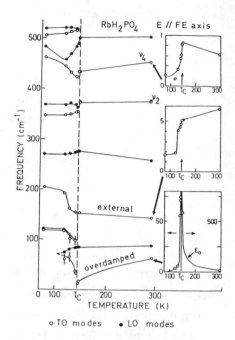

Figure 1 : TO-mode structure, obtained by fitting with the factorized form of the dielectric function.

Figure 2 : Temperature dependence of TO and LO modes of RDP. Inserts shows oscillator strengths of modes pointed by the arrows.

2. FERROELECTRICS.

The most interesting results are obtained for polarization of the electric field of the infrared electromagnetic wave oriented along the fourfold c-axis, which becomes the ferroelectric axis. The preceding works on this subject were concentrated on the low-frequency phenomena, particularly on the so-called ferroelectric fluctuation, overdamped at any temperatures, and on its coupling with the lowest-frequency lattice mode, which is the M - H_2XO_4 translation mode.[6] Figure 1 displays the imaginary part of the dielectric function (which corresponds to the TO-mode structure) for all compounds and Figure 2 the temperature dependence of TO and LO frequencies for RDP. Comparison of all spectra allows assignment of the vibration modes. The ν_{4c} PO_4 internal mode (near 430 cm^{-1}) exhibits a precursor effect of the phase transition : it is highly damped even at lowest temperatures, and its frequency softens and its oscillator strength increases upon approaching T_C. The eigenvectors of this mode are very similar to the static displacements at T_C, i.e., to the ferroelectric fluctuation component, and are coupled by electrostatic interaction to the intersite tunneling proton motions. The instability of mode ν_{4c} pushes down the lower-frequency components, especially the translation mode Rb--H_2PO_4, and this gives rise to the ferroelectric soft mode via a mixed mechanism between classical mode-mode and Fano-type discrete state-quasicontinuum interaction. The phase transition is then displacive, but with at the origin an order-disorder character of protons, consistent with neutron diffaction measurements.[2] The same argumentation can be applied to KDP and KDA too.[7]

3. ANTIFERROELECTRICS.

The antiferroelectric ammonium compounds are less known than their ferroelectric isomorphs, particularly about dynamic properties in the low-temperature phase, because the crystals invariably shatter at the phase transition. This has precluded spectroscopic investigations below T_N up to very recently, with low-frequency infrared reflectivity [8] and Raman [9] results. The present results are obtained in a larger spectral range, including internal PO_4 and ν_4 NH_4 modes. The TO-mode structure is displayed on Figure 1, for orientation of the IR electric field along the c-axis. The essential feature in these spectra is the occurence in the whole paraelectric phase of a highly damped mode at low frequencies (below 100 cm^{-1}), the frequency of which lowers upon approaching T_N from above, and which suddenly disappears on cooling just below T_N. The same is true, and even a little more marked, in the other polarization, corresponding to IR electric field oriented along the AFE axis. The disappearance of this mode explains the dielectric constant anomaly at T_N. The slight improper ferroelectric character of this lowest frequency mode originates from coupling with the zone-corner lattice instability. The other lattice mode are only very weakly affected by the phase transition. An other point must be noted, in the antiferroelectric phase of ADP : the weak damping of mode ν_{4c}. Its damping-to-frequency ratio is about 3%; whereas it is near 20% in RDP. This confirms the tendency to instability of this mode in ferroelectric compounds.

REFERENCES

1. Tokunaga, M. and Tatsuzaki, I., Phase Transitions 4, 97 (1984).
2. Nelmes, R.J., Kuhs, W.F., Howard, J.E., Tibballs, J.E. and Ryan ,T.W., J. Phys. C 18, L1023 (1985).
3. Blinc, R., Z. Naturforschung A 41, 249 (1986).
4. Courtens, E., Ferroelectrics 72, 229 (1987).
5. Gervais, F., in Infrared and Millimeter Waves (ed. K.J. Button), 8, 279 (Academic Press, New York, 1983).
6. Kobayashi, K. , J. Phys. Soc. Jpn. 24, 497 (1968).
7. Simon, P., Gervais, F. and Courtens, E., Phys. Rev. B 37, 1969 (1988).
8. Wyncke, B., Bréhat, F. and Arbouz, H., Phys. Stat. Sol. (b) 134, 523 (1986); J. Phys. C 19, 6893 (1986).
9. Kasahara, M., Tokunaga, M. and Tatsuzaki, I., J. Phys. Soc. Jpn. 55, 367 (1986).

NEUTRON SCATTERING STUDY OF THE PHASE TRANSFORMATION OF LaNi$_3$ INDUCED BY HYDRIDING[*]

J.H.RUAN, X.X.ZENG, S.W.NIU

(Institute of Atomic Energy, Box 275, Beijing, China)

and

J.M.ZHAO, K.D.YANG, C.H.DAI

(General Research Institute for Non-Ferrous Metals, Beijing)

1. INTRODUCTION

La-Ni hydrides are expected to be a new type of fuel in the near future for its high hydrogen storing capacity. LaNi$_3$ is a particular member of La-Ni family. Although some studies on its the property of physics and chemistry have been investigated using X-ray diffraction and differential thermal analysis[1], there are numerous unsolved problems, such as the influence of hydrogen on the structure is still an open question. The present paper is devoted to an further study of the phase transformation caused by hydiding and dehydiding using the neutron diffraction and neutron inelastic scattering methods at Institute of Atomic Energy in Beijing.

2. EXPERIMENT AND RESULTS

(1) Sample preparation:

Hydriding sample: The sample of LaNi$_3$ was loaded with hydrogen at room temperature under 40 H$_2$ atm. The hydrogen concentration was determined by both the excluding water and the pressure variation methods, corresponding to a composition LaNi$_3$H$_5$. The detailed procedure of preparation of sample and of hydrogenation are identical to that described in paper 1. The P-C isotherm of LaNi$_3$ hydrides measured at 50°C shows that no plateau appears in the P-C isotherm, i.e. there is only a α-phase in the hydriding sample.

Dehydriding sample: The dehydriding sample was produced by the hydriding sample dehydrided at 600°C for 10 minutes. The P-C isotherms measured at various temperatures show that two plateaus appear in the each of isotherm, i.e. there are two hydrides in the dehydriding sample.

(2) Neutron diffraction measurements:

In order to reducing the intense background caused by the large incoherent scattering cross section of hydrogen, the neutron diffraction experiments were performed on the tripe-axes neutron scattering spectrometer[2] at IAE in Beijing, using a wavelength of 1.227 Å.

The neutron diffraction pattern measured at room temperature are shown in figure 1, in which the abscissa and

[*] The project supported by National Science Foundation

the ordinate shows the diffracted intensities and 2θ (°) scattering angles respectively . The dotted line and the full line is the diffraction pattern of the hydriding sample and the dehydriding sample respectively. It can be seen from the dotted line that although a microcrystalline state of $LaNi_5$ was weakly exhibited, the pattern of amorphous type was revealed, that is, the hydriding sample of $LaNi_3$ hydrided at room temperature under 40 H_2 atm has been transformed from crystalline state of the $LaNi_3$ into amorphous state with a feature of microcrystalline of $LaNi_5$. In the diffraction pattern of the dehydriding sample shown the full line in figure 1 a number of the striking diffraction peaks composed by the crystalline states $LaNi_5$ and $LaH_{2.6}$ were exhibited (The notes of various crystal areas labeled on the top of the diffraction peaks), which show that the dehydriding sample behind the $LaNi_3-H_2$ dehydrided at 600°C for 10 minutes has been transformed from amorphous state into the new crystalline states.

(3) Neutron inelastic scattering experiment:

Neutron inelastic scattering experiments were performed on the Be filter detector spectrometer at IAE(Beijing) The details of spectrometer have been discribed in paper 3, which using the Ge(111) monochromator, a colded Be filter (15-cm long) placed between the sample and detector and 7-BF_3 detectors.

The local modes measured were shown in figure 2 , in which the abscissa indicates the neutron intensity detected by Be filter detector and normalized by the counts of monitor, which are proportional to the scattering function $S(Q,W)$, and the ordinate shows the energy of phonon in meV. The neutrons scattered were not corrected for multi-phonon and self-absorption. In figure 2, curve(1) and (2) shows the local modes of the hydriding sample and the dehydriding sample respectively, curve(3) shows that of the same sample as (2) but kept 50 days at ambient temperature and curve 4 shows that of the same sample as (3) but after heating to 50°C for 20 minutes.

We can see from figure 2 the procedure of evolution of local modes about the temperature and time stability of $LaNi_3-H_2$ system. The curve(1) presents a single peak with a much broader FWHM ($\sim$100meV) due to the local environment fluctuation in the amorphous state, and with the peak position around 64meV, which is nearly identical to that of the main peak of $LaNi_5-H_2$[4], i.e. The local environment of the hydrogen atoms in the hydriding sample of $LaNi_3$ is similar to that of $LaNi_5-H_2$. The curve(2) exhibit two peaks located around 63meV and 120meV respectively . The former is nearly the position of the main peak of $LaNi_5-H_2$, and the latter is that of local modes of La-H_2[5], this shows that the dehydriding sample has been decomposed into two compounds, $LaNi_5-H_2$ and La-hydrides , which is consistent with the result of the neutron diffr-

action mentioned above. In the curve (3) and (4), the structure of the two compounds are not exhibited because most
of the hydrogen atoms in LaNi$_5$ were released rapidly so
that the intensity of the peak around 64meV decrease down
to disappear and the position of the peak superposed by
both LaNi$_5$- and La-hydrides was shifted from 90meV to 120
meV, the latter is nearly a same position of the local
modes for La-hydrides.

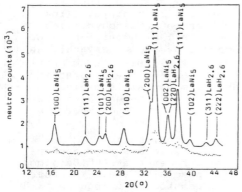

Fig.1, neutron diffraction patterns
——— :Hydriding sample, LaNi$_3$-H$_2$,obtained by
LaNi$_3$ activated at room temperature under
40 H$_2$ atm.
······ :Dehydriding sample, obtained by LaNi$_3$-H$_2$
dehydrided at 600°C for 10 minutes.

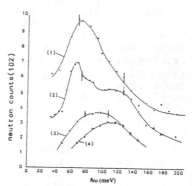

Fig.2. Evolution of the local modes
for LaNi$_3$-H$_2$ system.
(1): Hydriding sample, LaNi$_3$-H$_2$.
(2): Dehydriding sample.
(3): Same as (2) but kept 50 days
at room temperature.
(4): Same as (3) but after heating
to about 50°C for 20 minutes.

To summalize, in the recent study of LaNi$_3$-hydrides
all the experimental results presented here (neutron diffraction and inelastic scattering) show that LaNi$_3$-hydrides
has been transformed from crystalline state into amorphous
hydrides with a feature of microcrystalline for the LaNi$_5$,
and after the amorphous hydrides is heated (dehydrided)
at 600°C for 10 minutes it is decomposed and the new crystalline phase compounds, LaNi$_5$ and La-hydrides are produced. Perhaps, the feature of the phase transformation
induced by the hydriding and the dehydriding is closely
related to the special structure of LaNi$_3$, which is a
superstructure of the LaNi$_5$ type derived by an ordered
substitution of La for Ni, and to its thermodynamic
property.

REFERENCE

(1), Zhao Jian-Ming,etal., Journal of the Chinese Rare
     Earth Society, 6, 37(1988).
(2), Gou Cheng,etal., Atomic Energy Science and Techno-
     logy, 22, 151(1988).
(3), Ruan Jing-Hui,etal., Journal of High Energy and
     Nuclear Physics, 2, 441(1978).
(4), Cheng Gui-Ying,etal., Journal of the Chinese Rare
     Earth Society, 2, 168(1983).
(5), W.L.Korst,etal., Inorg. Chem., 5, 1719(1966).

A RAMAN STUDY OF THE PRESSURE INDUCED PHASE TRANSITIONS
OF THE ALKALINE-EARTH FLUORIDES

G.A. KOUROUKLIS
Physics Division, School of Technology, University of Thessaloniki
Thessaloniki GR-54006, GREECE

E. ANASTASSAKIS
Physics Department, Laboratory III, National Technical University
Zografou Campus, Athens GR-15773, GREECE

Raman measurements under hydrostatic pressure have allowed the study of the cubic—to—orthorhombic ($\beta \to \alpha$) phase transition in CaF_2, SrF_2, and BaF_2 and its hysterisis in $\alpha \to \beta$. Both transition pressures show a linear dependence on the inverse volume of the unit cell. Anharmonic contributions to the phonon frequency and linewidth in the cubic phase are also discussed.

We present the results of high-pressure Raman measurements in CaF_2, SrF_2, and BaF_2 at 300K for pressures covering the cubic—to—orthorhombic ($\beta \to \alpha$) phase transition. New optical phonons corresponding to the α-phase are observed above the transition pressure (P_c). The reverse ($\alpha \to \beta$) phase transition is also observed in CaF_2 and SrF_2 with decreasing pressure at $P'_c < P_c$, revealing a hysteresis. The β-phase pressure data are used to investigate the lattice anharmonic behavior. In addition, a correlation appears to exist between P_c, P'_c and the volume of the unit cell, namely both P_c and P'_c are proportional to the inverse volume of the unit cell.

The experimental set-up and procedure used in this study have been described elsewhere.[1,2] A diamond anvil cell was used to generate the pressure. The samples were platelet-like fragments from single crystals, ~ 50 μm in linear dimensions and $\sim 20\mu$m thickness.

In their cubic phase (O_h^5) the fluorite crystals exhibit one triply-degenerate optical phonon (F_{2g}) which is Raman active. At pressures $P > P_c$ new phonons appear indicating that a new phase of lower symmetry has set in. Typical spectra for CaF_2 and BaF_2 are shown in Fig. 1. The spectra of SrF_2 are similar.[1] In the case of CaF_2 and SrF_2 six Raman bands could be observed in the 200 - 450 cm^{-1} region. For BaF_2 seven Raman bands were observed in the 150 - 320 cm^{-1} region. We attribute these changes to the $\beta \to \alpha$ phase transition, where α is the orthorhombic phase (D_{2h}^{16}). More details about the experiment and the symmetry assignments of the phonons can be found elsewhere.[1,3]

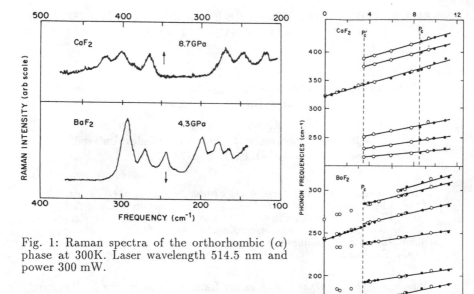

Fig. 1: Raman spectra of the orthorhombic (α) phase at 300K. Laser wavelength 514.5 nm and power 300 mW.

Fig 2: Phonon frequencies as a function of pressure. Solid (open) circles represent data for increasing (decreasing) pressure. Solid lines are linear least-squares fits to the experimental points.

The pressure dependence of the phonon frequencies is shown in Fig. 2. The P_c (P'_c) values in GPa for CaF_2 and SrF_2 are 8.5 (3.5) and 5.0 (1.7), all within 0.2. The results for CaF_2 are close to those reported from in-situ X-ray diffraction studies[4], i.e. $P_c \sim 5$ to 2. The $\beta \rightarrow \alpha$ phase transition in BaF_2 was observed at 2.3 $\pm$ 0.2 in agreement with earlier Raman results;[5,6] the transition pressure P_c could not be determined from X-ray diffraction techniques.[4] Furthermore, the $\alpha \rightarrow \beta$ phase transition did not seem to be totally completed at P = 0 since the strongest phonon of the α-phase always persisted even after the pressure was totally released.[4] Thus we can only say that the value of P'_c is at most 0 GPa.

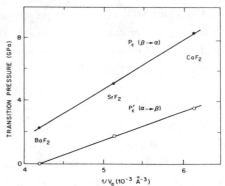

Fig. 3: Transition pressures P_c ($\beta \rightarrow \alpha$) and P'_c ($\alpha \rightarrow \beta$) as a function of the inverse volume of the unit cell.

In Fig. 3 the pressures P_c and P'_c are plotted as a function of $1/V_0$ where V_0 is the volume of the unit cell at ambient pressure and temperature. The relations appear to be linear. An attempt to incorporate into this scheme other materials with the fluorite structure, e.g. CdF_2[4] and PbF_2[5], did not produce consistent results, thus implying that the electronic structure of the cation is essential. (It is noted in this regard, that Ca, Sr, and Ba have similar electronic configurations, while in Cd and Pb the electronic configurations are different.) We believe that the above experimental facts may be of interest to theorists when trying to correlate the $\beta \to \alpha$ phase transitions to atomic radii and, accordingly, to electronic configurations and interatomic forces.

A detailed anharmonic study was undertaken based on the present results of $(d\omega/dP)_T$ and those of $(d\omega/dT)_P$ from the literature.[7] The Grüneisen parameters found here are 1.47, 1.51, 1.78 for CaF_2, SrF_2, and BaF_2, respectively. Details of the calculations can be found in Refs. 1-3. The results reveal similar behavior for all three materials, concerning the total ($\Delta\omega_t$), volume ($\Delta\omega_v$), and anharmonic ($\Delta\omega_a$) contributions to the T-dependent phonon frequency shifts: $\Delta\omega_v$ is essentially the same as $\Delta\omega_t$ for T> 100K allowing nearly zero contribution $\Delta\omega_a$. Below 100K the role of $\Delta\omega_v$ and $\Delta\omega_a$ is reversed. A more appropriate interpretation of these results is that in all three materials $\Delta\omega_t$ is largely due to volume expansion, at least for high temperatures.

Supported in part by the General Secretariat for Research and Technology, Greece.

REFERENCES

1. Kourouklis, G.A. and Anastassakis, E., Phys. Rev. B 34, 1233 (1986).
2. Liarokapis, E., Anastassakis, E., and Kourouklis, G.A., Phys. Rev. B 32, 8346 (1985).
3. Kourouklis, G.A. and Anastassakis, E., phys. stat. sol. (b) 152, 89 (1989).
4. Jamieson, C., and Dandekar, D., Tran. Amer. Crystallogr. Accoc. 5, 19 (1969).
5. Kessler, J.R., Monberg, E., and Nicol, M., J. Chem. Phys. 60, 5057 (1974).
6. Nicol, M., Kessler, J.R., Ebisuzaki, Y., Ellensen, W.D., Fong, M., and Grath, S., Dev. Appl. Spectroscopy 10, 79 (1972).
7. Mead, D.G. and Wilkinson, R.G., J. Phys. C 10, 1063 (1977).

FAR-IR STUDIES OF PHONONS AND PLASMONS
OF THE PHASE TRANSITION IN In$_2$Se$_3$

I.Lelidis [+][$],D.Siapkas [+],C.Julien[*],M.Balkanski[*].
[*]Laboratoire de Physique des Solides de l'Universite P.et M. Curie associe au CNRS,4,place Jussieu 75252 Paris Cedex 05,France.
[+]Physics Departement,Solid State Section,Aristotle University,54006 Thessaloniki,Greece.

INTRODUCTION

Among the III-VI family of chalcogenide compounds In$_2$Se$_3$ crystalizes with a layer-type structure and because its particular proparties,i.e energy gap of 1.356 eV,high anisotropy, defective character, disordering effects,etc,has received ingreasing attention.Recently,it has been demonstrated that lithium insertion is achieved with a moderate ion mobility[1], and its electrical proparties have been investigated as function of annealing and doping effects[2].In a previus paper[3,4], we have investigated the far-ir spectra of α-In$_2$Se$_3$ and analysed the dielectric function by both lattice vibrations and free-carrier contribution.The plasmon-phonon coupling and the effect of disorder on a variety of samples have been studied in relationship with their post-preparation treatment.

As it has already been reported by many authors[5-8], this material exist in at least three different crystalline modifications,denoted by α,β and γ with phase transition at 473K for α->β and at 923K for β->γ transition.The α-In$_2$Se$_3$ phase which exist at two differents polytypes rhomboedral (R3m), or exagonal (P6$_3$mmc), is transformed always, independent the polytype, to the rhomboedral (R3m) b-In$_2$Se$_3$.Although the phase transition is reversible a large hysterisis appears on cooling which has been attributed[6-8] to the formation of an intermediate β'-phase below 473K and above 333K.

The present work reports the first ir spectroscopic study of the stuctural phase transition α<->β of In$_2$Se$_3$ and the phonon and plasmon parameters are compared to those of α-In$_2$Se$_3$.

EXPERIMENTAL DETAILS

The samples have been prepared by direct fussion of elements in stoichiometric proportion.Two rates of cooling have been followed; very slow cooling rate to room temperature which provides us annealed samples (A) and rapid cooling to 77K which provides us quenched samples (Q).Some of the Q-type samples have been heated at 453K,a little below the α->β transition,over periods from a few hours to a week in order to get samples with different conductivity (from .1 to 100 Ω^{-1} cm^{-1}) depending upon the annealing process.This thermal treatment was done using a furnace under flowing argon atmosphere.

FIR reflectivity spectra have been recorded in E|c polarisation using a Bruker IFS113 vacoum interferometer.The measurments have been made on cleaved surfaces in the spectral range from 8 to 400 cm^{-1} with a resolution of 4 cm^{-1} using a Ge cooled bolometer as a detector,and each spectrum is the average of 100 scans.The phonons and the free carriers parameters have been determined by fitting the far-ir data to a dielectric model.Kramers Kronig analysis has also been done.

RESULTS AND DISCUSSION

Figure 1 shows the far-ir reflectivity spectra of α-In_2Se_3,β-In_2Se_3 and β'-In_2Se_3 at different temperatures during a cycle of heating-cooling the sample.The analysis has been made using the factor form formula[10]:

$$\varepsilon(\omega) = \varepsilon_{oo} \Pi(\omega^2-\omega^2_{li}+i\gamma_{li}\omega)/(\omega^2-\omega^2_{ti}+i\gamma_{ti}\omega) -\varepsilon_{oo} \omega^2_p/(\omega^2+i\gamma_p\omega)$$

where ω_{li}, γ_{li} are the LO mode frequency and the damping coefficient respectively of the i mode,ω_{ti} ,γ_{ti} are the TO mode frequency and the damping coefficient respectively of the i mode, ε_{oo} is the high frequency dielectric constant and ω_p ,γ_p are the plasma frequency and damping coefficient.At room temperatute (fig.1a) we observe in the spectrum of the α-phase three phonons features at ω_{to} about 90,168 and 187 cm^{-1} and a plasma edge at ω_p=370 cm^{-1} with a damping factor γ_p=260 cm^{-1}.After heating at 473K we get the spectrum of β-phase (fig.1b) which is characterized by only one phonon feature at ω_{to} = 162.5 cm^{-1} and a plasma edge at ω_p=26 cm^{-1} with a damping factor γ_p=64 cm^{-1}.

Figure 2 shows the real part ε_1, the imaginary part ε_2 of the dielectric function $\varepsilon=\varepsilon_1$-$\varepsilon_2$ and the loss energy function (-Im(1/ε)) of the α-phase (solid lines) and the β-phase (dushed lines) of In_2Se_3.

On cooling buck to R.T., below 473K,the β-phase is not transformed directly to the α-phase.As is shown at fig.1c an intermediate β'-phace is formed between 473 and 363K with the appearence of two new peaks in the low frequency region at 36.8 and 61.2 cm^{-1}.At 363K transformation to the original α-phase takes place.These structural phase transitions have also been observed by means of electron microscopy and electron diffraction methods[6,8].The observed extensive plasma frequency shift from ω_p=370 cm^{-1} to γ_p=26 cm^{-1} throw the phase transition results to approximately a 40 times decrease of conductivity justified by electrical measurments[6,9].

CONCLUSION

In this work,we have studied and analysed the far-ir spectra of In_2Se_3 over the transition temperature at 473K.Both the phonon multiplicity in α-phase and it's high conductivity are results of a reduction in the crystal symetry of β-phase associeted with the riversible distortion of the indium sublattice.The observed phonon and plasmon behavior supports a recently propoced model in which the α->β phase transition is driven by transverse acoustic phonons[6].

REFERENCES
$^\$$ Present address:Laboratoire de Physique des Solides de l'Universite P.et M. Curie
This work has been partly supported by the European Commisson Community under Stimulation program ST2P-0013-1F(CD).
The authors would like to thank the computing cender of University of Thessaloniki for using its IBM computing facilities.Also we would like to thank N.Hatzopoulos,J.Siapkas and S.Kotini ,for their help in the computation.
1.   Julien,C. and Samaras,I.,Solid State Ionics 27,101 (1988).
2.   Kambas,K.,Fotsing,J.,Julien,C.,Balkanski,M.,Mater.Sci.Engineer.B1,139 (1988).
3.   Lelidis,I.,Siapkas,D.,Julien,C.,Balkanski,M.,Mater.Sci.Engineer.B3,133 (1989).
4.   Lelidis,I.,Siapkas,D.,Julien,C.,Balkanski,M.,Phys.Scripta in press.
5.   Osamura,O.,Murakami,Y.,and Tomiie,Y.,J.Phys.Soc.Jpn.21,1848(1966).
6.   van Landuyt, J.,van Tendeloo, G.,and Amelinckx, S.,Phys.Stat. Sol. A30,299(1975).

1164

7.   Likforman,A.,Guittard,M.,Flahaut,J.,Poirier,R.and Szydlo,N.,J.Solid State Chem.33,91(1980).
8.   Manolikas,C.,J.Solid State Chem.74,319(1988).
9.   Miyazawa,H.and Sugaike,S.,J.Phys.Soc.Japan 12,312(1957).
10.   Gervais,F. and Baumard,F.,Solid Stats Commun.21 861(1977).

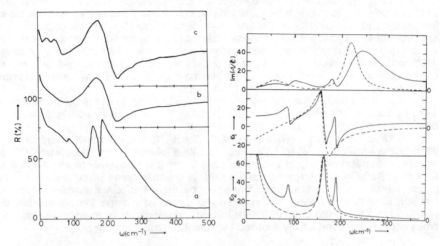

Figure 1.Reflexion spectra of α-In$_2$Se$_3$ (1a),β-In$_2$Se$_3$ (1b) and β'-In$_2$Se$_3$(1c).

Figure 2.The real part ε$_1$,the imaginary part ε$_2$ of the dielectric function and the loss energy function -Im(1/ε) of the α-In$_2$Se$_3$ (solid lines) and of the β-In$_2$Se$_3$ (dushed lines).

VIBRATIONAL SPECTROSCOPY OF UNUSUAL SEQUENCE OF STRUCTURAL PHASE TRANSITIONS IN LAYERED FERROELECTRICS

V.M.Burlakov, E.A.Vinogradov

Institute of Spectroscopy, USSR Ac. of Sciences, Troitsk, Moscow region, USSR, 142092

1. INTRODUCTION

The main interest to layered $TlGaSe_2$ and $TlInS_2$ crystals is connected with the unusual sequence of phase transitions (PT): the first PT is a transition into the incommensurate phase with $q_{inc}=(\delta,\delta,0.25)$ ($T_i=120$ K and 220 K respectively),and the second one ($T_c=110$ K and 200 K) into the ferroelectric phase with $q_c=(0,0,0.25)$ /1,2/.

There is a multiplication of unit cell (UC) volume which is an unusual phenomenon for proper ferroelectric PT through the phase with long period modulation of the structure /3/.

2. RESULTS AND DISCUSSION

2.1 Soft Mode and Intralayer Ferroelectric Instability

Low frequency Raman measurements of $TlGaSe_2$ and $TlInS_2$ allow to detect the presence of a ferroelectric soft mode interacting with some rigid modes below T_c. It was found that the temperature dependences of soft mode frequency and intensity are typical for proper ferroelectrics. To explain the multiplication of the UC volume in the ferroelectric phase we propose the intralayer nature for ferroelectric instability, i.e. the main contribution into the instability is made by intralayer interactions. Relatively weak interlayer forces provide a narrow band of unstable phonons and can easily lead to an incommensurate PT with $q_{inc} \cong (\delta, \delta, 0.25)$ before the ferroelectric PT occurs.

The full phonon band instability can lead to significant short range fluctuations near T_c. It means that the value of a fluctuation restricted in the volume $v \cong \xi_p^3$, where ξ_p is the order parameter (OP) correlation length may be sufficiently strong. The presence of such strong fluctuations is confirmed by some spectroscopic evidences described below.

2.2 Changes in Line Shapes of Rigid Modes Near T_c

Fig.1 represents the temperature dependence of line shape parameter $N = 100 \cdot I(\omega_0 + S) / I(\omega_0)$, where ω_0, S and I are the peak frequency, half width and intensity of IR E-band $\omega_1 = 310$ cm^{-1} and Raman Ag-line $\omega_2 = 20$ cm^{-1} in TlInS$_2$. The main feature in Fig.1 is a drop of N to gaussian below T_c and return to the initial level when temperature decreases. This behaviour can be explained by inhomogeneous broadening of vibrations in the field of slow OP fluctuations /4/, if the coherence length ξ_q of vibrations is of the order or less than ξ_p.

There is also a slow decrease in N(T) of the ω_1 line at low temperatures. Note that the behaviour of N(T) of the ω_2 line is identical in TlGaSe$_2$, Tl(InS$_2$)$_{0.99}$(GaSe$_2$)$_{0.01}$ and TlInS$_2$. It is natural to explain the low temperature decrease in N by the same mechanism of inhomogeneous broadening. Absence of such kind of N(T) change for ω_1 vibration indicates that the average dimensions of low temperature inhomogeneities are less than ξ_p near T_c. This may signify an additional PT into a partially chaotic phase.

2.3 Influence of Surface Electronic Density of States on the Ferroelectric PT Temperature.

Figure 2 represents the results of heat capacity

measurements of single crystalline (1) and well powdered (2) samples of $TlGaSe_2$. The IR reflection measurements from the cleaved surface (001) before and after grinding indicate the presence of a ferroelectric phase in the powdered sample. One can suppoze that the reason for such dramatic changes in Fig.2 is the surplus surface electronic states which lead to surface charge accumulation. The electric field of these surface charges induce the ferroelectric state at a distance of the order of Debye radius of screening.

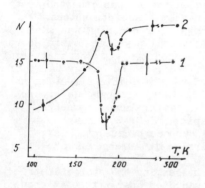 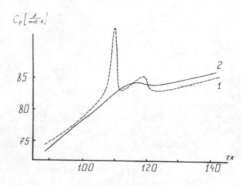

Fig.1   Fig.2

3. ACKNOWLEDGEMENTS

We thank Dr. M.Yakheev, Sh. Nurov, V. Rizak and M. Major for their help in experiments, N.M. Gasanly and Sh. Dzhuraev for providing the single crystalline samples.

4. REFERENCES

1. S.B. Vakhrushev et.al., Preprint No. 886, A.G. Ioffe Phys. Techn. Inst., Leningrad, USSR, 1984,(p.12).

2. A.A. Volkov et.al., Zh. eksper. teor. Fiz., Pisma, 1983 37, 517.

3. A.P. Levanyuk and D.G. Sannikov, Fiz. tverd. Tela, 1976, 18,1927.

4. V.M. Burlakov and A.G. Mitko, Fiz. tverd. Tela, 1988, 30, 3215.

EFFECT OF SPIN FLUCTUATION ON PHONON FREQUENCY IN METAMAGNETIC SEMICONDUCTOR $HgCr_2S_4$

Kunio Wakamura

Graduate School of Applied Physics, Okayama University of Science, 1-1 Ridai-cho, Okayama 700, Japan

The temperature dependence of infrared reflectivity spectra for a metamagnetic spinel $HgCr_2S_4$ has been measured through the Curie temperature Tc. The fundamental phonon frequency ω_j showed an anomalous variation below the temperature To higher than the Tc. The frequency shift below To is fitted enough by the curve proportional to the energy shift ΔE_G of absorption edge. This result will suggest the effect of spin fluctuation on the ω_j as found on ΔE_G.

The effect of long range spin ordering on the ω_j was observed as the rapid frequency shift at Tc for several magnetic semiconductors.[1-6] On the other hand, the effect of short range spin ordering was also observed as the appreciable variation of several physical values above Tc, for instance, the phonon damping constant γ_j,[1,2] the scattering intensity of spin disorder induced Raman band,[3,4] the energy shift of absorption edge[5], the electric resistivity[5], etc. This effect on the ω_j, however, has not been observed enough. In this paper, we try to observe such effect on ω_j by measuring the temperature dependence of infrared reflectivity spectra for a spinel $HgCr_2S_4$ through Tc(=60K). $HgCr_2S_4$ is appropriate for the purpose because the large spin fluctuation is expected for the spiral spin arrangement.[6]

$HgCr_2S_4$ was prepared in powdered form by heating the stoichiometric quantity of elements, HgS, Cr, and S in evacuated quartz tube at about 780 C for four days. The powder sample was sintered under the pressure of $4500Kg/cm^2$ at 620 C in vacuum for half an hour and further annealed at 600 C for ten days. The X-ray diffraction pattern showed almost the same feature as before the sintering. The density of sample is about 98% of ideal crystal. The reflectivity spectra were measured with a Fourier transform spectrometer for the entrance angle of 14° in the energy ranging from 40 cm^{-1} to $1000 cm^{-1}$. The resolution of the spectrometer is 1 cm^{-1}.

$HgCr_2S_4$ has the normal spinel structure. Four infrared active optical phonon modes are expected from the crystal symmetry of O_h^7. The phonon parameters were determined from the reflectivity R by the conventional procedure of fitting. The relations, $R=[(n+1)^2+k^2]/[(n-1)^2+k^2]$, $n^2-k^2=\varepsilon_1$, $2nk=\varepsilon_2$, where ε_1 or ε_2 reveals the real or imaginary part of dielectric function $\varepsilon(\omega)$, and n or k is that of refractive index, were applied in the analysis. The factorized form of $\varepsilon(\omega)$ is given by

$$\varepsilon(\omega)= \varepsilon_\infty \prod_j [(\omega_{Lj}^2-\omega^2-i\gamma_{Lj}\omega)/(\omega^2_{Tj}-\omega^2-i\gamma_{Tj}\omega)] . \qquad (1)$$

The ω_{Lj} or γ_{Lj} indicates the phonon frequency or damping constant of j-th LO-mode. The j takes all infrared active optical modes. Observed or fitted reflectivity is shown by the solid or dashed line in Fig.1. The agreements are excellent. Therefore, the determined parameters will be accurate enough as confirmed already.[1])

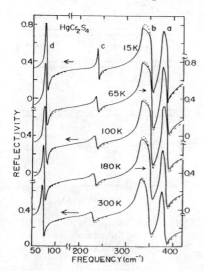

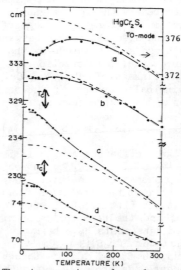

Fig.1 The reflectivity at several temperatures. The solid or dashed line represents the observed or fitted values.

Fig.2 The temperature dependence of ω_j for TO-mode. The dashed or solid line represents the fitted value with Eqs.(2),(3) or with Eqs.(2),(3),(4).

The ω_{TO} is shown as a function of temperature by the points in Fig.2. The appreciable shift was observed for all modes. It is, however, not rapid and begins at To($\fallingdotseq$120K). This deviation of To from Tc is the most different point from other spinels. Below To, the ω_j of the a- and b-bands or those of the c- and d-bands increase or decrease fairly with temperature. The ω_j above To exhibits a normal temperature dependence.

The frequency shift above Tc is attributed to the effect of thermal expansion and phonon-phonon interactions. Below Tc, the ω_j is further shifted by the spin-phonon coupling. The ω_j at temperature $T,\omega_j(T)$, is therefore represented as $\omega_j(T)=\omega_j(0)+\delta\omega_j(T)+\Delta\omega_j(T)+\Delta\omega_s(T)$, where $\omega_j(0)$ is the frequency extrapolated to 0 K. The $\delta\omega_j(T)$ is given by the linear thermal expansion coefficient α_v and the mode Gruneisen parameter γ_j as

$$\delta\omega_j(T)=\omega_j(0)[\exp(-3\gamma_j\int_0^T \alpha_v dT)-1] , \qquad (2)$$

At low temperature, only a cubic phonon-phonon term was adopted under the assumption which a phonon decays into two phonons having the equal energy $\omega_a(=\omega_j/2)$ and the opposite momentum. The frequency shift $\Delta\omega_j(T)$ is then written by the coupling coefficient C_j and ω_a as

$$\Delta\omega_j(T)=C_j[N(\hbar\omega_a)+0.5], \qquad (3)$$

The phonon population N is given by $N(\hbar\omega_a)=1/[\exp(\hbar\omega_a/kT)-1]$. The C_j and γ_j are roughly determined above To with Eqs.(2) and (3) by the same procedure as that in MCr_2S_4(M=Fe,Cd).[1,2])

In the spinels, the interaction energy between i-th and j-th spins is described by the ferromagnetic interaction J_{ij} for the linkage of Cr-S-Cr and also the energy between the i-th and k-th spins do by the antiferromagnetic interaction K_{ik}[7] for the linkage of Cr-S-M-S-Cr.[8] Following Baltensperger[8], the shift is then given in the form as[2]

$$\Delta\omega_s/\omega_j(0) = -A(S_1S_2) + B(S_1S_3), \quad (4)$$

where the coefficient A or B is the derivative of J_{ij} or K_{ik} with respect to the phonon displacements. The (S_1S_2) is a spin correlation function.

About the fitting of ω_j with Eqs.(2), (3),(4), and the reported value of α_v,[9] the good agreements have been obtained for NCr_2S_4(N=Fe,Cd)[1,2] and also MCr_2S_4(M=Hg,Cd)[10] though

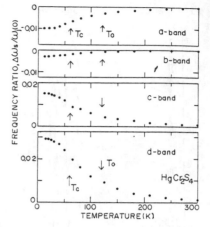

Fig.3 The value $\Delta\omega_s/\omega_j(0)$ as a function of temperature.

these compounds show the rapid shift at Tc. In $HgCr_2S_4$, the magnitude of (S_iS_j) has not been calculated. We now employ $\Delta E_G(T)$ in stead of (S_iS_j) since it is proportional to the spin correlation function.[6] The fitted results for ω_{TO} are shown by the solid line in Fig.2. The excellent agreements were obtained. The appreciable shift below To is also described enough.

To see more clearly the frequency shifts near T_o, the value$(\Delta\omega_s/\omega_j)$ (=r) is shown in Fig.3. The r shows near linear variation below To and the anomalous shift begins clearly at about To. Since the fitted vlaue below To agrees enough with the observed value , the origin of this shift is understood to be the same with that of ΔE_G, that is, the spin fluctuation effect as pointed out in the ΔE_G of $CdCr_2Se_4$ by the calculation of Erukhimov et al.[11] The spiral spin arrangement in $HgCr_2S_4$ may induce easily the spin fluctuation. The similar phenomena were also observed on the LO-phonon modes.

References
1) Wakamura,K. and Arai,T., J.Appl.Phys.63,5824(1988).
2) Wakamura,K., to be published in Solid State Commun.(1989).
3) Scagliotti,M., Jouanne,M., Balkanski,M., Ouvrard,G., and Benedek,G., Phys.Rev.B35,7097(1987).
4) Güntherodt,G. and Zeyhev,R., in Light Scattering in Solids IV, edited by Cardona,M. and Guntherodt,G.(Springer, Berlin,1984).
5) The data were listed in Stapele,R.P.van,in Ferromagnetic Materials, edited by Wohlfarth,E.P. (North-holland. Amsterdam,1982),Vol.3.
6) Lehmann,H.W. and Harbeke,G., Phys. Rev.B1,319(1970).
7) Baltzer,P.K., Wojtowicz,D.J., Robbins,M., and Lopantin,E., Phys. Rev.151,367(1966).
8) Baltensperger,W. and Helman,J.S., Helv.Phys.Acta.41,668(1968).
9) Slebarski,A.,Konopka,D.,and Chelkowski,A., Phys.Lett.50A,333(1974).
10) Wakamura,K.,Arai T., and Kudo K., J.Phys.Soc.Jpn.41,130(1976).
11) Erukhimov,M:M.Sh, Ovchinnikov,S.G., Gavrichkov,V.H.,and Ponomarev,I. Sov.Phys.Solid State 27,2685(1985).

OBSERVATION OF CORRELATION BETWEEN THE PHONON PARAMETERS AND THE

IONIC MOTION IN FAST IONIC CONDUCTOR Ag_3SBr

Kunio Wakamura and Kojiro Hirokawa

Graduate School of Applied Physics, Okayama University of Science,
1-1 Ridai-cho, Okayama 700, Japan

The infrared reflectivity spectra of Ag_3SBr have been measured in the normal and the superionic states. The phonon frequency ω_j and the phonon damping constant γ_j exhibit the similar temperature dependence to that of the ionic conductivity σ. This correlation was first observed and can be understood by the disorder arrangement of Ag ions above Tc and also the coupling of the zone boundary phonon (zb-phonon) with the mobile ions above 161 K.

We are interested in the coupling of the mobile ions with the phonons since it may relate predominantly to the mechanism of ionic motion. The possibility of coupling is expected from the energy of phonons nearly comparable to that of mobile ions in high temperature region and also from the fact that the mobile ions in the normal state construct a part of lattice vibrations.
The phenomenon will be observed by measuring the temperature dependences of ω_j and the γ_j at k=0. The infrared spectra, however, have been mainly remarked in the relation of the mobile ions,[1] which is observed as the increasing of reflectivity in the low frequency region. The detailed reports for the temperature dependence of phonon parameters in a superionic crystal are, therefore, not much. In this paper, we attempt to observe the correlation of ω_j and γ_j to the σ with the infrared reflectivity spectra in Ag_3SBr. The compound shows a clear structure of phonon band in infrared spectra of superionic state. Therefore, the compound is appropriate for the purpose.

The polycrystalline powder of Ag_3SBr was prepared by reacting the equimolar mixture of AgBr and Ag_2S powders at 300 C for 3 days. Obtained crystal was sintered under the pressure of 2000 Kg/cm^2 at 200 C for thirty minutes. The X-ray diffraction pattern for a sintered sample showed the same feature with that before the sintering. The infrared reflectivity spectra were measured by a Fourier transform spectrometer for the entrance angle of 14° with the resolution of 1 cm^{-1} in the energy ranging from 15 cm^{-1} to 500 cm^{-1}. The details for the measurements were described in elsewhere.[2]

The reflectivity R is analyzed by the factorized form of dielectric function $\varepsilon(\omega)$ as

$$\varepsilon(\omega)=\varepsilon_\infty \prod_j [(\omega_{Lj}^2 - \omega^2 - i\gamma_{Lj})/(\omega_{Tj}^2 - \omega^2 - i\gamma_{Tj}\omega] + S/(\Omega^2 - \omega^2 - i\Gamma\omega) \quad (1).$$

The first or the second term represents the contribution of phonons or the mobile ions. The latter term is derived by Brüesch et al.[3] The values ω_{Lj}, γ_{Lj}, ε_∞, Ω, and Γ reveal the frequency of j-th LO-mode and its damping constant, the optical dielectric constant, the averaged frequency and the damping constant of diffusive ions, respectively. The real and imaginary parts of refractive index n and k are respectively related to those of $\varepsilon(\omega)$, ε_1 and ε_2, as $\varepsilon_2 = 2nk$ and $\varepsilon_1 = n^2 - k^2$. The R for the normal incidence is represented as $R = [(n-1)^2 + k^2]/[(n+1)^2 + k^2]$. By using these parameters and the relations, the R is calculated. For the fitting to the observed R, the three phonon bands above 150 K, and the six bands between 115 K and 140 K were employed. The fitted results are shown in Fig.1. The good agreements were obtained. The R below 30 cm^{-1} was also fitted enough by the dominant contribution of second term in Eq.(1). The temperature dependencies of determined ω_j and γ_j are shown in Figs.2 and 3, respectively.

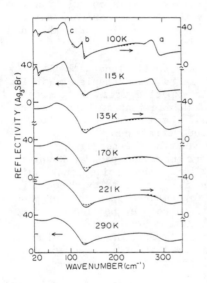

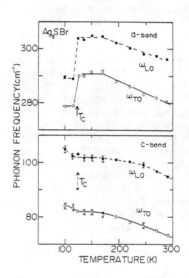

Fig.1 The reflectivity at several temperatures. The solid or the dashed line represents the observed or fitted value.

Fig.2 The temperature dependence of ω_j. The dashed or solid line represents the curve through the observed points.

About the ω_j, the drastic variation at Tc, nearly constant magnitude between Tc(= 128K) and 160 K, and also appreciable decreasing with temperature above 160 K were clearly observed for the a- and c-bands. The change of frequency shift at 160 K is not predicted from the volume thermal expansion and the anharmonic phonon-phonon interactions since the both elements give only a monotonic variation. Further, the shift for the a-band is about 3 % of ω_j between 160 K and 300 K and that for the c-band about 10 %. The latter is fairly large in comparison with those of normal ionic crystals.

About the γ_j, the drastic variation at Tc, the constant value between Tc and 160 K, and the appreciable increasing with temperature above about 160 K were found. The magnitude of γ_j above Tc is extremely large in comparison with that of normal ionic compounds as seen in

Fig.3. These variations can not be described by the usual phonon-phonon interactions.

Chiodelli et al. was observed the drastic variation of σ at Tc and the change of activation energy at 161 K.[4] The both temperatures equal to the temperatures arising the characteristic variation of ω_j and γ_j mentioned above. This correspondence of the temperature is understood by remarking the disorder arrangement of Ag ions above Tc and the coupling of zb-phonon to the mobile ions above 161 K.

The mobile ions begin to move near Tc. The arrangement of Ag ions is changed at Tc from the ordered state to the disordered state. From this result, the rapid increase of γ_j at Tc will be attributed to the effect of disordered Ag ions since the γ_j at the disordered state takes a large magnitude as found in disordered crystals.[5] Between Tc and 161 K, the disorder of Ag ions will not be much changed since the activation energy does not change.[4] The γ_j and the ω_j, therefore, will be small shifted by only the effect of anharmonic phonon-phonon interactions.

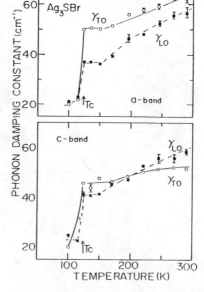

Fig.3 Temperature dependence of γ_j. The dashed and solid lines represent the curves through the observed points.

If the Ag ions move fluently above 161 K because of decreasing activation energy, the coupling of Ag ions will be dominantly arised on the phonons having the wavelength comparable to the length of short range ordering of Ag ions. The zb-phonon, therefore, will be strongly affected by the Ag ions. However, such an effect will be also appear on the zone center phonon because of the cubic phonon-phonon interactions which contains the zb-phonons. Then, the further decrease of ω_j or increase of γ_j with increasing σ will be arised. By these considerations, we can understand the variations of ω_j and γ_j above 160 K and conclude the predominant coupling of zb-phonon to the mobile ions.

References
1) For instance, Gras,B. and Funke,K., Solid State Ionics 2,341(1981), Brüesch,P., Beyeler,H.V., and Strässler,S., Phys.Rev.B25,541(1982),
2) Wakamura,K. and Arai,T.,J.Appl.Phys.63,5824(1988).
3) Brüesch,P., Strassler,S., and Zeller,H.R., Phys.Stat.Sol.a31,217 (1975).
4) Chiodelli,G., Magistris,A., and Schiraldi.A., Z.Phys.Chem.118,177 (1979).
5) For instance, Barker Jr.,A.S. and Sieverse,A.J., Rev.Mod.Phys.47, 141(1975).

12

Defect Scattering and
Defect Excitation

HIGH-FREQUENCY PHONONS IN RUBY

H. W. de Wijn and J. I. Dijkhuis

Faculty of Physics and Astronomy, University of Utrecht,
P.O. Box 80.000, 3508 TA Utrecht, The Netherlands

1. Introduction.

Much of the work to date on the dynamics of high-frequency phonons imprisoned by centers has been carried out for ruby (Al_2O_3:Cr^{3+}) held at low temperatures, in which 29-cm^{-1} phonons interact strongly with Cr^{3+} maintained in the $\overline{E}(^2E)$ state when resonant with the $\overline{E}(^2E) - 2\overline{A}(^2E)$ transition. A primary reason why ruby has provided a model study is that simple optical means permit to vary the concentration N^* of the active centers as well as the active volume. The first evidence for imprisonment of 29-cm^{-1} phonons by optically excited Cr^{3+} came from an elongated Orbach relaxation within the $\overline{E}(^2E)$ Kramers doublet.[1] Currently, the dynamics of the trapped 29-cm^{-1} phonons is usually registered via the phonon-induced R_2 luminescence emanating from $2\overline{A}(^2E)$.[2] The role of metastable Cr^{3+} in the resonant trapping was most strikingly revealed by inserting a second illuminated region, acting as an absorptive filter, in the flight path from a heater at the surface to the optical detection volume at a depth of several millimeters.[3] These experiments, as well as later ones employing optical phonon generation,[4] further pointed to a remarkable longevity of 29-cm^{-1} phonons.

For sufficiently high N^*, multiple absorption and emission of the phonons over the $\overline{E}(^2E) - 2\overline{A}(^2E)$ transition establish a dynamical equilibrium between, on the one hand, the phonons and, on the other hand, the spins (phonon bottleneck) well within the spontaneous decay time $T_d \sim 1$ ns of $2\overline{A}(^2E)$, i.e., fast compared to the dynamics of the phonons as coupled to the centers. This, in fact, permits to study the phonons via the spin system. It has been found (Sec. 3) that the effective relaxation time T_{eff} of the combined phonon-spin system, as viewed via the $2\overline{A}(^2E)$ population, increases with N^* and the typical dimension L of the excited zone,[5-7] but at higher N^* flattens out to a few microseconds.[6,7] At a low N^*L ($N^*L < 10^{15}$ cm^{-2}), the trapped phonons appear to escape by spatial diffusion.[5] At higher N^*L, however, spatial diffusion is too slow a process. Mechanisms involving spectral diffusion and transfer to such a distance from resonance that the mean free path exceeds L (spectral wipeout) apparently predominate.[6-8] The occurence of spectral diffusion at higher N^* was first demonstrated from an increase

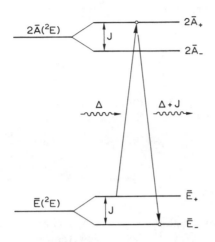

FIG. 1. One-site resonant-Raman mechanism of frequency shifting by Cr^{3+} in $\overline{E}(^2E)$ weakly exchange-coupled to Cr^{3+} in the $^4A_2$ ground state.

of the width of the nonthermal 29-cm^{-1} phonon spike upon gradual separation of the four $\overline{E}(^2E) - 2\overline{A}(^2E)$ Zeeman components with a magnetic field.[9] At low N^*, the resonant width is 0.017 cm^{-1}. While these experiments confirm the presence of spectral transport, they do not serve the purpose of observing the trapped phonon packet itself. The phonon spectrum may, however, be observed with a suitable form of fluorescence line narrowing (FLN) (Sec. 4).[10]

It has been realized that exchange-coupled Cr^{3+} pairs play a major role in the spectral dynamics of 29-cm^{-1} phonons. Strongly exchange-coupled pairs were shown to mediate in efficient spectral transport at high N^* from the ingrowth of the N_2 pair lines.[8] Spectroscopic evidence for the involvement of more weakly coupled pairs was derived from the development of their R_1 luminescence with time.[11] Further, monochromatic displaced phonons, generated at additional centers with a far-infrared laser, were indeed detected with metastable Cr^{3+}.[12] The first to suggest a mechanism for spectral transfer were Meltzer et al.,[6] who built on the concepts of resonant two-phonon-assisted energy transfer introduced by Holstein et al.[13] Here, the electronic energy of an optically excited Cr^{3+} ion is transferred to a nearby Cr^{3+} in the $^4A_2$ ground state via off-diagonal exchange, while the outgoing phonon takes up or supplies the balance in excitation energy. The mechanism requires microscopic broadening, i.e., energy mismatch of the $\overline{E}(^2E) - {}^4A_2$ distance among Cr^{3+} separated by less than 1 nm, and is infrequent according to explicit searches for optical energy transfer[14] and recent theoretical insights.[15] Furthermore, the mechanism of resonant phonon-assisted energy transfer fails to account for the observed dependence of the phonon loss on a magnetic field.[7] The second mechanism, due to Goossens et al.,[7] is based on one-site resonant Raman (Orbach) processes at Cr^{3+} weakly coupled to a nearby Cr^{3+} in the $^4A_2$ ground state by *diagonal* exchange. The model (Sec. 2) further relates the increased mean free path of frequency-shifted phonons to the dimensions of the active zone. Monte Carlo simulations based on the model are presented in Sec. 5.

2. Model.

To account for the spectral dynamics of trapped 29-cm^{-1} phonons, we rely on a model based on metastable Cr^{3+} in $\overline{E}(^2E)$ forming a pair with Cr^{3+} in the $^4A_2$ ground state by weak diagonal exchange.[7] The basic assumptions are the

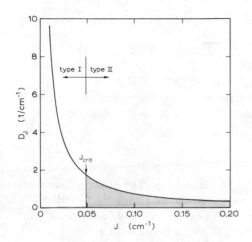

FIG. 2. D_J for 2500-at.ppm ruby. Optically excited Cr^{3+} of type II, experiencing an exchange larger than J_{crit}, are active in wipeout.

following: (i) Spectral contact between phonons of distinct frequencies is provided by one-site resonant Raman processes. Absorption of a phonon in a spin-nonflip transition from $\overline{E}(^2E)$ to $2\overline{A}(^2E)$ followed by emission in a spin-flip transition back to $\overline{E}(^2E)$, or vice versa, invokes a frequency shift of magnitude $\pm J$, with J the exchange parameter (Fig. 1); (ii) Phonons shifted from resonance are wiped out, i.e., do not interact with any other Cr^{3+} in the optically excited zone, if the shift is sufficiently large for the associated mean free path to exceed the typical zone dimension. It is important to note that diagonal exchange does only slightly affect the spin-nonflip frequencies between $\overline{E}(^2E)$ and $2\overline{A}(^2E)$.

For a genuinely random distribution of Cr^{3+} over the lattice and a J decaying with distance as $J = J_0 \exp(-ar)$, the distribution of the nearest-neighbor J is given by

$$D_J = \{4\pi N_0 [\ln(J/J_0)]^2 / a^3 J\} \exp\{4\pi N_0 [\ln(J/J_0)]^3 / 3a^3\} , \qquad (1)$$

in which N_0 is the total concentration of Cr^{3+} ions (Fig. 2). We have $a \sim 10$ nm^{-1} from theoretical estimates of the overlap integral,[16] and $J_0 \sim 330$ cm^{-1} from pair spectra.[17] The mean free path of a phonon shifted from the single-ion resonance by J, either to lower or higher frequency, equals

$$\Lambda = 2\rho T_d^{(f)} v / D_J N^* , \qquad (2)$$

in which ρ is the density of phonon states per unit of frequency, $D_J N^*$ is the density of excited Cr^{3+} pairs per unit of frequency, $T_d^{(f)}$ is the spontaneous spin-flip transition time from $2\overline{A}(^2E)$ to $\overline{E}(^2E)$, and v is the velocity of sound. The criterion for wipeout according to assumption (ii) now reads

$$\Lambda \geq L , \qquad (3)$$

in which L is the median flight path out of the zone. With Eqs. (1) and (2) we may thus define a "critical" exchange J_{crit}, beyond which wipeout is achieved. Integrating D_J over all frequencies larger than J_{crit}/h (Fig. 2), we find for the fraction of optically excited Cr^{3+} active in wipeout (type-II Cr^{3+})

$$N^{\mathrm{II}}/N^* = 1 - \exp\{4\pi N_0[\ln(J_{\mathrm{crit}}/J_0)]^3/3a^3\} \ . \tag{4}$$

Adopting $T_{\mathrm{d}}^{(f)} = 12$ ns, $v = 6.4 \times 10^5$ cm/s,[18] and $\rho = 7.3 \times 10^7$ Hz^{-1}cm^{-3}, we find, for 2500-at.ppm ruby, $J_{\mathrm{crit}} = 0.05$ cm^{-1} at a typical $N^* = 10^{18}$ cm^{-3} and $L = 200$ μm. Using Eq. (4), we then arrive at $N^{\mathrm{II}}/N^* = 0.27$.

A straightforward rate-equation calculation[7] further shows that, under the condition of strong bottlenecking, the combined $2\overline{A}(^2E)$-phonon system decays single-exponentially with the decay time

$$T_{\mathrm{eff}} = T_{\mathrm{d}}^{(f)}(N^* + \rho\Delta\nu)/\alpha N^{\mathrm{II}} \ . \tag{5}$$

Here, $\Delta\nu$ is the width of the $\overline{E}(^2E) - 2\overline{A}(^2E)$ transition. The quantity α expresses in a heuristic way that the coupling of resonant phonons with the type-II Cr^{3+} is modified from the coupling with isolated Cr^{3+} ions. For $N^* \gg \rho\Delta\nu \approx 4 \times 10^{16}$ cm^{-3}, T_{eff} is inversely proportional to the fraction of optically excited Cr^{3+} with $J > J_{\mathrm{crit}}$ as specified by Eq. (4). Equation (4), in turn, contains dependences on N^* and the dimension through J_{crit}, and a dependence on the ground-state concentration through N_0 and, to some extent, J_{crit}.

3. Optical generation and detection.

To measure T_{eff} averaged over the frequency, the decay of the R_2 luminescence, emanating from $2\overline{A}(^2E)$, was registered following removal of optical feeding into the broad bands. The optical excitation was provided by an argon laser operating at 514 nm, and the laser beam is repetitively switched off with an acousto-optical

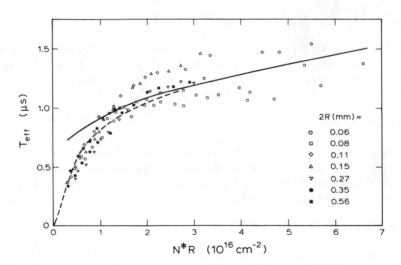

FIG. 3. T_{eff} of $2\overline{A}(^2E)$ vs N^*R for various diameters $2R$ of the excited zone in 700-at.ppm ruby. Solid line represents Eq. (5). Dashed line includes additional spatial diffusion as approximated by Eq. (6).

modulator. The R_2 luminescent intensity, as well as the R_1 intensity, which is a measure of N^*, were measured at right angles to the incident laser beam with a 0.85-m double monochromator followed by standard photon-counting equipment.

To examine the N^* and size dependences, we have plotted in Fig. 3 the results of the measurements for 700-at.ppm ruby as a function of N^* times the radius R of the optically excited cylinder. Note that $N^* \gg \rho \Delta \nu$, so that T_{eff} is inversely proportional to N^{II}/N^*. The latter quantity is in turn determined by J_{crit}, which according to Eqs. (2) and (3) depends on N^* and L in the combination N^*L only, or, upon identifying L with R, on N^*R. Indeed, the data in Fig. 3 collapse, within errors, on a single curve for R varying over as much as a decade. This observation is of significance as it proves quite directly that only a small fraction of the shifts result in phonon loss.

The solid curve in Fig. 3 represents T_{eff} vs N^*R as calculated from Eq. (5) with $\alpha = 6$. The model evidently accounts for the observed absolute magnitude of T_{eff} as well as its flattening out with increasing N^*L. At low N^*L, the model overshoots the measured T_{eff}. In this regime, however, there is a significant contribution from boundary-limited spatial diffusion of phonons resonant with Cr^{3+} ions of type I. We have[19]

$$T_{\text{eff}}^{\text{(dif)}} = 3(N^*L)^2/5.78(v\rho \Delta \nu^{\text{(f)}})^2 T_{\text{d}}' , \tag{6}$$

which in parallel with Eq. (5) yields the dashed curve in Fig. 3. The time constant $T_d' \sim 8$ ns, somewhere intermediate between $T_{\text{d}}^{\text{(f)}} \approx 12$ ns and $T_{\text{d}}^{\text{(n)}} \approx 1$ ns, expresses the average coupling of the diffusing phonons with type-I Cr^{3+}.

4. Fluorescence line narrowing.

The spectrum of nonequilibrium 29-cm^{-1} phonons in ruby as it evolves with time is, in the vicinity of the transition, accessible to observation with FLN.[10] FLN is a technique in laser spectroscopy in which the inhomogeneous broadening is eliminated by selective excitation of a *homogeneous* subset of the optical centers. For phonon detection, a three-level scheme is used, as in Fig. 4. A homogeneous subset of Cr^{3+} is excited by narrow-band optical excitation to the $\overline{E}(^2E)$ metastable state. The Cr^{3+} thus prepared act as luminescent phonon

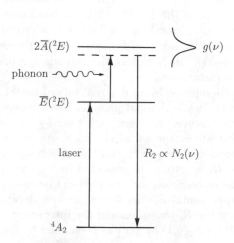

FIG. 4. Principle of frequency-selective phonon detection with FLN. The shape of the R_2 luminescence reflects the phonon spectrum.

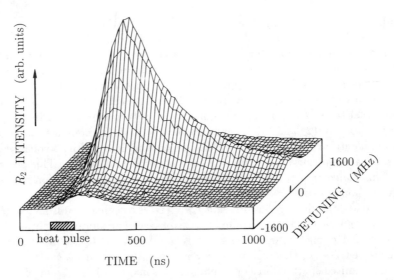

FIG. 5. Example of heat-pulse induced R_2 luminescence vs time and frequency for 700-at.ppm ruby at $N^* \approx 1.5 \times 10^{17}$ cm^{-3} and $2R = 200$ μm. Phonons are incident along the c axis.

detectors via phonon-assisted excitation to $2\overline{A}(^2E)$ and the subsequent luminescent return to $^4A_2$. The phonon spectrum, then, is directly reflected in the line shape of the R_2 luminescence, apart from corrections for the finite homogeneous width of $\overline{E}(^2E)$ and the instrumental resolution. When ignoring the width of $\overline{E}(^2E)$, the $2\overline{A}(^2E)$ population, $N_2(\nu,t)$, is in fact dictated by the bottlenecking condition

$$N_2(\nu,t) = N^*g(\nu)p(\nu,t) , \tag{7}$$

in which $p(\nu,t)$ is the time-dependent occupation number of the phonon modes with frequency ν, and $g(\nu)$ is the normalized shape of the $\overline{E}(^2E) - 2\overline{A}(^2E)$ transition. Note again that bottlenecking implies that dynamical equilibrium is established between the phonons and the spin system. So any modification of $p(\nu,t)$ goes associated with a similar modification of $N_2(\nu,t)$, and it is the combined spin-phonon system which reacts to an outside disturbance.

In the present experiments, phonons are injected by an external pulsed heater, providing an initially flat phonon spectrum, while thermal phonons are removed by cooling down to 1.5 K. The spectrum, however, modifies with time by virtue of scattering off excited Cr^{3+} and the ultimate escape out of the zone. An example of the measured R_2 intensity as a function of frequency and time, built up by repetitive stepwise scanning through the R_2 spectrum, is presented in Fig. 5. A Fabry-Perot interferometer in front of the monochromator provided the necessary resolution. To derive $p(\nu,t)$ from the experimental R_2 line shape, we first deconvolute for the minor homogeneous width of the R_1 transition and the instrumental profile, and subsequently divide the resultant $N_2(\nu,t)$ by $g(\nu)$ as prescribed by Eq.

(7). The line shape $g(\nu)$ is known to precision from far-infrared absorption spectroscopy[20] and FLN.[21] It contains a homogeneous Lorentzian part with a full width at half maximum (FWHM) $\Delta\nu_{hom} = 240 \pm 15$ MHz, and, for 700-at.ppm ruby, an inhomogeneous Gaussian part with a FWHM $\Delta\nu_{inh} = 295 \pm 15$ MHz.

Compelling evidence for spectral transport is derived from Fig. 6, in which the R_2 intensity is plotted vs the time for a selection of fixed distances $\Delta\nu$ from resonance. The four traces shown represent cross-sections of a complete data set like the one of Fig. 5, integrated over the ranges 0 ± 80, 375 ± 80, 675 ± 80, and 1000 ± 80 MHz from resonance. Figure 6 expresses quite directly the strong dependence of the phonon response on detuning. The most striking point is that near resonance the R_2 intensity, and thus $p(\nu, t)$, continues to rise after the heat pulse has been switched off. In the wings, by contrast, $p(\nu, t)$ starts to decline immediately after the pulse. The external supply of phonons having ceased, the continued ingrowth of $p(\nu, t)$ near resonance results from inelastic processes converting phonons in the wings to resonant ones.

An example of the temporal development of the complete phonon spectrum $p(\nu, t)$ around resonance is presented in Fig. 7 for 700-at.ppm ruby with $N^* \approx 1.5 \times 10^{17}$ cm^{-3}. The most salient feature here is that at the shorter times $p(\nu, t)$ exhibits a pronounced depression around resonance. The physical explanation for the dip is twofold: (i) initially the spin system preferentially absorbs resonant phonons, while the "heat capacity" of the combined system primarily resides with the spins $[N^*g(\nu) \gg \rho]$, and (ii) only part of the resonant phonons supplied by an outside source manage to penetrate into the excited zone because of diffusive motion back into free space. As to (i), we recall that under bottlenecking conditions the phonons maintain a dynamical equilibrium with the spin system, as expressed by Eq. (7). In connection with (ii), we shortly note that the penetration depth is of order $\sigma^{1/4}\Lambda$, with σ the bulk bottlenecking factor, or only 9 μm at

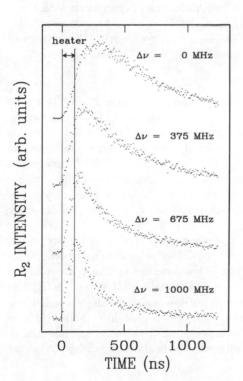

FIG. 6. Phonon-induced R_2 intensity vs time at various distances $\Delta\nu$ from the central frequency for 700-at.ppm ruby. $N^* \approx 1.5 \times 10^{17}$ cm^{-3} and $2R = 200$ μm.

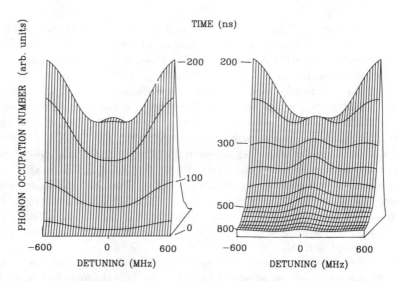

FIG. 7. Phonon spectrum vs time following a heat pulse of 100 ns duration for 700-at.ppm ruby at $N^* \approx 1.5 \times 10^{17}$ cm^{-3} and $2R \approx 200$ μm. At resonance $p(\nu, t)$ reaches about 3×10^{-2}.

the appropiate $\sigma \sim 500$. A second feature worth noticing in connection with Fig. 7 is that, similarly to what we have seen in Fig. 6, the decay of the phonon population at longer times becomes faster with the distance from resonance. This, evidently, expresses a drop of T_{eff} with detuning, in conformity with the model of Sec. 2.

5. Monte Carlo simulations.

For a more quantitative treatment of the spectral dynamics of 29-cm^{-1} phonons in interaction with the excited Cr^{3+} ions, we have set up Monte Carlo simulations based upon the model of Sec. 2, following the itinerary and life history of externally generated phonons until escape at the boundary of the excited zone. We define, for simplicity, a simple cubic lattice of *excited* Cr^{3+} ions, such that N^* lattice points are contained in a unit of volume. The phonons are assumed to travel with a unique velocity from one lattice point to the next along nearest-neighbor connections, and their frequencies and directions of travel are permitted to change by interaction with the excited Cr^{3+}.

Apart from computational shortcuts, a Monte Carlo step runs as follows. When the phonon has reached an excited Cr^{3+}, the latter is made constituent of an exchange-coupled pair by random drawing of an exchange splitting J out of the D_J distribution. Next, again randomly, one of the eight $\overline{E}(^2E)_\pm \rightarrow 2\overline{A}(^2E)_\pm \rightarrow \overline{E}(^2E)_\pm$ processes, or the most likely case of no interaction at all, is carried into effect. Here, due account is given of the transition probabilities and widths of the relevant transitions. In the event no interaction took place, the phonon is, of course, left to proceed on its course without modification and delay. When an

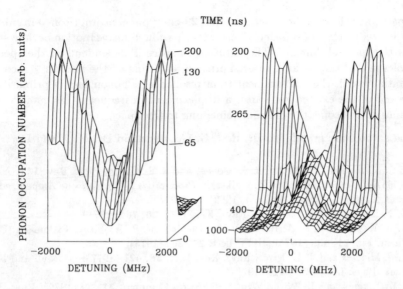

FIG. 8. Simulated phonon spectrum vs time for 700 at.ppm ruby with $N^* = 6 \times 10^{17}$ cm^{-3} in a volume of $0.2 \times 0.2 \times 10$ mm^3. Duration of the external phonon supply is 200 ns.

$\overline{E}(^2E)_\pm \rightarrow 2\overline{A}(^2E)_\pm \rightarrow \overline{E}(^2E)_\pm$ process is selected, however, the phonon frequency is adjusted if necessary, a new direction of travel is chosen out of the 6 allowed, and the age of the phonon is updated by adding the time T_d the excitation resides in $2\overline{A}(^2E)$. The procedure is repeated upon encountering the next Cr^{3+} a lattice constant away until the phonon manages to escape the excited zone. Each time the phonon has taken part in an $\overline{E}(^2E)_\pm \rightarrow 2\overline{A}(^2E)_\pm \rightarrow \overline{E}(^2E)_\pm$ process, its frequency and age are registered in a frequency-time histogram. The histogram, as it is, corresponds to the occupation $N_2(\nu, t)$ of $2\overline{A}(^2E)$, but may be converted to $p(\nu, t)$ by use of Eq. (7).

In Fig. 8, we present the simulated phonon spectrum for 700-at.ppm ruby as it develops with time after injecting phonons into a volume of $0.2 \times 0.2 \times 10$ mm^3. The parameters adopted are as discussed above, but, for best similarity with the experimental results, N^* has been increased from the experimental value within its range of uncertainty. The main features of the experimental phonon spectra (cf. Fig. 7), i.e., (i) the initial dip at resonance, and (ii) the faster decay of phonons further from resonance, both appear to be reproduced. Closer inspection of the simulated spectra further reveals that also the most direct evidence for spectral transfer, viz., the afterfeeding of resonant phonons after the external feeding has terminated (cf. Fig. 6), is satisfactorily accounted for.

6. Conclusions.

Experiments with wideband luminescent detection and FLN, and Monte Carlo simulations based on on-site resonant Raman scattering by weakly exchange-coupled

Cr^{3+} pairs, have lead to the conclusion that 29-cm^{-1} phonons imprisoned in ruby by optically excited Cr^{3+} of sufficient density take part in diffusive motion both in space and in the spectral domain. The diffusion in space is dependent on the spectral dynamics in that the distance covered prior to scattering off the next Cr^{3+} increases substantially with the displacement from resonance. Phonon loss is primarily by escape out of the excited zone after a displacement in frequency far enough from resonance to overcome the distance remaining to free space.

The authors are grateful to Dr. R. J. G. Goossens and Dr. M. J. van Dort.

1. S. Geschwind, G. E. Devlin, R. L. Cohen, and S. R. Chinn, Phys. Rev. **137**, A1087 (1965); S. Geschwind and L. R. Walker, in *Proceedings XVth Colloque Ampère*, edited by P. Averbuch (North- Holland, 1969), p. 460.

2. K. F. Renk and J. Deisenhofer, Phys. Rev. Lett. **26**, 764 (1971).

3. A. A. Kaplyanskii, S. A. Basun, V. A. Rachin, and R. A. Titov, Pis'ma Zh. Eksp. Theor. Fiz. **21**, 438 (1975) [JETP Lett. **21**, 200 (1975)].

4. R. S. Meltzer and J. E. Rives, Phys. Rev. Lett. **38**, 421 (1977); G. Pauli and K. F. Renk, Phys. Lett. **67A**, 410 (1978).

5. J. I. Dijkhuis and H. W. de Wijn, Solid State Commun. **31**, 39 (1979); Phys. Rev. B **20**, 1844 (1979).

6. R. S. Meltzer, J. E. Rives, and W. C. Egbert, Phys. Rev. B **25**, 3026 (1982).

7. R. J. G. Goossens, J. I. Dijkhuis, and H. W. de Wijn, Phys. Rev. B **32**, 7065 (1985).

8. S. A. Basun, A. A. Kaplyanskii, S. P. Feofilov, and V. L. Shekhtman, Fiz. Tverd. Tela, **25**, 2731 (1983) [Sov. Phys. Solid State **25**, 1570 (1983)].

9. J. I. Dijkhuis, A. van der Pol, and H. W. de Wijn, Phys. Rev. Lett. **37**, 1554 (1976).

10. M. J. van Dort, J. I. Dijkhuis, and H. W. de Wijn, J. Lumin. **38**, 217 (1987); and to be published.

11. S. A. Basun, A. A. Kaplyanskii, and S. P. Feofilov, Fiz. Tverd. Tela **28**, 3116 (1986) [Sov. Phys. Solid State **28**, 2038 (1986)].

12. U. Happek, T. Holstein, and K. F. Renk, Phys. Rev. Lett. **54**, 2091 (1985).

13. T. Holstein, S. K. Lyo, and R. Orbach, Phys. Rev. Lett. **36**, 891 (1976).

14. P. M. Selzer, D. S. Hamilton, and W. M. Yen, Phys. Rev. Lett. **38**, 858 (1977); P. M. Selzer and W. M. Yen, Opt. Lett. **1**, 90 (1977); P. E. Jessop and A. Szabo, Phys. Rev. Lett. **45**, 1712 (1980); S. Chu, H. M. Gibbs, S. L. McCall, and A. Passner, Phys. Rev. Lett. **45**, 1715 (1980).

15. S. Majetich, R. S. Meltzer, and J. E. Rives, Phys. Rev. B **38**, 11075 (1988).

16. Nai Li Huang, R. Orbach, E. Šimánek, J. Owen, and D. R. Taylor, Phys. Rev. **156**, 383 (1967); S. K. Lyo, Phys. Rev. B **3**, 3331 (1971).

17. P. Kisliuk, N. C. Chang, P. L. Scott, and M. H. L. Pryce, Phys. Rev. **184**, 367 (1969).

18. B. Taylor, H. J. Maris, and C. Elbaum, Phys. Rev. B **3**, 1462 (1971).

19. J. I. Dijkhuis and H. W. de Wijn, Phys. Rev. B **20**, 1844 (1979); Solid State Commun. **31**, 39 (1979).

20. N. Retzer, H. Lengfellner, and K. F. Renk, Phys. Lett. **A96**, 487 (1983).

21. M. J. van Dort, M. H. F. Overwijk, J. I. Dijkhuis, and H. W. de Wijn, Solid State Commun., in press.

BRILLOUIN SCATTERING AND THEORETICAL STUDIES
OF SUPERIONIC LANTHANUM FLUORIDE

P E NGOEPE[o*+], C R A CATLOW[o] AND J D COMINS[*]

[o] Department of Chemistry, University of Keele, Staffs, ST5 5BG U.K.

[*] Department of Physics, University of the Witwatersrand, Johannesburg, PO Wits 2050, South Africa

[+] Department of Physics, University of the North, Sovenga 0727, South Africa.

INTRODUCTION

The defect structures of the rare earth trifluorides have been receiving increased attention recently. This concerns their quite high ambient temperature ionic conductivity and the evidence for a diffuse phase transition to a superionic state above 1150K[1-3].

The present work concerns Brillouin scattering studies of LaF_3 in the temperature range 300 to 1500K in which a complete set of acoustic phonon frequencies associated with the elastic constants is determined. The elastic constants are calculated using computer modelling techniques: the effects of both lattice expansion and the presence of anion Frenkel and Schottky disorder are investigated. A comparison of the experimental and theoretical work is used to predict the dominant defect type and concentration in the high temperature superionic phase.

EXPERIMENTAL

Suitable single crystal specimens of LaF_3 were oriented, cut and polished to permit phonon propagation in appropriate directions, $(1,0,0)$, $(0,1,1)/\sqrt{2}$ and $(-1, \sqrt{2},1)/2$ using a $90°$ light scattering geometry. They were protected by an argon atmosphere within silica capsules and heated in an optical furnace. Brillouin spectra were excited by the 514.5 nm line of an argon–ion laser and analysed by a triple–pass Fabry–Perot interferometer. The temperature dependence of the elastic constants was extracted from the square of the acoustic phonon frequencies $(\Delta\omega_B)^2$ for the above phonon propagation directions. Birefringence effects are negligible and the hexagonal approximation to the trigonal structure is adequate. Accordingly elastic constants C_{11}, C_{13}, C_{33}, C_{44} and C_{66} were determined.

THEORETICAL

The P$\bar{3}$c1 space group[4] was used and the appropriate atomic coordinates were determined. Elastic constants were calculated using the THBREL computer code which is related to the PLUTO program[5]. The interatomic potentials were those of Jordan and Catlow[2]. The variation of the elastic constants with temperature was simulated by varying the lattice parameter according to known experimental behaviour. The presence of defects was incorporated using supercell procedures: a supercell comprises the defect and surrounding lattice repeated infinitely in space. The size of the chosen supercell (2,3,4,6,8 unit cells) permitted dilution of the defect concentration. Appropriate supercells were constructed for anion Frenkel and Schottky disorder.

RESULTS AND DISCUSSION

Fig. 1 shows the behaviour of the elastic constants as determined from the Brillouin scattering studies. The linear decreases in the elastic constants with temperature are satisfactorily accounted for by normal anharmonicity associated with lattice expansion, as determined by calculations of dC_{ij}/dT.

Above 1150K, there are abrupt reductions in the elastic constants, with the exception of C_{44} which tends to a nearly constant value. In this temperature region, there is also a substantial increase in the specific heat capacity.[6] This behaviour is attributed to the cooperative development of disorder. A rising defect interaction energy is calculated with increasing defect concentration which strongly supports this conclusion and is consistent with current thermodynamic models of the superionic transition.[7]

The experimental and computed percentage changes in the elastic constants (after subtraction of the anharmonic contribution) are given in table 2 for a temperature of 1400K.

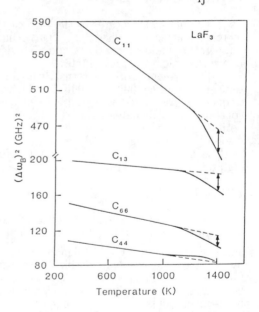

Fig.1 The temperature variation of $(\Delta\omega_B)^2$ corresponding to the individual elastic constants C_{11}, C_{13}, C_{66} and C_{44}.

It is seen that the experimental results are best reproduced by anion Frenkel disorder with a concentration of 2 mol % : in particular the behaviour of C_{44} is reproduced quite well. By contrast the latter feature is not displayed on the assumption of dominant Schottky disorder.

Table 1.   Compares the experimental percentage changes of elastic constants above 1150K and those calculated from anion Frenkel and Schottky models.

	Experimental 1400K	Frenkel model 2 mol %	Schottky model 2 mol %
	%	%	%
$\Delta C_{11}/C_{11}$	3.7	6.9	8.4
$\Delta C_{33}/C_{33}$	4.4	4.4	8.0
$\Delta C_{13}/C_{13}$	10.2	10.1	11.7
$\Delta C_{66}/C_{66}$	11.0	11.1	12.0
$\Delta C_{44}/C_{44}$	0.0	−3.0	15.8

The authors are grateful to R.A. Jackson and A.G. Every for useful discussions. PEN is grateful to the university of Keele for their hospitality and the British Council for a study award.

REFERENCES

1.   Chadwick, A.V., Hope, D.S., Jaroszkiewicz, G.A. and Strange, J.H., Fast Ion Transport in Solids, ed. P. Vashista, J.N. Mundy and G.K. Shenoy (Elsevier: North Holland, Amsterdam, 1979) p.683.
2.   Jordan, W.M. and Catlow, C.R.A., Cryst. Lattice Def. and Amorph. Mat. 15, 81 (1987).
3.   Ngoepe, P.E., Comins, J.D. and Every, A.G., Phys. Rev. B34, 8153 (1986).
4.   Zalkin, A., and Templeton, D.H., Acta Cryst. B41, 91 (1985).
5.   Catlow, C.R.A. and Norgett, M.J., AERE Harwell Report M.2763 (1978).
6.   Lyon, W.G., Osborne, D.W., Flotow, G.E., Grandjean, F., Hubbard, W. and Johnson G.K., J. Chem. Phys. 69, 167 (1978).
7.   Catlow, C.R.A., Comins, J.D., Germano, F.A., Harley, R.T. and Hayes, W., J. Phys. C11, 3197 (1978).

BRILLOUIN SCATTERING INVESTIGATIONS OF RARE EARTH DOPED CaF₂ IN THE HIGH TEMPERATURE SUPERIONIC REGION

P M MJWARA[*], J D COMINS[*], P E NGOEPE[+] AND A V CHADWICK[o]

[*] Department of Physics, University of the Witwatersrand, Johannesburg, P O Wits 2050, South Africa

[+] Department of Physics, University of the North, Sovenga 0727, South Africa.

[o] University Chemical Laboratory, University of Kent, Canterbury, Kent, CT2 7NH, U.K.

INTRODUCTION

A current model[1,2] of the diffuse transition to a superionic state in fluorites involves an attractive defect interaction term in the free energy. This causes in a reduction in the effective anion Frenkel defect formation energy and results in the co–operative generation of disorder. The defect interactions have been shown to result in defect clusters, both intrinsic[3] and extrinsic[4]. In the latter case, it has been shown by computational studies that the binding of thermally produced fluorine interstitials to the clusters can occur : this results in a reduction of the transition temperature T_c to the superionic state.

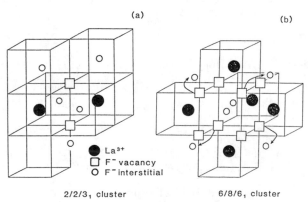

(a)   (b)

● La³⁺
□ F⁻ vacancy
○ F⁻ interstitial

2/2/3₁ cluster   6/8/6₁ cluster

Fig. 1a shows a 2/2/3₁ cluster.   Fig. 1b shows a 6/8/6₁ cluster.

The present work using measurements of the elastic constants by Brillouin scattering has been directed to provide further tests of the above concepts regarding extrinsic defect clusters. The system of CaF_2 doped with 10 mole % trivalent rare earth ions has been investigated recently by EXAFS techniques[5]. The results show there are two distinct situations : for the larger rare earth ions (La to Nd) smaller dimeric $2/2/3_1$ clusters dominate (Fig. 1a) while for the smaller ions (Gd to Yb) larger cubo–octahedral $6/8/6_1$ clusters (Fig 1b) are found. It would thus be expected that the transition temperature and the subsequent development of thermally generated disorder would depend on the type of cluster.

EXPERIMENTAL

Crystals of CaF_2 (10 mole % LaF_3) and CaF_2 (10 mole % TmF_3) were oriented, cut and polished to permit acoustic phonon propagation in the (1,0,0) direction using a 90° light scattering geometry. The were protected by an argon atmosphere within silica capsules and heated using an optical furnace. Brillouin spectra were produced using the 514.5 nm line of an argon–ion laser and analysed with a triple–pass Fabry–Perot interferometer. The temperature dependence of the elastic constant C_{11}, which is particularly sensitive to the presence of disorder in fluorites was determined from the temperature dependence of the square of the Brillouin frequency shifts $(\Delta\omega_B)^2$.

RESULTS AND DISCUSSION

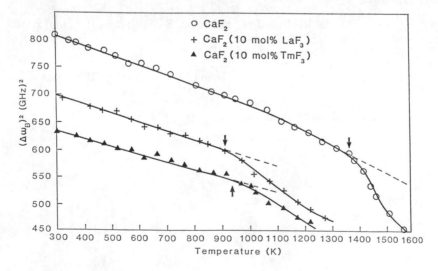

Fig. 2 shows the behaviour of $(\Delta\omega_B)^2$ associated with C_{11} as a function of temperature for pure and doped CaF_2.

In Fig 2, the behaviour of pure CaF_2, CaF_2 (10 mole % LaF_3) and CaF_2 (10 mole % TmF_3) is shown. It is observed that C_{11} decreases nearly linearly with temperature in all samples at first. This behaviour is attributed to normal anharmonicity and associated with lattice expansion.

In pure CaF_2 there is an abrupt reduction in C_{11} above a transition temperature T_c of 1375K. This behaviour has been observed previously and attributed to the co–operative generation of anion Frenkel disorder as discussed in the Introduction.

It is clear that the doping with the rare earth ions introduces substantial changes. In both doped samples there is a reduction in T_c by several hundred degrees. On examination of the results in more detail, it is clear that the values of T_c and the rate of change of C_{11} with temperature above T_c is significantly different in the case of CaF_2 (10 mole % LaF_3) in which there are $2/2/3_1$ clusters as compared with CaF_2 (10 mole % TmF_3) where $6/8/6_1$ clusters are present. This result, we believe, provides support for the influence of clusters on the processes controlling the superionic transition.

The results suggest that thermally induced fluorine interstitials are more strongly bound to the smaller dimeric clusters (lower T_c) than the larger cubo–octohedral clusters (higher T_c). The nature of the clusters would also result in different site limitation effects and defect–defect repulsions at the higher defect concentrations and temperatures, thus affecting the rate of defect generation above T_c.

REFERENCES

1. Catlow, C.R.A., Comins, J.D., Germano, F.A., Harley, R.T. and Hayes, W., J. Phys. C. 11 3197 (1978).
2. Hayes, W., Contemp. Phys. 27, 519 (1986).
3. Catlow, C.R.A. and Hayes, W., J. Phys. C. 15, L9 (1982).
4. Catlow, C.R.A., Comins, J.D., Germano, F.A., Harley, R.T., Hayes, W. and Owen, I.B., J. Phys. C. 14 329 (1981).
5. Catlow, C.R.A., Chadwick, A.V., Greaves, G.N. and Moroney, L.M., Cryst. Latt. Def. and Amorph. Mat. 12, 193 (1985).

THERMAL CONDUCTIVITY AND FERROELECTRIC
MICROREGIONS IN UNDOPED KTaO$_3$

B.SALCE, B.DAUDIN, J.L.GRAVIL

Département de Recherche Fondamentale, SBT/LCP

C.E.N. GRENOBLE, 85 X, 38041 GRENOBLE Cedex (France)

L.A.BOATNER

Solid State Division, O.R.N.L.

P.O. Box X, OAK RIDGE, Tennessee 37830 (USA)

The thermal conductivity K(T) of the cubic perovskyte KTaO$_3$ exhibits an unusual behavior for an insulating single crystal. At low temperature, K(T) follows a classical $\sim T^3$ law, in agreement with the calculated boundary scattering. Above 1 K, the thermal conductivity is usually dominated by intrinsic processes (dislocations, point defects,...) but, in the case of KTaO$_3$, a strong additional phonon scattering appears (figure 1). If one defines the scattering intensity as $I = W/W_o$ with $W = 1/K$ and $W_o = 1/K_o$, a calculated thermal resistance due to intrinsic processes, a sample dependent maximum I_m is observed around a fixed temperature $T_m \simeq 7$ K (figure 2).

The presence of extrinsic impurities, like paramagnetic ions, was invoked to account for this scattering, but it was shown that Cu, Ni, Co, Ag does not drastically change K(T)[1]. On the other hand, distorted microregions with temperature dependent size were detected in Raman scattering experiments[2]. By studying what we call an "ultra-pure" (UP) sample (grown by a flux technique with a specific procedure) we observed both the absence of microregions in Raman experiments and the lowest intensity I_m never measured in K(T). Then it can be concluded that phonon scattering in undoped KTaO$_3$ is in part due to symetry breaking defects (SBD). From recent fluorescence experiments, it was suggested that oxygen vacancies located near Ta^{3+} ions could be the origin of the microregions[3]. It should be noticed that the scattered phonon energies lie in the energy range of the soft transverse optic mode which gives to KTaO$_3$ the special nature of "incipient" ferroelectric. This mode is sensitive to polar impurities (Nb,Na,Li) which, due to the high lattice polarizability, induce phase transitions by setting its energy to zero. Then, the question of interaction between acoustic phonons and the soft optic mode arises, together with the possible role of Nb,Na,Li in the origin of microregions.

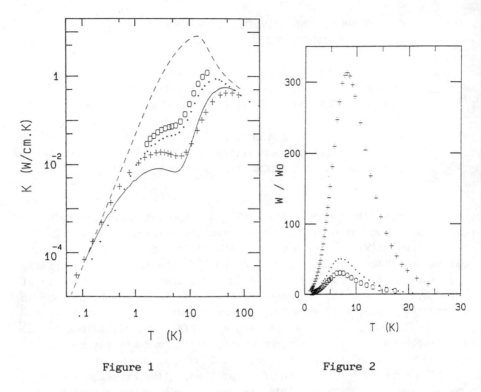

Figure 1   Figure 2

Thermal conductivity and phonon scattering intensity of undoped $KTaO_3$
o Sample (UP) ; • Sample (B) ; + Sample (P) ; − 0.15 % Nb doped $KTaO_3$
--- calculated conductivity

     To try to identify the nature of SBD, we have undertaken dielec-
tric measurements on several undoped crystals, together with K(T)
experiments. Both the real $\epsilon'(T)$ and imaginary $\epsilon''(T)$ parts of the
dielectric susceptibility were measured from 100 K down to 4 K. An
important result is shown in figure 3 : $\epsilon''(T)$ exhibits a peak located
around 40 K. Its magnitude Ip depends on the sample. It is worth
noticing that this peak is extremely weak in the case of the UP crystal. In figure 4 we plot Ip (measured at 1 kHz) versus I_m. A clear
relation can be seen for five different specimens. This shows that
the phonon scattering and dielectric losses are issued from the same
microscopic origin. One sample (J) does not fit so well into this
relation. Indeed, it was grown by a different method (top-seeded
solution growth) and $\epsilon''(T)$ exhibits additional peaks at T > 50 K. As
a comparison, we have studied one 0.15 % Nb doped $KTaO_3$. Although the
phonon scattering is stronger than in pure crystals, the losses peak
is smaller.

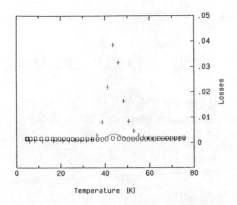

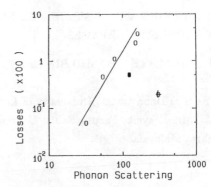

Figure 3 : Dielectric losses of
Samples (UP), (P) and Nb-doped

Figure 4 : I_m versus I_p
of 6 undoped Samples ($\bullet$ - (J))
$\oplus$ Nb - doped sample.

The general shape of the real part $\epsilon'(T)$ is quite the same for
all the undoped samples down to 20 K. Below this temperature, small
differences appear and the maximum value at 4 K cannot be related to
Ip. Introducing Nb results in a large increase in $\epsilon'(T)$, much more
larger than the corresponding change in Ip. Moreover, the temperature
T_m is decreased. These two effects are the signature of a long range
lattice polarisation leading to the optic mode softening. From this
set of results, it can be concluded that Nb (and presumably Na and
Li) is not the source of microregions in undoped $KTaO_3$.

To test the hypothesys of oxygen vacancies-Ta^{3+} complex as SBD,
we started low temperature electron irradiations in order to increase
the number of oxygen vacancies. Preliminary experiments show that
short irradiation at energy E ~ 500 kV does not modify the losses
peak at 40 K. A new very strong peak appears at T ~ 70 K. This peak
dissapears by annealing the sample at ~ 200 K. Taking into account
the electron energy, this peak could be attributed to oxygen displacements. In this case, the peak at 40 K could be due to oxygen vacancies stabilized by an other ion than Ta^{3+}. One possibility, based on
recent experiments, could be Fe^{3+}, but further investigation will be
necessary to support this hypothesys.

(1) Salce,B., De Goër,A.M. and Boatner,L.A.,
    J. Phys. Coll 42 C6, 424 (1981).
(2) Uwe,M., Lyons,K.B., Carter,M.L. and Fleury,P.A.,
    Phys. Rev. B 33, 6436 (1986).
(3) Grenier,P., Bernier,G., Jandl,S., Salce,B. and Boatner,L.A.,
    J. Phys. Condens. Matter, 1, 2515 (1989).

SCATTERING OF HEAT PULSES FROM ION IMPLANTATION DAMAGE AT SAPPHIRE SURFACES

J.K. WIGMORE[+], HAMID BIN RANI[*], S.C. EDWARDS[+] AND R.A. COLLINS[+]

[+] Physics Department, University of Lancaster, Lancaster, U.K.
[*] Jabatan Physik, Faculti Sains, Universiti Tecknologi Malaysia, 80990 Johor Bahru, Malaysia.

This work was motivated by the desire to understand the microscopic mechanism of phonon boundary scattering, and in particular to attempt to determine which out of surface roughness, point defects, dislocations, and amorphous two-level systems, all potentially present in a damaged surface layer, dominate the scattering[1]. In order to study well-defined defects, nominally perfect surfaces of sapphire single crystals were implanted with Al ions. The effects of such implantation are well understood and features of the resulting damage can be modelled quantitatively[2]. Heat pulses were used for time resolution and for the ability to control the frequency spectrum via heater excitation power.

The samples were 2 mm thick discs 2 cm in diameter of Crystal Systems HEMEX grade sapphire with faces parallel either to an A-plane ($10\bar{1}0$) or to an R-plane ($10\bar{1}2$). The faces were polished to the best "epi-finish" as supplied by Crystal Systems for silicon-on-sapphire technology. Implantation of Al$^+$ ions was carried out at AERE Harwell using a Lintott IV machine at energies between 10 and 200 keV and at doses between 3×10^{13} cm^{-2} and 5×10^{16} cm^{-2}. It is known that, at the lowest doses, the damage tracks are individual cascades of displaced atoms typically nm in cross section and extending tens of nm below the surface. At the highest doses the damaged regions overlap to form an amorphous layer of well-defined depth and thickness[2].

Details of the reflection heat pulse technique used in these experiments have been published earlier[3]. The heaters were of constantan 100 μm square, excited by 10 ns pulses, whilst the superconducting edge bolometers were of granular aluminium with transitions in the range 1.7 - 1.9 K. The transient changes in voltage occurring across the bolometers were pre-amplified with a B and H unit having a bandwidth of 3 GHz, and averaged by an EG and G boxcar integrator with a resolution of 350 ps, before being stored and processed by an ATARI 1040ST computer. Because the changes in signal took place on a time scale shorter than the response time of the bolometer ($\sim$ 20 ns), the raw data did not represent directly the phonon flux arriving at the bolometer. In order to determine the latter, it was necessary to deconvolute the bolometer signal numerically by recourse to the bolometer rate equation[4].

Typical results obtained for the deconvoluted phonon flux in an A-plane sample are illustrated in Figure 1, showing, unimplanted, low dose, and high dose data. The various specular reflection peaks from left to right are (polarisations L = longitudinal, FT and ST = fast and slow transverse): L-L; L-FT and FT-L coincident; L-ST and ST-L coincident; FT-FT; FT-ST and ST-FT coincident; ST-ST. Structure at later times related to focussed diffusive reflections. Each specular peak was followed by a discontinuous shift in baseline, due to diffusive reflection of the same polarisation character. The ramp before the L-L peak indicated the presence of bulk scattering also. Quantifying the shift from specular (S) to surface diffusive (D) reflection with increasing heater excitation power was the objective of the experiments.

The character and interpretation of the results are illustrated by Figure 2, showing the height of L-L reflections after low dose implantation as a function of heater power, plotted logarithmically. If the S > D scattering were frequency-independent, the slope of all such plots would be unity. Alternatively, if S > D conversion increased with frequency, the S gradient would be expected to be less than, and the D gradient greater than, unity. This is indeed observed; the data of Figure 2 yield values of 0.77 $\pm$ .07 and 1.13 $\pm$ .06 respectively.

This argument can be taken further, together with the dominant phonon approximation, to give an indication of the frequency dependence of the S > D scattering. The total specular flux arriving at the bolometer can be written as

$$W(t) \propto \int f^3 \exp\left(-af^n\right) df \propto \int \left(f^5 - af^{3+n} + ...\right) df$$

where f is dominant phonon frequency and weak scattering is assumed. The magnitude of the diffusive component will be directly proportional to the second

1198

term, neglecting higher ones, thus f^{4+n}. The slope of log W(t) against log (heater power) should therefore tend at low powers to $1 + \frac{n}{4}$. The consensus of our data taken on all samples is that n lies between $\frac{1}{2}$ and 1, but the presence of the bulk scattering increased the uncertainty of the figure. A value of 1 could indicate scattering from the static strain field of dislocations or from amorphous-type two-level systems[5],[6].

1) Wybourne M.N. and Wigmore J.K. Rep. Prog. Phys. __51__ 923 (1988).

2) Biersack J.P. and Ekstein W. Nucl. Instrum. Meth. __184__ 257 (1980).

3) Hamid bin Rani, Edwards S.C., Wigmore J.K. and Collins R.A. J. Phys. C: Solid State Phys. __21__ L701 (1988).

4) Edwards S.C., Hamid bin Rani and Wigmore J.K. J. Phys. E:Sci.Inst. (in press) (1989).

5) Nabarro F.R.N. Proc. Roy. Soc. __A209__ 278 (1951).

6) Kinder H. Physica __B107__ 549 (1981).

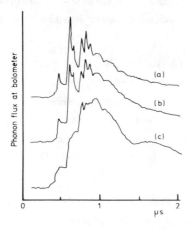

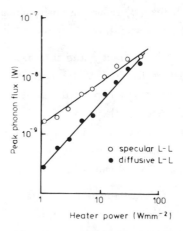

Figure 1: Deconvoluted phonon flux data for implantation doses a) 0, b) 1 x 10^{14} cm^{-2}, c) 5 x 10^{16} cm^{-2}.

Figure 2: Dependence of L–L specular and diffusive scattering on heater power.

THE STRESS DEPENDENCE OF THE SOUND VELOCITY IN MODERATELY Sb-DOPED Ge

T. UMEBAYASHI, F. AKAO[*], and K. SUZUKI

Department of Electrical Engineering, Waseda University, Shinjuku, Tokyo 169, Japan

[*]Department of Electronic Engineering, Okayama University of Science, Ridaicho, Okayama 700, Japan

1. INTRODUCTION

We have previously measured the stress(X) dependence of the change Δv in the sound velocity in Sb-doped Ge for the donor concentrations N near N_c where $N_c \sim 1.5 \times 10^{17}$ cm^{-3} is the critical concentration for the metal-nonmetal transition.[1] For $N \sim N_c$ the electrons are in the impurity band and the electron-phonon interaction for most of electrons is qualitatively not so different from that for the conduction band electrons. We have also calculated the quantity $\Delta \bar{v}(X) = \Delta v(X) - \Delta v(0)$ based on a donor-pair model.[2] The behavior of $\Delta \bar{v}(X)$ is qualitatively similar to that of an isolated donor system, though quantitatively different.

The purpose of this paper is to measure $\Delta \bar{v}(X)$ for the donor concentrations to which the donor-pair model seems to be applicable and to compare with a theory.

2. EXPERIMENT

Measurements have been done using the pulse-echo-overlap method with 60 MHz longitudinal waves at T=2.3, 3.0, and 4.2 K. The donor concentrations used are 1.1×10^{16} and 3.1×10^{16} cm^{-3} which belong to the lower end of the so-called intermediate concentration region. For the former q//[1$\bar{1}$1] and x//[110], while for the latter q//[1$\bar{1}$0] and X//[111] where q is the phonon wavevector and X is the uniaxial stress. The magnitude of X was varied from 0 to 3×10^8 dyn/cm^2.

3. RESULTS AND DISCUSSION

$\Delta \bar{v}(X)$ for N=1.1×10^{16} and 3.1×10^{16} cm^{-3} are shown by solid

circles in Figs. 1(a) and (b). The magnitude of $\Delta\bar{v}(X)$ is considerably smaller than the calculated values based on the donor-pair
model. The donor system in our samples is not to sensitive to the
external stress as the theory predicts. This is arised from that
the nealy degenerate ground state of Sb donors is easily modified by
internal strains and/or other internal fields existing in samples.
In fact the temperature dependence of $\Delta v(0)$ can be explained by
neither a calculation based on an isolated donor system nor that on
the donor-pair system provided that the idealized ground state of Sb
donors is assumed. In particular the calculations overestimate
considerably $|\Delta v(0)|$ at T<5 K.[2)]

(a)   (b)

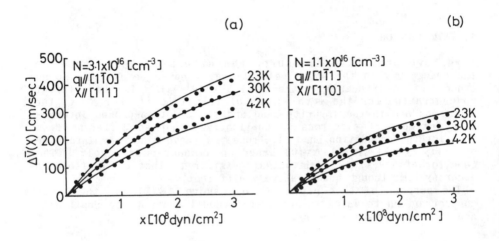

Fig. 1 $\Delta\bar{v}(X)$ versus X

The data of $\Delta\bar{v}(X)$ has been analyzed under the following assumptions. We take, for simplicity, the isolated donor system rather
than the donor-pair system. The system is under the internal uniaxial stress(x) whose magnitude is distributed according to a Gaussian
distribution. Here the varians σ_0 is determined so as to reproduce
the temperature dependence of $\Delta v(0)$. The expression for $\Delta v(X)$ is
given by

$$\Delta v(X) = (2\pi \sigma^2)^{-1/2} \int_{-\infty}^{\infty} \Delta v(x+X)\exp(-x^2/2\sigma^2)dx, \qquad (1)$$

$$\sigma = \sigma_0 + Ax. \qquad (2)$$

Here $\Delta v(x+X)$ is obtained using the expression given in Ref. 2 and the energy eigenvalues of the donor in Ref. 3.

The calculated $\Delta \bar{v}(X)$ for $N=1.1 \times 10^{16}$ and 3.1×10^{16} cm^{-3} are shown by solid lines in Figs. 1(a) and (b) where $\sigma_0(10^8$ $dyn/cm^2)=2.35$, 2.45 and $A=1.2$, 0.8, respectively. The values for σ_0 and A are too large. This is partly arised from that donor-pairs and other clusters existing in samples are not taken into account. The donor-pair system gives smaller values of $\Delta v(0)$ than the isolated donor system does.[2]

It has been pointed out that the dynamic Jahn-Teller effect plays an important role in the phonon attenuation by Sb donors.[4] We do not know whether this effect can explain more reasonably the experimental data of $\Delta v(0)$ and $\Delta \bar{v}(X)$. However, if the effect is important for shallow donors in Ge, a considerable reduction of the anisotropic part of the g-tensor and the shear deformation potential constant must occur. Experiments made so far seem to show that this is not the case.

In conclusion we do not yet get a complete understanding of acoustic properties of moderately Sb-doped Ge.

References

1. Kohno, M., Ogiwara, H., Asano, K., and Suzuki, K., J. Phys. C, 21, 4033(1988).
2. Yawata, M., Sota, T., Nakagawa, T., and Suzuki, K., J. Phys. C, 21, 1161(1988).
3. Suzuki, K., Phys. Rev. B11, 3804(1975).
4. Maier, J. and Sigmond, E., Phys. Rev. B34, 5562(1986).

THE MAGNETIC FIELD AND CONCENTRATION DEPENDENCE OF A PHONON DAMPING IN HEMATITE

V.S.Lutovinov, V.A.Murashov and T.A.Zharinova
Institute of Radioengineering, Electronics and Automation,
pr. Vernadskogo 78, Moscow, 117454, USSR.

The acoustic quality factor $Q=\omega/\gamma$ (ω is the sound frequency and γ is the relaxation rate) is one of the most essential characteristics of a magnetically ordered crystals. In ideal antiferromagnetic crystals with an easy plane type anisotropy (AFEP) the relaxation of both phonons and magnons is determined by the magnon-phonon interaction of relativistic origin which is enhanced by the exchange interaction [1-2]. The theoretical estimates of quasiparticle relaxation rates carried out for high-T_N antiferromagnets (α-Fe$_2$O$_3$, T_N=960K; FeBO$_3$, T_N=348K) show that magnon-phonon interaction of magnetoelastic origin cannot explain the values of experimentally observed relaxation rates in real AFEPs. As a rule the process of direct interaction of quasiparticles (magnons and phonons) with paramagnetic impurities cannot explain the observed values of relaxation rates also. Thus the problem of relaxation of quasiparticles in AFEP is still unsolved.This problem is important not only by itself; the knowledge of the relaxation mechanisms gives the possibility to reveal the main interaction processes between the quasiparticles in real (with defects) crystals and to predict the nonlinear properties of such crystals including their behavior in alternating fields, instability phenomena and so on.

In the present paper the damping of lowfrequency sound

waves ($\omega/2\pi=0.5$MHz) was investigated both theoretically and experimentally in hematite crystals. The magnetic field dependence of the acoustic quality factor in the series of Al -substituted hematite crystals α-$(Fe_{1-x}Al_x)_2O_3$ (0.015 < x < 0.082) was measured. The abrupt decreasing of the sound absorption (the increasing of acoustic quality) with the increasing of the external magnetic field shows that the magnetic subsystem of an antiferromagnet plays the dominant role in the formation of the relaxation characteristics of the acoustic subsystem. The sound damping which is quadratic in concentration of Fe^{++} ions (strong relaxators) is explained by the contribution of effective phonon-phonon and phonon-magnon anharmonic interactions induced by paramagnetic impurities [3] (divalent Fe ions inevitably appearing during the crystal grouth).

The Al- substituted hematite crystals α-$(Fe_{1-x}O_x)_2O_3$ (x=0.015;0.023;0.049;0.0699;0.078;0.082) was grown from the solution of Fe-oxide and Al-oxide in $NaBiO_3$. The content of Fe^{++} ions in nominally pure (without Al) and in Al-substituted crystals of hematite was found with the help of photo-calorimetric analysis. The dependence of the concentration of Fe^{++} ions c versus the Al concentration x is given in Fig.1. This curve obtained without annealing demonstrates two diffirent regions: the first one (0.015≤x≤0.023) with the strong dependence c(x) and the second plateau-like one (x≥0.023). The limiting concentration of Fe^{++} ions is $c\cong5.5\cdot10^{-3}$. The annealing decreases this value down to $c\approx3.6\cdot10^{-3}$. The first region in Fig.1 corresponds to the divalent

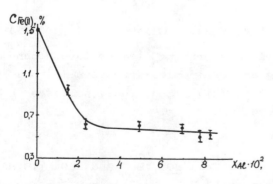

Fig.1.

Fe ions substituting trivalent Fe ones in the lattice and the region $x \geqslant 0.023$ corresponds to interstitial ions.

The damping of lowfrequency phonons in hematite at room temperature measured in a wide region of magnetic fields H strongly depends on the concentration of divalent Fe ions (see Fig.2). The acoustic quality factor Q changes by an order of magnitude when the concentration of Fe ions varies from $6.3 \cdot 10^{-3}$ to $1.5 \cdot 10^{-2}$. The logarithmic plot shows that the phonon damping is nearly quadratic in impurity (Fe^{++}) concentration in the range $c \geqslant 6.3 \cdot 10^{-3}$. Such a behaviour of a phonon damping is explained in the framework of a concept of effective 3-wave anharmonisms of an impurity origin [3]. The amplitude of these anharmonisms is proportional to the impurity subsystem magnetization $m \approx$ $c \cdot th(E/2T)$ (here c is the concentration, E is the energy spacing between the ground state and the first excited one of an impurity ion, T is the temperature). Calculations show that

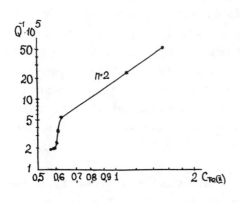

Fig.2.

$$Q^{-1} \approx c^2 \omega T th^2 (E/2T) f(H), \quad \omega \tau >> 1 . \qquad (1)$$

The additional experiments probing temperature dependence of a phonon damping are desirable. Expression (1) predicts that $Q^{-1}(T) \backsim T \cdot th^2(E/2T)$ and, thus, posesses a maximum at the temperature $T \approx E$.

REFERENCES

1. Lutovinov V.S., Fiz. Tverd. Tela 20, 1807 and 3294 (1978)
2. Lutovinov V.S., Preobrazhenski V.L. and Semin S.P., Zhur. Exp. Teor. Fiz., 74, 1159 (1978)
3. Lutovinov V.S., Acta Physica Polonica, A76, 45 (1989).

LOW-TEMPERATURE PHONON SCATTERING IN PLASTICALLY DEFORMED Nb AND Ta SINGLE CRYSTALS OF [001] ORIENTATIONS

W. WASSERBÄCH

MAX-PLANCK-INSTITUT für METALLFORSCHUNG, INSTITUT für
PHYSIK, D-7000 STUTTGART 80, HEISENBERGSTRASSE 1, FRG

1. INTRODUCTION

Phonons are scattered by mobile dislocations and/or by the elastic strain field around a sessile dislocation. The lattice thermal conductivity K of a plastically deformed crystal is influenced by both types of scattering mechanism unless one of them is suppressed. Multiple slip during plastic deformation of single crystals of [001] orientation leads to dislocation cell structures. These cell structures are observed at all stages of deformation[1,2] and consist of nearly dislocation-free regions ($N_c \leq 5 \cdot 10^{12} m^{-2}$) surrounded by dislocation walls of high density ($N_w \geq 10^{15} m^{-2}$). If the strain fields in a dense array of dislocations cancel, it is expected that the static scattering of phonons of wave lengths comparable with the dislocation spacings should be negligible[3]. The rather low density of dislocations in the cell interiors would give only a small "static" contribution to the lattice thermal resistivity. Thus it could be concluded that the predominant phonon scattering mode of such a dislocation arrangement should be of dynamic nature. We have investigated the influence of dislocations on the lattice thermal conductivity of [001] orientated Nb and Ta single crystals after tensile deformation.

2. RESULTS AND DISCUSSION

In fig. 1 the thermal conductivity K of the Ta specimen is plotted in the temperature range 0.3 - 20 K. The thermal conductivity decreases with increasing strain owing to the increase in the dislocation density. In addition, a change in the temperature dependence is observed. In order to analyse the influence of the plastic deformation, the lattice thermal resistivity W_d was calculated from the conductivity values. In fig. 2 the resitivity values of the Nb specimen are plotted in the temperature range 0.3 - 5 K. Surprisingly, for each deformation step a thermal resistivity W_d proportional to T^{-2} is found, indicating that the phonons must be scattered strongly by static strain fields rather than by dynamic phonon-dislocation processes: With dislocations fluttering freely[4] the interaction with phonons should give rise a thermal resistivity W_d proportional to $T^{-3.5}$ - T^{-4}.

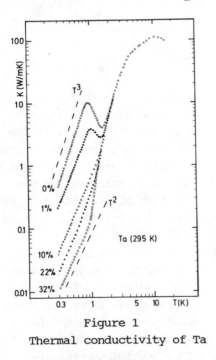

Figure 1

Thermal conductivity of Ta

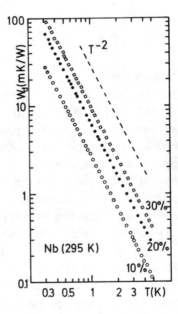

Figure 2

Thermal resistivity of Nb

Our experimental results of phonon scattering by static internal stressses can easily be explained with the composite model put forward by Mughrabi[5]. In crystals in which a heterogeneous dislocation distribution develops during deformation, substantial long-range internal stresses arise as a natural consequence of the compatibility requirements in the stress-applied state. The dislocation cell structure can be viewed as a composite of hard walls of high local dislocation density separated by soft cell interior regions. The dense walls act as obstacles to glide dislocations. As a consequence, a certain number of glide dislocations are held up at the interface between the walls and the cell interiors. The effect of the interface dislocations is that they provide tensile internal stresses in the cell walls and compressive internal stresses in the cell interiors. The tensile stress field σ_w extends into the cell interior with by about $0.1 \cdot d$ where d is the separation of two adjacent walls. The compressive internal stress field σ_c of the cell interior extends over a distance of $\approx 0.7d$. When the sample is unloaded, the stresses σ_w and σ_c remain frozen in the specimen. Because of these residual microstresses the phonons are scattered in a static sense which explains the temperature dependence of the observed thermal resistivities. With increasing strain the residual microstresses increase because of an increase in the dislocation density N_w according to

$$\sigma_w \approx \alpha Gbf_c/N_w \quad \text{and} \quad \sigma_c \approx -\alpha Gbf_w/N_w \qquad (1),$$

where α is a constant of ≈ 0.3, G is the shear modulus, b is the Burgers vector, f_w and f_c are the area fractions of the walls and the cell interiors. Since the spatial extensions of the residual microstresses are large compared with the mean phonon wavelength, the scattering of phonons by both residual microstresses are addi tive. Thus the phonons are scattered by an effective microstress

$$\sigma \approx |\tfrac{1}{2}\sigma_w| + |\sigma_c|. \qquad (2)$$

With the values[1,2] $f_w \approx 0.45$, $f_c \approx 0.55$, and $N_d \approx 0.4 \cdot N_w$ (N_d being the total dislocation density of the crystal) the effective microstress is given by $\sigma \approx 0.3Gb/N_d$. From the experimental values of the thermal resistivity W_d one obtains with the relationship $W_d = BN_dT^{-2}$ ($B \approx 1.4 \cdot 10^{-14}$ $m^3K^3W^{-1}$) and the measured values of the tensile stresses $\sigma \approx 0.2$ Gb/N_d which is in rather good agreement with the composite model.

3. REFERENCES

(1) Göttler, E., Phil. Mag. **28**, 1057 (1973)

(2) Ambrosi, P. Homeier, W., and Schwink, C. Scripta metall. **14**, 325 (1980)

(3) Ackerman, M. W. and Klemens, P. G., J. Appl. Phys. **42**, 968 (1971)

(4) Suzuki, T. and Suzuki, H., J. Phys. Soc. Jpn. **25**, 164 (1972)

(5) Mughrabi, H., Acta metall. **31**, 1367 (1983)

LOCAL DYNAMICS OF IMPURITIES IN PEROVSKITES WITH SOFT MODES

N. Kristoffel and M. Klopov
Institute of Physics, Estonian Academy of Sciences,
202400 Tartu, Estonian SSR, USSR

In the crystals with soft-mode driven structural phase transitions one can expect temperature dependent effects also in the local dynamics of defects. Paper [1] concludes that with $T \to T_c$ the frequencies of localized vibrations and resonances lower (further investigations on abstract models, see [2-6]).

Here some local dynamics of impurities in perovskite ferroelectrics $SrTiO_3$[7,8] and $BaTiO_3$[9] are presented. At $T_c = 108$ K $SrTiO_3$ shows a transition driven by the soft acoustic R_{25} zone corner mode; an optical polar Γ_{15} ferroelectric mode is pseudosoft (condenses formally at $T<0$). In $BaTiO_3$, the soft Γ_{15} mode drives the 120° C ferroelectric transition. We have described the lattice dynamics of these ideal crystals in a simple shell model. For $SrTiO_3$ charge and polarizability data have been taken from [10] and temperature-dependent short-range force constants have been found by using the data of [11] under the requirement of a correct description of both soft modes. For $BaTiO_3$ the same initial data were used and modified so that the observed behaviour of the soft mode was transmitted (here the variation of ionic charges only would suffice). Changes of other vibration branches with temperature are insignificant. We avoided the temperature region near T_c where the soft-mode description may become inadequate. On handling the perturbation matrix (only mass and force constant changes) in the usual dynamical Green's function scheme an effective interionic potential has been constructed. The defect region has been restricted to the impurity and its six neighbours with the symmetry O_h for the Ti-substitution and D_{4h} for the O-substitution.

The results of calculations consist in the following.

1. $SrTiO_3$. Localized vibrations may arise in the high-frequency region above 21.8 THz or in the gap (17.2 to 19.4 THz). In the case of a Ti-substituted defect the soft R_{25} mode is projected on the F_{1g} vibrations of the centre. To create a gap mode of this symmetry an increase of the noncentral force constant by $\Delta B = 2.4\ B$ is necessary. Further rise converts it in the long run into a high-frequency local mode. The dependence of these frequencies on temperature is weak and the same is valid for local vibrations of other symmetries. The temperature dependence of the conditions of localization is insigificant.

On the contrary, in the low-frequency region the soft-mode temperature dependence manifests itself in the behaviour of the F_{1g} resonances. A moderate weakening of noncentral forces leads to a pronounced peak in the acoustic region, which shifts monotonically to lower frequencies, becoming ever narrower as T approaches T_c.

The pseudosoft Γ_{15}-mode is projected into F_{1u} vibrations. For an isotopic defect the relative mass lowering (Q), necessary for the appearance of the gap mode, is 0.59 and 0.48, for the high-frequency local mode. A heavy impurity can now also generate temperature-dependent low-lying F_{1u} resonances. A very small enhancement of the central force constant $\Delta A/A$ suffices for a local A_{1g} mode to appear (gap mode requires 90% weakening). This reserved nature of mode-localization conditions is the result of the high weight of the totally symmetric vibrations in the region above the gap. For the E_g symmetry $\Delta A/A = 0.75$ is sufficient to generate simultaneous gap and high-frequency local modes.

In the case of a light isotopic defect on the O-site, high frequency polar A_{2u} modes can easily be created. The appearance of a E_u gap mode requires $Q = 0.31$ and that of a high-frequency local mode, $Q = 0.17$.

2. $BaTiO_3$. The top of the spectrum lies at 21.8 THz and the gap is between 17.2 and 18.9 THz. Concerning the local modes (Ti-substitution) analogous conclusions may be drawn like in the case of $SrTiO_3$. The appearance of A_{1g} local modes requires $\Delta A/A = 0.09$ and that of a gap mode, $\Delta A/A = -0.66$. Both gap and high-frequency E_g local modes may be present simultaneously ($\Delta A/A = 0.60$ required). The critical $\Delta B/B$ value for the F_{1g}

gap mode is 2.4 (3.9 for high-frequency local mode).

The ferroelectric soft mode is projected on the F_{1u} vibrations of the centre. In the isotopic model, the critical perturbations for the gap and high-frequency local modes are given by Q=0.5 and 0.43, respectively. The frequencies of these vibrations are practically independent of temperature. The soft mode manifests itself in the temperature behaviour of low-lying F_{1u} resonances, e.g. at Q=3.83 (W substituted for Ti) there is a well pronounced peak which shifts (becoming narrower) to zero frequency as $T \rightarrow T_c$. This behaviour is clearly observed in the case where the resonance lies at smaller frequencies than the soft mode peak in the full spectral distribution of F_{1u}-vibrations, see Fig.1, which represents the corresponding infrared spectrum. The lowering temperature

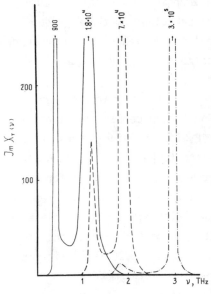

enhances markedly the resonance peak intensity. The position of this resonance is very sensitive to a small force constant weakening. In such a situation, the phase transition begins with local structural transformations around the impurities, reflecting the tendency of Ti-substituted impurities with softened bonds to become off-center.

Fig.1. The imaginary part of the transverse dielectric susceptibility of $BaTiO_3$ with Ti substituted defects (c= $=7.8 \cdot 10^{18}$ cm^{-3}) having Q=3.83, $\Delta A/A=0.1$, $\Delta B/B=-0.15$. For three curves from left to right T=430, 520, 700 K, respectively.

1. Oitmaa, J. and Maradudin, A.A., Solid State Commun. 7, 407 (1969).
2. Halperin, B.J. and Varma, C.M., Phys. Rev. B14, 4030 (1976).
3. Höck, K.-H. and Thomas, H., Z.Phys. B27, 267 (1977).
4. Schmidt, H. and Schwabl, F., Z.Phys. B30, 197 (1978).
5. Aksenov, V.L., Didyk, A.J., Fiz. Tverd. Tela 26, 2437 (1984).
6. Wiesen, B., Weyrich, K.M., Siems, R., Phys. Rev. B36, 3175 (1987).
7. Klopov, M. and Kristoffel, N., Fiz. Tverd. Tela 30, 3357 (1988).
8. Kristoffel, N. and Klopov, M., phys. stat. sol. 151b, K119 (1989).
9. Klopov, M. and Kristoffel, N., Fiz. Tverd. Tela 31, 321 (1989).
10. Stirling, W.G., J.Phys. C5, 2711 (1972).
11. Shirane, G. and Yamada, Y., Phys. Rev. 177, 858 (1969).

ISOTOPIC RANDOMNESS AND ISOTOPIC ORDERING IN PHONON PHYSICS

A.A. BEREZIN
Department of Engineering Physics, McMaster University,
Hamilton, Ontario, Canada, L8S 4M1

1. Introduction

Apart from some rare exceptions (e.g. NaI), almost all crystals have isotopic randomness. Some phonon-related studies of isotopicity done so far were recently reviewed in [1]. Most of the of work on isotopic effects based on comparative studies of hydrogenated and deuterated compounds. However, there have been relatively few studies (e.g. [2,3]) of the effects of isotopic replacements on thermal or elastic characteristics, melting and crystallization phenomena, structural phase transitions, or acoustic properties for isotopes heavier than hydrogen. Here we discuss some less-trivial possibilities of isotope effects in phonon physics.

2. Isotopic Correlations and Isotopic Ordering

Isotopic randomness affects heat conduction in a form of additional isotope scattering [4]. Much less pronounced are the variations in electrical conductivity due to subtle isotopic differences in electron and hole scattering cross-sections. In favorable cases (e.g. narrow bands, strong electron-phonon coupling) isotopic disorder can induce localization of electrons and/or holes [1]. Another possibility is electronic localization on extreme isotopic fluctuations in a regime when average-scale isotopic fluctuations leave the system in a sub-threshold state in terms of incipient localization. Also, in isotopically mixed lattices there could be a number of similarly structured isotopic microcomplexes. This might lead to low-frequency vibrational resonance effects and account for memory storage phenomena. For example, holographic-type memory effects in quartz crystals may be related to complexes involving minority isotopes of oxygen and/or silicon (e.g. ^{17}O and ^{29}Si) in a manner describable as a formation of "isotopic neural networks", similar to "neural networks" in spin glasses.

A commonly held (silent) assumption of perfectly random distribution of different stable isotopes among the corresponding lattice sites has been revised in [5] on the basis of spontaneous pattern formation in non-linear systems [6]. Figure 1 illustrates possible types of behavior of isotope correlation. Figure 2 sketches the anticipated regions of spontaneous isotope fractionation and ordering.

An artificial isotopic structuring can lead to light confinement (isotopic fiberoptics) [1]. A number of similar options could be tested for phonon transport phenomena in isotopically structured materials e.g., formation of phonon-focusing caustics (cusp and fold catastrophes) [7] due to phonon refraction on isotopic interfaces, phonon scattering on isotopically gradiented structures, etc.

3. Isotope Effects in Lattice Properties and Phase Transformations

Thermal expansion of crystals is a consequence of their lattice anharmonicity. It is well known that the freezing point of heavy water (D_2O) is +3.8°C. However, there are very few (if any) systematic studies of isotopic shifts of melting tempertures for isotopes heavier than hydrogen. A rough estimate of isotopic shift of melting temperature is provided by the dimensionality analysis, e.g. ΔT_{melt} = $(\Delta d/d)/\alpha$ (where α is the thermal expansion coefficient).

More refined experiments can be performed in "isotopic melting gap", i.e. in a temperature interval between the melting points of pure isotopes. For example, it could be instructive to study elastic properties, hardness, melting/crystallization under pressure and other bulk phonon-related properties of, say, natural copper (blend of ^{63}Cu and ^{65}Cu) between the melting points of ^{63}Cu and ^{65}Cu. In this narrow temperature interval ("no man's land") the T-behaviour of these properties could be largely affected by isotopic fluctuations. Also, hysteresic effects in phase transitions [8] could be enhanced in fluctuation-dominated regimes.

For most compounds the number of potentially available isotope combinations is quite significant - e.g. even such simple compound as SnTe exists in 10*8 = 80 isotopic versions. In each case the most contrasting (optimal) combinations can be chosen. Table of stable isotopes shows several especially "lucky" opportunities for comparative phonon spectra studies: e.g. SnTe can be studied in 3 equal-mass ("resonance") versions ($^{120}Sn^{120}Te$, $^{122}Sn^{122}Te$, $^{124}Sn^{124}Te$) against 2 "extreme" unequal-mass combinations ($^{112}Sn^{130}Te$ and $^{124}Sn^{120}Te$).

It is interesting to find out if differences in phonon spectra and phonon-controlled material characteristics (e.g. heat conductivity, lattice polarizability, etc.) for various isotopic versions of the same crystal are fully reducable to trivial mass scaling, or might exhibit some "singularities" (e.g. "cusp" behaviour of physical parameters when plotted as a function of participating isotopic masses - one-dimensional representation of "smooth" and "cusp" alternatives is given in Figure 3). In [9] some further options of "isotropic engineering" are listed while papers [10] and [11] contain more extended discussions of isotopicity for condensed matter physics and some biological aspects, respectively.

References

1. Berezin, A.A. and Ibrahim, A.M., Mat. Chem. and Phys., _19_, 407 (1988).
2. Hidaka, T. and Oka, K., Phys. Rev. B., _35_, 8502 (1987).
3. Collins, A.T. et al. J. Phys. C, _21_, 1363 (1988).
4. Klemens, P.G., Int. J. Thermophysics, 2, 323 (1981).
5. Berezin, A.A., Solid State Communs., _65_, 819 (1988).
6. Haken, H., "Synergetics", Springer-Verlag, 1978.
7. Chernosatonskii, L.A. and Novikov, V.V., Phys. Lett. A, _117_, 349 (1986).

8. De Sá, M.S. and Berezin, A.A., Bull. APS, <u>33</u>, 1198 (1988) [CF5].
9. Berezin, A.A., J. Phys. Chem. Solids, <u>50</u>, 5 (1989).
10. Berezin, A.A., Physics Letters A, <u>138</u>, 447 (1989).
11. Berezin, A.A., Biol. J. Linnean Soc. (London), <u>35</u>, 199 (1988).

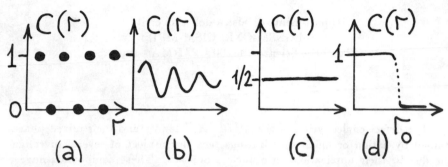

<u>Fig. 1</u> Positional isotopic correlation function C(r) for a 2-isotopic crystal with equal abundancies. (a) Perfect alternating ordering of isotopes.   (b) Partial microscopic isotopic ordering. (c) Perfectly random isotopic arrangement. (d) Complete isotope fractionation; r_0 is a grain size of a given "isotopic island".

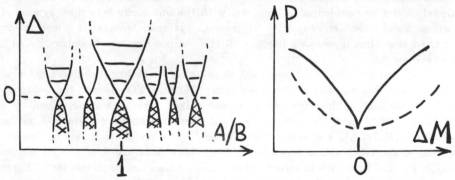

<u>Fig. 2</u> (left) Anticipated regions of spontaneous isotope fractionation (single horizontal hatching) and spontaneous isotopic ordering (double inclined hatching) for the simplest two isotope model. <u>Abscissa</u>: isotopic ratio A/B, <u>ordinate</u>:   Δ=E(AB)-E(AA); E is the corresponding bond energy. Outside the hatched cones ("Arnold toungues") isotopic distribution is essentially random, along the line corresponding to Δ=0 it is perfectly random.

<u>Fig. 3</u>   (right) Two possible ways a certain parameter p (e.g. phonon scattering length) for a 2-elemental lattice (SnTe) may depend on isotopic mass difference.

PECULIARITIES OF THE A-SYMMETRY MODES IN MIXED DEFECT CHALCOPYRITES

P.P. LOTTICI

Dipartimento di Fisica dell'Università
and GNSM-CNR, CISM-MPI
Viale delle Scienze - 43100 - PARMA - Italy

The defect chalcopyrites of the $A^{II}B_2^{III}X_4^{VI}$ family and their mixed phases obtained by cation or anion substitutions, are the subject of several investigations due to their applications in optoelectronics [1]. These compounds have a structure derived from that of the chalcopyrite, which in turn derives from that of the zincblende, by the inclusion of an ordered array of vacancies in cationic sites. Mixed crystals of the $CdGa_2(S_xSe_{1-x})_4$ and $ZnGa_2(S_xSe_{1-x})_4$ families have been previously investigated by Raman [2] and EXAFS [3] techniques. In these anion mixed compounds a generalized two-mode phonon behaviour for the vibrational modes was evidenced, in contrast with the one-mode behaviour typical of cation-mixed defect chalcopyrites [1]: in both systems, however, a conservation of the tetrahedral bonds as a function of the composition was found by EXAFS measurements.

A large attention has been paid to the zincblende-like (relative motion of the cation-anion sublattices) high frequency LO and TO modes: the composition dependence of their frequency in both mixed crystals was described by a MREI model [4,5] and the peculiar resonant Raman behaviour of the LO modes in $CdGa_2(S_xSe_{1-x})_4$ was pointed out in ref.6. On the other hand, no investigation of the composition dependence of the high intensity A-symmetry modes frequency and linewidth in mixed defect chalcopyrites has been reported. These modes are ascribed to the breathing motion of the chalcogen atoms around the vacancies [1] and occur at 141 cm^{-1} in $CdGa_2Se_4$, 221 cm^{-1} in $CdGa_2S_4$, 145 cm^{-1} in $ZnGa_2Se_4$ and at 230 cm^{-1} in $ZnGa_2S_4$. The fact that $\frac{141}{221} \approx \frac{145}{230} \approx (\frac{m_S}{m_{Se}})^{\frac{1}{2}}$, where m_S and m_{Se} are the S and Se atomic masses, respectively, is a striking evidence of the unique role of the anions in these modes.

The full Raman spectra at different compositions in $CdGa_2(S_xSe_{1-x})_4$ and in $ZnGa_2(S_xSe_{1-x})_4$ have been described elsewhere [2].

In Fig.1 it is shown the composition dependence of the frequency of the A-symmetry breathing modes in $CdGa_2(S_xSe_{1-x})_4$ and in $ZnGa_2(S_xSe_{1-x})_4$. The two-mode behaviour is evident, even if some interference between the S-type mode at low S content and a second A-symmetry mode at 188 cm^{-1} in $CdGa_2Se_4$ has to be expected [4].

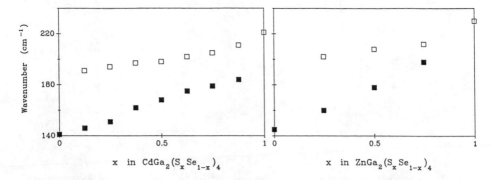

Fig.1 - Composition dependence of the A-symmetry S and Se-like modes (open and filled squares, respectively) in $CdGa_2(S_xSe_{1-x})_4$ and in $ZnGa_2(S_xSe_{1-x})_4$.

In $CdGa_2(S_xSe_{1-x})_4$ the frequency $\omega_S^{Cd}(x)$ of the S-type mode increases, in a nonlinear way, up to the end member value 221 cm^{-1}. From $x = 0$ to $x = 0.5$ it shows a slight downward curvature, whereas for $x \geq 0.5$ the curvature is upward. The frequency $\omega_{Se}^{Cd}(x)$ of the Se-type mode evolves to higher frequencies with increasing sulphur content, up to an extrapolated value at $x = 1$ of about 188 cm^{-1}. An interesting fact occurs at x=0.5: $\frac{\omega_{Se}(0.5)}{\omega_{Se}(0)} = (m_{Se}/\frac{m_{Se}+m_S}{2})^{\frac{1}{2}}$, as if the effective mass in the breathing like motion would be exactly the average anion mass.

As regards $ZnGa_2(S_xSe_{1-x})_4$, one observes a linear variation with increasing S content of the frequency $\omega_{Se}^{Zn}(x)$ of the Se-like mode, whereas $\omega_S^{Zn}(x)$ seems to change very slowly with composition x, apart from an abrupt change at low Se content.

From Fig.1 we notice that in both mixed crystals, the behaviour with decreasing Se content of the modes due to Se atoms is more "regular" than the vibrations involving the lighter S atoms.

This overall behaviour of the A-symmetry modes frequencies as a function of composition x can not be explained by a simple model involving only the anionic masses, as the simple relation between the end member frequencies could suggest. In addition to the stretching forces in the anion tetrahedra around the vacancies, one could consider also bond bending forces, which modify the effective frequencies of the anion-vacancy vibrations. This has to be done by an appropriate statistical counting, at each composition, of the different tetrahedra with 0,1,2,3,4 Se (or S) atoms at the corners. Preliminary calculations however show that the composition dependence of the breathing-like modes can not be reproduced.

The disorder introduced in the lattice by the anion mixing causes a broadening of the Raman peaks through a relaxation of the $\vec{q} = \vec{0}$ selection rule. The linewidths of the S-type and Se-type modes in $CdGa_2(S_xSe_{1-x})_4$ are reported in Fig.2.

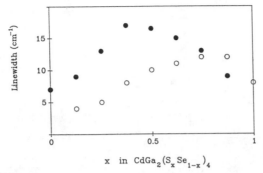

$$x \text{ in } CdGa_2(S_xSe_{1-x})_4$$

Fig.2 - Composition dependence of the linewidths (FWHM) of the A-symmetry S and Se-like modes in $CdGa_2(S_xSe_{1-x})_4$ (open and filled circles, respectively).

We notice that the maximum of the linewidth of the Se-like band is at about $x = 0.5$, where one should expect the maximum disorder, whereas for the S-like band the maximum is at low Se content. This unexpected result, whose origin is probably due to clustering effects [7], can be related to the corresponding abrupt change in the frequency of the S-like band at low Se concentrations. As regards $ZnGa_2(S_xSe_{1-x})_4$, one can estimate a similar behaviour of the linewidths as a function of composition, even if, due to a larger disorder, it is difficult to extract reliable linewidths from the Raman data.

In summary, we have evidenced that the composition dependence of the A modes in the anion-mixed defect chalcopyrites of the $A^{II}B_2^{III}X_4^{VI}$ family, in spite of their simple interpretation as breathing modes in the end member compounds, shows interesting peculiarities that need further theoretical investigation.

References

1. Razzetti, C., Lottici, P.P. and Antonioli, G., Prog. Crystal Growth and Charact. 15, 43 (1987).

2. Lottici, P.P., Parisini, A. and Razzetti, C., Prog. Crystal Growth and Charact. 10, 289 (1985).

3. Antonioli, G., Lottici, P.P., Parisini, A. and Razzetti, C., Prog. Crystal Growth and Charact. 10, 9 (1985).

4. Parisini, A. and Lottici, P.P., phys.stat.sol. (b) 129, 539 (1985).

5. Lottici, P.P., phys. stat. sol.(b) 146, 503 (1988).

6. Lottici, P.P., Parisini, A. and Razzetti, C., Proceed. Xth ICORS, Eugene 86 (Eugene, OR, USA: Univ. Oregon) pag.11/38.

7. Yamazaki, S., Ushirokawa, A. and Katoda, T., J. Appl. Phys. 51, 3722 (1980).

LOCALIZED MODE AND PHONON CONTRIBUTION
TO THE LUMINESCENCE OF KZnF$_3$:Cr^{3+}

Y. Vaills[*†], J.Y. Buzaré[*] and M. Rousseau[**]

[*]*Laboratoire de Spectroscopie du Solide, U.A. 807,*
Faculté des Sciences, Route de Laval, 72017 Le Mans Cedex, France
[**]*Laboratoire de Physique de l'Etat Condensé, U.A. 807,*
Faculté des Sciences, Route de Laval, 72017 Le Mans Cedex, France
[†]*Present address : C.N.R.S., Centre de Recherches sur la Physique*
des Hautes Températures, 1D, Av. de la Recherche Scientifique,
45071 Orléans Cedex 2, France

At room temperature the KZnF$_3$:Cr^{3+} luminescence spectrum is characterized by a 1500 cm^{-1} linewidth Gaussian band centred around 13000
cm^{-1}. As the temperature is decreased from room temperature to T = 10 K,
sharp lines and their vibronic replicas appear surimposed on the broad
band.

At low temperature the spectrum presents a lot of interesting
particularities not easily interpretable. Refering to the energy-level
scheme proposed by Brauch[1], at 10 K three zero lines corresponding to
the three Cr^{3+} sites are identified as follows :

c lines : σ = 14091 cm^{-1} : cubic site $^4T_{2g} \rightarrow {}^4A_{2g}$ transition

Δ lines : σ = 13765 cm^{-1} : trigonal site $^4E \rightarrow {}^4A_2$ transition

$\square$ lines : σ = 14016 cm^{-1} : tetragonal site $^4B_2 \rightarrow {}^4B_1$ transition

We have tried to reconstruct the 10 K emission band in considering the
multiphonon replicas of the zero-phonon lines with the continuous distribution of modes given by the phonon density of states. As previously used to calculate the emission band at 20000 cm^{-1} in KZnF$_3$:Ni, we
have worked with Pryce's model[2] assuming a linear coupling between
the electronic states and the crystal vibrations. In a first approach
we have taken the same coupling constant S to represent the modulation
of the three electronic states (cubic-tetragonal-trigonal) by the crystal vibration modes of the over all Brillouin zone.

In the Pryce's notation, the continuous emitted spectrum is
approximated by :
$$I(h\nu) = \frac{64\pi^4 \nu^4}{3c^3} \sum_{p=1}^{\infty} e^{-S} \frac{S^p}{p!} B_p(h\nu)$$

where $B_p(h\nu)$ defined from the normalized phonon density of states (Russi et $al^{3)}$) represents the p phonons contribution.

In the framework of this model, the broad emission band is well simulated with a coupling constant S = 3.7 ± 0.2. The comparison between the experimental emission band and the calculated one, both represented on figure 1, clearly shows that the sharp lines not attributed to zero-phonon lines cannot be interpreted by vibronic interactions with the phonons of the host lattice. In fact, most of these lines are shifted from the zero-phonon lines by an amount of multiples of 437 cm^{-1} or 574 cm^{-1}. When looking at the phonon density of states determined by Rousseau et $al^{4)}$ it is clear that 437 cm^{-1} lies in the vicinity of a gap in the pho-

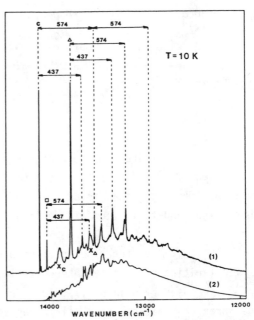

Figure 1 : KZnF$_3$:Cr^{3+} luminescence spectrum (1) experimental spectrum (pumpline λ = 4579 Å) (2) calculated multiphonon contribution.

non density of states whereas 574 cm^{-1} is greater than the upper cut off frequency of this material. Then the lines may be reasonably attributed to replicas of the purely electronic lines due to interactions with localized vibrations at 437 cm^{-1} and 574 cm^{-1}.

On the other hand, these two frequencies (574 cm^{-1} and 437 cm^{-1}) correspond to usual frequencies for normal vibrations of M^{III}F$_6$ octahedra in AAℓF$_4$ compounds (Bulou$^{5)}$). Furthermore two symmetric A$_{1g}$ vibration modes involving Cr^{3+} and F$^-$ ions have been reported at 586 cm^{-1} and 466 cm^{-1} in Rb$_2$CrF$_5$ (Boulard et $al^{6)}$).

In the framework of our description, the main features of the emission spectrum are interpreted with the exception of the two broad lines noted X$_c$ and X$_\Delta$ on figure 1. The intensities of these two lines are related to the relative intensities of the cubic and trigonal zero phonon lines respectively and they both lie at 200 cm^{-1} from these zero phonon lines. In fact, this frequency shift corresponds exactly to the frequency of the Brillouin zone center transverse optical mode having the largest oscillator strength (Ridou *et al*[7])). Then, a strong coupling between the electronic states and all the phonon modes having similar normal coordinates can produce a broad line in the luminescence spectrum even without any significant singularity in the phonon density of states.

Finally some particular features of the spectra have to be outlined even if it is not possible to draw definite conclusions about them. At 45 K (see figure 2), the trigonal line noticed Δ is flanked by a less intense line Δ'. Their vibronic replicas at 437 and 574 cm^{-1} are clearly apparent and so confirm the idea of

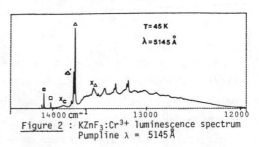

Figure 2 : KZnF$_3$:Cr^{3+} luminescence spectrum
Pumpline λ = 5145Å

localized modes. The splitting between Δ and Δ' is equal to 15 cm^{-1} which is a typical spin orbit coupling order of magnitude. As temperature decreases, the Δ' line intensity decreases whatever the pumpline is.

References

1) Brauch U August 1983 PhD Thesis University of Stuttgart.

2) Pryce MHL 1966 Phonon in perfect lattices and in lattices with point imperfections Ed by RWH Stevenson (Oliver and Boyd Edinburgh) 403.

3) Russi R, Barbosa GA, Rousseau M and Gesland JY 1984 J Phys 45 1773.

4) Rousseau M, Gesland JY, Hennion B, Heger G and Renker B 1981 Solid State Comm 38 45.

5) Bulou A 1985 Thèse d'Etat Université P. et M. Curie.

6) Boulard B, Rousseau M and Jacoboni C (in press) J Sol Stat Chem.

7) Ridou C, Rousseau M and Gervais F 1986 J Phys C 19 5757-5767.

LOCAL VIBRATIONAL MODES OF ISOLATED NITROGEN IMPURITIES IN DIAMOND

J. J. Sinai and S. Y. Wu
Department of Physics, University of Louisville
Louisville, Ky 40292

and

K. S. Dy
Department of Physics, University of North Carolina
Chapel Hill, North Carolina 27514

A nitrogen atom in diamond is a defect with many interesting properties. The main features[1] of its infrared absorption include a resonance band at 1130 cm^{-1} and a sharp peak at 1344 cm^{-1}. Both peaks are nitrogen-dependent. However, when ^{15}N substituted for ^{14}N, the 1130 peak shifted markedly, whereas no perceptible shift occurred for the 1344 peak. This suggests an indirect dependence of the 1344 peak on nitrogen. The substitutional nitrogen in diamond is also known to induce symmetry-breaking distortion[2]. The causes for the distortion are not well understood.

Ordinarily, a substitutional atom in a diamond lattice constitutes a five-atom cluster with T_d symmetry. However, nitrogen was observed to move off center towards three of its nearest neighbors. This distortion destroys the T_d symmetry. In fact, it was shown that[3], when T_d symmetry was maintained, it was not possible to reproduce the previously observed fine structure of the infrared absorption.

The local environment apparently plays an essential role in determining the vibrational properties of the system. To account for the observed lattice distrotion on the one hand, and to allow maximum flexibility for the theoretical fitting of the infrared absorption on the other, we considered an impurity configuration composed of a relaxed CN pair (the impurity 'molecule') with bonds of variable strengths connecting it to the neighboring carbon atoms in the lattice. The defect cluster is then an eight-atom cluster. The local vibrational modes were studied by calculating the local density of states (LDOS) of the defect cluster. The cluster LDOS is composed of the site LDOS's. Each site LDOS depends on the components of the displacement vector at that site. Thus each site LDOS consists of three components, one for each of the three directions x, y, or z. There may be equalities among these components depending on the symmetry of the subspace associated with the configuration at a given site for a given direction. For the eight-atom defect cluster, the components of the site LDOS associated with the nitrogen atom are equal and the same holds for the site LDOS associated with the carbon (labelled C2) in the relaxed CN pair. However, for the three carbon atoms (labelled C3 through C5) surrounding the nitrogen and those (labelled C6 through C8) surrounding C2, not all of the subspaces associated with the three directions are the same. It turns out that, for each such site, two of the three components are equal; which two depend on the location of the site. The equality is cyclically permuted among the three equivalent sites (C3 through C5 or C6 through C8). Thus, the cluster LDOS can be written as

$$\rho(\omega^2) = \frac{1}{8}[\rho_x(N)+\rho_x(C2)+\rho_z(C3)+\rho_z(C6)] + \frac{1}{4}[\rho_x(C3)+\rho_x(C6)] \qquad (1)$$

where $\rho_x(\alpha)$ is the x-component of the site LDOS at site α. The site LDOS itself is obtained as the limit of the imaginary part of the corresponding local Green's function, with the local Green's function calculated using the recursion method[4].

   In the calculation, only nearest neighbor interactions were taken into consideration. The force constant matrices are then defined by a two-parameter (α,β) model, with α and β denoting the central and non-central force constants respectively. The defect configuration under consideration allows the force constant matrices between N and its three neighbors and those between C2 and its three neighbors to vary separately. This procedure tends to decouple the dependence of the LDOS associated with C2 on the nitrogen isotopes, thus providing a scheme to recover the isotope-independent mode at 1344.

   There are three sets of parameters: (α_1',β_1') defines the relaxed interaction between the CN pair, (α_2',β_2') the interactions between N and its neighbors, and (α_3',β_3') the interactions between C2 and its neighbors. The interactions between the cluster and the host lattice are taken to be the parameters of the host, with $\beta/\alpha=0.65$.

The requirements on the variation process were: (i) to anchor the isotope-independent peak at 1344 and (ii) to obtain the best possible overall shape of the spectrum. The figure shows the spectrum obtained with $\alpha_1'=0.70$, $\beta_1'=0.45$, $\alpha_2'=1.01$, $\beta_2'=0.60$, $\alpha_3'=1.35$, and $\beta_3'=0.35$. The general shape of the spectrum agrees excellently with the experimental results[1], with one peak (1198) isotope-dependent while the other (1337) isotope-independent. The only deficiency is that the isotope-dependent peak occurs at 1198 rather than 1130. The peak could be moved closer to 1130 by reducing α_2', but it would have distorted the shape of the spectrum. Considering the fact that the fitting was accomplished with a nearest neighbor model, we feel that the result is excellent.

References

1. Collins, A. T. and Woods, G. S., Phil. Mag. 46, 77 (1982).
2. Smith, W. V., Sorokin, P. P., Gelles, I. and Lasher, G. J., Phys. Rev. 115, 1546 (1959).
3. Sinai, J. J. and Wu, S. Y., Bull. Am. Phys. Soc. 34, 993 (1989).
4. Sinai, J. J. and Wu, S. Y., Phys. rev. B39, 1856 (1989).

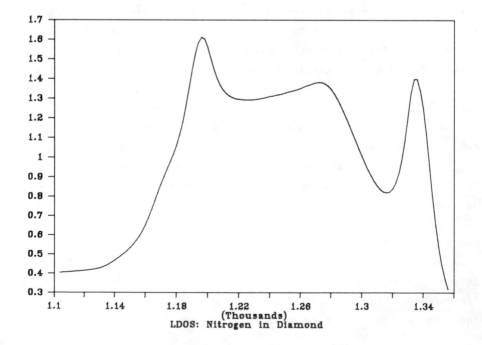

LDOS: Nitrogen in Diamond

13

Nonequilibrium Phonons

PHONON DYNAMICS IN HIGHLY
NONEQUILIBRIUM SYSTEMS

Y.B. Levinson

Institute of Microelectronics Technology
and High Purity Materials, USSR Academy
of Sciences, 142432 Chernogolovka,
Moscow District, USSR

The paper was stimulated by two recent experiments on phonon excitation. The first made by the group of J.P.Wolfe[1] deals
with ballistic phonons excited when a laser pulse is absorbed in
a Si crystal. In the second experiment made by the group of
W.Bron[2] nonequilibrium LO-phonons in GaP are probed by Anti-
Stokes Raman scattering. The aim of the paper is to discuss these two experiments and to show that the first experiment demo-
strates phonon hot spot explosion while the second demonstrates
a new type of collective phonon excitations (similar to the zero-
sound in Fermi-gas).

1. Phonon Hot Spot Explosion.

In this part of the paper I would like to discuss the ballistic phonon pulses excited in a semiconductor by laser pulses
absorbed in band-to-band transitions. Let us recall what processes occurr in this situation.

A laser pulse (typical pulse duration 10ps-10ns) is absorbed
near the crystal front surface. When a photon is absorbed, a
electron-hole pair is created, the electron and the hole being hot.
The carriers diffuse into the crystal bulk and relax via emission
of optical and intervalley high-energy phonons. The energy relaxation time is about 10ps, of the order of or shorter thah the laser pulse duration. As a result, during the laser pulse of about
one-half of the pulse energy is imparted into the phonon system.

(The rest of the laser pulse energy is transferred to the phonons in the course of electron–hole recombination). The high–energy relaxation phonons are created in a near–surface layer the thickness of which is of the order of a few microns. A bolometer or a tunnel–junction on the crystal rear surface is used to detect the nonequilibrium phonons. The distance L between the detector and the laser focal spot is about 1cm.

Let us emphasize that the frequency of the primary phonons created during the hot carrier relaxation is of the order of the Debye frequency ω_D irrespective of the excitation intensity. Therefore, the primary phonons are strongly scattered by isotopic impurities (elastic scattering) and due to the phonon anharmonicity (inelastic phonon–phonon scattering). In all crystals elastic scattering time $\tau'(\omega)$ is shorter than inelastic scattering time $\tau(\omega)$. As a result, during the time interval between two successive inelastic scattering events the motion of the phonons is diffusive. Due to the strong phonon scattering the crystal domain. occupied by the nonequilibrium phonons spreads slowly.

Nevertheless in some experiments with Ge[3] Si[1] and GaAs[4] sharp ballistic signals are detected. These signals are due to phonons arriving onto the detector without scattering. The absence of scattering is confirmed by two facts 1) the arrival time of the signal is exactly L/ s, where s is the sound velocity in the given direction; 2) the signals are very sensitive to phonon focusing. The ballistic signals are detected because a small number of low–frequency phonons with mean free part l>L are created in the excited crystal domain as a result of the decay of the high–frequency phonons. In other words, the nonequilibrium domain of the crystal is a source of ballistic phonons. The ballistic signal from this source contains information about the time evolution of the excited domain. It is this point that is discussed in what follows.

Two models are known to describe the time evolution of the excited crystal domain occupied by high–frequency phonons, namely the hot spot model (Hensel and Dynes[5]) and the phonon generation model (Kazakovtsev and Levinson[6]). These two models can be discriminated from the properties of the ballistic signal.

For simplicity we assume the primary phonons to be generated in a hemisphere with radius R_o and the laser pulse to be very short.

The phonon generation model is valid for low excitation levels. In this case the phonon occupation numbers are small, and because spontaneous phonon decay is the dominant phonon-phonon scattering process. Since on spontaneous decay the phonon energy is divided approximately on half the primary phonons with frequency $\omega_0 \simeq \omega_D$ give rise to a chain of phonon generation with frequencies ω_0, $\omega_1 = \omega_0/2$, $\omega_2 = \omega_1/2 = \omega_0/2^2$,... . The life-time of the generation with frequency ω_k is $\tau(\omega_k)$, where $\tau(\omega)$ is the phonon life-time against spontaneous decay. Each generation spreads by diffusion and the spreading length of the generation is of the order of the diffusion length $l(\omega_k)$, where

$$ l(\omega_{.}) = [D(\omega)\tau(\omega)]^{\frac{1}{2}}, \quad D(\omega) = \tfrac{1}{3}s^2\tau'(\omega). \tag{1} $$

For order-of-magnitude estimates one can use the long-wave approximation

$$ \tau(\omega) = \bar{\tau}x^{-5}, \quad \tau'(\omega) = \bar{\tau}'x^{-4}, \quad l(\omega) = \bar{l}x^{-9/2} , \quad x = \omega/\omega_D. \tag{2} $$

In Si the branch-averaged nominal values are: $\bar{\tau}$ =320ps, $\bar{\tau}'$ =13ps, $\bar{l}$=0.22 μm and $\omega_D/2\pi$ =13.4 THz[7]. Let R_o=5μm and assume the frequency of the primary acoustic phonons to be $\omega_0 = \omega_D/2$. From (2) one calculates $l(\omega_0)$=5μm and $l(\omega_1)$=110μm. This means that the second generation ω_1 escapes from the hemisphere R_o. The radius of the hemisphere, occupied by the nonequilibrium phonons grows with time. Let us find how this radius $R(t)$ depends on time.

The characteristic phonon frequency ω_t at the time t can be found from the relation $\tau(\omega_t)$ =t. If $l(\omega_t)<R_o$, then at the time t the phonons do not escape from the hemisphere R_o, where the primary phonons were excited, and hence, $R(t)=R_o$. If $l(\omega_t) > R_o$, then the radius of the excited hemisphere is of the order of the diffusion length of the last generation, i.e. $R(t)=l(\omega_t)$. As a result, we obtain

$$R(t) = R_o, \qquad t < t_o \qquad\qquad\qquad (3)$$
$$= R_o (t/t_o)^{9/10}, \quad t > t_o, \qquad t_o = \bar{\tau}(R_o/\bar{l})^{10/9}$$

where t_o is the time of escape from the initial hemisphere R_o. In the above mentioned numerical example $t_o = 10 ns$. Two important conclusions follow from the above consideration of the phonon generation model. 1) The size of the excited domain grows smoothly and unlimitedly. 2) This size and its grow rate do not depend on the laser power.

At higher excitation levels not only spontaneous decay but also phonon coalescence and induced phonon decay are important. When the number of phonons is so high, that the coalescence rate and the decay rate are of the same order of magnitude, Planckian phonon distribution is established due to phonon-phonon scattering. In this situation the excited domain of the crystal can be specified by a nonequilibrium temperature T which is higher than the bath temperature T_B. This is the phonon hot spot model.

In many papers the term "phonon hot spot" is used in all the cases, when phonons in the excited domain are strongly scattered, irrespective of whether the nonequilibrium phonon temperature is established, or not. We prefer to use the term "hot spot" only in the case when the nonequilibrium phonon temperature is indeed established. If it is not, we use the term "decay spot".

The initial nonequilibrium phonon temperature T_o can be calculated from energy balance

$$E/V_{R_o} = \varepsilon_{T_o} \qquad\qquad\qquad (4)$$

where E is the energy absorbed from the laser pulse, V_{R_o} is volume of the hemisphere R_o, and ε_T is the phonon energy per $1cm^3$ at temperature T. The Planckian with temperature T_o is established, if $R_o > l_T$, where l_T is the diffusion leugth $l(\omega)$ of thermal phonons with $\omega = 2.82T$. From the condition $R_o = l_T$, combined with (4) one can find the cross-over energy E^*. If $E > E^*$, the hot spot model is appropriate. If $E < E^*$, the decay model is appropriate. Let us adopt the long-wave approximation

$$\varepsilon_T = \bar{\varepsilon}(T/T_D)^4, \qquad T_D = \hbar\omega_D \qquad\qquad (5)$$

where $\bar{\varepsilon} = 2.5 \times 10^4$ J/cm^3 for Si. Then for a focal spot with $R_o = 5\,\mu m$ the cross-over energy $E^* = 10nJ$.

Now let discuss the question about the time evolution of the hot spot size and temperature. This evolution is governed by the equation

$$\varepsilon_T V_R = E, \qquad dR^2/dt = D_T \qquad (6)$$

the first being the energy conservation law and the second being the diffusion law. V_R is the volume of a hemisphere with radius R and D_T is $D(\omega)$ for thermal phonons with $\omega = 2.82T$. It follows from eqs. (6) in the case of the long-wave approximation, that

$$R(t) = R_o \left[1 - (E^*/E)(t/t_o)\right]^{-1},$$
$$T(t) = T_o \left[R(t)/R_o\right]^{-3/4}, \qquad (7)$$

where t_o has the same meaning as in (3). One can see from (7), that the hot spot dynamics depends on the imparted energy E. The higher this energy, the slower the hot spot spreading. This is in contrast to the decay spot, the spreading rate of which does not depend of the imparted energy E. Another pecularity of the hot spot is its exlosion. In a finite time interval $(E/E^*)t_o$ the hot spot becomes infinitely large, and its temperature becomes zero. To understand what actually happens, compare the hot spot radius $R(t)$ and the diffusion length $l_{T(t)}$ at various times. As the temperature $T(t)$ falls, the radius and the diffusion length grow: $R \sim T^{-4/3}$, $l_T \sim T^{-9/2}$, the diffusion length growing faster. Therefore, the condition $l_T < R$ is violated at some moment t^+ (Fig.1). At this moment the phonon-phonon scattering becomes too weak to support the Planckian distribution and the hot spot is destructed. In other words, eqs. (6) and (7) are valid only for $t < t^+$.

The radius and the temperature of the hot spot at the moment of its destruction R^+ and T^+ can be obtained from the condition $R^+ = l_T +$ giving

$$T^+ = T_o (l_T/R_o)^{6/19}, \qquad R^+ = R_o (l_T/R_o)^{-8/19}. \qquad (8)$$

Combining eqs. (8) with eqs. (7), we find the destruction moment

$$t^+ = t_o (E/E^*) \left[1 - (l_T/R_o)^{8/19}\right] \approx t_o (E/E^*). \qquad (9)$$

At the moment t^+ the hot spot is transformed to a decay spot and accordingly the slow spreading is transformed to fast spreading.

In the case of the above considered example in Si for
$E=100$nJ one obtais the initial temperature of the hot spot $T_O=$
230K. The thermalization length $l_{T_O}=0.2 \mu$m is short compaired
to $R_O=5 \mu$m. The hot spot is destructed at $t^+=67$ns, having at
the destruction moment the temperature $T^+=83$K and the radius
$R^+=19 \mu$m.

Now we compare the theoretical predictions about ballisitic
signal emitted from a hot spot and from a decay spot. The shape
of a ballistic signal emitted from a decay spot does not depend
on the laser pulse energy, the intensity of the signal being proportional to this energy. The decay spot can be considered as
a localized source of ballistic phonons until the nonequilibrium
high-frequency phonons are in the excitation domain determined
by the laser focal spot. When these phonons escape from the
excitation domain the size of the source grows rapidly. As a
result the decay spot is to be considered as a source of ballistic phonons with size R_O and life-time t_O. When ballistic phonons are emitted from a phonon hot spot not only the intensity
of the ballistic signal depends on the laser pulse energy, but
also the shape of the signal. The hot spot spreads slowly to
destruction. Heuce, the hot spot is to be considered as a source
of ballistic phonons with size R^+ and life-time t^+. Both these
source properties depend on laser energy: $R^+ \sim E^{9/19}$, $t^+ \sim E$, in
contrast to size R_O and life-time t_O of the ballistic phonon source in the case of a decay spot.

Now we are going to compare theory and experiment. First
we consider the recent experiments in Si performed with high resolution by the group of J.Wolfe[1]. The width of caustic Δx was
measured in a phonon focusing experiment. This width is of the
order of the size of the ballistic phonon source. The dependence
of Δx on E is shown in Fig.2. This dependence is strongly pronounced in the region $E > 10$nJ, where $\Delta x \sim E^{0.36}$. This high-energy region corresponds to phonon emission from a hot spot. The
theory predicts the size of the phonon source to be $R^+ \sim E^{0.47}$,
which is in a fairly good agreement with the experiment. In region

E<10 nJ the dependence of Δx on E is weak. This region corresponds to phonon emission from a decay spot. The experimental
cross-over energy E=10nJ agress with the calculated one.

It is also possible to discuss the source size. For E=100nJ
the theory predicts the size $2R^+=40\mu m$. The measured value $\Delta x=$
$100\mu m$. For E=1nJ the theory predicts the size to be the focal spot
diameter $2R_0=10\mu m$. The measured value $\Delta x=25\mu m$. Bearing in mind
that the long-wave approximation of the scattering rate (2)and of
the phonon energy density (5) is crude for high frequencies and
temperatures, we come to the theory argess with the experiment
quite satisfactorily.

Experiments with $Ge^{3)}$ measure the life-time of the ballistic phonon source. This life-time depends on pulse energy $\sim E^{0.7}$. The
theory predicts the life-time $t^+ \sim E$ for ballistic phonon emission from
a hot spot. This looks like a fairly good agreemnt, but it should
be remembered, that much higher energy excitation levels are used
in the experiments with Ge and Debye temperature for Ge crystal
is lower, so the long-wave approximations for Ge are much less
appropriate than for Si.

2. Collective Excitations in Nonequilibrium Phonon System.

First let us formulate the results obtained in the paper $^{8)}$ for
a more general situation. Consider a nonequilibrium phonon system
with acoustical phonons excited in a thin spherical shell in the k-
space. The distribution function of these phonons is assumed to be
a Lorenzian

$$N_K = N_0 \left(\Delta\omega/2\right)^2 \left[\left(\omega_K - \omega_0\right)^2 + \left(\Delta\omega/2\right)^2\right]^{-1}. \tag{10}$$

Here ω_0 is the frequency in the center of the distribution, N_0 is the
peak occupation number, $\Delta\omega$ is the width of the distribution N_K
$(\Delta\omega \ll \omega_0)$, $\omega_K=sk$, where s is the sound velocity.

Now we concentrate our attention on the two-particle Green
function K and find the poles of this function in the total energy
$\omega = \omega_1 + \omega_2$ at the total. momentum $k_1+k_2=0$. It follows from simple
calculations that K has a pole at $\omega = 2\omega_0 - i\Delta\omega$. This pole does not
represent a bound state of two acoustical phonon with momenta k
and -k, since the location of the pole does not depend on the
parameter responsible for coupling between acoustical

phonons. The pole $2\omega_0 - i\Delta\omega$ represents a collective excitation, similar to zero-sound in a nearly- ideal Fermi-gas. In both cases the collective excitation exists because the distribution function has some "singularity", i.e. the sharp Lorenzian peak or the Fermi step.

A special situation occurs when ω_0 is close to one- half of the longwave LO-phonon frequency Ω_0. In this situation the collective acoustic phonon excitation is in resonance with the optical phonon. The two modes strongly interact and mixed coupled modes arise in the total phonon (acoustical + optical) system. The higher the acoustical phonon occupation numbers N_0, the stronger the mixing. The corresponding mixing parameter is given by $\xi = N_0/N_0^*$, where $N^* = (\alpha-1)^2/8$, $\alpha = 2\Delta\omega/\Gamma_0$, Γ_0 being the spontaneous decay width of the LO-phonon.

We depict the coupled modes in the complex Ω- plane (Fig.3) For small ξ there are two poles, the first close to the LO-phonon pole $\Omega_1 = \Omega_0 - i\Gamma_0/2$ and the second close to the collective mode pole $\Omega_2 = \Omega_0 - i\Delta\omega$ .The oscillator strength of the first pole $f_1 \approx 1$, and of the second $f_2 \ll 1$. As parameter ξ increases, the two poles move in the vertical direction to meet half-way at $\xi = 1$. For $\xi > 1$ the poles go away in the horizontal direction. As a result, the phonon system is represented by two oscillators with the same frequency and different damping factors at $\xi < 1$, while at $\xi > 1$ it is represented by two oscillators with different frequencies and the same damping factor. The frequencies and damping factors are given

$$\xi < 1: \quad \Omega_1 = \Omega_2 = \Omega_0 \; , \quad \Gamma_{1,2} = \tfrac{1}{4}\Gamma_0 \left[(\alpha+1) \pm (\alpha-1)(1-\xi)^{\frac{1}{2}} \right] \quad (11)$$

$$\xi > 1: \quad \Omega_{1,2} = \Omega_0 \pm \tfrac{1}{4}\Gamma_0 (\alpha-1)(\xi-1)^{\frac{1}{2}}, \quad \Gamma_1 = \Gamma_2 = \tfrac{1}{2}(\tfrac{1}{2}\Gamma_0 + \Delta\omega)$$

It is my understanding that these coupled collective phonon modes were discovered in the experiments performed by W.Bron and coworkers[2]. LO- phonons in GaP were excited via nonlinear interaction of two picosecond laser pump pulses the difference in frequency of the laser pulses being equal the LO-phonon frequency Ω_0. The intensity of the Anti-Stoles scattering with a frequency shift Ω_0 from a probe pulse delayed in time with respect to the pump was measured. The dependence of the AS-signal on the delay time t tells us about the damping of the oscillations with frequency Ω_0 of the phonon system.

At low pumping levels the AS-signal decay with the time constant close to the LO-phonon life-time τ_{LO}. At high pumping levels the AS-signal displays two components (Fig.4). One component decay with the time constant $\tau_1 = \tau_{LO}$, while the other decays mucn faster, with the time constant depending on the pump intensity.

The LO-phonons generated by the pump decay into LA-phonons with frequency $\omega_0 = \Omega_0/2$. The decay time is $\tau_{LO} = 26ps$ ($\Gamma_0 = 0.20cm^{-1}$). The life-time of the LA-phonons is not known, yet, anyhow it is longer than the delay time $t < 250ps$. Hence, we have a stationary distribution of LA-phonons for $t > 26ps$. This distribution creates collective acoustic phonon excitations, which mix with LO-phonons giving two coupled excitations. If we assume ξ to be small even in the case of a high pumping level, both coupled excitations have same frequency Ω_0 and are detected by AS-scattering. The two components of the AS-signal correspond to different dampings Γ_1 and Γ_2 of these excitations.

Now let us make quantitative comparison between the theory and the experiment. It follows from [9] that $N_0 = (8/\pi)(J/J^*)^2$, where J is the irradiance of each of the two laser pulses and J^* is some relevant irradiance. To calculate J^* one can use (3,5) from [9] with the pulse length 8ps instead of the LA-phonon lifetime and th e Raman scattering cross-section for GaP from [10]. Taking into account the reflectiy of the laser (R=0.3), one obtains $J^* = 3.8GW/cm^2$. Hence, for experimental pumping levels $N_0 < 0.08$. The width of the LA-phonon distribution $\Delta\omega$ can be calculated from the measured values of the decay rates using the identity $\Gamma_1 + \Gamma_2 = \Gamma_0/2 + \Delta\omega$ valid for $\xi < 1$. For the highest pumping level one obtains $\Delta\omega = 0.55cm^{-1}$ and $N_0^* = 0.4$ which yield $\xi = 0.2$ confirming the assumpion $\xi < 1$. For $\xi = 0.2$ one obtains from (11) $2\Gamma_1 = 0.24cm^{-1}$ and $2\Gamma_2 = 0.98cm^{-1}$ wnich is in fairly good agreement with the experimental decay rates $1/\tau_1 = 0.18cm^{-1}$ and $1/\tau_2 = 1.1cm^{-1}$.

1234

REFERENCES

1. Shields J.A. and Wolfe J.P., Preprint (1988).
2. Bron W.E., Juhasz T. and Mehta S., Preprint (1988).
3. Greenstein M., Tamor M.A. and Wolfe J.P., Phys. Rev. B26, 5604 (1982).
4. Danilchenko B.A., Poroshin V.N., Slutskii M.I. and Asche M., Phys.stat. sol.(b) 136, 63(1986). Danilchenko B.A., Kazakovtsev D.V. and Slutskii M.I., Solid State Comm. (in press).
5. Hensel J.C. and Dynes R.C., Phys. Rev. Lett. 39, 969 (1977).
6. Kazakovtsev D.V. and Levinson Y.B., Sov. Phys. JETP 61, 1318 (1985).
7. Tamura S., Phys. Rev. B31, 2574 (1985).
8. Bulgadaev S.A. and Levinson Y.B. Sov. Phys. JETP 40, 1161 (1975).
9. Levinson Y.B., Sov. Phys. JETP 38, 162 (1974).
10. Calleja J.M., Vogt H. and Cardona M., Phil. Mag. A45, 239 (1982).

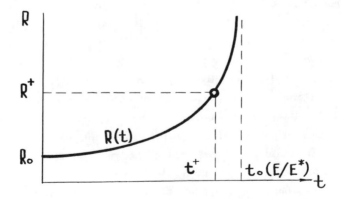

Fig. 1

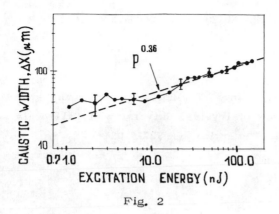

Fig. 2

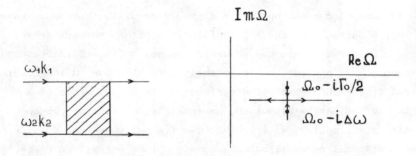

Fig. 3

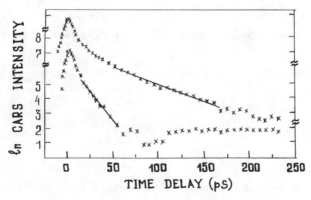

Fig. 4

A NEW NONEQUILIBRIUM PHONON STATE

W. E. Bron, T. Juhasz and S. Mehta

Department of Physics, University of California

Irvine, CA 92717 USA

Evidence for a new optically driven nonequilibrium phonon state has been observed. At thermal equilibrium optical phonons decay, with a single rate, into two acoustic phonons,[1,2,]. Bulgadaev and Levinson[3,4,5] predict that a strongly optically pumped longitudinal optical (LO) phonon spectral distribution, in the presence of a highly excited narrow-bandwidth, nonequilibrium acoustic phonon spectral distribution, will exhibit two (rather than one) decay rates below a critical nonequilibrium acoustic phonon occupation number N^*. We report the first observation of two decay rates for the dephasing of LO phonons.

In order to search for the nonequilibrium phonon state, we have performed picosecond time resolved anti-Stokes Raman scattering (TRCARS) measurements[1] using a recently developed[6] dual synchronously pumped and synchronously, strongly amplified, laser system.

The two decay rates of a strongly excited LO phonon mode can be shown[7] to be

$$\Gamma_{1,2} = \frac{\Gamma_o}{2} \left[(\alpha+1) \mp \left\{ (\alpha-1)^2 - \beta \right\}^{1/2} \right] \tag{1}$$

in which $\alpha = 2\Delta\omega/\Gamma_o$, $\beta = 8\alpha n_A$, $2\Delta\omega$ is the FWHM, and n_A occupation number at the peak of a nonequilibrium Lorentzian acoustic phonon distribution formed by the decay of the strongly excited LO phonon distribution with bandwidth Γ_o. Note that as n_A approaches zero, Γ_1 and Γ_2 approach Γ_o and $2\Delta\omega$, respectively. Thus, the solution with $\Gamma_1 = \Gamma_o$ is the one usually observed in spontaneous Raman measurements at relatively low laser intensities. For low values of $n_A \left(\lesssim 10^{-2} \right)$ a typical result of the TRCARS measurement is shown in Fig. 1a. As noted in reference 1, the TRCARS signal intensity has two readily

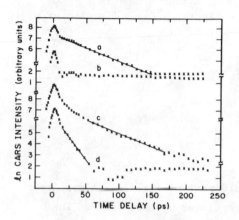

Fig. 1. TRCARS signal intensity as a function of the
delay, Δt, between the
formation of LO phonons near
and the time of the probe
pulse. For details see
reference 7.

distinguishable components. The first is attributable to the non-linear response of bound electrons in the GaP sample. The remaining signal for $\Delta t > 0$ reflects the dephasing of the LO phonons generated near $\Delta t = 0$. The inverse slope of this exponentially decreasing signal is by tradition referred to as $T_2/2$. Subtraction of the long-term, exponentially decreasing phonon part of the total signal, leaves only the response of bound electrons and a time independent residue ascribable to various sources of stray background radiation (see Fig. 1b). For values of $n_A \gtrsim 10^{-2}$, however, a second, additional (shorter) dephasing time is readily observed. The faster decay rate is, again, extracted by subtracting the long-term component from the total TRCARS signal (Figs. 1c and 1d).

Figure 2 (open circles) is a compilation of observed values of Γ_1 and Γ_2 (in cm^{-1}) for the range of n_A over which two different decay rates are observed. Moreover, separate evaluations of n_A for fixed values of α, lead to Γ_1 and Γ_2 as predicted through Eq. 1 (dashed lines). Note that α increase as the laser intensity (and,

1238

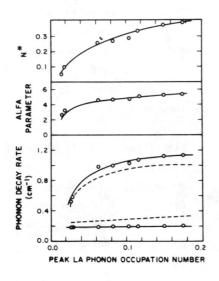

Fig. 2. (Lower figure) Compilation of observed (open circles) and theoretically predicted curves (dashed lines) of Γ_1 and Γ_2 (in cm^{-1}) as a function of n_A. (Middle figure) Alpha parameter as a function of n_A. (Upper figure) N^* as a function of n_A.

therefore, n_A) increases (see middle part of Fig. 2). A nonlinear dependence of α on n_A was indeed predicted by Levinson[3]. The difference between the observed and predicted values of Γ_1 and Γ_2 is not yet understood.

W.E.B. acknowledges funding through the NSF Grant No. DMR 86-03888 and through NATO Grant No. DO34188.

REFERENCES

1. Bron, W. E., Kuhl, J. and Rhee, B. K., Phys. Rev. B34, 6961 (1986).

2. See, e.g., Maradudin, A. A. and Fein, A. E., Phys. Rev. 128, 2589 (1962).

3. Levinson, I. B., Sov. Phys. JETP 38, 162 (1973).

4. Bulgadaev, S. E. and Levinson, I. B., Sov. Phys. JETP Lett. 19, 304 (1974).

5. Bulgadaev, S. A. and Levinson, I. B., Sov. Phys. JETP 40, 1161 (1974).

6. Juhasz, T., Kuhl, J. and Bron, W. E., Opt. Lett. 13, 577 (1988).

7. Bron, W. E., Juhasz, T., and Mehta, S., Phys. Rev. Lett. 62, 1655 (1989).

HIGH FREQUENCY ACOUSTIC PHONON RELAXATION

D.V.Kazakovtsev, A.A.Maksimov, D.A.Pronin, I.I.Tartakovskii

Institute of Solid State Physics, USSR Academy of Sciences,
142432, Chernogolovka, Moscow Region, USSR

One of the interesting problems in phonon physics is the relaxation of a nonequilibrium-phonon system in the course of the equilibrium temperature establishment in the crystal. Using optical method, proposed in[1], we have measured the occupation numbers of nonequilibrium phonons of several frequencies Ω at different time delays t_D after the intense laser pulse application and have compared them quantitatively with the model-based calculation results. To eliminate the effect of phonon propagation on the phonon energy relaxation, we applied, as in[2], the homogeneous surface excitation by a wide laser beam.

The experiment[3] can be summarized as follows. An anthracene single crystal of thickness d=2-5 μm was placed in an optical helium constant-temperature bath held at T_0=4.3K. Its front surface (the developed *ab*-plane) was uniformly excited with pulses from a nitrogen TEA laser (ω=28670 cm^{-1}, pulse length $\tau_p \approx$0.5 ns, repetition frequency of 100 Hz, maximum pump intensity of 100 kW/cm^2 at the sample). A fraction of the pulse, delayed by t_D=0-10 ns by means of an optical-delay line, excited a probe pulse of a tunable dye laser with frequency ω near the bottom of the lowest exciton *b*-band, ω_T=25096 cm^{-1} (half-width of the lasing band $\Delta\omega \approx$1 cm^{-1}, pulse length $\tau_L \approx$0.3 ns, polarization of light $E \| b$-axis). As was shown in[1], for $\omega \lesssim \omega_T$ the change in the absorption coefficient $\Delta k = k(\omega) - k_0(\omega)$ of anthracene crystals is proportional to the change $\Delta n(\Omega)$ in the occupation numbers of phonons with frequency $\Omega = \omega_T - \omega$. Measuring the intensity of the probe pulse passing

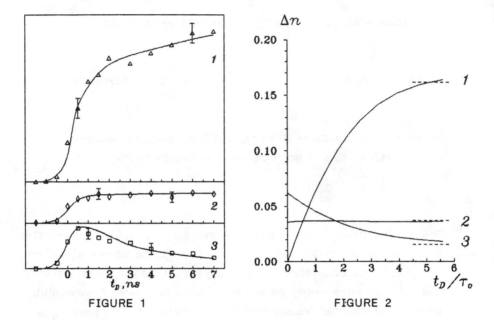

FIGURE 1   FIGURE 2

across the crystal plate at different times t_D for frequency ω, we obtain information on the kinetics of phonons with frequency $\Omega = \omega_T - \omega$.

Fig. 1 represents the kinetics of phonon occupation numbers for three frequencies Ω (1 – 14, 2 – 25, 3 – 32 cm^{-1}), obtained in the experiment described above. We see that pulsed laser pumping results in a generation of acoustic phonons in a broad frequency range, the principal excitation energy at $t_D = 0$ being concentrated in the high frequency region of the phonon spectrum. The occupation numbers for high phonon frequencies $\Omega > 25$ cm^{-1} decrease in the course of relaxation, while those for $\Omega < 25$ cm^{-1} increase.

We simulated numerically[3] the phonon kinetics for different frequencies. The corresponding kinetic equation, written in the model of one longitudinal and two degenerate transverse, dispersion-free, isotropic branches, was solved numerically on a uniformly spaced grid: the frequency interval from zero to the anthracene boundary acoustic frequency $\Omega_0 \approx 40$ cm^{-1} was partitioned into N levels, where $N = 128$. The matrix elements of the anharmonic three-phonon processes were assumed to be proportional to the product of the moduli of the phonon wave vectors. One of the variable parameters of the calculation was the lifetime of the boundary acoustic phonons, $\tau_0 = \tau_{decay}(\Omega_0)$. Comparing the calculations for various values of τ_0 with the experiment, we found, primarily on the basis of the behavior of phonons with $\Omega \approx 32$ cm^{-1}, the best agreement to be obtained at $\tau_0 \approx 1$ ns.

For the second adjustable parameter we chose the spectrum of acoustic phonons which were produced from the optical ones, i.e., the initial energy distribution of the acoustic phonons. Assuming it to have Gaussian shape, we changed the position of the maximum of this spectrum, Ω_{max}, and its half-width, $\Delta\Omega$. The total energy of the phonon system, which was assumed to be constant, corresponded to the finite temperature $T_c \approx 10.5$K, consistent with the experimental conditions. The kinetic curves at $\Omega_{max} \approx 0.7\Omega_0$ and $\Delta\Omega \lesssim \Omega_0/4$, yielding the best agreement with the experiment, are depicted in Fig. 2.

In contrast with the "generation" model[4] with the set of levels Ω_0, $\Omega_0/2$, $\Omega_0/4$, etc., a system of uniformly spaced in frequency phonon states makes it possible to follow the behavior of phonons with all intermediate frequencies and to quantitatively compare the calculations with experiment. The generation model, because of its crude frequency separation, puts the experimental kinetics dependences essentially in the range of a single phonon generation. The low-frequency phonon relaxation shows the strongest deviations from the results of the generation model, because in the model with a quasicontinuous phonon spectrum these phonons are produced in the first decay already, while in the generation model they appear only after several successive decays.

1) Maksimov, A.A., Tartakovskii, I.I., Pis'ma Zh. Tekh. Fiz. **12**, 112(1986) [Sov. Tech. Phys. Lett. **12**, 46(1986)]

2) Happek, U., Ayant, Y., Buisson, R., and Renk, K.F., Europhys. Lett. **3**, 1001(1987)

3) Kazakovtsev, D.V., Maksimov, A.A., Pronin, D.A., and Tartakovskii, I.I., Pis'ma Zh. Eksp. Teor. Fiz. **49**, 51(1989) [Sov. Phys. JETP Lett. **49**, 61(1989)]

4) Schaich, W.L., Solid State Commun. **49**, 55(1984)

LUMINESCENCE OF SEMICONDUCTORS IN THE PRESENCE OF NONEQUILIBRIUM PHONONS

A.V.Akimov and A.A.Kaplyanskii

A.F.Ioffe Physical-Technical Institute,
194021 Leningrad, USSR

Over the years luminescence has become a powerful technique for the detection of terahertz acoustic phonons in
crystals[1]. Used mainly in insulators the method is based
on the interaction of nonequilibrium phonons with electronic emitting states of "probe" impurities in photoexcited
crystals (such as ruby, doped fluorite, etc.). The technique has an advantage over the methods using superconducting
devices because of a much wider spectral range of phonon
frequencies and the lack of the boundary between crystal
and detector.

In this paper we present the results of our recent
studies of high-frequency phonons in semiconductor crystals
by measuring the effect of nonequilibrium phonons on the
edge luminescence in crystals at liquid helium temperatures.
In comparison with insulators, semiconductors have two specific features essential for luminescence detection of phonons, namely:

1. Existence in an excited semiconductor of "free" quasiparticles — free carries (FC) (electrons and holes) and
excitons (FE). Radiative transitions are due to different
types of electron — hole recombination (band-band, exciton,
band-impurity luminescence). Nonequilibrium phonons, injected into the crystal may change the concentration of FC and
FE, as well as the energy and momentum distribution of FC
and FE revealed in luminescence in different ways, the basic of these being: (i) the heating of FC and FE gas leading
to the broadening of their distribution function in the
energy band and of corresponding luminescence lines;

(ii) the phonon heating of FC often results in the quenching of band-band and band-impurity luminescence due to
competing processes of thermally activated FC capture by
nonradiative recombination centres; (iii) FC and FE drag by
directed phonon flux - a well known effect for electron-
hole droplets in Ge[2], which is accompanied by the spatial
shift of luminescence region in the sample.

2. Small binding energy of localized electronic states
in semiconductors - levels of shallow impurities and of bo-
und excitons (BE). Therefore THz acoustic phonons may induce ionization of shallow impurities and dissociation of
BE-FE. These processes lead to the increasing of FC or FE
luminescence and, correspondingly, to the quenching of luminescence lines due to localized states (donor - acceptor
recombination, BE luminescence, etc.).

The luminescence technique often posesses also frequency selectivity for phonons. Indeed, in the case of localized states phonon-induced ionization (dissociation) may
only take place if phonon energy is larger than threshold
being equal to the binding energy of shallow impurity or BE.
In the case of phonon interaction with free quasiparticles
(FC or FE), the phonon modes that appear active in the interaction, may be found from the well known energy and momentum conservation rules.

The exciton-phonon luminescence spectrum directly reflecting the FE energy distribution[3] was used to study spectral distribution of nonequilibrium phonons produced by relaxation of photoexcited carriers (phonon "hot spot") in
cuprous oxide crystals[4]. It was shown, that the low frequency part of phonon distribution is depleted in phonons,
in comparison with Planck equilibrium distribution. This result confirmed the assumption that the escape of relatively
low-frequency phonons does exist, while high-frequency ones
are trapped in the region of hot spot.

A study of the quenching of the conduction band - acceptor (carbon) emission in the spectrum of GaAs/AlGaAs
quantum wells induced by electron gas heating by nonequilibrium phonon pulses has revealed characteristic features in
the acoustic phonon interaction with 2D-electron gas in
quantum wells[5]. Comparison of the amplitude of phonon-induced "negative" luminescence pulses from 2D and bulk (3D)
electron gas revealed stronger heating of electrons by nonequilibrium phonons in quantum wells as compared with their

1244

heating in the bulk GaAs. The main conclusion from the experiment is that while for 3D gas phonons actually involved in electron-phonon interaction have threshold frequencies <0.1 THz, for 2D electrons in quantum well the value of permissible phonon frequency is increased to 0.3 THz.

The phonon heat pulses injected in Si sample affected strongly the intensity of free (FE) and bound (BE) exciton lines in the edge luminescence excited near the opposite side of the sample[6]. Two oppositely acting factors were included in the analysis of the effect of heat pulses on FE and BE luminescence: (i) exciton drag towards the surface by the flux of "low frequency" (<0.4 THz) phonons resulting in a FE luminescence quenching due to fast surface recombination of FE; (ii) bound exciton dissociation BE→FE, induced by "high-frequency"(>0.9 THz) phonons, enhanced the FE luminescence. In these experiments ballistic component has first been detected in the propagation of phonons with frequency as high as 1 THz in Si; strong reflection of phonons has been observed to occur from as-cleaved surface of Si in contact with superfluid He; frequency down-conversion for phonons has been observed in a-Ge films, evaporated on Si.

A study of the effect of heat pulses on the microsecond-range kinetics of the hole-neutral donor luminescence in epitaxial n-GaAs has shown this long kinetics to be due to the slow spontaneous influx of the free holes from some metastable localized states[7].

These experiments demonstrate some new possibilities of the use of nonequilibrium phonons in luminescence studies of semiconductors, intended to provide essential information on high-frequency acoustic phonons, electronic free and localized states in semiconductors, as well as on the coupling between the phonon and electron subsystems.

REFERENCES

1. Bron W.E. in "Nonequilibrium Phonons in Nonmetalic Crystals", ed. by W.Eisenmenger and A.A.Kaplyanskii. North--Holland, Amsterdam 227-275 (1986)
2. Keldysh L.V. and Sibeldin N.N., Ibid. 455-686.
3. Ivchenko E.L. and Takunov L.V., Fiz.Tverd.Tela $\underline{30}$ 1161 (1988) (in Russian)
4. Akimov A.V., Kaplyanskii A.A. and Moskalenko E.S., Sov. Phys.Solid State $\underline{29}$ 288 (1987)
5. Akimov A.V., Kaplyanskii A.A., Kopyev P.S., Kozub V.I. and Meltzer B.Y., Sov.Phys.Solid State $\underline{29}$ 1058 (1987)
6. Akimov A.V., Kaplyanskii A.A. Moskalenko E.S. and R.A.Titov JETP $\underline{94}$(11) (1988)
7. Akimov A.V., Kaplyanskii A.A., Krivolapchuk V.V. and Moskalenko E.S.,JETP Lett. $\underline{46}$ 42 (1987)

NUMERICAL SIMULATIONS AND EXPERIMENTAL DATA ON TRANSPORT OF 0.7 THZ ACOUSTIC PHONONS IN GALLIUM ARSENIDE

M. Fieseler[1], M. Schreiber[1], R.G. Ulbrich[2], M. Wenderoth[2], R. Wichard[1]

[1]*Inst. f. Phys., Univ. Dortmund,* [2]*IV. Phys. Inst., Univ. Göttingen, F. R. Germany*

The transport properties of acoustic phonons in GaAs are investigated by time-of-flight spectroscopy and by velocity-resolved focusing-technique. Time-of-flight signals and focusing patterns are explained by a Monte-Carlo calculation including isotope scattering.

At low temperature and in samples of sufficient perfection the transport of high-frequency acoustic phonons in GaAs is presumably limited by isotope scattering. Theoretical calculations [1] predict a mean free phonon path $\Lambda \approx 1.8$ mm at 0.7 THz. We present experiments and numerical investigations for sample lengths d comparable to the mean free path of the phonons, so that ballistic transport as well as isotope scattering are important.

The samples are $\langle 100 \rangle$-orientated slabs of undoped GaAs. A superconducting Pb tunnel junction is evaporated on one surface of the sample. The sharp onset of the detector at 0.65 THz enables us to separate dispersive phonons from the low frequency background. A pulsed 10 keV e-beam is focused on the opposite surface of the sample to generate non-equilibrium phonons. Time-of-flight signals and focusing structures can be measured without removing the sample. In the first case the detector signal is recorded at a fixed exitation position of the e-beam. In the other case the e-beam is scanned across the surface. A computer-controlled boxcar integrator selects the energy flux belonging to a constant group velocity. Fig. 1 shows time-of-flight signals in the main symmetry directions of a 1.5 mm thick sample. While the leading edges of the signals are due to ballistic phonons it has been a point of controversy whether the trailing edges are caused by scattering or dispersion. The corresponding focusing patterns in fig. 2 show sharp structures which are typical for ballistic propagation. In Fig. 2 a) the fast TA-velocity was selected. The

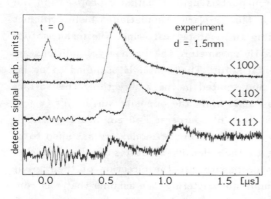

Fig. 1: *Experimental time-of-flight signal*

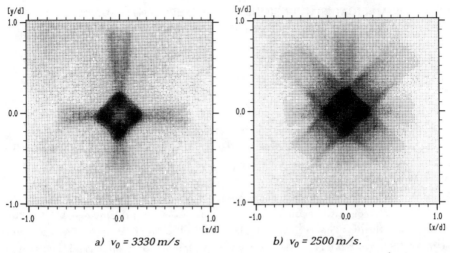

a) $v_0 = 3330\,m/s$ b) $v_0 = 2500\,m/s$.

Fig. 2: Experimental focusing patterns at constant group velocity

width of the v-shaped FTA ridge is used to determine the frequency of the detected phonons. Comparing the data with our numerical evaluations [2] of the adiabatic bond charge model [3] (which is known to be a reliable model for lattice dynamics in GaAs) we conclude that the detected phonons are within a narrow frequency range around 0.7 THz. The characteristic FTA structure does not broaden when the velocity is reduced (Fig. 2 b). Instead it remains in the focusing pattern although the corresponding phonons do not fit into the selected velocity interval. This behaviour can be explained by a "glowing" source with a time constant of 64 ns. But the long tails of the time-of-flight signals have to be explained mainly by scattering.

In the investigated regime transport is neither ballistic propagation nor diffusion. We therefore present a Monte-Carlo simulation which combines the previous calculations of focusing and isotope scattering. The mean isotope scattering time τ is regarded as a fit parameter. The boundary conditions are chosen according to the experimental setup. The crystal is assumed to be infinite in the x- and y- direction and limited in the z-direction. The surface $\vec{r} = (x,y,d)$ is regarded as a phonon sink while the opposite surface $\vec{r} = (x,y,0)$ scatters back all incident phonons. A point source at $\vec{r} = 0$ emits a large number $(>10^6)$ of phonons into the crystal. The acoustic modes are assumed to be occupied uniformly corresponding to their density of states in the frequency range 0.65 - 0.75 THz. The phonons propagate into the direction of their group velocity. On their path they may undergo scattering. We assume that τ is constant in the small frequency range and does not depend on the wavevector $\vec{q}$.

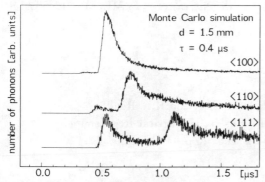

Fig. 3: Numerical time-of-flight curves

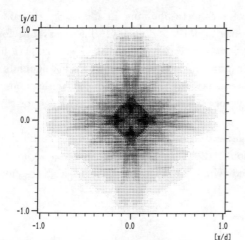

Fig. 4: Theoretical focusing pattern

For the scattering process the product of the polarization vectors is taken into account. Thus the phonons will be preferentially scattered into their old propagation direction or backward if the mode remains unchanged. Otherwise scattering will preferentially occur under large angles. Any phonon that reaches the surface at z = d is detected. Fig. 3 shows the resulting time-of-flight curves. In the $\langle 100 \rangle$-direction the shape of the curve is mainly determined by the temporal behaviour of the source whereas in the non-focusing directions it depends on the mean scattering time τ. We obtained best agreement with the experimental data using $\tau = 0.4 \pm 0.1 \, \mu s$. The corresponding time-integrated focusing pattern is shown in Fig. 4. It is significant that the structures remain comparatively sharp in spite of the scattering.

Our numerical simulation yields time-of-flight curves in the main symmetry directions as well as focusing patterns for different sample lengths, which are in excellent agreement with experiment. The obtained scattering time $\tau = 0.4 \pm 0.1 \, \mu s$ agrees with Tamura's calculations [1]. We conclude that the experimental features can be explained mainly by isotope scattering, while dispersive effects are found to be weak.

[1] Tamura, S., Phys. Rev. B27, 858 (1983)
[2] Schreiber, M. , Fieseler, M. , Mazur, A. , Pollmann, J. , Stock, B. and Ulbrich, R. G., in "Proceedings of the 18th International Conference on Physics of Semiconductors", edited by Engström, O. (Singapore, World Scientific 1987), p. 1373, and to be published
[3] Rustagi, K.C. and Weber, W., Sol. Stat. Com. 18, 673 (1976)

DECAY OF TERAHERTZ-PHONONS IN $Ca_{1-x}Sr_xF_2$ MIXED-CRYSTALS

U. Happek, W.W. Fischer, and K.F. Renk
Institut für Angewandte Physik, Universität Regensburg,
8400 Regensburg, Fed. Rep. of Germany

We have studied the decay of terahertz-phonons in $Ca_{1-x}Sr_xF_2$ mixed-crystals with Sr concentrations between x=0.01 and x=0.4. These crystals form homogeneous solid solutions, i.e. Sr ions randomely substitute Ca ions. We find that the decay rate varies as $\lambda\nu^5$ as in pure fluoride crystals but that λ depends on the concentration x. We attribute this behaviour to defect-induced anharmonicity. An analysis leads to an anharmonicity per Sr ion which is extremely large for small x bur decreases strongly for large x.

High frequency phonons were generated in the sample volumes via multiphonon absorption of CO_2 laser radiation at 28 THz[1]. By multiphonon absorption mainly optical phonons are generated which decay very fast into high-frequency acoustic phonons.

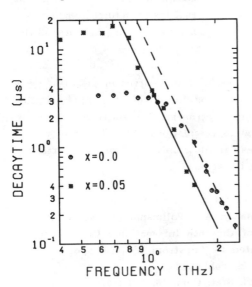

Fig. 1: Phonon decay time for pure and Sr-doped CaF_2.

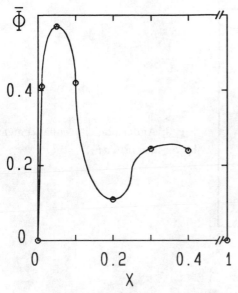

Fig. 2: Average anharmonicity change in $Ca_{1-x}Sr_xF_2$ mixed-crystals.

The crystals (typical size 5x5x5mm^3) were immersed in liquid helium and illuminated with pulsed radiation from a Q-switched CO_2 laser (pulse duration 100 ns, energy 10^{-4} J). For phonon detection we used the vibronic sideband spectrometer[2], for this purpose the crystals are weakly doped with Eu^{2+} ions (0.01 mol%). We measured phonon induced signals in a frequency range between 0.3 THz and 2.0 THz.

Fig. 1 shows the phonon decay time for a mixed-crystal with x=0.05 and a pure CaF_2 crystal. The decay time for the mixed crystal is by a factor of two shorter than for CaF_2, i.e. the Sr ions induce additional anharmonicity in the crystal lattice. For samples with different Sr concentration we found a ν^{-5} dependence of the decay time at high phonon frequencies that is typical for spontaneous phonon decay, where a phonon splits due to lattice anharmonicity into two phonons of lower frequency. We assume that the phonon decay is determined mainly by the decay of LA phonons and that the population of TA phonons and LA phonons are in equilibrium due to fast mode conversion by elastic scattering and describe the phonons in an isotropic model of a dispersionless solid. Then, the anharmonic decay time τ_a is given by[3] $\tau_a^{-1} = \lambda_a \cdot \nu^5$ with $\lambda_a = C \cdot \Phi^2 \cdot \rho^{-1} \cdot c^{-5} \cdot (1 + D_{TA}/D_{LA})^{-1}$, where Φ is an anharmonicity parameter, ρ the mass density, c the velocity of longitudinal acoustic phonons; D_{LA} and D_{TA} are the phonon density of states and C is a constant. For a discussion of the mixed crystals we interpolate ρ, c, D_{TA} and D_{LA} linearely with x from the values of CaF_2 and SrF_2. In the same way we determine an average anharmonicity parameter Φ_m from the experimental anharmonicity parameters for CaF_2 and SrF_2. On the other hand, we obtain an experimental anharmonicity parameter $\Phi(x)$ from the experimental $\lambda_a(x)$. To compare these values we have

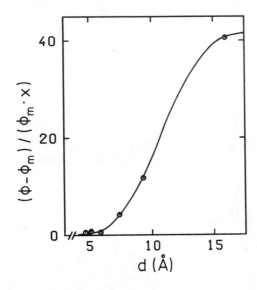

Fig. 3: Anharmonicity enhancement by Sr ions in CaF_2.

plotted in fig.2 the deviation of Φ from Φ_m, $\bar{\Phi} = \Phi/\Phi_m - 1$. We find a strong increase of the anharmonicity for small concentrations of Sr, showing a maximum around x=0.05, with the measured anharmonicity being about 1.6 times the interpolated anharmonicity Φ_m. For higher concentrations of Sr the anharmonicity decreases, reaching a value comparable to that of pure CaF_2 at x=0.2, and increases again for still higher Sr concentrations.

Fig. 3 shows the average parameter of anharmonicity per unit of concentration, $(\Phi/\Phi_m - 1)/x$ (points and guessed slope) as a function of distance d between Sr ions. There is a strong increase of the anharmonicity parameter with d: in the dilute case the anharmonicity for a Sr defect ion is nearly two orders of magnitude larger than for the substituted Ca ion. In the concentrated case (small d's) overlapping of disturbed regions seems to reduce the defect-induced anharmonicity almost to zero. We suggest that large elongations of the Sr ions relative to the neighbouring Ca ions are responsible for the strong anharmonicity enhancement in the dilute case.

In summary, the anharmonic decay time for high frequency acoustic phonons in $Ca_{1-x}Sr_xF_2$ mixed crystals has been measured for the first time. As in pure fluorid crystals, the phonon decay time shows a ν^{-5} dependence. The additional anharmonicity per defect-ion increases strongly with the distance of the defects and is for diluted mixtures almost two orders of magnitude larger than for the unperturbed crystal.

1) Happek,U., Ayant,Y., Buisson,R. and Renk,K.F.,Europhys. Lett.,3, 1001 (1987)
2) Bron,W.E. and Grill,W., Phys. Rev. B 21, 5303, 5313 (1977)
3) Baumgartner,R., Engelhardt,M. and Renk,K.F. Phys. Rev. Lett., 47, 1403 (1985)

POLARITON DEPHASING

T. Juhasz and W. E. Bron

Department of Physics, University of California, Irvine, CA 92717

Polariton dephasing rates in GaP have been directly measured, for the first time, in the time domain. The present experimental techniques differ from those previously used[1]. We use subpicosecond coherent Raman excitation (CRE)[2] to generate coherent polariton distributions and use subpicosecond time resolved coherent anti-Stokes Raman scattering (TRCARS)[2] to investigate the dephasing process. The laser system[3] contains synchronously pumped tunable picosecond and linear cavity femtosecond lasers. Typically the central polariton wavevector $q \sim 2 \times 10^3$ cm^{-1}, corresponds to a spectral distribution peaked at $\omega_\pi \sim 354$ cm^{-1}. However, a 120 mm focal length focusing lens focuses the laser beams into the sample, which produces a spread in the wavevector of the laser beam and, hence, in a spread in polariton wavevector of between $\sim 1.4 \times 10^3$ to $\sim 2.6 \times 10^3$ cm^{-1}. The corresponding range of ω_π is from ~ 330 cm^{-1} to ~ 360 cm^{-1}.

In order to avoid the generation of LO phonons in addition to the polaritons, the bandwidth of the femtosecond laser was limited to ~ 45 cm^{-1} which corresponds to a pulse duration of ~ 300 fs.

Figure 1a illustrates the typical temporal dependence of the TRCARS signal amplitude as a function of probe decay, Δt, and at an ambient temperature of 5K. The TRCARS signal shows three different components. The first component from ~ 0 ps $< \Delta t < \sim 2.5$ ps is the response of the bound electrons in GaP through the third order nonlinear electronic susceptibility [3]. The second component of the signal ($\sim$ 1.5 ps < Δt < $\sim$ 5ps), which occurs after polariton excitation has ceased indicates a curved structure implying that the dephasing does not correspond to a single decay constant. The third component ($\sim$ 5 ps < Δt < 12.5 ps) illustrates a clearly exponential decay corresponding to a polariton dephasing time $T_2/2$ = 1.72 $\pm$ or $\mp$ 0.16 ps. Subtracting this third component from the total signal,

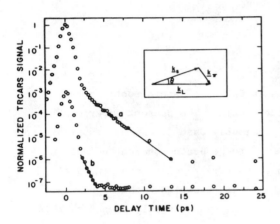

Fig. 1. a) Normalized TRCARS signal intensity as a function of the delay of the probe pulse. Insert: Wavevector conservation between the two pump laser $\vec{k}_\ell$, $\vec{k}_s$ and that for the polariton, $\vec{k}_\pi$.

uncovers a second exponential decay (see Fig. 1b) with a dephasing time $T_2/2$ = 620 $\pm$ 60 fs. The temperature dependence of $T_2/2$ and that of the polariton dephasing rate (in cm^{-1}) $\Gamma_\pi[= (2\pi c T_2/2)^{-1}]$ is indicated in Fig. 2.

We now set out to prove that in the present case there are two possible decay channels, namely,
$\pi(w_\pi, q\sim 0) \overset{\leftarrow}{\rightarrow} LA(w_\pi/2, q') + LA(w_\pi/2, -q')$, the "half-energy" channel, and $\pi(w_\pi, q\sim 0; \overset{\rightarrow}{\leftarrow} LA(\sim X) + TA(\sim X)$, the "X-point" channel, π refers to polaritons, and X refers to the X-point of the Brillouin zone.

For three-particle interactions, at temperature T, the polariton dephasing rate Γ, at frequency w_π, becomes $\Gamma(w_\pi, T) = \Gamma_0(w_\pi, T)$ $\times \left\{ n_1(w_1, T) + n_2(w_2, T) \right\}$, in which w_i and n_i represent, respectively, the frequency and thermal occupation numbers of the two phonons into which the polariton decays.

We have performed two-parameter, least χ^2 fits of the data of

Fig. 2 to the equation for $\Gamma(\omega_\pi, T)$. For the longer $T_2/2$ (Fig. 2a) the best fit is $\omega_1 = (118 \pm 16)$ cm^{-1} and $\omega_2 = (236 \pm 16)$ cm^{-1}. This

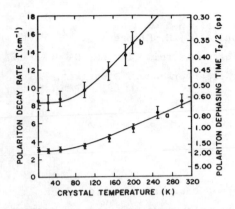

Fig. 2. Temperature dependence of the polariton dephasing rate $\Gamma(\omega_\pi)$, and of the dephasing time $(T_2/1)$. a) "X-point" decay channel, b) "half-energy" decay channel.

result, when compared to the phonon frequencies at the X-point as obtained from neutron scattering $\{\omega_{TA(X)} = (107 \pm 3)$ cm^{-1} and $\omega_{LA(X)} = (249 \pm 4)$ cm$^{-1}\}$ identifies this as the "X-point" decay channel. For the shorter $T_2/2$ (Fig. 3b) a best fit is obtained with $\omega_1 = (172 \pm 26)$ cm^{-1} and $\omega_2 = (172 \pm 26)$ cm^{-1}. This identifies the "half-energy" decay channel. The dispersion relations for GaP permit only excitation of LA phonons for these values of $\omega_\pi/2$. A detailed analysis of these results and then comparison to other work in this area will be presented elsewhere.

ACKNOWLEDGEMENT

W. E. B. acknowledges funding through NSF Grant No. DMR 86-03888. We also acknowledge helpful discussions with S. Mehta, A. Mayer and the technical assistance of K. Harris

REFERENCES

1. S. Ushioda, J. D. Mullen, and M. J. Delaney, Phys. Rev. B8, 4634 (1973).

2. W. E. Bron, J. Kuhl and B. K. Rhee, Phys. Rev. B34, 6961 (1986).

3. T. Juhasz, G. Smith, S. Mehta, K. Harris and W. E. Bron, to be published in the IEEE J. Quantum Electr. The femtosecond laser has a limiting temporal resolution for ~ 70 fs.

OPTICAL PHONONS-ONE-COMPONENT PLASMA INTERACTION

W. E. Bron[1], S. Mehta[1], J. Kuhl[1,2], M. Klingenstein[2]

(1) Dept. of Physics, University of California, Irvine, CA 92717 USA
(2) Max Planck Institute für Festkörperforschung, Stuttgart, FRG

The effect of a strongly damped, thermally excited, one-component plasma on near-zone-center, longitudinal, optical phonons in GaP has been studied in detail[1]. We show for n-doped GaP that the temperature dependence of the Raman scattering bandwidths, peak frequencies and other parameters of a coupled phonon-plasma system can be accounted for in terms of simple dielectric response theory without the introduction of fitting parameters. Specifically, we have investigated in some detail, the interaction of LO phonons in GaP with a thermally activated (electron) plasma. Samples of various nominal concentrations were held at various ambient temperatures. Figure 1a to 1d indicate a series of Raman scattering lines from the sample nominally doped with 1.7×10^{17} cm^{-3} of Silicon. Similar results are obtained for all other samples and temperatures.

We treat the coupled phonon-plasma system in terms of a dielectric response function, $\epsilon(\omega)$, given by

$$\epsilon(\omega) = \epsilon_\infty + \frac{\Omega^2}{\omega_T^2 - \omega^2 - i\omega\Gamma} - \frac{\omega_p^2}{\omega(\omega + i\gamma)}. \tag{1}$$

In Eq. (1) $\Omega^2 = \epsilon_\infty\left(\omega_L^2 - \omega_T^2\right)$, ω_L is the LO phonon frequency, ω_T is the transverse optical (TO) phonon frequency, Γ is the damping rate of the LO phonons in the absence of the plasma, γ is the damping rate of the plasma in the absence of the phonons, $\omega_p^2 = \epsilon_\infty\tilde{\omega}_p^2$, ϵ_∞ is the static dielectric constant, and $\tilde{\omega}_p$ is the plasma frequency in the

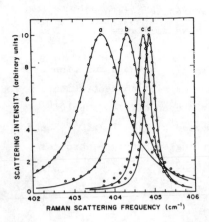

SCATTERING INTENSITY (arbitrary units)

RAMAN SCATTERING FREQUENCY (cm⁻¹)

Fig. 1. Observed incoherent Raman scattering lines from a sample held at an ambient temperature of 2) 250K, b) 175K, c) 100K, d) 50K. The solid lines are best fit Lorentzian line profiles. (See also reference 1.)

presence of the ion core background.

The real part of $\epsilon(\omega)$, set equal to zero, leads to the eigenfrequencies, $\omega_{\pm}$, of the coupled modes and the imaginary part of $(1/\epsilon(\omega))$ leads to a spectral function $S(\omega)$. Thus,

$$\omega_{\pm}^2 = \frac{1}{2}\left(\omega_L^2 + \tilde{\omega}_p^2 - \gamma^2\right) \pm \frac{1}{2}\Big[\left(\omega_L^2 + \tilde{\omega}_p^2 - \gamma^2\right)^2 \qquad (2)$$

$$+ 4\left(\omega_L^2\gamma^2 + \tilde{\omega}_p^2\gamma\Gamma - \tilde{\omega}_p^2\omega_T^2\right)\Big]^{1/2}$$

and

$$S(\omega) = \frac{\Omega^2\left(\omega_L^2-\gamma^2\right)}{\epsilon_\infty^2\left(\omega_L^2-\gamma^2-\tilde{\omega}_p^2\right)}\frac{1}{4\omega_+}\Big[\frac{\Gamma_{eff}}{\left(\omega_+-\omega\right)^2 + \left(\Gamma_{eff}/2\right)^2}. \qquad (3)$$

Here

$$\Gamma_{eff} = \Gamma + \Big[\frac{\tilde{\omega}_p^2\Omega^2\gamma}{\left(\omega_L^2+\gamma^2 - \tilde{\omega}_p^2\right)\omega_L^2\epsilon_\infty}\Big] \qquad (4)$$

where Γ_{eff} is the effective damping rate of the LO phonon coupled mode in the presence of the plasma.

Thus, there are two equations (2; ω_+ only) and (4), for the two unknown quantities $\tilde{\omega}_p$ and γ. Γ is obtained from the previously reported TRCARS measurements [2]. Exact solutions of Eqs. (2) and (4) for $\tilde{\omega}_p$ and γ, as a function of temperature, have been obtained.

Independent measurements of the Hall coefficient make it

possible to convert from ambient temperatures to carrier concentration n_c. Typical results are shown in Figure 2.

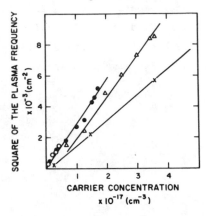

Fig. 2. $\tilde{\omega}_p$ as a function of the carrier concentration in cm^{-3} open circles 3.6×10^{16}, closed circles 1.7×10^{17}, open triangles 3.6×10^{17}, crosses 1.8×10^{18}.

The dependence of $\tilde{\omega}_p$ on n_c is used to obtain the carrier effective mass ratio, m^*/m_e, through the well-known relation $\tilde{\omega}_p^2 = n_c e^2/(\epsilon_o \epsilon_\infty m^*)$. The results for the lesser doped samples vary from 0.35 to 0.37, but ≈ 0.56 for the more highly doped sample. This result must be compared to $m^*/m_e \sim 0.3$ obtained from the literature.

It is clear that the dielectric response function given by Eq. (1) adequately accounts for the interaction between the damped LO phonon modes and strongly damped one-component plasma modes. More-over, various useful parameters such as the effective phonon damping rate, the plasma damping rate, the plasma frequency, the effective electronic mass, and the dependence of these parameters on temperature and carrier concentration, are readily recovered.

W.E.B. acknowledges funding through the National Science Foundation grant No. DMR-86 03888 and W.E.B. and J. K. acknowledge support through NATO grant No. 0034188.

[1] W. E. Bron, S. Mehta, J. Kuhl and M. Klingenstein, Phys. Rev. 39, 12642 (1989).

[2] W. E. Bron, J. Kuhl and B. K. Rhee, Phys. Rev. 34, 6962 (1986).

PECULARITIES OF PROPAGATION OF NONEQUILIBRIUM PHONONS
IN SINGLE CRYSTALS WITH ANISOTROPIC IMPURITIES

I. Sildos, I. Dolindo and G. Zavt

Institute of Physics, Estonian Acad. Sci., 202400 Tartu,
Estonian SSR, USSR

1. Introduction

In alkali halide crystals doped with molecular ions O_2^-, S_2^-, OH^-, NO_2^-, the librational-rotational motions and reorientation mechanisms of impurities have been well studied [1-4]. In the present paper, the possibilities of controlling the propagation of nonequilibrium phonons in the impurity crystal $KCl-NO_2$ have been studied at HeT by applying the uniaxial stress which effectively hinders rotational-reorientational motions strongly interacting with phonons [4,5]. Two original methods of optical detection of nonequilibrium phonons in impurity crystals have been developed.

2. Zero-phonon line (ZPL) of an impurity as a phonon detector

This method is easily applicable for impurity crystals with strong electron-phonon interaction that leads to sharp temperature dependence of ZPL intensity of the impurity. An absorption version of ZPL phonon detector has been developed, where the information about the phonons in crystal is obtained from the impurity absorption spectra [5]. The detection method was tested on $KCl-NO_2$ crystal at 5 K. Phonons were generated as a result of a rapid nonradiative relaxation following the laser-pulse excitation of NO_2^- impurities. Phonon-induced changes in the impurity absorption spectrum were recorded with a variable time delay by means of a probe pulse (Fig. 1,a). The probing light was detected by a linear CCD array that allows one to observe with high time resolution the spatial phonon propagation in the crystal (Fig. 1,b,c,d).

The spatial spreading of the nonequilibrium phonon area in the crystal in the absence of uniaxial stress can be described by phonon diffusion accompanied by anharmonic decay. The mean free path of phonons is limited by strong resonant scattering on the rotational-reorientational motions of impurities [5]. The relaxation of phonons proceeds in two stages (analogously to [6]): at the first stage the phonon distribution shifts to the low-frequency side. At the second stage the distribution is stabilized and the process as a whole can be described by the diffusion (dotted lines in Fig. 1,b).

On applying uniaxial stress the hindering of reorientations of NO_2^- impurities takes place. At sufficiently high stress (~ 6 kG/mm^2) the impuri-

ties line up on the (110) plane and rotation motions turn into libra-
tions [4]. The resonant scattering processes of phonons are suppressed
and acceleration of diffusion takes place. An essential portion of heat
is lead out of the hot track, probably, by subthermal phonons [7]. In
Fig. 1,c,d, the cooling of the hot track with a rate correlating with

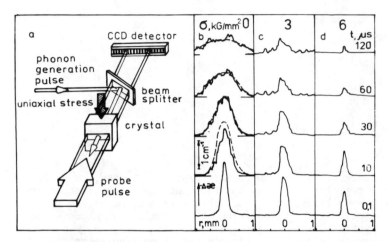

Fig. 1. a – the experimental scheme for spatial detection of phonon propagation in the impurity crystal; b,c,d – spreading of nonequilibrium
phonon area in $KCl-NO_2^-$ crystals at 5 K and the influence of uniaxial
stress on phonon propagation (dotted lines represent the approximation
by diffusion model).

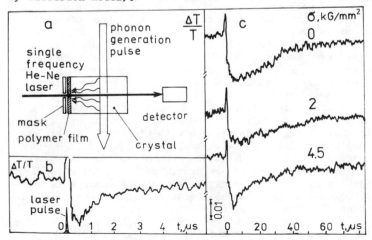

Fig. 2. a – the experimental scheme for phonon detection by chlorine-
doped spectral-hole-burned polymer film; b,c – time dependence of SHB
film transmission after the laser excitation generating phonons in
$KCl-NO_2^-$ crystal (distance between the film and the excitation beam ∿0.3
mm) and the influence of uniaxial stress.

the decay time of the long-lived component of NO_2^- triplet state [8] is depicted. It means that uniaxial stress has hindered the fast spin-rotational relaxation channel (4 µs) between the sublevels of the triplet state and now, due to radiationless transitions from the triplet state, NO_2^- impurities work as long-lived phonon generators.

3. Spectral hole-burned (SHB) film as a phonon detector

Perylene-doped SHB-prepared polymer films have been successfully used for phonon detection in sapphire [9], where luminescence modulations due to phonon-induced hole filling were recorded. We have modified this method, observing the phonon-induced modulations of optical density of SHB film (Fig. 2,a). For this purpose a thin layer (∿20 µm) of low-temperature Varnish glue doped with chlorine molecules (∿5 × 10^{-3} M) was applied to the surface of KCl-NO_2^- crystal.

At 5 K by means of single-frequency He-Ne laser (0.1 mW) a hole was burned into the films (for SHB specifications, see [10]). To detect the phonons the laser beam was attenuated by 10^2 times and, synchronously with phonon-generating laser pulses, the modulations of film transmission were detected (Fig. 2,b,c). It should be mentioned that the sharp peak at the moment of the laser pulse is caused by the scattered light occurring in the detector. One can see that the phonons propagating up to the film cause the decrease of film transmission, i.e. a temporary filling of the hole takes place. The modulations of transmission consist in fast (Fig. 2,b) and rather slow (Fig. 2,c) processes. Indeed, according to [5], fast quasiballistic phonons are followed by diffusely propagating ones. The time resolution of such phonon detector is determined mainly by the film thickness [11] to which in our case contributes also the nonideal (cylindrical) geometry of the phonon source. Both factors together determine the half-width (1 µs) of the fast response (Fig. 2,b). We have found that the fully reversible filling of spectral holes in the film occurs by heating up to ∿8 K. On applying to the crystal a uniaxial stress perpendicular to the laser beams the acceleration of diffusion processes takes place (Fig. 2,c), which is confirmed by the inferences of Section 2.

In conclusion, the application of uniaxial stress to crystals with anisotropic impurities (defects) leads to the hindering of low-energy rotational-reorientational motions, through which it is possible to control the modes of nonequilibrium phonon propagation in the crystal.

The authors are indebted to I. Renge for preparing polymer films.

1. Narayanamurti V. and Pohl R.O., Rev. of Modern Phys. 42, 201 (1970).
2. Hartel H. and Lüty F., Phys. Stat. Sol. 12, 347 (1965).
3. Treshchalov A. and Rebane L., Sov. Phys. Solid State, 20,272 (1978).
4. Dolindo I., Sild O. and Sildos I., Eesti NSV Tead. Akad. Toim. Füüs. Matem. (Proc. Estonian SSR Acad. Sci.) 36, 364 (1987).
5. Zavt G., Sildos I. and Dolindo I., Sov. Phys. Sol. State, 26, 864 (1984).
6. Baumgartner R., Engelhart M. and Renk R., Phys. Lett. 94A, 55 (1983).
7. Levinson Y., Sov. Phys. JETP, 79, 1394 (1980).
8. Sildos I., Rebane L. and Peet V. J. of Molec. Struct. 61, 67 (1980).
9. Beck K., Roska G., Bogner U. and Maier M. Sol. State Comm. 57, 703 (1986).
10. Burkhalter F., Suter G., Wild U., Samoilenko V., Rasumova V. and Personov R., Chem. Phys. Lett. 94, 483 (1983).
11. Beck K., Bogner U. and Maier M., Solid State Comm. 69, 73 (1989).

EXCITON DRAG BY HEAT PHONON PULSES IN SILICON

A.V.Akimov, A.A.Kaplyanskii and E.S.Moskalenko

A.F.Ioffe Physical-Technical Institute,
194021 Leningrad, USSR

Pure silicon is widely used in studies of nonequilibrium phonons in crystals. The present paper is concerned with the study of exciton drag by nonequilibrium acoustic phonon flux first observebly in Si. The phenomenon is used as a basis of a new experimental technique for the study of various properties of THz acoustic phonons in pure Si (see also[1]).

In typical experiments the excitonic cloud (EC) is formed near the crystal surface by cw-Ar-laser interband excitation, the phonon pulses being injected into the sample from metal film heater "h", deposited on the opposite face of the crystal, which is immersed in superfluid He (see Fig.1, inset). The effect of heat pulses on EC is studied by measuring the phonon-induced changes $\Delta I(t)$ in the intensity of lines of free excitons (FE) in luminescence spectrum of EC.

There are two principal processes responsible for phonon-induced changes of EC luminescence: (i) FE drag towards the surface by the flux of "low-frequency" phonons ($\omega < 0.4$ THz) which are active in FE - phonon interaction. The process results in FE luminescence quenching due to fast surface recombination of FE in Si. (ii) Dissociation of bound excitons BE→FE induced by "high-frequency" phonons with $\omega > \Delta E$, where $\Delta E = 0.9$ THz is the binding energy of BE. This obviously leads to the enhancement of FE-luminescence (and to the quenching of BE-luminescence).

Thus the above two contributions to the signal $\Delta I(t)$ have opposite signs and the resulting signal strongly de-

Fig.1 Phonon induced relative differential FE luminescence pulses for EC on freshly-cleaved (a) and oxidized (b) Si surface; lines-$\bar{q}\,\|<111>$, dots-$\bar{q}\,\|<100>$.

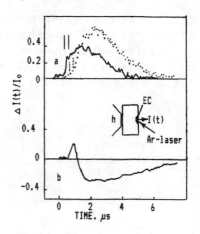

pends on experimental conditions. It was shown[1] that boundary conditions near the surface with EC are of great importance. In the case of surface-vacuum or surface-He bubble interfaces the observed phonon-induced signal $\Delta I(t)$ appears to be always positive, thus indicating the predominant role of BE→FE processes, the contribution of FE drag effect being negligible. The latter is attributed to the total reflection of phonons from interfaces. Indeed the reflected phonon flux compensates the incident one, as a result FE drag towards the surface disappears. Presented below are several qualitative results obtained with above technique.

1. Strong Reflection of Phonons from Freshly Cleaved Si Surface. The luminescence pulses $\Delta I(t)$ measured in the case of atomic pure (111) surface, cleaved in situ in liquid He are always positive (Fig.1a). This is due to the strong phonon reflection from perfect surface-liquid He boundary[2], which compensates the net phonon flux in EC region. Oxidation of previously cleaved (111) surface results in the reversal of the sign $\Delta I(t)$ which becomes negative indicating a strong FE drag contribution (Fig 1b). This is attributed to the well-known strong transmission of low-frequency phonons across oxidized Si surface-liquid He boundary (Kapitza anomaly) resulting in the occurrence of a directed phonon flux and FE drag in the EC region.

2. Ballistics and Bulk Scattering of 0.9 THz Acoustic Phonons in Si. Position of sharp leading edges of positive luminescence pulses (Fig.1) coincide with ballistic arrival

Fig.2 Phonon induced relative
differential FE luminescence
pulses for phonons generated
in the presence of a-Ge film
(1) and without it (2).

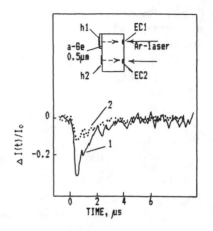

times of LA and TA phonons, which points to a ballistic propagation in pure silicon of a fairly large amount of high-
-frequency phonons with $\omega \geqslant 0.9$ THz involved in BE→FE dissociation. This is confirmed by the observation of phonon focusing for these phonons propagating in <100> direction
which reveals in the strong enhancement of corresponding
luminescence signal (compare the lines and dots in Fig.1a).
Experiments performed on samples with "ordinary" oxidized
surfaces in "back" geometry[1] have also demonstrated occurrence of strong bulk scattering of phonons $\omega >0.9$ THz in
pure Si. The theoretical estimation[3] of the mean free path
for 1 THz phonons for isotopic scattering in natural Si
gives l=4mm, which is in general accordance with the above
experimental results.

3. Phonon Frequency-Down Conversion in a-Ge Films.
Fig.2 shows the FE luminescence pulses from the heater (h1),
deposited on a-Ge film, evaporated on part of oxidized Si
surface and from similar heater (h2) on the same Si surface
without amorphous film. We attribute enhancement of negative signal amplitude in first case to phonon frequency
down-conversion of phonons in a-Ge, which leads to the increase of the low frequency phonon flux in the sample and
the FE drag.

REFERENCES

1. Akimov A.V., Kaplyanskii A.A., Moskalenko E.S. and
   Titov R.A. JETP, 94(11) (1988)
2. Weber J. , Sandman W. , Dietshe W. and Kinder H.,
   Phys.Rev.Lett., 40 1469 (1978)
3. Kazakovtsev D.V. and Levinson I.B., Phys.St.Sol.b.
   136 425 (1986)

NEW EFFECTS IN BOUND EXCITON RECOMBINATION
INDUCED BY NONEQUILIBRIUM ACOUSTIC PHONONS

B.L.Gelmont, N.N.Zinov´ev, D.I.Kovalev,

V.A.Kharchenko, I.D.Yaroshetskii and I.N.Yassievich

A.F.Ioffe Physico-Technical Institute,
Academy of Sciences of the USSR, 194021 Leningrad, USSR

Nonradiative Recombination

According to the Rashba model/1/ bound excitons(BE) should decay on a radiative way because of gigantic oscillator strength. But at the same time BE form a cloud of high local e-h concentration 10^{18} - 10^{20} cm^{-3}. This should lead to high probability of Auger nonradiative recombination/2/. To reconcile this divergence we have attempted to study a role of low lying BE energy state in nonradiative decay.

Stimulated Light Emission Induced With Acoustic Phonons

Stimulated emission of light at BE recombination was supposed to result from two-quantum (phonon + photon) radiative transition. In addition to this suggestion we propose that acoustic phonon generation with high occupation numbers might be detected at such multi-quantum radiative transition like having been discussed elsewhere/3/.

Results And Discussion

Quantum yield (QY) - η - of donor bound exciton luminescence in CdS has been tested under influence of nonequilibrium acoustic phonons (NAP). In turn, QY of acceptor bound exciton luminescence does not depend on phonon occupation numbers and therefore this luminescence band was used for detecting absolute values of NAP spectra with the method described elsewhere/4/. Experiments were performed at temperature T = 1.3K. Two luminescence spectra were recorded simultaneously: in the absence of NAP in the region of excitation(I_L) and differential spectrum characterizing an increment(δI_L) to the luminescence induced with the NAP emitted by metal heater. To get additional information temperature dependence of QY and excitation spectra have been studied. Data of (Fig.1) show strong quenching of the BE luminescence under increasing of the NAP flux intensity in the band with maximum energy near 2 meV. The measuring with NAP allowed to obtain the mean quenching energy and probability of the nonradiative decay process involving acoustic phonons(Fig.2). Symmetry of the BE energy state is characterized with a total angular momentum. According to the angular momentum selection rule Auger process for the BE ground state with total angular momentum J=1 is forbidden. The NAP absorption can transfer BE into the first excited state with J=0 where Auger transition is dominant. This type of local excitation of BE is related to the hole oscillations in potential produced by the donor core and two electrons. The quenching energy E_0=2.5 meV corresponds to the hole transition from the ground state into the excited one. Scrutinizing the BE nonradiative decay on the base of two level model we can obtain the following formula

$$\eta = (1 + C*N_{ph}(h\nu))^{-1} \text{ at } h\nu = E_0, \quad C = W_0\tau_r/(1 + W_0\tau_n) \qquad (1)$$

W_0 is probability of hole transition between ground and excited states, τ_n is a life time of Auger decay, τ_r is a life time of radiative decay according to the Rashba model, C_{exp} = 7, C_{theor} = 5 - 11.

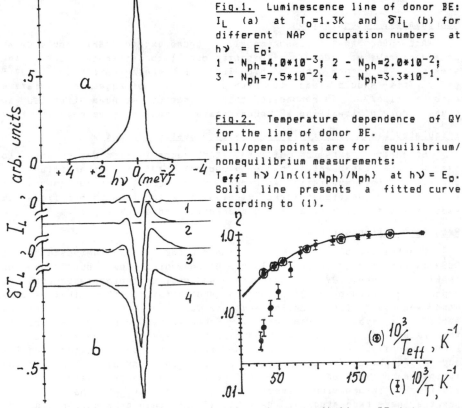

Fig.1. Luminescence line of donor BE: I_L (a) at T_0=1.3K and δI_L (b) for different NAP occupation numbers at $h\nu$ = E_0:
1 - N_{ph}=4.0*10^{-3}; 2 - N_{ph}=2.0*10^{-2}; 3 - N_{ph}=7.5*10^{-2}; 4 - N_{ph}=3.3*10^{-1}.

Fig.2. Temperature dependence of QY for the line of donor BE.
Full/open points are for equilibrium/nonequilibrium measurements:
T_{eff}= $h\nu$ /ln{(1+N_{ph})/N_{ph}} at $h\nu$ = E_0.
Solid line presents a fitted curve according to (1).

Now we turn to the experiments where radiative BE decay are prompted with NAP. We have compared evolution of the luminescence spectra I_L and δI_L as a function of the optical pumping(Fig.3). There is clear correlation in the behaviour of the M-band and the

δI_L responce, owing to the NAP. On account of the positive sign of

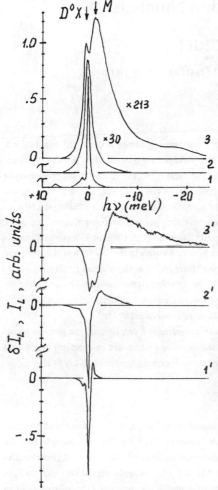

Fig.3. Luminescence spectra for different optical pumping:
$I_0 = 50W/cm^2$ - 1 and 1';
$50I_0$ - 2 and 2';
$10^3 I_0$ - 3 and 3'.
D^0X is luminescence line of exciton localized at neutral donor.

signal δI_L related to the M-band the mechanism of M-band luminescence should be attributed to the BE stimulated radiative recombination with emission of acoustic phonons. This fact is also supported with QY rapidly increase. Together with mentioned above, nonlinear growth in the spectral width of the δI_L is observed and exceeds the region of one-phonon states up to $\sim 2\nu_D$. This strong "expansion" of exciton-phonon band greatly exceeds the spectral width of the "probe" distribution, too. This behaviour directly indicates not only the electron-phonon nature of the M-band, but gives a proof of nonlinear dependence of luminescence spectrum on the phonon occupation numbers. We note that the observed nonlinearity can take place if in some spectral region occupation numbers of phonons generated at this stimulated radiative decay are significantly increased.

To summarize we note that the same impurity center was found to be transfered with internal excitation - NAP - from radiative state to nonradiative one. The new method of investigation of recombination induced with NAP, proposed in this work, can be fruitfully used to study nonradiative and radiative transition in crystals.

1. Rashba E.I., Gurgenishvili G.E., Fiz. Tverd. Tela(Leningrad), 4, 1029, (1962){Sov. Phys. Sol. State, 4, 759, (1962)}.
2. Landsberg P.T., Beattie A.R., J. Phys. Chem. Solids, 8, 73, (1959).
3. Grill W., Hirchbiegel L., Phys. Rev. B31, 8148, (1985).
4. Zinov'ev N.N., Kovalev D.I., Kozub V.I., Yaroshetskii I.D. this volume.

Dynamics of 29 cm^{-1}-Phonons in Ruby for High Phonon Occupation Numbers.

M. SIEMON, J. WESNER AND W. GRILL

Phys. Inst., Johann Wolfgang Goethe-Universität, 6000 Frankfurt, F. R. of Germany

SUMMARY

We investigated the trapping of resonant 29 cm^{-1} phonons (phonon bottleneck) in ruby for a wide range of phonon occupation numbers, up to 30. The phonons are generated by direct optical excitation of the $2\bar{A}(^2E)$ level and detected by fluorescence measurements and by a superconducting bolometer, using resonant scattering to identify the resonant 29 cm^{-1} portion. For high occupation numbers, more than 75% of the generated phonons can leave the excitation volume during the first 10 ns, comparable to the ballistic time of flight. Only the remaining portion becomes trapped for up to some 100 ns. We attribute this to a saturation of the $\bar{E}(^2E) \rightarrow 2\bar{A}(^2E)$ transition, leading to an increased mean free path similar to "bleaching" effects at high intensities in optics. To test this, we excited additional Cr^{3+} ions to the $\bar{E}(^2E)$ level with a second laser, prior to the phonon generating pulse. Here the whole pulse is trapped. By comparison with the background signal generated by de-tuning the generating laser, we can also identify the phonons which were frequency shifted due to spectral diffusion[1,2]. We find no evidence for a strong nonlinear behaviour as reported by Hu[3].

With a crossed beam geometry for the resonant scattering, we can identify the phonon signal from a volume of about $(50\ \mu m)^3$. Measurements of the focusing pattern are in excellent agreement with the calculations of Weiss[4] and with the measurements on thermal phonons of Every et al[5].

EXPERIMENTS

Fig. 1a shows the geometry for most of the measurements. The exciting dye laser with an intracavity etalon has a pulse power of max. 4 kW and a pulse width of about 7 ns. The two laser beams can be moved with a resolution of 1/100 mm under computer control. In most cases we used granular aluminium or tin bolometers at temperatures of about 1.8 K and 3 K. The ruby crystal of 1 cm^3 volume with a Cr concentration of 0.05% is mounted in an optical dewar in vacuum, one face in direct contact with liquid He.

Fig. 1b presents a typical signal for phonons traveling in a focusing direction: curve #1 is the raw detected signal with a peak at 0.1 μs due to scattered light and some electrical noise. The resonant signal is attenuated by the scattering argon laser (#2), the difference signal (#3) shows only the resonant phonons at 0.9 μs and reveals also the longitudinal phonons at 0.6 μs (arrow). This signal can be compared with the difference signal obtained by de-tuning the dye laser in Fig. 2a which generates only non-resonant phonons by the slow decay of high frequency optical phonons excited at impurities. This signal has a similar shape, showing that nearly all resonant phonons reach the bolometer after a travel of about 3 mm. A portion of about 7% of the generated phonons is not detected resonantly, in

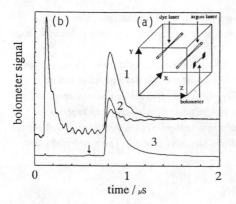

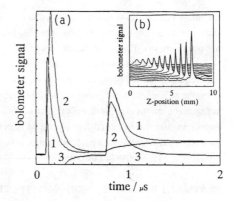

Figure 1

Figure 2

agreement with calculations including phonons from the wings of the emission line[6]. The inset 2b demonstrates the focusing of the resonant phonons, which can also be used for their identification.

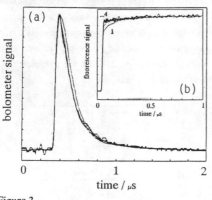

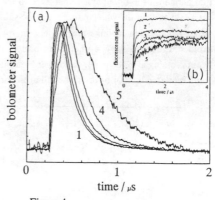

Figure 3

Figure 4

Fig. 3a shows the differential bolometer signal for different laser powers and phonon occupation numbers. The general shape of the signal can be fitted by considering geometry effects and by a two-time-constant model for the bolometer response[6]. Up to the highest power levels available (excitation density N^*: 0.1 to $5 \cdot 10^{18}$ cm^{-3}), no change of the leading edge is observed, apart from a small increase of the trapping time for the highest excitation. The fluorescence signal of the $\overline{E}(^2E)$ level (Fig. 3b) shows that always more than 75% of the phonons can escape in the first 10 ns. We attribute this to a "bleaching" of the phonon absorption in the first moment due to the limited number of excited ions and also by the escape of a considerable amount of phonons from the wings of the gaussian-shaped excitation profile, where the mean free path is greater.

In Fig. 4a, more ions in the $\overline{E}(^2E)$ level were generated by a second dye laser prior to the measurement pulse (100 kW, 1.5 μs). Here trapping is observed also for the first part of the curve, depending on the varying power of the second laser (N^*: 1.8 to $9.5 \cdot 10^{18}$ cm^{-3}). Curve #5 for the highest

1267

excitation density can be sufficiently fitted to a diffusion profile[6]. The fluorescence measurements also show the strong phonon trapping for some μs. The change in intensity of the total fluorescence signal has been used to calibrate the excitation density.

By comparison with the non-resonant excited background signal, we can identify the portion of the frequency shifted phonons (spectral diffusion). Fig. 5a shows that these phonons escape earlier from the volume than the resonant phonons, but they also show some delayed behaviour, giving evidence that even these phonons suffer some collisions before the final "wipe-out" step happens[1,2]. Figs. 5b and 5c show the measured dependence on the excitation density N^* of the frequency-shifted portion and the total normalized signal in comparison to the signal at low N^*, which decreased due to broadening of the beam by saturation.

SPATIALLY RESOLVED PHONON DETECTION

By modifying the basic set-up of Fig. 1a to a crossed beam geometry, shifting the argon laser beam near to the dye laser beam and using a very small bolometer, we can identify the resonant phonon signal emerging from a volume of about $(50 \ \mu m)^3$ (Fig. 6a). This allows the investigation of the focusing pattern (Fig. 6b). Excellent agreement with the calculations of Weiss and the measurements on thermal phonons of Every et al is found. Due to the extensive time to average the signals, no attempt was yet made to scan the whole pattern. Fig. 6c shows the signals for 3 different positions. Due to the restricted "excitation" volume they appear sharper than in Figs. 1 to 3, because geometric effects are avoided. Curve #1 shows a small contribution at the time of the maximum of #2, due to the dye laser beam crossing also the upper and lower high intensity spots similar to Pos. 2, which are modulated by some scattered light or by re-emitted fluorescence light and add to the signal.

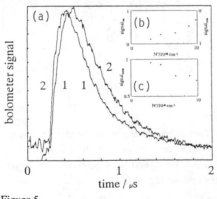

Figure 5

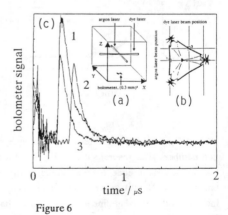

Figure 6

1. R. J. Goossens, J. I., Dijkhuis, H. W. de Wijn, Phys. Rev. B 32, 7065, (1985)
2. S. Majetich, R. S. Meltzer, J. E. Rives, Phys. Rev. B 38 11075 (1988)
3. P. Hu, Phys. Rev. Lett. 44, 417 (1980)
4. F. Roesch, O. Weis, Z. Physik B 25, 101+115 (1976)
5. A. G. Every, G. L. Koos, J. P. Wolfe, Phys. Rev. B 29, 2190 (1984)
6. M. Siemon, Thesis, University of Frankfurt (1989)

RESONANT TRAPPING OF 29-cm^{-1} PHONONS IN SUBMICRON RUBY LAYERS

R. J. van Wijk, P. J. M. Wöltgens, J. I. Dijkhuis, and H. W. de Wijn

Faculty of Physics and Astronomy, University of Utrecht,
P.O. Box 80.000, 3508 TA Utrecht, The Netherlands

Resonant trapping of 29-cm^{-1} phonons by optically excited Cr^{3+} ions in ruby has been studied extensively.[1] In bulk ruby, the removal of these phonons from the excited zone was found to occur through one-site inelastic scattering by weakly exchange-coupled Cr^{3+} pairs (spectral wipeout),[2,3] and elastic spatial diffusion to the zone border.[4] In both mechanisms, the product N^*L of the excited-ion concentration N^* and the typical zone dimension L determines the effective relaxation time T_{eff} of the combined spin-phonon system. At high N^*L, wipeout predominates over spatial diffusion. The minimum active-zone dimensions that can be achieved in bulk experiments is, however, limited to the diameter of the focused laser beam ($\gtrsim 50$ μm). In the experiments presented here, 29-cm^{-1} phonons are trapped in submicron excited zones, created by use of supported ruby films.

The ruby layers, having a thickness of 0.5 μm, were deposited on polished sapphire substrates with dimensions $10 \times 10 \times 0.5$ mm^3 by evaporation of ground 1-at.% ruby at a temperature of 2200 °C, and subsequently annealed in air at 1900 °C for several hours. The luminescent R_1 line from the ruby film is found to be broadened to 0.6 cm^{-1}, and stronger than the residual R_1 from the substrate by at least an order of magnitude. This corresponds to a Cr^{3+} concentration of about 0.4 at.ppm in the substrate. Furthermore, the $\overline{E}(^2E)$ Zeeman splitting in the film equals that in the substrate, indicative of a well-defined c axis collinear with the c axis of the substrate.

To study resonant trapping of 29-cm^{-1} phonons in these films, a sufficient concentration of Cr^{3+} was maintained in the $\overline{E}(^2E)$ state by optical pumping via the broad bands with an argon ion laser operating at all lines. The phonons were repetitively supplied by a constantan heater deposited onto the film. The trapped phonons were registered via the R_2 luminescence, which was detected by use of a 0.85-m double-monochromator, standard photon counting, and time-to-amplitude conversion to extract the development with time. To permit excitation and detection from above the ruby film, the focused laser beam was incident on the film near the heater at an angle of $\approx 40°$. To eliminate stray effects related to the optical pumping, the laser beam was interrupted by means of an acousto-optical modulator

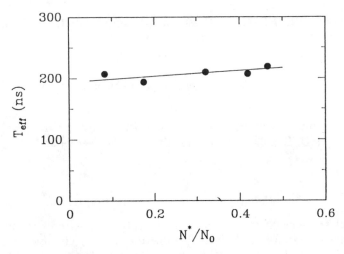

FIG. 1. Decay time of $2\overline{A}(^2E)$ following heat pulses of 1 μs vs N^*/N_0.

during the firing of a sequence of heat pulses.

For long heat pulses ($\sim$ 1 μs), the decay of the R_2 luminescence can adequately be described by a single exponential. The resultant T_{eff}, presented in Fig. 1 as a function of N^*/N_0 (total Cr^{3+} concentration $N_0 \approx 4 \times 10^{20}$ cm^{-3}), is seen to be approximately 200 ns, only marginally dependent on N^*. Indeed, T_{eff} of this order are expected from a calculation of spectral wipeout[2] in a 0.5-μm zone in 1-at.% ruby, and furthermore from extrapolation of the experimental dependence of T_{eff} at high N^*L on N_0.[2] We thus conclude that wipeout is the dominant mechanism of phonon removal.

Repeating the experiment with shorter heat pulses turnes out to result in a faster decay of the R_2 luminescence. Figure 2 shows T_{eff}, as obtained from fits to a single exponential, versus the duration of the ingrowth. Insight into the cause of this behavior is gained by considering the path via which the phonons are fed into the ruby film. We first note that the phonons emitted by the heater primarily travel ballistically into the substrate. Below the excited zone in the film, however, the sapphire substrate is excited to a typical $N^* \sim 10^{16}$ cm^{-3}. This implies a mean free path for 29-cm^{-1} phonons of order 20 μm. Resonant phonons thus are captured in the excited pencil in the substrate, and in part diffuse towards the film, maintaining a constant population of virtually *monochromatic* 29-cm^{-1} phonons at the film-substrate interface during the heat pulse. The phonons subsequently progress into the film, where their mean free path becomes substantially reduced, and they become subjected to spectral wipeout. Two processes are active in the decay after termination of the heat pulse. Phonons that reside close to the film-substrate interface may leave the zone by spatial diffusion back towards the substrate, but phonons having penetrated deeper into the film are removed by wipeout. The latter process will be dominant after long heat pulses.

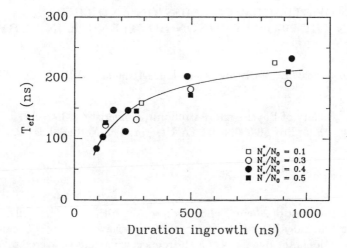

FIG. 2. Decay time of $2\overline{A}(^2E)$ from fits to a single exponential vs the duration of the ingrowth for the N^*/N_0 indicated.

Modeling the above mechanism of supply and removal of resonant phonons in the film, we find penetration profiles whose extent is, for long heat pulses, limited by wipeout. In the latter case, the profiles are exponential with a penetration depth given by $\sqrt{D\tau}$, with D and τ the diffusion constant and the wipeout time pertaining to the relevant N^*, respectively. For $N^* = 10^{20}$ cm^{-3}, $\sqrt{D\tau} \approx 0.2$ μm. Integrated over the zone, the decay of the $2\overline{A}(^2E)$ population after termination of the heat pulse reads

$$N_{2\overline{A}} = N_{2\overline{A}}^0 \sqrt{D\tau} \left\{ \mathrm{erfc}\left(\sqrt{t/\tau}\right) - \mathrm{erfc}\left(\sqrt{(t+T_h)/\tau}\right)\right\} , \qquad (1)$$

in which T_h is the duration of the heat pulse, and $N_{2\overline{A}}^0$ is the $N_{2\overline{A}}$ maintained at the interface during the pulse. This expression adequately accounts for the data. Fits of Eq. (1) with τ as adjustable parameter yield $\tau = 0.27 \pm 0.03$ μs, independent of the heat-pulse duration. Finally, according to Eq. (1) spatial diffusion towards the film-substrate interface plays a negligible role for heat-pulse durations over 0.5 μs, confirming the conclusion arrived at from Fig. 1.

This work was supported by the Netherlands Foundations FOM and NWO.

1. For a review, see K. F. Renk, in *Phonon Scattering in Condensed Matter*, edited by W. Eisenmenger, K. Laßmann, and S. Döttinger (Springer, 1984), p.10.
2. R. J. G. Goossens, J. I. Dijkhuis, and H. W. de Wijn, Phys. Rev. B **32**, 7065 (1985).
3. S. Majetich, R. S. Meltzer, and J. E. Rives,ves, Phys. Rev. B **38**, 11075 (1988).
4. J. I. Dijkhuis and H. W. de Wijn, Phys. Rev. B **20**, 1844 (1979).

EFFECTS OF STIMULATED EMISSION AND LINE BROADENING ON THE 29-cm⁻¹ PHONON BOTTLENECK IN RUBY

K. Z. Troost, E. van Brummelen, J. I. Dijkhuis, and H. W. de Wijn

Faculty of Physics and Astronomy, University of Utrecht,
P.O. Box 80.000, 3508 TA Utrecht, The Netherlands

The imprisonment of 29-cm⁻¹ phonons resonant with the $\overline{E}(^2E) - 2\overline{A}(^2E)$ transition of Cr^{3+} in ruby has been the subject of intensive studies[1-4] because of the insight it provides in the dynamics of high-frequency phonons scattered by centers in insulators. To a first approximation, the frequency band of interest to the bottleneck may roughly be divided into two classes: (I) a narrow, resonant, regime extending up to a few times the linewidth (0.02 cm⁻¹) at either side, and (II) frequencies sufficiently far from resonance that phonons are able to escape out of the active zone without being reabsorbed. Inelastic scattering of resonant phonons by one-site resonant Raman processes involving a spin-flip transition of weakly-coupled Cr^{3+} pairs, then, causes spectral transport between the two classes, resulting in phonon loss by wipe-out.[3] Conversely, an external supply of off-resonant phonons may, of course, lead to a feeding of the central phonons (wipe-in). The line shape, therefore, has a strong impact on the phonon dynamics, and already at low phonon occupations results in a frequency distribution of time constants of the coupled spin-phonon system.[5,6]

In the present paper phonons are fed to the combined spin-phonon system from a point contact driven to high temperatures by pulsed electrical excitation. It is shown that stimulated emission and concomitant line broadening induced by the high phonon densities thus achieved lead to a substantial modification of the dynamics. The spin system is maintained in the $\overline{E}(^2E)$ state by broad-band pumping with an Ar ion laser, and the dynamics of the spin-phonon system is monitored under bottlenecking conditions via the R_2 luminescence, which in turn is proportional to the frequency-integrated population $N_{2\overline{A}}(t)$ of $2\overline{A}(^2E)$. The resulting time traces for various power dissipations in the point contact are shown in Fig. 1. All traces are distinctly nonsingle-exponential, the more so the higher the power dissipation.

To account for the experimental results, we rely on a frequency-resolved description with explicit consideration of the line shape $g(\nu)$ of the transition, and with inclusion of a flat pulsed phonon feeding ϕ of duration τ_p. The quantity ϕ is proportional to the near-resonant phonon occupation p_{nr} supplied by the point contact and converted to resonant phonons by mediation of weakly-coupled Cr^{3+}

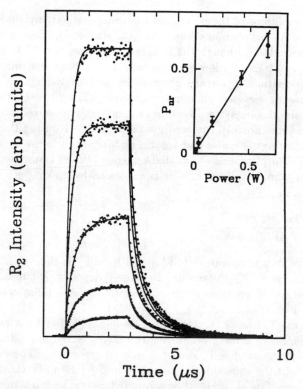

FIG. 1. Time traces of the point-contact induced frequency-integrated R_2 lumines-cence in 700-at.ppm ruby for various power dissipations. $N^* \approx 5 \times 10^{18}$. Inset shows the fitted near-resonant phonon occupation number p_{nr} averaged over the pumped region vs power.

pairs. The frequency-integrated $2\overline{A}(^2E)$ population can then be derived to read[6]

$$N_{2\overline{A}}(t) = \int_{-\infty}^{+\infty} \phi\, T_{\text{eff}}(\nu) \left(1 - e^{-t/T_{\text{eff}}(\nu)}\right) d\nu \,, \qquad (t \leq \tau_{\text{p}}) \qquad (1)$$

$$N_{2\overline{A}}(t) = \int_{-\infty}^{\infty} \phi\, T_{\text{eff}}(\nu) \left(1 - e^{-\tau_{\text{p}}/T_{\text{eff}}(\nu)}\right) e^{-(t-\tau_{\text{p}})/T_{\text{eff}}(\nu)} d\nu \,. \qquad (t > \tau_{\text{p}}) \qquad (2)$$

Here, the effective time constant of the coupled spin-phonon system at frequency ν is given by

$$T_{\text{eff}}(\nu) = \alpha g(\nu) T_{\text{d}}^{(\text{f})}/(2p_{\text{nr}} + 1) \,, \qquad (3)$$

with $T_{\text{d}}^{(\text{f})}$ the spontaneous spin-flip $2\overline{A}(^2E) \to \overline{E}(^2E)$ transition time, and α a pre-factor weakly dependent on the system parameters. The occupation of resonant phonons $p(\nu)$ is, in turn, determined by the condition of dynamical equilibrium

up of the contributions of the individual frequency packets, weighted according to the line shape. At low phonon occupation $[p_{nr}, p(\nu) \ll 1]$, these contributions still are single-exponential, although the integral, of course, is not. Indeed, a detailed analysis of $N_{2\overline{A}}(t)$ traces for low p_{nr} such as in Fig. 1 show that marked deviations from pure exponentials are already observed at the lowest electrical powers applied to the contact. At elevated phonon occupation $[p_{nr}, p(\nu) \approx 1]$, the dynamics of the coupled spin-phonon system becomes considerably more complex. Not only is the relaxation time, Eq. (3), shorter by a factor $2p_{nr} + 1$ during the pulse, but the $p(\nu)$ built up during the pulse and remaining afterwards keeps on to broaden the transition via absorption and stimulated emission. When assuming, for simplicity, a purely homogeneous transition, this is expressed by making $g(\nu)$ dependent on $p(\nu)$ according to

$$g(\nu) = \frac{2\pi}{\Gamma} \frac{(\Gamma/2)^2}{(\Gamma/2)^2 + \nu^2} , \qquad\qquad \Gamma = (2\overline{p} + 1)\,\Gamma_0 , \qquad\qquad (4)$$

where Γ_0 is the zero-temperature full linewidth, and $\overline{p}$ is the weighted sum of $p(\nu)$ over the line, or $\overline{p} = \int_{-\infty}^{+\infty} g(\nu)p(\nu)d\nu$. Because $\overline{p}$ depends on time, the broadening leads to larger deviation from single-exponentiality with point-contact power, as observed (cf. Fig. 1).

A precise calculation of $N_{2\overline{A}}(t)$ from Eqs. (1) – (4) has been accomplished by simultaneous numerical integration over the traces of the individual packets with continuous adjustment of $\overline{p}$, Γ, and $g(\nu)$ in a recursive way. The calculations also take into account the substantial depopulation of $\overline{E}(^2E)$ in favor of $2\overline{A}(^2E)$ in the center of the line. Fits of $N_{2\overline{A}}(t)$ thus computed appear to track the data in Fig. 1. The inset further shows that the p_{nr} derived from the fits indeed scale linearly with the applied power. This, in fact, provides considerable justification of the theory outlined above.

The work was supported by the Netherlands Foundations FOM and NWO.

1. K. F. Renk, in *Phonon Scattering in Condensed Matter*, edited by W. Eisenmenger, K. Lassmann, and S. Döttinger (Springer Verlag, 1984), p. 10.
2. A. A. Kaplyanskii and S. A. Basun, in *Nonequilibrium phonons in nonmetallic crystals*, edited by W. Eisenmenger and A. A. Kaplyanskii (North-Holland, 1986), p. 373.
3. R. J. G. Goossens, J. I. Dijkhuis, and H. W. de Wijn, Phys. Rev. B **32**, 7065 (1985).
4. S. Majetich, R. S. Meltzer, and J. E. Rives, Phys. Rev. B **38**, 11075 (1988).
5. H. Lengfellner, U. Werling, J. Hummel, H. Netter, N. Retzer, and K. F. Renk, Solid State Commun. **38**, 1215 (1981); H. Lengfellner, J. Hummel, H. Netter, and K. F. Renk, Opt. Lett. **8**, 220 (1983).
6. M. J. van Dort, J. I. Dijkhuis, and H. W. de Wijn, J. Lumin. **38**, 217 (1987); and to be published.
7. K. Z. Troost, J. I. Dijkhuis, and H. W. de Wijn, to be published.

PROPAGATION OF NONEQUILIBRIUM PHONONS IN QUARTZ

B. Sujak-Cyrul, T. Tyc
Institute for Low Temperature and Structural Research
Polish Academy of Sciences, Pl-50-950 Wrocław, P.O.Box 937
Poland

1. Introduction

Because of it's considerable industrial application quartz is one of the most interesting materials. Phonon propagation in α-SiO$_2$ was experimentally investigated by von Gutfeld at all [1], Eichele at all [2], Grill and Weis [3] whereas theoretical calculations of it have been made by Rosch and Weis [4,5], Koos and Wolfe [6].

2. Experiment

The characteristics of nonequilibrium phonon propagation along the x-axis in α-SiO$_2$ crystal have been measured at temperatures near 2K. The conventional heat-pulse method was used. Fig.1 presents a block diagram of the employed equipment.

The sample of the size 1.65 x 2.05 x 4.03 mm was cut from high-quality crystal α-SiO$_2$. Two faces of it (1.65x2.05) being perpendicular to the $(10\bar{1}1)$ direction were mechanically and chemically polished to the optical grade. Phonon generator and detector were evaporated on two opposite sides of the sample (fig.2). They had the same size of the active area smaller than 0.2 x 0.2 mm and large thin-film pads made of In. The generator-detector direction was parallel to the x-axis of the crystal with the accuracy of about 5°. Incoherent phonons were generated by heating of thin silver film (about 60 - 100 Ω in 300 K) passing short-duration (100ns) electric pulses with the frequency of 5 kHz. The power emitted during a pulse was approximately equal to 0.5W. Phonon pulses were detected after passing the crystal by superconducting Al-bolometer, maintained very close to its transition temperature and supplied by direct current (between 5 and 30 μA). Their resistance at 300K was contained between 100 and 300 Ω. For bolometers made in this way the N-S transition temperature was in range from 1.6K to 2.1K and the sensitivity was of the order of 1000K/Ω. The temperature was stabilized electronically with the accuracy better than 0.001K.

3. Results

Fig. 2 shows typical time-of-flight signals, received for α-SiO$_2$ crystal at a temperature close to 2K and for I_B = 20μA. The length of the path was equal to 5.1mm. Four peaks and a broad ramp are observed. The first peak is a mark of the phonon flight start. Maxima of the next three distinct singularities lay in turn at 1.0, 1.53 and 1.87 μs. Their broadnesses are larger than 100ns pulse durations. Intensities of the singularities decrease when the pulse amplitude and the bolometer current decrease. The singularities clearly demonstrate the ballistic nature of phonon propagation in quartz and confirm the existence of transverse (T_1, T_2) and obligate (OM) modes, which was theoretically calculated in [4]. The geometrical proper-ties of SiO$_2$ cause that the longitudinal mode propagating to the detector are not observed. The maxima positions yield the following values of the group velocities: $V_{T1} \cong 5.3$ km/s, $V_{T2} \cong 2.8$ km/s (see [3,4]).

References

[1] von Gutfeld,R.,Nethercot,A.,Phys.Rev.Let.12,641,(1964)

[2] Eichele,R. at all, Z.Phys. B48, 89, (1982)

[3] Grill,W.,Weis,O., Phys.Rev.Let., 35, 588, (1975)

[4] Rösch,F.,Weis,O., Z.Phys. B25, 101, (1976)

[5] Rösch,F.,Weis,O., Z.Phys. B27, 33, (1977)

[6] Koos,G.,Wolfe,J., Phys.Rev.B,30,3470 (1984)

[7] Hałaczek,T.,Tyc,T.,Zossel,R.,Chłodnictwo,7-8,8,(1987)
    - in polish

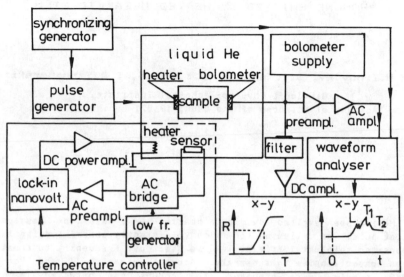

Fig.1. Block diagram of the employed equipment.
First version of it was presented in [7].

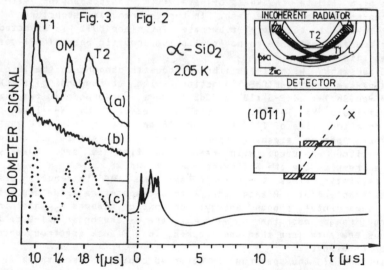

Fig.2. Typical time-dependent bolometer signal for quartz.
Ins.- momentary phonon position after phonon emission [4].
Fig.3. Singularities of bolometer signal from fig.2.
  a - signal, b - background, c - subtraction of a and b

SPECTRUM AND KINETIC OF NONEQUILIBRIUM ACOUSTIC PHONONS EMITTED BY HEATED METALLIC FILM

N.N.Zinov'ev, D.I.Kovalev, V.I.Kozub, I.D.Yaroshetskii

A.F.Ioffe Physico-Technical Institute,
Academy of Sciences of the USSR, 194021 Leningrad, USSR

Having been excited by Joule heating as a result of passage of current pulse or illumination with sufficiently powerful pulse laser a deposited thin metal film at $T=T_m$ is used most frequently to create a phonon nonequilibrium in substrate. Acoustic mismatch/1/ and Perrin-Badd/2/ models were supposed to account for observed in the experiments/3/ some deviation from a Planck distribution due to an effect of a finite thickness of metal film in comparison with the electron-phonon mean free path. In turn, granular structure of the metal layer at the metal-substrate interface(MSI) is expected to render a great influence on nonequilibrium acoustic phonon(NAP) spectra emitted into the sample.

Experiments with nonequilibrium acoustic phonons in the thin film metal(Sn)-semiconductor(CdS) junction with corresponding thickness $d\{2*10^{-5}cm(Sn)\} + L\{(2-5)*10^{-3}cm(CdS)\}$ were performed at temperature of superfluid ^4He $T=1.3K$. Sn-film was excited by N_2- laser pulse with duration time of $10^{-8}sec$. The NAP spectra in CdS were analysed by the "phonon sideband spectrometer" (luminescence of acceptor bound exciton was used) with resolution time $10^{-9}sec$. Summary of obtained results are the following - Fig.1 and 2:

- at ballistic times $t \sim L/s \sim 10^{-8}sec$ the NAP distribution differs considerably from a Planck one with a temperature T_E determined by the same total phonon energy; occupation numbers $N_{ph}(h\nu)$ in the energy range near the $h\nu \sim h\nu_{max}$ are of the order 0.1-0.5. These values are much more than one occuring in a Planck spectrum which is order of 0.067;

- evolution of the spectrum is observed beginning from the ballistic time of flight right up to $10^{-5}sec$; when delay time is increased the total energy of phonons decreases monotonously, but a non-Planck nature of the spectrum is conserved for too long delay time;

- there is no significant change in the energy $h\nu_{max}$, when pumping rate of metal is varied; the dependences of the total energy, T_E and η on the pumping intensity (I) are looking as following: $E \varpropto I$,

To understand the observed phenomena, the phonon relaxation processes were analysed in the details. Estimation and comparison with thermal conductivity data prove that in this case the most effective thermalization should be due to electron-phonon interaction in metal because of multiple return of phonons into the metal film. The relaxation time value in this case should not be more then 10^{-7} sec. To explain, why the non-Planck form is conserved, we took into account the role of the MSI structure. Even if metal evaporation were produced without additional gases like O_2, fine granules should be formed due to the nature of metal vapour condensation of on the substrate surface in the vacuum and the fact the semiconductor surface is not wetted with the metal. Therefore penetration of NAP into substrate should take place through the small region of "touch point" of spheroidal granules. So and the MSI looks like the random network of point contacts. Having in mind the dominant contribution of elastic phonon scattering on defects, dislocations, and defect clusters inside the metal film, the NAP spectrum transmitted through such boundary can be estimated in the following form:

$$N_{ph}(h\nu) = N_0(h\nu,T_m)\{D_{ph}(\nu)/\widetilde{D}_{ph}\}\{C(T_m)/\widetilde{C}\}\{\rho_m(\nu)/\rho_0(\nu)\}d/L$$

where ρ_0 and ρ_m - the phonon density of states in semiconductor and metal, waved $\widetilde{D}$ and $\widetilde{C}$ correspond to the mean values determining the thermal conductivity in the point contact, $C(T_m)$ is the metal specific heat at $T=T_m$, $D_{ph}(\nu)$ is the phonon diffusion coefficient, $N_0(h\nu,T_m)$ is a Planck distribution at $T=T_m$. We can see that the phonon distribution in the sample may differ considerably from a Planck spectrum and because of the strong frequency dependence $D_{ph}(\nu)$ we can expect significant shift of the phonon distribution towards lower frequencies. If $D_{ph}(\nu)\infty \nu^{-n}$, $dE/d\nu \infty \nu^{3-n} \exp(- h\nu /T_m)$ at $h\nu >T_m$, and $dE/d\nu \infty \nu^{2-n}$ at $h\nu <T_m$ in contrast to the dependence typical for a Planck formula. Low frequency cutoff of the evaluated frequency dependence occurs at $h\nu_{lim}$ determined with a size effect condition: $s\tau_{ph-i}(\nu_{lim}) \sim min(a,d)$, a - mean contact size. At $\nu < \nu_{lim}$ we have usual form like as $dE/d\nu \infty \nu^2$. If $n>2$, the frequency dependence of $dE/d\nu$ has a maximum at $h\nu_{lim}$. When $T_m > h\nu_{lim}/2.83$ the position of this maximum should not depend significantly on T_m or respectively on the pumping rate I. This is exactly the behaviour which is observed in Fig.1 and 2. It seems appropriate to refer to /4/ where reducing of the transmission efficiency with increasing phonon energy is observed for the Sn-Si interface and to /5/.

1. Wigmore J.K., Phys. Rev. B5, 700, (1972); Weis O. J., Physique Coll. 33, C4-107, (1972); Rosch F., Weis O. Z., Phys. B25, 101, (1976).
2. Perrin N., Badd H., Phys. Rev. Lett. 28, 1701, (1972).
3. Bron W.E., Grill W., Phys. Rev. B16, 5303, (1977).
4. Eisenmenger W. in "Nonequilibrium Superconductivity, Phonons, and Kapitza Boundaries", ed. by E.Gray, NATO ASI Series, B65, 73, (1981).
5. Kulik I.O., Omel'yanchuk A.N., Yanson I.K., Fiz. Nizk. Temp. 7, 263, (1981) {Sov. J.Low Temp. Phys. 7, 129, (1981)}.

$$T_E \infty I^{1/4}, \qquad \rho = N_{ph}(h\nu_{max})/(\exp h\nu_{max}/T_{max} - 1)^{-1} \infty I;$$
$$h\nu_{max} = 2.83 T_{max}.$$

- degree of the non-Planck destortion decreases with pumping lowering and delay time increasing.

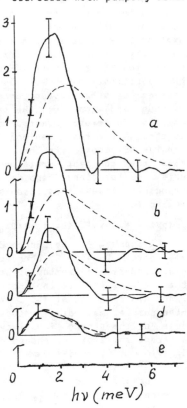

Fig.1. Spectral distributions of the bulk energy density for $I = 10^4 W/cm^2$, obtained for different delays t to the moment of NAP injection into a sample kept at $T_0 = 1.3K$:
a) t = 30 nsec, T_E = 9.3K;
b) t = 1 sec, T_E = 8.1K;
c) t = 2 sec, T_E = 7.6K;
d) t = 5 sec, T_E = 5.1K;
e) t = 9 sec.
The dashed curves represent the Planck spectra with $T = T_E$.

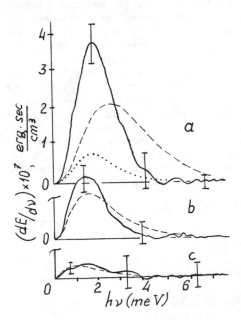

Fig.2. Spectral distribution of the bulk energy density measured for a fixed delay of $t \sim L/s$ ~ 30 nsec and different pumping intensity of Sn-film($I = 10^4 W/cm^2$ and $T_0 = 1.3K$):
a) I = I_0, T_E = 9.9K;
b) I = 0.141I_0, T_E = 6.5K;
c) I = 0.021I_0, T_E = 3.4K.
The dashed curves and dotted line represent the Planck spectra with $T = T_E$ and $T = T_{max}$ respectively.

ELECTRONS AND PHONONS IN NON-EQUILIBRIUM
IN DISORDERED THIN NARROW FILMS

G.Bergmann, Wei Wei and Yao Zou

University of Southern California, Physics Department,
University Park, Los Angeles CA. 90089-0484

R.M.Mueller

Institut für Festkörperforschung der Kernforschungsanlage,
D-5170 Jülich, West Germany

If an electrical current flows through a metal, the external power supply transfers energy into the electron system. The electron system then transfers its excessive energy into the phonon system by electron phonon processes and any temperature difference between the electron and phonon temperature relaxes as

$$dT_e/dt = -(T_e - T_{ph})/\tau_{ep} \qquad (1)$$

Since the electrical current increases the electron temperature according to $\rho j^2 = \gamma T_e dT_e/dt$ the resulting temperature difference is

$$[T_e - T_{ph}/2]^2 - T_{ph}^2/4 = \rho \tau_{ep}/\gamma \, j^2 \qquad (2)$$

where γ is the Sommerfeld constant $\gamma = \pi^2 k_B^2 2N_o/3$, $2N_o$ the density of states at the Fermi energy and ρ the resistivity of the film.

In this paper we report about recent experiments in which we use a thin narrow film with a thickness of 4 μ and a width of 3 nm. These films are quench condensed onto a quartz plate which is at He temperature. We send currents between 3 μA and 1 mA through the narrow film. The higher current corresponds roughly to a current density of 10^7 A/cm^2 and raises the electron temperature above the phonon temperature. However, since the total volume of our film is so small, the generated heat is still relatively small.

In Fig.1 the dependence of the resistance per square of a Au film is plotted versus the temperature of the quartz plate. The dashed curve gives the resistance in thermal equilibrium (for sufficient small current). From each equilibrium point a full curve branches off which is obtained by increasing the current from .003 μA to 1.3 mA in 13 steps. The equilibrium curve has a resistance minimum of about 956.8 Ω at 24.0 K. The non-equilibrium curves with higher currents yield considerably lower minima in the resistance. If the electron and phonon systems were in equilibrium then the minimum of the resistance would be independent of the current. Obviously the current density is high enough to heat the electron temperature considerably above the phonon system.

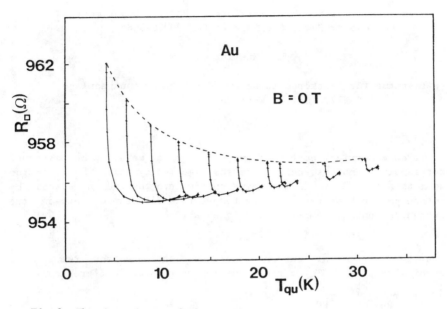

Fig.1: The dependence of the resistance per square of a Au film is plotted versus the temperature of the quartz plate for different currents through the film.

The electrical resistance in our film has the following contributions: i) the residual resistance of the film, ii) the contribution of weak localization, iii) the contribution of the Coulomb anomaly due to electron-electron interaction and iv) the (usual) thermal part of the electrical resistance due to electron-phonon processes. The Coulomb correction

$$\Delta R_{ee} = R_o^2 \ \frac{e^2}{2\pi^2\hbar} \ (1-F') \ \ln(T) \qquad (3)$$

depends only on the electron temperature and is therefore a suitable electron thermometer. (R_o is the resistance per square of the film.) We suppress the temperature dependence of weak localization by applying a magnetic field of 7 T. The contribution (iv) of the thermal part can be neglected for films with such a high resistance per square. At present time we approximate the phonon temperature

T_{ph} by the temperature of the quartz substrate T_{qu}.

From relation (2) we easily derive that a plot of j^2 versus $(T_e-T_{ph}/2)^2$ yields the electron-phonon relaxation time τ_{ep}. Such a plot is given in Fig.2 for the equilibrium temperature of 8.35 K.

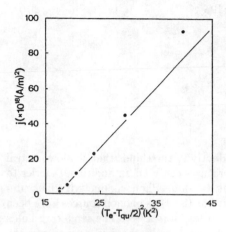

Fig.2: The square of the current density is plotted versus $(T_e-T_{ph}/2)^2$ for the equilibrium temperature of 8.35 K.

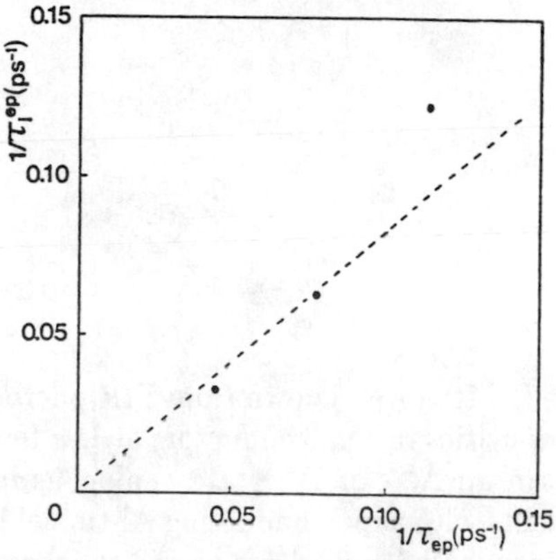

Fig.3: The two rates $1/\tau_{ep}$ and $1/\tau_i^{ep}$ are plotted as a function of each other.

We measure at the same time the magneto-resistance of the narrow film and determine with the theory of weak localization (for reference see for example [1]) the dephasing rate $1/\tau_i$ which is partially due to electron-phonon and electron-electron processes, i.e. $1/t_i=1/\tau_i^{ep}+1/\tau_i^{ee}$. We separate the two contributions. $1/\tau_i^{ep}$ obeys a T^2-law. This suggests that the Eliashberg-function $\alpha^2F(\omega)$ which describes the electron-phonon interaction is linear in the energy $\hbar\omega$. For such a dependence $1/\tau_{ep}$ and $1/\tau_i^{ep}$ should differ only by a factor of .83. In Fig.3 the two rates $1/\tau_{ep}$ and $1/\tau_i^{ep}$ are plotted as a function of each other at different temperatures. The theoretical ratio of .83 (for a linear $\alpha^2F(\omega)$) is shown as the dashed line.

Acknowledgement: The research was supported by NSF Grant Nr DMR-8521662 and by a grant from the NATO.

[1] Bergmann, G. Physics Reports 107, 1 (1984)

DEPOPULATION EFFECTS IN THE PHONOCONDUCTION RESPONSE OF A$^+$- STATES IN SILICON AND GERMANIUM

P. GROSS, M. GIENGER, AND K. LASSMANN
Universität Stuttgart, 1.Physikalisches Institut
Pfaffenwaldring 57, D-7000 Stuttgart 80, FRG

It is well known from FIR photoconductivity thresholds that shallow neutral impurities in semiconductors at low temperatures can bind an additional carrier to form an A$^+$- or D$^-$–state analogous to the H$^-$–ion. Phonoconductivity measurements with superconducting Al-tunnel junctions (STJ) as phonon sources have been shown to be a sensitive means to investigate these states [1, 2]. The same thresholds as with FIR photoconductivity were found, but the threshold is steeper and the signal shape beyond the threshold depends strongly on the intensity of illumination which is necessary to produce the charged states (Fig. 1a). Also the stress dependence of the A$^+$–conductivity threshold appears to be quite different for either FIR or phonon spectroscopy on the other hand [3, 4]. Haug et al [5] calculated the phonoconduction response for the cases of Si : B$^+$ and Si : In$^+$ and found good qualitative agreement with our experimental results. The point is that the short wavelength phonons are very sensitive to changes in the spatial extent of the A$^+$–wavefunction and that deviations from the parabolic form of the valence band near $k = 0$ are important. For these calculations it was assumed that the spectrum emitted by the STJ is monochromatic and that only a small part of the A$^+$–centers is depopulated by the phonons.

We show here that deviations from these assumptions are effective in the experiment and responsible in particular for the observed illumination dependence of the phonoconduction signal. The phonon spectrum of an Al – STJ at bias $V = eU > 2\Delta_{Al}$ consists of a continuum with a sharp threshold at $\Omega_m = V - 2\Delta_{Al}$ (Fig. 2a). In the differentiated spectrum (Fig. 2b) Ω_m becomes the most prominent bias-tunable feature, namely the quasimonochromatic line of phonon spectroscopy used for the approximate analysis of sharp phonon scattering resonances such as the O$_I$ line of Fig. 1a (which shows inversion of the center due to strong scattering). Since the calculated A$^+$–response to phonons has a width of several meV (Fig. 1bα) the complete differentiated spectrum has to be taken into account for comparison with theory. The low-frequency part of the differentiated spectrum may contribute to the signal at some bias $V - 2\Delta_{Al}$ even if the quasimonochromatic phonons at

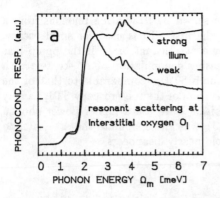
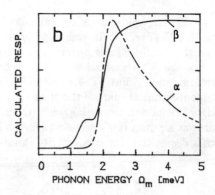

Figure 1: phonoconduction response a) of Ga$^+$ in Si at different illumination intensities b) calculated for α) monochromatic β) complete emission spectrum of the STJ for high illumination intensity (i.e. without depopulation effects)

$\Omega_m = V - 2\Delta_{Al}$ no more contribute to the signal because of the reduced interaction of high-frequency phonons with the A$^+$-state (Fig. 1bβ).

However, we have to keep in mind that with increasing $\Omega_m > E(A^+)$ there is an increasing amount of phonons with frequencies at $\Omega_m \gtrsim E(A^+)$ (Fig. 2a). This leads to a substantial depopulation of the A$^+$-centers if the illumination intensity (i.e. the A$^+$-concentration) is small and the A$^+$-phonon interaction is strong which is the case at the maximum of Fig. 1aβ. By such a reduction of the A$^+$-concentration the signal corresponding to strong illumination is more and more reduced beyond the threshold for weak illumination as shown in Fig. 1a.

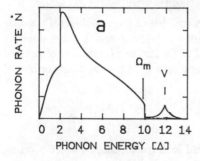
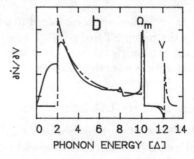

Figure 2: a) phonon emission spectrum [6] of an Al-STJ at bias $U = V/e > 2\Delta_{Al}$ ($2\Delta_{Al}$ = superconductor gap) b) differentiated spectrum (dashed line normal conductor approximation used for calculations)

We can roughly estimate the A$^+$-concentration from the measurement of the sample resistance to be at maximum $10^7 \ldots 10^9 \text{cm}^{-3}$. In the case of depopulation there is a change of the A$^+$-concentration by phonons in the same order of magnitude.

The depopulation effect can also be shown by irradiating phonons from a second STJ (Fig. 3). In this experiment, the first STJ is used for spectroscopy, while the second STJ is operated as a pump at constant bias. If the pump bias exceeds the A^+-threshold we get an increasing depopulation of A^+-centers with increasing pump bias. The effect of the pump STJ is comparable to that of the nonmonochromatic part of the emission spectrum of the spectroscopy STJ, namely a signal reduction increasing with pump bias. If we use a higher power phonon source as a pump (for example a NbAl-STJ), we can depopulate the A^+-states to such a degree that the conductivity threshold nearly disappears.

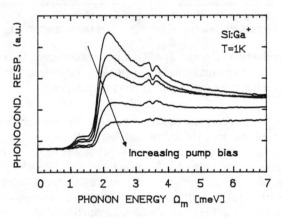

Figure 3: phonoconduction response of Si:Ga$^+$ at various biases of the second (pump) STJ

In principle such a combined irradiation could lead to non-thermal occupation by the pump STJ of possibly existing higher A^+-levels and a corresponding phonoconduction signal from the analyzing STJ. So far no population of excited bound states (as one might infer from the luminescence multiplet of acceptor bound excitons [7]) was found.

1) Burger, W. and Laßmann, K., Phys. Rev. Lett. 53, 2035 (1984)

2) Burger, W. and Laßmann, K., Phys. Rev. B 33, 5868 (1986)

3) Sugimoto, N., Narita, S., Taniguchi, M., and Kobayashi, M., Solid St. Comm. 30, 395 (1979)

4) Groß, P., Gienger, M., and Laßmann, K., Proc. 18th Int. Conf. on Low Temp. Phys., Kyoto 1987, pub. Jap. J. Appl. Phys. Vol. 26 (1987), suppl. 26-3

5) Haug, R. and Sigmund, E., accepted for publication in Phys. Rev. B 40, september 1989

6) Scheitler, W., thesis, Univ. Stuttgart 1989

7) Thewalt, M.L.W., Can. J. Phys. 55, 1463 (1977)

PHONON HOT SPOT IN Ge CRYSTAL

M. M. Bonch-Osmolovskii, T. I. Galkina, Yu. Yu. Pokrovskii, A. Yu. Blinov

P. N. Lebedev Physical Institute, USSR Academy of Sciences

117924, Moscow, Leninskii prosp., 53, USSR

Introduction. The establishment of local temperature (phonon hot spot[1,2]) after light pulse absorption was under study. The theory[1] gives no predictions of how the intrinsic hot spot (HS) processes are to be revealed in the remote detector signal. It is clear, that below the HS formation threshold, all the generated phonons will have independent fate and the detector signal will be linear in pulse energy E and independent of its density P. Above the HS formation threshold certain processes are supposed to depend on P. We have measured the phonon pulse amplitude and shape in Ge crystal at various E and P.

Experimental setup. Conventional heat pulse technique was used.

Sample: Ge [100] single crystal, $N_D + N_A \approx 5 \times 10^{+12} cm^{-3}$; $8 \times \varnothing 30 mm^3$ size. The sample was immersed in liquid helium, but the pumped surface was in vacuum, allowing to get rid of He boiling effects, that are known to have influence on heat pulse amplitude and shape[3].

Excitation: N_2 pulsed laser, $\lambda = 0.34 \mu m$, $\tau_P = 7.3 ns$. The excitation spot diameter, D, was changed through axial displacement of the lens.

Detector: superconducting bolometer of granular aluminum with broadened transition[4], $\Delta T \approx 0.1K$, $T_0 = 1.74K$; $1.3 \times 1.3 mm^2$ in size.

Results and discussion. As it was pointed above, at low P no dependence is expected of pulse shape and amplitude on P and this statement is supported by Fig. 1: At low P (E=const) all the curves are identical. The signal amplitude decrease at $P > P_0$ ($P_0 \approx 3 \mu J/mm^2$, can be found from Fig. 2) is supposed to be due to the HS formation. The estimated P_0 value[1] is 15 times greater. Bearing in mind, that[1] gives only "an order-of-magnitude" estimates, nevertheless, consider possible reasons for such discrepance. First: theory[1] utilized the anharmonic constant γ calculated in the isotropic approximation[5]; at the same time, the account of anisotropy of inelastic properties[6] (for Si) results in strong rise of estimated value of phonon anharmonic decay rate. Obtaining from[1] that $P_0 \propto \gamma^{-4/9}$ we see that enhancement of effective γ value by a factor of 200 will be enough to account for observed P_0 value. Strictly speaking, this makes the assumption[1] of quasidiffusive mode of phonon propagation questionable. Second: the existence of an additional anharmonism on the crystal surface might result in the decrease of observed P_0 value.

The dependence of phonon pulse amplitude on P obtained through changing D (Fig.2) has two plateaus: at "low" and at "high" P values. The former we have discussed above, so let us turn to the latter. Here the HS temperature can be very high (can even exceed Ge melting point), one could expect the heat pulse shape to change markedly. Consider, however, the model[1] for our experimental conditions. The HS is formed within the region where the energy is supplied. It expands then into the crystal interiour, initially in the plane geometry, its temperature going down to its minimum value, T^+, - the temperature of HS "death". At medium P values the "death" occures in the plane geometry. For the main HS parameters (the HS lifetime, t, and maximum thickness, L) we find to be extremely sensitive to the P value: $t \propto P^{10}$, $L \propto P^9$.

These estimates show, that the interval of P values, where the HS exists, but dies still in the plane geometry, is very narrow (not wider than P_0 to $2P_0$). From this moment on (with increasing P), the last phase of its life the HS will exist being three-dimensional (3D). Again, it is possible to obtain from[1], that here t and L are to depend on energy, E, only: $t \propto E^{10/19}$, $L \propto E^{9/19}$ and not on P. The latter is supported by the presence of the plateau at high P values, Fig.2. Thus, the "transition" region of each curve, reflects the process of setting of HS 3D-expansion geometry. For the point $\hat{P}$, where the plane geometry is to change into 3D-one, we find from[1]: $\hat{P} \propto E^{1/19}$, i.e. is almost independent of E. The experimental plot of $\hat{P}$ vs E, is given in the inset to Fig.2 ($\hat{P}$ was taken as a "transition" midpoint of Fig.2 curves), the slope being 0.55.

Two possible reasons for signal amplitude drop upon HS formation: 1. Energy trapping whithin the HS makes a delay in the time moment for low-frequency phonons to appear. Here, the total phonon signal is supposed to conserve and amplitude to diminish. 2. At high E and P values, the HS is so large, that makes a misalignement in the "source" - focussing pattern -detector geometry, causing phonon signal loss.

Conclusion. The threshold energy density of phonon hot spot formaton in Ge crystal was measured, $P_0 \approx 3\mu J/mm^2$. The phases in the HS history were identified: formation, transition from 1D to 3D expansion, 3D-expansion, depending on values of absorbed energy and its density. The phonon anharmonic decay rate is supposed to be substantially higher, than calculated in isotropic approximation.

Acknowledgement. We are grateful to D. Kazakovtsev for discussion.

References

1. Kazakovtsev D. V. and Levinson Y. B., Sov. Phys. JETP, **61**, 1318 (1985).

2. Greenstein M., Tamor M. A., Wolfe J. P., Phys. Rev. B, **26**, 5604 (1982).

3. Tsvetkov V. A. et al Sov. Phys. JETP Lett., **42**, 335 (1985).

4. Blinov A. Yu. et al, Kratkie Soobsch. Po Fizike FIAN, No7 (1989).

5. Tamura S., Phys. Rev. B., **31**, 2574, (1985).

6. Berke A., Mayer A. P., Wehner R. K., Solid State Comm., **54**, 395 (1985).

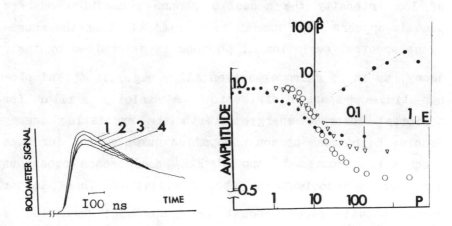

Fig. 1. STA-phonon pulse in Ge [100] at constant excitation energy, E=0,43 μJ, but various values of energy density P, $\mu J/mm^2$: 1- 11 and less; 2- 18; 3- 40; 4- 160 and more

Fig. 2. STA-phonon pulse amplitude (normalized to its value at low P) vs energy density P (obtained through changing the excitation spot diameter D) with energy E, μJ, as parameter o - 3.7×10^{-2}; ∇ - 2.7×10^{-1}; O - 1.1. Inset: Dependence of $\hat{P}, \mu J/mm^2$ vs E, μJ, ($\hat{P}$ was taken as P value, corresponding to the "transition" midpoint for curves like those in Fig. 2).

PHONON HOT SPOT FORMATION IN GaAs UNDER OPTICAL EXCITATION.

B.A.Danil'chenko, D.V.Kazakovtsev[*] and M.I.Slutskii

Institute of Physics, Ukrainian Academy of Sciences,
252008, Kiev, USSR

[*]Institute of Solid State Physics, USSR Academy
of Sciences, 142432, Chernogolovka, Moscow region, USSR

For phonon excitation by free carriers generated near the semiconductor surface by short interband light pulses of low intensity the acoustic phonon occupation numbers usually appears to be small, i.e., $n(\omega) \ll 1$. Then the subsequent spectral evolution of phonons is determined by their decay, as it is shown experimentally, e.g., in[1], and phonon time-and-space distribution is strictly similar for different absorbed energies. With the excitation energy density E increase phonon occupation numbers $n(\omega)$ for some frequencies become ~1, and at $E \gtrsim E_0$ coalescence processes also become important for phonon evolution. These latter competing with decays result in a hot spot formation[2], theoretical estimation giving for GaAs $E_0 \sim 80$ $\mu J/mm^2$. One can expect the nonequilibrium phonon time-of-flight spectra to change their form at $E \sim E_0$ compared to the $E \ll E_0$ case.

In[3] some changes in the nonequilibrium phonon time-of-flight spectra in GaAs has been noted at $E \sim 1$ $\mu J/mm^2$. However, as we have shown in[4], these could result not from the hot spot formation, but from nonstationary heat transfer from the excited surface of the sample into liquid helium.

We measured nonequilibrium phonon fluxes, generated by

interband absorption of nitrogen laser pulses of 10 ns du-

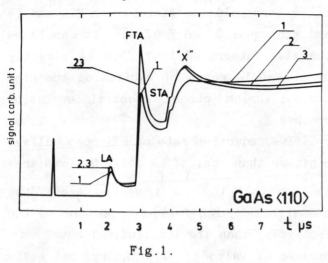

ration in vacuum placed GaAs crystal (0.86x 0.8x0.8cm³ sample of n-GaAs with impurity concentration $N_d = 5 \cdot 10^{15} cm^{-3}$, faced by (110) and (100) planes) by an In bolometer

Fig.1.

with transition temperature adjusted to 2K by an external magnetic field.

Fig.1 demonstrates the time-of-flight spectra for ⟨100⟩ direction normalized to the excitation energy. One can see that E growth from 0.5 to 3.2 µJ/mm² (curves 2,3) does not result in any change of spectrum form. Further increase of E to 22.6 µJ/mm² (curve 1) changes the spectrum form via

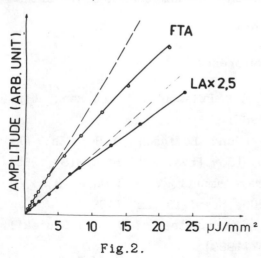

ballistic phonon components relative reducing and time-delayed phonon flux contribution increase.

The dependences on E of ballistic phonon pulse amplitudes for FTA and LA modes are represented in Fig.2. The linear dependences of low excitation ener-

Fig.2.

gy domain are seen to change into sublinear ones at larger energy densities. The threshold energies E_t for FTA and LA modes are different and equal 3 and 7 $\mu J/mm^2$. It should be mentioned that the time-integrated heat flux is strictly linear in E along the whole range of excitation energies used. So the nonequilibrium phonon spectral evolution changes as E approaches E_0.

Fig.2 shows relative reduction rate of FTA mode ballistic phonons to be higher than that of LA mode. Our explanation of this fact[4] is that the low frequency transverse phonons coalescence with the high frequency phonons of the hot spot more effectively, than the longitudinal ones.

Comparing experimental value E_t with theoretical estimation E_0, one must bear in mind that E_0 corresponds to the case of hot spot formation, when coalescence processes are comparable with decays. We observe at E_t the "precursors" of such situation, coalescence processes just beginning to compete with decays. We believe the gradual increase of the role of coalescence processes to explain the sublinearity of ballistic signals observed. Our semiquantitative estimations[4] show it is reasonable.

References

1) Happek, U., Ayant, Y., Buisson, R., and Renk, K.F., Europhys. Lett. **3**, 1001(1987)

2) Kazakovtsev, D.V., and Levinson, Y.B., Zh. Eks. Teor. Fiz. **88**, 2228 (1985) [Sov.Phys. JETP **61**, 1318 (1985)]

3) Ulbrich, R.G., Narayanamurti, V. and Chin, M.A., J. de Phys.(Paris), **42**, Col.C6, Suppl.13, 226 (1981)

4) Danil'chenko, B.A., Kazakovtsev, D.V. and Slutskii, M.I., Phys. Lett. **A138**, 77(1989)

ANOMALIES, NONEQUILIBRIUM PHONONS AND BACKGROUND OF THE MEASURED
POINT-CONTACT SPECTRA OF Cu AT 0.7 K

M. Reiffers, P. Samuely
Institute of Experimental Physics, Solovjevova 47,
043 53 Košice, Czechoslovakia

Anomalies moreover the known phonon non-linearities are observed
in the point-contact (PC) spectra of Cu. Their analysis including
the reabsorption of the nonequilibrium phonons in the PC region,
are reported.

The point-contact (PC) spectroscopy enables to obtain direct informations about the energy dependence of the electron-phonon interaction (EPI) in metals. The PC spectrum (the second derivative $d^2V/dI^2(V)$
of I-V characteristic of the metallic PC) is directly proportional to
the PC function of EPI - $G(\omega)$.

The PC spectra of polycrystalline Cu were measured at 0.7 K, e.g.
with the experimental smearing $\delta \sim 0.4$ meV. We observed the small humps
or breaks except the well-known transversal and longitudinal phonon
nonlinearities and zero-bias anomalies[1] (Fig.1 - marked by arrows).
The energy positions of these anomalies are in an agreement with the
measured characteristic energies of the U- and N-electron-phonon scattering processes (EPS) which are obtained from the noise[2] and the ballistic temperature[3] PC spectra of Cu.

The measurements were made on the polycrystalline samples and therefore the characteristic energies for all crystallographic orientations could be observed in the PC spectra because of the accidental
orientation of PC axe. According to that we can explain this nonlinearities as follows.

The anomalies at 5.2 meV; 9.5 meV and 12.4 meV can be connected
with the threshold energies of the U processes of EPS. The anomalies at
7.4 meV; 11.2 meV; 13.5 meV and 24 meV can be connected with the threshold energies of the N processes of EPS. The following at 14.6 meV;
22 meV and 26 meV can be connected with U processes (at 20.2 meV with
N processes) or with Van Hove singularities in the phonon density of
states.

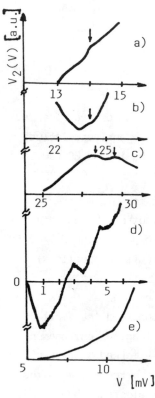

Fig.1

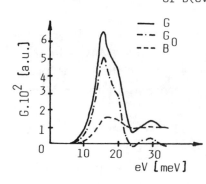

$\overline{\omega}$ = 6.06 meV; λ_{pc}^{K}=0.21

λ_{pc}^{N}= 0.18

Fig.2

Another explanation for the small humps at $\sim$10 meV; $\sim$13.5 meV and $\sim$20.5 meV is that they probably reflect the Kohn anomalies from the Cu phonon disperse curves observed by the neutron scattering[4].

The low energy anomaly at 2.9 meV can be connected with the non-Kohn anomaly which was . observed also by the neutron scattering[4]. This is the manifestation of the electron-wave-energy dependence probably.

The measured PC spectra contain the spectral part G_0(eV) and the background signal B(eV). It had been assumed the smooth shape of B(eV)[1]. The influence of the nonequilibrium phonons reabsorption on the nonlinear electrical conductivity of the metallic PC was studied theoretically[5,6] in the spectroscopic regime when the PC diameter d is much smaller than the energy relaxation length of electrons and phonons. If the impuls mean free path of phonons l_r is comparable with d, the non-smooth shape of B(eV) arises. If $l_r \lesssim$ d the incomplete reabsorption (IR) of the nonequilibrium phonons in the PC region arises and one introduces the quantity $\overline{\omega}$ - the individual PC characteristic connected with the nonequilibrium relaxation phonon frequency. If $l_r \rightarrow 0$ the complete reabsorption (CR) arises. Then $\overline{\omega}$= 0 and it is possible to determine the parameter $\langle q \rangle$ for phonons which depends on the PC geometrical form (e.g. orifice, channel, etc....).

In Fig.2 the obtained dependences of G_0(eV) and B(eV) of Cu are shown for

for IR ($\lambda_{pc}^{K,N,U}$ - the EPI parameter where K - usual background[1]; N - IR and U - CR; $\langle q \rangle_{orifice}$ = 0.29).

We obtained $\overline{\omega}$ = 0.3 - 32.4 meV in some our measured PC spectra of Cu. The $\overline{\omega}$ magnitude depends on the PC preparation way. The spectra obtained at the first touch of bulk electrodes performing the PC have $\overline{\omega} > 0$. Then $\overline{\omega}$ was decreasing after the contact pressure change without disconnection of PC. From that the $\overline{\omega}$ dependence on the PC region deformation (the magnitude l_r) follows. The background parameter γ were 0.1 - 0.3 in the case IR. We found $\gamma \sim \overline{\omega}^{-1/2}$. In Fig.2 the background nonlinearities are seen at the characteristic energies. Their intensities increase with the $\overline{\omega}$ decreasing. λ_{pc}^{N} values are smaller than λ_{pc}^{K} values in average 10% and they not depend on $\overline{\omega}$.

If $\overline{\omega}$ = 0 (CR) the γ was 0.3 - 0.7 and the intensities of background nonlinearities were higher. We obtained $\langle q \rangle$ changing from 0.3 to 0.6. We found that the influence of nonequilibrium phonons on the renormalization of electron spectrum increases with the $\langle q \rangle$ increasing. The change of $\lambda_{pc}^{U}/\lambda_{pc}^{K}$ is proportional to $\langle q \rangle^{-1/2}$. The obtained various values of $\langle q \rangle$ are in a contrast to the finding that PCs are formed in the orifice form. This is possible to explain by approximation of $\langle q \rangle$ [5].

This way of the background subtracting has no influence on the experimental ballistic regime criteria[1]. It is necessary to calculate with the reabsorption of the nonequilibrium phonons for the high γ PC spectra.

REFERENCES

1. Yanson, I.K., Kulik, I.O. and Batrak, A.G., J.Low Temp.Phys. 42, 527 (1981)

2. Akimenko, A.I., Verkin, A.B. and Yanson, I.K., J.Low Temp.Phys. 54, 247 (1984)

3. Reiffers, M., Flachbart, K. and Janos, S., Sov.JETP Lett. 44, 298 (1986)

4. Nilsson, G. and Rolandsson, S., Phys.Rev. B9, 3278 (1974)

5. Kulik, I.O., Sov.J.Low Temp.Phys. 11, 937 (1985)

6. Ickovitch, I.F. and Schechter, R.I., Sov.J.Low Temp.Phys. 11, 1176 (1985)

COOPERATIVE HOT-CARRIER DYNAMICS IN EXTRINSIC SEMICONDUCTORS SUPPORTED BY DIFFUSIVE PHONON COUPLING

J. Parisi[1], J. Peinke[1], B. Röhricht[2], and O.E. Rössler[3]

[1]Physikalisches Institut, Lehrstuhl Experimentalphysik II, Universität Tübingen, Morgenstelle 14, D-7400 Tübingen, Fed. Rep. Germany

[2]Laboratorium für Festkörperphysik, Eidgenössische Technische Hochschule Zürich, Hönggerberg, CH-8093 Zürich, Switzerland

[3]Institut für Physikalische und Theoretische Chemie, Universität Tübingen, Morgenstelle 8, D-7400 Tübingen, Fed. Rep. Germany

ABSTRACT

We report on long-range spatial correlation of self-generated current oscillations in electrically separated and diffusively coupled subsections of a single p-germanium crystal driven into low-temperature avalanche breakdown.

It is by now well established that complex nonlinear transport phenomena developing during low-temperature impact ionization breakdown of homogeneously doped semiconductor crystals can be attributed to the simultaneous presence of competing localized oscillation centers, arising spontaneously and interacting in a nonlinear way.[1]-[5] The study of the underlying spatial coupling mechanism is facilitated by the help of an experimental circuit configuration consisting of electrically separated sample subsections.[6][7] We report both experimental and numerical evidence that long-range spatial correlations between different crystal parts mainly result from diffusive energy coupling of autocatalytic hot-carrier subsystems via the rapid exchange of acoustical phonons emitted by the charge carriers. The concrete physical transport properties are finally discussed in the light of a generic reaction-diffusion model.

Our experiments were performed on single-crystalline p-doped germanium (with an acceptor impurity concentration of about 10^{14} cm^{-3} and typical dimensions of about $0.25 \times 2 \times 8$ mm^3), electrically driven into impact ionization breakdown at liquid-helium temperatures.[8] The semiconductor crystal was divided into separate parts by placing properly

spaced ohmic contacts on the specimen surface.[6)7)] The different subsections could then independently be operated in their hot-carrier breakdown regime, each being capable of generating inherent spontaneous current oscillations. In this way, the mutual coupling between the different oscillatory behavior of spatially and electrically separated subsystems can be investigated.

To probe for spatial correlation and long-range crosstalk between
different semiconductor subsections, one crystal part was biased in the
avalanche hot-carrier mode to act like an "emitter" of charge carriers,
phonons, and photons, while the other parts were kept in the sensitive
pre-breakdown mode to act like a "detector". We found that even for distances up to some millimeters the sensitivity of the passive detectors
was large enough to rigidly couple to the intrinsic current oscillations
of the active emitter. Moreover, if all subsections were gradually operated in their active oscillatory mode (under increase of the corresponding electric fields applied), we observed an abrupt resonance transition
from linearly correlated (phase-locked) to linearly anticorrelated
(phase-reversed) behavior. These experiments were carried out in such a
way that the extremely high lattice heat conductivity of germanium at
low temperatures remained as the only relevant spatial coupling mechanism. Further possible interactions via charge carriers, photons, or
magnetic fields could definitely be ignored. In fact, diffusive long-
range propagation of acoustical phonons between separate crystal parts
has been demonstrated experimentally by the help of superconducting bolometers attached to the surface of the semiconductor sample.

The above cooperative transport phenomena display a striking similarity to the behavioral characteristics of a generic phenomenological
model based on the well-known Rashevsky-Turing theory of morphogenesis.
The simplest version of Turing's model can be realized by a two-cellular
symmetrical reaction-diffusion system consisting of two cross-inhibitor-
ily coupled, potentially oscillating two-variable subsystems.[9)10)] This
4-D flow is capable of generating symmetry-breaking nonequilibrium phase
transitions from phase-locked coherent oscillations of both subsystems
to a phase lag of up to half a period (phase reversal). On the basis of
such an universal model approach, first attempts towards the development
of an adequate semiconductor transport theory involving the concrete

physical mechanisms of hot-carrier dynamics[11)12)] look highly promising. Here energy relaxation oscillations of two electronic subsystems are driven by the autocatalytic process of impurity impact ionization and coupled through heat diffusion via phonons emitted by the hot charge carriers.

1) Parisi, J., Peinke, J., Röhricht, B., and Huebener, R.P., "Turbulent Behavior of the Impact Ionization Induced Avalanche Breakdown in Germanium", in Proc. 18th Intern. Conf. Phys. Semicond., ed. O. Engström (World Scientific, Singapore, 1987) p. 1571.

2) Parisi, J., Peinke, J., Röhricht, B., and Mayer, K.M., "Self-Organized Formation of Spatial and Temporal Dissipative Structures in Semiconductors", Z. Naturforsch. 42 a, 329 (1987).

3) Peinke, J., Parisi, J., Röhricht, B., Wessely, B., and Mayer, K.M., "Quasiperiodicity and Mode Locking of Undriven Spontaneous Oscillations in Germanium Crystals", Z. Naturforsch. 42 a, 841 (1987).

4) Peinke, J., Parisi, J., Röhricht, B., Mayer, K.M., Rau, U., and Huebener, R.P., "Spatio-Temporal Instabilities in the Electric Breakdown of p-Germanium", Solid State Electron. 31, 817 (1988).

5) Stoop, R., Peinke, J., Parisi, J., Röhricht, B., and Huebener, R.P., "A p-Ge Semiconductor Experiment Showing Chaos and Hyperchaos", Physica 35 D, 425 (1989).

6) Röhricht, B., Wessely, B., Parisi, J., and Peinke, J., "Crosstalk of the Dynamical Dissipative Behavior Between Different Parts in a Current-Carrying Semiconductor", Appl. Phys. Lett. 48, 233 (1986).

7) Röhricht, B., Parisi, J., Peinke, J., and Huebener, R.P., "Spontaneous Resistance Oscillations in p-Germanium at Low Temperatures and their Spatial Correlation", Z. Phys. B - Condensed Matter 66, 515 (1987).

8) Parisi, J., Rau, U., Peinke, J., and Mayer, K.M., "Determination of Electric Transport Properties in the Pre- and Post-Breakdown Regime of p-Germanium", Z. Phys. B - Condensed Matter 72, 225 (1988).

9) Röhricht, B., Parisi, J., Peinke, J., and Rössler, O.E., "A Simple Morphogenetic Reaction-Diffusion Model Describing Nonlinear Transport Phenomena in Semiconductors", Z. Phys. B - Condensed Matter 65, 259 (1986).

10) Parisi, J., Peinke, J., Röhricht, B., Rau, U., Klein, M., and Rössler, O.E., "Comparison Between a Generic Reaction-Diffusion Model and a Synergetic Semiconductor System", Z. Naturforsch. 42 a, 655 (1987).

11) Schöll, E., Parisi, J., Röhricht, B., Peinke, J., and Huebener, R.P., "Spatial Correlations of Chaotic Oscillations in the Post-Breakdown Regime of p-Ge", Phys. Lett. 119 A, 419 (1987).

12) Schöll, E., Naber, H., Parisi, J., Röhricht, B., Peinke, J., and Uba, S., "Resonance Transition of the Spatial Correlation Factor of Self-Generated Oscillations in the Post-Breakdown Regime of p-Ge", Phys. Lett. A (to be published).

NOVEL EVIDENCE OF QUASIPARTICLE TRAPPING

C. PATEL, D. J. GOLDIE, N. E. BOOTH and G. L. SALMON

Department of Physics, University of Oxford,
Nuclear Physics Laboratory, Keble Road, Oxford OX1 3RH, United Kingdom

ABSTRACT Superconducting tunnel junctions (STJ) have been used to study the non-equilibrrium state established in a single crystal of In by pulsed laser irradiation. Using thin Cu films, in intimate contact with the crystal, a pronounced change is observed in the time-evolution of the excess quasiparticle density which indicates that the quasiparticles are trapped.

The magnitude of the signal current from a STJ, when biased below the sum of the energy gaps and exposed to a source of excitation, is inversely proportional to the volume of the films of the junction. To overcome this effect, Booth has recently proposed the idea of quasiparticle trapping (Fig. 1), where the excess of quasiparticles created in a superconductor of energy gap Δ, for example by a nuclear particle interaction, can be trapped into a smaller volume of a smaller gap superconductor.[1]

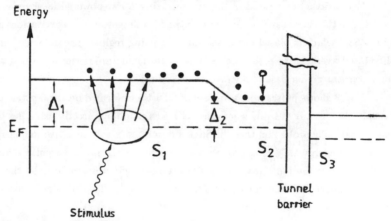

Figure 1 Quasiparticle trapping. An external stimulus breaks Cooper pairs in the superconductor S_1. The excitations diffusing into the smaller energy gap material S_2 scatter by phonon emission and are thus trapped.

Here we show that by a suitable choice of trap material, the magnitude and timing of the signal detected by the STJ can be improved.

The trapping time is proportional to the scattering time τ_s for a quasiparticle of energy E ($\approx \Delta$) to relax to close to the energy gap of the trap Δ_{trap} by phonon emission. This scattering time varies approximately as $(E/\Delta_{\text{trap}})^{-3}$ and for various superconductors scales as the characteristic time τ_0.[2]

To investigate this phenomenon we have performed a series of experiments to establish the time evolution of the quasiparticles and phonons in a 2 g single crystal of

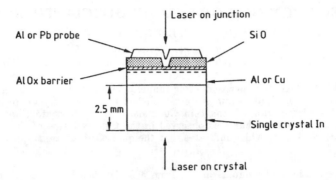

Laser on junction

Al or Pb probe

Si O

Al Ox barrier

Al or Cu

2.5 mm

Single crystal In

Laser on crystal

Figure 2 Experimental arrangement to investigate trapping from single crystal In into Al and Cu thin films.(Not to scale)

indium both with and without trapping. Fig. 2 shows a schematic of the experimental arrangement.

Excitations are created in the crystal using a low-powered laser pulse which is directed onto the crystal surface opposite to the STJ detectors. The excitations are detetcted by the STJ which is biased in the thermal tunneling region. The detected signals are amplified and averaged to reduce noise. The experiments are performed using a commercial He-3 cryostat giving base temperatures of 360 mK.

Fig. 3 shows pulses from a variety of STJ types formed on the crystal surface. Trace (a) was observed at 491 mK where the STJ was made by directly oxidizing the surface of the crystal. The probe film was Pb. Such a detetctor is only sensitive to quasiparticles and phonons of energy $\Omega \geq 2\Delta_{In}$. We obtain good fits to the observed pulse shape and to the leading edge of the pulse if we use a finite Green's function to describe the quasiparticle diffusion and a diffusion coefficient D_{qp} of 10 m^{2}s^{-1}. This implies a mean-free-path of order 100 μm for the quasiparticles.

Trace (b) was obtained from an Al-Al tunnel junction deposited on the crystal but where a surface layer impedes the transmission of the quasiparticles into the Al. This device thus detects phonons and has a threshold of $2\Delta_{Al}$. The decay time of this pulse and its temperature dependence are the same as for the quasiparticle pulse of trace (a). We find that we can account for this pulse shape using a simple model to describe the production of sub-$2\Delta_{In}$ phonons by the decay of recombination phonons into the detector range and subsequent decay below threshold.

We have also investigated the effect of trapping the quasiparticles into a normal metal. Here the proximity effect induces an energy gap in the Cu film used. Trace (c) shows a pulse observed at 369 mK. Although accurate results are only obtained by using a full Green's function to include the effects of diffusion on the pulse shape, the reduction in the quasiparticle decay time is evident.

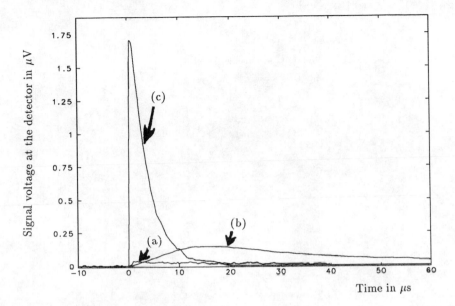

Figure 3 Pulses from STJ detectors on In crystal. See text for details .

The decay time of the quasiparticles is given by

$$\tau_{qp}^{-1} = \tau_r^{*-1} + \tau_{\text{trap}}^{-1}, \tag{1}$$

where τ_r^* is the effective recombination time in the superconductor and τ_{trap} is the trapping time. The trapping time is given by

$$\tau_{\text{trap}} = \frac{V_{\text{crystal}}}{V_{\text{trap}}} \tau_s, \tag{2}$$

where V_{crystal}, V_{trap} are the volumes of the crystal and trap respectively.

We find $\tau_{qp} = 3.4$ μs for this 250 nm Cu film implying a trapping time of (3.8 ± 0.1) μs. We have performed a series of experiments to investigate the effect of film thickness and energy gap on the pulse decay time.[3] We find a value of τ_0 of (1.0 ± 0.2) ns for Cu with proximity effects independently of film thickness and energy gaps in the range $0 < \Delta_{\text{Cu}} \leq 180$ μeV.

In conclusion we have studied the time-evolution of the quasiparticles and sub-gap phonons produced in superconducting In by laser irradiation. Using Cu as a trap reduces the quasiparticle decay time in the crystal and increases the amplitude of the detected signal.

REFERENCES

1] Booth N.E., Appl. Phys. Lett. **50**, 293 (1987)

2] Kaplan S.B. *et al.* Phys. Rev. B **14**, 4854 (1976)

3] Goldie D.J., Booth N.E., Patel C., Salmon G. *submitted to* Phys. Rev. Lett. (1989)

14

Transport Phenomena

PICOSECOND OPTICS STUDIES OF THE KAPITZA RESISTANCE BETWEEN SOLIDS AT HIGH TEMPERATURE

R. Stoner, D.A. Young, H.J. Maris, J. Tauc
Department of Physics, Brown University, Providence, RI 02912, USA.

H. T. Grahn
Max-Planck Institut fur Festkorperforschung, 7000 Stuttgart 80, FRG.

In this paper we report on the use of picosecond optics techniques to measure the Kapitza conductance between metal films and alkali halide crystals at temperatures up to room temperature. The experiment is shown schematically in Fig.1. A metal film (Au or Al in these experiments) of thickness 65-330Å is heated by a light pulse (the "pump" pulse) of duration about 1psec. The change in the temperature ΔT of the film produces a change in the optical reflectivity $\Delta R(t)$ which is measured by a time-delayed probe pulse. In the experiments reported here the film was heated by pump pulses of energy about 0.1nJ focussed onto an area of radius about 20μ. Under these conditions ΔT is typically 10K or less, and so we can assume that $\Delta R(t)$ is proportional to $\Delta T(t)$. Thus the measurement of the transient change in the reflectivity enables us to study the cooling of the thin film with extremely good time resolution. Results obtained in this way for the cooling of a 95Å Al film on a LiF substrate at 300K are shown in Fig.2.

From these cooling curves it is possible to determine the Kapitza resistance, R_K at the film-substrate interface. The simplest method is to use the fact that the initial rate of cooling must be unaffected by the thermal properties (specific heat and thermal conductivity) of the substrate. Thus, initially

$$\frac{d \log\Delta T(t)}{dt} = \frac{-\sigma_K}{C_F d} \qquad (1)$$

where σ_K is the Kapitza conductance, C_F is the specific heat of the film per unit volume and d is its thickness. Using Eq.(1) we have obtained the results for the Kapitza conductance σ_K for Au films shown in Fig.3. These results were for films evaporated in a vacuum of about 10^{-6} Torr onto air-cleaved (111) surfaces of CaF_2 and BaF_2 and (110) surfaces of LiF. The data exhibit the following features:

1) The Kapitza conductance has only a small variation with temperature in the range studied (70-300K). This is not unexpected since the Debye temperature of Au is only 170K, and model calculations[1] predict that for $T \geq 0.3\theta_D$, σ_K should be almost constant .

2) From the data one can calculate an average phonon transmission coefficient $<t>$ using the relation[2] $<t>=4\sigma_K/C_F d$. The results for $<t>$ lie in the range 0.05 to 0.15.

3) It is expected[1] that the conductance will be larger for combinations of materials in which the phonon densities of states match. Since θ_D is 170K for Au, and 282K, 514K, and 737K for BaF_2, CaF_2 and LiF respectively, one should have σ_K decreasing going from BaF_2 to CaF_2 to LiF. This is in agreement with the experimental results, but the variation of σ_K is probably smaller than expected considering the large variation of θ_D amongst the three substrates.

To understand the complete cooling curve (instead of just the initial exponentially decaying part) it is necessary to consider the heat transport processes in the substrate. The simplest approach is to use Fourier's law together with literature values for the thermal conductivity. This gives a fairly good description of the entire cooling curve at room temperature, but does not work well for all samples at lower temperatures (eg. 100K). Some of these difficulties may be due to the presence of a damaged surface layer of the substrate (produced in the process of cleaving) with a lowered thermal conductivity. In addition, in the lower part of the temperature range studied the Umklapp collision time τ_U is comparable to the time scale of the experiment. In this regime it is not permissible to treat heat flow in the substrate as a simple diffusive process, and a more sophisticated theory allowing for the separate rôles of N and U-processes is required. For Al films on LiF and CaF_2 we have observed cooling curves in which the temperature does not decrease monotonically with time, and these may possibly be related to such "second sound" effects. We will describe these data in a subsequent paper.

This work was supported in part by the US DOE under grant DEFG0286ER45267.

References

1. Young, D.A., and Maris, H.J., in this volume.
2. See, for example, Challis, L.J. , J. Phys. C 7, 481 (1974).

Fig.1 Schematic diagram of the experiment.

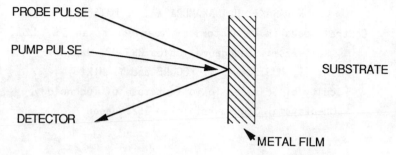

PROBE PULSE

PUMP PULSE

SUBSTRATE

DETECTOR

METAL FILM

Fig.2 Cooling of Al film on LiF at 300K.

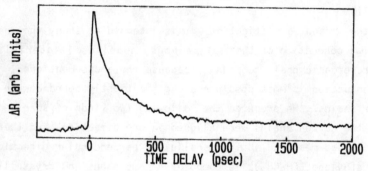

Fig.3 Kapitza conductance as a function of temperature.

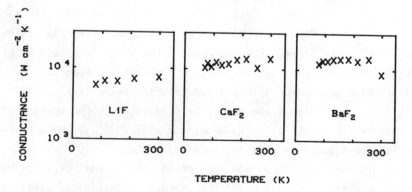

PHONON CONDUCTION IN ÓRIENTED ORGANIC MATERIALS

K. KAWASAKI, H. NAKAMURA and M. MATSUURA
Central Research Laboratories, Idemitsu Kosan Co., Ltd.,
Kami-izumi, Sodegaura, Chiba 299-02, Japan
T. MUGISHIMA, Y. KOGURE and Y. HIKI
Faculty of Science, Tokyo Institute of Technology,
Oh-okayama, Meguro-ku, Tokyo 152, Japan

INTRODUCTION

In the present investigation, it is intended to study the one-dimensional conduction of thermal phonons in solids. Materials chosen are organic chain compounds, because the above-mentioned phonon conduction is most possible along the tightly bound carbon molecular chains. We prepared the following two kinds of specimens in which the carbon chains were aligned in one direction, and thermal diffusivity was measured in the parallel or perpendicular direction.

Polyethylene $(CH_2CH_2)_n$ is composed of amorphous and crystalline parts, and the crystallites are randomly oriented. By extruding the material, the crystallites are oriented, and carbon chains crossing crystalline and amorphous parts are formed in the extruded direction.

The Langmuir-Blodgett (LB) film is a stacking of monolayers of organic molecules formed on a substrate. The molecule should be of amphiphilic, namely, having the polar end (head) and nonpolar chain (tail), which are hydrophilic and hydrophobic, respectively. The tails are aligned almost accurately perpendicular to the surface of the substrate. The LB films of arachidic acid $CH_3(CH_2)_{18}COOH$ formed on glass substrates are used in the present study.

As a measure of the phonon conduction, the thermal diffusivity D (= $\kappa/\rho C$; where κ, C, and ρ are the thermal conductivity, specific heat, and material density) can be chosen. By considering the gas kinetics relation, $\kappa = (1/3)Cvl$; where v and l are the velocity and mean free path of phonons, the simple relation $D \propto l$ can be utilized.

ORIENTED POLYETHYLENE

Polyethylene material was molded, and extruded through a conical die. Disk—shaped specimens 1 mm in thickness with their axes parallel or perpendicular to the extruded direction were prepared. The thermal diffusivities of the specimens $D_\shortparallel$ and $D_\perp$, and of an unextruded one D_o, were measured at low temperatures. The measurements have been made by the laser—flash method: the front face of specimen is heated by a laser pulse, the time (t) variation of back face temperature (ΔT) is recorded, and the diffusivity is determined from the ΔT—vs—t curve.[1]

The experimental data are shown in Fig. 1. The diffusivities $D_\shortparallel$, $D_\perp$, and D_o decrease with increasing temperature; the diffusivity in extruded specimen is strongly anisotropic (e.g. $D_\shortparallel/D_\perp$ = 4.7 and 45 at T = 2 and 300 K); the diffusivity D_o is between the two extremes $D_\shortparallel$ and $D_\perp$; detailed temperature dependences of three diffusivities are different with each other. These results have thoroughly been analyzed,[2] and the essential points are as the following.

In the case of heat conduction perpendicular to the extrusion direction, the crystalline and amorphous parts are connected in series arrangement; while for the parallel conduction, a long carbon—chain part is added in parallel arrangement to the above two parts. The origins of the thermal resistances in the crystalline and amorphous parts are the usual phonon Umklapp process and the scattering due to some localized low—energy phonons, respectively. Because of the two scattering mechanisms, the thermal diffusivity decreases with temperature. If one—dimensional phonon conduction is occurring along the long carbon chains, the decrease of the diffusivity is weakened, since no Umklapp process can occur in the conduction. Such behavior can really be seen in the case of the parallel thermal conduction.

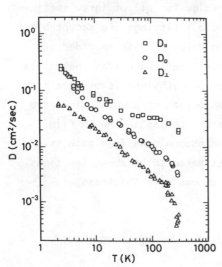

Fig. 1. Temperature dependence of thermal diffusivity.

LANGMUIR–BLODGETT FILMS

The specimens have been prepared as follows. On a glass plate (1 mm in thickness), an indium film thermometer (30 nm) is sputtered. The plate is kept vertically and dipped into a monolayer of arachidic acid spread on a water, and is moved up and down repeatedly to form a thick LB film on the indium surface. The produced LB film is of the Y–type: the stacking of the molecules is head to head or tail to tail. The thickness of a monolayer determined by x rays is 2.7 nm, being in accordance with the molecular length of the material. An indium film heater (100 nm) is further evaporated onto the LB film.

The room–temperature thermal diffusivity of the LB films has been measured by the pulse heating method, by applying a voltage pulse (32 nsec in width) to the heater. The ΔT–vs–t data for three specimens with different thicknesses are shown in Fig. 2. The data for the 2000–layer specimen were analyzed with considering the heat–loss effect,[1] and also the finite pulse–time effect since the specimen was very thin compared with that of polyethylene. The thermal diffusivity was determined as $D = 0.40$ cm^2/sec. The lines in the figure are the fitted curves with using the above D value for all of three specimens, and all fittings are acceptable. The value of D is an order of magnitude larger than that of $D_{\parallel}$ for polyethylene, showing a high conduction of phonons along the carbon chains in the LB film. The phonon mean free path is estimated to be about ten times the monolayer thickness.

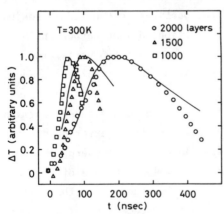

Fig. 2. Recorded data of temperature variation.

REFERENCES
1) Kogure, Y., Mugishima, Y. and Hiki, Y., J. Phys. Soc. Jpn. 55, 3469 (1986).
2) Mugishima, T., Kogure, Y., Hiki, Y., Kawasaki, K. and Nakamura, H., J. Phys. Soc. Jpn. 57, 2069 (1988).

ELECTRON-PHONON COUPLING IN RHODIUM THIN FILMS*

E.T. Swartz** and R.O. Pohl
Laboratory of Atomic and Solid State Physics
Cornell University, Ithaca, N.Y. 14853

The coupling between electrons and phonons in bulk metals at low temperatures is quantitatively understood only in the case where the metal is a perfect single crystal. To our knowledge, there has been no systematic study of the effects of surfaces on the coupling between electrons and phonons. Surfaces and interfaces could well affect the coupling between electrons and phonons in thin metallic films at low temperatures, because significant inelastic[1] scattering of both electrons and phonons is likely to occur at the metal-substrate interface and at the free surface. (Elastic scattering will not couple energy between the two systems.) We have measured the thermal coupling between electrons and phonons in thin (0.5 μm) films of Rh:Fe(0.5 at%) in the temperature range 1 K to 5 K. We find evidence for a correlation between scattering of phonons at the metal-substrate interface (indicated by the thermal boundary resistance at the interface) and the electron-phonon scattering (indicated by the phenomenological electron-phonon thermal resistance). It appears that the greater the scattering of phonons at the interface, the greater is the electron-phonon coupling. The interface appears to affect the value of the electron-phonon coupling by a factor of two or more. Therefore the electron-phonon coupling in thin films at low temperatures may well be as much a property of the surfaces as it is a property of the film itself.

The experimental arrangement is shown in Fig. 1. Details of the technique are published elsewhere.[2] The temperature difference between the electrons in the film and the lattice of the substrate is due to the electron-phonon phenomenological thermal resistance in the film and to the thermal boundary resistance between the film and the substrate. The thermal boundary resistance has a temperature dependence of $R_{bd} \propto T^{-3}$, and the electron-phonon thermal resistance in pure, single crystal metals has a temperature dependence of $R_{ep} \propto T^{-4}$. We naively assume that R_{ep} in the film has a T^{-4} temperature dependence, and that R_{bd} and R_{ep} simply add (only approximately true). Therefore, in a plot of the total thermal resistance multiplied by T^4 versus T, the data should fall on a straight line, the slope should be R_{bd}, and the intercept should be R_{ep}. In Fig. 2 are such a plots for Rh:Fe deposited on sapphire. The polishing treatment of the substrate is shown for each measurement. The data for each sample fall remarkably well on a straight line, apparently allowing independent assignment of R_{bd} and R_{ep} as suggested above. In Table 1 is a summary of the values of R_{bd} and R_{ep} for the various substrates and substrate polishing treatments. For Rh:Fe on sapphire in this temperature range, the lower the boundary resistance, the less is the phonon

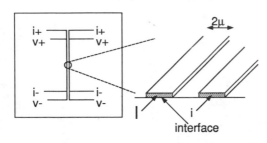

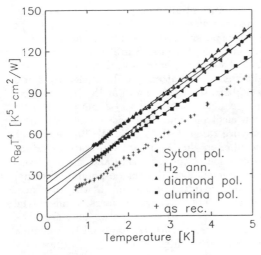

Figure 1. Experimental geometry. Two parallel, thin (0.4 μm thick) 4-lead resistors are patterned on an insulating substrate. One resistor (the left one) is Joule heated and its resistance (and therefore the temperature of the electrons) is measured. The resistance of the second resistor is measured using negligible excitation current. Its temperature is thus the same as that of the phonons of the underlying substrate.

Figure 2. Total thermal resistance plotted as a boundary resistance times T^4 versus temperature for Rh:Fe deposited on five different sapphire substrates with differing surface preparation. The intercept multiplied by the film thickness is the electron-phonon phenomenological thermal resistance. The slope is the thermal boundary resistance. All substrates were carefully cleaned immediately before deposition. For the as-received samples the substrate surface is as received from the manufacturer (an unspecified and proprietary treatment which utilizes Syton at some point). For the data marked alumina, the substrate was polished with a sequence of alumina powders, finishing with 500 Å. For the data marked Syton, the substrate was given the same sequence of alumina polishes, and then Syton polished for 1 hour. For the data marked H$_2$, substrates given the same alumina and Syton polishing steps were subsequently annealed in a hydrogen atmosphere at 1470 K for 30 min. For the data marked diamond, substrates given the same alumina and Syton polishing steps were subsequently polished with 2000 Å diamond polishing paste. The surface of the quartz substrate (data not shown) was Syton polished.

Table 1. Values of $R_{bd}T^3$ and $R_{ep}T^4$ for the various substrates and substrate polishing treatments.

Substrate material	surface treatment	R_{bd} K^4/(W/cm^2)	$10^4 R_{ep}$ K^5/(W/cm^3)
sapphire	hydrogen	21	5.9
sapphire	alumina	20	5.9
sapphire	diamond	23	4.9
sapphire	Syton	24	4.0
sapphire	as-received	21	2.5
quartz	as-received	13	2.7

scattering at the surface. The boundary resistance to quartz is lower than to sapphire because of the difference in the densities and velocities of sound in the two substrates. It is found that rough surfaces (diamond polished) are strong scatterers of phonons. Syton polished surfaces also scatter phonons very strongly, whereas hydrogen polished and alumina polished surfaces are poor scatterers of phonons in this temperature range. These conclusions are discussed elsewhere, see ref. 2; similar conclusions have been drawn by others.[3] As shown in Table 1, not only does the Syton polished surface scatter phonons strongly, it also lowers R_{ep}. The diamond-polished surface scatters phonons and also promotes electron-phonon coupling. In contrast, the hydrogen and alumina polished surfaces (poor phonon scatterers) do not appear to promote electron-phonon coupling. The Syton polished substrates are much smoother than the diamond polished substrate, yet the electron-phonon coupling is stronger on the Syton polished substrates. This suggests that roughness alone may not be effective in promoting electron-phonon coupling, that the type of defect abundant at Syton polished surfaces (whatever that is?) is quite effective in promoting electron-phonon coupling, and that much of the scattering at this type of defect is inelastic.

Thus, we have seen evidence that electron-phonon coupling in thin metal films may not be an intrinsic property of the film, but may depend on the properties of the interface between the film and the substrate on which it is deposited. We also conclude that damaged interfaces cause a significant amount of *inelastic* scattering. The magnitude of the electron-phonon phenomelogical thermal resistance can vary by about a factor of two, depending on the surface preparation. Films deposited on substrates with damaged surfaces have stronger electron phonon coupling than films deposited on substrates with near-ideal surfaces. The as-received substrates (which have been Syton polished in some manner by the manufacturer) both apparently have defects which are very effective in promoting electron-phonon thermalization. The pattern of low R_{bd} implying high R_{ep} is broken for the as-received sapphire, indicating there may be effects which we have not considered. However, an experimental error of as little as 10% could appear to break the pattern; the as-received sapphire substrate was the first substrate studied, and the experimental technique may not have been fully developed.

* This work was supported by Semiconductor Research Corporation contract 82-11-001.

** Present address: RMC-*Cryosystems*, Tucson, Arizona

1. Inelastic phonon scattering has been seen at surfaces and interfaces. For example, see L. J. Challis, A. A. Ghazi, and M. N. Wybourne, Phys. Rev. Lett. **48**, 759 (1982); and A. F. G. Wyatt, in *Nonequilibrium Superconductivity, Phonons, and Kapitza Boundaries,* edited by K. E. Gray (Plenum, NY, 1981), p. 31.

2. E.T. Swartz and R.O. Pohl, Rev. Mod. Phys. **61**, 605 (1989).

3. Diffuse scattering at Syton polished interfaces and specular scattering at alumina polished interfaces have been seen by S. Burger, W. Eisenmenger, and K. Lassmann, LT17 **1**, 659 (1984). Specular scattering at hydrogen polished sapphire surfaces has been seen by C. G. Eddison and M. N. Wybourne, J. Phys. C **18**, 5225 (1985).

RELAXATION OF INITIALLY SPATIALLY UNHOMOGENEOUS STATES OF PHONON GASES SCATTERED BY POINT MASS DEFECTS EMBEDDED IN ISOTROPIC MEDIA

Cz. Jasiukiewicz and T. Paszkiewicz, Institute of Theoretical Physics, Wrocław University, ul. Cybulskiego 36, Pl-50-205 Wrocław, Poland

1. INTRODUCTION

Suppose that we deal with a massive dielectric specimen at temperatures much lower than the Debye temperature. Suppose further, that this specimen is ideal, apart of presence a small concentration, g, of point mass defects (e.g. isotopic impurities). An initial spatially unhomogeneous state of a gas of long-wave acoustic phonons is created (e.g. in the form of a heat pulse). Then, the main mechanism providing the relaxation of this initial state is scattering of phonons by point imperfections. The scattering events are elastic, so the collisions homogenize an initially unhomogeneous gas, isotropize an initial distribution of the wave vectors directions $\hat{k}$, and mixe their polarizations. Since the phonon energy $\hbar\omega$ is conserved in scattering events, the gas does not reach the complete thermodynamic equilibrium. Only collisions with other quasiparticles (being in contact with a thermal bath) provide the mechanism of relaxation towards this complete equilibrium.

The purpose of this paper is to describe the evolution in time of an initial phonon distribution function (PFD).

2. THE EXPLICIT TIME DEPENDENCE OF PFD

Each acoustic phonon is characterized by the frequency ω, the wave vector $k=(k=|k|, \hat{k}=k/k)$ and polarization index j. For simplicity we restrict ourselves to the case of isotropic elastic media. Then, there are (D-1) types of purely transversal phonons (D=2,3 is the dimensionality of system): $j=t_1, t_2$ (D=3); $j = t$ (D=2), and one type of longitudinal $- j=\ell$. Since we deal with long-wave acoustic phonons in an isotropic medium

$$\omega = c_j k \qquad (D=3: \ c_{t_1} = c_{t_2} = c_t; \ \text{for both D:} \ 0 < c_t \leq c_l)$$

Introduce the ratio of velocities $p = c_t/c_l$ ($0 < p \leq 1$) and the Debye velocity c_d.

PDF obeys the Boltzmann-Peierls equation (BPE) with the collision integral related to the scattering on point mass defects suplemented with an initial condition

$$f(\omega, \hat{K}, r, t=0) = F(\omega, \hat{K}, r),$$

where $F(\omega, \hat{K}, r)$ is a given function of ω, $\hat{K} = (\hat{k}, j)$ and r.

Solving BPE we obtained the explicit expression for the Fourier transform of PFD

$$f'(\omega, \hat{K}, q, t) = \int d^D r \ e^{iqr} f(\omega, \hat{K}, r, t). \qquad (D=2,3)$$

Next, we shall introduce the dimensionalless variable $\varkappa$

$$\varkappa = qc_d\tau \sim l_{pi}/\lambda, \qquad (q = |q|)$$

where $\tau \sim \omega^{-4} g$ is an average time lapsing between succesive collisions, l_{pi} – is the mean free path and $\lambda \sim q^{-1}$ is the length characterizing the spatial unhomogeneity related to the q-th Fourier harmonic.

The explicit formula for $f(\omega, \hat{K}, \hat{q}, \varkappa, t) \equiv f'(\omega, \hat{K}, \ \hat{q}q(\varkappa))$ reads:

$$f(\omega, \hat{K}, \hat{q}, \varkappa, t) = f_o(\omega) + e^{-t/\tau} \Phi_c(\omega, \varkappa, q, t) + \sum_{n=1}^{2} e^{\zeta_n(p, \varkappa)t/\tau} \Phi_n(\omega,$$

$$\varkappa, \hat{q}, t)\theta(\varkappa_c^{(n)} - \varkappa). \qquad (1)$$

where $\theta(x)$ is the Heaviside step function. The distribution function of the incomplete equlilibrium, $f_o(\omega)$, depends on the initial state (e.g. on an initial density of phonons of frequency ω). The exponential functions are universal, i.e. they do not depend on the initial condition. The preexponential functions Φ_c, Φ_n are nonuniversal.

Inspecting the formula (1) we found that the third and fourth terms of it are nonzero if respectively: $0 < \varkappa < \varkappa_c^{(1)}$, $0 < \varkappa < \varkappa_c^{(2)}$. The parameters $\varkappa_c^{(n)}$ (n=1,2) depend on the ratio of velocities p and dimensionality D

$$x_c^{(1)} = \frac{\pi}{2} g_4^{(3)}(p) \quad (D=3), \quad x_c^{(1)} = g_4^{(2)} = \sqrt{2}\frac{(p^3 + 1)}{(p^2 + 1)^{3/2}} \quad (D=2),$$

where $\quad g_n^{(3)}(p) = 3^{(n-3)/2}(p^n + 2)/(p^3 + 2) \quad (n=1,4).$ (2)

Let us notice that $g_4^{(2)} \geq g_4^{(3)}$ ($0 < p \leq 1$). Similarly

$$x_c^{(2)} = \frac{\pi}{5} g_4^{(3)}(p) \quad (D=3), \quad x_c^{(2)}(p) = \frac{1}{2} g_4^{(2)}(p). \quad (D=2)$$

Thus, for both two and three dimensional systems: $x_c^{(1)}(p) > x_c^{(2)}(p)$ ($0 < p \leq 1$).

Consider two asymptotic regimes. In the collision dominated regime $x \ll 1$ (or $l_{pi} \ll \lambda$) and

$$\zeta_1^{(D)}(x,p) = \frac{1}{D} x^2 g_1^{(D)}(p), \qquad \qquad \zeta_2 = \begin{cases} -3/5 & D=3 \\ -1/2 & D=2 \end{cases}.$$

where $g_1^{(2)}(p) = 1$ and $g_1^{(3)}(p)$ is given by eq. 2. Introduce
the diffusion constant $\mathcal{D}$:

$$\mathcal{D} = \frac{1}{D} c_d^2 \tau \, g_1^{(D)}(p).$$

Since $0 < g_1^{(3)}(p) \leq 1$, the diffusion coefficients obey the
following inequality: $\mathcal{D}_2 \geq \mathcal{D}_3$. We see that in the
collision dominated regime for long lapses of time ($t \gg \tau$)

$$f(\omega, x, \hat{q}, t) - f_o(\omega) \sim \exp(-\mathcal{D}q^2 t). \quad (D=2,3)$$

For x $x_c^{(n)}$ the suitable $\zeta_n^{(D)}$ -1 ($n= 1,2$ and $D = 2,3$).
Hence we observe the crossover from kinetic to diffusive
regime. In this respect the considered system resembles
the Lorentz gas [2].

In the collisionless regime (i.e. for $x > 1$ or $l_{pi} > \lambda$)
the distribution function relaxes essentially with the
characteristic time τ.

REFERENCES

[1] Gurevich,V.L., Transport in Phonon Systems,
North-Holland, Amsterdam, 1986 (sect. 7,9).

[2]Jasiukiewicz Cz.,Paszkiewicz T.,Physica 145A, 239(1984)

ELASTIC SCATTERING OF PHONONS ON POINT MASS DEFECTS: SPECTRUM OF COLLISION INTEGRAL FOR LONG - WAVE ACOUSTIC PHONONS IN TRANSVERESELY ISOTROPIC MEDIA

T. Paszkiewicz, Institute of Theoretical Physics, Wrocław University, ul. Cybulskiego 36, Pl-50-205 Wrocław, Poland

M. Wilczyński, Institute of Low Temperature Physics and Structural Research, Polish Acad. Sci., ul. Okólna 2, Pl-50-422 Wrocław, Poland

1. INTRODUCTION

We discuss relaxation of an initially spatially homogeneous state of a gas of long-wave acoustic phonons scattered by point mass defects embedded in transversely isotropic media (TIMs). It is assumed that the concentration of these imperfections, g, is small and one deals with an otherwise perfect massive dielectric specimens at temperatures much lower than the Debye temperature T_D. Then, the mean free path for the phonon-impurity scattering, l_{pi}, mean free path for phonon-phonon collisions, l_{pp}, and the length d characterizing a specimen fulfil the inequalities

$$l_{pp}, d \ll l_{pi}.$$

At such conditions the main machanism of the initial state relaxation is the scattering of phonons on imperfections.

For homogeneous state of massive specimens the spectrum of collision operator determines solely the relaxation toward the equilibrium state. Therefore, we shall study the spectrum of collision integral operator related to phonon-imperfection collisions, B_{pi}.

Obviously, the elastic properties of media affect the spectrum of B_{pi}. One can study this influence in the whole region of elastic stability. We found that changes of elastic constants may substantially modify the spectrum of collision integral. However, here we shall confine ourselves only to the case of the existing materials. As a reference source of elastic constant values for TIMs we chose McCurdy's paper [1].

2. CHARACTERISTICS OF PHONON GASES

Acoustic phonons are characterized by the frequency ω, wave vector $\mathbf{k} = (k = |\mathbf{k}|, \hat{\mathbf{k}} = \mathbf{k}/k)$ and polarization $j, (j=0,1,2)$. For small k the dispersion law is linear:

$$\omega(k,j) = c(\theta_k, j)k \quad (\hat{k}\hat{c} = \cos\theta_k).$$

An homogeneous state of this phonon gas is described by the phonon distribution function (PDF), which depends on the frequency ω, direction of propagation $\hat{\mathbf{k}}$, polarization index j and time t: $f = f(\omega, \hat{K}, t)$, $(\hat{K} = (\hat{k}, j))$. PDF obeys the Boltzmann-Peierls equation, which for an homogeneous state of a gas filling a massive specimen, reads [2]

$$\frac{\partial f(\omega, \hat{K}, t)}{\partial t} = B_{pi} f\omega, \hat{K}, t).\qquad(1)$$

Eq.1 should be supplemented with an initial condition: $f(\omega, \hat{K}, t=0) = F(\omega, \hat{K})$, where $F(\omega, \hat{K})$ is a given function of ω and $\hat{K}$. The collision integral operator B_{pi} consist of two terms

$$B_{pi} = -\nu(\Theta_k)I + B_1, \quad (\Theta_k \equiv \theta_k, j), \quad \text{where}\qquad(2)$$

B_1 is an integral operator. The kernel of it is essentially the scattering cross section for the phonon-defect collisions [3]. I is the unit operator, and

$$\nu(\Theta_k) = \varkappa_3 + \frac{1}{3}(1 - 3\varkappa_3)\sin^2\theta_e(\Theta_k),$$

$\theta_e^{(j)}$ being the angle between polarization vector $e(\Theta_k)$ and the vector $\hat{c}$. Above, $\varkappa_3$ is a constant characterizing TIMs.

The spectrum of B_{pi} consists of a discrete and a continuous part. A general discussion of the spectrum structure will be given in [3]. Below, we describe the spectrum of B_{pi} for the materials mentioned above [1]. For them the continuous part of the spectrum makes up an interval $\langle -\nu_{max}, -\nu_{min} \rangle$, where

$$-\nu_{max} = \min[-\varkappa_3, \frac{1}{2}(\varkappa_3 - 1)], \quad -\nu_{min} = \max[-\varkappa_3, \frac{1}{2}(\varkappa_3 - 1)].$$

The discrete part contains always the eigenvalue $\nu_o^{(M)} = 0$ corresponding to the only conserved quantity - the total number of phonons of energy $\hbar\omega$. Additionally there are three negative eigenvalues : $-\nu_1^{(F)}$, $-\nu_1^{(G)}$ and $-\nu_1^{(M)}$. For some materials $\nu_1^{(G)}$, or/and $\nu_1^{(F)}$ are absent or the interval $<-\nu_{max}, -\nu_{min}>$ is very narrow. The results of calculation arranged according to a growing anizotropy parameter α [3] are collected in the **Table 1**. Let us add that the time variable is scaled with the factor $\Lambda^{-1} = (3v_0/4\pi)g\omega^{-4}c_D^{-3}$ (v_0 is the elementary cell volume), i.e. $t \to \tau = t\Lambda^{-1}(\omega,g)$.

REFERENCES

[1] McCurdy, A.K., Phys. Rev. **B9**, 466,(1974).

[2] Gurevich, V.L., Transport in Phonon Systems,Amsterdam, North - Holland, 1986.

[3] Paszkiewicz, T., Wilczyński, M., in preparation.

Table 1

Material	$-\nu_{max}$	$-\nu_{min}$	$-\nu_G$	$-\nu_F$	$-\nu_M$	α
Hafnium	−0.342	−0.315	−0.261	−0.186	−0.100	.042
Magnesium	−0.339	−0.320	−0.267	−0.184	−0.093	.052
Cadmium sulfide	−0.343	−0.313	−0.274	−0.183	−0.078	.070
Titanium	−0.351	−0.297	−0.259	−0.187	−0.094	.070
Yttrium	−0.337	−0.325	−0.270	−0.183	−0.091	.078
Zirconium	−0.333	−0.332	−0.276	−0.181	−0.085	.078
Beryllium	−0.351	−0.297	−0.241	−0.191	−0.127	.085
Cadmium selenide	−0.347	−0.305	−0.273	−0.184	−0.075	.087
Rhenium	−0.342	−0.315	−0.274	−0.183	−0.081	.094
Cobalt	−0.359	−0.281	−0.264	−0.187	−0.070	.146
Thallium	−0.419	−0.161	none	none	−0.068	.151
Helium 4[a]	−0.368	−0.263	−0.256	−0.189	−0.057	.221
Cadmium	−0.514	−0.242	none	−0.151	−0.093	.280
Zinc	−0.525	−0.237	none	−0.155	−0.112	.294
Helium 4[b]	−0.395	−0.209	none	−0.199	−0.063	.307

[a] Elastic const. measured at a molar volume of 20.37 cm^3
[b] Elastic const. measured at a molar volume of 20.5 cm^3

HYDRODYNAMIC EFFECTS IN SOLIDS

R.N.GURZHI

Physico-Technical Institute of Low Temperatures,
Acad.Sci. UkSSR, Kharkov,

L.P.MEZHOV-DEGLIN

Institute of Solid State Physics Acad. Sci. USSR,
142432 Chernogolovka, Moscow distr, USSR.

Under certain conditions the motion of a guasiparticlegas (phonon, magnon or electron gases) in large nearly perfect crystals may be analogous to space nonuniform flow of a viscous gas or liquid along a tube. This may result in significant changing of temperature and dimension dependence of the proper kinetic coefficients (thermal or electrical conductivity, thermopower, etc.). Actually, the onset of Poiseuille flow of the phonon gas and Knudsen minimum were observed in thermal conductivity of helium and bismuth crystals. Phonon second sound was observed in ^{4}He, ^{3}He, NaF and Bi crystals. Recently observed the minimum in electrical resistance of potassium samples can also be explained by manifestation of weak hydrodynamics under conditions that the electron-impurity scattering predominates over the normal electron-phonon scattering.

The review papers on hydrodynamic phenomena in solids were published earlier by R.N.Gurzhi, Usp.Fiz.Nauk (USSR),1968,94,689 and H.Beck,P.F.Meier and A.Thellung, Phys.Stat.Sol.,1974,24,11. The main purpose of this report is to sum up the results of the recent theoretical and experimental investigations and to discuss some of the predicted effects not revealed up to date.

ABSTRACT:

1. Introduction.
2. Transport phenomena in dielectrics.
   A. Theory.
   2.1. Hydrodynamic effect in heat conductance of phonon or spin wave systems:
      a. Poiseuille flow.
      b. Knudsen minimum.
      c. Nonlocal hydrodynamics.
   2.2. Second sound.
   2.3. High order anharmonic effects in thermal conductivity of ferrodielectrics.
   B. Experiments.
   2.4. Hydrodynamic effects in thermal conductivity of nearly perfect crystals.
   2.5. Phonon-phonon drag thermopower in Bi crystals.
   2.6. Phonon second sound.
3. Transport phenomena in metals.
   3.1. Hydrodynamic effects in the electrical conductivity:
      a. Uniform flow of electron gas in thick crystals.
      b. Nonuniform flow.
      c. Weak hydrodynamics in impurity limit and minimum in electrical resistivity of potassium samples.
      d. Hydrodynamic effects in two-dimensional metallic systems.
   3.2 Thermal conductivity and electron second sound in compensated metals.
4. Conclusion.

1322

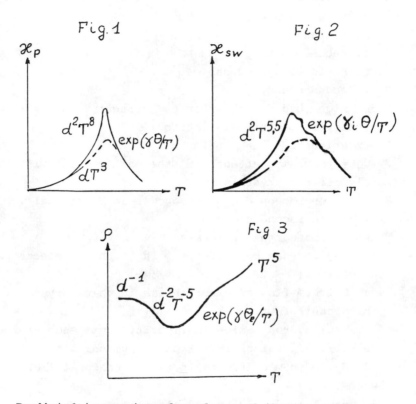

Fig. 1

Fig. 2

Fig 3

Predicted temperature dependence of the thermal conductivity of dielectric (Fig.1) and ferrodielectric crystals (Fig.2) and electrical resistivity of noncompensa-
ted metal (Fig.3) the hydrodynamic conditions are fulfilled.

SCATTERING OF PHONONS BY FINE PARTICLES

Y. HIKI and Y. KOGURE

Faculty of Science, Tokyo Institute of Technology,

Oh-okayama, Meguro-ku, Tokyo 152, Japan

H. NAKAMURA and K. KAWASAKI

Central Research Laboratories, Idemitsu Kosan Co., Ltd.,

Kami-izumi, Sodegaura, Chiba 299-02, Japan

INTRODUCTION

Phonons are scattered at an interface between two materials. The scattering can most effectively be observed when fine particles are used for the scatterers due to their high surface/volume ratio. By investigating the phonon scattering, we are able to obtain useful information on the characteristics of the interface, and also on the interactions of the phonons with other phonons or electrons. We have experimentally studied the scattering of thermal phonons by metallic and non-metallic fine particles embedded in a non-metallic medium.

EXPERIMENTAL METHOD

The thermal diffusivity D and specific heat C of polyethylene (PE) specimens containing dispersed fine particles of SiO_2, Ni, Fe, and Cu have been measured in the temperature range of 4.2 - 300 K. The diameters of the particles are: $2a$= 7, 16, 40 nm, 1, 5, 10 μm for SiO_2; 16, 27, and 53 nm for Ni, Fe, and Cu. The volume contents of the particles are X = 1 - 18 % for SiO_2; 10 % for Ni, Fe, and Cu. The PE material is of high purity, with the specific gravity of 0.954.

Both measurements were made dynamically. The thermal diffusivity was measured by the laser-flash method,[1] by heating the front face of a disk-shaped specimen by a pulse of laser and observing the time variation of back face temperature. The specific heat was measured by the thermal relaxation method,[2] by applying a step-like voltage to a heater on the front face and observing the back face temperature.

RESULTS AND ANALYSIS

The thermal diffusivity D of PE is lowered by the inclusion of non-metallic and metallic particles, especially at low temperatures. Examples of the data are shown in Fig. 1. For the specific heat, an additivity has been found, namely, $C'(X) = (1-X)C'_{PE} + XC'_{particle}$, where C' is the heat capacity per unit volume, and X is the volume content of the particles. For the SiO_2 inclusion, the dependences of the diffusivity on the volume content X and on the particle radius a were precisely studies. It was shown that the thermal conductivity κ ($= C'D$) was lowered with increasing X, and also with decreasing a. Analyses were made by considering the phonon scattering by particles.

The thermal resistivity of PE containing particles (radius a, and volume content X) is represented as $\kappa^{-1} = (1-X)\kappa_{PE}^{-1} + XR$, where R is the thermal resistivity due to the particles. Further, R is a sum of the surface resistivity and the bulk resistivity of the particles. By considering the usual gas kinetics formula, thermal conductivity $\kappa = (1/3)C'vl$; where v and l are the velocity and the mean free path of phonons, it can be shown that the surface resistivity is represented as $9\gamma/4avC' = A/a$, where γ is the normalized scattering cross section (cross section divided by πa^2). The bulk resistivity (B) is of course independent of particle radius. Thus, the total thermal resistivity is $\kappa^{-1} = (1-X)\kappa_{PE}^{-1} + X(A/a + B)$. Experimental κ-vs-X (a = const.) and κ-vs-a (X = const.) data for the SiO_2 inclusion are well fitted to the above relation. From the curve fitting, we can determine the above two quantities A and B at various temperatures, and finally obtain the temperature dependence of the scattering cross section γ. The above procedure is for the case of non-metallic particles. For the metallic particles, the thermal resistivity is represented as $\kappa^{-1} = (1-X)\delta\kappa_{PE}^{-1} + XA/a$. In this case, the heat carriers in the particles

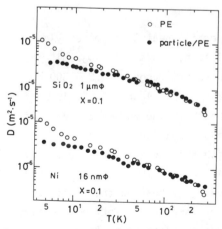

Fig. 1. Thermal diffusivity containing fine particles.

are electrons which are highly conductive, and the bulk resistivity term B can be ignored. The coefficient δ (< 1) comes from the Maxwell correction, considering the modification of the heat flow lines near a particle of high thermal conductivity.[3] The final results, the temperature dependences of the scattering cross section γ for the non-metallic and metallic scattering particles, are shown in Fig. 2.

DISCUSSION

The experimental results show that the scattering cross section increases linearly with temperature for SiO_2 particles, while the temperature dependences are not distinct for Ni, Fe, and Cu particles.

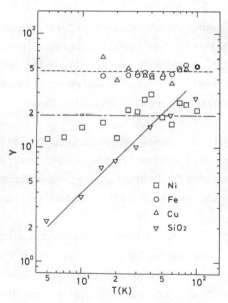

The above temperature dependences, and also the magnitude of the cross section, cannot be explained by the usual acoustic mismatch mechanism. We consider the following things.
(i) Non-metallic scatterer – Extra localized phonons exist near the surface of the particle, and they interact with the thermal phonons. The surface phonons have low Debye temperature, and the number of the phonons, and the cross section, are proportional to the temperature.
(ii) Metallic scatterer – Thermal phonons interact with the electrons in the particle. The number of the electrons, and the cross section, are independent of temperature.

Fig. 2. Temperature dependence of scattering cross section.

REFERENCES
1) Kogure, Y., Mugishima, T. and Hiki, Y., J. Phys. Soc. Jpn. 55, 3469 (1986).
2) Bachmann, R. et al., Rev. Sci. Instrum. 43, 205 (1972).
3) Maxwell, J. C., A Treatise on Electricity and Magnetism (Clarendon Press, Oxford, 1904).

The Kapitza Resistance between Liquid ^{3}He and Solid ^{3}He.

F. Graner[1,2], S. Balibar[1], R. Bowley[2], P. Nozières[2], E. Rolley[1].

(1) Laboratoire de Physique Statistique - Ecole Normale Supérieure,
24 rue Lhomond, 75231 Paris Cedex 05, France.
(2) Institut Laue-Langevin, BP 156X, 38042 Grenoble Cedex, France.

We study the dissipative coefficients which control the growth kinetics of ^{3}He crystals. The experimental value of the Kapitza resistance between liquid ^{3}He and solid ^{3}He is much smaller than expected. A theoretical calculation explains this good heat transmission by taking into account the existence of transverse sound modes in Fermi liquids. Our measurements of the growth resistance also agrees with existing theories.

The interface between a helium crystal and its melt is free of impurities and in a well known surface state; it also quickly reaches its thermodynamic equilibrium. It is then possible to measure and theoretically interpret the dissipative Onsager coefficients which relate the heat and mass fluxes across such an interface to the chemical potential and temperature differences $\mu_L - \mu_S$, $T_L - T_S$.

A similar study was already performed in ^{4}He, which improved the understanding of the crystal to superfluid interface[1]. We now observe the growth kinetics of ^{3}He crystals at 0.32 Kelvins and link our measurements of Onsager coefficients with various properties of the ^{3}He Fermi liquid[2,3]. In particular, we measure a weak value of the Kapitza resistance R_K (thermal resistance of the interface). This means a high heat transmission across the interface, which can be explained from the appearance of transverse sound in the liquid at low temperature. Including this mode in an acoustic mismatch calculation leads to a good agreement with our experimental results.

1. EXPERIMENTAL METHOD AND RESULTS

In an optical cryostat which has been described elsewhere[4,2], we grow very pure ^{3}He crystals from their melt. A slightly deformed monocrystal relaxes towards thermodynamic equilibrium. Its shape evolves by local melting and recrystallization under the effect of gravity (since a slight difference in density $\rho_S - \rho_L$ exists between solid and liquid) and the surface tension γ. We observed that the growth velocity v is a linear function of the applied force and defined an effective growth resistance by

$$R_{eff} = - [(\rho_S - \rho_L) \, g \, (z - z_0) + \gamma C] / \rho_S v \qquad (1)$$

$C = 1/r_1 + 1/r_2$ being the mean surface curvature and z_0 the equilibrium height of a flat interface. R_{eff} comes from two dissipation sources: first, the intrinsic resistance R_{int} due to a friction on the Fermi liquid excitations; second, a term due to the necessary evacuation of the latent heat away from the interface. This thermal term can be related to the Kapitza resistance, for the following argument. During the growth, when an atom from the liquid sticks to the solid, it can radiate only 10^{-5} of its energy as phonons in the solid or in the liquid; while the "quasiparticles" of the Fermi liquid are better coupled to this atom and can take away its energy[3]. We thus think that the latent heat L is released on the liquid side. Since the thermal conductivity of the liquid is poor (4400 times less than in the solid around 0.32 K), the heat crosses the interface and is evacuated through the solid thermal resistance z_S. According to this model, one should have[2]:

$$R_{eff} = R_{int} + \rho_S.(R_K + z_S).L^2 / T_m \qquad (2)$$

R_{eff} has been measured for different temperatures between 0.30 and 0.34 Kelvins, i.e. near the minimum cf the ^{3}He melting curve at $T_m = 0.32$ K; at T_m, the latent heat L is zero and, as expected, R_{eff} goes through a minimum value. Around this minimum our measurements can be fitted by a parabola, giving the first direct measurement of R_{int}:

$$R_{eff} = 5.5 + 3.9 \times 10^4 \, (T - T_m)^2 \qquad (units: m/s) \qquad (3)$$

As z_S is negligible here, we also deduce $R_K = 1.3$ cm^2.K.W^{-1}, which is smaller than previously obtained in a criticable experiment by Castaing et al [5].

2. INTERPRETATION

R_K can be determined from the "acoustic mismatch" (difference of acoustic impedances) between the solid and the liquid, and a calculation of the energy transmission rate for phonons across the interface. A simple evaluation including only the longitudinal sound in the liquid, longitudinal and transverse modes in the solid would yield to a value about 5 times higher than our experiment[5].

However, in a Fermi liquid like ^{3}He, a collective mode of frequency ω undergoes a transition between two propagation regimes. For low ω or high T, $\omega\tau \ll 1$ (here τ is the characteristic relaxation time, varying as T^{-2}) and the acoustic impedance of the liquid is the same as in a normal viscous liquid, allowing only longitudinal sound. In the opposite limit $\omega\tau \gg 1$, this "first" (normal) sound is replaced by a "zero-sound" in which a deformation of the Fermi surface propagates; the interactions between the fermions act as a restoring force around the equilibrium state. Transverse modes thus appear, in addition to the longitudinal one, and the liquid behaves somehow like an elastic medium.

The longitudinal modes in both regimes have a velocity $c \approx 402$ m/s, close to the velocities in the solid. The transverse zero-sound modes are 10 times slower. Their transmission rate is smaller, due to acoustic mismatch; but they contribute more to the phonon specific heat (varying as $1/c^3$) and hence to the heat flux across the interface (as $1/c^2$). Considering this new transmission channel leads to a decrease in R_K by a factor 5.

Indeed, detailed numerical calculations (including the anisotropy of the sound velocity in the solid and the growth resistance of the interface) allowed us to determine R_K as a function of the transverse acoustic impedance of the liquid. In our experiments around 0.32 K, the thermal frequencies are in the cross-over regime between first and zero-sound. We find an agreement with our experimental value.

We also interpret the value of R_{int}, which is found in good agreement with a recent theory[C,B] if the mean sticking probability of the atoms on the solid is about 10%. We finally predict that when the thermal phonons get in the zero-sound regime (T $\leq$ 0.1K), their transmission rate reaches a high value independent of T; $R_K T^3$ remains then constant and as low as 0.025 cm^2.K^4.W^{-1}. We can compare this behaviour with the case of a rough ^{4}He crystal-superfluid interface[6], where the very mobile interface acts as a pressure node; it hinders the low-frequency phonon transmission, leading to $R_K T^3 \sim T^{-2}$.

3. CONCLUSION

We have measured the Kapitza resistance between crystalline ^{3}He and liquid ^{3}He at 0.32 K by observing the relaxation kinetics of crystals shapes. A weak value was found which is explained by taking into account the transverse zero-sound in the liquid. However, more direct measurements of R_K should be useful to check this indirect value, as well as our predictions at lower temperatures. We also measured and interpreted the intrinsic growth resistance of the interface.

REFERENCES :
[1] Wolf, P.E., Edwards, D.O., Balibar, S., J. Low Temp. Phys. 51, 489 (1983).
[2] Graner, F., Balibar, S., Rolley, E., J. Low Temp. Phys. 75, 69 (1989).
[3] Bowley, R., Graner, F., Nozières, P., to be published.
[4] Rolley, E. et al., Europhys. Lett. 8 (6), 523 (1989).
[5] Castaing, B. et al., private communication and J. Low Temp. Phys. 62, 315 (1986).
[6] Puech, L. et al, J. Physique - Lettres 43, L-809 (1982).

THEORY OF KAPITZA CONDUCTANCE BETWEEN SOLIDS AT HIGH TEMPERATURES

D. A. YOUNG and H. J. MARIS

Department of Physics, Brown University, Providence, RI 02912, USA

The Kapitza resistance R_K arises because phonons attempting to cross a boundary from one material to another may be reflected due to the different properties of the two materials. Most experimental studies of R_K have been performed at low temperatures where the thermal phonons have long wavelengths. Under these conditions the reflection can be calculated using continuum elasticity. Recently, new techniques have made possible the measurement of R_K at high temperatures[1-3]. To interpret these experiments it is necessary to have a theory of phonon reflection based on a lattice dynamical model for the two materials, so as to correctly allow for the effects of phonon dispersion, structure in the phonon densities of states, etc.

In this article we describe results obtained for the phonon reflection and Kapitza resistance at an interface between two fcc lattices. The model is shown in Fig. 1 and the calculations are described in detail in ref. 4. We consider two lattices in registry with the interface parallel to the (001) plane. The springs and masses in the interior of the lattices are K, K', M and M'. The springs connecting the nearest neighbors on opposite sides of the interface are of strength K''. It is straightforward to show that the conductance $\sigma(\equiv 1/R_K)$ can be written

$$\sigma(T) = \frac{1}{2} \int C(\omega, T) < v_z(\omega) >< t(\omega) > d\omega \tag{1}$$

where $C(\omega, T)$ is the heat capacity per unit volume of the phonons of frequency ω, $< v_z(\omega) >$ is the average group velocity, and $< t(\omega) >$ is the average transmission coefficient for phonons of frequency ω. The average $< t(\omega) >$ is weighted by the component of the phonon group velocity normal to the interface. We computed these averages using a random sampling ($> 10,000$) of phonon wavevectors in the Brillouin zone, and forming histograms with bin widths 0.01 of the maximum phonon frequency. We were able to make overall checks on the calculations using detailed balance conditions that the transmission coefficients from the two sides of an interface must obey.

1330

We have investigated a broad range of the ratios of the spring constants and masses in the two materials. It is convenient to express the results in terms of the low-temperature θ_D which is $2.97\hbar/k_B\sqrt{K/M}$, and the quantity $\sigma_o = k_B a^2 \sqrt{K/M} = 0.34 k_B^2 \theta_D /\hbar a^2$ which has the dimensions of Kapitza conductance. Here we summarize the aspects of the results that appear to be general features of the phonon transmission at high temperatures:

Temperature Dependence of σ. The temperature dependence of σ is similar to that of the specific heat, but σ rises somewhat less rapidly with increasing T. This is due to the decrease of the average group velocity at high temperatures. As an example, Fig. 2 shows the reduced conductance for pairs of lattices with $M'/M = 2$.

Phonon Density of States. For a series of pairs of lattices with a given ratio of M' to M the largest conductance occurs when

$$K'/K = M'/M \qquad (2)$$

An example of this effect can be seen in Fig. 2. When this condition is satisifed the conductance is maximized because the density of states is the same in both lattices.

Frequency-Dependence of the Transmission. $< t(\omega) >$ is in most cases fairly independent of ω, except when ω approaches the upper frequency limit of the lattice into which transmission occurs. Typical results for $< t(\omega) >$ are shown in Fig. 3, which is for $M = M' = 1$ and $K = 1$. The existence of only modest variations of $< t(\omega) >$ with ω is somewhat surprising considering that the match of the phonon density of states is important in determining the Kapitza conductance.

Interface Spring Constants. Provided K'' lies in the range between K and K' the results are almost independent of K''. For K'' having any value which lies in the range above K and K' the change in σ is also fairly small. If K'' is decreased below the smaller of K and K' the conductance decreases, and for sufficiently small K'' goes as K''^2.

We thank J. Tauc for helpful discussions. This work was supported in part by the U.S. DOE under grant DE-FG02-86ER45267.

REFERENCES

1. Young, D.A., Thomsen, C., Grahn, H.T., Maris, H.J., and Tauc, J., in Phonon Scattering in Condensed Matter V, edited by A.C. Anderson and J.P. Wolfe (Springer, Berlin, 1986), p.49.

2. Swartz, E.T., and Pohl, R.O., Appl. Phys. Lett. **51**, 2200 (1987).

3. Stoner, R.A., Young, D.A., Maris, H.J., Tauc, J., and Grahn, H.T., in this volume.

4. Young, D.A., and Maris, H.J., Phys. Rev. B to appear.

5. Young. D.A., Ph.D. thesis, Brown University, 1989.

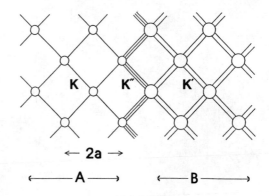

Fig.1. A section through the interface.

Fig.2. Kapitza conductance as a function of reduced temperature for various values of K'.

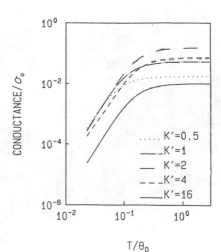

Fig.3. Frequency dependence of phonon transmission from a lattice with M=1 and K=1 into a lattice with M'=1.

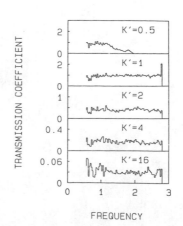

15

Phonon Imaging

CURRENT TOPICS IN PHONON IMAGING

J.P. Wolfe

University of Illinois at Urbana-Champaign
Urbana, IL 61801 USA
and
Technische Universität München
Physics E 10, 8046 Garching, FRG

1. INTRODUCTION

This Phonon Conference marks a decade since phonon-imaging experiments were first introduced at the Third International Phonon Conference on Phonon Scattering in Condensed Matter[1]. A decade earlier, Taylor, Maris and Elbaum[2], introduced the idea of phonon focusing, upon which the phonon-imaging technique is based. In this paper I will highlight some of the progress in the last decade and describe selected experiments which indicate the future directions of research in this area.

Following the pioneering demonstration of ballistic heat-pulse propagation in solids by von Gutfeld and Nethercot[3], it was soon observed that the intensities of longitudinal and transverse heat pulses were not isotropic. Taylor et al.[2] recognized that the anisotropy in heat-flux from a point source could be explained in terms of the basic elastic anisotropy of the crystal. A small angular variation in sound velocity (say, 30% in a typical crystal) leads to immense variations in ballistic heat flux.

A recent experiment on InSb demonstrates these ideas[4]. Figure 1a shows the heat pulses generated by a pulsed laser beam exciting a metallized surface of this semiconductor. The detector is a PbTl tunnel junction and the temperature is 1.6 K. The focused laser beam can be translated to vary the angle of ballistic phonon propagation. At position A, three pulses associated with longitudinal (L), fast transverse (FT), and slow transverse (ST) modes are observed. The time of flight for these pulses corresponds to the propagation distance divided by the three sound velocities along this direction. If the laser beam is translated a small distance to position B, the FT pulse almost completely vanishes! The difference of these two traces shows only the FT pulse.

This large anisotropy in energy flux is displayed by "sitting" on the FT peak and continuously scanning the laser position, as shown in

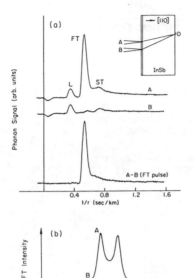

Figure 1. (a) Heat pulses in InSb produced by exciting a 250 nm copper film with a focused laser beam. As the beam is translated slightly a dramatic change in the FT pulse is seen due to phonon focusing. Subtraction of the two traces shows only the FT pulse. The detector is a PbTl tunnel junction. (b) Peak intensity of the FT heat-pulse intensity as the laser beam is continuously translated in the ($\underline{1}$10) plane. The maxima correspond to caustics in phonon flux. (Hebboul and Wolfe[4])

Figure 1b. The FT heat flux is beamed into a narrow channel centered on the (110) plane: The full extent of this phonon-focusing effect is shown as a phonon image, Figure 2, which converts the heat-flux intensity into brightness on a video screen. (Northrop and Wolfe[5]) Figure 1b is a horizontal scan across the vertical ridge at the center of the image. The rounded diamonds above and below this ridge are centered on <100> axes and are due to phonon focusing of ST phonons. The cusped structures at the left and right sides of the image are centered on the <111> axes and are also due to ST phonons.

The bright lines in the phonon image are actually mathematical singularities, or "caustics", in phonon flux. They correspond to lines of zero Gaussian curvature (parabolic lines) on the slowness surface. One can also construct a group-velocity (or "wave") surface defined by all the normals to the slowness surface. Parabolic lines on the slowness surface produce folds in the wave surface, and it is the projection of these folds onto the experimental plane which produces the caustics in the phonon image[5].

2. PROGRESS IN THE THEORY OF PHONON FOCUSING

The complex topology of these acoustic surfaces--a direct result of the fourth-rank elasticity tensor--can be displayed graphically using the concepts of parabolic lines and caustics. Figure 3 shows a three dimensional plot of the ST slowness surface of Si. Heavy curves are parabolic lines which separate convex, saddle (shaded), and concave regions. Group-velocity normals to the surface at these lines give the caustic pattern similar to that shown in Figure 2 for InSb.

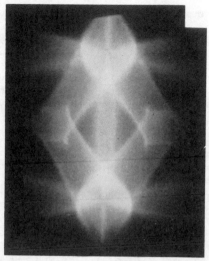

Figure 2. Phonon image of InSb taken with an aluminum bolometer. A horizontal line-scan through the center quarter of the image corresponds to Fig. 1b. Bright lines are caustics in ballistic heat flux due to phonon focusing. (Hebboul and Wolfe[4])

Systematic classification of acoustic surfaces and caustic patterns in cubic crystals have been made by Hurley and Wolfe[6] and by Every et al.[7] Given the two ratios of elastic parameters (C_{11}/C_{44} and C_{12}/C_{44}), the positions of phonon caustics can be quickly ascertained. Systematic studies of known tetragonal crystals have been made by Every and McCurdy.[8]

A key concept in understanding the topology of the slowness surface is degeneracy. In general, the L, FT, and ST sheets of the slowness surface are degenerate only at a few discrete points. The existence of conical or tangential degeneracies produces concave and saddle regions on the surfaces, and, therefore, caustics. By concentrating on the degeneracies in the slowness surface it is possible to see the relationship between the focusing pattern of different crystals. For example, the focusing pattern of trigonal crystals such as quartz and sapphire are distortions of those for higher symmetry hexagonal crystals[9].

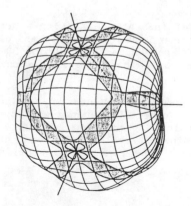

Figure 3. ST slowness surface of Si, which is similar to that of InSb. Normals to this surface along the heavy parabolic lines give the real-space directions of caustics, such as those observed in the phonon image of Fig. 2. Notice the corresponding three and fourfold axes in the two figures. (Hurley and Wolfe[6])

The symmetry of a phonon-focusing pattern is higher than that of the crystal itself. For example, all slowness surfaces are centro-symmetric, even for crystals lacking a center of inversion. In general, the "acoustic" symmetry determined by the elastic tensor is even higher than the Laue symmetry observed in a diffraction pattern of the crystal. The addition of piezoelectricity stiffens the elastic constants and lowers the acoustic symmetry, in some cases to that of the crystal point group. Systematic studies of these ideas have been made by Every[10] and Every and McCurdy[11]. The effect of piezoelectricity on ballistic heat propagation was shown experimentally by Koos and Wolfe[12] for lithium niobate.

3. DISPERSIVE PHONON FOCUSING AND LATTICE DYNAMICS

The discussion so far has been concerned with the continuum limit, where the slowness surface (and, hence, focusing pattern) is independent of phonon frequency. As the frequency of the phonon is increased, its wavelength cannot become shorter than twice the atomic spacing. Therefore, the constant-frequency surfaces distort as they approach the Brillouin zone of the crystal. Along with this distortion come radical shifts in the caustic pattern. For InSb the ST phonon frequency at the <111> zone boundary is about 1 THz, so dispersive effects occur above a few hundred GHz.

The experimental problem is to observe the ballistic propagation of phonons with such short wavelength. Scattering from mass defects--analogous to Rayleigh scattering of light--increases rapidly as ω^4. Hence, even the existence of naturally occuring isotopes greatly reduces the mean free path of large-wavevector phonons. Also, one must select phonons of within a narrow frequency range.

These problems were solved in pure crystals of InSb by using relatively thin samples (0.4 mm) and tiny ($10 \times 10 \ \mu m^2$) superconducting tunnel junctions (Hebboul and Wolfe[4]). A range of lead-alloy concentrations were chosen to systematically shift the onset frequency of the junction ($2\Delta/h$), which corresponds to the minimum energy required to break a cooper pair. Isotope scattering provided a natural high-frequency cutoff, so a bandwidth $\Delta\nu \cong 0.2 \ \nu_{onset}$ was obtained.

Figure 4 shows a phonon image of InSb at 1.6 K using a detector with onset at about 690 GHz. The phonon-focusing caustics are clearly shifted from those shown in Figure 2, corresponding to about 300 GHz. The shift in the caustic pattern is due to a change in the shape of the slowness surface due to dispersion.

Hebboul et al.[4] made quantitative measurements of the angular dimensions of the caustic patterns. These measurements were compared to the dispersive phonon-focusing predictions of various lattice-dynamics models: rigid, dipole, shell and bond charge. Although it had the fewest parameters, the bond-charge model agreed best with the data.

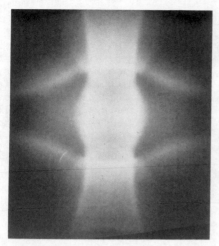

<u>Figure 4</u>. Phonon image of
InSb using a PbBi tunnel-
junction detector with 690
GHz onset frequency. Caustics are shifted from those
of the low frequency image
of Fig.2 due to dispersion.
Quantitative comparison of
such images with those cal-
ulated using lattice-dynamics models has been made
by Hebboul and Wolfe[4].

4. BULK SCATTERING OF PHONONS

Phonon imaging can also give direct information about scattering
processes.   Phonons undergoing just a few scattering events retain
some of the ballistic anisotropy[13].   This channeling effect is shown
graphically in the experiment of Shields, Wolfe and Tamura[14], Figure
5.   A cube of Si is slotted so that about half of the ballistic
focusing pattern is blocked from the detector.   (The complete
ballistic pattern is similar to that of Ge, shown in Fig. 6.)   The
right side of the image contains ballistic plus scattered phonons and
the left side (behind the slot) contains only flux from scattered
phonons. The continuation of the FT ridges behind the slot shows that
the scattered phonons retain some of the ballistic flux.   This effect
is confirmed by Monte Carlo simulations.

Other experiments have examined the scattering from impurities
and defects.   Held et al[15] measured the scattering from Cr impurities
in GaAs by extracting a diffusive background from the ballistic flux.
Previously, highly anisotropic scattering from dislocations in LiF was
observed, demonstrating the "fluttering string" motion of these
extended defects (Northrop et al.[16]).   Anisotropic scattering from
point defects and impurities has not yet been demonstrated but is a
natural future direction for the imaging technique.

Another   experimentally   unexplored   area   is   the   anisotropy
associated with phonon-phonon interactions.   Tamura and Maris[17] have
conducted a theoretical study on NaF which shows that phonon-phonon
scattering depends significantly on phonon mode and propagation
direction.

Related to phonon-phonon interactions is the so-called phonon hot
spot which appears at high excitation levels.   Recently the size of
the excitation source has been measured by high resolution imaging of
the phonon caustics in Si (Shields and Wolfe[18]).   The sharpness of the
caustics depends on the excitation level, as is expected for a heated

Figure 5. Image of heat flux through a half-slotted crystal of silicon. A slot cut in the crystal parallel to the detection surface prevents ballistic flux from reaching the left side of the image. The scattered phonons on the left side are channeled along continuations of the FT caustics to form the "x" structure. (Shields et al.[14])

region which expands with power. Still, much needs to be determined for this phenomenon, such as the emitted frequency distribution and the temporal evolution.

5. PHONON INTERACTIONS WITH INTERFACES

Like a photon, the phonon is expected to obey Snell's Law when it is specularly reflected or transmitted at a crystal interface. But the situation is much more complicated and interesting for a phonon, which can undergo mode conversion at the interface. A phonon image of the phonon flux reflected from a sapphire boundary (Northrop and Wolfe[19]) contains distinct caustic patterns for each of the mode-conversion processes (L → FT, FT → ST, etc.). These patterns are replaced by a more diffusive distribution when the surface is roughened. With this information, one can determine the ratio of specular to diffusive scattering at the surface, a quantity which is related to the anomalous Kapitza conductance.

Recently, imaging experiments by Höß et al.[20] have demonstrated the refraction of high frequency phonons at a solid-solid interface. A film of MgO is deposited on a "clean" surface of Ge. The sound velocity in MgO is higher than that in Ge, so phonons incident on the film from the Ge crystal should display a critical cone beyond which there is total internal reflection.

Figure 6 shows the experimental focusing pattern with and without an MgO film. With the 250 nm film, the FT phonons display a critical angle for total internal reflection of about 32°. The critical angle for ST phonons is smaller. If a film with thickness comparable to the mean phonon wavelength ($\cong$ 10 nm) is deposited on the Ge, a significant rise in transmission beyond the critical angle is observed. This transmission of the phonons is analogous to frustrated total internal refraction in optics.

Phonons at crystal/metal interfaces have displayed anomalous

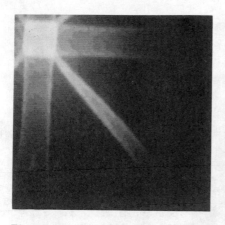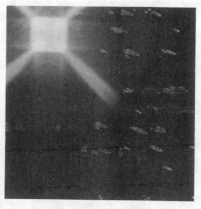

<u>Figure</u> <u>6</u>. Experiment showing specular refraction of phonons at an MgO/Ge boundary. Left is the phonon image of Ge without MgO film and right is with a 260 nm film. Cutoffs in phonon flux correspond to total internal reflection beyond the critical cones for the two transverse modes. (Höß et al[20])

reflectivity, attributed to diffuse scattering by defects or "dirt" at the interface[21]. Using phonon imaging, Every et al.[22] discovered an increased transmission at the critical cone for L-to-T conversion. This critical-cone channeling effect occurs when there is a "soft" buffer layer of atoms between the metal and crystal, and suggests that such imaging experiments may help to characterize the interface..

The propagation of phonons through the multiple interfaces of a synthetic superlattice has lead to interesting interference effects and the discovery of intramode stopbands. This subject is reviewed by Tamura in these Proceedings.

6. ELECTRON-PHONON INTERACTIONS

One of the experiments leading to the development of phonon-imaging techniques was the discovery that electron-hole droplet clouds in Ge were highly anisotropic[23], an effect produced by the ballistic phonon wind emanating from the optical excitation point.

Recently, phonon imaging has been used to study the two-dimensional electron gas (2DEG) formed in a GaAs/AlGaAs heterostructure[24]. See the review paper by Dietsche in these Proceedings. The phonons transfer their momentum to the electrons in the conducting layer and produce a voltage in the device -- i.e. a phonon-drag voltage. By scanning a laser heat source across the opposite side of the crystal, the angle of the phonon wavevector is changed. The resulting phonon image, shown in Figure 7, is a

1342

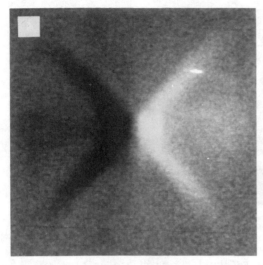

Figure 7. Phonon image of GaAs using a 2DEG detector. The 2-dimensional electron gas at the surface of the device absorbs the ballistic phonons, which induce a voltage according to the direction of the momentum transfer. Light and dark are + and - voltages. The pattern is a result of a phonon focusing pattern similar to Fig. 6 and the electron phonon coupling. (Karl et al[24])

convolution of the GaAs phonon-focusing pattern and the probability of absorption by the electrons in the 2-dimensional layer. Light and dark areas correspond to positive and negative voltages (grey is zero voltage). The sign of the phonon-drag voltage depends upon whether the phonon wavevector parallel to the channel is directed to the right or left. The shape of the pattern corresponds to piezoelectric rather than deformation potential coupling.

7. APPLICATIONS TO PARTICLE DETECTORS

Superconducting phonon detectors have recently been applied to the detection of elementary particles, and phonon focusing may potentially be used to localize the event. (See the papers by Cabrera and Zehnder in these Proceedings.) The detection of single nuclear events has been demonstrated using tunnel junctions and bolometers cooled to sub-Kelvin temperatures. While the heat pulses from a nuclear event are generally not time resolved, due to the use of charge-integration amplifiers, time-correlation studies suggest that the heat pulses in Si propagate with velocities several times slower than the sound velocity.[25]

This is similar to the observation of broad heat pulses produced by direct photoexcitation of semiconductors. The hot carriers thermalize by emitting mainly optical phonons, which quickly down convert to high frequency acoustic phonons. Much of the resulting thermal energy propagates diffusively because the down-conversion rate of acoustic phonons decreases rapidly with frequency (ω^5), and the mean free path of these bottlenecked high frequency phonons can be smaller than the crystal dimension, due to elastic scattering.

Yet, following a spatially restricted absorption of β particles, a significant anisotropy in phonon flux is observed by employing several detectors.[25] Diffusive heat pulses yet ballistic anisotropy! The answer to this apparent contradiction is probably that even a small fraction of low, frequency ballistic phonons can have a significant flux along strong focusing directions. Clearly some phonon imaging and spectroscopy studies are needed to measure the basic down-conversion and elastic scattering processes of phonons at very low temperatures. (For example, the hot-spot experiment mentioned previously suggests a rapid down conversion process at sufficient photoexcitation levels.)

8. CONCLUSION

Basically, phonon imaging is useful when a property to be studied depends on the direction of the phonon wavevector, e.g., ballistic propagation or phonon scattering. The utility of this method is expanding with technical advances: for example, fabrication of tiny frequency-tunable detectors[4], development of spatially scanable tunnel-junction detectors[27], and phonon imaging by electron-beam scanning[15]. It also seems likely that phonon-imaging experiments could employ generators and detectors based on scanning tunneling microscopy (STM) and optical spectroscopy.

In addition to scanning propagation direction, phonon imaging can also be used to image sub-surface or surface structures. Two examples include the shadow imaging of sub-surface laser-drilled holes in quartz[27], and the micro-imaging of semiconductor devices[28].

At present, there remains a surprising number of basic phonon properties which are not quantitatively known. Acoustic phonon down-conversion processes have been measured only in a few crystal systems[29]. The lifetimes and dynamics of large-wavevector phonons remain to be studied. Isotope scattering rates are generally not known experimentally. Anisotropic phonon scattering from impurities and defects is a broad area which has hardly been investigated. Phonon interactions with surfaces and interfaces still contain many puzzles. Spectroscopy and imaging techniques hold the potential for resolving many of these issues.

ACKNOWLEDGEMENTS

I wish to thank H. Kinder and his research group at the Technische Universität München for a stimulating sabbatical year. The Alexander von Humboldt Foundation made this possible. Illinois work described here was supported under MRL grant NSF DMR 86-12860.

REFERENCES

1. Maris, H.J., Phonon Scattering in Condensed Matter (Plenum, New York, 1980).
2. Taylor, B., Maris, H.J. and Elbaum, C., Phys. Rev. Lett. 23, 416 (1969).
3. Von Gutfeld, R.J. and Nethercot Jr., A.H., Phys. Rev. Lett 12,

641 (1964).

4. Hebboul, S.E. and Wolfe, J.P., Z. Phys. B74, 35 (1989).

5. Northrop, G.A. and Wolfe, J.P., Proc. of the NATO Advanced Study Institute of Nonequilibrium Phonon Dynamics, Les Arcs, France 1984,. ed. by W.E. Bron, Plenum, New York (1985).

6. Hurley, D.C. and Wolfe, J.P., Phys. Rev. B32, 2568 (1985).

7. Every, A.G., Phys. Rev. B24 3456 (1981); Every, A.G. and Stoddart, A.J., ibid 32, 1319 (1985).

8. Every, A.G., Phys. Rev. B37, 9964 (1988); McCurdy, A.K., Phys. Rev. B9, 466 (1974).

9. Every, A.G., Phys. Rev. B34, 2852 (1986).

10. Every, A.G. J. Phys. C: Solid State Phys. 20, 2973 (1987).

11. Every, A.G. and McCurdy, A.K., Phys. Rev. B36, 1432 (1987).

12. Koos, G.L., and Wolfe, J.P., Phys. Rev. B30, 3470.

13. Ramsbey, Wolfe J.P. and Tamura S., Z. Phys. B73, 167 (1988); see also Fieseler, M., Wenderoth, M., and Ulbrich, R.G., Proc. 19th Int. Conf. Physics Semiconductors (1989, Warsaw).

14. Shields, J.A., Wolfe ,J.P. and Tamura, S., Z. Phys., to be published.

15. Held, E., Klein W. and Huebener, R.P., Z. Phys. B75, 17 (1989).

16. Northrop, G.A., and Wolfe, J.P., Phys. Rev. B27, 6395 (1983).

17. Tamura, S. and Maris, H.J., Phys. Rev. B31, 2595 (1989).

18. Shields, J.A. and Wolfe, J.P., Z. Phys. B75, 11 (1989).

19. Northrop, G.A. and Wolfe, J.P., Phys. Rev. Lett. 52, 2156 (1984).

20. Höß, C., Schreyer, H., Dietsche, W., Wolfe J.P. and Kinder, H., these Proceedings.

21. Marx, D. and Eisenmenger, W., Z. Phys. B48, 277 (1982).

22. Every, A.G., Koos, G.L. and Wolfe, J.P., Phys. Rev. B29, 2190 (1984).

23. Greenstein, M. and Wolfe, J.P., Phys. Rev. B24, 3318 (1981).

24. Karl, H.,Dietsche, W., Fischer, A. and Ploog, A., Phys. Rev. Lett. 61, 2360 (1988).

25. Peterreins, Th., Prbst, F., von Feilitzsch, F., Mäbauer, R.L., and Kraus, H., Phys. Lett B 202, 161 (1988).

26. Schreyer, H., Dietsche, W. and Kinder, H., LT17-Contributed papers, ed. by U. Eckern et al. (North-Holland, 1984), p. 665.

27. Huebener, R.P., Held, E., Klein, W., and Metzger, W, Phonon Scattering in Condensed Matter V, ed. by A.C. Anderson and J.P. Wolfe (Springer Verlag, Berlin, 1986), p. 305.

28. Schreyer, H., Dietsche W. and Kinder, H., Phonon Scattering in Condensed Matter V, ed. by A.C. Anderson and J.P. Wolfe (Springer Verlag, Berlin, 1986), p. 315.

29. Baumgartner, R., Engelhardt, M. and Renk, K.F., Phys. Rev. Lett. 47, 1403 (1981).

Phonons and Scanning Tunneling Microscopy
(INVITED)

by

H.K.Wickramasinghe, J.M.R.Weaver and C.C.Williams
IBM T.J.Watson Research Center,
P.O.Box 218,
Yorktown Heights,
New York, N.Y.10598

1. Introduction

The invention of the Scanning Tunneling Microscope (STM) {1} has stimulated the development of a number of novel scanned probe microscopes {2} which are capable of measuring various physical properties of surfaces on a scale of several nanometers. However, in spite of this, little work has been done on the study of phonons on this scale. Some initial work has been reported on the observation of phonon spectra in the tunneling electrodes (tip and sample) of a STM by measuring steps in junction conductance (caused by creation of phonon assisted tunneling channels) as a function of applied voltage {3}. Further work from the same group has discussed the measurement of vibrational spectra of molecules within the tunnel gap using the same technique {4}. However, the results and interpretation are not definitive and further work remains to be done.

In this paper we will review some of our work on the study of phonons (both coherent and incoherent) using the STM and related techniques and present some of our recent results in this area. In section 2 we discuss the work on the detection of incoherent phonons using a) a thermal probe and b) a novel tunneling thermometer and demonstrate how these techniques can be used to perform spectroscopy on a nanometer scale. Section 3 discusses the generation of coherent phonons in the 0.1 THz range and presents our initial results on a new phonon absorption microscope. Finally, in section 4 we present some brief concluding remarks.

2. Detection of Incoherent Phonons on a Nanometer Scale

In 1985, we started a project on a Scanning Thermal Probe with the initial aim of profiling insulating surfaces for the semiconductor industry (which is not possible with the STM) and the broader aim of studying thermal interactions on the scale of tens of nanometers. Fig. 1 shows a schematic of the thermal probe used in our initial experiments. It consists of a tungsten tip that is etched to a radius of 20 nm at the very end and then coated with an insulating film everywhere except the very end. The probe is then overcoated with a second metal (nickel) - which produces a thermocouple junction only a few hundred angstroms in dimension at the tip. We can use such a thermal probe to accurately control the tip over a conducting or insulating surface. We do so by first passing a DC current across the juction to raise the temperature of the tip 10 K or so above ambient. As the tip approaches a sample surface, heat conducts away to the surface thereby cooling it. If we vibrate the tip up and down (10 Å or so) towards and away from the surface, typically at 1 KHz, we can detect an AC thermoelectric voltage which increases as the average tip-sample distance decreases. The amplitude of this signal can then be used in a feedback loop - just like in a STM - to stabilize the tip at a fixed distance over the surface using piezoelectric controls. In our initial experiments, we plotted the average up and down (z) motion of the tip in order to profile the surface of a resist pattern as we scanned the tip in X and Y {5}. In later experiments, we were able to simultaneously stabilize the tip over a surface and measure AC heating due to a second, modulated heating source (either electrical or optical) by chosing the heating frequency to be outside the bandwidth of the gap control feedback loop (see Fig.2). Using this system, we were able to map the temperature variations across a uniformly heated, 100nm thick 200nm period resist grating on a chromium film {6}. In this case, the chromium film was thick enough so that all the pump light was absorbed and the film acted as a uniform heater. The spatial resolution achieved was around 30nm and we were able to observe spatial variations in the thermal resistance across the chrome-resist system. The temperature sensitivity of our system was around 0.1mK in 1Hz bandwidth.

When the tip-sample spacing is greater than 70 nm (i.e. the mean free path of the air molecules), classical conduction mechanisms through the air accounts for the heat transfer between tip and sample. However, as we further decrease the tip-sample spacing, we observe that the AC thermoelectric signal no longer increases - it reaches a plateau until the tip approaches very close (10 nm) to the sample at which point it rapidly increases all the way until the tip actually crashes into the sample. In this regime, the mechanisms of heat transfer are not well understood. Clearly, the conduction via gas molecules alone cannot account for the rapid variation. In fact, Dransfeld has pointed out that one should not expect any variation with gap spacing in this near-field regime if the heat tranfer were purely due to thermal conduction via gas molecules {7}. He has theorised that in the case of ionic samples, this rapid variaton in signal can be accounted for by surface charge density fluctuations caused by optical phonon modes in the solid. These fluctuations give rise to an evanescent electromagnetic field which decays with a decay length equal to the phonon wavelength. These localised electric fields can be very strong and when a conducting tip enters into this region considerable energy can be transferred via Joule heating of the tip.

This strong near-field distance dependence we observed gave us the hope that one might be able to build a thermal probe with a resolution far better than 30nm. Unfortunately, it proved to be a technological challenge to build thermocouple tips which were significantly smaller than 30nm in dimension. However, recent developments in our group has changed this situation. We have developed a novel tunneling thermometer which is capable of monitoring thermal variations down to the atomic scale {8}. In the tunneling thermometer, we bring a conducting tip (e.g. tungsten) within tunneling range of a second conducting surface (e.g. gold) using any feedback control mechanism. In our case we have used tunneling microscopy for control although atomic force or capacitance microscopy would be just as good. We then switch off the tunneling control loop and measure the thermoelectric potential across the tip-sample electrode system (Fig. 3). The thermoelectric potential can be understood in the following way. As the tip and sample electrodes come within tunneling range (5 Å or less) of each other, their electronic states become strongly coupled and their Fermi levels tend to equalize due to two-way electron tunneling across the gap. If the tip-sample system is at a different temperature with respect to the measurement amplifier, a thermoelectric voltage will be developed across the tip-sample system which can be measured by the amplifier. Under conditions of local thermodynamic equilibrium, and in the case where the spatial extent of the thermal variations in the sample are large as compared with the electron scatter lengths, this voltage becomes the classical thermoelectric voltage between tip and sample electrode and the system acts as a regular thermocouple junction with the junction resistance replaced by the tunneling resistance. When the tip and sample are not in thermodynamic equilibrium, we can still measure a thermoelectric voltage although the thermoelectric coefficient is different to the classical one.

In our initial experiments, we have measured the absorption spectrum of a 1nm region of a 20nm thick annealed gold film on mica using the tunneling thermometer to measure the periodic heating caused by a 1 KHz modulated and tunable optical beam incident on the film via the mica. Fig.4 shows the recorded spectrum of gold (dots) compared with the absorption curve derived from published values of the optical parameters. Fig.5 shows a comparison of the tunneling topography image of the gold surface with the thermal image taken at a single wavelength. The usual atomic steps are clearly resolved in the tunneling image. Some of these steps are also seen in the thermal image in addition to contrast that is not present in the tunneling image. This contrast arises from sub-surface topography and adhesion variations in the mica substrate which will not be detected in the tunneling image. We believe these results pave the way for the study of infra-red and Raman spectra of surfaces on a nanometer scale.

3. Phonon Absorption Microscope

In this section we discuss our initial work on a Phonon Absorption Microscope. The basic idea here is to replace the modulated optical pump described in the previous section with a modulated coherent phonon pump and use the tunneling thermometer as before to record the spatial variations in the temperature caused by the absorption of the phonons by a sample. Photoacoustics has been used to generate acoustic waves above 2 GHz as far back as 1967 {9}. A photoacoustic microscope operating at 1GHz was demonstrated in 1978 {10} whereby photoacoutically generated acoustic waves from a scanned sample were detected using a conventional acoustic microscope lens. More recently, Thomsen et. al.{11} have shown that it is possible to photoacoustically generate THz phonons in sub-micron thick metal films using picosecond optical pulses and detect the echo's using a second ps probe pulse - which is simply a delayed and attenuated version of the pump. The probe pulse detects the echo's though the strain induced reflectivity changes in the film.

In our experiments, we have used the scheme of Thomsen et.al. to generate and detect picosecond phonons in thin copper films. Fig.6 shows a typical result obtained

1348

using this pump-probe technique; 100 GHz phonons can be readily generated and detected. We find that by adjusting the energy per pulse to the point where the transient (ps) temperature rise is close to the melting point of the film, we are able to generate average phonon powers in the 0.5 mW range - quite comparable to the power levels used in our optical absorption experiments described in section 2.

Our first demonstration of phonon absorption microscopy was carried out using the sample configuration shown in Fig.7. A 60nm copper film was deposited onto a c-axis sapphire plate which acted as a buffer rod. On the other surface of the sapphire we deposited a 5nm thick resist absorber film which was then overcoated with 100nm of gold in order to be able to record the temperature variations caused by the phonon absorption using the tunneling thermometer. Using ps pulses focused to a 4 micron spot, we were able to generate 50 GHz phonon pulses in the buffer rod. At room temperature, the phonon absorption in the buffer rod is about 6 dB and the absorption in the resist film is about 12 dB. Thus with a gold overlayer film, most of the energy incident on the resist film will be converted into heat. Fig.8 shows the temperature variation across the resist film mapped out using the tunneling thermometer. We believe this represents the phonon beam profile generated photoacoustically by the copper transducer. In order to make sure that we were only observing the phonon absorption and not any other spurious signal, we chopped the ps pulse train simultaneously at two frequencies (1 MHz and 1.1 MHz). A phonon absorption signal would be expected to produce a mixing signal at 0.1 MHz whereas other signals such as optical breakthrough would be expected to provide a linear response. The nonlinear mixing term is what we have plotted in Fig.8. The chopping frequencies were chosen to be high (MHz) so that any residual average heating at the transducer surface will produce a corresponding temperature modulation at the gold film which is well below our limit for temperature detection.

4. Conclusion

In conclusion, we have reviewed the current status of experiments involving both coherent and incoherent phonons with the STM. We have described some new sensors which open up the possibility of studying phonon interactions on a nanometer scale. With further work, we believe these new techniques will find applications ranging from phonon spectroscopy to microscopy of nanometer structures.

References

1. Binnig,G., Rohrer,H., Gerber,Ch.,& Weibel,E., *Phys.Rev.Lett.* **49,** 57 (1982)

2. Martin,Y., Williams,C.C.,& Wickramasinghe,H.K., *Scanning Microscopy.* **2 (1), 3** (1988)

3. Smith,D.P.E., Binnig,G.,& Quate,C.F., *Appl.Phys.Lett.,* **49 (24),** 1641 (1986)

4. Smith,D.P.E., Kirk,D.,& Quate,C.F., *J.Chem.Phys.,* **86 (11),** 6034 (1987)

5. Williams,C.C.,& Wickramasinghe,H.K., *Appl.Phys.Lett.,* **49 (23),** 1587 (1986)

6. Williams,C.C.,& Wickramasinghe,H.K., *Proc. SPIE,* **897,** 129 (1988)

7. Dransfeld,K.,& Xu,J., *Journal of Microscopy*, **152 (1)**, 35 (1988)

8. Weaver,J.M.R., Walpita,L.M. & Wickramasinghe,H.K., *Nature*, (1989) - in press

9. Brienza,M.J.,& De Maria,A.J., *Appl.Phys.Lett.*, **11**, 44 (1967)

10. Wickramasinghe,H.K., Bray,R.C., Jipson,V., Quate,C.F., & Salcedo, J.R., *Appl.Phys.Lett.*, **33 (11)**, 923 (1978)

11. Thomsen,C., Grahn,H.T., Maris,H.J., & Tauc,J., *Phys.Rev.B*, **34 (6)** 4129 (1986)

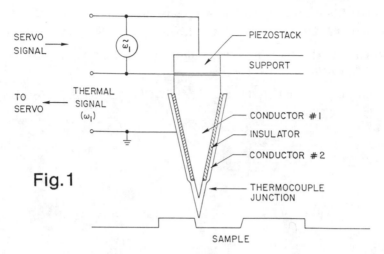

Fig.1

Scanning Thermal Profiler

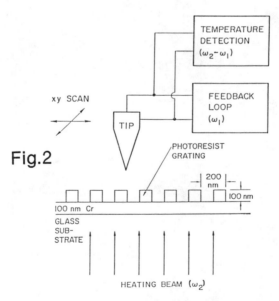

Fig.2

**Photothermal
Temperature Mapping**

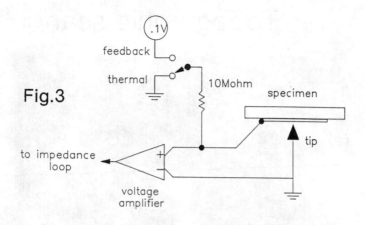

Fig.3

STM feedback loop switching arrangement

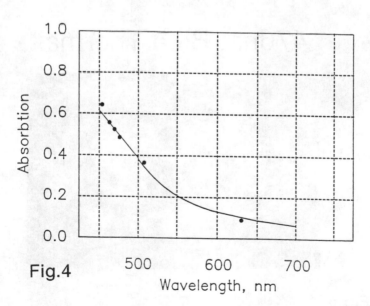

Fig.4

Absorption Spectrum of 20nm Gold Film

Topographic scan

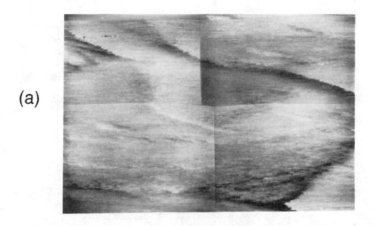

(a)

476nm Blue Thermal

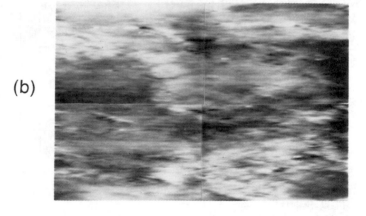

(b)

Fig.5

10nm

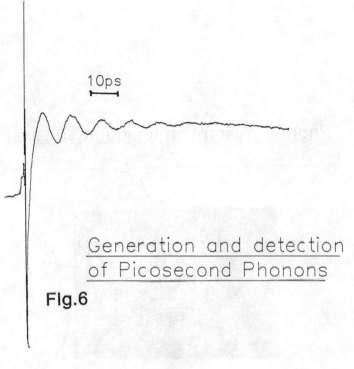

10ps

Generation and detection
of Picosecond Phonons

Fig.6

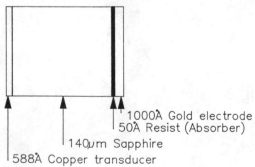

1000Å Gold electrode
50Å Resist (Absorber)
140μm Sapphire
588Å Copper transducer

Fig.7
Specimen construction for
Phonon Absorption Microscope

Phonon Absorption Image at 50 GHz

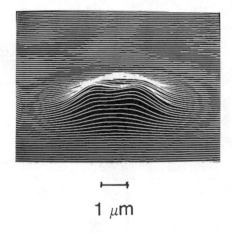

├───────┤

1 μm

Fig.8

ACOUSTIC SYMMETRY AND FIRST ORDER SPATIAL DISPERSION EFFECTS IN PHONON FOCUSING

A.G. Every
Physics Department, University of the Witwatersrand,
P O Wits 2050, South Africa.

1. THE SYMMETRY OF PHONON FOCUSING PATTERNS

In the absence of magnetism, phonon focusing patterns are centrosymmetric and possess crystallographic point group symmetry. Generic focusing patterns thus conform to the 11 Laue symmetry groups. If the phonon temperature is sufficiently low only long wavelength acoustic phonons are excited. These are governed by continuum elasticity theory and their focusing patterns thus possess the symmetry of the elastic constant tensor, or acoustic symmetry. Stiffening of the elastic constants in piezoelectric crystals leads in certain crystal classes to a reduction in the acoustic symmetry[1].

For Laue TII group crystals a suitable rotation about the four-fold axis can reduce the seven independent constant form of the elastic constant matrix to the six constant form for TI group crystals. The acoustic symmetry of all non–piezoelectric tetragonal crystals is thus 4/mmm. Piezoelectric stiffening however lowers the acoustic symmetry for the crystal classes 4 and $\bar{4}$ to 4/m. In a similar way the acoustic symmetry of all non–piezoelectric trigonal crystals is $\bar{3}$m, but piezoelectric stiffening reduces this to $\bar{3}$ for the crystal class 3. The acoustic symmetry for all non–piezoelectric hexagonal crystals is transverse isotropy, ∞/mm. Piezoelectric stiffening reduces this to 6/mmm in the case of the crystal classes $\bar{6}$ and $\bar{6}$m2.

2. FIRST ORDER SPATIAL DISPERSION EFFECTS IN PHONON FOCUSING.

The onset of dispersion can be treated by expanding the wave–vector–dependent elastic moduli $C_{ij\ell m}(\vec{k})$ in power series in $\vec{k}$ and retaining only leading terms of low order. The Christoffel characteristic equation thereby takes the form[2,3]

$$| \Gamma_{i\ell}^{(0)} + \Gamma_{i\ell}^{(1)} + ... - \rho\omega^2 \, \delta_{i\ell}| = 0, \qquad (1)$$

where $\Gamma_{i\ell}^{(0)} = C_{ij\ell m} k_j k_m$ and $C_{ij\ell m}$ are the limiting elastic moduli for $\vec{k} \to 0$, piezoelectrically stiffened if required, and $\Gamma_{i\ell}^{(1)} = id_{ij\ell mn} k_j k_m k_n$ etc. The coefficients $d_{ij\ell mn}$ comprise the acoustic gyrotropic tensor. It is non–vanishing only for the non–centrosymmetric crystal classes[4] and leads, with certain exceptions[2], to the removal of branch degeneracy along acoustic axes, and to the phenomenon of acoustical activity.

Well away from acoustic axes $\Gamma^{(1)}$ contributes only in second order of perturbation to the dispersion relation, while the subsequent term $\Gamma^{(2)} = f_{ij\ell mnr} k_j k_m k_n k_r$ contributes in first order. As a result both terms yield corrections to the phonon velocities which are quadratic in $|k|$. In the vicinity of acoustic axes however $\Gamma^{(1)}$ leads, by degenerate perturbation theory, to splitting of the branches, elliptically/circularly polarized modes, and to linear dependence of velocities on $|k|$, while the effects of $\Gamma^{(2)}$ are still quadratic in $|k|$ and hence negligible by comparison.

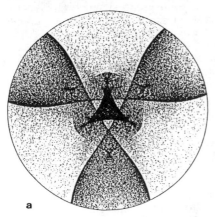

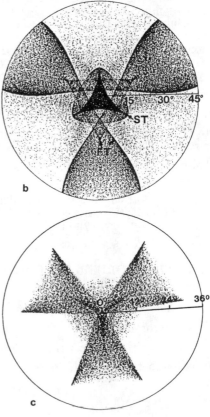

Figure 1. Calculated transverse phonon focusing patterns in quartz centred on c–axis.
(a)   No dispersion,
(b)   Modes with $|k| = 2 \times 10^8 m^{-1}$,
(c)   Modes with $|k| < 1.162 \times 10^9 m^{-1}$ and $V_Z > 4800$ m/s.

T = 10 K.

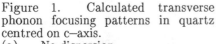

Particular attention has been paid by a number of investigators to acoustical activity associated with the c–axis conical point in quartz. Only one component $d_{543} = d_{13233}$ plays a significant role here, and measurements, based on a variety of techniques, have established that $d_{543} \approx 13.4$ N/m.

Marked discrepancies exist between the experimental phonon focusing pattern of quartz[6] and the predictions of non–dispersive continuum elasticity theory, and to a large extent these can be resolved by taking first order spatial dispersion into account. Fig. 1(a) shows the calculated phonon focusing pattern of quartz in the region of the c–axis, neglecting dispersion. It differs in a number of details from the corresponding experimental phonon images obtained by Koos and Wolfe[6].

Fig.1(b) shows the calculated focusing pattern for transverse modes of $|k| = 2 \times 10^8$ m^{-1} ($\nu \approx 150$ GHz), taking account of dispersion. A new triangular feature has appeared and there has been some displacement of existing caustics, mostly so near the centre of the pattern, where the c–axis is located. When a number of such images are superposed to simulate the effects of a thermal distribution of phonon frequencies, the caustics are transformed into broadened features. Fig. 1(c) shows the focusing pattern corresponding to a phonon temperature $T = 10$K. The calculation has been restricted to FT modes with $|k| < 1.162 \times 10^9$ m$^{-1} = 0.1$ k_{BZ} and $V_z > 4800$ m/s. This focusing pattern is in good agreement with the experimental image obtained by Koos and Wolfe[6] with an early setting (800ns) of their time gates.

3.   CONCLUSIONS

The symmetry group classification of phonon focusing patterns differs depending on whether piezoelectric stiffening of the elastic constants and dispersion are present or not.

In acoustically active crystals spatial dispersion has pronounced effects on phonon focusing in the neighbourhood of acoustic axes, even at relatively low phonon temperatures. The onset of dispersion can be satisfactorily treated by use of the acoustic gyrotropic tensor. Well away from acoustic axes dispersion in phonon velocities is quadratic in $|k|$ and only becomes significant at relatively high temperatures.

REFERENCES

1.   Every, A.G., J. Phys. C20, 2973 (1987).
2.   Portigal, D.L. and Burstein E., Phys. Rev. 170, 673 (1968).
3.   Every, A.G., Phys. Rev. B36, 1448 (1987).
4.   Kumaraswamy, K. and Krishnamurthy, N., Acta Crystallogr. A36, 760 (1980).
5.   DiVincenzo, D.P., Phys. Rev. B34, 5450 (1986).
6.   Koos, G.L. and Wolfe, J.P., Phys. Rev. B30, 3470 (1984).

THE ROLE OF PHONON FOCUSING IN THE SURFACE TEMPERATURE PROFILE OF A SILICON WAFER HEATED BY A 2DEG

A.G. Every[o], N.P. Hewett[+*], L.J. Challis[+] and J. Cooper[+]

[o]Department of Physics, University of the Witwatersrand, Johannesburg, P O Wits 2050, South Africa

[+]Department of Physics, University of Nottingham, Nottingham, NG7 2RD, UK

[*]British Telecom Research Labs, Martlesham Heath, Ipswich, UK

In the ballistic phonon regime the temperature profile of the surface opposite a heated two–dimensional electron gas (2DEG) provides information on the angular distribution of the phonons emitted by the 2DEG. In this paper we describe simulations that we have carried out on a silicon wafer containing a heated 2DEG, and it is planned to extend this to GaAs. The integral equation approach of Klitsner et al[1] is adapted for our wide slab–shaped samples, in which the temperature can be regarded as independent of the large dimension transverse to the direction of heat flow. These simulations are being carried out in parallel with experimental investigations in an effort to determine the directional dependence of the phonon emission from a 2DEG in zero field and in quantising magnetic fields at low power densities where pulse techniques cannot be used.

The conventional approach to boundary limit phonon conduction[2] assumes a uniform thermal gradient along the axis of a rod–shaped sample. We have followed the more general approach of Klitsner et al[1] by establishing an integral equation for the surface temperature profile of our samples based on the condition of local radiative equilibrium. A typical sample in our experiments has the form of a rectangular wafer 0.38 mm thick, 5mm wide and 10 (or 5)mm long as shown in Figure 1. The 2DEG is a narrow strip (1 mm wide) located inside one face and extending across much of the width (3 mm). The sample sits in a vacuum, heat is generated by passing an electrical current through the 2DEG, and one end of the sample acts as a heat sink with the temperature fixed at around 1K. Silicon and GaAs of this thickness are transparent to phonons of frequency less than 1500 GHz. Specular reflections are not expected to be important under the operating conditions we are concerned with, and we assume that the phonons are thermalised at the surface.

We have considered various models for the phonon emission by the 2DEG, and have investigated the influence of phonon focusing. In these various models we assume the heat sink end of the sample to be at a fixed temperature T_c and that the surface temperature of the sample is independent of z (see Figure 1), being a function only of s, the distance measured around the sample.

In one model we take the heating power to be distributed uniformly over the area of the 2DEG. Phonon emission from a hot 2DEG is known to be

Figure 1

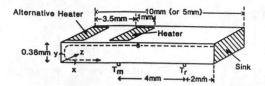

Figure 2

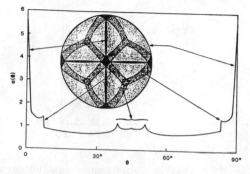

largely restricted to a cone of directions close to the normal by the need to conserve momentum in the plane of the 2DEG[3,4]. We have simulated this effect by introducing a cut–off to the energy flux from the 2DEG for $\theta > \theta_c$ while keeping the total heat output fixed. The model further assumes that phonons which are subsequently emitted by all surface elements (including those of the Si/SiO$_2$ interface above the 2DEG) pass through the 2DEG without being absorbed.

Under steady state conditions the temperature profile T(s) of the surface (excluding the sink) is such that each surface element radiating as a black body emits the same total energy flux as it receives from all other surface elements together with that from the 2DEG. This leads to an integral equation for $\delta T = T - T_c$, which to first order in δT takes the form

$$\delta T(s) = \frac{1}{2} \int_{\text{all } s'} \delta T(s') \, \alpha\left(\theta(s')\right) \cos\left(\theta(s')\right) \frac{d\theta}{ds'} ds' + K_s$$

where K_s is a source term representing the heat received from the 2DEG, and $\alpha(\theta)$ is a factor that incorporates the effects of phonon focusing. A Monte Carlo calculation for a silicon sample with faces parallel to the cube planes and taking account of all three phonon branches with suitable weighting yields the result shown in Figure 2. Interestingly, in integrating out the one angular coordinate to obtain $\alpha(\theta)$ the intense caustics, which are a prominent feature of phonon focusing patterns, are either eliminated entirely or softened into logarithmic divergences. Figure 2 shows the correspondence between certain caustics and features in the $\alpha(\theta)$ curve.

For computational purposes we have converted the integral equation into a set of linear equations by partitioning the surface into N = 208 parallel strips of approximately equal width, and have employed an interative method to solve these equations.

1360

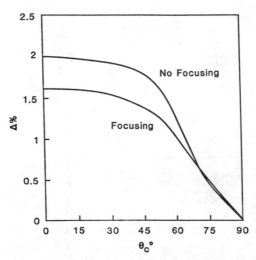

Figure 3

The calculations show that in the region midway between the heat source and sink a uniform thermal gradient exists, as expected, and the two faces of the sample are at the same temperature. The magnitude of this thermal gradient agrees, to within a few percent, with the average gradient obtained by dividing the temperature difference between source and sink by the distance Δs between their midpoints. Some other features of the temperature profile are, however, more surprising. Sizeable temperature discontinuities occur at the edges of the heat source and sink, and nearby the thermal gradients are non–uniform. Near to the heat source the temperatures of the opposite faces differ appreciably. In the region of the slab remote from the heat sink, there is a finite thermal gradient, even though the net heat flux in the x direction there is zero.

As θ_c is reduced the maximum temperature δT_m on the back face at x = 4 mm increases as a result of the increasingly concentrated phonon intensity in the forward direction. Figure 3 shows the fractional increase Δ in δT_m as a function of θ_c. Also shown are results obtained by ignoring phonon focusing (taking $\alpha(\theta) = 1$). Reducing θ_c, as can be seen, has a slightly smaller effect in the case of focusing.

The maximum change in temperature (relative to the end of the sample) seen experimentally is ~ 2% so there is quite good agreement, but the effect of phonon focusing remains to be confirmed.

References

1. Klitsner, T., VanCleve, J.E., Fischer, H.E. and Pohl, R.O., Phys. Rev. B38, 7576 (1988).
2. Casimir, H.B.G., Physica 5, 495 (1938).
3. Rothenfusser, J.M., Foster, L. and Dietsche, W., Phys. Rev. B34, 5518 (1986).
4. Challis, L.J., Toombs, G.A. and Sheard, F.W., Physics of Phonons ed T. Paskiewiez (Berlin: Springer) p389.

Phonon Imaging of the Critical Cone and Frustrated Total Internal Reflection at Solid-Solid Interfaces

C. Höss, H. Schreyer, W. Dietsche, J.P.Wolfe[*] and H. Kinder

Physik Department E 10, Technische Universität München, 8046 Garching, FRG

[*]Physics Dept., University of Illinois at Urbana–Champaign, Urbana, IL 61801, USA

Although Snell's law of refraction is very well known in optics, few experiments have dealt with the specular refraction of phonons at solid–solid interfaces. Due to the elastic anisotropy of dielectric crystals and the existence of waves with longitudinal polarisation as well as transverse, refraction of phonons is much more interesting than for photons. In this paper we present measurements which display for the first time the effects of total internal reflection and frustated total internal reflection of high frequency phonons at solid–solid interfaces. The technique we use is phonon imaging[1] – –a scanning heat pulse experiment– – which is well suited for this type of measurement.

Fig.1 shows a cross section of the sample and the associated geometry for phonon transmission through it. We have chosen Ge (the softer material) as a substrate material and MgO (the harder material) for the thin layer[2]. The ratio of the phase velocity in MgO to that in Ge is about 2. The MgO film was covered with a thin Al layer, which when photoexcited with a focused He–Ne–laser (spot size ≈30 μm) acts as a broad band phonon source. The phonons transmitted through the film and substrate are detected by a superconducting Al–O–Al junction having a gap of $2\Delta/h \approx 140$ GHz and dimensions of 30x30 μm². The sample

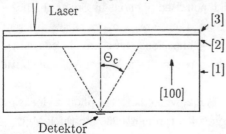

Laser
[3]
[2]
Θ_c
[1]
[100]
Detektor

Fig: 1: Cross section of the sample geometry and a critical cone for phonon transmission. The substrate [1] is a 3 mm (100) oriented Ge crystal covered with a polycrystalline MgO layer [2]. On top of it is a 100 nm Al layer [3] in which phonons are generated. The detector, a Al–O–Al junction, is placed on the opposite side of the Ge sample.

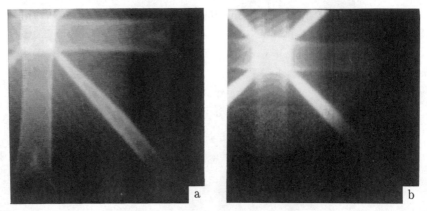

Fig. 2: a) Phonon—focusing pattern of Ge in the (100)—plane. The central 'box' and broad ramps are formed by ST phonons, whereas FT intensity appears as the ridges rotated by 45° from the ST ramps. L phonons are nearly isotropic in this plane. b) The spatial intensity distribution for phonon transmission through a 230 nm MgO film on (100) Ge. As described in the text, phonon transmission is only possible inside the critical cones of total internal reflection.

is placed in an optical cryostat and operated at a temperature of 1 K.

We assume that the phonons generated in the Al layer are radiated isotropically into the MgO layer, where they propagate according to the elastic properties of polycrystalline MgO. If one assumes pure specular transmission, i.e., that the acoustic mismatch model is valid, the transmission of phonons from MgO to Ge is determined by Snell's law $\sin\Theta_1/v_1 = \sin\Theta_2/v_2$ (for L, FT, ST polarisations separately), which states that the wavevector component parallel to the interface is conserved. The velocities v_1 and v_2 are the phase velocities in MgO and Ge, which for an anisotropic solid, depend on wavevector direction and are different for each of the three modes.

From Snell's law the maximum angle of transmission into the Ge substrate occurs when $\Theta_1 = 90°$, which defines the critical angle $\sin\Theta_c = v_2/v_1$ different for each mode. An additional feature is the possibility of mode conversion, i. e., of L or T polarised phonons in MgO to L, FT or ST polarised phonons in Ge, at the interface. Taking all these possibilites into account, there are as many as six different Θ_c's for a given propagation plane. In three dimensions there are six concentric critical cones for total internal reflection, distorted by the anisotropy of the elastic medium.

The experimental result is shown in Fig. 2, where both the phonon—focusing patterns of (100) Ge (no film) and the transmission through a 230 nm thick MgO film are displayed. Transmitted intensity only occurs inside the critical cones,

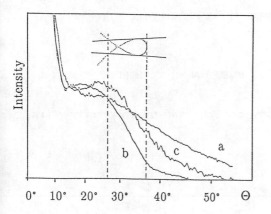

Fig. 3: Three diagonal raster scans, starting from the (100) point in the focusing pattern, which show the transmitted intensity along the FT ridges. Scan **a)** is taken from Fig.2a, which is the no film case. Scan **b)** is from Fig.2b), which is the thick film case. The dropoff in intensity, caused by total internal reflection, is clearly seen. Scan **c)** is from a measurement with a 12 nm MgO film and shows a rise in intensity beyond Θ_c due to frustrated total internal reflection of phonons.

which, transformed into group–velocity space, form complex structures. So ST intensity is present inside the $T_1 \rightarrow ST_2$ critical cone, which contains the central 'box' and only small parts of the horizontal and vertical ramps. FT intensity is transmitted inside the $T_1 \rightarrow FT_2$ critical cone, which causes the FT ridges to be cut off at about 35°, compared to 30° predicted by the above equation. A third critical cone is the $L_1 \rightarrow L_2$ one at about 40° and is nearly circular.

A very interesting situation is obtained when the MgO layer is reduced in thickness to one which is comparable to the penetration depth of the evanescent wave. The phonons no longer obey total internal reflection, instead they 'tunnel' through the MgO at angles beyond Θ_c. Fig. 3 shows this case in comparison to the no–film and thick–film cases. The inset shows the angular dependence of the Θ_c condition for FT phonons within the FT ridge, i. e., there is no sharp step. The results shown here are in good agreement with continuum–limit calculations which we have carried out. In spite of total internal reflection, there is significant intensity beyond Θ_c for each mode, due to a diffuse part in the phonon signal, caused by surface scattering of phonons at the Ge–MgO interface. From this it is possible to determine the ratio of diffuse to specular transmission through the interface, which, in our experiments, gives about 85% specular transmission.

1. G.A. Northrop and J.P. Wolfe, in <u>Nonequilibrium Phonon Dynamics</u>, ed. by W.E. Bron (Plenum, New York 1985)

2. Polycrystalline MgO films were deposited on highly polished Ge substrates, which are thermally heated before the evaporation to remove any sort of contamination. Isotropic L and T modes for MgO are assumed with elastic constants given by $c_{11} = 3.4$ Mbar, $c_{44} = 1.345$ Mbar and density $\rho = 3.595$ g/cm^3.

A SPATIALLY RESOLVING HEAT PULSE PHONON DETECTOR

A.J. KENT, G.A. HARDY, P. HAWKER AND D.C. HURLEY

Department of Physics, University of Nottingham, University Park, Nottingham NG7 2RD U.K.

Techniques which allow us to study the spatial anisotropy of ballistic heat pulse propagation in crystalline solids have developed rapidly in the past decade[1] . Standard 'phonon imaging' techniques use a pulsed laser as a heat source and a superconducting bolometer as a detector. A 2-D map of the phonon intensity can be built up by raster scanning the laser over the sample. Another technique, which is useful if the phonon source cannot be moved, uses an extended tunnel junction as a detector[2] . A chosen spot on the junction is made sensitive to phonons by illuminating it with a focused laser beam. Scanning the laser enables the phonon image to be built up.

A disadvantage of the above techniques is that, because they both rely on superconducting principles, they cannot be used in the presence of an applied magnetic field. We have developed a scanned detector phonon imaging system based on an extended cadmium sulphide (CdS) thin film bolometer which is able to operate in magnetic fields of up to at least 11T[3] . It is being used to study phonon emission by the 2DEG in Si MOSFET and GaAs HEMT devices in quantising magnetic fields.

The property of CdS bolometers which makes them suitable for use in this way is that they are sensitive to phonons only after illumination. This promotes electrons to shallow traps below the conduction band. The phonon pulse liberates these electrons thus increasing the conductivity and with a small bias current applied a corresponding voltage signal may be detected.

The CdS film (1-10μm thick) is vacuum evaporated onto the sample, a large area (1cm x 1cm) interdigital copper electrode comprising 100μm fingers spaced by 100μm is fabricated on the CdS by photolithography. The detector has a resistance of 100Ω at 300K, when it is cooled the electrons freeze out and the resistance rises to a few MΩ at 4.2K. The bolometer is insensitive to phonons in this state. A selected spot on the bolometer is sensitized by illuminating with an Ar$^+$ laser beam focused to 200μm diameter, when it reaches a few KΩ the bolometer is ready and the laser beam is switched off. The sensitive state is persistent and so following data acquisition the bolometer must be reset before another position can be sensitized. To do this the active spot is heated by passing a short current pulse (≈50mA), on recooling to 4.2K the bolometer is in its high resistance state again. The rest of the imaging system comprises an acousto-optic modulator for laser control, a pair of computer controlled galvanometer mirrors for X-Y positioning of the laser and the associated electronics.

The bolometer signal is pre-amplified and then fed to a digital storage oscilloscope which is used in conjunction with a microcomputer for signal processing. Using this arrangement two-dimensional maps and line-scans of the phonon intensity can be obtained, under computer control, to a maximum resolution of 200μm.

Figure 1 shows heat pulse signals for three different propagation directions in silicon. The phonon source was an electrically heated metal film on the (001) surface of the 5mm thick crystal opposite the bolometer. In both cases the heat pulse is propagating in the (110) plane and θ is the angle between the propagation direction and <001> .

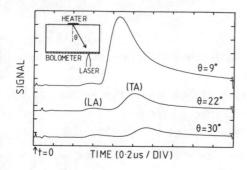

Fig 1. Heat pulse signals for different propagation directions θ in Si. Transverse (TA) and Long-itudinal (LA) modes can be identified.

These results demonstrate that the sensitization/reset process works well there being no retention of signal from previous positions. Figure 2 shows a 2D image of the phonon intensity on the (001) surface to a resolution of 0.3mm. The 22x22 points were each obtained by integrating signals like in fig. 1 over a narrow time window chosen to select the transverse modes. The characteritic focusing patterns for transverse modes in Si are clearly visible. Figure 3 shows the phonon intensity as a function of angle for a single line corresponding to propagation in the (110) plane. Quantitative comparison with theoretical monte-carlo simulations of the phonon propagation[4] confirms that the bolometer response is uniform over an extended area.

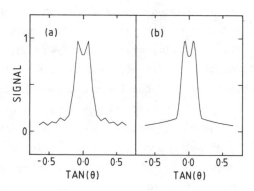

Fig.2 Two dimensional image of transverse mode intensity opposite the heater.

Fig.3 (a) Line scan of phonon intensity in the (110) plane, (b) Theoretical simulation.

This work is supported by a grant from the Science and Engineering Research Council of the UK.

REFERENCES

1) Northrop G A and Wolfe J P (1985) Nonequilibrium Phonon Dynamics ed. W E Bron (New York:Plenum) p165.
2) Dietche W (1986) Phonon Scattering in Condensed Matter vol. V, ed. A C Anderson and J P Wolfe (Berlin:Springer) p366
3) Ishiguro T and Morita S, Appl. Phys. Lett. $\underline{25}$,533 (1974).
4) Northrop G A, Comp. phys. Commun. $\underline{28}$, 103 (1982).

PHONON IMAGING: COMPARISON OF ENERGY AND QUASIMOMENTUM FOCUSING IN TRANSVERSELY ISOTROPIC MEDIA

Cz. Jasiukiewicz, T. Paszkiewicz, Institute of Theoretical Physics, University of Wrocław, ul. Cybulskiego 36, 50-205 Wrocław, Poland

1. Introduction

In most of phonon imaging experiments the focusing of energy is studied (cf. [1], [2]). However, recent progress in semiconductor devices technology allows to study also the *quasimomentum* focusing [3]. Here we compare the focusing patterns for both above quantities. We restrict ourselves to to the simplest case of long-wave acoustic phonons in transversely isotropic media (TIMs)(cf. [4]).

2. Properties of transversaely isotropic media

Elastic properties of TIMs are characterized by a symmetry axis (specified by an unit vector $\hat{c}$) and five independent elastic constants. For TIMs the Christoffel equation allows an explicit solutions for the phase velocities and polarization vectors of three modes ($j=0,1,2$) [4]. For explaining results of imaging experiments both sets of above quatities are indispensable. Having phase velocities one can calculate the group velocities $V(k,j)$. The quasimomentum focusing pattern is additionally afected by a phonon-electron interaction matrix element, which involves the polarization vectors of bulk phonons. The influence of this matrix element on observed focusing pattern was studied by Karl et al [3]. In our paper we compare only the densities of energy and quasimomentum.

3. Densities of energy and quasimomentum currents

Suppose that one deals with a massive perfect specimen at temperature much lower than the Debye temperature. Then injected phonons move ballistically and the influence of the boundary scattering on their propagation is negligible.

As a natural idealization of such conditions we consider half-space filled with a TIM. At an arbitrary point of a plane, being the only boundary of this medium, there is a point source generating infinitely short Planckian heat pulses. Assume that the source is located at r=0. The phonon distribution function $f(k,j;r,t)$ obeys the Boltzmann – Peierls equation (BPE) with a source term [5]

$$\frac{\partial f(k,j;r,t)}{\partial t} + V(k,j)\nabla f(k,j;r,t) = Af_o\left[\frac{\hbar\omega}{k_B T_H}\right]\Theta(V_z)\delta(r)\delta(t),$$

where A is a constant, ω -frequency of a phonon (k,j), T_H- the source temperature, $\Theta(x)$- Heaviside step function, $\delta(x)$ – is Dirac δ-function. For simplicity we assume that the symmetry axis is perpendicular to the medium boundary, the z axis of the Cartesian coordinate system is parallel to the symmetry axis, x and y axes are oriented arbitrarily. The focusing pattern is created on a surface parallel to the medium boundary. The solution of BPE has a simple form

$$f(k,j;r,t) = Af_o\left[\frac{\hbar\omega}{k_B T_H}\right]\Theta(t)\Theta(V_z)\delta[r - V(k,j)t]. \qquad (1)$$

Next we shall calculate the components of energy and quasimomentum currents densities

$$j_{E\alpha}'(r,t) = \sum_{j=0}^{2} \frac{1}{(2\pi)^3} \int d^3k \; \hbar\omega(k,j)V_\alpha(k,j)f(k,j;r,t) \qquad (2a)$$

$$\Pi_{\alpha\beta}'(r,t) = \sum_{j=0}^{2} \frac{1}{(2\pi)^3} \int d^3k \; \hbar k_\alpha V_\beta(k,j)f(k,j;r,t). \qquad (2b)$$

Introduce two new quantities $j_E = j_E' B^{-1}$, $\Pi = \Pi' B^{-1}$, where $B = A\pi^2(k_B T_H)^4(60\hbar^3)^{-1}$, and calculate quantities characterizing the focusing of energy and quasimomentum.

$$I(r) = \int_0^\infty dt \hat{z} j_E(r,t), \qquad P_\tau(r) = \int_0^\infty dt \; \hat{\tau} \; \Pi(r,t)\hat{z}, \qquad (3a,b)$$

where $\hat{\tau}$ is an unit vector tangent to a strip containing 2D electron gas.

The results of calculation of P_τ (for z=1) for He^4 and Zn are shown in Figs. 1,2. In the same figures there are shown cross-sections of calculated I(r), obtained with the help of the plane in which the vectors $\hat{c}$ and $\hat{\tau}$ are lying. It is seen that generally the energy focusing pattern is different than the quasimomentum focusing pattern. Only the main characteristics - the caustics are identical. Thus, apart from electron-phonon interaction the different nature of focused quantities may lead to the additional differece of focusing patterns studied by Karl et al [3].

References

1.Maris H.J, in: Nonequilibrium Phonons, ed. by W.Eisen-menger A.A. Kaplyanskii, North-Holland, Amsterdam, 1986.
2.Northrop G.A.,Wolfe J.P., in: Nonequilibrium Phonon Dynamics, ed. by W.E.Bron, Plenum, New York, 1985.
3.Karl H., Dietsche W., Fischer A., Ploog K., Phys. Rev. Lett. 61, 2360, (1988).
4.Fedorov F.I.,Theory of Elastic Waves in Crystals, Plenum Press, New York, 1985.
5.Kwok P.C., Phys. Rev. 175, 1208 (1968).

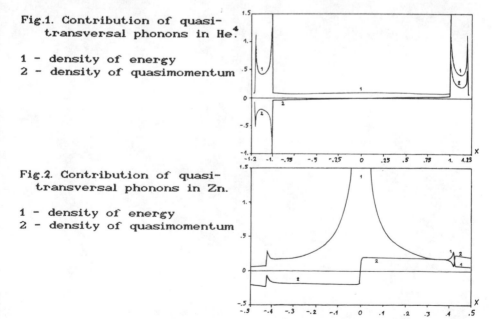

Fig.1. Contribution of quasi-
    transversal phonons in He^4.

1 - density of energy
2 - density of quasimomentum

Fig.2. Contribution of quasi-
    transversal phonons in Zn.

1 - density of energy
2 - density of quasimomentum

16

Particle Detection

PHONON-MEDIATED DETECTION OF ELEMENTARY PARTICLES

B. Cabrera

Physics Department, Stanford University
Stanford, California 94305

1. INTRODUCTION

Semiconductor diode particle detectors now provide the highest energy resolution (≈ 3 keV FWHM for 1 kg) and the lowest thresholds available (≈ 4 keV) for large mass detectors [1]. In the keV range, less than 30% of the deposition energy is converted directly into the electron-hole pair signal, the rest forming phonons. The characteristic energies of these phonons is ≈ 1 meV, 10^3 less than the excitation energy for an electron-hole pair in a semiconductor (≈ 1 eV). Thus in principle, energy resolutions over an order of magnitude better are possible if the phonon signal is used.

The desire for higher resolution and lower threshold detectors is motivated by several exciting experiments in weak interaction physics. These include a search for a hypothetical flux of weakly interacting massive particles which may make up the dark matter around our galaxy, a first measurement at a reactor of the predicted coherent neutrino-nucleus elastic scattering process, a self-normalizing reactor neutrino oscillation measurement for detecting a finite neutrino mass, and a future solar neutrino observatory capable of measuring the flux and energy spectrum of solar neutrinos.

Research efforts now underway are aimed at developing a new class of elementary particle detector based on phonon-production in insulating crystals such as silicon at temperatures below 1 K. A number of groups are pursuing the idea of using insulating crystals such as silicon for bolometers [2]. The Debye heat capacity becomes so small at tens of mK that the change in temperature caused by a particle interaction in the few keV range is detectable in a large silicon crystals [3]. Currently, the best high-efficiency x-ray spectrometer uses a 10^{-5} g silicon crystal micro-bolometer at ≈ 100 mK and has achieved a resolution of 12 eV FWHM for the ≈ 6 keV x-rays from ^{55}Fe [4]. In addition several groups [3] have operated larger silicon bolometers and a resolution of ≈ 25 keV FWHM has been demonstrated in a 100 g silicon crystal [5].

More recently a few groups including our group at Stanford are investigating detectors in which the high energy phonons which travel ballistically from the particle interaction region within the crystal are detected before they are down-converted to the quasi-thermal phonons used in the bolometers [6,7]. The advantage of such a detector over the simple bolometer is that imaging of each event is possible. The additional information substantially improves the background suppression. In this paper, we discuss the current efforts on this newest type of detector which we call a SiCAD (silicon crystal acoustic detector).

2. Phonons from Local Events in Silicon Crystals

Within a single crystal of silicon, energy depositions of a few keV are contained within a sphere a few μm in diameter for electron recoils and a few tens of nm in diameter for nuclear recoils. A thermal-like spectral phonon distribution is generated with a characteristic temperature of 10-20 K. This spectral distribution arises from the rapid decay of electron-hole excitations to the band edge, first generating very short wavelength phonons, which quickly relax to longer wavelength phonons within less than $\approx$ 10 ns. The decay rates are very strongly dependent on phonon energy ($\propto v^5$) and for wavelengths greater than several hundred lattice spacings further decays are negligible. These longer wavelength phonons propagate throughout the crystal with little scattering and no dispersion and are called the the ballistic phonons [8].

2.1 Ballistic Phonons

An interesting and important aspect of ballistic phonon propagation is that strong focussing effects occur within the crystal [9]. These focussing patterns permit three dimensional reconstruction of event locations within the crystal and can resolve tracks or multiple scattering events. We have performed Monte Carlo calculations on these effects [7]. Fig 1a shows the result of such a calculation for the phonon energy density arriving at the [100] surface of a silicon crystal from a point energy deposition within the crystal. The number of phonons used in the calculation is for a 1 keV energy deposition at a point within the crystal. Roughly 12 % of the total phonon energy is beamed in 2 % of the solid angle. For comparison, in Fig 1b we have shown the energy distribution resulting from ballistic propagation in an isotropic medium. The two figures are normalized to the same total energy deposition. The actual energy distributions have a ballistic component like that shown in Fig 1a and a more diffuse component resulting from scattered phonons. For sufficiently pure silicon, the scattering is determined by the intrinsic isotope concentration which in turn is affected by the phonon energy distribution produced by anharmonic decay.

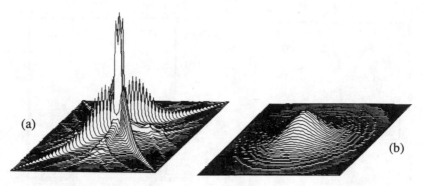

Fig 1. (a) Calculated ballistic phonon energy density incident on surface of [100] silicon crystal. (b) For isotropic medium.

2.2 Anharmonic Decay

The effects of the anharmonic decay process are shown in Fig 2. We have run a Monte Carlo on a simple model where the lifetime for decay is assumed to be of the form $\tau_A = (3~\mu s)~(1~THz~/~v)^5$ for silicon crystals [10]. Here we assume that all phonon modes as well as all wavevectors directions have the same decay rates. In fact, the lifetime depends on the mode and the magnitude and direction of $\mathbf{k}$, nevertheless, the results are qualitatively correct. For the calculation, the anharmonic lifetime τ_A is used to determine whether a phonon decays during each time step and a simple spherical density of states argument is used to determine the partitioning of the phonon energy into two phonons when a decay does occur. The probability peaks for decay phonons with half of the original energy and is proportional to $E_1^2 E_2^2$ where $E_1 + E_2 = E$ is the energy of the parent phonon. We have assumed a linear dispersion relation so that E is proportional to k and assume the density of states to be uniform in k-space. In Fig 2a, we begin the calculation with a delta function of phonons at 4 THz. Then each of the nine curves follows the time evolution of the Monte Carlo from 0.01 μs to 2.56 μs with each curve from right to left representing a factor of two increase in the time. In Fig 2b, we proceed exactly the same as in Fig 2a except that we begin the calculation with the initial energy uniformly distributed over a frequency interval. Remarkably, after a short time the resulting phonon energy distribution is the same in Fig 2a and 2b. Thus the decay process quickly forgets the initial distribution and depends only on the phonon physics below about 1 THz where the approximations that we have made are most applicable. After several μs the phonons have traveled about 1 cm and distribution has the same average phonon energy as a Planck distribution at 10-20 K, although the latter is skewed to lower energy with a high energy tail.

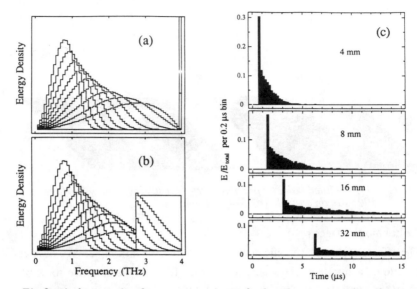

Fig 2. *Anharmonic down-conversion of phonon energy distribution starting from (a) delta function, and (b) from a uniform distribution. (c) Affect of isotope scattering on time dispersion of signal.*

2.3 Isotope Scattering

As the propagation distance increases through an intrinsic silicon crystal, a smaller fraction of the energy arrives ballistically and a larger fraction arrives at delayed times. Again in the spirit of our qualitative tour, we can estimate the fraction of ballistic energy transport as a function of crystal thickness with another simple Monte Carlo. We take a sphere of silicon and assume that it is an isotropic medium (no focussing) with both anharmonic decay and isotope scattering. We begin each of a thousand 4 THz phonons at the center of the sphere and follow each out towards the surface keeping track of anharmonic decay events and isotope scattering events. For the anharmonic decay events we use the same lifetime as in Sec 2.2 and assume that both lower energy phonons continue along the same path as the parent phonon. For the isotope scattering, we assume that the scattering is isotropic and that the lifetime is given by $\tau_I = (0.2\ \mu s)\ (1\ \text{THz}\ /\ v)^4$ for silicon crystals, again the same for all phonon modes and for all wavevectors directions [10]. The results of the calculation are shown in Fig 2c where for each of four thicknesses, we keep track of the arrival times of the phonon energy in 0.2 μs time bins. If we identify the first bin with the ballistic component, then roughly 30 % of the phonon energy arrives ballistically through 4 mm of silicon and about 8 % through 32 mm.

3. SUPERCONDUCTING PHONON SENSORS

To detect the phonons at the crystal surfaces two sensor designs have been developed both utilizing superconductivity. A third technique utilizes neutron transmutation doped germanium thermistors and will not be discussed here [11].

3.1 Superconducting Transition Edge Sensors

The first, called a transition edge sensor, is a single thin-film patterned into a series meandering circuit and biased with a constant current. A phonon flux incident on the film will drive those portions normal where the phonon energy density exceeds a critical value. Then a voltage is seen across the circuit providing a signal which is proportional to the length of the circuit driven normal. If the current is below a latching critical current, then self-extinguishing pulses are seen, otherwise the resistive heating of the normal state is sufficient to expand the normal region across the whole circuit. Such a system is straightforward to manufacture over large crystal surfaces with photolithographic techniques, but the physics governing the response is non-linear. These types of devices, often called superconducting bolometers in the phonon physics literature, have been used for many years [9]. However, in the phonon experiments the devices were biased in temperature at the midpoint of the resistive transition and have a roughly linear response for small changes about the bias point. On the other hand, for our application better performance is obtained by biasing at the foot of the transition on the superconducting side, and our signal comes from those portions of the film that were driven fully into the normal state. To indicate this distinction, we prefer the term transition edge sensor.

We estimate the threshold energy density E_ρ necessary for this superconducting to normal state transition, by integrating the heat capacity $C_{es}(T)$ of the superconductor from the bias temperature T_b to the critical temperature T_c. The electronic heat capacity below T_c is given by the BCS theory [12] and shown in Fig 3a. We assume that the time constants within the superconductor are sufficiently short to allow quasi-thermal equilibrium. The critical energy density E_ρ as a function of bias temperature T_b is shown in Fig 3b. For a film thickness d, the asymptotic form near T_c of the surface energy density E_σ given by E_ρ d is $E_\sigma \approx 5.0$ N(0) Δ_0^2 d (1-T/T_c), where Δ_0 is the gap at T = 0 and N(0) is the density of states at the Fermi surface in the normal metal. In Fig 3c we plot the surface critical energy density for 40 nm thick films in eV/μm^2 as a function of temperature for the elemental superconductors Al, Ti, Ir, and W.

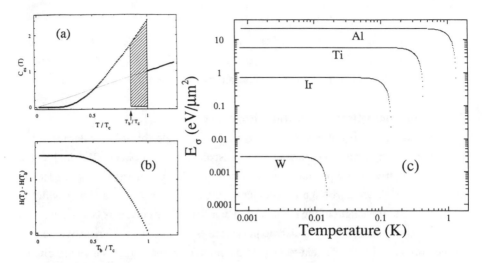

*Fig 3. (a) BCS heat capacity versus temperature in units $(2\pi^2/3)N(0)k^2T_c$.
(b) Differential enthalpy between T_b & T_c in units $(\pi^2/3)N(0)k^2T_c^2$.
(c) Critical surface energy density for 40 nm films.*

3.2 Superconducting Tunnel Junctions

A second device, the superconducting tunnel junction, possess the opposite properties. These are more difficult to manufacture, but the physics is nearly linear. Such a device consists of two superconducting films separated by a thin oxide barrier (typically 1-2 nm). This device is biased with a constant voltage given by Δ/e, where Δ is the superconducting energy gap. Phonons from an event within the crystal reach the surface and enter the superconducting film. Once inside the superconductor, those phonons with energies greater than 2Δ (most) are strongly absorbed by breaking Cooper pairs and forming electron-like excitations called quasiparticles. In fact for phonon energies well above threshold, several Cooper pairs will be broken. These quasiparticles will tunnel across the oxide barrier providing a signal if the tunneling times are short compared to the quasiparticle recombination time. The condition is satisfied for $T \leq T_c/10$, where quasiparticle recombination with thermal quasiparticles is negligible. For these devices the tunneling current is proportional to the energy absorbed by the film as long as the quasiparticle density produced by the phonon flux remains dilute so that little self-recombination occurs before tunneling.

4. EXPERIMENTS

As examples of the current status of the research, we will discuss two recent experiments in more detail. The first uses aluminum superconducting tunnel junctions to image the location of alpha particle interactions within a large silicon crystal and the second uses titanium transition edge sensors to obtain spectral information for x-rays down to several keV. Several other efforts are reported in these proceedings, including the recent work at S.I.N. by the Zehnder group [13].

4.1 Detection of Alpha Particles with Tunnel Junctions

A series of elegant experiments were performed by Peterreins, von Feilitzsch, et al, of the Munich group [14], and are based on the development of high sensitivity superconducting tunnel junctions [15]. Fig 4a shows a schematic of the experiment. Three Al-Al oxide-Al junctions with $\approx$ 100 μm diameters were deposited on 5 mm centers along one face of a 3 mm-thick silicon crystal. On the other side, a copper sheet was positioned between the crystal and the ^{241}Am alpha emitter. This Cu sheet acted as a mask with five small 0.2 mm diameter holes drilled in the positions shown, so that alpha particles could strike the crystal only at the hole locations. These $\approx$ 5 MeV alpha particles are not very penetrating and deposit all of their kinetic energy within $\approx$ 20 μm of the silicon crystal surface. The experiment was operated in a vacuum at a temperature of $\approx$ 380 mK. Fig 4b shows data in which the pulse height (charge sensitive amplifiers were used) seen in junction 1 is plotted against the pulse height seen coincidently in junction 3. The alphas from each of the five holes are easily distinguished in the five dark clusters. The tails correspond to events where the alphas passed through the edges of the holes in Cu sheet loosing part of their energy before hitting the silicon. Another way of presenting the data (Fig 4c) is to plot the difference in arrival time of the signal between junctions 1 and 3. The five alpha locations are again clearly visible. However, the delay times are roughly ten times longer than expected from the ballistic time of flight, indicating that the largest part of the signal arrives after multiple scatterings either within the crystal or at the crystal surfaces. These experiments represent the first demonstration of phonon-mediated imaging of particle events through silicon crystals.

4.2 Detection of X-Rays with Transition Edge Sensors

Our Stanford group has demonstrated the transition edge sensor technique using aluminum films with alpha particle sources [16], and most recently we have obtained a factor of one hundred improvement in energy resolution and threshold using titanium films in x-ray experiments [17]. These most recent sensors are made by depositing 40 nm of titanium on crystalline silicon wafers which are 1 mm thick. The polished wafer

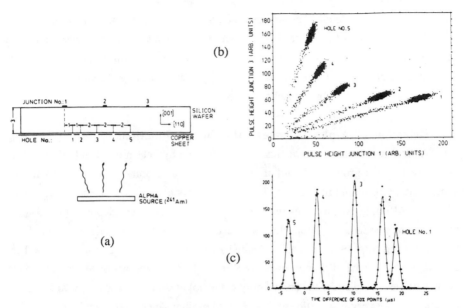

Fig 4. (a) Schematic of phonon-mediated imaging with alpha particles.
(b) Integrated charge from junction 1 versus from junction 3.
(c) Arrival time difference between junctions 1 and 3.

faces are perpendicular to the [100] axis. The meander pattern consists of 299 parallel lines each 2 μm wide on 5 μm centers. The active area of the pattern is 4.5 mm long and 1.5 mm wide. The normal resistance just above $T_c \approx 312$ mK is ≈ 18 kΩ per line.

Here we describe several x-ray experiments. These were performed with a 24 μC source of ^{241}Am. The decay spectrum of ^{241}Am contains several alpha lines around $\approx$ 5.5 MeV, a nuclear gamma at 60 keV and two atomic x-rays at 14 and 18 keV. Foils of Pb or Sn which are 125 μm thick are placed between the source and detector and stop all but the more penetrating 60 keV gamma rays which are attenuated in number by $\approx$ 0.5. As shown in Fig 2c, our Ti films which are biased at $T_b / T_c \approx 0.95$ have a critical surface energy density $E_\sigma \approx 1$ eV/μm². We use cryogenic GaAs MESFET voltage-sensitive amplifiers with $\Delta V_{rms} \approx 1$ nV/$\sqrt{Hz}$ at 1 MHz [18].

Fig 5a shows typical single pulses, of $\approx$ 5-10 μs duration, resulting from the photoabsorption of the 60 keV gamma rays in the Si substrate. These gammas have a 30 mm absorption length, much longer than the crystal thickness, so that they interact at a nearly uniform rate throughout the interior of the crystal. In Fig 5b we show the leading edge of each pulse with electronics-limited risetimes of $\approx$ 140 ns. Fig 5d is a pulse height spectrum obtained using the 125 μm Sn absorber. The prominent peak at

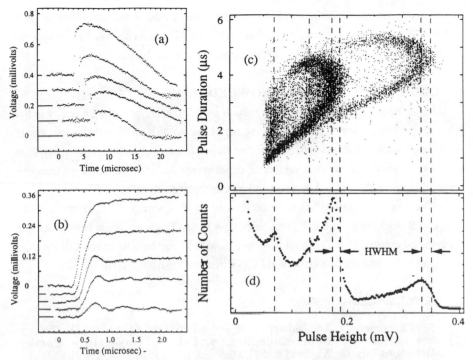

Fig 5. (a) Pulses from 60 keV gammas showing (b) fast leading edges.
(c) Plot of pulse height versus pulse duration for pulses form (a).
(d) Pulse height spectrum of 60 keV, 25 keV and 8 keV.

the upper end of the spectrum is the 60 keV photopeak, the sharp feature at the lower end is consistent with secondary 8 keV x-rays produced in the surrounding Cu and then striking the front side of the detector, and the central peak around 25 keV is due to secondary emission of K_α x-rays from the Sn. The central peak disappears when the Sn absorber is replaced by Pb which has its K-edge above 60 keV.

Our first direct data on the distribution of signal amplitudes as a function of distance into the silicon is shown in Fig 5c. The graph is a two dimensional plot of the height of each pulse versus its duration. The 60 keV and 25 keV branches are clearly visible and we estimate an energy resolution of ~ 2.5 keV (FWHM). This technique is effective in separating two branches for each energy, because events close to the titanium drive a small area normal, but greatly exceed threshold in this normal region. This area takes a longer time to return to equilibrium since the excess phonon energy must leave the film. On the other hand, events with the same peak height which are further from the film, just exceed threshold, so that the relaxation to equilibrium is faster. Thus the upper portions in Fig 5c correspond to events near the titanium film and the lower branches to events further from the films. At the 60 keV peak, we estimate that the event location is ≈ 200 μm from the crystal surface.

5. CONCLUSIONS AND ACKNOWLEDGEMENTS

Imaging phonon-mediated detectors such as SiCADs are becoming a reality in the laboratory. The prospects look promising for obtaining the additional improvements necessary to achieve thresholds of a keV or better in large crystals. Such a detector would be of great interest for a number of experiments including dark matter searches for weakly interacting neutral particles and reactor neutrino mass experiments.

The work at Stanford has been performed by B. Cabrera, C. J. Martoff, A. T. Lee, and B. A. Young. Also, B. Neuhauser now at San Francisco State University has participated extensively. This work has been funded in part by a Research Corporation Grant, a Lockheed Research Grant and DOE Contract DE-AM03-76-SF00-326.

6. REFERENCES

[1]   See for example: D. O. Caldwell, et al, Phys. Rev. Lett. **61**, 510 (1988).
[2]   See for example: **Superconducting and Low-Temperature Particle Detectors**, eds. G. Waysand and G. Chardin, (North-Holland, 1989).
[3]   B. Cabrera, L. Krauss and F. Wilzcek, Phys. Rev. Lett. **55**, 25 (1985).
[4]   S.H. Moseley, private communication; and S.H. Moseley, et al, J. Appl. Phys. **56**, 1257 (1984); and D. McCammon, et al, J.Appl. Phys. **56**, 1263 (1984).
[5]   N. Coron, et al, op. cit. [2].
[6]   B. Cabrera, **Massive Neutrinos in Astrophysics and in Particle Physics**, eds. O. Fackler and J.Tran Thanh Van, p. 423 (Editions Frontieres, 1986). A detector of this type was independently proposed by F. V. Feilitzsch, L. Stodolsky, and A. Drukier (1983, unpublished).
[7]   B. Cabrera, J. Martoff and B. Neuhauser, Nucl. Instr. & Meth. **A275**, 97(1989).
[8]   See for example: **Nonequilibrium Phonon Dynamics**, ed. W.E. Bron, NATO ASI Series **B124**, Plenum Press, N.Y., 1985.
[9]   See for example: G. A. Northrop and J. P. Wolfe, Phys. Rev. **B22**, 6196 (1980).
[10] S. Tamura, Phys. Rev. **B31**, 2574 (1985).
[11] N. Wang, et al, UC, Lawrence Berkeley Laboratory, Preprint LBL-26884.
[12] See for example: M. Tinkham, **Introduction to Superconductivity**, (Krieger Publishing, 1975).
[13] See paper by Zehnder, et al, in these proceedings.
[14] Th. Peterreins, et al, op. cit. [2]; and paper by Th. Peterreins in these proceedings.
[15] D. Twerenbold, Europhys. Lett. **1**, 209 (1986); H. Kraus, et al, Europhys. Lett. **1**, 161 (1986); and D. Twerenbold and A. Zehnder, J. Appl. Phys. **61**, 1 (1987).
[16] B. Neuhauser, B. Cabrera, C.J. Martoff and B.A. Young, *Jap. J. of Appl. Phys.* **26**, 1671 (1987).
[17] B. A. Young, B. Cabrera, A. T. Lee and C. J. Martoff, IEEE Trans. on Mag **25**, 1347 (1989).
[18] A. T. Lee, Stanford Preprint No. BC73-88.

Phonon-mediated Detection of Particles

Report of the Round Table Discussion

B. Sadoulet[1, 2, 3]; B. Cabrera[4,1]; H. J. Maris [5]; J. P. Wolfe[6]

1)   Center for Particle Astrophysics, University of California, Berkeley, CA 94720
2)   Department of Physics, University of California, Berkeley, CA 94720
3)   Lawrence Berkeley Laboratory, University of California, Berkeley, CA 94720
4)   Department of Physics, Stanford University, Stanford, CA 94305
5)   Department of Physics, Brown University, Providence, RI 02912
6)   Department of Physics and Material Research Laboratories, University of Illinois, at Urbana-Champaign, Urbana, IL 61801

Over the past five years particle physicists, nuclear physicists and astrophysicists have been increasingly interested in using phonons to detect particle interactions. In these detection attempts it is obviously critical to integrate the understanding that the phonon physicists have accumulated on the mechanisms governing the production, propagation and detection of those phonons. Vice versa, some of the issues raised by the particle detection problem may be of significant interest and the high sensitivity methods being developed may become important for phonon physics investigations. These were the motivations for a round table discussion between members of the two communities. This report attempts to summarize the themes of a very interesting discussion.

1. The Detection Problem

1.1 *Low Rate, Small Energy Deposition Experiments*

In the past few years, new fascinating questions have emerged which require breakthroughs in instrumentation. For instance, it is possible that weakly interacting particles could account for the dark matter which constitutes more than 90% of the mass of the universe and could be detected with suitable detectors[1]. The detection of coherent neutrino interactions[2] and the study of solar neutrinos[3] are other examples of considerable interest. Those fundamental experiments raise two instrumentation challenges:

The expected rates are extremely low requiring both massive detectors (total mass $\geq$ 1kg, or even tons for solar neutrinos) and exceptionally high rejection of the radioactive background. It is therefore necessary to have maximum redundancy. The experimentalist would like to have access to the **energy spectrum** (to check compatibility with expectations and identify, for instance, X-ray photo-peaks), the **position** of its interaction (to reject surface contamination and veto low efficiency regions in the detector), some **directionality** (a powerful signature, since the direction of the incident particle is usually known) and if possible some indication of the **nature of the recoil** (nucleus or electron). In many applications, electron recoils are radioactive background contamination.

At the same time, in most applications (e.g. dark matter, coherent ν scattering) the energy deposition is low (≈ 1 keV) and in the form of nuclear recoil which have very little ionization[4]. Conventional semiconductor ionization detectors have to be pushed to their ultimate limits[5] and, working near the electronic noise, they cannot provide the redundancy necessary for background rejection.

1.2 High Resolution Spectroscopy

Simultaneously, other fascinating problems require spectroscopic resolution unattainable by semiconductors. The measurement of the mass of the neutrinos, especially with neutrinoless double β decay[6] is a typical example. Another one is high efficiency (approximately 100%), high resolution spectroscopy (~1 eV) of X-rays in the 1 keV to 10 keV region. In particular the iron line at ~6 keV is important in astrophysics[7].

1.3 The Hopes Raised by "Cryogenic Detectors"

These challenging but fundamental problems have recently triggered a wide interest in the development of "cryogenic detectors"[8]. The main idea is to use for the detection, systems in which very low energy quanta can be created. Instead of using electron-ion pairs (typical binding energy ~10 eV), as in a gaseous proportional courrier, or electron-hole pairs (~1 eV) as in a semiconductor ionization detectors, it is proposed to use Cooper pairs (~1 meV) in a superconductor or phonons in an insulator (~ 1μeV at 10mK). If these quanta could be **efficiently** detected, very low threshold, and possibly very high redundancy or very high resolution could be achieved. Obviously, since thermal generation of excitations must be avoided, such detectors have to work at low temperature (≤ 4K).

2 . Phonon-mediated Particle Detection

Given the nature of this conference, the discussion was limited to phonon mediated detectors. After all, between 2/3 and 9/10 of the total energy of a particle interaction appears in phonons and at low energy, it is natural to attempt to measure this component, in order to get maximum sensitivity. Two basic ideas are being developed.

On one side calorimetry where one measures the temperature rise produced by the particle interaction, can be very sensitive at low temperature and may be interpreted as the measurement of **thermalized** phonons. Recent results of the McCammon, Moseley and coworkers[9] are impressive. (FWHM ~ 11.5eV for a 6 keV X-rays!) and progress has been made in the extrapolation to high mass, in particular by Coron[10] and Fiorini *et al.,*[11].

On the other side, it may be argued that it is more advantageous to attempt to measure the phonons **before** they thermalized. As remarked first by Maris[12] the thermalization time will be long at low temperature and as advocated by Cabrera[13] the "ballistic" phonons carry a much larger amount of information. In this context "ballistic" has to be interpreted loosely since, as we will see, the phonon propagation may be quasi-diffusive. A special case of this idea is being developed by the Brown University group[16] which attempts to use rotons in liquid helium. This medium has the advantages of being cheap and very pure.

The round-table discussion concentrated mainly on the second concept of trying to use phonons out of equilibrium. In order to focus the ideas, it may be useful to sketch a generic detector using out of equilibrium phonons (Fig. 1).

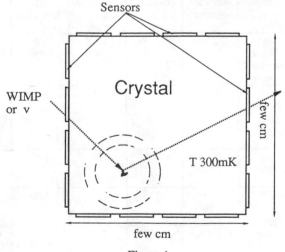

Figure 1

A generic detector

We choose the case described under 1.1 of a low rate detector. The detector (here represented as cubic, but which may be in some cases a thin wafer) is made of an insulating crystal (Si, Ge, LiF, B) maintained at very low temperature ($T \leq 300mK$). Typical dimensions of a few centimeters are needed to reach detector unit cells of at least a few grams. A (weakly interacting) particle interacts somewhere inside the crystal and the goal is to detect the resulting phonons by suitable sensors placed on the surfaces of the crystal.

We therefore discuss three regions: the phonon production region close to the interaction (Section 3), the propagation (Section 4) and the phonon sensing at surfaces (Section 5).

3. The Phonon Production Region

3.1 *The Initial Interaction*

In the discussion of the phonon production processes it is important to distinguish between the various particle interactions and energy regions (Table 1). While nuclei deposit their energies very locally leading to a high energy density, electrons deposit theirs in a very diffuse fashion. Except for the α particles, these conditions are quite different from those obtained in typical heater or laser beam excitation in phonon propagation experiments.

Table 1: Typical parameters of particle interactions

Recoiling Object	Energy	Physics Interest	Energy deposition volume (Si)	Energy Density (J/μ^3)	Initial Temperature	Amount of Ionization	Remarks
Nucleus	1-10 keV	Dark Matter Coherent ν scattering	$(0.01\mu)^3$	10^{-10} -10^{-9}	160K -290K	5% -10%	Local disruption of crystal
Electron	6 keV	X ray Spectroscopy	$(0.5\mu)^3$	10^{-14}	16K	33%	Stored ionization in metastable states
Electron	100 keV	ν-e scattering	$60\mu \times 100\mu^2$	3×10^{-18}	2K	33%	-//-
Electron	1 MeV	Double β, Nuclear Physics	$1.8mm \times 100\mu^2$	10^{-18}	1.5K	33%	-//-
α particle	6 MeV	Convenient test particle	$10\mu \times 1\mu^2$	10^{-13}	30K	33%, 2×10^6 e-hole pairs	Close to formation of e-hole droplet?

3.2 *The Phonon Generation Mechanism*

The detailed mechanism of phonon generation was the subject of many discussions during (and after) the round table. It is clear that both phonons and electronic excitations (electrons, holes and possibly excitons) are simultaneously produced.

But what is the proportion of optical and acoustic phonons? Are the phonons produced "coherently" (through the mechanical movement of the recoiling particle)? Is there some memory (e.g., through momentum conservation or the geometry of the source) of the direction of the recoil?

The LO phonons clearly decay very rapidly into acoustic phonons. How is this influenced by the above effect (e.g. by interference)? A sizeable non sphericity or asymmetry in the phonon flux would be of course a very important discrimination tool in particle detection!

Are the densities of excitations and the volume of the excited region large enough for a statistical equilibrium to establish itself? This could possibly lead to local down-conversion of a fair amount of the phonons to an energy low enough for the isotopic scattering in the bulk to be negligible. This question is intimately related to the question of the "hot spot" very much discussed in the conference[15]. And the spectrum for the phonons emerging from the interaction region is clearly important for the bulk propagation discussed in the next section. Taking Levinson's[15] analysis at face value, it does not seem that a hot spot is formed, except may be for nuclear recoil, but this has to be checked.

Another important issue is whether it is possible to distinguish between nuclear recoil and electron recoil. The ratio between the energy stored in ionization and the total energy may be a possible handle. But the ionization energy may be difficult to measure: drifting it out produces a phonon flux which may exceed the total initial phonon flux[16] and this method is practical only for germanium and silicon. The observation of the luminescence coming from the exciton decay may not be efficient enough (few 10^{-2}) for small energy deposition. In pure materials, the exciton lifetime (few microseconds) may give rise to a larger pulse width which may be detectable (for the ballistic component). Finally, if there is a hot spot effect for the recoiling nuclei, the spectrum may be different from the case of a recoiling electron and give rise to a different ratio of ballistic to diffuse components.

4. The Propagation Region

4.1 *The bulk propagation*

The spectrum of the phonons emerging from the interaction region is critical to gauge the effect of anharmonic decays and isotopic scattering[17]. In the round table discussion Wolfe presented new data which may be very relevant to this problem. When a laser pulse (λ=904nm) is sent on to metal film evaporated onto the surface of a 3 mm-thick silicon crystal, very efficient down-conversion is achieved and the detected phonon appear mainly ballistic. However, if the laser is incident directly onto the crystal (<111>-face), a very substantial diffusive component appears and only at high intensity are small ballistic peaks apparent. The average arrival time of the phonons is about 3 times the ballistic time of flight. A natural explanation is that there is no efficient down conversion in the illuminated region (except to some extent at high intensity where a hot spot may be formed) and that the phonons which stay at high energy suffer severe isotopic scattering. This conclusion is substantiated by preliminary phonon spectra obtained at high intensity in the TA ballistic peak (to be reported by Wolfe, Berberich, and Koester). For thicker crystals the diffusive contribution may extend over a longer period of time and not appear so dominant in the heat pulse. This diffusive component however may well contain most of the energy of the event!

If this is indeed the situation with particles, most of the phonons will be at high frequency and will arrive quasi-diffusively at the sensors. This may explain both the reduced ballistic component and the long delays observed in experiments by Peterreins *et al.*[18] and Cabrera and coworkers[19]. A small fraction of ballistic phonons can still provide signigicant focusing along strong focussing directions (e.g. <100>). The expected strong isotopic scattering would tend to spoil any memory of the particle initial direction.

With suitable material (e.g. NaF which is isotopically pure) it may be possible to suppress the isotopic scattering and thus decrease the diffuse component. Without a large industry dedicated to such materials, it may be difficult however, to obtain pure enough materials with low enough concentration of impurities and defects.

4.2 *Interfaces and boundaries*

Boundaries and interfaces are notoriously difficult to understand in terms of their observed reflection, transmission and down-conversion properties and very few data are of sufficient quality to be able to guide the work of the detector physicists (otherwise than to advise to limit the number of interfaces).

Provocative results has been presented by the PSI detector group[20] which interpret their data as evidence for the excitation of surface waves! This hypothesis has clearly to be checked further!

5. The Sensor Technology

Figure 2 attempts to summarize the sensor technologies which are presently studied, or have been suggested.

5.1 *Semiconductor thermistors*

The Berkeley group has indications[21] that Neutron Transmutation Doped (NTD) germanium thermistors may provide excellent coupling between the charge carriers and the high energy phonons which may be absorbed within a distance of a few hundred microns. However, in order for the detector to be fast, it is essential to develop an interface which does not thermalize too drastically the phonons. An attempt is currently being made with a Au - Ge eutectic (Fig. 2a) for bonding.

5.2 *Transitio n edge superconducting film*

The Stanford group is developing a very simple method[19] based on a superconducting narrow and thin strips of Ti (Fig. 2b). A large enough flux of phonons of energy above twice the gap, destroys the superconductivity of the strip producing locally a normal region. This is sensed by a current flowing through the strip. Biasing at the foot of the transition allows a faster and more uniform detector but introduces a threshold. The detector response is then dependent on the depth of the interaction. In order to obtain spectral resolution, it is necessary to measure this depth, e.g., by two films on opposite sides of the crystal.

5.3 *Superconducting film and tunnel junctions*

Two detector groups have used small tunnel junctions[18,20] to detect phonons generated by particle interaction. However, in order to cover large areas without having large junctions (which will have too large leakage current) and compromising on the speed, it is necessary to adopt more complex structures: junction arrays or probably better, large superconducting films equipped with a quasiparticle trapping structure[22] (a lower gap material) immediately below a small tunnel junction (Fig. 2c). In that way, the tunneling time may stay small in spite of the large ratio between the collecting and tunnelling areas.

1390

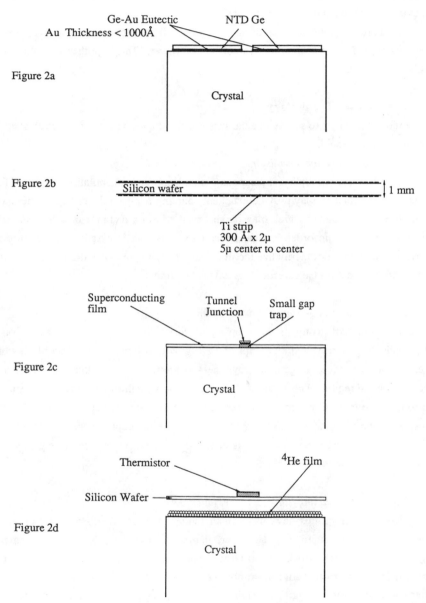

Figure 2 Various phonon detector schemes (see text)

5.4 *Desorption of film atoms*

Following the work of Wyatt and Goodstein [23], Maris has proposed using the fact that helium atoms can be ejected with high efficiency ($\geq 10\%$) from the free surface of liquid or a thin ^{4}He film on a solid crystal (Fig. 2d). These atoms can be detected by absorption on a silicon wafer. A relatively large rise in temperature is expected because of the release of the Van der Waals binding energy and the very small heat capacity of the wafer. This is the scheme that Maris and coworkers[14] would like to use to detect neutrino-electron interactions through the production of rotons in liquid helium.

Kinder and coworkers[24] have demonstrated a method for greatly amplifying the number of helium atoms desorbed from the surface of a cold (~ 100mK) silicon crystal. The desorbed atoms gain kinetic energy from a 50 K plate located only microns from the cold He film. Above a critical power thresfold the entire He film is desorbed, resulting in a Geiger-Counter like giant heat pulse. It is hoped that this multiplication process can be triggered by quantum desorption of single He atoms; however, this requires improvement of the present sensitivity by several orders of magnitude.

Acknowledgements

The authors would like to acknowledge support from the U.S. Department of Energy. [Contracts #DE-AC03-76.SF00-98 (B. Sadoulet), DE-AM03-76-SF00-326 (B. Cabrera), DE-FG02-88ER-40452 (H. Maris)] and the U.S. National Science Foundation [Office of the Science and Technology Centers under Cooperative Agreement No. AST-8809616- (B. Sadoulet) and MRL Grant-DMR-86-12860 (J. Wolfe)].

References.

1)   See for instance Primack, J. R., Seckel, D. and Sadoulet, B., Detection of Cosmic Dark Matter, Ann.Rev.Nucl.Part.Sci., **38** (1988) 751, and references therein.
2)   Drukier, A. K. and Stodolski, L., Phys. Rev., **D30** (1984) 2295.
    Cabrera, B., Krauss, L. M. and Wilczek, F., Bolometric Detection of Neutrinos, Phys. Rev. Lett., **55** (1985) 25.
3) See for instance: Davis, R., AIP Conference Proceedings No. 126, "Solar Neutrinos and Neutrino Astronomy", (1986)
4)   For a recent review see: Sadoulet, B. et al., Testing the WIMP explanation of the solar neutrino puzzle with conventional silicon detectors, Astrophys. J., **324** (1988) L75.
5)   Ahlen, S. P.et al., Limits on Cold Dark Matter Candidates from an Ultralow Background Germanium Detector, Phys. Lett. B, **195** (1987) 603.

Caldwell, D. O., Eisberg, R. M. and Grumm, D. M.et al., Laboratory Limits on Galactic Cold Dark Matter, Phys. Rev. Lett., **61** (1988) 510.

Luke, P. N., Goulding, F. S., Madden, N. W. and Pehl, R. H., Low Capacitance Large Volume Shaped-Field Germanium Detector, IEEE Trans. On Nucl. Sc., **NS-36** (1989) 976.

6)   Avignone, F. T. et al., Ultra-Low Background Study of Neutrinoless Double Beta Decay of 76-Ge; New Limit on the Majorana Mass Of ve, Phys. Rev. Lett., **54** (1985) 2309.

Caldwell, D. O., Phys.Rev.Lett., **59** (1987) 419.

7)   See e.g. Koyama, K. et al., Intense 6.7 keV iron line emission from the Galactic Centre, Nature, **339** (1989) 603.

8)   See for instance: the Proceedings of the Workshop on Low Temperature Devices for the Detection of Low Energy Neutrinos and Dark Matter, Ringberg Castle (Bavaria), (1987), Editors K. Pretzl, N. Schmitz and L. Stodolsky (Springer-Verlag: Berlin, Heidelberg), of the European Workshop on Low Temperature Devices for the Detection of Low Energy Neutrinos and Dark Matter, Annecy (1988) and of the Conference on Superconducting and Low Temperature Particle Detectors, Strasbourg(1988), G. Waysand and G. Chardin edit. (Elsevier Sc.: Amsterdam:). A shorter review is given by Sadoulet, B., Cryogenic Detectors of Particles: Hopes and Challenges, IEEE Trans. on Nucl. Sci., **NS-35** (1988) 47.

9)   Moseley, S. H.et al., Advances toward High Spectral Resolution Quantum X-ray Calorimetry, IEEE Trans. on Nucl. Sci., **NS-35** (1988) 59.

10)   N. Coron et al.,1989. in Proceedings of the Conference on Superconducting and Low Temperature Particle Detectors, G. Waysand and G. Chardin. Elsevier Sc.: Amsterdam.

11) Alessandrello, A.et al., Construction and operation of low temp erature bolometers for detectors of gamma-rays, IEEE Trans. on Nucl. Sci., **NS-36** (1989) 141.

12)   Maris, H. J., Design of Phonon Detectors for Neutrinos, Fifth International Conference on Phonon Scattering in Condensed Matter, Urbana, Illinois, June 2-6, (1986) 404.

13)   Cabrera, B., Martoff, J. and Neuhauser, B., Acoustic Detection of Single Particles, Nuclear Instrumentation and Methods, **A275** (1989) 97.

14) R. E. Lanou, H. J. Maris and G. M. Seidel,1987. in Proceeding of the Workshop on Low Temperature Detectors for Neutrinos and Dark Matter, Editors K. Pretzl, N. Schmitz and L. Stodolsky. p. 150. Springer-Verlag: Berlin, Heidelberg.

15) Wolfe, J. P., Current Topics in Phonon Imaging, this volume, Levinson, Y. B., Phonon Dynamics in highly non equilibrium systems, this volume.

16)   Luke, P. K., Voltage-Assisted Bolometric Ionization Detector, Submitted to Appl. Phys. LBL preprints (1988).

Lanou R., Maris H. and Seidel G., Superfluid Helium as Dark Matter Detector, VIIIth Moriond Astrophysics Workshop on Dark Matter, (1988).

17)   Klemens, P. G., in "Solid State Physics" , Seitz, F. and Turnbull, D., Edit., **Vol 1** (1958).

Klemens, P. G., Phys. Rev., **148** (1966) 845.

Klemens, P. G., J.Appl. Phys., **38** (1967) 4573.

Tamura, S., Phys. Rev., **B31** (1985) 2574.

18) Peterreins, T. H.et al., A new detector of Nuclear Radiation Based on Ballistic Phonon Propagation in a Single Crystal at Low Temperatures, Phys. Lett. B, **202** (1988) 161 and this volume.

19) Young, B. A. et al., Phonon-Mediated Detection of X-rays in Silicon Crystals Using Superconducting Transistion Edge Phonon Sensors, IEEE Trans. on Mag, **25** (1989) 1347.

Cabrera, B., Phonon-Mediated Detection of Elementary Particles, This Volume.

20) Hagen, C., Rothmund, W. and Zehnder, A., Slow Surface Phonons in Silicon, this volume,

21) Wang, N.et al., Electrical and Thermal Properties of Neutron Transmutation Doped Germanium at 20mK, Submitted for publication, (1989).

Wang, N.et al., Particle Detection with Semiconductor Thermistors at Low Temperatures, IEEE Trans. on Nucl. Sci., NS-36 (1989) 852.

22) Booth, N. E., Quasiparticle trapping and the quasiparticle multiplier, Apl. Phys.Lett., **50** (1987) 293.

C. Patel et al., Quasiparticle Trapping from a Single Crystal Superconductor into a Normal Film via the Proximity Effect., This volume.

23) Goodstein, D. L. et al., Phys. Rev. Lett., **54** (1985) 2034..

Wyatt, A. F. G., in: Phonon Scattering in Condensed Matter V, Anderson, A. C., Wolfe, J. P., Springer Verlag, (1986) 196 and this volume.

24) Wurdack, S., Gunzel, P. and Kinder, H., Impact Evaporation of Helium Films as a Phonon Multiplier for Particle Detection, this volume.

IMPACT EVAPORATION OF HELIUM FILMS AS A PHONON MULTIPLIER FOR PARTICLE DETECTION

S. Wurdack, P. Günzel, and H. Kinder

Physik Department E 10, Technische Universität München, 8046 Garching, FRG

Particle detection by phonon methods is of topical interest. An ambitious goal is the detection of neutrinos by their coherent interactions with entire nuclei.[1] This process has a much higher probability than the inverse beta decay. But the recoil energy is so small that the coherent scattering has not yet been verified experimentally. Therefore, we seek a new phonon detector which should in fact be sensitive to single phonon quanta. Moreover, the time resolution should be as high as possible for an effective suppression of background events.

The principle of the new phonon detection scheme is depicted in the inset of Fig. 2. A substrate crystal of Si is covered by a helium film of several monolayers and held at 0.1 to 0.2 K by a small dilution refrigerator. At this temperature, there

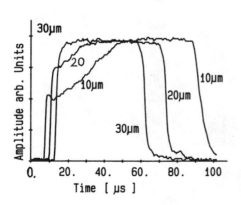

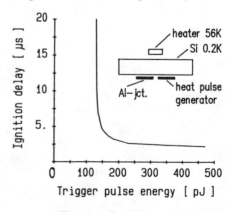

Fig. 1. Giant phonon pulses triggered by a heat pulse at time t=0 for several gap widths.

Fig. 2. Ignition time vs. energy of trigger pulse. Inset shows sample arrangement.

is no helium gas above the film. If phonons are created inside the substrate crystal by impact of particles, they will travel to the surface where they will release single helium atoms by the phonoatomic effect.[2] These neutral atoms must be multiplied before they can be detected. This is accomplished by a heater plate which is placed closely above the surface, at a distance of about 20 μm. The heater plate is suspended by a thin glass fiber which serves as an effective thermal insulator so that the plate can be heated to 70 K. If the helium atoms impinge upon the heater plate, they are reflected with much higher energy back into the helium film on the substrate. If this impact leads to secondary evaporation of more than one helium atom, a chain reaction develops which eventually evaporates the total helium film. The resulting hot gas generates a giant phonon pulse which we detect by a tunnel junction on the bottom of the substrate crystal. The accompanying temperature dip of the heater plate is monitored by a thermometer.

In the preliminary experiments presented here, we did not use a particle source but rather a pulsed heater to generate the phonons which then trigger the avalanche. Fig. 1 shows the resulting giant phonon pulses for the substrate at 0.2 K and the heater plate at 56 K. There is a distinct time delay of about 10μs between the trigger pulse and the "ignition" of the giant pulse. We have varied the gap between substrate and heater plate by a piezo drive. As expected, the ignition delay increases with the gap width due to the increased time of flight of each individual round trip of the atoms. All giant pulses appear to have the same height because the Al junction detector was saturated. The pulse duration depends also on the gap width, probably because the hot gas escapes more quickly from the larger gaps.

The ignition delay depends also on the power of the triggering heat pulse. This is shown in Fig. 2 for a gap width of 15 μm, trigger pulses of 50 ns duration, and substrate and heater temperatures again of .2 K and 56 K, respectively. Below about 200 pJ, the delay time increases dramatically, and below 130 pJ the ignition of giant pulses is not possible at all. This phenomenon seems to indicate that the gain of helium atoms at each round trip depends on the number of helium atoms already present inside the gap. At a small pulse energy, only a small number of atoms is released and the gain is less than unity, i.e. no giant pulses can be formed. At the minimum energy of 130 pJ, the initial gain becomes unity. Slightly above this energy the gain is still small so that many round trips are necessary to produce the giant pulse. This leads to the large ignition delay. For larger numbers of helium atoms in the gap, corresponding to larger trigger pulse energies, the gain saturates.

1396

In this limit of a large number, the gain can be estimated from the rise time of the giant pulses. We find 1.8 if we assume a typical round trip time of 300 ns.

For the detection of single phonon quanta, the gain should be greater than unity even for a few atoms in the gap. So it is important to understand why the gain depends on the number. We have set up a model where the energy of an individual "hot" atom first thermalizes and then relaxes by Kapitza conduction, by evaporation, and by lateral heat conduction along the superfluid film. Unfortunately, the latter effect dominates under the present experimental conditions. This dilutes the energy quickly so that the evaporation probability is low. Only when there are enough particles per unit time impinging in the vicinity, then the "warm" regions around individual impacts overlap and the outflow of heat fram a particular region is compensated by the influx of heat from neighboring regions. As a result, the probability of evaporation increases, until its cooling effect dominates over the lateral conduction. Correspondingly, the gain increases and eventually saturates.

To achieve atomic sensitivity we must therefore try to reduce the lateral heat conduction, e. g. by reducing the film thickness, or enhancing surface roughness and probably also the amount of surface contamination. We also plan to try mixtures of ^{3}He and ^{4}He. In its present state, our detector is about as sensitive as the tunnel junction used to read out the giant pulses.

The temperature dip of the heater plate during the ignition is plotted in Fig. 3 as a function of time. The initial drop is not resolved, but it is seen that the recovery of the device takes about 30 s under present conditions.

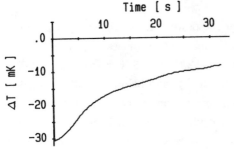

Fig. 3. Temperature drop of heater plate during ignition. This may provide a more favourable read–out in applications for particle detection, because it leaves the substrate silicon crystal free of any attachements.

References

[1] Drukier, A. and Stodolsky, L., Phys. Rev. D30, 2295 (1984)

[2] Goodstein, D. L., Maboudian, R., Scaramuzzi, F., Sinvani, M., Vidali, G., Phys. Rev. Lett. 54, 2034 (1985).
Wyatt, A. F. G., in: Phonon Scattering in Condensed Matter V, ed. by A. C. Anderson and J. P. Wolfe (Springer 1986), p. 196.

DETECTION OF PHONONS CREATED BY SINGLE α-PARTICLES

TH.PETERREINS, F.PRÖBST, J.JOCHUM, F.v.FEILITZSCH, R.L.MÖSSBAUER, and H.KRAUS (work supported by BMFT)

Physik-Department, Technische Universität München, James-Franck-Str., D-8046 Garching, Federal Republic of Germany

Abstract and References:
The absorption of single α-particles with an energy of less than 10^{-12} J in a Silicon wafer of dimensions 20 x 10 x 3 mm^3 was detected by signals in three superconducting Al/Al$_2$O$_3$/Al-junctions evaporated onto the surface of the wafer. The high sensitivity is largely due to the low operating temperature of T=0.37 K. In certain crystal directions strong phonon focusing effects were registered, proving the presence of a large percentage of ballistic phonons in the signals. Scattered and reflected phonons however may dominate in other directions. Position information with a resolution of fractions of a millimeter can be gained from investigating the correlations between coincident pulses in different junctions. The aim of this work is the development of a detector for neutrinos and certain dark matter candidates. The reader should consult the following references for more detailed information:

1) Th.Peterreins et al., Physics Letters B 202, 161 (1988).
2) F.Pröbst et al., Nucl. Instr. Meth. A 280, 251 (1989).
3) Th.Peterreins, F.Pröbst, F.v.Feilitzsch, and H.Kraus, "Tests of a Nonequilibrium Phonon Detector", in: G.Waysand and G.Chardin (eds.): "Superconducting and Low-Temperature Particle Detectors", Proceedings of Symposium C of the 1988 E-MRS Fall Conference, Strasbourg, November 8-10, Elsevier (North Holland) Amsterdam 1989, p.127.
4) Th.Peterreins, Ph.D. Thesis Technische Universität München 1989 (unpublished).

The absorption of nuclear radiation in a single crystal creates a variety of primary excitations. These excitations subsequently decay into phonons with time constants mostly in the ns - μs range. In a highly pure crystal at low temperatures, the phonons can propagate over several cm, sometimes ballistically (without scattering and at the speed of sound), to the crystal surface. It has therefore been suggested to construct a new detector for nuclear and particle physics which should consist of a single crystal serving as absorber, onto which an array of superconducting tunneling junctions acting as phonon detectors is evaporated. A detector of this type does not depend on ionization and should therefore be able to register non-ionizing events. One special application could be the detection of neutrinos and several dark matter candidates. Certain interactions of these particles with the nuclei of the absorber material have favourable cross sections, but can hardly be detected with conventional detectors relying on ionization, because energies of only 1 keV or less are deposited in the detector material. A further advantage of devices based on detection of nonequilibrium phonons would be the possibility to determine the position of interaction within the crystal by investigating the timing of the pulses or exploiting phonon focusing patterns.

The propagation of phonons in pure single crystals at low temperatures is governed by the rates for anharmonic decay ($\propto \omega^5$) and for elastic scattering off the randomly distributed isotopic atoms of the detector material ($\propto \omega^4$). Truly ballistic propagation dominates only if the frequency spectrum of the phonons peaks at values well below 1 THz. This seems to hold when phonons are created by the metal film heater technique or by laser pulses, as can be seen by the sharp focusing patterns observed in imaging experiments. In the case of nuclear radiation, the energy per quantum is much higher, but the total energy per pulse is usually orders of magnitude smaller than in laser experiments, because every pulse is caused by a single quantum. Details of the frequency spectrum created after absorption of nuclear radiation are unknown,

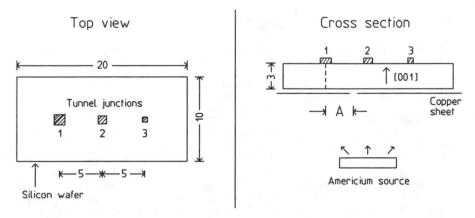

Fig.1: Simplified view of the experimental set-up (not fully to scale). Dimensions are given in mm. A Si single crystal with a volume of 10 x 20 x 3 mm³ has three superconducting Al tunnel junctions (labeled 1,2, and 3) evaporated onto its top surface. The wafer is illuminated from the back with an ^{241}Am α-source. Distinct points of absorption on the Si wafer are defined by holes in a Copper mask which lie in various distances A from the main focusing direction [001] relative to junction 1. Only one hole is shown for simplicity. The Cu sheet itself is sufficiently thick to absorb the α's.

though it is probable that very high frequency phonons are generated in the first place which then decay to lower frequency phonons.

We have carried out several experiments with a prototype device consisting of a silicon wafer of dimensions 20 x 10 x 3 mm³. Three superconducting Al/Al$_2$O$_3$/Al junctions were evaporated onto the top surface (Fig.1). The back of the wafer was illuminated with an ^{241}Am-source emitting α-particles with an energy of 5.5 MeV. This is less than 10^{-12} J. The high sensitivity of the tunnel junctions largely results from the low operating temperature of T=0.37 K. Small holes in a copper sheet define distinct spots where the α's hit the wafer surface. They deposit their energy within a depth of about 25 μm. Because the wafer is much thicker, only a transferring mechanism can lead to pulses in the junctions. By using α-particles and the mask arrangement, the location of energy deposition is well known.

The main focusing direction in Si is [001]. The holes were located at various distances A from the point lying in [001]-direction relative to junction 1. Well resolved pulses were obtained in all three junctions. Pulses in junctions 1,2, and 3 were recorded in digitized form and the time differences between correlated pulses in junctions 1,2, and 3 caused by the same α-particle were determined (see below).

The absorption of phonons in the superconducting films comprising the junctions leads to the breaking up of Cooper pairs and a subsequent enhancement of the single-particle tunneling current. Because the excess tunneling current pulses $I(t)$ were rather noisy, we often used the charge sensitive mode where the current pulses are integrated up ($Q = \int I(t)dt$). This mode allows to obtain high sensitivities, though the time resolution is worse than in tunneling current measurements. Typical rise times of the charge pulses are some tens of microseconds. The different phonon modes (L, FT, ST) could be resolved in neither case.

In a first experiment we tried to determine how large the amount of ballistic phonons in the signals is. A unique signature of ballistic phonons is the occurrence of phonon focusing. Because [001] is the main focusing direction in Si, it is expected (Fig.1) that the signal in junction 1 when A is chosen to be $A=0$ is much higher than for neighbouring directions (we chose $A=0.75$ mm, outside the main focusing spot). Indeed, the pulse height ratio we observed in junction 1 for α-absorption at the two positions was about 2.5:1. In the hypothetical case of isotropic phonon propagation we would expect a pulse height ratio of only 1.1:1. From the pulse height ratio observed experimentally, one can estimate a percentage of about 70 % of ballistic phonons in the signals of junction 1 for $A=0$. The amplitude of the current pulses for $A=0$ indicates an almost complete conversion of α-energy into nonequilibrium, nonthermal phonons within the time scale of our measurements.

Pulses in the junctions are however also detectable when the distance from α-absorption to detecting junction is more than 10 mm and if the junction lies out of focusing directions (e.g. pulses in junction 3 when $A=0$). This means that besides ballistic phonons, there are other components in the signal. This point becomes even clearer when looking at the time differences.

Extreme care was taken in evaluating the time differences between correlated pulses in different junctions caused by the same α-particle. For determining the onset of the pulses, we did not rely on a simple hardware trigger, but fitted each pulse separately to a model function. Because of the large focusing effect observed in the first experiment, one is tempted to assume that ballistic phonons should also contribute a significant portion of the signal in junction 2 for $A=0$. In the case of a well defined, dominating ballistic phonon front arriving at the diode, the charge pulse should be described by a rising exponential (because of the integration of the tunnel current). The onset of the charge pulse is defined by the arrival time of the ballistic phonon front, the rise time is essentially given by the life time of the excess quasiparticle density in the junction. The rise time was a free parameter in the fit procedure.

The time differences between pulse onsets in junctions 1 and 2 for $A=0$ as obtained from the exponential fit were much longer than one can calculate for ballistic phonon propagation to both junctions. This does not mean that the phonons propagate at a velocity below that of sound, but simply that the assumption that the pulses in junction 2 are dominated by a single wave front is wrong. It seems that for $A=0$, the portion of ballistic phonons in the signal of junction 2 is negligible. The most natural explanation is that the signal is dominated by phonons that were scattered off impurities and isotopes and/or reflected several times at the surfaces. It is conceivable that the frequency spectrum of the detected phonons peaks at about 1 THz. In this case isotope scattering would prevent the first, ballistic phonon front from reaching junction 2 on a straight path (the mean free path against isotope scattering for 1 THz transverse phonons is about 2 mm).

By investigating the correlated pulse heights in two junctions or time differences between pulses in those junctions we could show that points of α-absorption A spaced only fractions of a mm apart can be distinguished. This way of obtaining position resolution does not rely on focusing patterns and can be achieved with only a few junctions evaporated onto the crystal surface.

From the amplitude of current pulses, the amount of phonon energy absorbed in the junctions could be calculated. Together with the observed signal-to-noise ratio of the charge pulses (up to 100:1 in focusing directions), it can be estimated that an amount of phonon energy absorbed as low as 20 eV $= 3.2 \cdot 10^{-18}$ J would lead to a measureable signal in the charge sensitive mode. The deposition of 250 keV anywhere in the detector volume would be detected in at least one junction. In focusing directions, the detector threshold lies at about 20 keV. To achieve a threshold below 1 keV, the junction sensitivity can be increased by still lowering the operating temperature.

Using Doped Semiconductors for a Cryogenic Phonon Detector

Ning Wang[1,3], B.Sadoulet[1,3], F.C. Wellstood[1,3], E.E. Haller[2,3], J. Beeman[3]

[1] Department of Physics, University of California, Berkeley CA 94720

[2] Department of Material Science and Mineral Engineering, University of California, Berkeley CA 94720

[3] Lawrence Berkeley Laboratory, Berkeley CA 94720

ABSTRACT

We are developing a cryogenic phonon detector for the direct detection of dark matter particles in the laboratory. For this purpose, we have studied neutron transmutation doped Ge (NTD Ge) near 20 mK for its possible application as a sensor of both thermal and high energy ballistic phonons. We have investigated the thermal and electrical properties of NTD Ge. We find that results of both DC and AC measurements can be well explained by the decoupling of thermal phonons from charge carriers, an effect presumably due to a vanishing phase space of the thermal phonons at this low temperature. Such an effect is expected to be less profound for high energy phonons. This has been confirmed by the sample's responses to X-rays. We are currently developing a method to attach such a sensor to a single crystal, which is needed in the final construction of a dark matter detector. In order to detect ballistic phonons generated in the crystal, the interface between the sensor and the crystal has to allow high energy phonons to pass through from the crystal to the sensor with the least down-conversion.

INTRODUCTION

A particle interacting with a crystal generates optical phonons, which down-convert in a few nanoseconds to acoustic ballistic phonons. These in turn down-convert to thermal phonons after a few hundred microseconds. The conventional way to detect ballistic phonons is to let them thermalize and measure the temperature rise. Direct detection of these ballistic phonons can provide position and directionality of the incoming particles, which can be very crucial for background rejection. Therefore the critical problem is how to build a high energy ballistic phonon sensor. For this purpose, we have studied neutron transmutation doped Ge[1] near 20 mK.

PROPERTIES OF NTD GERMANIUM

The electrical conduction of neutron transmutation doped Ge is dominated by variable range hopping[2),3)]. The I-V curves of these thermistors are however very nonlinear. One theory is that the resistance at finite bias depends not only on the temperature of phonons, but also on the electrical field across the sample[4]. However, we find that our data is incompatible with this theory. We propose a model, based on the decoupling of the charge carriers from the phonons, to interpret the nonlinear I-V curves. We find that this model not only agrees well with the I-V curves, but also gives good fits to voltage pulses observed in AC measurements[5] (Fig. 1).

The decoupling between the charge carriers and the phonons reduces the expected sensitivity of such thermistors; however, a NTD Ge thermistor is still sensitive enough that a very low energy threshold can be achieved with the use of a low noise FET preamplifier. Pulses have been observed from 18 and 60 KeV X-rays from [241]Am with a fast rise time, indicating that the high-energy phonons are well coupled to the charge carriers. Therefore NTD Ge can be used as a high-energy phonon detector, provided it can be attached to a pure single crystal with an interface that passes ballistic phonons. We are developing a eutectic bond to achieve this aim.

EUTECTIC INTERFACE

A eutectic bond is formed by putting two surfaces, each coated with metal, in contact with each other and heating to above the eutectic temperature. In particular, we are interested in Au-Ge or Au-Si eutectic bonds.

To make a Au-Si or Au-Ge eutectic bond, one takes two chips of Ge or Si, evaporates Au on each chip, and places the Au-coated surfaces in contact with each other. The chips are pressed together and then heated to above the eutectic temperature in vacuum. An SEM picture of a Au-Si eutectic is shown in Fig. 2.

Au-Si and Au-Ge eutectics can be made repeatably, with good mechanical properties. We are preparing an experiment to measure the transmission of phonons through the eutectic bond at dilution refrigerator temperatures.

REFERENCES

1) Haller, E.E., Infrared Phys. 25, 257 (1985)

2) Wang,N., Sadoulet, B., Shutt, T., Beeman, J., Haller, E.E., et al.," A 20 mK Temperature Sensor", IEEE Trans. in Nuclear Science, NS 35, 55-58 (1988).

3) Wang, N., et al., "Particle Detection with Semiconductor Thermistors at Low Temperatures", IEEE Trans. in Nuclear Science, NS 36, 852-856 (1989).

4) Rosenbaum,T.F., et al., Solid State Commun. 35, 663-666 (1980).

5) Wang, N., et al., submitted to Phys. Rev. B (1989).

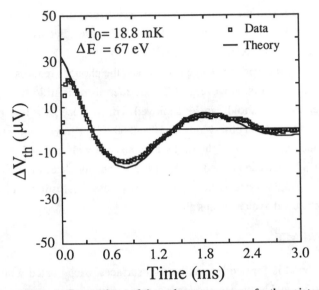

Fig. 1 Comparison of the voltage response of a thermistor with the decoupling theory.

Fig. 2 Au-Si eutectic. Au thickness is 4,800 Å. Magnification is 7,000 X.

17

New Experimental Methods

MEASUREMENTS OF PHONON DISPERSION CURVES BY X-RAYS WITH VERY HIGH ENERGY RESOLUTION

B. Dorner[+], E. Burkel[++], Th. Illini[++] and J. Peisl[++]

[+] Institut Laue-Langevin, 156X, 38042 Grenoble Cedex, France
[++] Sektion für Physik der Ludwig Maximilians Universität,
8000 München 22, W. Germany

ABSTRACT

Very high energy resolution of $\delta E = 17$ meV for 13.8 keV X-rays has been achieved by backscattering at monochromator and analyser. Longitudinal and transverse phonon dispersion curves could be determined in Beryllium and in Diamond. The observed frequencies and the intensities are in good agreement with neutron data and the scattering function.

1. INTRODUCTION

The availability of synchrotron radiation (SR) stimulates new techniques in many areas of physics. Originally SR was considered as the unwanted but unavoidable loss of energy in closed circuit accelerators, thus limiting the highest achievable energy. Now storage rings with high brightness dedica⁺ed to SR are being built. SR from highly relativistic electrons or positrons is strongly collimated in the horizontal plane and has high intensity with a white spectrum from visible light to hard X-rays. The upper limit in photon energy depends mainly on the particle energy.

At HASYLAB on the storage ring DORIS in HAMBURG, we constructed the instrument "INELAX" for inelastic X-ray scattering /1-6/. An energy resolution near to $\delta E/E \simeq 10^{-6}$ could be achieved. Inelastic scattering requires an energy determination before and after the scattering process together with a technique to vary at least one energy continuously in a controlled way.

This technique is complementary to inelastic neutron scattering in concentrating on high energy transfers, because the energy resolution of actually 17 meV is still large compared to inelastic scattering with thermal or cold neutrons. Working with X-rays has the advantage that the momentum transfer can be kept small at large energy transfers up to several eV, while the resolution does not vary with energy transfer. Inelastic X-ray scattering is expected to be applied to different areas of research such as lattice dynamics of hard materials, as presented in the following, and materials containing light elements; molecular spectroscopy; collective excitations in liquids; dynamics of surfaces; excitation of electronic states etc.

2.   THE BASIC PRINCIPLE OF THE MONOCHROMATOR

It is common knowledge, that in the case of Bragg reflection from a thick single crystal the radiation has limited penetration depth. For a perfect crystal a penetration depth can be expressed by a number of interfering reflecting planes. Using the optical principal of interference we conclude that the resolution $\delta E/E$ should be proportional to the reciprocal of the number of participating planes.

We shall now derive the reflectivity per single plane following the ideas of Darwin /7/. A plane wave with amplitude A_0 incident under the angle θ_i onto a planar array of scattering particles is partly reflected as the specular beam under the outgoing angle θ_f. Note that for an infinite plane and a large distance R of the observer from the scattering plane the angle $\theta_f = \theta_i$. To obtain the amplitude of the reflected wave at the observer we use the Fresnel construction. Solving the Fresnel integrals we find that essentially only the innermost Fresnel zone contributes. Its size is $\lambda \cdot R$, where λ is the wavelength of the radiation. If M is the number density of particles on the plane with scattering power (atomic form factor) $f(Q)$ ($Q = 4\pi\sin\theta/\lambda$) then the amplitude A arriving at the observer is

$$A = A_0\ \lambda \cdot R \cdot \frac{Mf(Q)}{\sin\theta} \cdot \frac{1}{R} \tag{1}$$

The $1/\sin\theta$ comes in, because the reflecting plane is inclined with respect to the Fresnel reference plane and $1/R$ accounts for the decrease of the amplitude over the distance R. The reflectivity q of one plane is then

$$q = \frac{A}{A_o} = \lambda \frac{M \cdot f(Q)}{\sin\theta} \tag{2}$$

As far as one plane is concerned, there is no correlation between λ and θ. But for a three dimensional crystal we obtain the Bragg equation

$$2d' \sin\theta = \lambda \tag{3}$$

where d' is the distance between reflecting planes. For an n-th order reflection (n-1) of these, planes are virtual and $d' = d/n$, d being the plane spacing in real space.

The number of particles per plane is for a fundamental reflection (n = 1) $M = N \cdot d$, where N is the three dimensional density. If we average the number of particles over real and virtual planes in the case of higher order reflections, then we get

$$M = N \cdot d' \tag{4}$$

Now we rewrite the reflectivity per plane with help of eqn. (3) and (4)

$$q = 2 \ d'^2 \ N \cdot f(Q) \tag{5}$$

The number of participating planes is proportional to $1/q$ and in turn the resolution is proportional to q :

$$q \ \sim \frac{\delta\lambda}{\lambda} = \frac{\delta E}{E} = \frac{4}{\pi} \ d'^2 \ N \ f(Q) \tag{6}$$

Eqn. (6) quotes the exact formula for an infinitely thick perfect crystal as derived by dynamical theory /8/.

In backscattering ($\sin\theta = 1$) the energy of the reflected X-rays is

$$E \ \sim \frac{1}{\lambda} = \frac{1}{2d'} \tag{7}$$

Therefore the absolute energy resolution is

$$\delta E \ \sim \frac{2}{\pi} \ d' \ N \ f(Q) \ \sim \frac{N \cdot f(Q)}{\pi} \ \frac{1}{E} \tag{8}$$

and we conclude that the absolute energy resolution can be improved using higher order reflections and this means higher X-ray energies. Note that this does not hold for neutrons, where $E \sim 1/\lambda^2$.

Considering the penetration depth we easily realize that it should be proportional to $1/d'$ and not to $1/d'^2$ because the increase of the number of virtual planes does not contribute. But a depth to be measured in mm is difficult to define, because the exact expression (eqn. (6)) is derived for the infinitely thick crystal. In Fig. 1 we

show the resolution in backscattering from an infinitely thick crystal.
If the thickness of perfect material is only once Δ_0 /8/

$$\Delta_0 = \frac{\pi}{2d' \cdot N \cdot f(Q)} \tag{9}$$

then side wiggles appear in the distribution of E around E_I, see Fig.
1. For the (777) reflection from Silicon, see next chapter, it turnes
out $\Delta_0 = 0.25$ mm, while the resolution calculated with help of eqn. (6)
is $0.36 \cdot 10^{-6}$.

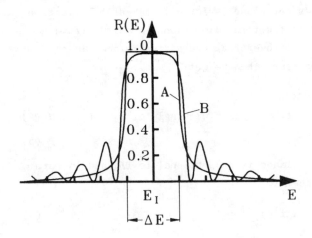

Fig. 1) Reflectivity of perfect crystals in backscattering; A for the infinitely thick crystal, B for a thickness of Δ_0, eqn. (9).

3. THE INSTRUMENT "INELAX"

The main features of the instrument (Fig. 2) are described in /3/
and /6/. Despite the high intensity of synchrotron radiation emitted by
the hard X-ray wiggler at HASYLAB we need focussing optics on the
monochromator and the analyser to obtain measurable signals. A
perfectly curved crystal, 1:1 imaging in nearly backscattering, could
collect the intensity from a large solid angle without degrading the
resolution. But the real devices, not imaging 1:1, and the fact, that
true backscattering is not possible, contribute to the resolution in an
amount of the order of 10^{-6}. Remark that the Bragg angle on the
analyser is about $0.08°$ away from $90°$.

To measure the energy resolution we need a sample which scatters
elastically into a large solid angle to fully illuminate the analyser.
Amorphous fused silica served very well. The best resolution obtained
so far experimentally is $\delta E = 17$ meV with $E_I = 13.8$ keV /5/. This

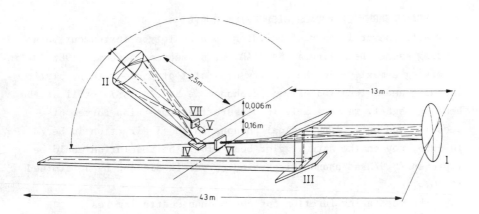

Fig. 2) Schematic presentation of the instrument "INELAX". I and II are the spherically curved monochromator and analyser crystals. III is the two crystal premonochromator, IV the sample and V the detector. VI and VII are diaphragms to reduce the background and to avoid a broadening of the resolution due to imperfect curvature of monochromator and analyser.

resolution contains probably a high contribution from the curved focussing crystals by means of a reduced penetration depth. The crystal disks are crosswise groved to create cubes of 1x1x1 mm^3 which are held together by the rest material on the bottom of the groves of about 0.2 mm thickness. The main deformation takes place in these thin parts, but nevertheless some strain mounts into the cubes. We have doubts, whether the upper part of 0.25 mm, or better more, from the surface really remains strain free.

At the same time as we struggle to get more intensity on the sample, we suffer from the heat load on the first optical element. To avoid the heat load on the main monochromator (I) we installed a double crystal premonochromator (III). This double monochromator has a strongly reduced transmission because the first crystal heats up under radiation leading to a mismatch of the lattice parameter between the two crystals.

To perform a scan in energy transfer we vary the temperature of the analyser crystal with respect to the monochromator, while the mechanical setup stays at rest. The lattice parameter of the analyser, as a function of temperature, defines the reflected wavelength. Such a constant scattering angle scan is always a constant Q scan.

4. MEASUREMENT OF PHONON DISPERSION CURVES

After several years of building up a prototype instrument on a bending magnet beam line at HASYLAB, we succeeded in summer 1986 /3/ in observing a maximum in the density of states of phonons in pyrolytic graphite and a longitudinal optic phonon in Be. In 1987, still at the bending magnet, we were able to measure the dispersion curves of acoustic and optic longitudinal phonons in [00 ξ] direction in Be /4/. In 1988, now on the hard wiggler beam line, the resolution could be improved /5,9/ and phonon dispersion curves in diamond be determined /5,10/.

The scattering function for one phonon scattering reads

$$S(\vec{Q},\omega) = G(\vec{Q},\vec{q},j) \cdot F(\omega,T) \qquad (10)$$

where the dynamical structure factor G is

$$G(\vec{Q},\vec{q},j) = \left| N \cdot \sum^{\text{unit cell}} f_d(Q) \cdot e^{-W}d(\vec{Q}\cdot\vec{e}_d(\vec{q},j))M_d^{-1/2}e^{i\vec{Q}\vec{d}} \right|^2 \qquad (11)$$

Here f_d is the atomic form factor of atom d at position $\vec{d}$ and $\vec{e}_d$ is the component for atom d of the normalized phonon eigenvector in the mode j with phonon wavevector $\vec{q}$. $e^{-W}d$ is the Debye–Waller factor of atom d and M_d its mass. N is the number of atoms in the unit cell. The response funciton F for undamped phonons is given by

$$F(\omega,T) = \frac{\langle n \rangle + 1/2 \pm 1/2}{\omega_j} \delta(\omega_{\pm}\omega_j). \qquad (12)$$

Here the upper sign holds for energy loss and the lower one for energy gain by the X-rays. $\langle n \rangle$ is the Bose occupation factor and ω_j the frequency of mode j. Note that the scattering law for X-rays can be obtained from the one for neutrons by simply replacing the Fermi scattering length by the atomic form factor. Here f_d already includes the classical electron radius. The validity of eqn. (10) has been proved by analysing measured intensities /4/.

For both samples, Be and diamond, we selected for our investigation symmetry directions in which the acoustic and optic modes are "pure" either longitudinal or transverse. The scattering geometry was then used in the way, that the dynamical structure factor (eqn. (11)) was zero (at least almost) for one kind of polarisation. Both materials have structures, hcp for Be and fcc for diamond, which contain two

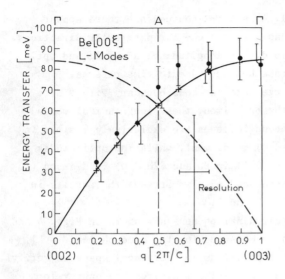

Fig. 3) Phonon dispersion curve of Be in the extended zone scheme. The Brillouin zone boundary is at A (q = 0.5). o results from inelastic X-ray scattering /4/; + results from inelastic neutron scattering /11/. The dashed branch is invisible in the [001] direction.

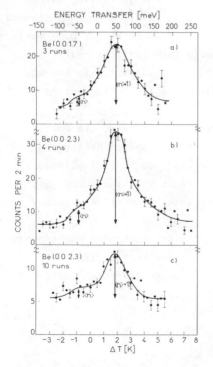

Fig. 4) Constant Q-scans in Be at q = 0.3. The temperature difference between monochromator and analyser is plotted as the bottom abscissa, while the energy transfer is given on the upper abscissa. Obviously there is good agreement between a) and b). c) is performed with reduced apertures of the diaphragms and slightly improved resolution to observe the energy gain peak.

atoms per primitive unit cell. As a first consequence there appear acoustic and optic modes and as a second one for particular directions, [00ξ] in Be and [ξ00] in diamond, the longitudinal acoustic and optic modes are degenerate at the zone boundary due to time reversal symmetry. The two branches merge into the zone boundary with finite slope opposite in sign. For the experiment these longitudinal modes have the advantage that the acoustic modes are exclusively visible in zones (0 0 21) in Be and (4h 0 0) in diamond, while the optic modes have only intensity in zones (0 0 1) in Be and (2h 0 0) in diamond (1 and h being odd integers). Correspondingly we present the results in the extended zone scheme, Figs. 3 and 5.

Particular scans for longitudinal phonons are shown in Figs. 4 and 6. In the case of Be (Fig. 4) we measured for the same phonon wavevector q = 0.3 at different positions in reciprocal space to verify that the signal stems from a phonon excitation. From the energy loss signal we derived <n> and indicated the corresponding expected intensity on the energy gain side /4/. This interpretation is not in contradiction with the data (δE = 55 meV) but not convincing either. After improvement in resolution (δE = 17 meV) /5,9/, a scan in diamond at q = 0.2 (phonon frequency close to the one in Be at q = 0.3) very clearly shows the separation of the energy gain and loss peaks. The absence of an increased background at elastic scattering is somewhat astonishing (at least for somebody coming from inelastic neutron scattering).

The transverse modes in [ξ00] direction in diamond are doubly degenerate all along. No additional degeneracy appears at the zone boundary and both the acoustic and optic branches have simultaneously intensity near the zone boundary. Fig. 7 shows that the energy gap between the two modes is resolved by the resolution of δE = 17 meV /10/.

The results on Be showed just that the new technique works because the dispersion curve has been measured with neutrons earlier /11/. The case is different for diamond where almost no experimental neutron data /12/ exist on the longitudinal optic mode.

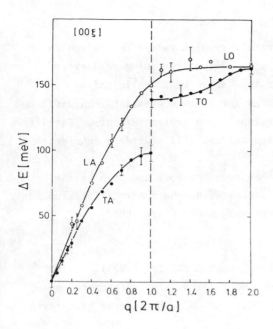

Fig. 5) Phonon dispersion curves in diamond as measured with inelastic X-ray scattering. In the extended zone scheme the Brillouin zone boundary is at q = 1.

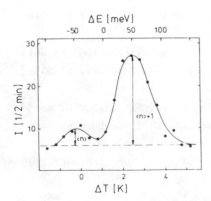

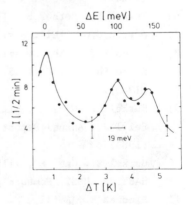

Fig. 6) Constant Q-scan in diamond at q = 0.2 in longitudinal geometry. The abscissae are as in Fig. 5. The energy loss ($\sim\{\langle n \rangle+1\}$) and the energy gain ($\sim\langle n \rangle$) peaks are well separated.

Fig. 7) Constant Q-scan in diamond at q = 1 (Brillouin zone boundary) in transverse geometry. The TA and TO modes are clearly separated.

5. CONCLUSION

Inelastic X-ray scattering with very high energy resolution is feasible, but still suffering from low intensity. Going to heavier elements the scattering volume of the sample will be limited by absorption. Some components, such as the premonochromator, can still be improved to gain intensity. The European Synchrotron Radiation Facility (ESRF) with undulators as radiation sources will provide a considerable increase in flux.

We wish to acknowledge the continuous support and help by the staff of HASYLAB. The project is supported by the Bundesministerium für Forschung und Technologie under project number 03 PE 1LMU2.

REFERENCES

1) Dorner B. Peisl J., Nucl. Instr. Meth. 208, 587 (1983).
2) Dorner B, Benda T., Burkel E. and Peisl J., Festkörperprobleme (Advances in Solid State Physics) XXV, 685, Ed. Grosse P., Vieweg, Braunschweig (1985).
3) Burkel E., Peisl J. and Dorner B., Europhys. Lett. 3, 957 (1987).
4) Dorner B., Burkel E., Illini Th. and Peisl J., Z. Phys. B Condensed matter 69, 179 (1987).
5) Burkel E., Illini Th. and Peisl J., Proc. SRI Conference, Japan (1988).
6) Burkel E., Dorner B., Illini Th. and Peisl J., to be published.
7) Darwin C.G., Phil. Mag. 27, 315, 675 (1914).
8) James R.W., "The Dynamical Theory of X-ray Diffraction" in Solid State Physics, Vol. 15, Eds. F. Seits and D. Turnbull, p. 53 (1963).
9) Burkel E., Gaus S., Illini Th., Löhnert K., Strohmeier M. and Peisl J., Annual Report-1988 of HASYLAB, DESY, Hamburg p. 441 (1989).
10) Burkel E., Gaus S., Illini Th. and Peisl J., Annual Report-1988 of HASYLAB, DESY, Hamburg p. 417 (1989).
11) Stedman R., Amilius Z., Pauli R. and Sundin O., J. Phys. F6, 157 (1976).
12) Warren J.L., Yarnell J.L., Dolling G. and Cowley R.A., Phys. Rev. 158, 805 (1976).

DETECTION OF BALLISTIC PHONONS IN MAGNETIC FIELDS

U.Valbusa, C.Boragno

Dipartimento di Fisica and Unità GNSM/CISM

Via Dodecaneso 33, 16146 GENOVA, ITALY

1. INTRODUCTION

Phonon spectroscopy in presence of a magnetic field is of great interest in several fields of solid state physics. We should mention the phonon studies of the 2DEG of a Si MOSFET recently done by Rampton et al. [1] which observed the modification of emission and adsorption of phonons by a 2DEG induced by the presence of a magnetic field and the capabilities offered by the new phonon spectroscopic technique of Eisenstein et al. [2]; both methods represent important studies of the electron-phonon scattering in the quantum hall regime and indicate that this one will be a fruitful area for future work. In addition it is worthwhile to remember the experiments of scattering from impurities in magnetic field which have been extensively performed in the past by using the heat-pulse technique [3],[4],[5].

From the experimental point of view the major problem arising from the use of a magnetic field is that the ballistic phonon detector cannot be neither a superconductor bolometer nor a Josephson junction, since these detectors cannot operate in an external magnetic field. For this use a semiconductor bolometer is more convenient: this operates in magnetic field and in addition can change the operation temperature. On the other hand, the use of semiconductor bolometers for detection of ballistic phonons is limited by the need of having a fast response time associate with an high sensitivity.

In the past several authors developed semiconductor bolometers with the aim of use them in heat-pulse experiments in magnetic field. Some of this detectors will be reviewed in this paper and particular emphasis will be given to an integral bolometer recently fabricated by ion implantation of silicon with phosphorus.

2. SEMICONDUCTOR BOLOMETERS AS HEAT-PULSE DETECTORS

The first bolometer used in presence of a magnetic field was a carbon bolometer [6]. This detector is sensitive enough to detect heat pulses but its response time is too slow for ballistic modes of duration of less than few microseconds. For this reason it has not been used frequently in ballistic phonon experiments.

In 1968, Wigmore developed the avalanche bolometer [7],[8]. This is a small chip of doped material biased to the point where impact ionization takes place. In this situation free carries obtain enough energy to ionize the neutral impurities with which they collide. In this way the small pulse of current generated by the heat pulse will be amplified by impact ionization. The avalanche phenomenon should take place in almost all doped semiconductors in which the dopant concentration is low enough to avoid impurity conduction. The avalanche bolometer has a typical response time $\tau = 2 \cdot 10^{-7}$ sec, a sensitivity of $5 \cdot 10^{-2}$ Volt/W and a noise equivalent power NEP = 10^{-7} watt/$\sqrt{Hz}$. The bolometer has been tested in magnetic field up to 7 Tesla. The avalanche bolometer should be bonded to the sample where one whishes to study the phonon propagation with a bonding agent which disturbes the detection of phonons.

The CdS bolometer has been first developed by Ishiguro and

Morita [9] and later on used by other groups [10],[3]. The device is a thin film of CdS of 0.5 - 1 μm thickness able to respond fast enough to the heat pulses as shown in Fig. 1.

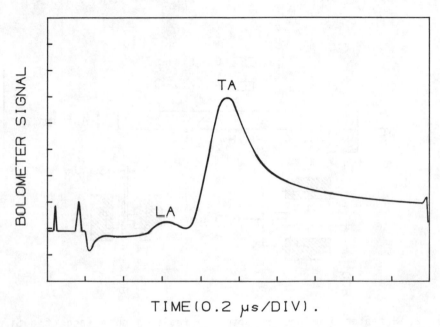

TIME (0.2 μs/DIV).

Fig. 1. Typical response of a CdS bolometer to an heat-pulse as function of phonon time of flight. The excitation power is 1W mm^{-2} and the pulse length τ = 100 nsec. From ref. 10.

Recently a Ge:Au bolometer has been fabricated [11]. The doping concentration is close to the metal insulator transition. The authors developed an empirical method which allows to produce bolometers with desired responsivity in the temperature range of interest. This detector has an high impedence and needs to be coupled with FETS or PMODFETS which work at low temperature and in magnetic field [12].

3. INTEGRAL BOLOMETER

An integral bolometer has been recently fabricated and used in magnetic field [13]. The device is described in figure 2.

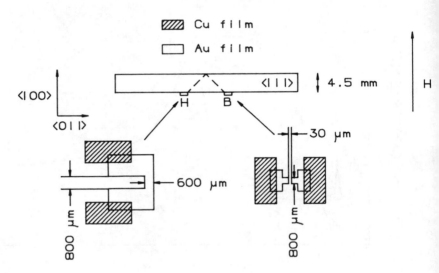

Fig. 2. H heater B bolometer. Both are realized on a face of a Si crystal of 50 mm diameter.

The bolometer is realized on a small zone of the crystal of 200mm^2 which has been doped by ion implantation with phosphorus in order to obtain a thin layer of 560 nm thickness homogeneously doped at a net donor centration close to the metal insulator transition. The electrical contacts to the bolometer are made by two thin gold films and two large copper pads as described in figure 2.

This device has a resistance R which as a function of crystal temperature T has as behaviour which follows

$$R = R_0 \exp(T_0/T)^{1/n} \qquad (1)$$

indicating a variable range hopping conduction. The bolometer has a magnetoresistance R(H) which in first approximation behaves as

$$R(H) = R(0) + aH^2 \qquad (2)$$

Figure 3 reports the plot of R(H) vs H^2. The behaviour of magnetoresistance is in agreement with the theoretical one predicted in the region of variable-range hopping conduction [14] in the weak field approximation.

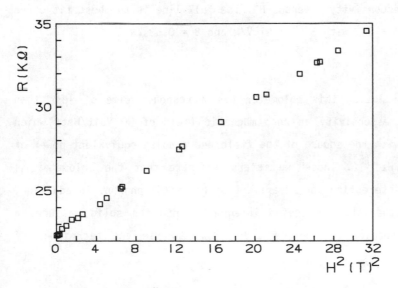

Fig. 3. R(H) versus H^2. The best fit gives R(0)=22.5KΩ and a=3.86 10^{-1} KΩ T^{-2}

The dependence of the resistivity on H affects the responsivity S of the bolometer which shows a similar behaviour as reported in figure 4.

The responsivity has been determined by the loading curves.

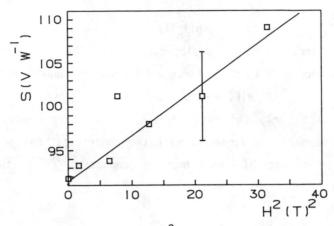

Fig. 4. Responsivity S versus H^2. The full line is the best fit curve $S(H) = S(0) + BH^2$ with $S(0) = 91$ V/W and $B = 0.53$ $VW^{-1}T^{-2}$.

In conclusion this bolometer has a response time of less then 50nsec, a responsivity at zero magnetic field of 90 Volt/Watt which increases with the square of the field and a noise equivalent power of 10^{-12} W/$\sqrt{Hz}$ [15]. These parameters indicate that the bolometer is extremely interesting for experiment of ballistic phonon. In addition, its charactheristic of beeing integrated into the solid, assures a correct detection of phonons and allows to use the techniques of microelectronics.

4. SPECTROSCOPY OF BALLISTIC PHONONS IN MAGNETIC FIELD

As application of this bolometer, it is interesting describe the experiment, recently performed [16], for studying the reflection of ballistic phonons from a silicon surface covered by a Pb film.

The experimental configuration is the same as the one described in figure 2 where on the other face it has been deposited a thin film of lead. The experiment has been performed at a temperature (kept constant) of 1.3K and it has been recorded the time of flight spectrum of the ballistic phonons reflected by the Pb film. It has been observed an increase in the area of the time of flight spectrum as function of the applied magnetic field for values below the critical field and a saturation for magnetic fields above the critical one as reported in fig. 5.

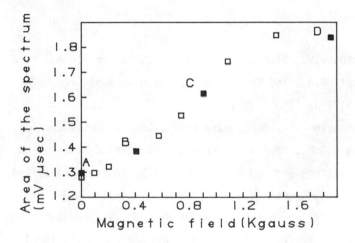

Fig. 5. Area of the time of flight spectra as function of the magnetic field H. It is evident a smooth increase in the area and a saturation reached at H≈1000 gauss which remains constant up to 3000 gauss.

The phenomenon can be explained by a two step process. In the first step the phonons are either reflected from the interface, either adsorbed into the film. The adsorbed phonons excite the electron gas which reemits incoherently phonons back into the silicon crystal. In our case we assume that the phonon adsorption occurs in the normal regions and then the signal increase when the normal phases involve the whole film.

In this kind of application the bolometer has to be able to work in presence of a magnetic field. We should point out that at these low fields the responsivity of the bolometer increase of less than 0.1% while the area of the spectrum changes of 30% at the same field.

REFERENCES

1)   Rampton V.W., Newton M.I., Kent A.J., Carter P.J.A., Hardy G.A.,
     Russell P.A., and Challis L.J., Japan J. Appl. Phys. 26,
     Suppl.26-3, 1755 (1987)

2)   Eisenstein J.P., Narayanamurti V., Stormer H.L., Cho A.Y. and
     Hwang J.C.M., Phonon Scattering in Condensed Matter V ed.
     A.C.Anderson and J.P.Wolfe (Berlin: Springer), 401 (1986)

3)   Kent A.J., Rampton V.W., Miyasato T., Challis L.J., Newton M.I.,
     Russel P.A., Hewett N.P. and Hardy G.A., Phonon Scattering in
     Condensed Matter V ed. A.C.Anderson and J.P.Wolfe (Berlin:
     Springer), 135 (1986)

4)   Miyasato T., Tokumura M., Toguchi M. and Akao F., Journal of the
     Physical Society of Japan 50, 1986 (1981)

5)   Miyasato T., Tokumura M., Suzuki K., Phonon Scattering in
     Condensed Matter ed. W.Eisenmenger, K.Lassmann and S.Döttingen
     (Berlin:Springer), 307 (1983)

6)   Cannon W.C. and Chester M., Rev. Sci. Instrum. 38, 318 (1967)

7)   Wigmore J.K., Appl. Phys. Lett. 13, 73 (1968)

8) Wigmore J.K., J. of Appl. Phys. 41, 1 (1970)

9) Ishiguro T. and Morita S., Appl. Phys. Lett. 25, 533 (1974)

10) Kent A.J., Rampton V.W., Newton M.I., Carter P.J.A., Hardy G.A.,
    Hawker P., Russel P.A. and Challis L.J., Sur. Sci. 196, 410
    (1988)

11) Chin M.A., Baldwin K.W., Narayanamurti V. and Stormer H.L.,
    Phonon Scattering in Condensed Matter V ed. A.C.Anderson and
    J.P.Wolfe (Berlin: Springer) 395 (1986)

12) Stormer H.L., Baldwin K., Gonard A.C., Wiegmann W., Appl. Phys.
    Lett. 44, 1062 (1984)

13) Boragno C., Ferdeghini C. and Valbusa U., J. Phys. E: Sci.
    Instrum. 22, 150 (1989)

14) Shklovski B.I. and Efros A.L., Electrical Properties of doped
    semiconductors (Berlin:Springer) 211 (1984)

15) Boragno C., Valbusa U. and Pignatel G., Appl. Phys. Lett. 50, 583
    (1988)

16) Boragno C. and Valbusa U., Solid State Comm. 65, 267 (1988)

OPTICAL DETECTION OF NONEQUILIBRIUM TERAHERTZ PHONONS IN DISORDERED SOLIDS USING SITE-SELECTIVE FLUORESCENCE EXCITATION

S.A.Basun, P.Deren*, S.P.Feofilov,
A.A.Kaplyanskii, and W.Strek*

A.F.Ioffe Physical-Technical Institute,
194021, Leningrad, USSR
* — Institute for Low Temperature and Structure Research,
P.O. Box 937, PL-53529, Wroclaw, Poland

The optical methods of high frequency acoustic phonon detection are widely used in application to doped insulators [1]. In the present paper (see also [2],[3]) we suggest a new type of "phonon spectrometer" for frequency selected time resolved phonon detection in doped disordered insulators.

Most of our experiments were performed in disordered $MgAl_2O_4$:0.1% Cr^{3+} spinel crystals in which Cr^{3+} impurity ions occupy distorted octahedral sites and the lower lying Cr^{3+} excited state is ^2E doublet [4]. The ^2E state energy shift and its splitting Δ site to site variations are of the same order of magnitude, the $^4A_2 - {}^2E$ transition forming in the region 680 - 710 nm a single inhomogeneously broadened (~ 150 cm^{-1}) R-line contour both in emission and absorption. If a relatively narrow band (~ 1 cm^{-1}) laser excitation is performed inside this contour the light is absorbed either by transition to the lower lying ^2E sublevel $\bar{E}$ or by transition to the upper sublevel $2\bar{A}$ (Fig.1). In a luminescence spectrum the transitions resonant with the excitation frequency ν_{exc} and nonresonant ones are observed. Resonant transitions are $\bar{E} - {}^4A_2$ in ions excited via $\bar{E}$ while $2\bar{A} - {}^4A_2$ in ions excited via $2\bar{A}$. Nonresonant transitions are $2\bar{A} - {}^4A_2$ in ions excited via $\bar{E}$ (anti-Stokes) whereas $\bar{E} - {}^4A_2$ in ions excited via $2\bar{A}$ (Stokes). Nonresonant transitions in different ions form a continious spectrum close to the resonantly narrowed line.

At helium temperatures resonant and Stokes transitions exclusively can be observed owing to fast $2\bar{A} \rightarrow \bar{E}$ relaxation. The heat pulse technique was used for the injection of pho-

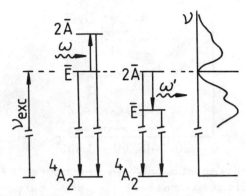

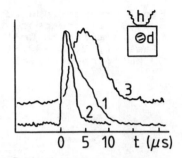

Fig.2 Heat pulse induced
anti-Stokes fluorescence
pulses in spinel at fre-
quency $\nu_{exc}+\omega$. l –
heater-detector distance.
1 – ω =20 cm^{-1}, l=0
2 – ω =60 cm^{-1}, l=0
3 – ω =20 cm^{-1}, l=1.5 mm.

Fig.1 The scheme of selective
excitation of Cr^{3+} ions ^{2}E-^{4}A$_2$
fluorescence inside the inhomogeneously broadened contour.

nons into a crystal immersed in liquid helium (inset in
Fig.2). If nonequilibrium phonons of frequency ω enter the
excited volume they induce $\bar{E}$ – $2\bar{A}$ transitions in Cr^{3+} ions
with Δ =ω, which results in anti-Stokes $2\bar{A}$ – ^{4}A$_2$ luminescence. The intensity and kinetics of this luminescence at
frequency ν_{exc} +ω reflect temporal evolution of phonons
with frequency ω in the excited volume and can thus serve
as a "phonon spectrometer" [1].

Fig.2 shows the phonon induced luminescence pulses at
different frequencies ω and different heater-detector distances. The long pulse decay time, as well as the absence
of a sharp front edge in the pulses apart from the heater,
evidence for the diffusive regime of phonon propagation.
The increase of pulse decay time with heat pulse energy and
its observed decrease with phonon frequency are typical
for the case of strong phonon-phonon interaction and "hot
spot" creation. A surprizing result is that the phonon propagation regime in disordered crystal does not differ much
from that in "ordered" crystals, such as fluorite[1] and YAG.

Optically excited phonons generated in the course of
radiationless relaxation of electronic excitations (for instance, $2\bar{A}$ → $\bar{E}$ in the ions excited via $2\bar{A}$ – Fig.1) were
also observed by anti-Stokes fluorescence in our experi-
ments with pulsed selective laser excitation.

We have also applied the suggested technique (Fig.1)
for the detection of phonon excitations in glasses. In most
of the experiments 0.2% Cr^{3+} doped silicate glass was used.
Two lasers were used in the experiment (inset in Fig.3):

<u>Fig.3</u> Phonon-induced anti-Stokes fluorescence kinetics in silicate Cr^{3+} doped glass; 1, 2, 3 – ω =10, 30, 50 cm^{-1}, respectively.

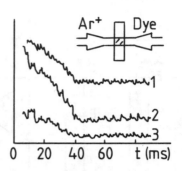

chopped Ar-laser light generating phonons in the course of radiationless electronic excitation relaxation and selective excitation ν_{exc} inside $^4A_2$ – ^2E inhomogeneously broadened contour serving for the phonon detection in the same sample volume. The anti-Stokes phonon signals are shown in Fig.3. The phonon excitation decay times appear to be very long (tens of ms), increasing with Ar-laser power. A surprizing feature of the observed phonon-induced anti-Stokes luminescence decay time is that it is frequency-independent. Another feature is also worthy of note here, i.e. the generality of the observed phonon-induced anti-Stokes luminescence relaxation regime for different glasses; similar results were obtained in Cr^{3+} doped borate glasses and glassy $Pr(PO_3)_3$. These preliminary results of the experiments with glasses are not fully understood and call for further studies.

Finally mention must be made of the earlier papers on organic crystal films [5] and ruby [6] where the selective laser excitation was used for phonon detection.

REFERENCES

1. Renk K.F. in "Nonequilibrium Phonons in Nonmetallic Crystals", ed. by W.Eisenmenger and A.A.Kaplyanskii. North-Holland, Amsterdam, 277-316 (1986). Bron W.E., Ibid. 227-275.
2. Basun S.A., Deren P., Kaplyanskii A.A., Strek W., Feofilov S.P., Fiz. Tverd. Tela 31 no.3, 199 (1989).
3. Basun S.A., Deren P., Feofilov S.P., Kaplyanskii A.A., Strek W., Proc. of the 7th Int. Conf. on Dynamical Processes in Excited States of Solids (Athens, USA, 1989) (to be published).
4. Mikenda W., Preisinger A., J.Lumin. 26 53 (1981). Strek W., Deren P., Jezowska-Trzebiatowska B., J. de Phys. 48 Suppl. no.12, C7-475 (1987).
5. Akimov A.V., Basun S.A., Kaplyanskii A.A., Titov R.A., Fiz. Tverd. Tela 20 no.1, 220 (1978). Bogner U., Phys. Rev. Lett., 37 no.14, 909 (1976).
6. Van Dort M.J., Dijkhuis J.I., de Wijn H.W., J. Lumin. 38 no.1-6, 217 (1987).

MONOCHROMATIC GENERATION OF HIGH-FREQUENCY PHONONS BY DEFECT-INDUCED ONE-PHONON ABSORPTION OF FAR-INFRARED RADIATION

U.Happek, P.T.Lang and K.F.Renk
Institut für Angewandte Physik
Universität Regensburg
8400 Regensburg, Fed. Rep. of Germany

We report monochromatic generation of high-frequency acoustic phonons by defect-induced one-phonon absorption of far-infrared radiation. In a recent study, one-phonon absorption due to defects has been used to generate high-frequency phonons in diamond[1]; however, only decay products of the monochromatically generated phonons could be observed. Here we report direct observation of phonons produced by far-infrared laser excitation, demonstrating for the first time that defect-induced one-phonon absorption leads to generation of monochromatic phonons.

Defect-induced one-phonon absorption occurs because the translational symmetry is lost at a defect and the conservation rule of the quasimomentum is no longer valid. The absorption coefficient for one-phonon absorption is proportional to the square of the dipole moment of the perturbed phonon modes, the concentration of defects and the density of states[2].

We generated phonons in a CaF_2 crystal making use of yttrium- induced infrared absorption. Fig.1 shows the absorption coefficient for a Y^{3+} doped CaF_2 crystal as a function of frequency[3]. At our generation frequency ($\nu_{FIR} = 0.89$ THz) the absorption coefficient is about 1 cm^{-1} for a crystal doped with 0.1 mole%YF_3. For phonon detection, the crystal was additionally doped with Eu^{2+} ions in a

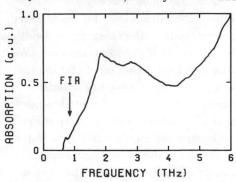

Fig.1:
Defect-induced far-infrared absorption in $CaF_2 : Y^{3+}$.

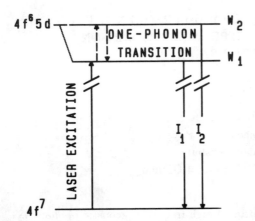

Fig.2:
Principle of phonon detection.

small concentration (0.003%) and fluorescence radiation was observed from the first excited (Γ_8^+) level. Applying uniaxial stress to the crystal, the Γ_8^+ splits into two sublevels W_1 and W_2 (Fig.2), with an energy separation Δ that is variable by stress. Phonons at a frequency $\nu = \Delta/h$ are in equilibrium with the relative population of the W_1 and W_2 levels. The phonon occupation number n can be determined from the ratio of the populations in the W_2 and W_1 levels, monitored by the corresponding fluorescence intensities I_2 and I_1.

The crystal (size $5 \cdot 5 \cdot 8$ mm^3) was fixed in a mechanical press and immersed in liquid helium at a temperature of 1.8 K. Radiation of a continuous wave HCN laser was focused on the crystal, about 10 mW power was absorbed in the bulk. The electronic W_1 level of the Eu^{2+} ions were excited resonantly with radiation of a continuous wave dye laser. Fluorescence radiation was selected with a double-grating monochromator and detected with a photomultiplier tube. A phonon induced signal was detected by lock-in technique.

Experimental results are presented in Fig. 3. The phonon occupation number shows a peak at the laser frequency and additionally a broad distribution at lower frequencies. The width of the peak (inset) corresponds to the detector bandwidth (110 GHz), determined by the width of the $W_1 \longrightarrow W_2$ phonon transition. We note that the observed peak was not caused by direct transition due to far-infrared absorption between the W_1 and W_2 levels of the Eu^{2+}-ions since this transition is forbidden; also experimentally we could not find evidence for far-infrared absorption by $W_1 \longrightarrow W_2$ transition. Using the technique of vibronic sideband spectroscopy[4] for phonon detection, a peak of the phonon occupation number at the laser frequency could be observed, too. In this case, the signal was very weak because the vibronic sideband spectrometer is about a factor of 10 less sensitive than the detector described above. Thus, we attribute the peak in the phonon distribution to phonons generated by yttrium-induced one-phonon absorption.

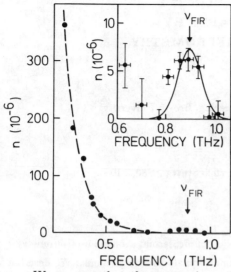

Fig.3:
Experimental occupation number of phonons generated by defect-induced one-phonon absorption of laser radiation at frequency ν_{FIR}.

We suggest that the occupation number of phonons at the laser frequency was larger than the experimental one ($n_{exp} \simeq 6 \cdot 10^{-6}$) and estimated from the ratio of the detector bandwidth and the spectral width of the far-infrared laser radiation ($\Delta\nu \simeq 5$ MHz) that the actual phonon occupation number had a value of the order of 0.1.

We attribute phonons below 650 GHz to anharmonic decay products of the originally generated phonons. Due to the lattice anharmonicity, the generated phonons decay fast into lower frequency phonons before they can leave the crystal into the helium bath, giving rise to the large phonon occupation numbers at lower frequencies.

In conclusion, we presented experimental results showing that defect-induced one-phonon absorption allows to generate monochromatic phonons. Since defect-induced one-phonon absorption is not limited to a fixed frequency but is a broadband absorption mechanism and occurs in principle in each crystal – and also in amorphous solids – this new technique should be applicable for generation of phonons in the whole phonon band of solids. Thus, a new method is available for the study of nonthermal phonons.

The work was supported by the Deutsche Forschungsgemeinschaft.

References
1)   Schwartz,H. and Renk,K.F., Solid State Commun. 54, 925 (1985).
2)   Genzel,L., in *Festkörperprobleme* 6 ed. Madelung,O., (Vieweg 1966), 52.
3)   Hayes,W., Wiltshire,M.C.K., Berman,R. and Hudson,P.R.W., J.Phys.C: Solid State Phys. 6, 1157 (1973).
4)   Bron,W.E. and Grill,W., Phys. Rev. B 21, 5303, 5313 (1977).

PHONON ATTENUATION AND VELOCITY MEASUREMENTS IN FUSED QUARTZ BY PICOSECOND ACOUSTIC INTERFEROMETRY

H.-N. LIN, H. J. MARIS, AND J. TAUC

Department of Physics and Division of Engineering, Brown University, Providence, RI 02912, USA

H. T. GRAHN

Max-Planck-Institut für Festkörperforschung, 7000 Stuttgart 80, FRG

In this article, we report the first application of picosecond acoustic interferometry[1] to measure the attenuation and velocity of longitudinal acoustic phonons. We describe the experimental technique, discuss the uncertainties of the measurement, and present data obtained in amorphous SiO_2.

The experiment is shown schematically in Fig.1. The sample in these experiments was a-SiO_2[2] with a 100 Å film of Al sputtered on the surface. A picosecond light pulse (pump) passed through the sample and was absorbed in the Al film. Thermal expansion of the Al launched a longitudinal strain pulse into the a-SiO_2. A second light pulse (probe) time-delayed by t relative to the pump is reflected at the Al film with angle of incidence θ and also at the strain pulse (see Fig.1). These three reflected beams interfere and, since the beams reflected at the strain pulse have travelled a distance $2vt\cos\theta$ more or less than the main beam (a), the reflected intensity has an oscillatory component of frequency $2nv\cos\theta/\lambda$ (λ light wavelength and n index of refraction) as can be seen in Fig.2. In our experiment the transient temperature rise of the Al film is a few K (at room temperature), and the strain amplitude and the amplitude of the oscillations in reflectivity are in the range 10^{-4} to 10^{-5}.

The spatial form of the propagating strain pulse is expected to be complicated and dependent on the details of the Al transducer film and its coupling to the sample. It can be shown[1], however, that the rate of damping of the oscillations in reflectivity is equal to the rate of attenuation of a longitudinal acoustic phonon at the same frequency as the observed oscillations. We have used this to obtain the temperature dependence of the attenuation at a frequency of 30 GHz as shown in Fig.3. We include in the figure results obtained by Vacher et al.[3] at a frequency of 35 GHz with Brillouin scattering. Allowing for possible differences in the sample material, the agreement between the two experiments is good.

We mention some of the conditions that have to be satisfied in order for this technique to give an accurate attenuation measurement.[4] The coherence time of the laser must be sufficiently long that there is negligible loss of coherence between the three reflected beams. With 1.7 psec FWHM pulses, we estimate this to be a negligible error for delay times of 1.6 nsec or less. Secondly, the pulse propagation is assumed to be approximately one-dimensional during the experimental time. This is a good approximation provided that the linear dimensions of the region illuminated (about 20 μm diameter in the present experiment) by the pump light are sufficiently large. Finally, it is important that the probe beam be at nearly normal incidence (0.06 rad in our experiment) so that it always interacts with the same part of the acoustic wavefront, regardless of the delay time t. These three possible sources of error are all expected to lead to a non-exponential decay of the oscillatory part of the reflectivity. A detailed analysis of the experimental data, however, showed that the decay of the oscillations was exponential, thereby confirming that the three effects discussed above lead to small errors under the conditions of the experiment.

From the period τ of the oscillations, we can calculate the sound velocity using $v = \lambda/2n\tau\cos\theta$. Results are shown in Fig.4 together with the data of Vacher et al[3]. Again, the data are in good agreement considering possible differences in sample material.

We acknowledge helpful discussions with C.D. Zhu and T.R. Kirst. This work was supported by the US Department of Energy through grant DE-FG02-86ER45267.

REFERENCES

1) Thomsen, C., Grahn, H.T., Maris, H.J., and Tauc, J., Opt. Commun. 60, 55 (1986).

2) ESCO Products Inc., Oak Ridge, N.J., USA

3) Vacher, R., and Pelous, J., Phys. Rev. B14, 823 (1976).

4) We will discuss these points and present more extensive data in a subsequent paper.

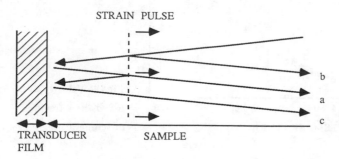

STRAIN PULSE

TRANSDUCER
FILM

SAMPLE

b

a

c

Fig. 1. Geometry of the experiment.

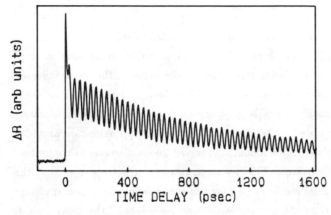

Fig. 2. Typical result for the change in the optical reflectivity as a function of time.

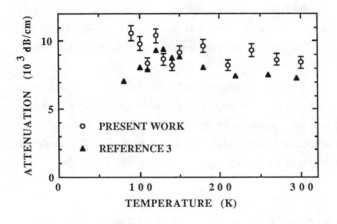

Fig. 3. Attenuation in a-SiO₂ as a function of temperature.

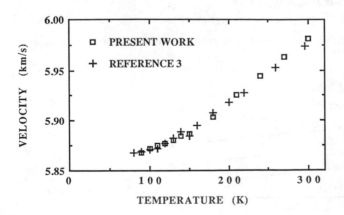

Fig. 4. Velocity in a-SiO₂ as a function of temperature.

PIEZOLECTRIC GENERATION AND DETECTION OF EXTREM NARROW, LATERAL MOVABLE SOUND BEAMS BY MEANS OF PLANAR ELECTROMAGNETIC RESONANCE STRUCTURES*

Thomas Aeugle and Olaf Weis

Abteilung Festkörperphysik, Universität Ulm

Oberer Eselsberg, D-7900 Ulm, FRG

We report on piezoelectric pulse-echo experiments performed in connection with sound-beam topography [1] using new developed planar electromagnetic resonators in the frequency range up to 35 GHz. H-slot and Hertzian resonators [2] can easily be produced by photolithography at millimeter-wave frequencies where the conventional reentrant cavities are no longer practicable due to production difficulties.

The spatial resolution in sound-beam topography is determined by the width of the sound beam which is at present only dependent on the experimentally achievable minimum extension of the exciting electrical field at the piezoelectric crystal face. Using a reentrant cavity of the kind shown in Fig. 1, we were able to manifacture coupling holes as small as 1 mm diameter at 24 GHz and 0.6 mm diameter at 35 GHz [1] and obtained with these electromagnetic structures for longitudinal phonons

a beam width of   300 μm at 24 GHz and 180 μm at 35 GHz at a

wavelength of   238 nm at 24 GHz and 163 nm at 35 GHz.

Hence, the beam width is by a factor 10^3 larger than the wavelength. This involves a beam collimation better than that of a HeNe Laser. To increase spatial resolution in sound-beam topography further, a reduction of the beam width to a diameter of 20...30 μm seems desirable at these frequencies.

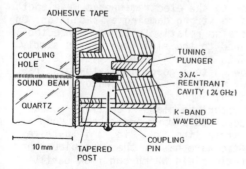

Fig. 1: Excitation and detection of a lateral movable sound beam at 24 GHz by means of a reentrant cavity [1]

*   Supported by Deutsche Forschungsgemeinschaft

Having this aim in mind, we began |2| to investigate planar reso-
nator structures at 1, 3 and 9 GHz. Based on the gained experiences,
we are now able to report in this paper on our first experiments with
planar H-slot and Hertzian resonators performed at 24 GHz and 35 GHz.

Analogous to the H-slot structure used earlier at 9 GHz, we varied systematically the H geometry and obtained the optimum resonator
dimensions, the electromagnetic coupling to the resonator, and the pho-
non-echo train shown in Fig. 2. The coupling between resonator and in-

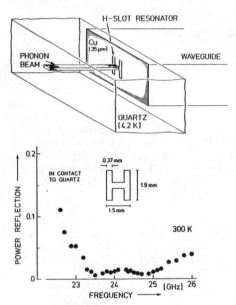

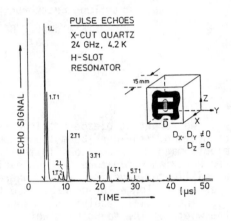

Fig. 2: Experimental results
using a H-slot resonator on
Teflon substrate at 24 GHz

cident electromagnetic wave is due to the continuation of the surface
current at the short by a dielectric displacement current in the H
slot. The measured power reflection as a function of frequency shows a
very broad resonance dip with a flat peak in the center which indicates too strong coupling. In principle, this coupling can be reduced by
placing the H slot in a more sideward position in the waveguide short.
But the main disadvantage of the H-slot resonator consists in its relatively large dimensions in comparison to the electromagnetic wavelength
in the crystal substrate which leads to strong damping by radiation. On
the other hand, the beam diameter is also relatively large, since the
exciting electric field is not confined to the slot area -as is simply
assumed in Fig. 2- but extends far into the metallized regions [2].
The observed phonon-pulse echoes show quite a good signal-to-noise ratio and demonstrate the usefulness of planar H-slot resonators for ap-
lications where the beam diameter is not important.

Therefore, we concentrated our effort on planar Hertzian resonators for which it was necessary to develope a more practicable coupling
than the earlier used coupling to a strip line |2|. Fig. 3 illustrates
the final solution. The electromagnetic wave has at the waveguide short
an especially strong horizontal magnetic field which can flow partly
through the inductive loop of the tiny Hertzian resonator. The coupling
can easily be modified by changing the included magnetic flux either by

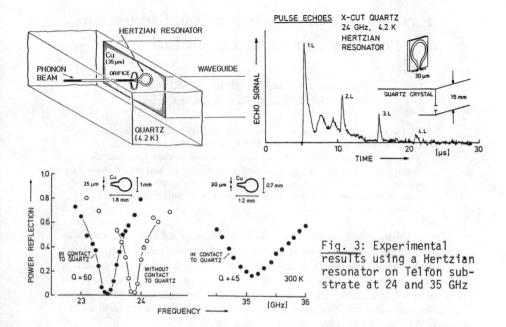

Fig. 3: Experimental results using a Hertzian resonator on Telfon substrate at 24 and 35 GHz

turning the loop by a certain angle or by changing the loop area or its position, respectively. According to the measured power reflection of Fig. 3, the resonator quality is about 60 at 24 GHz and 45 at 35 GHz. These Hertzian resonators were etched in a 35 µm thick copper film on a 200 µm thick Teflon substrate. A pulse-echo train observed with such a resonator is also shown in Fig. 3. According to our experiences at lower frequencies, the beam width should be 30 to 40 µm in these experiments. Using evaporated thin film structures, it should be possible to reduce the dimensions further. Of course, the signal-to-noise ratio is reduced too, since the maximum excitation power is limited by electric spark-over.

1. H. Edel, H. Bialas, and O. Weis: Z.Phys.B-Condensed Matter 64, 437 (1986)
2. Th. Aeugle, O. Weis: Z.Phys.B-Condensed Matter 71, 149 (1988)

First Results of Inelastic X-Ray Studies on Polycrystalline and Liquid Lithium

E.Burkel, S.Gaus, Th. Illini and J.Peisl

Sektion Physik Universität München, FRG

Inelastic measurements with x-rays can be performed with high energy resolution at the instrument INELAX at HASYLAB, Hamburg. The high intensity of the HARWI wiggler allows in the meantime to perform even inelastic x-ray studies of polycrystalline and liquid materials. Lithium was chosen for the first studies because of its low absorption cross-section for x-rays. The results on polycrystalline and liquid lithium are presented.

The first direct observation of phonon excitations in solids with x-rays was carried out in 1986 [1] by using the inelastic x-ray scattering spectrometer INELAX at HASYLAB in Hamburg. The very high energy resolution in the range of meV is achieved by backscattering at perfect single crystals used as monochromator and analyser. Complete descriptions of the method, the development and the latest instrumental set-up as it is installed at the HARWI wiggler beamline are given in [2,3,4,5,6].

Inelastic scattering has been to this point essentially the domain of thermal neutron scattering. However some research fields open now, in which the new technic of inelastic scattering with x-rays can give complementary valuable informations, not available with neutrons. Scattering experiments at small momentum transfer can be performed with x-rays at any desired energy transfer in contrast to neutron scattering experiments [7,8] , where the velocity of the incoming neutrons has to be larger than the velocity of sound in the sample material for the detection of a phonon loss peak.

This is an important range in the energy - momentum space for the determination of the frequency spectra for excitations in polycrystalline and liquid materials. Only in the first Brillouin zone around $Q = 0$ there exist distinct vibrational maxima allowing clear interpretation. They correspond to longitudinal modes averaged over all lattice directions for polycrystalline materials and to collective density fluctuations in liquids.

For the x-ray studies lithium was chosen, because of its low absorption cross-section for x-rays and its high velocity of sound, which makes it unaccessable for neutrons. It was investigated in polycrystalline form at room temperature and as liquid at 600 K.

Fig. 1 shows the scattered x-ray intensity as a function of the temperature difference of monochromator and analyser and the corresponding energy transfer for polycrystalline lithium at $Q = 1.0 \ \mathring{A}^{-1}$. Despite the large error bars one can detect the energy loss peak at $\delta E = (33 \pm 5) \ meV$ due to

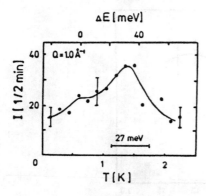

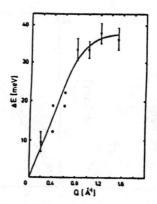

Figure 1: Scattered x-ray intensity of a constant Q-scan for $Q = 1.0 \, \text{Å}^{-1}$ of polycrystalline Li shown as a function of the temperature(energy)difference

Figure 2: Phonon dispersion of longitudinal modes in polycrystalline Li in the first Brillouin zone

longitudinal vibrations. Fig. 2 gives the information on energy transfers available so far as a function of the momentum transfer, thus indicating the dispersion branch for polycrystalline lithium in the first Brillouin zone.

Fig 3 and 4 show the scattered intensity for liquid lithium at the Q values $1.4 \, \text{Å}^{-1}$ and $2.0 \, \text{Å}^{-1}$ as a function of the energy transfer. Strong quasielastic scattering intensity and additional side peaks or shoulders are observed, as it is expected from Brillouin scattering and as it was observed with neutrons for liquid Rb [9,10] as well.

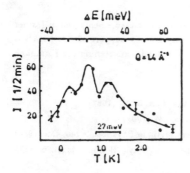

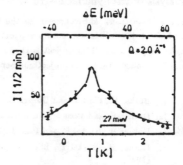

Figure 3: Scattered x-ray intensity of a constant Q-scan for $Q = 1.4 \, \text{Å}^{-1}$ for liqiud Li shown as a function of the temperature(energy)difference

Figure 4: Scattered x-ray intensity of a constant Q-scan for $Q = 2.0 \, \text{Å}^{-1}$ for liquid Li shown as a function of the temperature(energy)difference

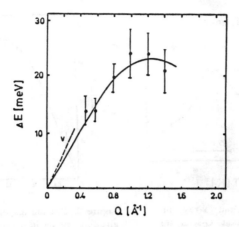

Figure 5: Dispersion relation for the collective excitations for liquid Li

Extracting the energy values for the observed distinct excitation maxima for the measured Q values up to 1.4 $\mathring{A}^{-1}$ leads to fig.5 in a first attempt of a dispersion relation of the collective excitations in liquid lithium. The slope of the velocity of sound from literature [11] is indicated as well. The position of the maximum of the observed dispersion relation corresponds to $Q_0/2$ with Q_0 being the position of the first correlation peak of the static structure factor of liquid lithium.

After these promising first results on polycrystalline and liquid lithium further measurements with improved energy resolution will lead to detailed analysis of the dynamical structure factors.

The authors thank M.Strohmeier for the help during the measurements and B.Dorner for fruitful discussions and steady support. The very friendly support of HASYLAB is highly appreciated. This project is supported by the Bundesministerium für Forschung und Technologie under the project number 03 PE 1LMU2.

[1] Burkel E., Peisl J. and Dorner B.: Europhys. Lett. 3,957 (1987)

[2] Dorner B. and Peisl J.: Nucl.Inst.Meth. 208, 587 (1983)

[3] Dorner B., Burkel E. and Peisl J.: Nucl.Inst.Meth. in Phys.Res. A246, 450 (1986)

[4] Dorner B., Burkel E., Illini Th. and Peisl J.: Z. Phys. B Cond. Matt. 69, 179 (1987)

[5] Burkel E., Illini Th. and Peisl J.: J.Proc. SRI Conf. Japan (1988)

[6] Dorner B., Burkel E.,Illini Th. and Peisl J., Proc. of Phonons Conf.(1989)

[7] Weinstock R.: Phys.Rev. 65, 1 (1944)

[8] Dorner B.: KFA Report Jül-412-NP (1966)

[9] Copley J.R.D. and Rowe J.M.: Phys.Rev.Lett. 32, 49 (1974)

[10] Copley J.R.D. and Rowe J.M.: Phys.Rev. A 9, 1656 (1974)

[11] Ruppersberg H. and Speicher W.: Z.Naturforsch. 31a, 47 (1976)

PHASE TRANSITION TEMPERATURES DETERMINATION WITH ELECTRODELESS CRYSTALS IN A LARGE RANGE OF THICKNESSES.

X. GERBAUX

Laboratoire Infrarouge Lointain

Université de Nancy 1,

B.P. 239 - F 54506   Vandoeuvre-les-Nancy Cedex,

France.

Far infrared absorption is very sensitive to small changes in the crystal structure, e.g. appearance of new bands in the low temperature phase, but also a broad increase in transmission. The measurement of such an increase does not need a high spectral resolution. A large spectral width can be used, hence a much higher energy can be sent through the sample. Hence, phase transitions temperatures can be measured with a good accuracy on electrodeless crystals either thin (95% transmission), or thick (5% transmission).

As an example, Sodium ammonium tartrate phase transition temperature has been determined (Fig. 1), for two crystal plates cut perpendicular to the c axis: Tc = 113.8 ± .2K, significantly higher than quoted in the literature from measurements of dielectric constant with metal coated samples (Tc = 109 K) [1]. The dashed line is corresponding to a 45 μm thick plate and the full line to a 680 μm thick one. In that case the effect of thickness is negligible.

Well known ferroelectric cystals such as KDP [2], and TGSe [3], have been used to look at the method efficiency (Fig.2 and 3).

REFERENCES

1 - Jona F. & Pepinsky R., Phys. Rev. 92, 1577 (1953).
2 - Busch G. & Scherrer P., Naturwiss. 23, 737 (1935).
3 - Matthias B.T., Miller C.E. & Remeika J.P. Phys. Rev. 104, (1956).

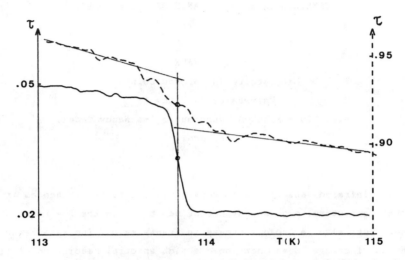

Fig. 1: Sodium ammonium tartrate: 45 µm thick crystal plate (---), and 680 µm (——). Phase transition temperature determined at 113.8 K.

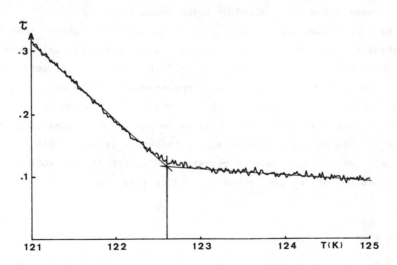

Fig. 2: KDP: 25 µm thick crystal plate parallel to the polar axis: phase transition temperature located at 122.6 K.

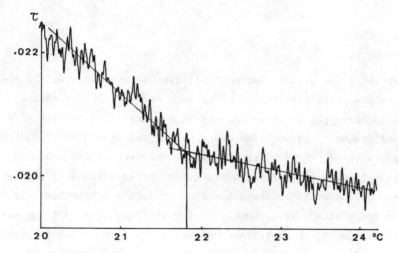

Fig. 3: TGSe: 500 μm thick crystal plate parallel to the polar axis: transition temperature at 21.8 °C.

STUDY OF DYNAMICAL BEHAVIOR OF INTERFACE IN SOLID-LIQUID PHASE
TRANSITION OF METALS BY PHOTOACOUSTIC MEASUREMENTS

S. KOJIMA

Institute of Applied Physics, University of Tsukuba,

Tsukuba, Ibaraki 305, Japan

1. INTRODUCTION

Recently the photoacoustic investigations have been widely applied to the phase transitions of various materials. The advatages
of the gas-microphone method are its high sensitivity and the wide
range of frequency between 1Hz and 200kHz. The method is therefore
very suitable for the studies of a dynamical process below 200kHz.
The present work is the first application of wide-frequency photoacoustic measurements to the dispersion of such a phenomenon. To
analyze the photoacoustic anomaly in the vicinity of a melting point,
the oscillation of the interface between a solid and a liquid phase
is predicted [1]. So this paper reports the experimental results of
the photoacoustic dispersion due to the dynamical behavior of the
interface.

2. EXPERIMENTAL

All photoacoustic measurements were done by the following hand-
made setup. The prototype was already reported in detail in ref.2,
and altered items are enumerated briefly. The photoacoustic effect
is excited by the 7884A light from a laser diode with maximum power
of 30mw. The intensity is electrically modulated in the frequency
range from DC to 5MHz. The photoacoustic signals are detected by the
condencer microphones with diameter 1/2 inch. The connection of a
microphone to a photoacoustic cell is done by the use of a probe tube
with an inner diameter of 1mm and a length of 200mm.

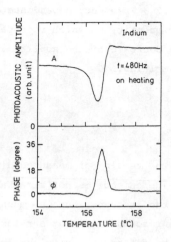

Fig.1 : Temperature dependence of photoacoustic signals in the vicinity of a melting point of an indium specimen of 99.9% purity.

3. TEMPERATURE DEPENDENCE OF PHOTOACOUSTIC SIGNALS ON MELTING

The first observation of photoacoustic anomaly of a pure metal was reported on the melting of a gallium specimen by Florian et al.[3]. According to their measurements the photoacoustic amplitude decreased remarkably in the vicinity of the melting point. Secondly Korpiun and Tilgner developed a theoretical model on the photoacoustic anomaly at a first order phase transition on the assumption that the interface between a solid and a liquid phase oscillates with the modulation frequency [1]. In the present study, the temperature dependence of both the photoacoustic amplitude and phase was accurately determined on melting for gallium, indium and tin specimens of various purities. The photoacoustic anomaly on melting for the indium specimen of 99.9% purity is shown in Fig.1. The similar behavior was also observed in other specimens [4]. Therefore it is concluded that these anomalies are essentially the intrinsic phenomena of pure metals.

4. DISPERSION OF PHOTOACOUSTIC ANOMALY ON MELTING

To clarify the mechanism of the anomaly, the changes of both amplitude and phase were measured as a function of the modulation frequency in the range 1Hz and 200kHz. In order to characterize the anomaly, the quantity $\Delta A = (A_s - A_{min})/A_s$ is introduced, where A_s and A_{min} denote the amplitude far below the melting point and the minimum amplitude, respectively. The quantity ΔA is proportional to the maximum

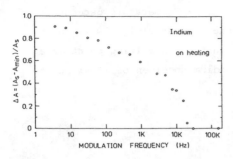

Fig.2 : Frequency dependence of
the quantity ΔA which represents
the decrease of the amplitude
on melting.

amplitude of the interface oscillation. The result of measured dispersion is shown in Fig.2. The frequency dependence is not remarkable , however, ΔA decreases monotonically as the frequency becomes much higher. This type of dispersion is not a resonant type but a relaxational type. Such a gentle slope cannot be fitted by a single relaxation, and the distribution of the relaxation time should be introduced to analyze the dispersion quantitatively. This multi-relaxational process is considered to be caused by certain types of friction which react to the motion of the interface between a solid and a liquid phase.

5. CONCLUSION

The photoacoustic anomaly of pure metals has been studied in detail on melting. The temperature dependences have been investigated by changing the modulation frequency of the incident beam in the wide range between 1Hz and 200kHz. The obvious dispersion of the photoacoustic anomaly has been found in the vicinity of a melting point. Such a dispersion strongly suggests that a multi-relaxational mechanism exists in the dynamics of the interface.

REFERENCES

1. Korpiun, P. and Tilgner, R.,J. Appl. Phys. 51, 6115 (1980)
2. Kojima, S., Proc. Tyohashi Int. Conf. Ultrasonic Technology, 141 (1987)
3. Florian, R., Pelzl, J., Rosenberg, M., Vargas, H. and Wernhardt, R. ,Phys. Stat. Solidi. A48, K35 (1978)
4. Kojima, S., Jpn. J. Appl. Phys. 27-1, 226 (1988)

Raman Scattering of Semiconductor Layers Grown by Laser
Atomic Layer Epitaxy

Tadaki MIYOSHI[*], Yasufumi IIMURA, Sohachi IWAI, Yusaburo
SEGAWA, Yoshinobu AOYAGI and Susumu NAMBA

[*]Technical College, Yamaguchi University, Ube, Yamaguchi
755, Japan ; RIKEN (The Institute of Physical and Chemical
Research), Wako, Saitama 351-01, Japan

ABSTRACT: Raman spectra were measured at 300 K in layers
grown by laser atomic layer epitaxy in order to
characterize thin layers of semiconductors. Two Raman
lines were observed in $Ga_{1-x}Al_xAs$ layers. The molar
fraction x of Al was determined with the frequency shift
of the Raman lines.

1. INTRODUCTION

Atomic layer epitaxy (ALE) of the GaAs/GaAlAs system
is an attractive method, since this method seems to be a
promising candidate for producing thin epitaxial layers
and abrupt interfaces controlled in one atomic layer
scale. Ideal ALE of GaAs would be achieved by successively
depositing a monolayer of Ga atoms followed by a monolayer
of As atoms. To achieve the ALE, switched laser
metalorganic vapour phase epitaxy (SL MOVPE) has been used
for GaAs crystal growth.[1] Application of the SL MOVPE to
a line patterning is possible by scanning laser beam.
However, characterization of the narrow epitaxial line is
difficult. In this paper, we describe the
characterization of thin and small region of GaAlAs
epitaxial layers using Raman scattering.

2. EXPERIMENTAL PROCEDURE

Epitaxial layers were grown on (100) oriented Si
doped (n = 10^{18} cm^{-3}) GaAs substrates in a low pressure
MOVPE system. Triethylgallium/triethylaluminium (TEG/TEA)
and AsH_3 were switched on and off alternatively. A laser
beam from an Ar laser (NEC GLG-3300, λ = 488.3 nm and 514.5
nm) was also switched by a shutter and was introduced into
the reactor for irradiating a substrate surface. One
cycle for epitaxial growth consists of the supply of
TEG/TEA for 1 s followed by the purge for 1 s and supply
of AsH_3 for 1 s followed by the purge for 1 s. Thus, one
cycle was completed in 4 s. Laser irradiation was
performed concurrently with the introduction of TEG/TEA.
TEG and TEA were introduced simultaneously. The growth
rate of the GaAlAs epitaxial layer was 0.28 nm/cycle (one
monolayer/cycle), so that the layer with thickness about
0.4 μm was grown with 1500 cycle. The growth area was

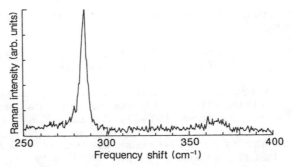

Fig. 1. Raman spectrum of $Ga_{1-x}Al_xAs$ grown by laser atomic layer epitaxy.

about 0.7 mm^2. The growth condition of the GaAlAs epitaxial layer is as follows: laser power density is 200 W/cm^2, growth temperature is 350-355 $^{\circ}$C, flux of TEG is 2×10^{-7} mol/cycle, flux of TEA is 1.2×10^{-7} mol/cycle, flux of AsH$_3$ is 3×10^{-5} mol/cycle and total flow rate of H$_2$ carrier gas is 2.8×10^{-3} m^3/min.

Raman spectra have been measured at room temperature in the backscattering geometry, using an Ar laser (Spectra Physics 168B, λ = 514.5 nm). The diameter of the laser beam was 0.1 mm. The scattered light was analyzed by a 1 m double monochromator ended by a photomultiplier (Hamamatsu R464S). The spectra were measured with a photon counter, a computer and an X-Y plotter.

3. RESULTS AND DISCUSSION

Figure 1 shows Raman spectrum of GaAlAs epitaxial layer. Two lines are observed: GaAs-type LO mode (287 cm^{-1} line) and AlAs-type LO mode (365 cm^{-1} line). The dependence of frequency shifts on the molar fraction x of Al in $Ga_{1-x}Al_xAs$ has been reported by Abstreiter et al.[2] The value of x = 0.15 is obtained with the frequency shifts of the Raman lines.

Raman spectra of different parts of the sample were measured in order to study the spatial profile of the Al concentration. Figure 2 shows cross-sectional profile of the sample and molar fraction of Al as a function of the distance from the central part of the epitaxial layer. They are related with spatial profile of the laser beam. GaAlAs layer becomes thinner with increase in the distance from the central part. This decrease is attributable to the reduction of growth rate caused by lower laser power as shown in Fig. 3.

Figure 3 shows that TEG is decomposed with light and not decomposed thermally at about 350 $^{\circ}$C,[1] while TEA is decomposed thermally.[3] In the central part of the epitaxial layer (region 1 in Fig. 2), TEG is decomposed

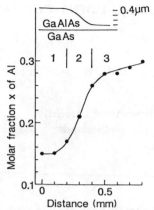

Fig. 2. Cross-sectional profile of the sample and molar fraction x of Al as a function of the distance from the central part of the laser-irradiated region.

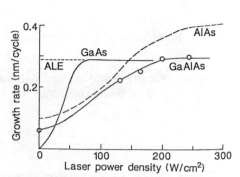

Fig. 3. Growth rate of GaAlAs, GaAs and AlAs[4] as a function of laser power density.

rapidly with light and Ga atoms are deposited prior to the deposition of Al atoms, so that the Al concentration is low. In region 3 (without laser irradiation), TEG is not decomposed, while TEA is decomposed. Thus the Al concentration is expected to be 100 %. However, experimental value of the Al concentration is about 30 %. This result suggests that TEG decomposition is assisted with TEA. In region 2, decomposition of TEG is reduced, so that the Al concentration becomes higher. Since the Al concentration depends on laser power, Al/Ga ratio of epitaxial layer is controllable with change in the laser power.

In summary, Raman spectra were measured in thin $Ga_{1-x}Al_xAs$ layers grown by laser atomic layer epitaxy. The molar fraction x of Al was determined with the frequency shift of the Raman lines. The Raman scattering experiments give information on the properties of thin and small crystalline layers.

REFERENCES

1) Doi, A., Aoyagi, Y. and Namba, S., Appl. Phys. Lett., 49, 785 (1986).
2) Abstreiter, G., Bauser, E., Fischer, A. and Ploog, K., Appl. Phys. 16, 345 (1978).
3) Ozaki, K., Meguro, T., Yamamoto, Y., Suzuki, T., Okano, Y., Hirata, A., Aoyagi, Y. and Namba, S., Int. Conf. Electronic Materials ICEM '88, W41 (1988).
4) Although apparent growth rate of AlAs does not show the ALE, laser-enhanced growth rate shows the ALE.

SPATIALLY TUNABLE SUPERCONDUCTING POINT-CONTACT BOLOMETER

K. Z. Troost, M. Barmentlo, J. I. Dijkhuis, and H. W. de Wijn

Faculty of Physics and Astronomy, University of Utrecht,
P.O. Box 80.000, 3508 TA Utrecht, The Netherlands

High-resolution ballistic phonon imaging requires small-sized phonon generators and detectors.[1] Earlier, a whisker-film metallic point contact has been employed as a micrometer-sized source of intense phonon pulses.[2] Here, we demonstrate the use of a thin superconducting tungsten film, forming part of a whisker-film point contact, as a bolometer at 1.5 K. Under appropriate conditions, the bolometer achieves subnanosecond time resolution, and a spatial resolution better than 10 μm. An interesting feature further is that the sensitive area can be tuned away from the point contact itself by increasing the bias current.

Thin tungsten films deposited by dc sputtering with a noble gas exhibit substantially enhanced superconducting transition temperatures T_c and a high specific resistivity above T_c,[3] and so these films are well suited for use as a superconducting bolometer. When the film is made part of the circular symmetric point-contact configuration, the bias current density drops inversely with the distance from the contact. With proper adjustment of the current, therefore, the sensitive area associated with the transition from the superconducting to the normal state consists of a small ring around the contact. The diameter of this ring is in fact given by $D_c = I/\pi d J_c$, in which I denotes the bias current, d is the thickness of the film, and J_c is the critical current density. The present superconducting point-contact bolometer consists of a 500-Å thick Kr-sputtered tungsten film ($J_c = 10^4$ A/cm^2, $T_c = 3.5$ K, and $\rho = 10^{-3}$ Ωcm), deposited onto a 0.4-mm thick sapphire crystal. Electrical contact is established with an electrochemically etched tungsten whisker, having a tip of approximately 1-μm radius.

We have investigated the intrinsic response time of the bolometer by direct laser-induced heating with the focused beam of a mode-locked Ar ion laser delivering pulses with a duration of 500 ps at a repetition rate of 82 MHz. Response times as short as 0.4 ns were found, comparable to the fastest large area film bolometers.[5]

For an application, we have performed phonon propagation experiments across the sapphire crystal, i.e., along the twofold crystalline axis. The phonon pulses were generated by laser heating of a second film, consisting of a 1-μm thick, totally opaque Au-layer deposited onto the crystal face opposite to the bolometer. Again, a

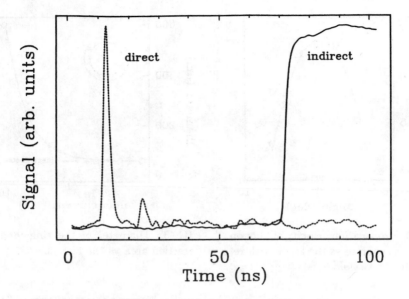

FIG. 1. Signal from a phonon pulse traveling across the sapphire crystal, and, for comparison, the signal from direct excitation.

focused Ar ion laser beam was used, except that for the sake of avoiding integration effects the repetition rate was reduced to 800 kHz by cavity-dumped mode-locked operation. Figure 1 shows the time trace of the phonon-induced bolometer signal, as well as, for timing purposes, the response to direct excitation. From the arrival time the phonon propagation velocity along the twofold axis is determined to be $(6.67 \pm 0.05) \times 10^3$ m/s. This result agrees remarkably well with the ballistic propagation velocity of fast transverse (T_1) acoustic phonons in sapphire, as calculated from elastic constants.[4]

By virtue of the small size of the point-contact bolometer, its combination with a second laser-heated film also permits to study phonon focusing. In the present case of propagation along the twofold axis of sapphire, both the arrival time and signal amplitude have been recorded as a function of the angle from the axis by varying the position of incidence of the laser beam. The dependence of the propagation velocity on angle is presented in Fig. 2. It appears to be in excellent agreement with the group velocity in the relevant direction as taken from the complete group-velocity surface of the T_1 mode calculated by Weis.[4] Similarly, the decrease of the signal amplitude, not shown here, appears to be in agreement with the calculated intensity surface for this mode.[4]

We finally demonstrate that the area of detection itself may be tuned away from the contact simply by increasing the bias curent. This was accomplished by scanning the laser beam across the tungsten film and monitoring the induced voltage as a function of the lateral displacement. The results for the outer diameter

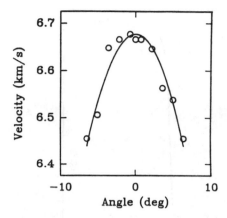

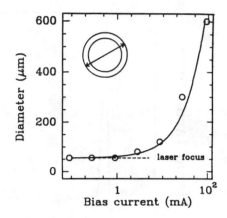

FIG. 2. Propagation velocity of phonon pulses in sapphire vs the angle with respect to the twofold axis.

FIG. 3. Diameter of the ring-shaped detection area vs the bias current.

of the sensitive area are given in Fig. 3. For smaller bias currents, the extent of the signal is, of course, determined by the laser focus, which measures 50 μm across. For larger currents, the diameter of the ring has risen to 0.6 mm.

In conclusion, we have shown that a superconducting whisker-film point contact may serve as a small, fast, and spatially tunable bolometer. In combination with a second laser-heated film, the bolometer further allows one to perform phonon propagation experiments with high spatial and temporal resolution.

The authors thank C.R de Kok and A.J. Michielsen for invaluable assistance. The work was supported by the Netherlands Foundations FOM and NWO.

1. R. P. Huebener, E. Held, W. Klein, and W. Metzger, in *Phonon Scattering in Condensed Matter V*, edited by A. C. Anderson and J. P. Wolfe (Springer, 1986), p. 305.
2. R. J. G. Goossens, J. I. Dijkhuis, H. W. de Wijn, A. G. M. Jansen and P. Wyder, Physica **127B**, 422 (1984).
3. K. L. Chopra, Phys. Lett. **25A**, 451 (1967); W. W. Y. Lee, J. Appl. Phys. **42**, 4366 (1971); P. H. Schmidt, R. N. Castellano, H. Barz, A. S. Cooper, and E. G. Spencer, J. Appl. Phys. **44**, 1833 (1973).
4. F. Rösch and O. Weis, Z. Physik **B25**, 101 (1976).
5. J. P. Maneval, J. Desailly and B. Pannetier, in *Phonon Scattering in Condensed Matter*, edited by W. Eisenmenger, K. Laßmann, and S. Döttinger (Springer, 1984), p. 43.

OPTIMISATION AND NOISE PERFORMANCE OF CONSTANT TEMPERATURE BOLOMETRIC PHONON DETECTORS

C D H WILLIAMS

Department of Physics, University of Exeter, Exeter, EX4 4QL, United Kingdom.

ABSTRACT

This paper gives a brief description of the conditions under which a constant temperature bolometric detector performs optimally and and compares the signal bandwidth and quality attainable with different types of device commonly used for the detection of phonons at low temperatures.

1. INTRODUCTION

Although designs for bolometer detectors rather more sensitive than the constant temperature detector CTD circuit (figure 1) have been proposed [1] a CTD based on a self-balancing bridge has several features that make it attractive as a detector of phonon pulses: it is simple, cheap and reliable; it can cover a frequency range from DC to several MHz; it eliminates detector nonlinearity by operating at a fixed sensor temperature, and offers absolute calibration.

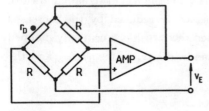

Figure 1. The CTD circuit.

The dynamic behaviour of a noise-free CTD is treated in the literature [2]. This paper gives a brief description of the unbiased, symmetric CTD in the presence of noise and compares the performance that might be expected in practice from superconducting transition and graphite film phonon detectors used in this configuration. Limited space precludes presentation of the detailed derivations of some of the results but these are given in another paper to appear elsewhere [3].

2. CTD PARAMETERS

The symmetric, unbiased bridge can be described by relatively few parameters. The detector is characterised by its temperature coefficient of resistance α, heat capacity C_{th}, and the thermal conductance to its surroundings σ_i. The bridge comes to equilibrium when the heating effect of the current I_D flowing through the detector has increased its temperature, and hence resistance r_D, to the point where $r_D \approx R$. Thermal noise from the bridge resistors can be eliminated, either by keeping R less than about 500Ω or by cooling, leaving the amplifier input noise voltage n_1 the dominant system noise source. The amplifier is assumed to have a large DC gain G_A, be fully compensated, and have a gain bandwidth product $G_A\omega_0$ which can be adjusted with an external compensation capacitor.

The change in output voltage V_e for a given small, low frequency change in the net power loss from the detector W_h is given by the responsivity $H_{th}(0) = 1/I_D$. The product of the useful (undistorted) signal bandwidth ω_s and the square of the low frequency responsivity by the mean square noise at the output defines the system sensitivity S which has the SI units [rad s^{-1}W^{-2}] and is given by

$$S = \frac{\omega_s |H_{th}(0)|^2}{\int_0^\infty n_1^2(\omega) |H_n(-j\omega)|^2 \, d\omega}. \tag{1}$$

The optimum value of $G_A\omega_0$ turns out to be quite critical and occurs when $G_A\omega_0 = \sigma_i^2/2\alpha^2 I_D^4$: below this value the useful signal bandwidth is reduced whilst the sensitivity stays almost constant; above this value the useful signal bandwidth stays nearly constant but the sensitivity is seriously reduced by the increased output noise. When $G_A\omega_0$ is optimally adjusted the CTD has a critically damped response to thermal inputs, a useful signal bandwidth $\omega_s = 0.2\sigma_i/C_{th}$, and the sensitivity is given by the simple formula

$$S = \frac{0.10\alpha^2 I_D^2}{\overline{n_1^2}\sigma_i^2}. \tag{2}$$

A related quantity is the system resolution: this is the smallest energy pulse that can be detected against the noise background and is given by

$$R = \left(\frac{\overline{n_1^2} \sigma_i C_{th}}{0.05 \alpha^2 I_D^2} \right)^{\frac{1}{2}}.$$

(3)

in fact the 'optimum' value of $G_A \omega_0$ given earlier actually minimises this quantity.

3. PHONON DETECTORS

In the following examples it is assumed that the bolometers are of area $1\,\text{mm}^2$ and at 1K in liquid ^{4}He have Kapitza boundary conductances estimated [4,5] to be $\sigma_i \approx 5 \times 10^{-4}\,\text{WK}^{-1}$ in each case. A reasonable operational amplifier might have white input noise n_1 equivalent to about $3\,\text{nV}/\sqrt{\text{Hz}}$.

3.1. Graphite Film Bolometers

A thin film of colloidal graphite has semiconducting properties at low temperatures and hence a large coefficient of resistance, typically in the range 10^3–$10^4\,\Omega\text{K}^{-1}$. A small bias current, in the range 1–$10\,\mu\text{A}$, is needed to avoid self-heating and the heat capacity, inferred from the frequency response, is expected to lie in the range 10^{-7}–$10^{-9}\,\text{JK}^{-1}$. So, typical values [5] of $\alpha = 3 \times 10^3\,\Omega\text{K}^{-1}$, $I_D = 3\,\mu\text{A}$ and $C_{th} = 10^{-8}\,\text{JK}^{-1}$ yield estimates for the sensitivity $S \approx 4 \times 10^{18}\,\text{HzW}^{-2}$ and resolution $R \approx 3 \times 10^{-12}\,\text{J}$.

3.2. Zinc Film Bolometers

Superconducting transition-edge bolometers have typical values of α and C_{th} in the same range as the graphite film devices described above. However the much lower electrical resistance permits a higher bias current: $I_D = 1\,\text{mA}$ being typical. This gives a big increase in the sensitivity and resolution with $S \approx 4 \times 10^{23}\,\text{HzW}^{-2}$ and $R \approx 10^{-14}\,\text{J}$.

4. REFERENCES

[1]   McDonald DG 1987 *Appl. Phys. Lett.* **50** 775–7
[2]   Sherlock RA 1984 *J. Phys. E.* **17** 386–93
[3]   Williams CDH 1989 *submitted to J. Phys. E.*
[4]   Challis LJ 1974 *J. Phys. C* **7** 481–495
[5]   Sherlock RA and Wyatt AFG 1983 *J. Phys. E.* **16** 669–672

AUTHOR INDEX